44,90
68I

Theory of Microwave
Remote Sensing

Wiley Series in Remote Sensing
Jin Au Kong, Editor

Tsang, Kong, Shin *Theory of Microwave Remote Sensing*
Hord *Practical Topics in Digital Image Processing of Remotely Sensed Data* (in preparation)

Theory of Microwave Remote Sensing

Leung Tsang
Jin Au Kong
Robert T. Shin

A Wiley-Interscience Publication
JOHN WILEY & SONS
New York • Chichester • Brisbane • Toronto • Singapore

Copyright © 1985 by John Wiley & Sons, Inc.

All rights reserved. Published simultaneously in Canada.

Reproduction or translation of any part of this work
beyond that permitted by Section 107 or 108 of the
1976 United States Copyright Act without the permission
of the copyright owner is unlawful. Requests for
permission or further information should be addressed to
the Permissions Department, John Wiley & Sons, Inc.

Library of Congress Cataloging in Publication Data
Tsang, Leung
    Theory of microwave remote sensing.

    "A Wiley-Interscience publication."
    Bibliography: p.
    Includes index.
    1. Remote sensing – Equipment and supplies. 2. Microwave devices. I. Kong, Jin Au, 1942-   . II. Shin, Robert T. III. Title.
G70.6.T76   1985   621.36'78   84-17397
ISBN 0-471-88860-5

Printed in the United States of America

10 9 8 7 6 5 4 3 2 1

To our parents

PREFACE

Remote sensing of the Earth has become a reality with the advent of the space age. The use of microwaves was largely prompted by their ability to penetrate clouds and to provide day and night coverage. It is important to develop theoretical models in conjunction with the collection of experimental data from various passive and active remote sensors.

During the past decade, progress has been made in the development of theoretical models for the interpretation of experimental data, and controlled field experiments have been performed to test such models. Our initial research interest was in the area of remote sensing of snow, ice, and soil moisture. This was later extended to vegetation cover, forestry, clouds, and rainfall. For such earth terrain, it is known that volume scattering plays an important role in the electromagnetic response of radiometers in passive remote sensing and radars in active remote sensing.

When volume-scattering effects were not included, the theoretical model has emphasized wave scattering and emission from stratified media and rough surfaces. As our first topic, we shall discuss in Chapter 2 the electromagnetic response for such models with particular application to the remote sensing of soil moisture.

In our study of the volume-scattering effects, we realized that terrain media are very often mixtures of substances with very different electric properties. Initial success was realized with the use of the random medium model, in which earth terrains such as snow-ice and vegetation are described by a correlation function, with the variance characterizing the strength of the permittivity fluctuation of the medium and correlation lengths corresponding to the scales of the fluctuation. To simulate the discrete nature of the scattering particles, a theoretical model consisting of discrete scatterers imbedded in a homogeneous layered medium was also studied.

The radiative transfer theory has been used extensively in the interpretation of experimental data obtained from field as well as aircraft and spacecraft measurements. In Chapter 3, vector radiative transfer equations for nonspherical particles are developed for both active and passive remote sensing. The extinction matrix, phase matrix, and emission source vector for the Stokes parameters are derived. Numerical illustrations of solutions of the radiative transfer equations with application to remote sensing problems are given in Chapter 4. Simple models of single scattering using point and Rayleigh scatterers are first illustrated, and multiple scattering effects for more complicated scatterers are then discussed.

To study the limitations of the radiative transfer theory, the analytic wave theory is examined in Chapter 5 by using Dyson's equation and the Bethe-Salpeter equation. Backscattering enhancement effects that are not accounted for by the radiative transfer theory are also illustrated. A strong permittivity fluctuation theory has also been used to account for the large variance in the permittivity fluctuations that are common in geophysical media. For the cases where a reflective boundary is present, a set of modified radiative transfer equations is derived that includes correlation of downward- and upward-going waves.

As demonstrated in controlled laboratory experiments, the assumption of independent scattering in the radiative transfer theory is not valid for a medium with an appreciable fractional volume of scatterers. As most geological materials are mixtures of constituents with significant volume fractions, it has become necessary to develop a dense-scatterer medium model. In Chapter 6, we study electromagnetic wave scattering by dense distributions of discrete scatterers, with attention given to the pair-distribution functions. The attenuation constants and backscattering coefficients are calculated and illustrated for dense media.

This book contains the results of research in active and passive microwave remote sensing of earth terrain. Contributions from many other sources and investigators are cited in the reference list, which is at best representative but by no means exhaustive. The manuscript of this book has been used as a graduate text. Many of our colleagues and students have contributed to the research. We would like to thank Michael Zuniga, Tarek Habashy, Boucar Djermakoye, Eni Njoku, Soon Yun Poh, Jay Kyoon Lee, Shun-Lien Chuang, Weng Cho Chew, Yaqiu Jin, Boheng Wen, and Yunsoo Choe for their helpful input. Special thanks are due to Soon Poh for his assistance throughout the preparation of this manuscript. We are indebted to Professors Akira Ishimaru, Irene Peden, and James Meditch of the University of Washington; Jonathan Allen, Joel Moses, David Staelin, and Gerald Wilson of the Massachusetts Institute of Technology; Andrew Blanchard, William Jones and Richard Newton of Texas A & M University for their support and encouragement. We would also like to acknowledge the financial support from the National Science Foundation, the National Aeronautics and Space Administration, Jet Propulsion Laboratory, Office of Naval Research, Schlumberger-Doll Research Center, and the Joint Services Electronics Program. The manuscript was prepared by Cindy Kopf,

Preface

Linna Wu, and Huoy-Ming Yeh. Finally, we would like to thank our wives Hannah Tsang, Wen-Yuan Kong and Sangmi Shin for their patience and encouragement during the preparation of this book.

Leung Tsang
Jin Au Kong
Robert T. Shin

Seattle, Washington
Cambridge, Massachusetts
Cambridge, Massachusetts
January 1985

CONTENTS

CHAPTER 1

INTRODUCTION   1

1. Active Remote Sensing   1
2. Passive Remote Sensing   4
3. Reciprocity Relation   10
4. Kirchhoff's Law   14
5. Characteristics of Particles in Earth Terrain   16
   Problems   21

CHAPTER 2

SCATTERING AND EMISSION BY LAYERED MEDIA   23

1. Introduction   24
2. Reflection and Transmission   25
3. Dyadic Green's Function   32
4. Fluctuation-Dissipation Theorem for Remote Sensing   41
5. Scattering and Emission by Periodic Rough Surfaces   53
6. Scattering and Emission by Random Rough Surfaces   70
   Problems   115

CHAPTER 3

RADIATIVE TRANSFER THEORY – EXTINCTION MATRIX, EMISSION VECTOR, AND SCATTERING PHASE MATRICES   119

1. Introduction   120
2. Stokes Parameters   121
3. Vector Radiative Transfer Equation: Constituents, Reciprocity, and Energy Conservation   128
4. Phase Matrix for Simple Objects   155

5. Phase Matrix and Extinction Matrix for General Nonspherical Particles with Prescribed Orientation and Size Distribution: T-Matrix Approach — 168
6. Boundary Conditions for Radiative Transfer Equations — 200
Problems — 208

CHAPTER 4
SOLUTIONS OF RADIATIVE TRANSFER EQUATIONS WITH APPLICATIONS TO REMOTE SENSING — 219

1. Introduction — 220
2. Iterative Method — 220
3. Discrete Ordinate–Eigenanalysis Method — 258
4. Method of Invariant Imbedding Applied to Problems with Inhomogeneous Profiles — 291
Problems — 312

CHAPTER 5
ANALYTIC WAVE THEORY FOR SCATTERING BY LAYERED RANDOM MEDIA — 317

1. Introduction — 318
2. Scattering by Layered Random Media – Born Series — 319
3. Analytic Wave Theory — 337
4. Strong Permittivity Fluctuations — 375
5. Modified Radiative Transfer Equations for Volume Scattering in the Presence of Reflective Boundaries — 390
Problems — 410

CHAPTER 6
SCATTERING BY RANDOM DISCRETE SCATTERERS — 425

1. Introduction — 427
2. Simple Model for Scattering from a Dense Medium — 430
3. Multiple Scattering Equations and Derivations — 439
4. Approximations of Multiple Scattering Equations — 455
5. Pair-Distribution Functions — 479
6. Scattering of Electromagnetic Waves from a Half-Space of Dielectric Scatterers – Normal Incidence — 490

7.	Scattering of Electromagnetic Waves from a Half-Space of Dielectric Scatterers – Oblique Incidence	506
8.	Nonspherical Particles	525
9.	Dispersion Relations Based on Coherent Potential	542
10.	Multiple Scattering of Second Moment	548
	Problems	563

BIBLIOGRAPHY 575

INDEX 603

Theory of Microwave Remote Sensing

1

INTRODUCTION

1.	Active Remote Sensing	1
2.	Passive Remote Sensing	4
	2.1 Thermal Radiation	4
	2.2 Planck's Radiation Law	4
	2.3 Specific Intensity	8
	2.4 Brightness Temperature	10
3.	Reciprocity Relation	10
4.	Kirchhoff's Law	14
5.	Characteristics of Particles in Earth Terrain	16
	Problems	21

1   ACTIVE REMOTE SENSING

For active remote sensing, a radar consisting of a transmitter system and a receiver system is utilized. Bistatic radar has the transmitting system and the receiving system situated at different locations, whereas in monostatic radar, these systems are located at the same place, usually sharing the same antenna system. The transmitter sends out a signal to the target and the scattered signal in a specified direction is measured by the receiver. The received power P_r can be summarized by the radar equation (Ruck et al.,

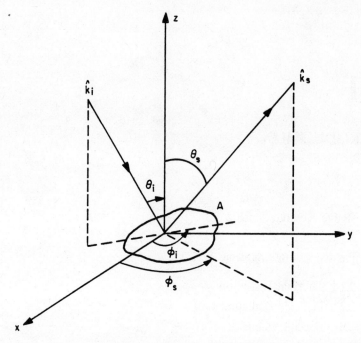

Fig. 1.1 Incident and scattered directions in calculating bistatic scattering coefficients.

1970; Skolnik, 1962, 1970).

$$P_r = \frac{P_t G_t}{L_t} \frac{1}{4\pi r_t^2 L_{mt}} \sigma_A \frac{1}{4\pi r^2 L_{mr}} \frac{G_r \lambda^2}{4\pi L_r} \frac{1}{L_p} \qquad (1)$$

The parameters of the above equation are defined as

P_t = transmitter power in watts

G_t = gain of the transmitting antenna in the direction of the target

L_t = numerical factor to account for losses in the transmitting system

L_r = a similar factor for the receiving system

r_t = range between the transmitting antenna and the target

σ_A = radar cross section

L_{mt}
L_{mr} = numerical factors allowing propagation loss in the medium

r = range between the target and the receiving antenna

G_r = gain of the receiving antenna in the direction of the target

λ = radar wavelength

L_p = numerical factor to account for polarization losses

The target is characterized by the quantity σ_A which has the dimension of area and is dependent on direction and polarization. It is defined as

$$\sigma_A = 4\pi \lim_{r\to\infty} r^2 \frac{\overline{E}^s \cdot \overline{E}^{s*}}{\overline{E}^i \cdot \overline{E}^{i*}} \tag{2}$$

where $\overline{E}^i$ and $\overline{E}^s$ are the incident and scattered electric fields, respectively. The scattered field is defined as the difference between the total field $\overline{E}$ (with the target present) and the incident field $\overline{E}^i$ (the field in the absence of the target).

$$\overline{E}^s = \overline{E} - \overline{E}^i \tag{3}$$

For the case of terrain and sea return, the cross section is normalized with respect to the area, A, illuminated by the radar. The bistatic scattering coefficient is defined as

$$\gamma_{\beta\alpha}(\theta_s, \phi_s; \theta_i, \phi_i) = \lim_{r\to\infty} \frac{4\pi r^2 |E_\beta^s|^2}{|E_\alpha^i|^2 A \cos\theta_i} \tag{4}$$

Notice that $A\cos\theta_i$ is the illuminated area A projected on to the plane normal to the incident beam. Thus, the fractional power that is scattered into polarization β and the angle (θ_s, ϕ_s) within solid angle $d\Omega_s$ is

$$\frac{1}{4\pi}\gamma_{\beta\alpha}(\theta_s, \phi_s; \theta_i, \phi_i) d\Omega_s \tag{5}$$

In (4) and (5), (θ_i, ϕ_i) and (θ_s, ϕ_s) are, respectively, the incident and scattered angles and α and β are respectively the incident and scattered polarizations (Figure 1.1).

From the figure, the incident and scattered directions $\hat{k}_i$ and $\hat{k}_s$ can be written as follows

$$\hat{k}_i = \sin\theta_i \cos\phi_i \hat{x} + \sin\theta_i \sin\phi_i \hat{y} - \cos\theta_i \hat{z} \tag{6a}$$

$$\hat{k}_s = \sin\theta_s \cos\phi_s \hat{x} + \sin\theta_s \sin\phi_s \hat{y} + \cos\theta_s \hat{z} \tag{6b}$$

In the backscattering directions $\theta_s = \theta_i$ and $\phi_s = \pi + \phi_i$, the monostatic (backscattering) coefficient is defined as

$$\sigma_{\beta\alpha}(\theta_i, \phi_i) = \cos\theta_i \gamma_{\beta\alpha}(\theta_s = \theta_i, \phi_s = \pi + \phi_i; \theta_i, \phi_i) \tag{7}$$

2 PASSIVE REMOTE SENSING

2.1 Thermal Radiation

All substances at a finite absolute temperature radiate electromagnetic energy. This electromagnetic radiation is measured in passive remote sensing. According to quantum theory, radiation corresponds to the transition from one energy level to another. When the transition is between levels with energies $\mathcal{E}_1$ and $\mathcal{E}_2$, the frequency of the emitted photon is given by Planck's equation

$$f = \frac{\mathcal{E}_1 - \mathcal{E}_2}{h}$$

where h is Planck's constant and is equal to 6.634×10^{-34} joule-sec.

There are different kinds of transition, and they include electronic, vibrational, and rotational transitions. For complicated systems of molecules with an enormous number of degrees of freedom, the spectral lines are so closely spaced that the radiation spectrum becomes effectively continuous, emitting photons of all frequencies.

Thermal emission of radiation by a substance is generally caused by collisions as a result of random motion of molecules. The collision rate depends on the kinetic energy of the random motion and hence increases with the absolute temperature of the substance.

2.2 Planck's Radiation Law

To derive the relation between temperature and radiated power, consider an enclosure in thermodynamic equilibrium with the radiation field it contains. First we shall find the energy density spectrum of the radiation field at equilibrium, after which it will be easy to find the intensity spectrum. The appropriate model for the radiation in the enclosure is an ideal gas of photons. Photons are governed by Bose-Einstein statistics. Thus the photons are indistinguishable and any number can occupy any allowed mode or state. The procedure for finding the energy density spectrum of the radiation field consists of three parts: (1) finding the allowed modes of the enclosure, (2) finding the mean energy in each mode, and (3) finding the energy in a volume V and frequency interval $d\nu$.

Without loss of generality, we shall choose a rectangular metal cavity because its modes are easiest to analyse. Let the cavity be filled with a material of permittivity ϵ and of dimensions $a, b,$ and d. The resonating

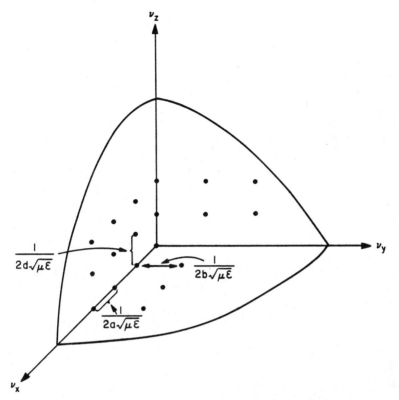

Fig. 1.2 Rectangular cavity resonating modes in ν space.

condition is

$$\nu^2 = \frac{1}{\mu\epsilon}\left[\left(\frac{l}{2a}\right)^2 + \left(\frac{m}{2b}\right)^2 + \left(\frac{n}{2d}\right)^2\right] = \nu_x^2 + \nu_y^2 + \nu_z^2 \qquad (1)$$

where $l, m,$ and $n = 0, 1, 2, \ldots$. The number of modes in a frequency interval $d\nu$ can be determined using (1) and a simple diagram (Figure 1.2). Each point in the diagram corresponds to a value for $l, m,$ and n and thus to a specific cavity mode. Thus, the volume of one mode in ν space is $1/8abd(\mu\epsilon)^{3/2} = 1/8V(\mu\epsilon)^{3/2}$ with V being the physical volume of the resonator. If a quarter-hemispherical shell has a thickness $d\nu$ and radius ν, then the number of modes contained in the shell is

$$N(\nu)d\nu = \frac{4\pi\nu^2 d\nu}{8} \times 8V(\mu\epsilon)^{3/2} \times 2 = 8\pi\nu^2 V d\nu (\mu\epsilon)^{3/2} \qquad (2)$$

where the factor of 2 accounts for the existence of TE and TM modes. If there are n photons in a mode with frequency ν, then the energy

$E = nh\nu$. Using the Boltzmann probability distribution, the probability of a state with energy ν is

$$P(E) = B e^{-E/KT} \tag{3}$$

where B is a normalization constant, K is Boltzmann's constant (1.38×10^{-23} joule/K), and T is temperature in Kelvin. Thus the average energy $\overline{E}$ in a mode with frequency ν is

$$\overline{E} = \frac{\sum_{n=0}^{\infty} E P(E)}{\sum_{n=0}^{\infty} P(E)} = \frac{\sum_{n=0}^{\infty} nh\nu \, e^{-nh\nu/KT}}{\sum_{n=0}^{\infty} e^{-nh\nu/KT}} = \frac{h\nu}{e^{h\nu/KT} - 1} \tag{4}$$

The total amount of radiation energy per unit frequency interval and per unit volume is $w(\nu) = N(\nu)\overline{E}/V$. Hence,

$$w(\nu) = \frac{8\pi h \nu^3 (\mu\epsilon)^{3/2}}{e^{h\nu/KT} - 1} \tag{5}$$

To compute radiation intensity, consider a slab of area A and infinitesimal thickness d. Such a volume would contain radiation energy

$$W = 8\pi A d (\mu\epsilon)^{3/2} \frac{h\nu^3}{e^{h\nu/KT} - 1} \tag{6}$$

per unit frequency interval. The radiation power emerging in direction θ within solid angle $d\Omega$ is $2I \cos\theta A d\Omega$, where I is the specific intensity per polarization and the radiation pulse will last for a time interval of $d\sqrt{\mu\epsilon}/\cos\theta$. Thus,

$$W = \int d\Omega \, 2AI \cos\theta \frac{d\sqrt{\mu\epsilon}}{\cos\theta} = I d\sqrt{\mu\epsilon} 8\pi A \tag{7}$$

Equating (6) and (7),

$$I = \mu\epsilon \frac{h\nu^3}{e^{h\nu/KT} - 1} \tag{8}$$

In the Rayleigh-Jean's approximation $h\nu/KT \ll 1$; this gives for a medium with permeability μ and permittivity ϵ

$$I = \frac{KT}{\lambda^2} \frac{\mu\epsilon}{\mu_o \epsilon_o} \tag{9}$$

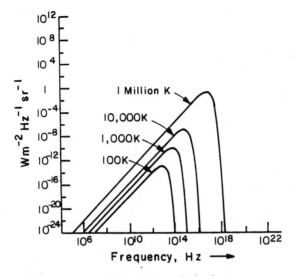

Fig. 1.3 Planck radiative-law curves.

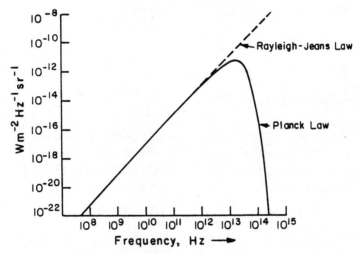

Fig. 1.4 Comparison of Planck's law with Rayleigh-Jean's law.

where $\lambda = c/\nu$ is the free-space wavelength. In free space

$$I = \frac{KT}{\lambda^2} \qquad (10)$$

for each polarization. The specific intensity given by (10) has dimension watts m^{-2} Hz^{-1} Sr^{-1} (power per unit area per unit frequency interval per unit solid angle).

A family of curves of I as a function of frequency for different T is shown on Figure 1.3. The Rayleigh-Jean's law (9) is a low frequency

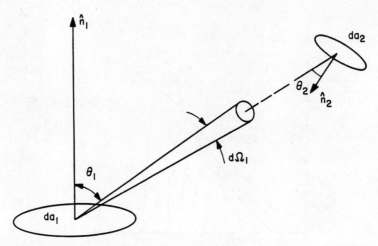

Fig. 1.5 Specific intensity: Transmitting element da_1 and receiving element da_2.

approximation to Planck's law. In Figure 1.4, we compare Planck's law with Rayleigh-Jean's law at $T = 300$ K. The difference is less than 1% if

$$f < 1.2 \times 10^{11} \text{ Hz} \tag{11}$$

Thus the Rayleigh-Jean's law can be used in the microwave region.

2.3 Specific Intensity

The fundamental quantity in the radiative transfer theory is the specific intensity. The amount of power dP flowing within a solid angle $d\Omega_1$ through an elementary area da_1 in a frequency interval $(\nu, \nu + d\nu)$ in terms of specific intensity $I(\bar{r}, \hat{s})$ is given by

$$dP = I(\bar{r}, \hat{s}) \cos\theta_1 da_1 d\Omega_1 d\nu \quad \text{(watts)} \tag{1}$$

where θ_1 is the angle which the direction $\hat{s}$ makes with outward normal $\hat{n}_1$ of da_1 (Figure 1.5). Specific intensity has unit of watts m^{-2} Sr^{-1} Hz^{-1} and, in general, will vary from point to point and also with direction through every point. A radiation field is said to be isotropic if the intensity is independent of direction and homogeneous if the intensity is the same at all points.

Consider a receiving surface with area da_2 in the direction θ_1 at a distance r from da_1 with its normal $\hat{n}_2$ at an angle θ_2 with respect to direction $\hat{s}$ of propagation. Then the angle subtended by da_2 at da_1 is $d\Omega_1 = \cos\theta_2 da_2/r^2$. The power intercepted by da_2 is, using (1)

$$dP_2 = I(\bar{r}, \hat{s}) \cos\theta_1 da_1 \frac{da_2}{r^2} \cos\theta_2 d\nu \tag{2}$$

Brightness Temperature

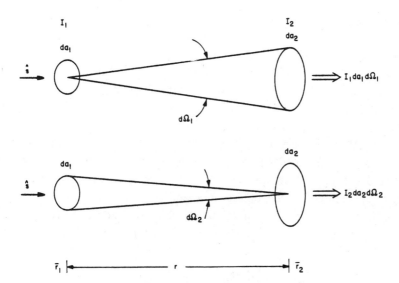

Fig. 1.6 Invariance of propagation of specific intensity.

However,

$$\frac{da_1}{r^2} \cos \theta_1 \equiv d\Omega_2$$

is the angle subtended by da_1 at da_2. Therefore,

$$dP_2 = I(\bar{r}, \hat{s}) d\Omega_2 da_2 \cos \theta_2 d\nu \tag{3}$$

Thus, the expression for power

$$dP = I(\bar{r}, \hat{s}) \cos \theta \, d\Omega \, da \, d\nu \tag{4}$$

is applicable to both transmission and reception of radiation.

Invariance of Specific Intensity in Free Space

Consider intensities I_1 and I_2 at two points $\bar{r}_1$ and $\bar{r}_2$ separated by a distance r along the direction $\hat{s}$ and two areas da_1 and da_2 perpendicular to $\hat{s}$ (Figure 1.6). We can express the power received by da_2 in two different ways. In terms of I_1 it is $I_1 da_1 d\Omega_1 d\nu$. However, in terms of I_2 it should be $I_2 da_2 d\Omega_2 d\nu$. Thus, the two should be equal

$$I_1 da_1 d\Omega_1 d\nu = I_2 da_2 d\Omega_2 d\nu \tag{5}$$

since $d\Omega_1 = da_2/r^2$ and $d\Omega_2 = da_1/r^2$. Therefore, $I_1 = I_2$, which shows that the specific intensity is invariant along the ray path in the free space.

2.4 Brightness Temperature

The absorptivity of a body is defined as the ratio of the total thermal energy absorbed by the surface to the total thermal energy incident upon it. A black body has absorptivity $a = 1$. According to Kirchhoff, the emissivity e of a body is equal to its absorptivity a. Thus, a black body has $e = 1$ and is the most efficient radiator.

Real materials emit less than a black body and the specific intensity emitted generally depends on the direction and the polarization. In passive remote sensing, the radiometer acts as a receiver of the specific intensity $I_\beta(\theta, \phi)$ emitted by the object under observation where β denotes the polarization and (θ, ϕ) denotes the angular dependence. An equivalent radiometer temperature called the brightness temperature $T_{\beta B}(\theta, \phi)$ can be defined as follows:

$$T_{\beta B}(\theta, \phi) = I_\beta(\theta, \phi) \frac{\lambda^2}{K} \tag{1}$$

If the body has uniform physical temperature T, then the emissivity $e_\beta(\theta, \phi)$ is defined as

$$e_\beta(\theta, \phi) = \frac{T_{\beta B}(\theta, \phi)}{T} \tag{2}$$

For a black body, we can substitute (10), Section 2.2, in (1) and (2) and get back the result $e_\beta(\theta, \phi) = 1$. Using (1), Section 2.3, and (1) and (2), the power emitted in a particular direction with polarization β per unit frequency interval is

$$dP_\beta(\theta, \phi) = e_\beta(\theta, \phi) \frac{KT}{\lambda^2} \cos\theta \, da \, d\Omega \, d\nu \tag{3}$$

3 RECIPROCITY RELATION

The reciprocity relation will be used in the derivation of Kirchhoff's law in Section 4. In this section, we will derive the reciprocity relation for a half-space of inhomogeneous and anisotropic medium with rough interface S (Figure 1.7).

Case a

Consider a time harmonic source $\overline{J}_a$ placed in region 0 producing fields $\overline{E}_a$ and $\overline{H}_a$ in region 0 and fields $\overline{E}_{1a}$ and $\overline{H}_{1a}$ in region 1. The time harmonic dependence of $e^{-i\omega t}$ is assumed. The Maxwell's equations are

$$\nabla \times \overline{H}_a = -i\omega\epsilon\overline{E}_a + \overline{J}_a \tag{1}$$

Reciprocity Relation

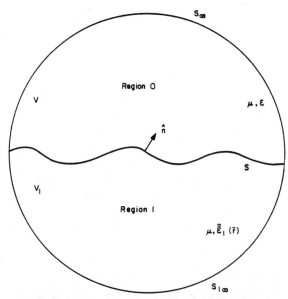

Fig. 1.7 Reciprocity relation for half space of inhomogeneous anisotropic medium with rough interface.

$$\nabla \times \overline{E}_a = i\omega\mu\overline{H}_a \tag{2}$$

$$\nabla \times \overline{H}_{1a} = -i\omega\overline{\overline{\epsilon}}_1(\overline{r}) \cdot \overline{E}_{1a} \tag{3}$$

$$\nabla \times \overline{E}_{1a} = i\omega\mu\overline{H}_{1a} \tag{4}$$

with boundary conditions on S

$$\hat{n} \times \overline{H}_a = \hat{n} \times \overline{H}_{1a} \tag{5}$$

$$\hat{n} \times \overline{E}_a = \hat{n} \times \overline{E}_{1a} \tag{6}$$

Case b

Consider a source $\overline{J}_b$ placed in region 0. Maxwell's equations with source b are

$$\nabla \times \overline{H}_b = -i\omega\epsilon\overline{E}_b + \overline{J}_b \tag{7}$$

$$\nabla \times \overline{E}_b = i\omega\mu\overline{H}_b \tag{8}$$

$$\nabla \times \overline{H}_{1b} = -i\omega\overline{\overline{\epsilon}}_1(\overline{r}) \cdot \overline{E}_{1b} \tag{9}$$

$$\nabla \times \overline{E}_{1b} = i\omega\mu\overline{H}_{1b} \tag{10}$$

with boundary conditions on surface S

$$\hat{n} \times \overline{H}_b = \hat{n} \times \overline{H}_{1b} \tag{11}$$

$$\hat{n} \times \overline{E}_b = \hat{n} \times \overline{E}_{1b} \tag{12}$$

Take the difference of (1) dot multiplied by $\overline{E}_b$ and (8) dot multiplied by $\overline{H}_a$, and the difference of (2) dot multiplied by $\overline{H}_b$ and (7) dot multiplied by $\overline{E}_a$. Also, take the difference of (3) dot multiplied by $\overline{E}_{1b}$ and (10) dot multiplied by $\overline{H}_{1a}$, and the difference of (4) dot multiplied by $\overline{H}_{1b}$ and (9) dot multiplied by $\overline{E}_{1a}$. Then, we arrive at

$$\nabla \cdot (\overline{E}_b \times \overline{H}_a) = i\omega\epsilon\overline{E}_a \cdot \overline{E}_b + i\omega\mu\overline{H}_a \cdot \overline{H}_b - \overline{J}_a \cdot \overline{E}_b \tag{13}$$

$$\nabla \cdot (\overline{E}_a \times \overline{H}_b) = i\omega\mu\overline{H}_a \cdot \overline{H}_b + i\omega\epsilon\overline{E}_a \cdot \overline{E}_b - \overline{E}_a \cdot \overline{J}_b \tag{14}$$

$$\nabla \cdot (\overline{E}_{1b} \times \overline{H}_{1a}) = i\omega\mu\overline{H}_{1a} \cdot \overline{H}_{1b} + i\omega\overline{E}_{1b} \cdot \overline{\overline{\epsilon}}_1(\overline{r}) \cdot \overline{E}_{1a} \tag{15}$$

$$\nabla \cdot (\overline{E}_{1a} \times \overline{H}_{1b}) = i\omega\mu\overline{H}_{1a} \cdot \overline{H}_{1b} + i\omega\overline{E}_{1a} \cdot \overline{\overline{\epsilon}}_1(\overline{r}) \cdot \overline{E}_{1b} \tag{16}$$

Integrate the difference of (13) and (14) over volume V of the upper half space and apply divergence theorem and radiation condition over S_∞, and integrate the difference of (15) and (16) over volume V_1 of the lower half space, and apply divergence theorem and radiation condition over $S_{1\infty}$. Also assume (Kong, 1975)

$$\overline{\overline{\epsilon}}_1^t(\overline{r}) = \overline{\overline{\epsilon}}_1(\overline{r}) \tag{17}$$

where t denotes the transpose of the matrix. We obtain the results

$$\int_S (\overline{E}_a \times \overline{H}_b - \overline{E}_b \times \overline{H}_a) \cdot \hat{n} \, dS = \int dV \, (\overline{E}_a \cdot \overline{J}_b - \overline{J}_a \cdot \overline{E}_b) \tag{18}$$

$$\int_S (\overline{E}_{1a} \times \overline{H}_{1b} - \overline{E}_{1b} \times \overline{H}_{1a}) \cdot \hat{n} \, dS = 0 \tag{19}$$

In view of the boundary conditions (5), (6), (11) and (12), the left-hand sides of (18) and (19) are equal. Hence,

$$\int dV \, (\overline{E}_a \cdot \overline{J}_b - \overline{J}_a \cdot \overline{E}_b) = 0 \tag{20}$$

Thus, the system is reciprocal if the reciprocity condition in (17) is satisfied. We can express the total field as a sum of incident and scattered fields

$$\overline{E}_a = \overline{E}_a^i + \overline{E}_a^s \tag{21}$$

$$\overline{E}_b = \overline{E}_b^i + \overline{E}_b^s \tag{22}$$

Reciprocity Relation

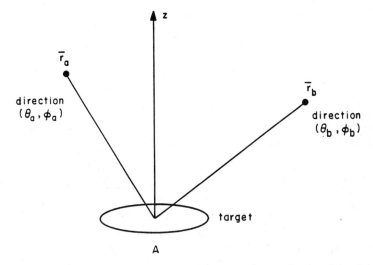

Fig. 1.8 Reciprocity relation for bistatic scattering coefficients in directions (θ_a, ϕ_a) and (θ_b, ϕ_b).

The previous analysis can be repeated for the case when the half-space medium is absent. Hence,

$$\int dV \, (\overline{E}_a^i \cdot \overline{J}_b - \overline{J}_a \cdot \overline{E}_b^i) = 0 \tag{23}$$

Subtracting (23) from (20) gives the reciprocity theorem for the scattered field

$$\int dV \, (\overline{E}_a^s \cdot \overline{J}_b - \overline{J}_a \cdot \overline{E}_b^s) = 0 \tag{24}$$

To derive the reciprocity relation for the bistatic scattering coefficients, we let $\overline{J}_a$ and $\overline{J}_b$ be Hertzian dipoles located at $\overline{r}_a$ and $\overline{r}_b$, respectively, and currents pointing in directions $\hat{\alpha}$ and $\hat{\beta}$, respectively. Thus,

$$\overline{J}_a = \hat{\alpha} Il \, \delta(\overline{r} - \overline{r}_a) \tag{25}$$

$$\overline{J}_b = \hat{\beta} Il \, \delta(\overline{r} - \overline{r}_b) \tag{26}$$

We also let $\overline{J}_a$ and $\overline{J}_b$ be at the far field region in the directions (θ_a, ϕ_a) and (θ_b, ϕ_b), respectively, and at distances r_a and r_b, respectively, from the target (Figure 1.8). Then in the vicinity of the target, the incident field will be in the form of plane waves

$$\overline{E}_a^i = \hat{\alpha} \left[-\frac{i\omega\mu Il}{4\pi r_a} e^{ikr_a} e^{-ik\hat{r}_a \cdot \overline{r}} \right] \tag{27}$$

and

$$\overline{E}_b^i = \hat{\beta}\left[-\frac{i\omega\mu Il}{4\pi r_b} e^{ikr_b} e^{-ik\hat{r}_b\cdot\overline{r}}\right] \tag{28}$$

Substituting (25) and (26) in (24), the reciprocity theorem gives

$$\overline{E}_a^s(\overline{r}_b) \cdot \hat{\beta} = \overline{E}_b^s(\overline{r}_a) \cdot \hat{\alpha} \tag{29}$$

From (4), Section 1, and making use of (27)

$$\gamma_{\beta\alpha}(\theta_b, \phi_b; \theta_a, \phi_a) = \frac{4\pi r_b^2 |\hat{\beta} \cdot \overline{E}_a^s(\overline{r}_b)|^2}{\left(\dfrac{\omega\mu Il}{4\pi r_a}\right)^2 A\cos\theta_a} \tag{30}$$

while using (4), Section 1, and (28)

$$\gamma_{\alpha\beta}(\theta_a, \phi_a; \theta_b, \phi_b) = \frac{4\pi r_a^2 |\hat{\alpha} \cdot \overline{E}_b^s(\overline{r}_a)|^2}{\left(\dfrac{\omega\mu Il}{4\pi r_b}\right)^2 A\cos\theta_b} \tag{31}$$

Hence, in view of (29) through (31), we arrive at the reciprocity relation for bistatic scattering coefficients

$$\cos\theta_a\, \gamma_{\beta\alpha}(\theta_b, \phi_b; \theta_a, \phi_a) = \cos\theta_b\, \gamma_{\alpha\beta}(\theta_a, \phi_a; \theta_b, \phi_b) \tag{32}$$

With slight modification, the method can be used to derive reciprocity relations for multilayered media with rough interfaces separating the layers (Problem 6). Thus, reciprocity holds if the permittivity tensors at all points are symmetric. This is true even in the presence of volume and rough surface scattering.

4 KIRCHHOFF'S LAW

Consider a plane wave with polarization α incident onto a scattering medium with area A. Then the power intercepted by the surface area A is $|E_\alpha^i|^2 A\cos\theta_i/(2\eta)$. The power scattered into the upper hemisphere is equal to

$$\sum_{\beta=v,h} \int_{\text{upper hemisphere}} r^2 \frac{1}{2\eta} |E_{\beta s}|^2 d\Omega_s \tag{1}$$

Thus the fractional power that is absorbed by the surface is given by one minus the fractional power that is scattered back into the upper hemisphere

Kirchhoff's Law

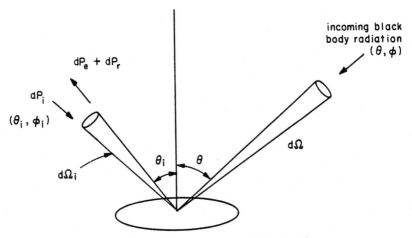

Fig. 1.9 Geometry for derivation of Kirchhoff's law.

and is known as the absorptivity of the surface.

$$a_\alpha(\theta_i, \phi_i) = 1 - \frac{\sum_{\beta=v,h} \int_0^{\pi/2} d\theta_s \sin\theta_s \int_0^{2\pi} d\phi_s\, r^2 |E_{\beta s}|^2}{|E_\alpha^i|^2 A \cos\theta_i}$$

$$= 1 - \frac{1}{4\pi} \sum_{\beta=v,h} \int_0^{\pi/2} d\theta_s \sin\theta_s \int_0^{2\pi} d\phi_s\, \gamma_{\beta\alpha}(\theta_s, \phi_s; \theta_i, \phi_i)$$

(2)

Kirchhoff's law describes the relation between the emissivity and the absorptivity of the body. Consider the body to be in temperature equilibrium with black-body radiation in the half-space above it (Figure 1.9). Under this equilibrium condition, it will be assumed that just as much energy of a given polarization leaves the surface in a given direction as falls upon it from the same direction with the same polarization.

The power incident on the surface with polarization β from direction (θ_i, ϕ_i) is, from (3), Section 2.4,

$$dP_i = \frac{KT}{\lambda^2} d\Omega_i\, A \cos\theta_i\, d\nu \qquad (3)$$

The power leaving the surface in the same direction is the sum of two parts, the thermal emission from the surface, and the external black-body radiation scattered by the surface in direction (θ_i, ϕ_i). The first part is given by

$$dP_e = e_\beta(\theta_i, \phi_i) \frac{KT}{\lambda^2} \cos\theta_i\, A\, d\Omega_i\, d\nu \qquad (4)$$

The second part arises from incoming radiation from all directions (θ, ϕ) that are scattered into the direction (θ_i, ϕ_i) with polarization β. Thus, the power scattered into solid angle $d\Omega_i$ with polarization β is

$$dP_r = \int_{\text{upper hemisphere}} \frac{KT}{\lambda^2} d\Omega \cos\theta \, A$$
$$\times \left\{ \gamma_{\beta v}(\theta_i, \phi_i; \theta, \phi) + \gamma_{\beta h}(\theta_i, \phi_i; \theta, \phi) \right\} \frac{1}{4\pi} d\Omega_i \, d\nu \qquad (5)$$

Since the body is in thermodynamic equilibrium with the half-space above it

$$dP_i = dP_e + dP_r \qquad (6)$$

Equating (3) to the sum of (4) and (5), and further making use of the reciprocity relation in (32), Section 3, we have (Peake, 1959)

$$e_\beta(\theta_i, \phi_i) = 1 - \frac{1}{4\pi} \sum_\alpha \int_0^{\pi/2} d\theta \sin\theta \int_0^{2\pi} d\phi \, \gamma_{\alpha\beta}(\theta, \phi; \theta_i, \phi_i)$$
$$= a_\beta(\theta_i, \phi_i) \qquad (7)$$

Equation (7) is a useful formula that calculates the emissivity from the bistatic scattering coefficient γ. It also relates active and passive remote-sensing measurements.

To illustrate the usefulness of (7), let us consider the case of scattering by a plane dielectric interface. We note that the power reflected will be in the specular direction and has the same polarization as the incident wave and the fractional reflected power is $|R_\beta(\theta_i)|^2$ with $R_\beta(\theta_i)$ the Fresnel reflection coefficient for polarization β and incident direction θ_i. Using (7) we have

$$|R_\alpha(\theta_i)|^2 = \frac{1}{4\pi} \sum_\beta \int d\Omega_s \, \gamma_{\beta\alpha}(\theta_s, \phi_s; \theta_i, \phi_i) \qquad (8)$$

Therefore,

$$\gamma_{\beta\alpha}(\theta_s, \phi_s; \theta_i, \phi_i) = |R_\alpha(\theta_i)|^2 \, 4\pi \, \delta(\cos\theta_s - \cos\theta_i) \, \delta(\phi_s - \phi_i) \delta_{\alpha\beta} \qquad (9)$$

5 CHARACTERISTICS OF PARTICLES IN EARTH TERRAIN

Geophysical media are very often mixtures of different types of particles. They can be characterized by the physical parameters of the particles, such as size, concentration or fractional volume, shape, orientation, and the dielectric property. Recently, because of the applications of microwave remote sensing to earth terrain, extensive theoretical and experimental studies have been made on the physical parameters of the particles and their dielectric properties at microwave frequencies. In subsequent chapters the physical parameters will be used as input parameters in the theoretical models. In the following we will give a brief description of the particles of the geophysical media that are dealt with in the book and refer the readers to references for detailed statistical data and experimental measurements.

Soils

Soil can be regarded as a mixture of water and the bulk material which is a mixture of sand, silt, and clay. The dielectric constant of the bulk material is approximately equal to 3 which is generally much smaller than that of water. The complex relative permittivity of free water with no salinity can be described by the Debye equation

$$\frac{\epsilon}{\epsilon_o} = \epsilon_\infty + \frac{\epsilon_0 - \epsilon_\infty}{1 - i2\pi f \tau} \qquad (1)$$

where ϵ_0 is the static dielectric constant, ϵ_∞ is the optical limit of dielectric constant, τ is the relaxation time, and f is the frequency. For details, refer to Lane and Saxton (1952), Saxton and Lane (1952), and Ray (1972). The dielectric constant of water depends on whether the water is *bound* to the soil particle or is *free*. Free water has a much higher dielectric constant than bound water.

According to the United States Department of Agriculture, the particle diameter of clay is less than $2\,\mu$, the particle diameter of silt is between $2\,\mu$ and $50\,\mu$, and the particle diameter of sand is between $50\,\mu$ and $2000\,\mu$. Thus, for the same volumetric content of moisture, there is more free water in sand than in clay. Hence, the dielectric constant of soil is also dependent on the soil texture. For example Yuma sand is almost exclusively made up of sand and Long Lake clay consists primarily of silt and clay. At 1.4 GHz, the real part of dielectric constant can vary from 3 to 25 and the imaginary part can vary from 0 to 5 with increasing soil moisture (Lundien, 1971; Newton, 1976; Njoku and Kong, 1977; Wang and Schmugge, 1980; Schmugge, 1980; Newton and Rouse, 1980)

The measurements of dielectric constants as a function of moisture content in soils have been carried out over a wide microwave frequency range in the past several years. Some of these measurements were made for many soil samples with widely different textures and thus provided systematic studies on the variation of the dielectric constant with soil texture. As a result of these studies, two distinct features associated with the relation between the soil dielectric constant and moisture content have emerged. First, for all soil samples the dielectric constant increases slowly with moisture content initially. After reaching a transition moisture value, the dielectric constant increases steeply with moisture content. Secondly, the transition moisture is found to vary with soil type or texture, being smaller for sandy soils than for clayey soils. Because of this variation of the transition moisture with soil types, the plots of the measured dielectric contents versus moisture content show differences for soils with different textures. These differences persist whether the moisture content is determined in percent by dry weight or by volume basis.

Because of irrigation, the surfaces of the soils are generally not flat. The surfaces can be classified as smooth, medium rough and rough (Newton, 1976; Newton et al., 1982). The surface can also have row structures (Wang et al., 1980). Thus, study of scattering by periodic surfaces and random rough surfaces are important problems in microwave remote sensing of soils.

Ice

The dielectric properties of ice can be found in Cumming (1952), Evans (1965) and Ray (1972). At microwave frequencies, the real part of the relative permittivity of ice is about 3.2 and the imaginary part vary from 10^{-4} to 0.05. Thus ice is not an absorptive media and depending on the frequency and the size of the particles in ice, it can be a strong scattering media. The dielectric properties of ice is also dependent on its temperature (Ray, 1972). Sea ice is a mixture of ice, salt, brine pockets and air bubbles. The loss tangent of sea ice varies with the ice type depending on whether it is pure ice, first-year sea ice or multi-year sea ice. The dimensions of brine pockets in sea ice are between 0.5 to 3 mm and air bubbles are between 0.1 and 2 mm. Thus volume scattering effects are important.

Snow

Snow is classified as a dense medium because each of the constituents forming snow can occupy an appreciable fractional volume. Dry snow is a mixture of ice and air and wet snow is a mixture of ice, air and water. The amount of water in wet snow is given in terms of snow wetness which is the fractional volume of water f_w in snow. The amount of ice in snow can be calculated from snow density M (g/cm^3). Suppose that the specific density of ice is 10% less than that of water, the fractional volume of ice in snow f_i in terms of M is approximately given by $f_i = M/0.9$. Typical values of f_w run between 0% and 10%. The value of M usually falls between 0.10 g/cm^3 and 0.30 g/cm^3. Particle diameters of ice and water are between

0.1 mm and 2 mm. For hydrological applications, it is desirable to infer the water equivalent from the remote sensing measurements. The water equivalent W (in cm) is the amount of water that remains when the snow is melted. If the snow layer is of thickness d, then the water equivalent W is governed by the approximate relation $W = Md$. Particles do not scatter independently when they are densely distributed. To study the scattering from snow requires a theory of scattering from dense media. At very low frequencies, when the particle sizes are very much smaller than a wavelength and scattering can be ignored, the several constituents of snow contribute to an effective permittivity of snow which can be described by mixture formulas (Maxwell-Garnett, 1904; Polder and van Santern, 1946; Bottcher, 1952) and verified by experimental measurements (Cumming, 1952; Evans, 1965; Linlor, 1980; Ambach and Denoth, 1980; Colbeck, 1982; Tiuri, 1982). For dry snow, the real part of relative permittivity varies from 1.2 to 2.8 depending on the snow density and the imaginary part varies from 10^{-4} to 10^{-2}. For wet snow, the real part of relative permittivity varies from 2 to 6 and imaginary part from 10^{-3} to 1 depending on temperature, wetness and frequency. Thus it is desirable that the theory of scattering by dense media should reduce to a good mixture formula at very low frequency. A study of the geometry and grain structure of snow has been done by Colbeck (1972, 1979, 1982).

Vegetation

Vegetation consists of leaves and stalks embedded in air, such as alfalfa, sorghum, corn, soy beans, wheat, etc. The fractional volume occupied by leaves and stalks per unit volume of vegetation including the air space is between 0.1% and 1%. The particles in vegetation are non-spherical in shape with large aspect ratios. Leaves have the shape of thin disks and stalks assume the form of long slender cylinders. They also have preferred orientation distribution. Scattering by non-spherical particles can give strong depolarization return. Depending on the type of vegetation the thickness of leaves are of the order of 0.1 mm to 1 mm and the surface area can vary from 1 cm^2 to 10^3 cm^2. Stalk diameters vary from 1 cm to 5 cm and the length can vary from 5 cm to 100 cm. For geometry and inclination characteristics of leaves and stalks in sorghum one can refer to Havelka (1971).

Very little theoretical and experimental study has been done on the variation of the permittivity of vegetation as a function of moisture and frequency. One permittivity model that has been used is the following (de Loor, 1968; Fung and Ulaby, 1978).

$$\frac{\epsilon}{\epsilon_o} = \epsilon'_r + i\epsilon''_r \qquad (2)$$

$$\epsilon'_r = 5.5 + \frac{\epsilon_m - 5.5}{1 + (1.85/\lambda)^2} \tag{3}$$

$$\epsilon''_r = (\epsilon_m - 5.5)\frac{1.85/\lambda}{1 + (1.85/\lambda)^2} \tag{4}$$

$$\epsilon_m = 5 + 51.56 V_w \tag{5}$$

$$V_w = \frac{M}{d_w/d_s + M(1 - d_w/d_s)} \tag{6}$$

where λ is the free space wavelength in centimeters, V_w is the volume filling factor of the dispersed granules in leaves, M is the moisture contents by weight, d_s is the density of the solid material, and d_w is the density of water.

Atmosphere

In weather radar, the important quantities are the size, composition, shape, and orientation of aerosols and hydrometeors in the atmosphere. In this section we give a brief summary of the characteristics of aerosols and hydrometeors.

Aerosols are particulate matter suspended in the atmosphere. Examples include smog, smoke, haze, clouds, fog, and fine soil particles. Their sizes are generally under 1 μm in radius. Hydrometeors are the water partricles in solid or liquid form in the atmosphere. Some examples are mist, rain, freezing rain, ice pellets, snow, hail, ocean spray, clouds, and fog, whose sizes are generally 1 μm or more in radius.

The size distribution of rain droplets (Chu and Hogg, 1968) depends on the precipitation rate (rain rate) p, which is normally expressed in millimeters per hour. The spherical droplet with a given diameter $2a$ falls with a certain terminal velocity $v(a)$. Let $n(p,a)\,da$ be the number of droplets per unit volume having radius between a and $a + da$ at the precipitation rate p. Then the precipitation rate p in millimeters per hour is given by

$$p = 1.51 \times 10^7 \int_0^\infty v(a) n(p,a) a^3 \, da \tag{7}$$

where v is measured in meters per second, $n\,da$ in reciprocal cubic meters, a in meters, and $1.51 \times 10^4 = 3600 \times 4\pi/3$. Typical values of p are 0.25 mm/hr (drizzle), 1 mm/hr (light rain), 4 mm/hr (moderate rain), 16 mm/hr (heavy rain), and 100 mm/hr (extremely heavy rain).

The size distribution n can be represented by an empirical expression obtained by Marshall and Palmer (1948)

$$n(p,a) = n_o e^{-\alpha a} \tag{8}$$

where $n_o = 8 \times 10^6 \text{ m}^{-4}$, $\alpha = 8200 \, p^{-0.21}$ (m^{-1}), and p is in millimeters per hour. Note that even though there are more particles with smaller radii, these small particles have a relatively small effect on wave propagation and scattering.

The terminal velocities of raindrops depend on the drop radius. It has also been shown that over the diameter range 1 through 4 mm, the terminal velocity v (m/sec) can be approximated by

$$v = 200.8 \, a^{1/2} \quad (a \text{ in meters}) \tag{9}$$

Another parameter that describes rain is the liquid water content M (g/m^3) which is related to particle size distribution by

$$M = \frac{4\pi}{3} \times 10^6 \int a^3 n(a) \, da \tag{10}$$

The liquid water content is the mass of water contained in one cubic meter.

Cloud droplets are water particles, and their radii are generally smaller than 100 μm. The median radii are typically 2.5 to 5 μm and the number density may vary from 10^6 to 10^9 m^{-3} with a typical value of 10^8 m^{-3}. Typical liquid water content may vary from 0.03 g/m^3 to 1 g/m^3. Thus, generally, the fractional volume occupied by particles is much less than 1 and the atmosphere is a medium with sparse concentration of particles.

References for rain and cloud are Laws and Parson (1943), Marshall and Palmer (1948), Medhurst (1965), Blanchard (1972), Fraser et al. (1975). The shapes of cloud and rain droplets are discussed in Pruppacher and Pitter (1971). The emission and absorption of other atmospheric gases can be found in Waters (1976).

PROBLEMS

1.1 In (4), Section 2.2, we have made use of the relation

$$\frac{\sum_{n=0}^{\infty} nh\nu \, e^{-nh\nu/KT}}{\sum_{n=0}^{\infty} e^{-nh\nu/KT}} = \frac{h\nu}{e^{h\nu/KT} - 1}$$

Prove this relation by making use of the summation of geometric series.

1.2 Consider a rectangular resonator of dimensions a, b, and d in x, y, z directions, respectively. Write down the field solutions for the

TM_{lmn} and TE_{lmn} modes. The T stands for transverse with respect to the z direction. Show that the two modes TM_{lmn} and TE_{lmn} have degenerate resonating frequencies.

1.3 By using (7) and (8), Section 4, show that the emissivity of a half space with a smooth surface is

$$e_\beta(\theta_i, \phi_i) = 1 - |R_\beta(\theta_i)|^2$$

with $\beta = v, h$.

1.4 Consider a dielectric medium with $\epsilon_1 = 2\epsilon_o$ and with temperature $T = 300$ K. Plot the brightness temperature $T_{Bv}(\theta)$ and $T_{Bh}(\theta)$ as a function of θ for θ between 0 deg and 90 deg. At what value of θ will $T_{Bv}(\theta)$ be maximum? Why?

1.5 Consider a two-layer medium with rough interfaces S_1 and S_2. Regions 1 and 2 consist of inhomogeneous and anisotropic media characterized by permittivity and permeability tensors $\bar{\bar{\mu}}_1(\bar{r})$, $\bar{\bar{\epsilon}}_1(\bar{r})$ and $\bar{\bar{\mu}}_2(\bar{r})$, $\bar{\bar{\epsilon}}_2(\bar{r})$, respectively. Derive the reciprocity condition for such a configuration.

1.6 Consider an n-layer medium with adjacent layers separated by rough interfaces. Denote the anisotropic and inhomogeneous permittivity of the lth layer by $\bar{\bar{\epsilon}}_l(\bar{r})$. Derive the reciprocity condition for the n-layer medium.

2

SCATTERING AND EMISSION BY LAYERED MEDIA

1.	Introduction	24
2.	Reflection and Transmission	25
	2.1 Formulation	26
	2.2 Illustrations	29
3.	Dyadic Green's Function	32
	3.1 Integral Representation of Free-Space Dyadic Green's Function	32
	3.2 Dyadic Green's Function for a Two-Layer Medium	34
	3.3 Dyadic Green's Function for Stratified Medium	38
4.	Fluctuation-Dissipation Theorem for Remote Sensing	41
	4.1 Fluctuation-Dissipation Theory	42
	4.2 Brightness Temperature of Stratified Medium	45
5.	Scattering and Emission by Periodic Rough Surfaces	53
	5.1 Formulation	54
	5.2 Coupled Matrix Equations	60
	5.3 Emissivity of Periodic Surfaces	65
6.	Scattering and Emission by Random Rough Surfaces	70
	6.1 Kirchhoff Approach	70
	6.2 Small Perturbation Method	98
	Problems	115

1 INTRODUCTION

In the geophysical application of remote sensing, one is faced with the problem of characterizing a medium in terms of its scattering properties, i.e., scattering cross sections, or its thermal emission spectrum. At microwave frequencies, the scattering and emission from natural surfaces are largely determined by the surface roughness and the inhomogeneous profile of the dielectric constant, temperature, and the volume scattering property. In the remote sensing of soil moisture at microwave frequencies, the nonuniform moisture and temperature profiles in the near-surface region and the rough surface play a dominant role. The effect of subsurface volume scattering plays a secondary role because of high absorption due to soil moisture. In this chapter, the theoretical models applicable to the remote sensing of soil moisture will be presented. Scattering and emission from a layered medium with a nonuniform temperature profile has been studied with the propagating matrices and the fluctuation-dissipation approach. The effect of rough surfaces has been studied by modeling the natural surfaces as a periodic rough surface or as a random rough surface.

In Section 2, the reflection and transmission of an electromagnetic wave incident on a stratified medium is solved using the propagating matrices. For a medium at a uniform temperature, the brightness temperature equals the emissivity times the physical temperature. However, if the temperature of the medium is not uniform, the definition of emissivity loses meaning and other approaches must be sought. In Section 3, the dyadic Green's function for stratified medium will be derived and it will be used in Section 4, together with the fluctuation-dissipation approach, to solve for the microwave thermal emission from a layered medium with nonuniform temperature profiles. It will also be used in Chapter 5 for scattering by layered random medium.

In Section 5 the scattering of electromagnetic waves from a dielectric periodic rough surface is studied for an incident wave with arbitrary incident wave vector. The sinusoidal profile is used to model plowed vegetation fields with row structures. The components of the electric and magnetic fields along the row direction are used as the unknown scalar functions to reduce the vector nature of the problem to a scalar one. Then, the extended boundary condition (EBC) approach with the Fourier series expansion for the surface fields are used to obtain the matrix equations governing the scattered field amplitudes. In general, we find that the E waves, which are characterized by the components of the electric fields along the row direction, and the H waves, which are characterized by the components of the magnetic fields along the row direction, are coupled. Results are illustrated with sinusoidal profiles. The scattered power calculated is shown to satisfy the principles of reciprocity and energy conservation. The emissivity of a periodic rough surface is also calculated from one minus the reflectivity.

In Section 6 the scattering of electromagnetic waves from a randomly rough surface is studied with the Kirchhoff approach (KA) and small perturbation method (SPM). The KA approximates the surface fields using the tangent plane approximation. Under the tangent plane approximation, the fields at any point of the surface are approximated by the fields that would be present on the tangent plane at that point. The SPM assumes that the surface variations are much smaller than the incident wavelength and the slopes of the rough surface are relatively small. The bistatic scattering coefficients for the reflected and transmitted waves are derived using both the KA and SPM. The Kirchhoff-approximated diffraction integral for a dielectric rough surface is still difficult to evaluate analytically and further approximations are usually made. We first expand the integrand, which is a function of local slopes, around the zero slope and then apply integration by parts discarding the edge effect. The integrals can then be evaluated by keeping only a few terms of the expansion. In the high frequency limit, the geometrical optics solution can be obtained using the method of stationary phase. The geometrical optics solution is independent of frequency and states that the scattered intensity is proportional to the probability of the occurrence of the slopes which will specularly reflect or transmit the incident wave into the direction of observation. The SPM is used to calculate the scattered fields up to the second order. The zeroth-order solutions are just the reflected and transmitted fields from a flat surface. The first-order solutions give the lowest-order incoherent transmitted and reflected intensities. However, the first-order solution does not give the depolarization effect in the backscattering direction. The second-order solution gives the lowest-order correction to the coherent reflection and transmission coefficients. The depolarization of the backscattering intensity is also exhibited.

2 REFLECTION AND TRANSMISSION

Reflection and transmission by a layered medium for an incident plane wave is important in both active and passive remote sensing. In this section, we solve for the reflected and transmitted wave amplitudes by a layered media. The thickness and the electromagnetic properties of each layer are variables subject to specification. Any incident polarization can be expressed as a linear combination of the horizontal and vertical polarizations. The horizontally polarized wave is also called a transverse electric (TE) wave because its electric field vector is parallel to the surface of the layer medium

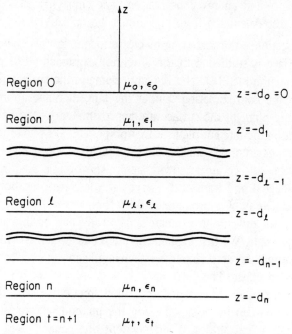

Fig. 2.1 Configuration of a $n+1$ layered medium.

or in optics language, perpendicular to the plane of incidence. The vertically polarized wave is also called a transverse magnetic (TM) wave because its magnetic field vector is perpendicular to the plane of incidence. The derivations are only shown for the TE case, since the results for the TM case can be obtained by duality. The reflection coefficient is expressed in a recurrence relation form that is easy to program in a computer, and the amplitudes of waves in each layer are obtained using the propagating matrices.

2.1 Formulation

Consider a stratified medium with boundaries at $z = -d_0, -d_1, \ldots -d_n$, with $d_0 = 0$ (Figure 2.1). The $(n+1)$-th region is semi-infinite and is labelled t, $t = n + 1$. In any region l, the medium is characterized by permeability μ_l and permittivity ϵ_l which can be complex. A plane wave is incident upon the stratified medium. The polarization of the incident wave is decomposed into horizontal and vertical polarizations which are treated separately. The plane of incidence is determined by the z axis and the incident $\overline{k}$ vector. We will concentrate on the horizontally polarized incident wave case since the solutions for the vertically polarized incident

Formulation

wave can be obtained by duality with the replacements $\overline{E} \to \overline{H}$, $\overline{H} \to -\overline{E}$, and $\epsilon \leftrightarrow \mu$.

For a horizontally polarized incident wave, the electromagnetic field components can be expressed in terms of a single component H_z. Maxwell's equations yield, in any region l,

$$(\nabla_\perp^2 + k_l^2 - k_{lz}^2) H_{lz} = 0 \tag{1}$$

$$\overline{H}_{l\perp} = \frac{1}{k_l^2 - k_{lz}^2} \nabla_\perp \left[\frac{\partial}{\partial z} H_{lz} \right] \tag{2}$$

$$\overline{E}_{l\perp} = \frac{i\omega\mu_l}{k_l^2 - k_{lz}^2} \nabla_\perp \times \overline{H}_{lz} \tag{3}$$

where $k_l = \omega(\mu_l \epsilon_l)^{1/2}$, $k_{lz} = (k_l^2 - |\overline{k}_\perp|^2)^{1/2}$ and $\overline{k}_\perp$ is the component of the incident wave vector $\overline{k}$ in the xy plane, $\nabla_\perp$ and $\nabla_\perp^2$ are the two-dimensional del and Laplacian operators in the xy plane. For a TE wave with $H_z = H_0 e^{-ik_z z} e^{i\overline{k}_\perp \cdot \overline{r}_\perp}$ incident on the stratified medium, the total field in region l can be written as

$$H_{lz} = \left(A_l e^{ik_{lz}z} + B_l e^{-ik_{lz}z} \right) e^{i\overline{k}_\perp \cdot \overline{r}_\perp} \tag{4}$$

There is no subscript l for $\overline{k}_\perp$ because by phase matching, $\overline{k}_\perp$ values in all regions are equal. We note that in region 0 where $l = 0$,

$$A_0 = R H_0 \tag{5}$$
$$B_0 = H_0 \tag{6}$$

In region t where $l = n + 1 = t$, we have

$$A_t = 0 \tag{7}$$
$$B_t = T H_0 \tag{8}$$

We now determine R from the boundary conditions. At the interface $z = -d_l$ which separates l and $l+1$ the tangential $\overline{E}_\perp$ and $\overline{H}_\perp$ fields must be continuous. We obtain

$$k_{lz} \left(A_l e^{-ik_{lz}d_l} - B_l e^{ik_{lz}d_l} \right) = k_{(l+1)z} \left(A_{l+1} e^{-ik_{(l+1)z}d_l} \right.$$
$$\left. - B_{l+1} e^{ik_{(l+1)z}d_l} \right) \tag{9a}$$

$$\mu_l \left(A_l e^{-ik_{lz}d_l} + B_l e^{ik_{lz}d_l} \right) = \mu_{l+1} \left(A_{l+1} e^{-ik_{(l+1)z}d_l} \right.$$
$$\left. + B_{l+1} e^{ik_{(l+1)z}d_l} \right) \tag{9b}$$

There are $n+1$ boundaries and at each boundary there are two equations such as (9). We therefore have $2n+2$ equations to solve for the $2n+2$ unknowns A_l and B_l, $l = 1, 2, \ldots, n$ and A_0 and B_t. Expressing A_l and B_l in terms of A_{l+1} and B_{l+1}, we find from (9)

$$A_l e^{-ik_{lz}d_l} = \frac{1}{2}\left[\frac{\mu_{l+1}}{\mu_l} + \frac{k_{(l+1)z}}{k_{lz}}\right]\left[A_{l+1}e^{-ik_{(l+1)z}d_l} + R_{l(l+1)}B_{l+1}e^{ik_{(l+1)z}d_l}\right] \tag{10a}$$

$$B_l e^{+ik_{lz}d_l} = \frac{1}{2}\left[\frac{\mu_{l+1}}{\mu_l} + \frac{k_{(l+1)z}}{k_{lz}}\right]\left[R_{l(l+1)}A_{l+1}e^{-ik_{(l+1)z}d_l} + B_{l+1}e^{ik_{(l+1)z}d_l}\right] \tag{10b}$$

where

$$R_{l(l+1)} = \frac{\mu_{l+1}k_{lz} - \mu_l k_{(l+1)z}}{\mu_{l+1}k_{lz} + \mu_l k_{(l+1)z}} \tag{11}$$

is the reflection coefficient for horizontal polarization in region l at the interface separating regions l and $l+1$. Forming the ratio of (10a) and (10b), we obtain

$$\frac{A_l}{B_l}e^{-i2k_{lz}d_l} = \frac{\dfrac{A_{l+1}}{B_{l+1}}e^{-i2k_{(l+1)z}d_{l+1}}e^{i2k_{(l+1)z}(d_{l+1}-d_l)} + R_{l(l+1)}}{\dfrac{A_{l+1}}{B_{l+1}}e^{-i2k_{(l+1)z}d_{l+1}}R_{l(l+1)}e^{i2k_{(l+1)z}(d_{l+1}-d_l)} + 1} \tag{12}$$

This equation is a recurrence relation that expresses $(A_l/B_l)e^{-i2k_{lz}d_l}$ in terms of $(A_{l+1}/B_{l+1})e^{-i2k_{(l+1)z}d_{l+1}}$, which can in turn be expressed in terms of $(A_{l+2}/B_{l+2})e^{-i2k_{(l+2)z}d_{l+2}}$ and so on until the transmitted region t is reached where $A_t/B_t = 0$. We can obtain the reflection coefficient for the stratified medium $R_h = A_0/B_0$ by using the following procedure. Since $A_t/B_t = 0$ and $t = n+1$, we first compute $(A_n/B_n)e^{-2ik_{nz}d_n}$ by using (12). Next we calculate $(A_{n-1}/B_{n-1})e^{-2ik_{(n-1)z}d_{n-1}}$ and so on until the zeroth region is reached and $R_h = A_0/B_0$ is calculated.

A dual procedure applies to the derivations for vertically polarized waves. The results are simply obtained by the following replacements: $\epsilon_l \to \mu_l$, $\mu_l \to \epsilon_l$, $\overline{E}_l \to \overline{H}_l$, and $\overline{H}_l \to -\overline{E}_l$. For a layered media with uniaxial permittivity and permeability tensors, a similar closed form solution is available (Kong, 1975) when the optic axes are perpendicular to the interfaces. The reflection coefficient for a few inhomogeneous profiles can be calculated using special functions (Wait, 1970).

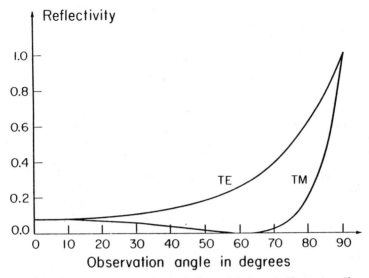

Fig. 2.2 Reflectivity as a function of observation angle for a half-space medium with $\epsilon_t = 3.2(1 + i0.1)\epsilon_o$.

2.2 Illustrations

The reflectivity r is related to emissivity e by

$$e = 1 - r \tag{1}$$

and reflectivity is given by

$$r = |R|^2 \tag{2}$$

A. Reflection from a One-Layer (Half-Space) Medium

The reflection coefficients for the horizontal and vertical polarizations are given by

$$R_h = \frac{\mu_t k_{oz} - \mu_o k_{tz}}{\mu_t k_{oz} + \mu_o k_{tz}} \tag{3a}$$

for horizontal polarization and

$$R_v = \frac{\epsilon_t k_{oz} - \epsilon_o k_{tz}}{\epsilon_t k_{oz} + \epsilon_o k_{tz}} \tag{3b}$$

In Figure 2.2 we illustrate $r_h = |R|^2$ and $r_v = |R_v|^2$ as a function of the incident angle θ_i for $\mu = \mu_o$ and $\epsilon_t = 3.2(1 + i0.1)\epsilon_o$ which corresponds to permittivity of ice (Cumming, 1952; Evans, 1965). The Brewster angle effect is exhibited for the case of vertically polarized waves.

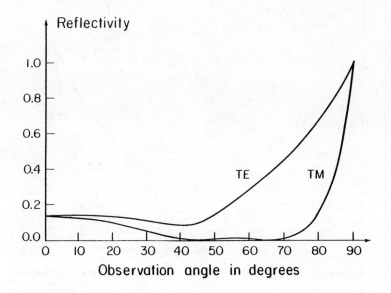

Fig. 2.3 Reflectivity as a function of observation angle for a two-layered medium with $\epsilon_1 = 3.2(1 + i0.1)\epsilon_o$, $d_1 = 50\text{cm}$, $\epsilon_t = 80\epsilon_o$, and frequency at 1 Ghz.

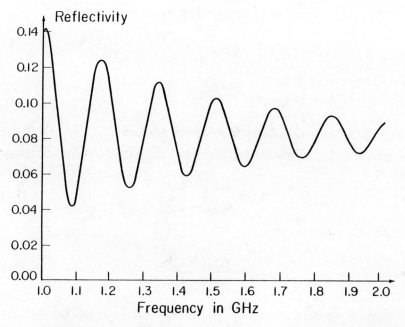

Fig. 2.4 Reflectivity as a function of frequency for a two-layered medium with $\epsilon_1 = 3.2(1 + i0.1)\epsilon_o$, $d_1 = 50$ cm, and $\epsilon_t = 80\epsilon_o$.

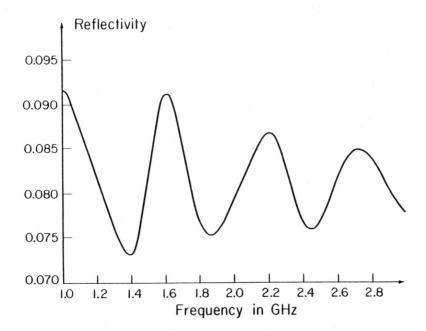

Fig. 2.5 Reflectivity as a function of frequency for a six-layer medium with $d_0 = 0$ cm, $\epsilon_1 = 3.2(1 + i0.1)\epsilon_o$, $d_1 = 50$ cm, $\epsilon_2 = 3.5(1 + i0.2)\epsilon_o$, $d_2 = 25$ cm, $\epsilon_3 = 3.8(1+i0.3)\epsilon_o$, $d_3 = 35$ cm, $\epsilon_4 = 4.0(1+i0.4)\epsilon_o$, $d_4 = 45$ cm, $\epsilon_5 = 4.2(1+i0.5)\epsilon_o$, $d_5 = 50$ cm, and $\epsilon_t = 5.0(1 + i0.6)\epsilon_o$.

B. Reflection from a Two-Layer Medium

$$R = \frac{R_{01} + R_{1t}\, e^{i2k_{1z}d_1}}{1 + R_{01} R_{1t}\, e^{i2k_{1z}d_1}} \tag{4}$$

In Figure 2.3, we illustrate r_h and r_v as a function of observation angle for $\mu_t = \mu_1 = \mu_o$, $\epsilon_1 = 3.2(1 + i0.1)\epsilon_o$, $d_1 = 50$ cm, and $\epsilon_t = 80\epsilon_o$ which corresponds to permittivity of fresh water at the frequency $f = \omega/2\pi = 10^9$ Hz.

C. General Case

We illustrate in Figure 2.4 a plot of reflectivity as a function of frequency for a two-layer stratified medium and in Figure 2.5 a six-layer stratified medium. The interference pattern is a result of constructive and destructive interferences of the upward and downward traveling waves.

3 DYADIC GREEN'S FUNCTION

3.1 Integral Representation of Free-Space Dyadic Green's Function

In this section, the integral representation of the free-space dyadic Green's function is derived. The dyadic Green's function shall be expressed in terms of the basis vectors for TE and TM polarizations, which we apply to stratified medium in the next section. The dyadic Green's function in free space (Tai, 1971, 1973) satisfies

$$\nabla \times \nabla \times \overline{\overline{G}}(\bar{r}, \bar{r}') - k_o^2 \overline{\overline{G}}(\bar{r}, \bar{r}') = \overline{\overline{I}} \delta(\bar{r} - \bar{r}') \tag{1}$$

where $k_o^2 = \omega^2 \mu_o \epsilon_o$ and $\overline{\overline{I}}$ is the unit dyad. The expression for $\overline{\overline{G}}(\bar{r}, \bar{r}')$ is

$$\overline{\overline{G}}(\bar{r}, \bar{r}') = \left[\overline{\overline{I}} + \frac{1}{k_o^2} \nabla \nabla \right] g(\bar{r}, \bar{r}') \tag{2}$$

where

$$g(\bar{r}, \bar{r}') \equiv g(\bar{r} - \bar{r}') = \frac{e^{ik_o|\bar{r} - \bar{r}'|}}{4\pi |\bar{r} - \bar{r}'|} \tag{3}$$

$$(\nabla^2 + k_o^2) g(\bar{r}, \bar{r}') = -\delta(\bar{r} - \bar{r}') \tag{4}$$

Since $g(\bar{r}, \bar{r}')$ depends on $\bar{r} - \bar{r}'$ only, we can temporarily choose $\bar{r}'$ to be the origin and calculate $g(\bar{r})$ as well as $\overline{\overline{G}}(\bar{r})$.

The Fourier transformation gives

$$g(\bar{r}) = \frac{1}{(2\pi)^3} \int d^3\bar{k}\, e^{i\bar{k}\cdot\bar{r}} g(\bar{k}) \tag{5}$$

$$\delta(\bar{r}) = \frac{1}{(2\pi)^3} \int d^3\bar{k}\, e^{i\bar{k}\cdot\bar{r}} \tag{6}$$

Substituting (5) and (6) into (4) with $\bar{r}' \equiv 0$, we obtain

$$g(\bar{k}) = \frac{1}{k^2 - k_o^2} \tag{7}$$

We next integrate (5) over k_z first, by noticing that for $z > 0$ we deform the contour upward such that $Im[k_z] > 0$, and pick up the contribution at the pole $k_z = (k_o^2 - k_\rho^2)^{1/2}$, with $k_\rho^2 = k_x^2 + k_y^2$. Similarly, for $z < 0$, we deform downward, picking up the pole contribution at $-(k_o^2 - k_\rho^2)^{1/2}$. We obtain

$$g(\bar{r}) = \frac{i}{(2\pi)^2} \int d^2\bar{k}_\perp \frac{e^{i\bar{k}_\perp \cdot \bar{r}_\perp + ik_z|z|}}{2k_z} \tag{8}$$

Integral Representation

To find the expression for $\overline{\overline{G}}(\bar{r})$, we note that there is a discontinuity in the derivative with respect to z at $z = 0$.

$$\frac{\partial}{\partial z} g(\bar{r}) = - \int \frac{d^2 \bar{k}_\perp}{(2\pi)^2} e^{i\bar{k}_\perp \cdot \bar{r}_\perp + ik_z|z|} \epsilon(z) \qquad (9)$$

where

$$\epsilon(z) = \begin{cases} 1/2 & z > 0 \\ -1/2 & z < 0 \end{cases}$$

Further differentiation with respect to z will give a dirac delta function in addition to regular terms.

$$\frac{\partial^2}{\partial z^2} g(\bar{r}) = -\delta(\bar{r}) - \frac{i}{(2\pi)^2} \iint dk_x \, dk_y \, \frac{e^{i\bar{k}_\perp \cdot \bar{r}_\perp + ik_z|z|}}{2} k_z \qquad (10)$$

With the extraction of the singular term, the expression for $\overline{\overline{G}}(\bar{r})$ is:

$$\overline{\overline{G}}(\bar{r}) = -\hat{z}\hat{z}\frac{\delta(\bar{r})}{k_o^2} + \begin{cases} \dfrac{i}{8\pi^2} \displaystyle\int d^2\bar{k}_\perp \frac{1}{k_z}\left[\overline{\overline{I}} - \frac{\bar{k}\bar{k}}{k_o^2}\right] e^{i\bar{k}\cdot\bar{r}} & \text{for } z > 0 \\[1em] \dfrac{i}{8\pi^2} \displaystyle\int d^2\bar{k}_\perp \frac{1}{k_z}\left[\overline{\overline{I}} - \frac{\bar{K}\bar{K}}{k_o^2}\right] e^{i\bar{K}\cdot\bar{r}} & \text{for } z < 0 \end{cases} \qquad (11)$$

where

$$\bar{k} = k_x \hat{x} + k_y \hat{y} + k_z \hat{z} \qquad (12a)$$

$$\bar{K} = k_x \hat{x} + k_y \hat{y} - k_z \hat{z} \qquad (12b)$$

We recognize that $\bar{k}/k_o$ is the unit vector $\hat{k}$. Next, we form an orthonormal system of unit vectors $\hat{k}$, $\hat{h}(k_z)$, and $\hat{e}(k_z)$ as follows.

$$\hat{e}(k_z) = \frac{\hat{k} \times \hat{z}}{|\hat{k} \times \hat{z}|} = \frac{1}{k_\rho}(\hat{x} k_y - \hat{y} k_x) \qquad (13a)$$

$$\hat{h}(k_z) = \frac{1}{k_o}\hat{e} \times \bar{k} = \frac{-k_z}{k_o k_\rho}(\hat{x} k_x + \hat{y} k_y) + \frac{k_\rho}{k_o}\hat{z} \qquad (13b)$$

With the orthonormal system, then

$$\overline{\overline{I}} - \hat{k}\hat{k} = \hat{e}\hat{e} + \hat{h}\hat{h} \qquad (14)$$

The dyadic Green's function $\overline{\overline{G}}(\bar{r},\bar{r}')$, by translating the origin to $\bar{r}'$, is given by

$$\overline{\overline{G}}(\bar{r},\bar{r}') = -\hat{z}\hat{z}\frac{\delta(\bar{r},\bar{r}')}{k_o^2}$$

$$+ \begin{cases} \dfrac{i}{8\pi^2}\int d^2\bar{k}_\perp \dfrac{1}{k_z}\left[\hat{e}(k_z)\hat{e}(k_z) + \hat{h}(k_z)\hat{h}(k_z)\right]e^{i\bar{k}\cdot(\bar{r}-\bar{r}')} \\ \qquad\qquad\qquad\qquad\qquad\qquad z > z' \\ \dfrac{i}{8\pi^2}\int d^2\bar{k}_\perp \dfrac{1}{k_z}\left[\hat{e}(-k_z)\hat{e}(-k_z) + \hat{h}(-k_z)\hat{h}(-k_z)\right]e^{i\overline{K}\cdot(\bar{r}-\bar{r}')} \\ \qquad\qquad\qquad\qquad\qquad\qquad z < z' \end{cases}$$

(15)

where

$$\hat{e}(-k_z) = \hat{e}(k_z) \tag{16}$$

$$\hat{h}(-k_z) = \frac{1}{k}\hat{e} \times \overline{K} \tag{17}$$

3.2 Dyadic Green's Function for a Two-Layer Medium

A model for the subsurface of the earth is a two-layer medium. Thus, in this section, we shall derive expressions for the two-layer dyadic Green's function. Let the source be located in region 0 (Figure 2.1). In writing the dyadic Green's function $\overline{\overline{G}}_{lj}(\bar{r},\bar{r}')$, the first subscript l is used to denote the region of the observation point $\bar{r}$ and the second subscript j is used to denote the region of the source $\bar{r}'$. In view of (15), Section 3.1, we have

$$\overline{\overline{G}}_{00}(\bar{r},\bar{r}') = \frac{i}{8\pi^2}\int d\bar{k}_\perp \frac{1}{k_z}\Big\{\left[R^{TE}\hat{e}(k_z)e^{i\bar{k}\cdot\bar{r}} + \hat{e}(-k_z)e^{i\overline{K}\cdot\bar{r}}\right]$$

$$\cdot \hat{e}(-k_z)e^{-i\overline{K}\cdot\bar{r}'} + \left[R^{TM}\hat{h}(k_z)e^{i\bar{k}\cdot\bar{r}} + \hat{h}(-k_z)e^{i\overline{K}\cdot\bar{r}}\right]$$

$$\cdot \hat{h}(-k_z)e^{-i\overline{K}\cdot\bar{r}'}\Big\} \quad \text{for} \quad z < z' \tag{1}$$

$$\overline{\overline{G}}_{10}(\bar{r},\bar{r}') = \frac{i}{8\pi^2}\int d\bar{k}_\perp \frac{1}{k_z}\Big\{\left[A_1\hat{e}_1(k_{1z})e^{i\bar{k}_1\cdot\bar{r}} + B_1\hat{e}_1(-k_{1z})e^{i\overline{K}_1\cdot\bar{r}}\right]$$

$$\cdot \hat{e}(-k_z)e^{-i\overline{K}\cdot\bar{r}'} + \left[C_1\hat{h}_1(k_{1z})e^{i\bar{k}_1\cdot\bar{r}} + D_1\hat{h}_1(-k_{1z})e^{i\overline{K}_1\cdot\bar{r}}\right]$$

$$\cdot \hat{h}(-k_z)e^{-i\overline{K}\cdot\bar{r}'}\Big\} \tag{2}$$

Dyadic Green's Function

$$\overline{\overline{G}}_{20}(\bar{r},\bar{r}') = \frac{i}{8\pi^2} \int d\bar{k}_\perp \frac{1}{k_z} \Big\{ T^{TE} \hat{e}_2(-k_{2z}) e^{i\overline{K}_2 \cdot \bar{r}} \hat{e}(-k_z) e^{-i\overline{K}\cdot\bar{r}'}$$

$$+ T^{TM} \hat{h}_2(-k_{2z}) e^{i\overline{K}_2\cdot\bar{r}} \hat{h}(-k_z) e^{-i\overline{K}\cdot\bar{r}'} \Big\} \tag{3}$$

where

$$\bar{k}_l = k_x \hat{x} + k_y \hat{y} + k_{lz} \hat{z} \tag{4}$$

$$\overline{K}_l = k_x \hat{x} + k_y \hat{y} - k_{lz} \hat{z} \tag{5}$$

$$k_{lz} = (k_l^2 - k_x^2 - k_y^2)^{1/2} \tag{6}$$

In view of phase matching, we have

$$\hat{e}(k_z) = \hat{e}(-k_z) = \hat{e}_1(k_{1z}) = \hat{e}_1(-k_{1z}) \tag{7}$$

We then apply the boundary conditions that $\hat{z} \times \overline{\overline{G}}$ and $\hat{z} \times \nabla \times \overline{\overline{G}}$ are continuous at $z = 0$ and $z = -d$. By comparing the coefficients, we obtain

$$R^{TE} + 1 = A_1 + B_1 \tag{8a}$$

$$k_z(R^{TE} - 1) = k_{1z}(A_1 - B_1) \tag{8b}$$

$$k(R^{TM} + 1) = k_1(C_1 + D_1) \tag{8c}$$

$$\frac{k_z}{k}(R^{TM} - 1) = \frac{k_{1z}}{k_1}(C_1 - D_1) \tag{8d}$$

and

$$A_1 e^{-ik_{1z}d} + B_1 e^{ik_{1z}d} = T^{TE} e^{ik_{2z}d} \tag{9a}$$

$$k_{1z}(A_1 e^{-ik_{1z}d} - B_1 e^{ik_{1z}d}) = -k_{2z} T^{TE} e^{ik_{2z}d} \tag{9b}$$

$$k_1(C_1 e^{-ik_{1z}d} + D_1 e^{ik_{1z}d}) = k_2 T^{TM} e^{ik_{2z}d} \tag{9c}$$

$$\frac{k_{1z}}{k_1}(C_1 e^{-ik_{1z}d} - D_1 e^{ik_{1z}d}) = -\frac{k_{2z}}{k_2} T^{TM} e^{ik_{2z}d} \tag{9d}$$

From the above equations, the constants can all be defined.

Solving (8) and (9) gives the same result as in (20), Section 2.2,

$$R^{TE} = \frac{R_{01} + R_{12} e^{i2k_{1z}d}}{1 + R_{01}R_{12} e^{i2k_{1z}d}} \tag{10a}$$

$$R^{TM} = \frac{S_{01} + S_{12} e^{i2k_{1z}d}}{1 + S_{01}S_{12} e^{i2k_{1z}d}} \tag{10b}$$

where R_{ij} and S_{ij} are respectively the TE and TM reflection coefficients between medium i and j with

$$R_{ij} = \frac{k_{iz} - k_{jz}}{k_{iz} + k_{jz}} \qquad (11a)$$

$$S_{ij} = \frac{\epsilon_j k_{iz} - \epsilon_i k_{jz}}{\epsilon_j k_{iz} + \epsilon_i k_{jz}}. \qquad (11b)$$

If we further define

$$D_2(k_\perp) = 1 + R_{01} R_{12}\, e^{i2k_{1z}d} \qquad (12a)$$

$$F_2(k_\perp) = 1 + S_{01} S_{12}\, e^{i2k_{1z}d} \qquad (12b)$$

and

$$X_{10} = 1 + R_{10} = \frac{2k_{1z}}{k_z + k_{1z}} \qquad (13a)$$

$$Y_{10} = 1 + S_{10} = \frac{2\epsilon_o k_{1z}}{\epsilon_1 k_z + \epsilon_o k_{1z}} \qquad (13b)$$

the following results are obtained

$$B_1 = \frac{2k_z}{k_z + k_{1z}} \frac{1}{D_2(k_\perp)} = \frac{k_z}{k_{1z}} \frac{X_{10}}{D_2(k_\perp)} \qquad (14a)$$

$$A_1 = B_1 R_{12}\, e^{i2k_{1z}d} \qquad (14b)$$

Similarly,

$$D_1 = \frac{k_1}{k} \frac{k_z}{k_{1z}} \frac{Y_{10}}{F_2(k_\perp)} \qquad (15a)$$

$$C_1 = D_1 S_{12}\, e^{i2k_{1z}d} \qquad (15b)$$

and T^{TE} and T^{TM} can be calculated from (9), (14), and (15).

Making use of the symmetric property of the dyadic Green's function (Tai, 1971; Problem 4)

$$\overline{\overline{G}}_{01}(\bar{r},\bar{r}') = \overline{\overline{G}}^{\,t}_{10}(\bar{r}',\bar{r}) \qquad (16)$$

we can calculate $\overline{\overline{G}}_{01}(\bar{r},\bar{r}')$ with source in region 1. With a change of variables, $k_x, k_y \to -k_x, -k_y$, we obtain

$$\overline{\overline{G}}_{01}(\bar{r},\bar{r}') = \int d\bar{k}_\perp\, \overline{\overline{g}}_{01}(\bar{k}_\perp, z, z')\, e^{i\bar{k}_\perp \cdot (\bar{r}_\perp - \bar{r}'_\perp)} \qquad (17a)$$

Dyadic Green's Function

where

$$\bar{\bar{g}}_{01}(\bar{k}_\perp, z, z') = \frac{i}{8\pi^2} \frac{1}{k_{1z}} \left\{ \frac{X_{10}}{D_2(k_\perp)} \hat{e}(k_z) \left[R_{12} e^{i2k_{1z}d} e^{ik_{1z}z'} \hat{e}_1(-k_{1z}) \right. \right.$$

$$\left. + e^{-ik_{1z}z'} \hat{e}_1(k_{1z}) \right] + \frac{k_1}{k_o} \frac{Y_{10}}{F_2(k_\perp)} \hat{h}(k_z)$$

$$\left. \cdot \left[S_{12} e^{i2k_{1z}d} e^{ik_{1z}z'} \hat{h}_1(-k_{1z}) + e^{-ik_{1z}z'} \hat{h}_1(k_{1z}) \right] \right\} e^{ik_z z} \tag{17b}$$

For microwave remote sensing, the observation point $\bar{r}$ in region 0 is in the far field. It is useful to have a far field approximation for $\bar{\bar{G}}_{01}(\bar{r}, \bar{r}')$ with $r \gg r'$. We may evaluate $\bar{\bar{G}}_{01}(\bar{r}, \bar{r}')$ by the stationary-phase method. First cast $\bar{\bar{G}}_{01}(\bar{r}, \bar{r}')$ into the form

$$\bar{\bar{G}}_{01}(\bar{r}, \bar{r}') = \int d\bar{k}_\perp e^{i\bar{k}\cdot\bar{r}} \bar{\bar{F}}(\bar{k}_\perp, z, z') e^{-i\bar{k}_\perp \cdot \bar{r}'_\perp} \tag{18}$$

with

$$\bar{\bar{F}}(\bar{k}_\perp, z, z') = e^{-ik_z z} \bar{\bar{g}}_{01}(\bar{k}_\perp, z, z') \tag{19}$$

Next apply a two-dimensional stationary phase method to the double integral in (18) (Born and Wolf, 1975). The exponent is

$$k_x x + k_y y + (k^2 - k_x^2 - k_y^2)^{1/2} z$$

Then the stationary point is at

$$k_x = k \sin\theta \cos\phi \tag{20a}$$

$$k_y = k \sin\theta \sin\phi \tag{20b}$$

$$k_z = k \cos\theta \tag{20c}$$

where

$$x = r \sin\theta \cos\phi \tag{21a}$$

$$y = r \sin\theta \sin\phi \tag{21b}$$

$$z = r \cos\theta \tag{21c}$$

Hence the far field solution for $\bar{\bar{G}}_{01}(\bar{r}, \bar{r}')$ is

$$\bar{\bar{G}}_{01}(\bar{r}, \bar{r}') = \frac{e^{ikr}}{4\pi r} \left\{ \bar{\bar{H}} e^{-i\bar{k}_1 \cdot \bar{r}'} + \bar{\bar{F}} e^{-i\bar{K}_1 \cdot \bar{r}'} \right\} \tag{22}$$

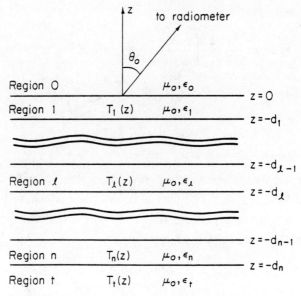

Fig. 2.6 Geometrical configuration of problem.

$$\overline{\overline{H}} = \frac{X_{01}}{D_2} \hat{e}(k_z)\hat{e}_1(k_{1z}) + \frac{k_o}{k_1}\frac{Y_{01}}{F_2} \hat{h}(k_z)\hat{h}_1(k_{1z}) \qquad (23)$$

$$\overline{\overline{F}} = \left[\frac{X_{01}}{D_2} R_{12}\hat{e}(k_z)\hat{e}_1(-k_{1z}) + \frac{k_o}{k_1}\frac{Y_{01}}{F_2} S_{12}\hat{h}(k_z)\hat{h}_1(-k_{1z})\right] e^{i2k_{1z}d} \qquad (24)$$

where the terms $\overline{\overline{F}}, \overline{\overline{H}}, \overline{k}_1$ and $\overline{K}_1$ are to be evaluated at the stationary point given by (20).

3.3 Dyadic Green's Function for Stratified Medium

The approach in Section 3.2 can be readily extended to derive the dyadic Green's function for a general n-layered stratified medium (Tsang et al., 1975). Consider a point source located above a stratified medium (Figure 2.6). The medium in region l is characterized by permittivity ϵ_l. Let region 0 be free space. For $z < z'$, the field in region 0, must be composed of both upward- and downward-going waves. Thus, the dyadic Green's function

Dyadic Green's Function

in this region must assume the following form,

$$\overline{\overline{G}}_{00}(\bar{r},\bar{r}')$$
$$= \frac{i}{8\pi^2}\int d\bar{k}_\perp \frac{1}{k_z}\left\{\left[R^{TE}\hat{e}(k_z)e^{i\bar{k}\cdot\bar{r}}+\hat{e}(-k_z)e^{i\bar{K}\cdot\bar{r}}\right]\hat{e}(-k_z)e^{-i\bar{K}\cdot\bar{r}'}\right.$$
$$\left.+\left[R^{TM}\hat{h}(k_z)e^{i\bar{k}\cdot\bar{r}}+\hat{h}(-k_z)e^{i\bar{K}\cdot\bar{r}}\right]\hat{h}(-k_z)e^{-i\bar{K}\cdot\bar{r}'}\right\} \text{ for } z < z' \quad (1)$$

where R^{TE} and R^{TM} are reflection coefficients to be determined from the boundary conditions. In order that the boundary conditions can be matched at $z = -d_l$, $l = 0, 1, \ldots, n$ for all $\bar{r}'$, the dyadic Green's function in other regions must assume the following form

$$\overline{\overline{G}}_{l0}(\bar{r},\bar{r}')$$
$$= \frac{i}{8\pi^2}\int d\bar{k}_\perp \frac{1}{k_z}\left\{\left[A_l\hat{e}_l(k_{lz})e^{i\bar{k}_l\cdot\bar{r}}+B_l\hat{e}_l(-k_{lz})e^{i\bar{K}_l\cdot\bar{r}}\right]\hat{e}(-k_z)e^{-i\bar{K}\cdot\bar{r}'}\right.$$
$$\left.+\left[C_l\hat{h}_l(k_{lz})e^{i\bar{k}_l\cdot\bar{r}}+D_l\hat{h}_l(-k_{lz})e^{i\bar{K}_l\cdot\bar{r}}\right]\hat{h}(-k_z)e^{-i\bar{K}\cdot\bar{r}'}\right\} \quad (2)$$

in region l, and

$$\overline{\overline{G}}_{t0}(\bar{r},\bar{r}') = \frac{i}{8\pi^2}\int d\bar{k}_\perp \frac{1}{k_z}\left\{T^{TE}\hat{e}_t(-k_{tz})e^{i\bar{K}_t\cdot\bar{r}}\hat{e}(-k_z)e^{-i\bar{K}\cdot\bar{r}'}\right.$$
$$\left.+T^{TM}\hat{h}_t(-k_{tz})e^{i\bar{K}_t\cdot\bar{r}}\hat{h}(-k_z)e^{-i\bar{K}\cdot\bar{r}'}\right\} \quad (3)$$

The coefficients R^{TE}, R^{TM}, A_l, B_l, C_l, D_l, T^{TE} and T^{TM} are related through boundary conditions.

The boundary conditions for the dyadic Green's functions at each boundary are continuity of $\hat{z}\times\overline{\overline{G}}$ and of $\hat{z}\times\nabla\times\overline{\overline{G}}$. Using (2), we find that

$$A_l e^{-ik_{lz}d_l}+B_l e^{ik_{lz}d_l}=A_{l+1}e^{-ik_{(l+1)z}d_l}+B_{l+1}e^{ik_{(l+1)z}d_l} \quad (4)$$

$$k_{lz}\left[A_l e^{-ik_{lz}d_l}-B_l e^{ik_{lz}d_l}\right]=k_{(l+1)z}\left[A_{l+1}e^{-ik_{(l+1)z}d_l}-B_{l+1}e^{ik_{(l+1)z}d_l}\right] \quad (5)$$

$$\frac{k_{lz}}{k_l}\left[C_l e^{-ik_{lz}d_l}-D_l e^{ik_{lz}d_l}\right]=\frac{k_{(l+1)z}}{k_{(l+1)}}\left[C_{l+1}e^{-ik_{(l+1)z}d_l}-D_{l+1}e^{ik_{(l+1)z}d_l}\right] \quad (6)$$

$$k_l\left[C_l e^{-ik_{lz}d_l}+D_l e^{ik_{lz}d_l}\right]=k_{(l+1)}\left[C_{l+1}e^{-ik_{(l+1)z}d_l}+D_{l+1}e^{ik_{(l+1)z}d_l}\right] \quad (7)$$

with $A_0 = R^{TE}$, $B_0 = 1$, $A_t = 0$, $B_t = T^{TE}$, $C_0 = R^{TM}$, $D_0 = 1$, $C_t = 0$, and $D_t = T^{TM}$.

The wave amplitudes can now be determined using the propagating matrices. The reflection coefficients R^{TE} and R^{TM} are dual of each other and can be calculated by the recurrence relation of (12), Section 2.1. However, because we are deriving the electric dyadic Green's function, the amplitudes C_l, D_l, and T^{TM} are not dual of A_l, B_l, and T^{TE}. We define

$$R_{l(l+1)} = \frac{k_{lz} - k_{(l+1)z}}{k_{lz} + k_{(l+1)z}} \tag{8a}$$

$$S_{l(l+1)} = \frac{\epsilon_{l+1} k_{lz} - \epsilon_l k_{(l+1)z}}{\epsilon_{l+1} k_{lz} + \epsilon_l k_{(l+1)z}} \tag{8b}$$

$R_{l(l+1)}$ and $S_{l(l+1)}$ are the reflection coefficients for TE and TM waves, respectively, between regions l and $l+1$. The wave amplitudes in regions $l+1$ and l are related by the TE and TM propagating matrices.

$$\begin{bmatrix} A_{l+1} e^{-ik_{(l+1)z} d_{l+1}} \\ B_{l+1} e^{ik_{(l+1)z} d_{l+1}} \end{bmatrix} = \overline{\overline{V}}^{TE}_{(l+1)l} \begin{bmatrix} A_l e^{-ik_{lz} d_l} \\ B_l e^{ik_{lz} d_l} \end{bmatrix} \tag{9a}$$

and

$$\begin{bmatrix} C_{l+1} e^{-ik_{(l+1)z} d_{l+1}} \\ D_{l+1} e^{ik_{(l+1)z} d_{l+1}} \end{bmatrix} = \overline{\overline{V}}^{TM}_{(l+1)l} \begin{bmatrix} C_l e^{-ik_{lz} d_l} \\ D_l e^{ik_{lz} d_l} \end{bmatrix} \tag{9b}$$

where $\overline{\overline{V}}^{TE}_{(l+1)l}$ is called the TE forward propagation matrix and is given by

$$\overline{\overline{V}}^{TE}_{(l+1)l} = \frac{1}{2}\left(1 + \frac{k_{lz}}{k_{(l+1)z}}\right) \begin{bmatrix} e^{-ik_{(l+1)z}(d_{l+1}-d_l)} & R_{(l+1)l} e^{-ik_{(l+1)z}(d_{l+1}-d_l)} \\ R_{(l+1)l} e^{ik_{(l+1)z}(d_{l+1}-d_l)} & e^{ik_{(l+1)z}(d_{l+1}-d_l)} \end{bmatrix} \tag{10a}$$

and, similarly

$$\overline{\overline{V}}^{TM}_{(l+1)l} = \frac{1}{2}\frac{k_l}{k_{(l+1)}}\left(1 + \frac{\epsilon_{(l+1)}}{\epsilon_l}\frac{k_{lz}}{k_{(l+1)z}}\right) \begin{bmatrix} e^{-ik_{(l+1)z}(d_{l+1}-d_l)} & S_{(l+1)l} e^{-ik_{(l+1)z}(d_{l+1}-d_l)} \\ S_{(l+1)l} e^{ik_{(l+1)z}(d_{l+1}-d_l)} & e^{ik_{(l+1)z}(d_{l+1}-d_l)} \end{bmatrix} \tag{10b}$$

The symmetry relation of dyadic Green's function (Problem 4) is

$$\overline{\overline{G}}_{0l}(\bar{r},\bar{r}') = \overline{\overline{G}}^t_{l0}(\bar{r}',\bar{r}) \tag{11}$$

In view of (3), we thus have

$$\overline{\overline{G}}_{0l}(\vec{r},\vec{r}') = \frac{i}{8\pi^2} \int d\vec{k}_\perp \frac{1}{k_z} e^{i\vec{k}\cdot\vec{r}}$$
$$\left\{ \hat{e}(k_z) \left[A_l \hat{e}_l(-k_{lz}) e^{-i\vec{K}_l\cdot\vec{r}'} + B_l \hat{e}_l(k_{lz}) e^{-i\vec{k}_l\cdot\vec{r}'} \right] \right.$$
$$\left. + \hat{h}(k_z) \left[C_l \hat{h}_l(-k_{lz}) e^{-i\vec{K}_l\cdot\vec{r}'} + D_l \hat{h}_l(k_{lz}) e^{-i\vec{k}_l\cdot\vec{r}'} \right] \right\} \quad (12)$$

A change of variable of integration from k_x, k_y to $-k_x$ and $-k_y$ is made in arriving at the above equation. The above equation will be used to derive the brightness temperature for a stratified medium in Section 4.

4 FLUCTUATION-DISSIPATION THEOREM FOR REMOTE SENSING

It has been shown in Chapter 1 that the intensity of radiation emitted from a medium can be expressed in terms of brightness temperature T_B. The brightness temperature is related to the physical temperature T of the medium and its emissivity as follows:

$$T_B = e(\theta, \phi) T \quad (1)$$

The emissivity is expressed in terms of the reflectivity $r(\theta, \phi)$, defined as the fraction of the power incident on the surface from direction (θ, ϕ) that is re-scattered:

$$e(\theta, \phi) = 1 - r(\theta, \phi) \quad (2)$$

The reflectivity in general involves an integration over the bistatic scattering coefficients of the surface, but for a uniform medium bounded by a smooth surface it simply reduces to the specular Fresnel reflectivity. However, if the temperature of the medium is nonuniform, the definition of emissivity loses meaning and other approaches must be sought. An alternative approach is that of radiative transfer which does not take into account coherent effects of layering.

An approach that takes into account the coherent nature of thermal emission from medium with nonuniform temperature is based on the fluctuation-dissipation theory (Landau and Lifshitz, 1960). In this section, the fluctuation-dissipation approach is developed for application to the remote sensing of areas with nonuniform moisture and temperature profiles. In this development, surface and subsurface scattering will be neglected.

4.1 Fluctuation-Dissipation Theory

Fluctuating electromagnetic fields occur in any dissipative body due to spontaneous local electric and magnetic moments arising from thermally induced random motions of its constituent charges. These thermally generated fields radiate power into the surrounding medium, and in turn power from the surrounding medium will be dissipated in the body and converted into thermal motion of the charges. To compute the radiated power, it is necessary to determine the time and space correlation functions of the fluctuating sources.

The frequency spectrum of the fluctuations can be related to the generalized susceptibility. This is a special case of a general theorem due to Callen and Welton (1951), in which a relation is established between the *impedance* of a general linear system and the fluctuations of appropriate generalized *forces*. This is the so-called fluctuation-dissipation theorem. The derivation of the theorem is well documented in literature (Landau and Lifshitz, 1960), so only the outline and results will be presented here. Let $F(t)$ represent the generalized external force acting on a system, resulting in a change in the energy Hamiltonian of the system:

$$H = AF(t) \qquad (3)$$

The quantity A can be considered as the response of the system and is a function of its position and momentum coordinates, and hence an implicit function of time. It is related to $F(t)$ by

$$A = \int_{-\infty}^{\infty} \chi(\tau) F(t-\tau) d\tau \qquad (4)$$

where $\chi(\tau) = 0$ for $\tau < 0$. The time average rate of change of energy is then

$$\left\langle \frac{\partial H}{\partial t} \right\rangle = \lim_{T \to \infty} \frac{1}{T} \int_{-T/2}^{T/2} A \frac{dF(t)}{dt} \, dt \qquad (5)$$

The functions $A(t)$ and $F(t)$ are considered to be stationary random processes and strictly speaking do not have Fourier transforms. This is circumvented in the usual manner by defining truncated functions that approximate the true functions as time T becomes large, for example:

$$A_T(t) = \begin{cases} A(t) & |t| \leq T/2 \\ 0 & |t| > T/2 \end{cases} \qquad (6)$$

The Fourier transforms can now be defined in the usual manner:

$$A_T(\omega) = \frac{1}{2\pi} \int_{-\infty}^{\infty} A_T(t) e^{i\omega t} \, dt \qquad (7)$$

In what follows, the subscript T will be understood when Fourier transforms are concerned. We may then write

$$A(\omega) = \chi(\omega) F(\omega) \tag{8}$$

where $\chi(\omega)$ is the general susceptibility.

By inserting the Fourier transform into (5), we find

$$\left\langle \frac{\partial H}{\partial t} \right\rangle = \int_{-\infty}^{\infty} G(\omega) \, d\omega \tag{9}$$

where the real part of the spectral power density

$$Re\, G(\omega) = \lim_{T \to \infty} \left[\omega |F(\omega)|^2 \, Im\, \chi(\omega) \right] \left(-\frac{4\pi^2}{T} \right) \tag{10}$$

is a measure of the power dissipation in the medium per $d\omega$.

By a quantum-mechanical procedure, Callen and Welton (1951) obtained another relation for $<\partial H/\partial t>$ in terms of the spontaneous mean-square fluctuations $<A^2>$ of the system in thermal equilibrium. By comparing the two relations, one obtains the result

$$<A^2> = \frac{\hbar}{2\pi} \int_{-\infty}^{\infty} Im\, \chi(\omega) \coth \frac{\hbar \omega}{2KT} \, d\omega \tag{11}$$

where T stands for temperature and $\hbar = h/2\pi$. The derivation can also be carried out classically to obtain the result, valid when $\hbar \omega \gg KT$:

$$<A^2> = \frac{KT}{\pi} \int_{-\infty}^{\infty} \frac{Im\, \chi(\omega)}{\omega} \, d\omega \tag{12}$$

By using the Fourier transform of $A(t)$, (12) can be expressed in terms of the spectral correlation function:

$$<A(\omega) A^*(\omega')> = \frac{KT}{\omega \pi} Im\, \chi(\omega) \delta(\omega - \omega') \tag{13}$$

This result is derived for the case of thermodynamic equilibrium. However, the results can still be applied to a body at nonuniform temperature by considering each differential volume element as a separate absorber and emitter of power at local thermodynamic equilibrium.

Application to Electromagnetic Fluctuations

The presence of fluctuating electric and magnetic moments in a medium necessitates the addition of source terms into Maxwell's equations. We may write these as

$$\overline{\overline{\nabla}} \cdot \overline{E} \equiv \nabla \times \overline{E} = i\omega(\mu \overline{H} + \mu_o \overline{M})$$

$$\overline{\overline{\nabla}} \cdot \overline{H} \equiv \nabla \times \overline{H} = -i\omega(\epsilon \overline{E} + \overline{P}) \qquad (14)$$

where $\overline{M}$ and $\overline{P}$ are the fluctuating magnetic and electric dipole moments whose time average is zero. Rearranging (14), we obtain

$$\begin{bmatrix} \overline{P} \\ \mu_o \overline{M} \end{bmatrix} = \begin{bmatrix} -\epsilon \overline{\overline{I}} & (i/\omega)\overline{\overline{\nabla}} \\ -(i/\omega)\overline{\overline{\nabla}} & -\mu \overline{\overline{I}} \end{bmatrix} \cdot \begin{bmatrix} \overline{E} \\ \overline{H} \end{bmatrix} \qquad (15)$$

To use the general result of (13), the relationship between the electromagnetic quantities under consideration and the quantities A, F, and χ must be established. To do this we may consider an external electromagnetic field (force F) that gives rise to induced dipole moments in the medium (response A). The time average rate of change of energy can be obtained from Poynting's theorem

$$\left\langle \frac{dW}{dt} \right\rangle = \lim_{T \to \infty} \frac{1}{T} \int_{-T/2}^{T/2} \int_V \left(\overline{H} \cdot \frac{\partial \overline{B}}{\partial t} + \overline{E} \cdot \frac{\partial \overline{D}}{\partial t} \right) dV\, dt \qquad (16)$$

In frequency domain, let

$$\overline{B} = \mu \overline{H} + \mu_o \overline{M}, \qquad \overline{D} = \epsilon \overline{E} + \overline{P}$$

By Fourier transform of (16), one obtains the power dissipation analogous to (10)

$$Re\, G(\omega) = \int_V \lim_{T \to \infty} \frac{2}{T} \pi \left[-\omega\, Im\, \{\mu_o \overline{H} \cdot \overline{M}^* + \overline{E} \cdot \overline{P}^* \} \right] dV \qquad (17)$$

By dividing up the medium into small but finite portions ΔV and taking mean values of the fields in each portion, we may express the integral as a sum:

$$\int_V (\,) dV \to \sum_V (\,) \Delta V$$

Comparing (8) with (15) and (10) with (17), and making the generalization of (8),

$$\overline{A} = \overline{\overline{\chi}} \cdot \overline{F}$$

we can now identify the following

$$\overline{A} = \begin{bmatrix} \overline{P} \\ \mu_o \overline{M} \end{bmatrix} \quad (18a)$$

$$\overline{F} = \Delta V \begin{bmatrix} \overline{E} \\ \overline{H} \end{bmatrix} \quad (18b)$$

$$\overline{\overline{\chi}} = \frac{1}{\Delta V} \begin{bmatrix} -\epsilon \overline{\overline{I}} & (i/\omega)\overline{\overline{\nabla}} \\ -(i/\omega)\overline{\overline{\nabla}} & -\mu \overline{\overline{I}} \end{bmatrix} \quad (18c)$$

Thus, for example, the components of $\overline{E}$ and $\overline{H}$ in each separate volume element ΔV make up the components of the vector $\overline{F}$ for the whole medium.

We now make the assumption that the fluctuations are uncorrelated between neighboring volume elements, and taking the limit $\Delta V \to 0$ we obtain the generalization of (13)

$$<\overline{P}(\overline{r},\omega)\overline{P}^*(\overline{r}',\omega')> = -\frac{iKT}{2\pi\omega}(\epsilon - \epsilon^*)\delta(\omega - \omega')\delta(\overline{r} - \overline{r}')\overline{\overline{I}} \quad (19a)$$

$$\mu_o^2 <\overline{M}(\overline{r},\omega)\overline{M}^*(\overline{r}',\omega')> = -\frac{iKT}{2\pi\omega}(\mu - \mu^*)\delta(\omega - \omega')\delta(\overline{r} - \overline{r}')\overline{\overline{I}} \quad (19b)$$

$$\mu_o <\overline{P}(\overline{r},\omega)\overline{M}^*(\overline{r}',\omega')> = 0 \quad (19c)$$

The limit is taken in the macroscopic sense, so that the fluctuations in fact become uncorrelated over distances of a few molecular radii. We shall be concerned henceforth with media that have no magnetic loss. Thus equation (19b) also becomes zero.

Therefore, the thermal radiation is generated by an equivalent current source $\overline{J}(\overline{r},\omega) = -i\omega \overline{P}(\overline{r},\omega)$, with expectation value

$$<\overline{J}(\overline{r},\omega)\overline{J}^*(\overline{r}',\omega')> = \frac{4}{\pi}\omega\epsilon''(\overline{r})KT(\overline{r})\overline{\overline{I}}\delta(\omega - \omega')\delta(\overline{r} - \overline{r}') \quad (20)$$

where $\epsilon''(\overline{r})$ is the imaginary part of the permittivity and only positive frequencies are considered in (20).

4.2 Brightness Temperature of Stratified Medium

In passive remote sensing with microwaves, the measured quantity from a radiometer is the brightness temperature which is determined by the geometrical configuration, the medium properties, and the temperature distributions of the area under observation. With the model of a vertically structured medium, Stogryn (1970) used the fluctuation-dissipation theorem and formulated the solution for the brightness temperature in terms of a two-point boundary value problem by solving a second-order differential equation, together with the evaluation of an integral. In this section, the propagation matrix formulation is used to solve for the brightness temperature of a stratified medium (Tsang et al., 1975; Njoku and Kong, 1977). A similar approach has also been employed by Wilheit (1978). The problem is an important one because inhomogeneous permittivities and nonuniform temperature profiles can be conveniently approximated by a stratified medium with many layers. On the other hand, analytical results can only be obtained for very few media with simple profiles (Wait, 1970). The solutions for the stratified medium are expressed in closed forms and numerical examples are given for different temperature and permittivity profiles.

Consider a stratified medium consisting of n layers as shown in Figure 2.6. From (20), Section 4.1, thermal radiation in region l is generated by an equivalent current source $\overline{J}_l(\overline{r}, \omega)$ with expectation values

$$<\overline{J}_l(\overline{r},\omega)\overline{J}_l^*(\overline{r}',\omega')> = \frac{4}{\pi}\omega\epsilon_l'' K T_l \overline{\overline{I}} \delta(\omega-\omega')\delta(\overline{r}-\overline{r}') \qquad (1)$$

where K is the Boltzmann constant and $<>$ denotes ensemble average.

The expected value of the energy density of the radiation with polarization $\hat{p}$ is, in free space,

$$U(\overline{r}) = \frac{1}{2}\epsilon_o <|\hat{p}\cdot\overline{\mathcal{E}}|^2> \qquad (2)$$

where

$$\overline{\mathcal{E}}(\overline{r},t) = \int_o^\infty d\omega \int d^3k\, \overline{E}(\overline{k},\omega)\, e^{i(\overline{k}\cdot\overline{r}-\omega t)} \qquad (3)$$

The energy density is related to the specific intensity, in free space, by

$$U(\overline{r}) = \int_o^\infty \frac{d\omega}{2\pi} \int d\Omega\, \frac{I(\overline{r})}{c} \qquad (4)$$

where $I(\overline{r})$ is specific intensity and c is the speed of light in free space. Combining (1) through (4), we have, since $d^3k = dk\, k^2 d\Omega$ and brightness temperature $T_B = I\lambda^2/K$,

$$T_{Bp}(\hat{k},\omega) = \frac{(2\pi)^3}{K}\left(\frac{c}{\omega}\right)^2 \frac{1}{2}c\epsilon_o \int_o^\infty d\omega' \int_o^\infty k^2\, dk \int_{-\infty}^\infty d^3k'$$

$$\times \left\{ \left(\hat{p}\cdot <\overline{E}(\overline{k},\omega)\overline{E}^*(\overline{k}',\omega')>\cdot\hat{p}\right) e^{i(\overline{k}-\overline{k}')\cdot\overline{r}-i(\omega-\omega')t} \right\} \qquad (5)$$

for $\hat{p}$ polarization waves.

Once the electric field $\overline{E}(\overline{k},\omega)$ in free space has been determined, the brightness temperature can be found from (5). The determination of $\overline{E}(\overline{k},\omega)$ in the presence of the current source given by (1) and the medium properties poses a well-defined electromagnetic boundary value problem.

Using dyadic Green's function, the electric field in region 0 is given by

$$\overline{E}(\overline{r},\omega) = \sum_{l=1}^{t} \int d\overline{r}'_\perp \int_{-d_l}^{-d_{l-1}} dz'\, i\omega\mu\, \overline{\overline{G}}_{0l}(\overline{r},\overline{r}') \cdot \overline{J}_l(\overline{r}') \qquad (6)$$

where $t = n+1$ and $d_t \to \infty$. The dyadic Green's function for a stratified medium is derived in Section 3.3 and we have on introducing (12) (Section 3.3), into (6)

$$\overline{E}(\overline{k},\omega) = -\frac{\omega\mu}{8\pi^2} \sum_{l=1}^{t} \int d\overline{r}'_\perp \int_{-d_l}^{-d_{l-1}} dz'\, \delta(k_z - \sqrt{\omega^2\mu\epsilon - k_x^2 - k_y^2})\frac{1}{k_z}$$

$$\times \left\{ \hat{e}(k_z)\left[A_l\, \hat{e}_l(-k_{lz})\, e^{-i\overline{K}_l \cdot \overline{r}'} + B_l\, \hat{e}_l(k_{lz})\, e^{-i\overline{k}_l \cdot \overline{r}'}\right] \right.$$

$$\left. + \hat{h}(k_z)\left[C_l\, \hat{h}_l(-k_{lz})\, e^{-i\overline{K}_l \cdot \overline{r}'} + D_l\, \hat{h}_l(k_{lz})\, e^{-i\overline{k}_l \cdot \overline{r}'}\right]\right\} \cdot \overline{J}_l(\overline{r}')$$

$$(7)$$

Using (1) and (5), we find the brightness temperature to be

$$T_{Bh}(\hat{k},\omega) = k^3 \cos\theta_o \sum_{l=1}^{t} \frac{\epsilon_l''}{\epsilon_o} \int_{-d_l}^{-d_{l-1}} dz'\, T_l(z')$$

$$\times \left|\frac{1}{k_z}\left[A_l\, \hat{e}_l(-k_{lz})\, e^{ik_{lz}z'} + B_l\, \hat{e}_l(k_{lz})\, e^{-ik_{lz}z'}\right]\right|^2 \qquad (8)$$

$$T_{Bv}(\hat{k},\omega) = k^3 \cos\theta_o \sum_{l=1}^{t} \frac{\epsilon_l''}{\epsilon_o} \int_{-d_l}^{-d_{l-1}} dz'\, T_l(z')$$

$$\times \left|\frac{1}{k_z}\left[C_l\, \hat{h}_l(-k_{lz})\, e^{ik_{lz}z'} + D_l\, \hat{h}_l(k_{lz})\, e^{-ik_{lz}z'}\right]\right|^2 \qquad (9)$$

In the derivation of (8) and (9), the relation $\delta\left(k_z - \sqrt{\omega^2\mu\epsilon - k_x^2 - k_y^2}\right) = U(k_z)\cos\theta_o\delta(k - \omega\sqrt{\mu\epsilon})$ is used, where $U(k_z) = 1$ for $k_z > 0$ and $U(k_z) = 0$ for $k_z < 0$.

The calculation of the integral in (8) and (9) is straightforward. Without loss of generality, we let the plane of observation be the x-z plane and set $k_y = 0$. We allow the temperature $T_l(z')$ in region l to be a constant in that region. Carrying out the integrations in (9), we find the brightness temperature as observed from a radiometer at an angle θ_o to be

$$T_{Bh} = \frac{k}{\cos\theta_o} \sum_{l=1}^{n} \frac{\epsilon_l'' T_l}{2\epsilon_o} \left\{ \frac{|A_l|^2}{k_{lz}''} \left(e^{2k_{lz}''d_l} - e^{2k_{lz}''d_{l-1}}\right) \right.$$

$$- \frac{|B_l|^2}{k_{lz}''} \left(e^{-2k_{lz}''d_l} - e^{-2k_{lz}''d_{l-1}}\right)$$

$$+ \frac{iA_l B_l^*}{k_{lz}'} \left(e^{-i2k_{lz}'d_l} - e^{-i2k_{lz}'d_{l-1}}\right)$$

$$\left. - \frac{iA_l^* B_l}{k_{lz}'} \left(e^{i2k_{lz}'d_l} - e^{i2k_{lz}'d_{l-1}}\right) \right\}$$

$$+ \frac{k}{\cos\theta_o} \frac{\epsilon_t'' T_t}{2\epsilon_o k_{tz}''} |T^{TE}|^2 e^{-2k_{tz}''d_n} \qquad (10a)$$

for horizontal polarization, and

$$T_{Bv}(\theta_o) = \frac{k}{\cos\theta_o} \sum_{l=1}^{n} \frac{\epsilon_l'' T_l (|k_{lz}|^2 + k_x^2)}{2\epsilon_o |k_l|^2}$$

$$\times \left\{ \frac{|C_l|^2}{k_{lz}''} \left(e^{2k_{lz}''d_l} - e^{2k_{lz}''d_{l-1}}\right) - \frac{|D_l|^2}{k_{lz}''} \left(e^{-2k_{lz}''d_l} - e^{-2k_{lz}''d_{l-1}}\right) \right.$$

$$+ \frac{|k_{lz}|^2 - k_x^2}{|k_{lz}|^2 + k_x^2} \frac{C_l D_l^*}{ik_{lz}'} \left(e^{-i2k_{lz}'d_l} - e^{-i2k_{lz}'d_{l-1}}\right)$$

$$\left. - \frac{|k_{lz}|^2 - k_x^2}{|k_{lz}|^2 + k_x^2} \frac{C_l^* D_l}{ik_{lz}'} \left(e^{i2k_{lz}'d_l} - e^{i2k_{lz}'d_{l-1}}\right) \right\}$$

$$+ \frac{k}{\cos\theta_o} \frac{\epsilon_t'' T_t(|k_{tz}|^2 + k_x^2)}{2\epsilon_o k_{tz}''|k_t|^2} |T^{TM}|^2 e^{-2k_{tz}''d_n} \qquad (10b)$$

for vertical polarization where $k_x = k\sin\theta_o$. In the derivation of (10), we made use of the identities $2k_{tz}' k_{tz}'' = \omega^2 \mu \epsilon_t''$ and $|k_{tz}|^2 + k_x^2 = \omega^2\mu(\epsilon_t' k_{tz}' + \epsilon_t'' k_{tz}'')/k_{tz}'$. The procedure for evaluating these expressions is as follows: (1) the surface reflection coefficients R^{TE} and R^{TM} are evaluated by the recurrence relation method as given in (12), Section 2.1; (2) the propagation matrix formalism of (9), Section 3.3, is used to calculate the upward and downward wave amplitudes A_l, B_l, C_l, and D_l in each layer, and the transmitted wave amplitudes in the bottom layer T^{TE} and T^{TM}. Since $A_o = R^{TE}$, $B_o = 1$, $C_o = R^{TM}$ and $D_o = 1$ are known, we can use (9),

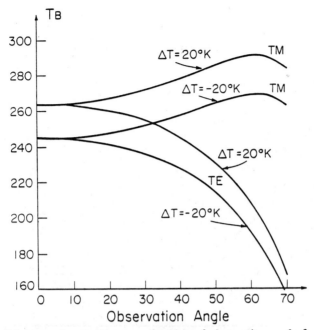

Fig. 2.7 Brightness temperature as a function of observation angle for horizontal polarized (TE) and vertical polarized (TM) waves.

Section 3.3, to calculate A_1, B_1, C_1 and D_1 and then A_2, B_2, C_2 and D_2, etc. (3) the temperature T_l and permittivity ϵ_l in each layer are used to perform the summation with the wave amplitudes previously obtained, as done in (10a) and (10b).

When the medium is of constant temperature T, then the brightness temperatures are given simply by

$$T_{Bh}(\theta_o) = (1 - |R^{TE}|^2) T \qquad (11a)$$

$$T_{Bv}(\theta_o) = (1 - |R^{TM}|^2) T \qquad (11b)$$

The solution of a stratified medium is an important one because inhomogeneous permittivity and temperature profiles can be conveniently approximated by a stratified medium by including a sufficient number of layers. Thus with the formalism in this section, brightness temperatures for a stratified and/or continuous profile can be calculated. The results of the coherent approach presented here and those of the radiative transfer approach (Newton, 1976; Burke et al., 1979) have been compared (Schmugge and Choudhury, 1981).

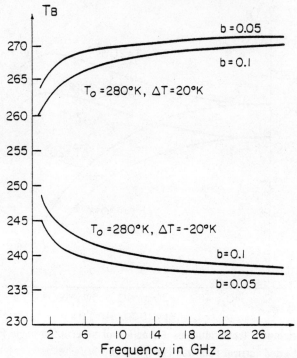

Fig. 2.8 Brightness temperature as a function of frequency.

Illustrations

In Figure 2.7 we examine the angular dependence of the brightness temperature. At 1 GHz, T_{Bh} and T_{Bv} are plotted as functions of observation angle for the following profile:

$$\frac{\epsilon_1(z)}{\epsilon_o} = 9.0(1 + i0.3) - (5.5 + i2.5)\, e^{az}$$

$$\epsilon_t = \epsilon_1(-d)$$

$$T(z) = T_o + \Delta T\, e^{bz}$$

$$T_o = 280 \text{ K}$$

$$T_t = T(-d)$$

with $a = 0.02\,\text{cm}^{-1}$, $b = 0.05\,\text{cm}^{-1}$, $d = 30\,\text{cm}$, and $\Delta T = \pm 20\text{K}$. It is interesting to observe a maximum for vertical polarization similar to the Brewster angle for a uniform half-space medium.

In Figure 2.8, we plot the brightness temperature as observed from nadir as a function of frequency for the following permittivity and temper-

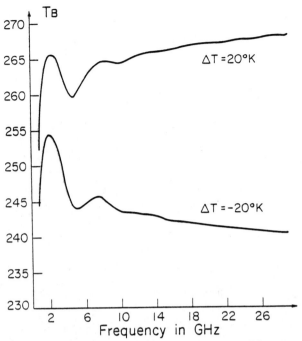

Fig. 2.9 Brightness temperature as a function of frequency.

ature profile:

$$\frac{\epsilon_1(z)}{\epsilon_o} = 9.0(1 + i0.3) - (5.5 + i2.5)\,e^{az}$$

$$\epsilon_t = \epsilon_1(-d)$$

$$T = T_o + \Delta T\,e^{bz},$$

$$T_t = T(-d)$$

with $a = 0.02\,\text{cm}^{-1}$, $d = 30\,\text{cm}$, $T_o = 280\,\text{K}$, $\Delta T = \pm 20\,\text{K}$, and $b = 0.05\,\text{cm}^{-1}$ and $0.1\,\text{cm}^{-1}$. In the calculations, the temperature and the permittivity profiles are stratified into 350 layers from 0 to $z = -d$. We see that at low frequencies, the subsurface temperature affects T_B more than at high frequencies. At very low frequencies the brightness temperatures for different values of parameters a and b approach the same value dictated by T_o and the effective emissivity of the medium.

In Figure 2.9 the brightness temperature as observed from nadir is plotted as a function of frequency for the following profiles:

$$\frac{\epsilon_1(z)}{\epsilon_o} = 9.0(1 + i0.1) - (5.5 + i0.83)\,e^{az}$$

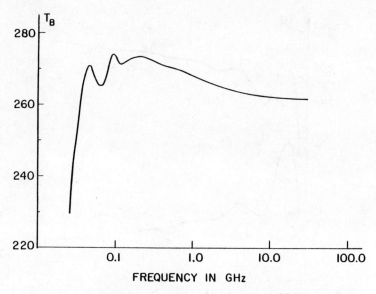

Fig. 2.10 Brightness temperature as a function of frequency

$$\frac{\epsilon_t(z \leq -20\,\text{cm})}{\epsilon_o} = 9.0(1 + i0.05)$$

$$T(z) = T_o + \Delta T\, e^{bz}$$

$$T_t = T(z = -20\,\text{cm})$$

with $a = 0.02\,\text{cm}^{-1}$, $b = 0.1\,\text{cm}^{-1}$, $T_o = 280$ K, and $\Delta = \pm 20$ K. Note that there is an abrupt change in the permittivity profile at $z = -20$ cm. At low frequencies, we see interference effects. For higher frequencies, the brightness temperature is determined by $T_o \pm \Delta T$ times the effective surface emissivity of the medium.

The interference effect at low frequencies is not characteristic of the abrupt change of the permittivity profile. For a continuous permittivity profile the interference is illustrated in Figure 2.10, where

$$\frac{\epsilon_1(z)}{\epsilon_o} = (2.88 + i0.34)\, e^{-az}$$

$$\epsilon_t = \epsilon_1(z = -1\,\text{m})$$

$$T(z) = 300 - 20\, e^{bz} \qquad -\infty < z < 0$$

$a = 2\,\text{m}^{-1}$ and $b = 3\,\text{m}^{-1}$. In the calculations, the region from $z = 0$ to $z = -1$ m is stratified into 350 layers. The temperature profile from $z = -1$ m to $-\infty$ is not stratified. For this profile, the exact solution can be obtained and is in terms of Bessel functions (Tsang et al., 1975). The result calculated with the stratified model is found to agree very well with that calculated with the exact formulas.

The model discussed in this section can be readily applied to calculate the brightness temperatures of soils. In recent years, extensive radiometer measurements of soil emission have been made with radiometers mounted on trucks or aircraft (Schmugge et al., 1974; Newton, 1976; Njoku and Kong, 1977; Newton and Rouse, 1980; Ulaby et al., 1981; Jackson and Schmugge, 1981; Njoku and O'Neill, 1981; Burke and Schmugge, 1982; Wang et al., 1983). The permittivities of soils as a function of soil texture and moisture content have also been studied (von Hippel, 1954; Hipp, 1974; Hoekstra and Delaney, 1974; Wang, 1980; Wang and Schmugge, 1980; Newton and Rouse, 1980).

5 SCATTERING AND EMISSION BY PERIODIC ROUGH SURFACES

The scattering of waves from a periodic surface has been studied extensively. The theory is used in the study of the laser light scattering from optical gratings (Hutley and Bird, 1973; Toigo et al., 1977; Chuang and Kong, 1981), the acoustic wave scattering from hard or soft boundaries (DeSanto, 1975), and the scattering of atoms from crystal surfaces (Masel et al., 1975; Goodman, 1977; Garcia and Cabrera, 1978). For the application to remote sensing of plowed vegetation fields with row structures, the model of the scattering of electromagnetic waves from a dielectric periodic rough surface for an incident wave with arbitrary incident wave vector can be used. In this section, the theory for the solution to such problem is illustrated. The components of the electric and magnetic fields along the row direction are used as unknown scalar functions to reduce the vector nature of the problem to a scalar one. Then, the extended boundary condition (EBC) (Waterman, 1975) approach with Fourier series expansion for the surface fields is used to obtain the matrix equations governing the scattered field amplitudes. In general, the E waves, which are characterized by the components of the electric fields along the row direction, and the H waves, which are characterized by the components of the magnetic fields along the row direction, are coupled together. Results are illustrated with sinusoidal profiles. The scattered power calculated is shown to satisfy the principles of reciprocity and energy conservation. The emissivity of a periodic rough surface is also calculated from one minus the reflectivity and compared with experimental data (Kong et al., 1984).

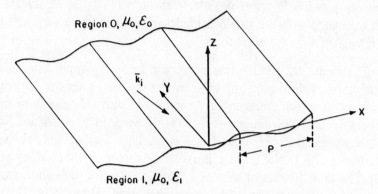

Fig. 2.11 Geometrical configuration of the problem.

5.1 Formulation

Consider a plane wave incident upon a periodic surface described by $f(x) = f(x+P)$, with P denoting the period of the surface in the $\hat{x}$ direction (Fig. 2.11). The electric field of the incident wave is given by

$$\overline{E}_i = \hat{e}_i E_o e^{i\overline{k}_i \cdot \overline{r}} \qquad (1)$$

where $\overline{k}_i$ denotes the incident wave vector and is equal to $\hat{x}k_{xi} + \hat{y}k_{yi} - \hat{z}k_{zi}$ and $\hat{e}_i$ is the polarization of the electric field vector.

Since the structure is uniform in the $\hat{y}$ direction, all the field components in both region 0 and 1 will have the same $\exp(ik_{yi}y)$ dependence. With this dependence, we can replace $\partial/\partial y$ in Maxwell's equation by ik_{yi}. It is possible to express all field components transverse to $\hat{y}$ in terms of longitudinal field components parallel to the y axis. This is in complete analogy to waveguide theory (Kong, 1975). Unless otherwise specified, we will suppress the $\exp(ik_{yi}y)$ dependence in our subsequent theoretical development.

In terms of the longitudinal components, the transverse components are

$$\overline{E}_{js}(\overline{r}) = \frac{i}{k_j^2 - k_{yi}^2}\left[k_{yi}\nabla_s E_{jy}(\overline{r}) + \omega\mu\nabla_s \times \overline{H}_{jy}(\overline{r})\right] \qquad (2a)$$

$$\overline{H}_{js}(\overline{r}) = \frac{i}{k_j^2 - k_{yi}^2}\left[k_{yi}\nabla_s H_{jy}(\overline{r}) - \omega\epsilon_j\nabla_s \times \overline{E}_{jy}(\overline{r})\right] \qquad (2b)$$

where $j = 0, 1$ signify regions 0 and 1, respectively, ∇_s is the transverse gradient operator that is transverse to $\hat{y}$ direction

$$\nabla_s = \hat{x}\frac{\partial}{\partial x} + \hat{z}\frac{\partial}{\partial z} \qquad (3)$$

Formulation

and $\overline{E}_{js}$ and $\overline{H}_{js}$ denote the transverse components of the electric and magnetic fields for region j. The subscript $j = 0$ is sometimes suppressed. The E waves are described by $H_{jy} = 0$ and H waves are described by $E_{jy} = 0$. The longitudinal components E_{jy} and H_{jy} satisfy the partial differential equations

$$\left(\nabla_s^2 + k_j^2 - k_{yi}^2\right) \left\{ \begin{array}{c} E_{jy} \\ H_{jy} \end{array} \right\} = 0 \qquad (4)$$

We let

$$\overline{\rho}_s = \hat{x}x + \hat{z}z \qquad (5)$$

to be the position vector that is transverse to the $\hat{y}$ direction. We recognize that E_{jy} and H_{jy} satisfy a two-dimensional wave equation. Thus, we shall use a two-dimensional Green's function to formulate Huygen's principle and the extinction theorem. The Green's function is

$$G_j(\overline{\rho}_s, \overline{\rho}_s') = \frac{i}{4} H_o^{(1)}(k_{js}|\overline{\rho}_s - \overline{\rho}_s'|)$$

$$= \frac{i}{4\pi} \int_{-\infty}^{\infty} dk_x \, \frac{1}{\sqrt{k_{js}^2 - k_x^2}}$$

$$\times \exp\left[ik_x(x - x') + i\sqrt{k_{js}^2 - k_x^2}\,|z - z'|\right] \qquad (6)$$

where $j = 0, 1$, and

$$k_{js} = \left(k_j^2 - k_{yi}^2\right)^{1/2} \qquad (7)$$

Using scalar Green's theorem, the extinction theorem (Waterman, 1975) can be derived and making use of Green's function of region 0, we have

$$E_{yi}(\overline{\rho}_s) - \int_{-\infty}^{\infty} d\sigma' \left\{ G(\overline{\rho}_s, \overline{\rho}_s') \hat{n} \cdot \nabla_s' E_y(\overline{\rho}_s') - E_y(\overline{\rho}_s') \hat{n} \cdot \nabla_s' G(\overline{\rho}_s, \overline{\rho}_s') \right\}$$

$$= \begin{cases} \overline{E}_y(\overline{\rho}_s) & z > f(x) \\ 0 & z < f(x) \end{cases} \qquad (8)$$

where the y-dependence $\exp(ik_{yi}y)$ has been suppressed and

$$d\sigma' \hat{n} = \left[\hat{z} - \hat{x} \frac{df(x')}{dx'}\right] dx' \qquad (9)$$

The surface integral in (8) is over infinite domain. However, it can be condensed into a single period. We first recognize that the surface fields have the property

$$\omega(\overline{\rho}_s + \hat{x}nP) = \omega(\overline{\rho}_s) e^{ik_{xi}nP} \qquad (10)$$

where $\omega(\overline{\rho}_s)$ can be $E_y(\overline{\rho}_s)$, $H_y(\overline{\rho}_s)$ or their normal derivatives, n is an integer, and P is the period of the surface.

From the theory of Fourier series, we note that a periodic train of Dirac delta functions can be represented by an infinite summation of complex exponentials (Goodman, 1968).

$$\sum_{m=-\infty}^{\infty} e^{i(k_{xi}-k_x)mP} = \sum_{m=-\infty}^{\infty} \frac{2\pi}{P} \delta\left(k_x - k_{xi} - \frac{2\pi m}{P}\right) \quad (11)$$

Making use of (6) and (11), we have

$$\sum_{m=-\infty}^{\infty} G_j(\overline{\rho}_s, \overline{\rho}'_s + mP\hat{x}) e^{ik_{xi}mP} = G_{jP}(\overline{\rho}_s, \overline{\rho}'_s) \quad (12)$$

where

$$G_P(\overline{\rho}_s, \overline{\rho}'_s) = \frac{-1}{2ik_s P} \sum_n \frac{1}{\beta_n} \exp[ik_s\alpha_n(x-x') + ik_s\beta_n|z-z'|] \quad (13)$$

$$k_s \alpha_n = k_{xi} + n\frac{2\pi}{P} \quad (14)$$

$$\beta_n = \begin{cases} (1-\alpha_n^2)^{1/2} & \alpha_n^2 < 1 \\ +i(\alpha_n^2 - 1)^{1/2} & \alpha_n^2 > 1 \end{cases} \quad (15)$$

$$G_{1P}(\overline{\rho}_s, \overline{\rho}'_s) = \frac{-1}{2ik_{1s} P} \sum_n \frac{1}{\beta'_n} \cdot \exp[ik_{1s}\alpha'_n(x-x') + ik_{1s}\beta'_n|z-z'|] \quad (16)$$

$$k_{1s}\alpha'_n = k_s\alpha_n \quad (17)$$

$$\beta'_n = \begin{cases} (1-\alpha_n'^2)^{1/2} & \alpha_n'^2 < 1 \\ +i(\alpha_n'^2 - 1)^{1/2} & \alpha_n'^2 > 1 \end{cases} \quad (18)$$

Making use of (10) and (12) in (8), we find that

$$E_{yi}(\overline{\rho}_s) - \int_P d\sigma' \left\{ G_P(\overline{\rho}_s, \overline{\rho}'_s) \hat{n} \cdot \nabla'_s E_y(\overline{\rho}'_s) - E_y(\overline{\rho}'_s) \hat{n} \cdot \nabla'_s G_P(\overline{\rho}_s, \overline{\rho}'_s) \right\}$$

$$= \begin{cases} E_y(\overline{\rho}_s) & z > f(x) \quad (19a) \\ 0 & z < f(x) \quad (19b) \end{cases}$$

where the integration $d\sigma'$ is over one period P, and similarly

$$H_{yi}(\overline{\rho}_s) - \int_P d\sigma' \left\{ G_P(\overline{\rho}_s, \overline{\rho}'_s) \hat{n} \cdot \nabla'_s H_y(\overline{\rho}'_s) - H_y(\overline{\rho}'_s) \hat{n} \cdot \nabla'_s G_P(\overline{\rho}_s, \overline{\rho}'_s) \right\}$$

$$= \begin{cases} H_y(\overline{\rho}_s) & z > f(x) \quad (20a) \\ 0 & z < f(x) \quad (20b) \end{cases}$$

Formulation

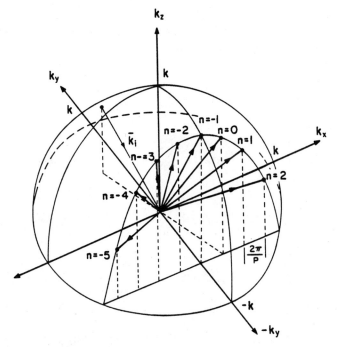

Fig. 2.12 Conical diffraction of the reflected wave in region 0.

Making use of periodic Green's function of Region 1, we have

$$\int_P d\sigma' \{ G_{1P}(\bar{\rho}_s, \bar{\rho}_s') \hat{n} \cdot \nabla_s' E_{1y}(\bar{\rho}_s') - E_{1y}(\bar{\rho}_s') \hat{n} \cdot \nabla_s' G_{1P}(\bar{\rho}_s, \bar{\rho}_s') \}$$

$$= \begin{cases} 0 & z > f(x) \quad (21a) \\ E_{1y}(\bar{\rho}_s) & z < f(x) \quad (21b) \end{cases}$$

$$\int_P d\sigma' \{ G_{1P}(\bar{\rho}_s, \bar{\rho}_s') \hat{n} \cdot \nabla_s' H_{1y}(\bar{\rho}_s') - H_{1y}(\bar{\rho}_s') \hat{n} \cdot \nabla_s' G_{1P}(\bar{\rho}_s, \bar{\rho}_s') \}$$

$$= \begin{cases} 0 & z > f(x) \quad (22a) \\ H_{1y}(\bar{\rho}_s) & z < f(x) \quad (22b) \end{cases}$$

From the representations of the Green's functions in (13) and (16), we see that the waves are propagating in discrete Floquet modes forming a cone shape when the observation point is either above the highest point or below the lowest point of the surface (Figure 2.12).

Equations (19) and (20) can be simplified by allowing z to be larger than f_{max} or z less than f_{min} where f_{max} and f_{min} are respectively the maximum and minimum values of the surface profile $f(x)$. For $z > f_{max}$, then the $|z - z'|$ term in (13), can be replaced by $z - z'$ while for $z < f_{min}$, the $|z - z'|$ term is replaced by $-(z - z')$. Thus

$$E_y(\bar{\rho}_s) = E_{yi}(\bar{\rho}_s) + \sum_n b_n \frac{e^{i\bar{k}_n^+ \cdot \bar{\rho}_s}}{\sqrt{\beta_n}} \quad z > f_{max} \tag{23a}$$

$$0 = E_{yi}(\bar{\rho}_s) - \sum_n a_n \frac{e^{i\bar{k}_n^- \cdot \bar{\rho}_s}}{\sqrt{\beta_n}} \quad z < f_{min} \tag{23b}$$

$$H_y(\bar{\rho}_s) = H_{yi}(\bar{\rho}_s) + \sum_n b_n^{(h)} \frac{e^{i\bar{k}_n^+ \cdot \bar{\rho}_s}}{\sqrt{\beta_n}} \quad z > f_{max} \tag{24a}$$

$$0 = H_{yi}(\bar{\rho}_s) - \sum_n a_n^{(h)} \frac{e^{i\bar{k}_n^- \cdot \bar{\rho}_s}}{\sqrt{\beta_n}} \quad z < f_{min} \tag{24b}$$

where

$$\bar{k}_n^\pm = \hat{x} k_{xn} \pm \hat{z} k_{zn} = \hat{x} k_s \alpha_n \pm \hat{z} k_s \beta_n \tag{25}$$

denote the propagation vectors of Floquet modes. We recognize that b_n and $b_n^{(h)}$ are scattered field amplitudes. The coefficients a_n, b_n, $a_n^{(h)}$, and $b_n^{(h)}$ are related to the surface fields by the following integrals,

$$b_n = \frac{1}{2ik_s P} \int_P d\sigma' \left\{ \frac{e^{-i\bar{k}_n^+ \cdot \bar{\rho}_s(x')}}{\sqrt{\beta_n}} \hat{n} \cdot \nabla'_s E_y(\bar{\rho}'_s) \right.$$
$$\left. - E_y(\bar{\rho}'_s) \hat{n} \cdot \nabla'_s \frac{e^{-i\bar{k}_n^+ \cdot \bar{\rho}_s(x')}}{\sqrt{\beta_n}} \right\} \tag{26}$$

and $b_n^{(h)}$ is the same expression as b_n with E_y replaced by H_y.

$$a_n = \frac{-1}{2ik_s P} \int_P d\sigma' \left\{ \frac{e^{-i\bar{k}_n^- \cdot \bar{\rho}_s(x')}}{\sqrt{\beta_n}} \hat{n} \cdot \nabla'_s E_y(\bar{\rho}'_s) \right.$$
$$\left. - E_y(\bar{\rho}'_s) \hat{n} \cdot \nabla'_s \frac{e^{-i\bar{k}_n^- \cdot \bar{\rho}_s(x')}}{\sqrt{\beta_n}} \right\} \tag{27}$$

and $a_n^{(h)}$ is the same expression as a_n with E_y replaced by H_y, where

$$\bar{\rho}_s(x') = \hat{x} x + \hat{z} f(x') \tag{28}$$

Formulation

is a point on the periodic surface. Similarly, making use of (21) and (22),

$$0 = -\sum_n B_n \frac{e^{i\vec{k}_{1n}^+ \cdot \vec{p}_s}}{\sqrt{\beta_n'}} \quad z > f_{max} \tag{29a}$$

$$E_{1y}(\vec{p}_s) = \sum_n A_n \frac{e^{i\vec{k}_{1n}^- \cdot \vec{p}_s}}{\sqrt{\beta_n'}} \quad z < f_{min} \tag{29b}$$

$$0 = -\sum_n B_n^{(h)} \frac{e^{i\vec{k}_{1n}^+ \cdot \vec{p}_s}}{\sqrt{\beta_n'}} \quad z > f_{max} \tag{30a}$$

$$H_{1y}(\vec{p}_s) = \sum_n A_n^{(h)} \frac{e^{i\vec{k}_{1n}^- \cdot \vec{p}_s}}{\sqrt{\beta_n'}} \quad z < f_{min} \tag{30b}$$

where

$$B_n = \frac{1}{2ik_{1s}P} \int_P d\sigma' \left\{ \frac{e^{-i\vec{k}_{1n}^+ \cdot \vec{p}_s(x')}}{\sqrt{\beta_n'}} \hat{n} \cdot \nabla_s' E_{1y}(\vec{p}_s') \right.$$
$$\left. - E_{1y}(\vec{p}_s') \hat{n} \cdot \nabla_s' \frac{e^{-i\vec{k}_{1n}^+ \cdot \vec{p}_s(x')}}{\sqrt{\beta_n'}} \right\} \tag{31}$$

and $B_n^{(h)}$ is the same expression as B_n with E_{1y} replaced by H_{1y},

$$A_n = \frac{-1}{2ik_{1s}P} \int_P d\sigma' \left\{ \frac{e^{-i\vec{k}_{1n}^- \cdot \vec{p}_s(x')}}{\sqrt{\beta_n'}} \hat{n} \cdot \nabla_s' E_{1y}(\vec{p}_s') \right.$$
$$\left. - E_{1y}(\vec{p}_s') \hat{n} \cdot \nabla_s' \frac{e^{-i\vec{k}_{1n}^- \cdot \vec{p}_s(x')}}{\sqrt{\beta_n'}} \right\} \tag{32}$$

and $A_n^{(h)}$ is the same expression as A_n with E_{1y} replaced by H_{1y}, and

$$\vec{k}_{1n}^\pm = \hat{x} k_{1xn} \pm \hat{z} k_{1zn} = \hat{x} k_{1s} \alpha_n' \pm \hat{z} k_{1s} \beta_n' \tag{33}$$

are the propagation vectors of the Floquet modes in Region 1. Thus, we have the fields represented in terms of space harmonics in (23), (24), (29), and (30). By letting the observation point be outside the trough region of the periodic surface, we see from (23b) and (24b) that

$$E_{yi}(\vec{p}_s) = \sum_n a_n \frac{e^{i\vec{k}_n^- \cdot \vec{p}_s}}{\sqrt{\beta_n}} \quad z < f_{min} \tag{34a}$$

$$H_{yi}(\vec{p}_s) = \sum_n a_n^{(h)} \frac{e^{i\vec{k}_n^- \cdot \vec{p}_s}}{\sqrt{\beta_n}} \quad z < f_{min} \tag{34b}cr$$

Comparing the incident plane wave in (1) with (34) gives the following solutions for the coefficients of the Floquet modes

$$a_n = \delta_{no}\sqrt{\beta_o}\, E_o\, (\hat{e}_i \cdot \hat{y}) \tag{35a}$$

$$a_n^{(h)} = \delta_{no}\sqrt{\beta_o}\, \frac{E_o}{\omega\mu}\, (\overline{k}_i \times \hat{e}_i \cdot \hat{y}) \tag{35b}$$

Also, from (29a) and (30a),

$$B_n = B_n^{(h)} \equiv 0 \quad \text{for all } n \tag{36}$$

Knowing a_n, $a_n^{(h)}$, B_n, and $B_n^{(h)}$, we can solve for the unknown fields on the surface using the relations of a_n, $a_n^{(h)}$, B_n and $B_n^{(h)}$ with the surface fields as represented by (27) and (31). After the surface fields are solved from the matrix equation as given by (27) and (31), then the scattered field amplitudes b_n, $b_n^{(h)}$, A_n, and $A_n^{(h)}$ can be obtained using (26) and (32).

5.2 Coupled Matrix Equations

Boundary Conditions

Applying the boundary conditions for the tangential fields on the periodic surface S we obtain, by noting that the normal vector $\hat{n}$ has no y-component,

$$E_y = E_{1y} \tag{1a}$$

$$H_y = H_{1y} \tag{1b}$$

$$\hat{n} \times \overline{E}_s = \hat{n} \times \overline{E}_{1s} \tag{2a}$$

$$\hat{n} \times \overline{H}_s = \hat{n} \times \overline{H}_{1s} \tag{2b}$$

where $\overline{E}_s$, $\overline{H}_s$, $\overline{E}_{1s}$, and $\overline{H}_{1s}$ are related to $\overline{E}_y$, $\overline{H}_y$, $\overline{E}_{1y}$, and $\overline{H}_{1y}$ by (2), Section 5.1. Since $\hat{n}$ is in the x-z plane,

$$\hat{n} \times (\nabla_s \times \overline{A}_y) = -\hat{y}\,\hat{n} \cdot \nabla_s A_y \tag{3}$$

where $\overline{A}_y = \hat{y} A_y(\overline{\rho}_s)$ and $A_y(\overline{\rho}_s)$ can be $E_y(\overline{\rho}_s)$, $E_{1y}(\overline{\rho}_s)$, $H_y(\overline{\rho}_s)$ or $H_{1y}(\overline{\rho}_s)$.

Now, (1a) can be written explicitly as

$$E_y[x, z = f(x)] = E_{1y}[x, z = f(x)] \tag{4}$$

Coupled Matrix Equations

Differentiating (4) with respect to x gives

$$\frac{dE_y}{dx} = \frac{\partial E_y}{\partial x} + \frac{\partial E_y}{\partial z}\frac{df(x)}{dx} = \frac{\partial E_{1y}}{\partial x} + \frac{\partial E_{1y}}{\partial z}\frac{df(x)}{dx} = \frac{dE_{1y}}{dx} \quad (5)$$

with $z = f(x)$. Since $\hat{n} = [-df/dx\,\hat{x} + \hat{z}]/[1 + (df/dx)^2]^{1/2}$, (5) can be put into the form,

$$\hat{n} \times \nabla_s E_y = \frac{dE_y/dx}{[1 + (df/dx)^2]^{1/2}}\hat{y}$$

$$\hat{n} \times \nabla_s E_{1y} = \frac{dE_{1y}/dx}{[1 + (df/dx)^2]^{1/2}}\hat{y} \quad (6)$$

with $z = f(x)$. Similarly,

$$\frac{dH_y/dx}{[1 + (df/dx)^2]^{1/2}}\hat{y} = \hat{n} \times \nabla_s H_y$$

$$= \hat{n} \times \nabla_s H_{1y} = \frac{dH_{1y}/dx}{[1 + (df/dx)^2]^{1/2}}\hat{y} \quad (7)$$

on surface S. Using (3), (6), and (7) and the relation of the transverse components of the fields to the y components as in (2), Section 5.1, we can deduce the following boundary conditions from (2)

$$\hat{y}(\hat{n} \cdot \nabla_s E_y) = c_o \hat{n} \times \nabla_s H_{1y} + c_2 \hat{y}(\hat{n} \cdot \nabla_s E_{1y}) \quad (8a)$$

$$\hat{y}(\hat{n} \cdot \nabla_s H_y) = -d_o \hat{n} \times \nabla_s E_{1y} + d_2 \hat{y}(\hat{n} \cdot \nabla_s H_{1y}) \quad (8b)$$

where

$$c_o = \left[\frac{k_s^2}{k_{1s}^2} - 1\right]\frac{k_{yi}}{\omega\epsilon_o} \quad (9a)$$

$$c_2 = \frac{\epsilon_1}{\epsilon_o}\frac{k_s^2}{k_{1s}^2} \quad (9b)$$

$$d_o = \left[\frac{k_s^2}{k_{1s}^2} - 1\right]\frac{k_{yi}}{\omega\mu_o} \quad (9c)$$

$$d_2 = \frac{\mu_1}{\mu_o}\frac{k_s^2}{k_{1s}^2} \quad (9d)$$

Equations (1a), (1b), (8a), and (8b) are the four boundary conditions needed for solving the unknown surface fields.

Surface Field Expansion

Noting that the surface fields are dependent on x only, we adopt the Fourier series expansions for the four unknown fields on the surface. With $\bar{\rho}_s(x) = \hat{x}x + \hat{z}f(x)$, we let

$$E_{1y}[\bar{\rho}_s(x)] = \sum_n 2\alpha_n^s \exp\left[ik_{xi}x + in\frac{2\pi}{P}x\right] \quad (10a)$$

$$d\sigma\hat{n} \cdot \nabla_s E_{1y}[\bar{\rho}_s(x)] = ik_{1s}dx \sum_n 2\beta_n^s \exp\left[ik_{xi}x + in\frac{2\pi}{P}x\right] \quad (10b)$$

$$H_{1y}[\bar{\rho}_s(x)] = \sum_n 2\gamma_n^s \exp\left[ik_{xi}x + in\frac{2\pi}{P}x\right] \quad (10c)$$

$$d\sigma\hat{n} \cdot \nabla_s H_{1y}[\bar{\rho}_s(x)] = ik_{1s}dx \sum_n 2\delta_n^s \exp\left[ik_{xi}x + in\frac{2\pi}{P}x\right] \quad (10d)$$

This choice of basis functions using the Fourier series expansion is appropriate because the surface fields, when multiplied by the term $\exp(-ik_{xi}x)$, are periodic functions of x (van den Berg, 1971; Chuang and Kong, 1981). Other basis functions, such as plane harmonics evaluated on the surface (Waterman, 1975) or a modified physical optics expansion (DeSanto, 1975; Whitman and Schwering, 1977), were used in the study of the classical case where the incident wave vector is perpendicular to the row direction.

Substitute (10) into (31), Section 5.1, and the similar expression for $B_n^{(h)}$. Using (36), Section 5.1, and defining $Q_{D_2}^\pm$, $Q_{N_2}^\pm$ as the Dirichlet and Neumann matrices with elements

$$\left[\overline{\overline{Q}}_{D_2}^\pm\right]_{mn} \equiv \frac{-1}{P}\int_P dx \frac{e^{-i\bar{k}_{1m}^\pm \cdot \bar{\rho}_s(x)}}{\sqrt{\beta_m'}} \exp\left[ik_{xi}x + in\frac{2\pi}{P}x\right]$$

$$= \frac{-1}{P\sqrt{\beta_m'}}\int_P dx \exp\left[-i(m-n)\frac{2\pi}{P}x - ik_{1s}(\pm\beta_m')f(x)\right] \quad (11a)$$

$$\left[\overline{\overline{Q}}_{N_2}^\pm\right]_{mn} \equiv \frac{1}{ik_{1s}P}\int_P d\sigma\,\hat{n} \cdot \nabla_s \frac{e^{-i\bar{k}_{1m}^\pm \cdot \bar{\rho}_s(x)}}{\sqrt{\beta_m'}} \exp\left[ik_{xi}x + in\frac{2\pi}{P}x\right]$$

$$= \frac{-1 + \alpha_m'\alpha_n'}{\pm\beta_m'\sqrt{\beta_m'}P}\int_P dx \exp\left[-i(m-n)\frac{2\pi}{P}x - ik_{1s}(\pm\beta_m')f(x)\right] \quad (11b)$$

where the integrations dx and $d\sigma$ are over one period, we obtain the two matrix equations

$$-\overline{\overline{Q}}_{D_2}^+\bar{\beta}^s - \overline{\overline{Q}}_{N_2}^+\bar{\alpha}^s = \overline{B} = 0 \quad (12a)$$

Coupled Matrix Equations

$$-\overline{\overline{Q}}^+_{D_2}\overline{\delta}^s - \overline{\overline{Q}}^+_{N_2}\overline{\gamma}^s = \overline{B}^{(h)} = 0 \tag{12b}$$

In deriving the second equality in (11b), we have performed an integration by parts. Here the vectors $\overline{B}$, $\overline{B}^{(h)}$, $\overline{\alpha}^s$, $\overline{\beta}^s$, $\overline{\gamma}^s$, and $\overline{\delta}^s$ contain the elements B_n, $B_n^{(h)}$, α_n^s, β_n^s, γ_n^s, δ_n^s, respectively. Similarly, (27), Section 5.1, and the similar expression for $a_n^{(h)}$, on using the boundary conditions of (1) and (8) and (6) through (7), can be written in the following matrix form

$$\overline{a} = c_o \overline{\overline{Q}}^-_{hy1} \overline{\gamma}^s + c_2 \frac{k_{1s}}{k_s} \overline{\overline{Q}}^-_{D_1} \overline{\beta}^s + \overline{\overline{Q}}^-_{N_1} \overline{\alpha}^s \tag{13a}$$

$$\overline{a}^{(h)} = -d_o \overline{\overline{Q}}^-_{hy1} \overline{\alpha}^s + d_2 \frac{k_{1s}}{k_s} \overline{\overline{Q}}^-_{D_1} \overline{\delta}^s + \overline{\overline{Q}}^-_{N_1} \overline{\gamma}^s \tag{13b}$$

where $\overline{\overline{Q}}_{hy1}$ denotes the hybrid matrix which couples α^s and γ^s to a and $a^{(h)}$, and

$$[Q^\pm_{D_1}]_{mn} = \frac{-1}{P\sqrt{\beta_m}} \int_P dx \exp\left[-i(m-n)\frac{2\pi}{P}x - ik_s(\pm\beta_m)f(x)\right] \tag{14a}$$

$$[Q^\pm_{N_1}]_{mn} = \frac{-1+\alpha_m\alpha_n}{\pm\beta_m\sqrt{\beta_m}P} \int_P dx \exp\left[-i(m-n)\frac{2\pi}{P}x - ik_s(\pm\beta_m)f(x)\right] \tag{14b}$$

$$[Q^\pm_{hy1}]_{mn} = -\alpha_n [Q^\pm_{D_1}]_{mn} \tag{15}$$

Equations (12a), (12b), (13a), and (13b) can be put into the following matrix equation.

$$\begin{bmatrix} Q^-_{N_1} & c_2\frac{k_{1s}}{k_s}Q^-_{D_1} & c_o Q^-_{hy1} & 0 \\ -d_o Q^-_{hy1} & 0 & Q^-_{N_1} & d_2\frac{k_{1s}}{k_s}Q^-_{D_1} \\ Q^+_{N_2} & Q^+_{D_2} & 0 & 0 \\ 0 & 0 & Q^+_{N_2} & Q^+_{D_2} \end{bmatrix} \begin{bmatrix} \alpha^s \\ \beta^s \\ \gamma^s \\ \delta^s \end{bmatrix} = \begin{bmatrix} a \\ a^{(h)} \\ 0 \\ 0 \end{bmatrix} \tag{16}$$

The unknown vectors α^s, β^s, γ^s, and δ^s, for the surface field expansions are obtained by solving the above matrix equation. Note that the quantities on the right hand side of (16) have been calculated and are given by (35), Section 5.1. Next, the scattered field amplitudes are derived from (26) and (32), Section 5.1, and similar expressions for $b_n^{(h)}$ and $A_n^{(h)}$. The upward going field amplitudes are

$$b = -c_o Q^+_{hy1} \gamma^s - c_2 \frac{k_{1s}}{k_s} Q^+_{D_1} \beta^s - Q^+_{N_1} \alpha^s \tag{17a}$$

$$b^{(h)} = d_o Q^+_{hy1}\alpha^s - d_2 \frac{k_{1s}}{k_s} Q^+_{D_1}\delta^s - Q^+_{N_1}\gamma^s \qquad (17b)$$

and the downward-going field amplitudes are

$$A = Q^-_{D_2}\beta^s + Q^-_{N_2}\alpha^s \qquad (18a)$$

$$A^{(h)} = Q^-_{D_2}\delta^s + Q^-_{N_2}\gamma^s \qquad (18b)$$

From the matrix equations we see that the E and H waves are generally coupled. An incident wave of E type ($a^{(h)} \equiv 0$) will be diffracted from the dielectric periodic surface into waves that have both E wave and H wave components. Both types of waves will coexist to satisfy the boundary conditions. When the incident wave vector is perpendicular to the row direction ($k_{yi} = 0$), then $c_o = d_o = 0$ and we can easily see that the scattered waves arising from $\overline{E}_{yi}$ are decoupled from those arising from $\overline{H}_{yi}$.

Sinusoidal Surface

For a sinusoidal rough surface with

$$f(x) = -h\cos\left(\frac{2\pi}{P}x\right) \qquad (19)$$

the $Q^\pm$ matrices can be calculated by carrying out the integrations in (11), (14), and (15) and expressed in terms of Bessel functions. We have

$$[Q^\pm_{D_1}]_{mn} = \frac{-1}{\sqrt{\beta_m}}(\pm i)^{|m-n|} J_{|m-n|}(k_s h \beta_m) \qquad (20a)$$

$$[Q^\pm_{N_1}]_{mn} = \frac{-1 + \alpha_m \alpha_n}{\pm\beta_m\sqrt{\beta_m}}(\pm i)^{|m-n|} J_{|m-n|}(k_s h \beta_m) \qquad (20b)$$

$$[Q^\pm_{D_2}]_{mn} = \frac{-1}{\sqrt{\beta'_m}}(\pm i)^{|m-n|} J_{|m-n|}(k_{1s} h \beta_m) \qquad (20c)$$

$$[Q^\pm_{N_2}]_{mn} = \frac{-1 + \alpha'_m \alpha'_n}{\pm\beta'_m\sqrt{\beta'_m}}(\pm i)^{|m-n|} J_{|m-n|}(k_{1s} h \beta_m) \qquad (20d)$$

$$[Q^\pm_{hy1}]_{mn} = \frac{\alpha_n}{\sqrt{\beta_m}}(\pm i)^{|m-n|} J_{|m-n|}(k_s h \beta_m) \qquad (20e)$$

where $J_{|m-n|}$ denotes the Bessel function.

Using the steps outlined, the scattered field amplitudes can be obtained from (17) and (18), after the surface fields expansion amplitudes are obtained by solving the matrix equation (16). This approach, which makes use of Green's theorem to derive the extended boundary conditions, is exact. The results can be shown to satisfy the principles of energy conservation and reciprocity. However, the matrices used may become ill-conditioned when the surface corrugation is deep or when the corrugation depth divided by the period is large (Garcia et al., 1978). When the surface corrugation is deep, we generally increase the number of basis functions and check the energy conservation. When ill conditioning occurs in the matrix the numerical errors may be large. This limits the applicable regime of this method unless one can minimize the errors in the surface field expansions. For a general profile of periodic surface defined by a single-valued function $z = f(x)$, the method can be applied by numerically integrating (11), (14), and (15) to calculate the elements of the $Q^{\pm}$ matrices.

5.3 Emissivity of Periodic Surfaces

The emissivity of a periodic surface can be calculated from one minus the reflectivity by using the principle of reciprocity and energy conservation. The reflectivity consists of the sum of all scattered power. The scattered fields from a periodic surface were derived in the previous sections using the extended boundary conditions approach. The results satisfy both the principles of reciprocity and energy conservation. In this section a sinusoidal surface is used to model a row-structured plowed field and the theoretical results are illustrated and compared with the experimental data obtained from field measurements.

The reflectivity can be obtained by calculating the sum of all the propagating power densities of the reflected waves and normalizing it to the incident power density. We define reflected power P_r for one period as follows:

$$P_r = \frac{1}{2P} \int_0^P Re[(\overline{E}_r \times \overline{H}_r^*) \cdot \hat{z}] dx \qquad (1)$$

and incident power as

$$P_{inc} = \frac{1}{2P} \int_0^P Re[(\overline{E}_{inc} \times \overline{H}_{inc}^*) \cdot (-\hat{z})] dx \qquad (2)$$

For the reflected E mode, it can quickly be shown by using (2) and (23a), Section 5.1, that for $z > f_{max}$

$$\overline{E}_r = \sum_n \frac{b_n}{\sqrt{\beta_n}} e^{i\overline{k}_n^+ \cdot \overline{\rho}_s} \begin{bmatrix} ik_{yi}k_{xn}/k_s^2 \\ 1 \\ ik_{yi}k_{zn}/k_s^2 \end{bmatrix} \qquad (3)$$

$$\overline{H}_r = \sum_n \frac{\omega \epsilon b_n}{\sqrt{\beta_n}} e^{i\overline{k}_n^+ \cdot \overline{\rho}_s} \begin{bmatrix} -k_{zn}/k_s^2 \\ 0 \\ k_{xn}/k_s^2 \end{bmatrix} \qquad (4)$$

where $k_{xn} = k\alpha_n$ and $k_{zn} = k_s \beta_n$ and the column vector denotes $\hat{x}$, $\hat{y}$ and $\hat{z}$ components. By substituting (3) and (4) in (1), we find that on integrating, the Floquet modes are orthogonal and for the E mode

$$P_r = \sum_n \frac{1}{2} Re \frac{\omega k_{zn} \epsilon}{k_s^2} \frac{|b_n|^2}{|\beta_n|} \qquad (5)$$

Similarly, the reflected power for H mode can be obtained and the total reflected power is

$$P_r = \frac{1}{2} \sum_n Re \frac{\omega k_{zn}}{k_s^2} \left(\epsilon \frac{|b_n|^2}{|\beta_n|} + \mu \frac{|b_n^{(h)}|^2}{|\beta_n|} \right) = \sum_n P_n^r \qquad (6)$$

where P_n^r can be interpreted as the reflected power for the n-th Floquet mode. In (5) and (6), Re means the real part of the expression. When the mode is evanescent, k_{zn} is purely imaginary. We have $P_n^r = 0$ for the evanescent mode. The incident power is obtained by using (2) and (34) and (35), Section 5.1,

$$P_{inc} = \frac{1}{2} \frac{\omega \epsilon}{k_s} \left(|a_o|^2 + \frac{\mu}{\epsilon} |a_o^{(h)}|^2 \right) \qquad (7)$$

Thus, the reflectivity for a wave with horizontal or vertical polarization is

$$r_\alpha = \frac{1}{P_{inc}} \sum_n P_n^r = \sum_n \frac{|b_n|^2 + |\eta b_n^{(h)}|^2}{|a_o|^2 + |\eta a_o^{(h)}|^2} \qquad (8)$$

where $\alpha = v, h$ polarization, and the summation in (8) is over the propagating modes and a_o and $a_o^{(h)}$ are obtained from (35), Section 5.1, by setting

$$\hat{e}_i = \hat{\alpha}_i \qquad (9)$$

The transmitted power per period of the nth mode passing through the surface defined by $z = z_{min}$ is

$$P_n^t = \frac{1}{2} Re \frac{\omega k_{1zn}}{k_{1s}^2} \left(\epsilon_1 \frac{|A_n|^2}{|\beta_n'|} + \mu_1 \frac{|A_n^{(h)}|^2}{|\beta_n'|} \right) e^{2Im[k_{1zn}]z_{min}} \qquad (10)$$

for all modes, where $k_{1zn} = k_{1s}\beta_n'$, and $Im[k_{1zn}] > 0$ should be used. When medium 1 is lossless, we have (letting $\eta_1 = \sqrt{\mu_1/\epsilon_1}$)

$$P_n^t = \frac{\omega \epsilon_1}{2k_{1s}} \left(|A_n|^2 + |\eta_2 A_n^{(h)}|^2 \right) \qquad (11)$$

for the nth propagating mode and $P_n^t = 0$ for the evanescent mode. The transmissivity is given by

$$t_\alpha = \sum_n \frac{\epsilon_1}{\epsilon} \frac{k_s}{k_{1s}} \frac{|A_n|^2 + |\eta_1 A_n^{(h)}|^2}{|a_o|^2 + |\eta a_o^{(h)}|^2} \qquad (\alpha = v, h) \tag{12}$$

Thus, for the lossless case, we have the following power conservation relation

$$r_\alpha + t_\alpha = 1 \tag{13}$$

The emissivity of a periodic rough surface is given by

$$e_\alpha = 1 - r_\alpha \tag{14}$$

We note that there is a considerable saving in computational effort in summing up the reflected power to calculate the emissivity compared with summing up the transmitted power. This is because the number of basis functions required to calculate the reflected modes is considerably less than the number required to calculate the transmitted modes to the same accuracy.

The theoretical results are illustrated for a sinusoidal surface in Figures 2.13, 2.14, and 2.15 at a frequency of 1.4 GHz. The effect of the row structure on the microwave emission from a bare agricultural field has been reported (Wang et al., 1980) together with the soil moisture contents for the measured data. The periodic surface has a height $h = 10$ cm and a period $P = 95$ cm and can be approximated by a sinusoidal function.

In Figure 2.13 we illustrate the comparison between the theoretical results and the experimental data for both the vertical and horizontal polarizations when the radiometer observation angle is along the row direction ($\phi = 90$ deg). The reported soil moisture content varies from 26% by dry weight at top 0 to 1 cm to 21.4% at 9 to 15 cm. In the theoretical results, we take $\epsilon = (5.5 + i1.2)\epsilon_o$ which corresponds to a soil moisture content of approximately 18% (Newton, 1976; Wang and Schmugge, 1980) which is reasonable since the soil becomes drier as the depth increases. In the same figure, we also show the theoretical curves for the flat surface case. It is seen that the brightness temperature for the periodic rough surfaces for the horizontal polarization is higher than that for the flat surface, whereas for the vertical polarization the brightness temperature is lower. For the flat surface, both polarizations have the same brightness temperature value when viewed from nadir, whereas for the periodic surface, the values for the horizontal polarization are higher than the vertical polarization at near-nadir angles and become lower at larger angles of observation.

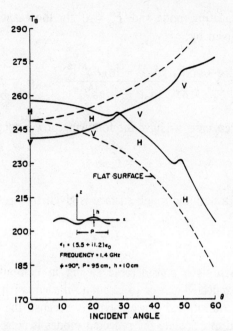

Fig. 2.13 Brightness temperature as a function of viewing angle. Radiometer observation plane is parallel to the row direction ($\phi = 90$ deg).

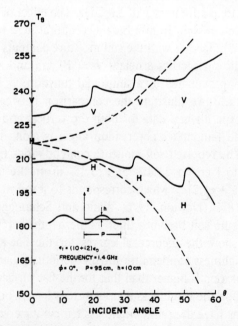

Fig. 2.14 Brightness temperature as a function of viewing angle. Radiometer observation plane is perpendicular to the row direction ($\phi = 0$ deg).

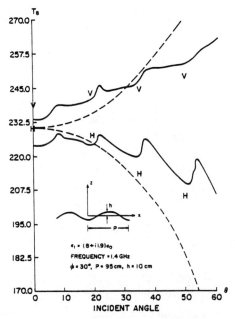

Fig. 2.15 Brightness temperature as a function of viewing angle. Radiometer observation plane is slanted with respect to the row direction ($\phi = 30$ deg).

In Figure 2.14, the radiometer observation angles are perpendicular to the row direction ($\phi = 0$ deg). The soil moisture content is 29% by dry weight at the top 0 to 5 cm and becomes drier with depth. We use $\epsilon = (10 + i2)\epsilon_o$, which is smaller than the corresponding permittivity at the top surface region. We see that at near-nadir angles, as compared with the flat surface cases, the brightness temperatures for the horizontal polarization are lower and for the vertical polarization higher. The effect of the rough surface as compared with the flat surface appears to bring both the horizontal and vertical polarization results closer together at higher incident angles. The case of observation plane at $\phi_i = 30$ deg is illustrated in Figure 2.15.

In Figures 2.13, 2.14, and 2.15, we observe that the T_B curves for the periodic rough surface are not smoothly varying. For instance, in Figure 2.14, there are kinks appearing at observation angles θ near 6 deg, 19 deg, 34 deg, and 51 deg. The corresponding change in T_B may be as high as 10 deg. Such a phenomenon can be explained by the appearance and disappearance of Floquet modes at various threshold angles (Bolotoviskii and Lebedev, 1968). The kinks are caused by the redistribution of the scattered power during the course of the disappearance and appearance of the modes.

6 SCATTERING AND EMISSION BY RANDOM ROUGH SURFACES

The scattering of electromagnetic waves from a randomly rough surface has been studied extensively for many years (Beckmann and Spizzichino, 1963; Bass and Fuks, 1979). Two basic analytical approaches have been the Kirchhoff approach (KA) and the small perturbation method (SPM). The KA approximates the surface fields using the tangent plane approximation. Under the tangent plane approximation, the fields at any point of the surface are approximated by the fields that would be present on the tangent plane at that point. Thus, the tangent plane approximation requires a large radius of curvature relative to the incident wavelength at every point on the surface. The SPM assumes that the surface variations are much smaller than the incident wavelength and that the slopes of the rough surface are relatively small.

In this section, we derive the bistatic scattering coefficients for the reflected and transmitted waves using both the KA and SPM. The Kirchhoff-approximated diffraction integral for a dielectric rough surface is still difficult to evaluate analytically and further approximations are usually made. The integrands, which depend on the local surface slopes, can be expanded in slope terms about zero slopes, and can then be integrated by parts discarding the edge effect. In the high frequency limit, the geometrical optics solution can be obtained using the method of stationary phase. The geometrical optics solution is independent of frequency and states that the scattered intensity is proportional to the probability of the occurrence of the slopes that will specularly reflect or transmit the incident wave into the direction of observation. The SPM is used to calculate the scattered fields up to the second order. The zeroth-order solutions are just the reflected and transmitted fields of a flat surface. The first-order solutions give the lowest-order incoherent transmitted and reflected intensities. However, the first-order solution does not give any depolarization effect in the backscattering direction. The second-order solution gives the lowest-order correction to the coherent reflection and transmission coefficients and also exhibits depolarization in the backscattering direction.

6.1 Kirchhoff Approach

Consider a plane wave incident upon a random rough surface (Figure 2.16). The electric field of the incident wave is given by

$$\overline{E}_i = \hat{e}_i\, E_o\, e^{i\overline{k}_i \cdot \overline{r}} \qquad (1)$$

Kirchhoff Approach

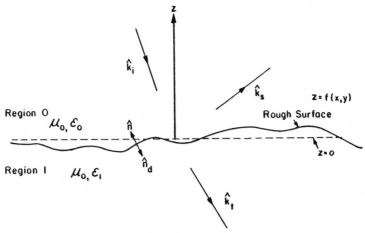

Fig. 2.16 Geometrical configuration of the problem.

where $\bar{k}_i$ denotes the incident wave vector and $\hat{e}_i$ the polarization of the electric field vector. The rough surface is characterized by a random height distribution $z = f(\bar{r}_\perp)$ where $f(\bar{r}_\perp)$ is a Gaussian random variable with zero mean, $<f(\bar{r}_\perp)> = 0$. From Huygen's principle, which expresses the field at an observation point in terms of fields at the boundary surface, the following expressions are obtained for the scattered fields in region 0 and the transmitted fields in region 1 (Kong, 1975).

$$\bar{E}_s(\bar{r}) = \int_{S'} dS' \left\{ i\omega\mu_o \bar{\bar{G}}(\bar{r}, \bar{r}') \cdot [\hat{n} \times \bar{H}(\bar{r}')] \right. \\ \left. + \nabla \times \bar{\bar{G}}(\bar{r}, \bar{r}') \cdot [\hat{n} \times \bar{E}(\bar{r}')] \right\} \quad (2a)$$

$$\bar{E}_t(\bar{r}) = \int_{S'} dS' \left\{ i\omega\mu_o \bar{\bar{G}}_1(\bar{r}, \bar{r}') \cdot [\hat{n}_d \times \bar{H}(\bar{r}')] \right. \\ \left. + \nabla \times \bar{\bar{G}}_1(\bar{r}, \bar{r}') \cdot [\hat{n}_d \times \bar{E}(\bar{r}')] \right\} \quad (2b)$$

where S' denotes the rough surface on which the surface integration is to be carried out, $\hat{n}$ and $\hat{n}_d$ are the unit vectors normal to the rough surface and pointing into the reflected and transmitted regions, respectively (Figure 2.16). The dyadic Green's function for homogeneous space of the region 0 and 1, $\bar{\bar{G}}(\bar{r}, \bar{r}')$ and $\bar{\bar{G}}_1(\bar{r}, \bar{r}')$, are

$$\bar{\bar{G}}(\bar{r}, \bar{r}') = \left[\bar{\bar{I}} + \frac{\nabla\nabla}{k^2} \right] \frac{e^{ik|\bar{r} - \bar{r}'|}}{4\pi |\bar{r} - \bar{r}'|} \quad (3a)$$

and

$$\bar{\bar{G}}_1(\bar{r}, \bar{r}') = \left[\bar{\bar{I}} + \frac{\nabla\nabla}{k_1^2} \right] \frac{e^{ik_1|\bar{r} - \bar{r}'|}}{4\pi |\bar{r} - \bar{r}'|} \quad (3b)$$

where $k = \omega\sqrt{\mu_o \epsilon_o}$ and $k_1 = \omega\sqrt{\mu_o \epsilon_1}$. If the observation point is in the far field region, then the dyadic Green's functions simplify to

$$\overline{\overline{G}}(\bar{r}, \bar{r}') \simeq (\overline{\overline{I}} - \hat{k}_s \hat{k}_s) \frac{e^{ikr}}{4\pi r} e^{-i\bar{k}_s \cdot \bar{r}'} \qquad (4)$$

$$\overline{\overline{G}}_1(\bar{r}, \bar{r}') \simeq (\overline{\overline{I}} - \hat{k}_t \hat{k}_t) \frac{e^{ik_1 r}}{4\pi r} e^{-i\bar{k}_t \cdot \bar{r}'} \qquad (5)$$

where $\hat{k}_s$ and $\hat{k}_t$ denote the scattered and transmitted directions in region 0 and region 1, respectively.

Substituting (4) and (5) into the diffraction integral (2), we obtain, in the reflected direction $\hat{k}_s$ and transmitted direction $\hat{k}_t$,

$$\overline{E}_s(\bar{r}) = \frac{ik\, e^{ikr}}{4\pi r}(\overline{\overline{I}} - \hat{k}_s \hat{k}_s)$$
$$\cdot \int_{S'} dS' \left\{ \hat{k}_s \times [\hat{n} \times \overline{E}(\bar{r}')] + \eta[\hat{n} \times \overline{H}(\bar{r}')] \right\} e^{-i\bar{k}_s \cdot \bar{r}'} \qquad (6a)$$

$$\overline{E}_t(\bar{r}) = \frac{ik_1 e^{ik_1 r}}{4\pi r}(\overline{\overline{I}} - \hat{k}_t \hat{k}_t)$$
$$\cdot \int_{S'} dS' \left\{ \hat{k}_t \times [\hat{n}_d \times \overline{E}(\bar{r}')] + \eta_1[\hat{n}_d \times \overline{H}(\bar{r}')] \right\} e^{-i\bar{k}_t \cdot \bar{r}'} \qquad (6b)$$

where η and η_1 are the wave impedances in the regions 0 and 1, respectively.

Tangent Plane Approximation

In the Kirchhoff approach, an approximate expression for the surface fields is obtained using the tangent plane approximation. The fields at any point of the surface are approximated by the fields that would be present on the tangent plane at that point. Thus, the tangent plane approximation requires a large radius of curvature relative to the incident wavelength at every point on the surface (Beckmann and Spizzichino, 1963).

First, an orthonormal system $(\hat{p}_i, \hat{q}_i, \hat{k}_i)$ is formed at the point $\bar{r}'$, with

$$\hat{q}_i = \frac{\hat{k}_i \times \hat{n}}{|\hat{k}_i \times \hat{n}|} \qquad (7)$$

$$\hat{p}_i = \hat{q}_i \times \hat{k}_i \qquad (8)$$

where, $\hat{n}(\bar{r}') = -\hat{n}_d(\bar{r}')$, is the normal to the surface at the point $\bar{r}'$ pointing into the region 0. The unit vectors $\hat{q}_i$ and $\hat{p}_i$ are the local perpendicular and parallel polarization vectors at the point $\bar{r}'$. In applying the

Kirchhoff Approach

tangent plane approximation, we solve the boundary value problem for the TE and TM polarization of a wave incident onto an infinite planar interface with the tangent plane as the interface. The incident field is decomposed into locally perpendicular and parallel polarization fields.

The perpendicular component of the incident field is

$$(\hat{e}_i \cdot \hat{q}_i)\hat{q}_i \, E_o \, e^{i\overline{k}_i \cdot \overline{r}'}$$

so that the local reflected field is

$$(\hat{e}_i \cdot \hat{q}_i)\hat{q}_i \, E_o R_h \, e^{i\overline{k}_i \cdot \overline{r}'}$$

where R_h is the local Fresnel reflection coefficient for perpendicular polarization.

$$R_h = \frac{k \cos \theta_{l_i} - \sqrt{k_1^2 - k^2 \sin^2 \theta_{l_i}}}{k \cos \theta_{l_i} + \sqrt{k_1^2 - k^2 \sin^2 \theta_{l_i}}} \tag{9a}$$

with θ_{l_i} as the local angle of incidence at the point $\overline{r}'$. The magnetic fields associated with the incident and reflected fields are, respectively,

$$\frac{1}{\eta}\hat{k}_i \times (\hat{e}_i \cdot \hat{q}_i)\hat{q}_i \, E_o \, e^{i\overline{k}_i \cdot \overline{r}'}$$

and

$$\frac{1}{\eta}\hat{k}_r \times (\hat{e}_i \cdot \hat{q}_i)\hat{q}_i \, E_o R_h \, e^{i\overline{k}_i \cdot \overline{r}'}$$

where $\hat{k}_r$ is the local reflected direction and is related to the incident direction by

$$\hat{k}_r = \hat{k}_i - 2\hat{n}(\hat{n} \cdot \hat{k}_i)$$

Hence, the tangential electric field of this perpendicular component at the point $\overline{r}'$ is

$$\hat{n} \times \overline{E} = (\hat{n} \times \hat{q}_i)(\hat{e}_i \cdot \hat{q}_i)(1 + R_h) \, E_o \, e^{i\overline{k}_i \cdot \overline{r}'}$$

and the associated magnetic field is

$$\hat{n} \times \overline{H} = \frac{1}{\eta}(\hat{e}_i \cdot \hat{q}_i)\hat{n} \times \left[(\hat{k}_i \times \hat{q}_i) + R_h(\hat{k}_r \times \hat{q}_i)\right] E_o \, e^{i\overline{k}_i \cdot \overline{r}'}$$

$$= -(1 - R_h)(\hat{n} \cdot \hat{k}_i)\frac{(\hat{e}_i \cdot \hat{q}_i)}{\eta}\hat{q}_i \, E_o \, e^{i\overline{k}_i \cdot \overline{r}'}$$

where we have made use of the relations $\hat{n} \cdot \hat{q}_i = 0$ and $\hat{n} \cdot \hat{k}_r = -\hat{n} \cdot \hat{k}_i$. The calculations can be repeated for local parallel polarized component with local reflection coefficient

$$R_v = \frac{\epsilon_1 k \cos\theta_{l_i} - \epsilon_o \sqrt{k_1^2 - k^2 \sin^2\theta_{l_i}}}{\epsilon_1 k \cos\theta_{l_i} + \epsilon_o \sqrt{k_1^2 - k^2 \sin^2\theta_{l_i}}} \tag{9b}$$

Summing up the local parallel and perpendicular polarized components, we obtain

$$\hat{n} \times \overline{E}(\bar{r}') = E_o \Big\{ (\hat{e}_i \cdot \hat{q}_i)(\hat{n} \times \hat{q}_i)(1 + R_h)$$
$$+ (\hat{e}_i \cdot \hat{p}_i)(\hat{n} \cdot \hat{k}_i)\hat{q}_i(1 - R_v) \Big\} e^{i\bar{k}_i \cdot \bar{r}'} \tag{10a}$$

$$\hat{n} \times \overline{H}(\bar{r}') = \frac{E_o}{\eta} \Big\{ -(\hat{e}_i \cdot \hat{q}_i)(\hat{n} \cdot \hat{k}_i)\hat{q}_i(1 - R_h)$$
$$+ (\hat{e}_i \cdot \hat{p}_i)(\hat{n} \times \hat{q}_i)(1 + R_v) \Big\} e^{i\bar{k}_i \cdot \bar{r}'} \tag{10b}$$

The local angle of incidence can be calculated from the formula

$$\cos\theta_{l_i} = -\hat{n} \cdot \hat{k}_i \tag{11}$$

The normal vector at the point $\bar{r}'$ is given by

$$\hat{n}(\bar{r}') = \frac{-\alpha\hat{x} - \beta\hat{y} + \hat{z}}{\sqrt{1 + \alpha^2 + \beta^2}} \tag{12}$$

where α and β are the local slopes in the x and y directions,

$$\alpha = \frac{\partial f(x', y')}{\partial x'} \tag{13a}$$

$$\beta = \frac{\partial f(x', y')}{\partial y'} \tag{13b}$$

Substituting (10) into (6), we obtain, after some algebraic manipulations,

$$\overline{E}_s(\bar{r}) = \frac{ik\, e^{ikr}}{4\pi r} E_o (\overline{\overline{I}} - \hat{k}_s \hat{k}_s) \cdot \int_{A_o} d\bar{r}'_\perp \overline{F}(\alpha, \beta)\, e^{i(\bar{k}_i - \bar{k}_s) \cdot \bar{r}'} \tag{14a}$$

Kirchhoff Approach

Similarly, for the transmitted field,

$$\bar{E}_t(\bar{r}) = -\frac{ik_1 e^{ik_1 r}}{4\pi r} E_o(\bar{\bar{I}} - \hat{k}_t \hat{k}_t) \cdot \int_{A_o} d\bar{r}'_\perp \overline{N}(\alpha, \beta) e^{i(\bar{k}_i - \bar{k}_t) \cdot \bar{r}'} \quad (14b)$$

In (14),

$$\overline{F}(\alpha, \beta) = (1 + \alpha^2 + \beta^2)^{1/2} \Big\{ -(\hat{e}_i \cdot \hat{q}_i)(\hat{n} \cdot \hat{k}_i)\hat{q}_i(1 - R_h)$$
$$+ (\hat{e}_i \cdot \hat{p}_i)(\hat{n} \times \hat{q}_i)(1 + R_v) + (\hat{e}_i \cdot \hat{q}_i)(\hat{k}_s \times (\hat{n} \times \hat{q}_i))(1 + R_h)$$
$$+ (\hat{e}_i \cdot \hat{p}_i)(\hat{n} \cdot \hat{k}_i)(\hat{k}_s \times \hat{q}_i)(1 - R_v) \Big\} \quad (15a)$$

$$\overline{N}(\alpha, \beta) = (1 + \alpha^2 + \beta^2)^{1/2} \Big\{ -\frac{\eta_1}{\eta}(\hat{e}_i \cdot \hat{q}_i)(\hat{n} \cdot \hat{k}_i)\hat{q}_i(1 - R_h)$$
$$+ \frac{\eta_1}{\eta}(\hat{e}_i \cdot \hat{p}_i)(\hat{n} \times \hat{q}_i)(1 + R_v) + (\hat{e}_i \cdot \hat{q}_i)(\hat{k}_t \times (\hat{n} \times \hat{q}_i))(1 + R_h)$$
$$+ (\hat{e}_i \cdot \hat{p}_i)(\hat{n} \cdot \hat{k}_i)(\hat{k}_t \times \hat{q}_i)(1 - R_v) \Big\} \quad (15b)$$

The orthonormal systems for the incident, scattered, and transmitted fields are given respectively by $(\hat{v}_i, \hat{h}_i, \hat{k}_i)$, $(\hat{v}_s, \hat{h}_s, \hat{k}_s)$, and $(\hat{v}_t, \hat{h}_t, \hat{k}_t)$ with

$$\hat{k}_i = \hat{x} \sin \theta_i \cos \phi_i + \hat{y} \sin \theta_i \sin \phi_i - \hat{z} \cos \theta_i \quad (16a)$$
$$\hat{h}_i = -\hat{x} \sin \phi_i + \hat{y} \cos \phi_i \quad (16b)$$
$$\hat{v}_i = -\hat{x} \cos \theta_i \cos \phi_i - \hat{y} \cos \theta_i \sin \phi_i - \hat{z} \sin \theta_i \quad (16c)$$

$$\hat{k}_s = \hat{x} \sin \theta_s \cos \phi_s + \hat{y} \sin \theta_s \sin \phi_s + \hat{z} \cos \theta_s \quad (17a)$$
$$\hat{h}_s = -\hat{x} \sin \phi_s + \hat{y} \cos \phi_s \quad (17b)$$
$$\hat{v}_s = \hat{x} \cos \theta_s \cos \phi_s + \hat{y} \cos \theta_s \sin \phi_s - \hat{z} \sin \theta_s \quad (17c)$$

$$\hat{k}_t = \hat{x} \sin \theta_t \cos \phi_t + \hat{y} \sin \theta_t \sin \phi_t - \hat{z} \cos \theta_t \quad (18a)$$
$$\hat{h}_t = -\hat{x} \sin \phi_t + \hat{y} \cos \phi_t \quad (18b)$$
$$\hat{v}_t = -\hat{x} \cos \theta_t \cos \phi_t - \hat{y} \cos \theta_t \sin \phi_t - \hat{z} \sin \theta_t \quad (18c)$$

We note that except for the phase factors, the expressions in the integrands of the diffraction integral (14) are not explicit functions of $\bar{r}'$. They are

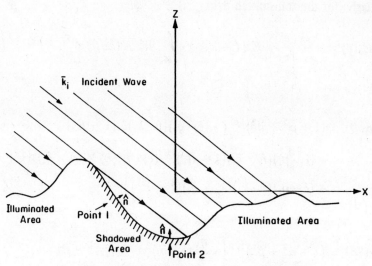

Fig. 2.17 The shadowed area on the rough surface.

explicit functions of the slopes α and β which are functions of $\bar{r}'$. The tangent plane-approximated diffraction integrals, as expressed in (14), do not take into account the effects of shadowing and multiple scattering.

When the angle of incidence is not normal to the x-y plane, some points on the rough surface will not be illuminated directly (Figure 2.17). For some points (point 1, Figure 2.17), the local angle of incidence θ_{l_i} is not defined since

$$\cos \theta_{l_i} = -\hat{n} \cdot \hat{k}_i < 0 \tag{19}$$

All the points on the rough surface with such local slopes will not be illuminated directly. Some other points (point 2, Figure 2.17) are not directly illuminated, even though the local angle of incidence is well defined, because of the height of rough surface at that point relative to the heights of the surrounding points. However, even without the complication of shadowing, the diffraction integrals for the scattered fields are difficult to evaluate analytically. This is because the local reflection coefficients R_v and R_h are functions of the surface slopes. In the limiting case of a perfectly conducting random rough surface, the local reflection coefficients R_v and R_h are 1 and -1, respectively, and do not depend on the local surface slopes. Then, by neglecting the shadowing effect so that at all points on the rough surface, $\hat{n} \times \overline{E} = 0$ and $\hat{n} \times \overline{H} = 2\hat{n} \times \overline{H}_i$, the diffraction integrals can be cast into a well-defined integral.

For the dielectric random rough surface, various approximations have been applied to the Kirchhoff-approximated diffraction integrals. The integrands that depend on the local surface slopes can be expanded about zero

Kirchhoff Approach

slopes, and, then integrated by parts with the edge effect discarded (Leader, 1971). Usually only the first few terms of the expansion are kept. In the high-frequency limit, the geometrical optics solution can be obtained from (14) with the stationary-phase method. The geometrical optics solution is independent of frequency and states that the scattered intensity is proportional to the probability of the occurrence of the slopes which will specularly reflect or transmit the incident wave into the direction of observation (Barrick, 1968).

In the calculation of the scattered fields, the expression for the diffraction integral (14a) contains the total field (incident and scattered) on the surface. An alternative is to use the scattered field. The scattered field evaluated from (14a) with the exact $\overline{F}(\alpha, \beta)$ is the same whether one uses the total field or the scattered field on the surface. However, when the integrand $\overline{F}(\alpha, \beta)$ is approximated, the results using total or scattered surface fields may not give the same result (Holzer and Sung, 1978). This is also true when shadowing is present. On the other hand, the geometrical optics solution is independent of whether total or scattered field is used since the integrand $\overline{F}(\alpha, \beta)$ is evaluated at the stationary phase points α_o and β_o.

Coherent and Incoherent Scattering Coefficients

The scattered intensities from a random rough surface can in general be decomposed into coherent and incoherent components. Coherent components only contribute in the specular reflected or transmitted directions whereas incoherent components contribute in all directions. In the limiting case of a flat surface, the scattered intensity consists of only the specularly reflected and transmitted coherent intensities. In the other limiting case of a very rough surface, the coherent components almost vanish and only incoherent components remain. In this section, the coherent and incoherent scattered intensities are solved by further approximating the integrands in the Kirchhoff diffraction integrals.

A commonly used approximation is to expand the integrands $\overline{F}(\alpha, \beta)$ and $\overline{N}(\alpha, \beta)$ about zero slopes and to keep only the first few terms (Leader, 1971). Expanding $\overline{F}$ and $\overline{N}$, we obtain

$$\overline{F}(\alpha, \beta) = \overline{F}(0,0) + \alpha \frac{\partial \overline{F}}{\partial \alpha}\bigg|_{\alpha,\beta=0} + \beta \frac{\partial \overline{F}}{\partial \beta}\bigg|_{\alpha,\beta=0} + \ldots \qquad (20a)$$

$$\overline{N}(\alpha, \beta) = \overline{N}(0,0) + \alpha \frac{\partial \overline{N}}{\partial \alpha}\bigg|_{\alpha,\beta=0} + \beta \frac{\partial \overline{N}}{\partial \beta}\bigg|_{\alpha,\beta=0} + \ldots \qquad (20b)$$

where $\overline{F}(0,0)$ is $\overline{F}(\alpha, \beta)$ evaluated at $\alpha = \beta = 0$, etc. For angles of incidence near normal and for surfaces with small root mean square (rms) slope, the Fresnel reflection coefficients only vary slightly with the change of local angle of incidence. Only the first terms in (20a) and (20b) are retained in our subsequent calculations. Thus, from (14), we have

$$\overline{E}_s = \frac{ik\, e^{ikr}}{4\pi r} E_o(\overline{\overline{I}} - \hat{k}_s \hat{k}_s) \cdot \overline{F}(0,0) I \qquad (21a)$$

$$\overline{E}_t = -\frac{ik_1\, e^{ik_1 r}}{4\pi r} E_o(\overline{\overline{I}} - \hat{k}_t \hat{k}_t) \cdot \overline{N}(0,0) I_t \qquad (21b)$$

where the integrals I and I_t are given by

$$I = \int_{A_o} e^{i(\overline{k}_i - \overline{k}_s)\cdot \overline{r}'} \, d\overline{r}'_\perp \qquad (22a)$$

$$I_t = \int_{A_o} e^{i(\overline{k}_i - \overline{k}_t)\cdot \overline{r}'} \, d\overline{r}'_\perp \qquad (22b)$$

The scattered and transmitted fields are next separated into a mean field and a fluctuating part of the field

$$\overline{E}_s(\overline{r}) = \overline{E}_{sm}(\overline{r}) + \overline{\mathcal{E}}_s(\overline{r}) \qquad (23)$$

$$\overline{E}_t(\overline{r}) = \overline{E}_{tm}(\overline{r}) + \overline{\mathcal{E}}_t(\overline{r}) \qquad (24)$$

with

$$<\overline{\mathcal{E}}_s(\overline{r})> = <\overline{\mathcal{E}}_t(\overline{r})> = 0 \qquad (25)$$

and $\overline{E}_{sm}$ and $\overline{E}_{tm}$ denote the mean scattered and transmitted fields, respectively. The total scattered intensity is then a sum of coherent and incoherent intensities

$$<|\overline{E}_s(\overline{r})|^2> = |E_{sm}|^2 + <|\overline{\mathcal{E}}_s(\overline{r})|^2> \qquad (26a)$$

$$<|\overline{E}_t(\overline{r})|^2> = |E_{tm}|^2 + <|\overline{\mathcal{E}}_t(\overline{r})|^2> \qquad (26b)$$

In view of (21) and (22), we have

$$|\overline{E}_{sm}(\overline{r})|^2 = \frac{k^2 |E_o|^2}{16\pi^2 r^2} \left\{ \left|[\hat{v}_s \cdot \overline{F}(0,0)]\right|^2 + \left|[\hat{h}_s \cdot \overline{F}(0,0)]\right|^2 \right\} |<I>|^2 \qquad (27a)$$

$$<|\overline{\mathcal{E}}_s(\overline{r})|^2> = \frac{k^2 |E_o|^2}{16\pi^2 r^2} \left\{ \left|[\hat{v}_s \cdot \overline{F}(0,0)]\right|^2 + \left|[\hat{h}_s \cdot \overline{F}(0,0)]\right|^2 \right\} D_I \qquad (27b)$$

$$|\overline{E}_{tm}(\bar{r})|^2 = \frac{k_1^2|E_o|^2}{16\pi^2 r^2}\left\{\left|[\hat{v}_t \cdot \overline{N}(0,0)]\right|^2 + \left|[\hat{h}_t \cdot \overline{N}(0,0)]\right|^2\right\}|<I_t>|^2 \quad (28a)$$

$$<|\overline{\mathcal{E}}_t(\bar{r})|^2> = \frac{k_1^2|E_o|^2}{16\pi^2 r^2}\left\{\left|[\hat{v}_t \cdot \overline{N}(0,0)]\right|^2 + \left|[\hat{h}_t \cdot \overline{N}(0,0)]\right|^2\right\} D_{I_t} \quad (28b)$$

where

$$D_I = <|I|^2> - |<I>|^2 \quad (29a)$$

$$D_{I_t} = <|I_t|^2> - |<I_t>|^2 \quad (29b)$$

Next, we specify further the height distribution $f(\bar{r}_\perp)$. The rough surface is assumed to be a stationary Gaussian process. The probability for $f(\bar{r}_\perp)$ is independent of the position $\bar{r}_\perp$ on the rough surface and has the Gaussian distribution

$$p(f(\bar{r}_\perp)) = \frac{1}{\sqrt{2\pi}\sigma} e^{-f^2/2\sigma^2} \quad (30)$$

where σ is the standard deviation of the surface height. For two points on the surface, $\bar{r}_{\perp 1}$ and $\bar{r}_{\perp 2}$, the joint probability density (Davenport and Root, 1958) is

$$p(f_1(\bar{r}_{\perp 1}), f_2(\bar{r}_{\perp 2})) = \frac{1}{2\pi\sigma^2\sqrt{1-C^2}} \exp\left[-\frac{f_1^2 - 2Cf_1 f_2 + f_2^2}{2\sigma^2(1-C^2)}\right] \quad (31)$$

where C is the correlation coefficient between the two points and is a function of $\bar{r}_{\perp 1}$ and $\bar{r}_{\perp 2}$. For a statistically homogeneous isotropic surface, it is only a function of $\rho = \sqrt{(x_1-x_2)^2 + (y_1-y_2)^2}$.

$$<f(\bar{r}_{\perp 1})f(\bar{r}_{\perp 2})> = \sigma^2 C(\rho) \quad (32)$$

$$C(0) = 1 \quad (33a)$$

$$C(\infty) = 0 \quad (33b)$$

It can be easily shown that the characteristic function is

$$<e^{i\nu f(\bar{r}_\perp)}> = \int_{-\infty}^{\infty} df\, p(f) e^{i\nu f} = \exp\left[-\frac{1}{2}\sigma^2\nu^2\right] \quad (34)$$

Similarly,

$$<e^{i\nu(f_1(\bar{r}_{\perp 1}) - f_2(\bar{r}_{\perp 2}))}> = \int_{-\infty}^{\infty}\int_{-\infty}^{\infty} df_1\, df_2\, p(f_1, f_2)\, e^{i\nu(f_1-f_2)}$$

$$= \exp\left[-\sigma^2\nu^2\left(1 - C(\rho)\right)\right] \quad (35)$$

The expressions for $|<I>|^2$, D_I, $|<I_t>|^2$ and D_{I_t} can now be derived in terms of the statistical moments of the height distribution.

The integral I is given by

$$I = \int_{A_o} e^{i\overline{k}_{d\perp} \cdot \overline{r}'_\perp} e^{ik_{dz} f(\overline{r}'_\perp)} d\overline{r}'_\perp \tag{36}$$

where

$$\overline{k}_d = \overline{k}_i - \overline{k}_s = k_{dx}\hat{x} + k_{dy}\hat{y} + k_{dz}\hat{z} \tag{37}$$

The ensemble average of I is given by

$$<I> = \int_{A_o} e^{i\overline{k}_{d\perp} \cdot \overline{r}'_\perp} <e^{ik_{dz} f(\overline{r}'_\perp)}> d\overline{r}'_\perp \tag{38}$$

In view of (34), we obtain, after carrying out the $d\overline{r}'_\perp$ integration,

$$<I> = 4L_x L_y \exp\left[-\frac{1}{2}k_{dz}^2 \sigma^2\right] \operatorname{sinc}(k_{dx} L_x) \operatorname{sinc}(k_{dy} L_y) \tag{39}$$

where $\operatorname{sinc} x = \sin x / x$, $2L_x$ and $2L_y$ are the lengths of rough surface illuminated in the x and y directions, respectively, so that

$$A_o = 4L_x L_y \tag{40}$$

By allowing L_x and L_y to approach infinity in the above expression, we obtain

$$|<I>|^2 = 4\pi^2 A_o \exp\left[-k_{dz}^2 \sigma^2\right] \delta(k_{dx}) \delta(k_{dy}) \tag{41}$$

where δ is the Dirac delta function and we made use of the identity (Goodman, 1968)

$$\lim_{L_x, L_y \to \infty} \frac{L_x L_y}{\pi^2} \operatorname{sinc}(k_{dx} L_x) \operatorname{sinc}(k_{dy} L_y) = \delta(k_{dx}) \delta(k_{dy}) \tag{42}$$

The integral for $<II^*>$ is given by

$$<II^*> = \int_{A_o} d\overline{r}_\perp \int_{A_o} d\overline{r}'_\perp e^{i\overline{k}_{d\perp} \cdot (\overline{r}_\perp - \overline{r}'_\perp)} <e^{ik_{dz}(f(\overline{r}_\perp) - f(\overline{r}'_\perp))}> \tag{43}$$

Using (35) and making the usual change of variables to the difference and half the sum of coordinates, we obtain

$$<II^*> = \int_{-2L_x}^{2L_x} dx \int_{-2L_y}^{2L_y} dy \, (2L_x - |x|)(2L_y - |y|)$$
$$\times \exp(ik_{dx}x + ik_{dy}y) \exp\left[-k_{dz}^2 \sigma^2 \left(1 - C(\rho)\right)\right] \tag{44}$$

Kirchhoff Approach

The correlation function $C(\rho)$ is assumed to have a Gaussian form (Barrick, 1970)

$$C(\rho) = e^{-\rho^2/l^2} \qquad (45)$$

where l is the correlation length for the random variable $f(\bar{r}_\perp)$ in the transverse plane. The expression for the standard derivation of the integral I can now be evaluated in closed form. We first note that $|<I>|^2$ can be also be expressed as

$$|<I>|^2 = \int_{-2L_x}^{2L_x} dx \int_{-2L_y}^{2L_y} dy \, (2L_x - |x|)(2L_y - |y|)$$
$$\times \exp(ik_{dx}x + ik_{dy}y)\exp[-\sigma^2 k_{dz}^2] \qquad (46)$$

Combining (44) and (46) and in view of (45), we note that the contribution of the integral of $<II^*> - |<I>|^2$ comes from $|x|$ and $|y|$ of the same order of l and the integrand is practically zero for $\rho = (x^2 + y^2)^{1/2}$ larger then a few l's. Assuming the illuminated rough surface contains many correlation lengths $L_x, L_y \gg l$, we obtain

$$D_I = <II^*> - |<I>|^2$$
$$= A_o \int_{-\infty}^{\infty} dx \int_{-\infty}^{\infty} dy \left\{ \exp\left[-\sigma^2 k_{dz}^2 (1 - C(\rho))\right] \right.$$
$$\left. - \exp\left[-\sigma^2 k_{dz}^2\right] \right\} \exp\left[ik_{dx}x + ik_{dy}y\right] \qquad (47)$$

Converting the integral in (47) to cylindrical coordinates and carrying out the integral in $d\phi$ give a Bessel function $J_o(k_\rho \rho)$ where

$$k_\rho = (k_{dx}^2 + k_{dy}^2)^{1/2} \qquad (48)$$

in the integrand. We further make a power series expansion

$$\exp\left[-\sigma^2 k_{dz}^2 \left(1 - C(\rho)\right)\right] - \exp\left[-\sigma^2 k_{dz}^2\right]$$
$$= \exp[-\sigma^2 k_{dz}^2] \sum_{m=1}^{\infty} \frac{(\sigma^2 k_{dz}^2)^m}{m!} \exp\left[-\frac{m\rho^2}{l^2}\right] \qquad (49)$$

and make use of the integral identity (Gradshteyn and Ryzhik, 1963)

$$\int_0^\infty d\rho \, \rho \, J_o(k_\rho \rho) e^{-m\rho^2/l^2} = \frac{l^2}{2m} \exp\left[-\frac{k_\rho^2 l^2}{4m}\right] \qquad (50)$$

Using (49) and (50) in (47), we obtain

$$D_I = <II^*> - |<I>|^2$$

$$= \pi A_o \sum_{m=1}^{\infty} \frac{(k_{dz}^2 \sigma^2)^m}{m!m} l^2 \exp\left[-\frac{(k_{dx}^2 + k_{dy}^2)l^2}{4m}\right] \exp\left[-\sigma^2 k_{dz}^2\right] \quad (51)$$

In a similar manner, the expressions for $|<I_t>|^2$ and D_{I_t} may be derived. They are

$$|<I_t>|^2 = 4\pi^2 A_o \exp\left[-\sigma^2 k_{tdz}^2\right] \delta(k_{tdx}) \delta(k_{tdy}) \quad (52)$$

and

$$D_{I_t} = \pi A_o \sum_{m=1}^{\infty} \frac{(k_{tdz}^2 \sigma^2)^m}{m!m} l^2 \exp\left[-\frac{(k_{tdx}^2 + k_{tdy}^2)l^2}{4m}\right] \exp\left[-\sigma^2 k_{tdz}^2\right] \quad (53)$$

where

$$\overline{k}_{td} = \overline{k}_i - \overline{k}_t = k_{tdx}\hat{x} + k_{tdy}\hat{y} + k_{tdz}\hat{z} \quad (54)$$

The bistatic scattering coefficients for the reflected intensities are defined as

$$\gamma_{ba}^r(\hat{k}_s, \hat{k}_i) = \frac{4\pi r^2 (S_r)_b}{A_o \cos\theta_i (S_o)_a} \quad (a, b = v, h) \quad (55)$$

where subscript a represents the polarization of the incident wave, subscript b the polarization of the scattered wave, S_o the Poynting power density of the incident wave, S_r the Poynting density of the scattered wave, A_o the area of the rough surface projected onto the x-y plane, and θ_i the incident angle. From (21) and (27), we calculate the vertically and horizontally polarized coherent and incoherent scattered intensities for the cases of vertically and horizontally polarized incident fields. Let

$$\overline{F}_b(0,0) = \overline{F}(0,0)\big|_{\hat{e}_i = \hat{b}_i} \quad (56)$$

$\overline{F}(0,0)$ can be calculated by setting $\alpha = \beta = 0$ in (12) and (15a). Next we take the dot product with $\hat{v}_s$ and $\hat{h}_s$. Thus,

$$\hat{h}_s \cdot \overline{F}_h(0,0) = \left[(1 - R_{ho})\cos\theta_i - (1 + R_{ho})\cos\theta_s\right]\cos(\phi_s - \phi_i) \quad (57a)$$

$$\hat{v}_s \cdot \overline{F}_h(0,0) = \left[(1 - R_{ho})\cos\theta_i \cos\theta_s - (1 + R_{ho})\right]\sin(\phi_s - \phi_i) \quad (57b)$$

$$\hat{h}_s \cdot \overline{F}_v(0,0) = \Big[(1+R_{vo}) - (1-R_{vo})\cos\theta_i\cos\theta_s\Big]\sin(\phi_s - \phi_i) \quad (57c)$$

$$\hat{v}_s \cdot \overline{F}_v(0,0) = \Big[-(1+R_{vo})\cos\theta_s + (1-R_{vo})\cos\theta_i\Big]\cos(\phi_s - \phi_i) \quad (57d)$$

The R_{vo} and R_{ho} of the above equations are respectively the Fresnel reflection coefficients of a smooth surface for vertically and horizontally polarized incident waves and are equal to the expressions in (9a) and (9b) with θ_{li} replaced by θ_i.

In view of (26), the bistatic scattering coefficients γ_{ab}^r can be decomposed into a coherent part $\gamma_{c_{ab}}^r$ and an incoherent part $\gamma_{i_{ab}}^r$.

$$\gamma_{ab}^r(\hat{k}_s, \hat{k}_i) = \gamma_{c_{ab}}^r(\hat{k}_s, \hat{k}_i) + \gamma_{i_{ab}}^r(\hat{k}_s, \hat{k}_i) \quad (58)$$

where

$$\gamma_{c_{ab}}^r(\hat{k}_s, \hat{k}_i) = \frac{k^2}{4\pi A_o \cos\theta_i} \left|\hat{a}_s \cdot \overline{F}_b(0,0)\right|^2 |<I>|^2 \quad (59a)$$

$$\gamma_{i_{ab}}^r(\hat{k}_s, \hat{k}_i) = \frac{k^2}{4\pi A_o \cos\theta_i} \left|\hat{a}_s \cdot \overline{F}_b(0,0)\right|^2 D_I \quad (59b)$$

The coherent scattered wave only exists in the specular directions. Thus, using (41) and (57), and since

$$\delta(k_{dx})\delta(k_{dy}) = \frac{\delta(\theta_s - \theta_i)\delta(\phi_s - \phi_i)}{(k^2 \sin\theta_i \cos\theta_i)}$$

we find that

$$\gamma_{c_{ab}}^r(\hat{k}_s, \hat{k}_i) = \frac{4\pi |R_{bo}|^2}{\sin\theta_i} \exp(-4k^2\sigma^2\cos^2\theta_i)\delta(\theta_s - \theta_i)\delta(\phi_s - \phi_i)\delta_{ab} \quad (60)$$

where δ_{ab} is the Kronecker delta.

By the same token, we define the bistatic transmission coefficients to be,

$$\gamma_{ba}^t(\hat{k}_t, \hat{k}_i) = \frac{4\pi r^2 (S_t)_b}{A_o \cos\theta_i (S_o)_a} \quad (61)$$

where S_t is the Poynting power density of the transmitted wave. Following the same procedure we obtain

$$\gamma_{ab}^t(\hat{k}_t, \hat{k}_i) = \gamma_{c_{ab}}^t(\hat{k}_t, \hat{k}_i) + \gamma_{i_{ab}}^t(\hat{k}_t, \hat{k}_i) \quad (62)$$

where

$$\gamma^t_{c_{ab}}(\hat{k}_t, \hat{k}_i) = \frac{k_1^2}{4\pi A_o \cos\theta_i} \frac{\eta}{\eta_1} \left|\hat{a}_t \cdot \overline{N}_b(0,0)\right|^2 |<I_t>|^2 \quad (63a)$$

$$\gamma^t_{i_{ab}}(\hat{k}_t, \hat{k}_i) = \frac{k_1^2}{4\pi A_o \cos\theta_i} \frac{\eta}{\eta_1} \left|\hat{a}_t \cdot \overline{N}_b(0,0)\right|^2 D_{I_t} \quad (63b)$$

with

$$\overline{N}_b(0,0) = \overline{N}(0,0)\Big|_{\hat{e}_i = \hat{b}_i} \quad (64)$$

and

$$\hat{h}_t \cdot \overline{N}_h(0,0) = \left[\frac{\eta_1}{\eta}(1 - R_{ho})\cos\theta_i + (1 + R_{ho})\cos\theta_t\right]\cos(\phi_t - \phi_i) \quad (65a)$$

$$\hat{v}_t \cdot \overline{N}_h(0,0) = \left[-\frac{\eta_1}{\eta}(1 - R_{ho})\cos\theta_i\cos\theta_t - (1 + R_{ho})\right]\sin(\phi_t - \phi_i) \quad (65b)$$

$$\hat{h}_t \cdot \overline{N}_v(0,0) = \left[\frac{\eta_1}{\eta}(1 + R_{vo}) + (1 - R_{vo})\cos\theta_i\cos\theta_t\right]\sin(\phi_t - \phi_i) \quad (65c)$$

$$\hat{v}_t \cdot \overline{N}_v(0,0) = \left[\frac{\eta_1}{\eta}(1 + R_{vo})\cos\theta_t + (1 - R_{vo})\cos\theta_i\right]\cos(\phi_t - \phi_i) \quad (65d)$$

Again, the coherent component only exists in the specular transmission direction, and we find

$$\gamma^t_{c_{ab}}(\hat{k}_t, \hat{k}_i) = 4\pi \frac{\cos\theta_{1i}}{\cos\theta_i} \frac{|1 + R_{bo}|^2}{\sin\theta_{1i}} g_a \exp\left[-(k_1\cos\theta_{1i} - k\cos\theta_i)^2\sigma^2\right]$$
$$\times \delta(\theta_t - \theta_{1i})\delta(\phi_t - \phi_i)\delta_{ab} \quad (66)$$

where $g_a = \eta_1/\eta$ for $a = v$ and $g_a = \eta/\eta_1$, for $a = h$ and θ_{1i} is related to θ_i by Snell's law

$$k_1 \sin\theta_{1i} = k\sin\theta_i \quad (67)$$

We note that the coherent component is only nonzero in the specular direction. Also as $k\sigma$ increases, the coherent component diminishes exponentially because scattered fields from different parts of the rough surface have a large variance in phase fluctuation.

Kirchhoff Approach

Geometrical Optics Solution

Under the geometrical optics limit as $k \to \infty$, the asymptotic solution to the Kirchhoff-approximated diffraction integrals can be derived using the method of stationary-phase. The coherent component of the scattered fields will vanish in this limit and only the incoherent component will remain. The bistatic scattering coefficients for the reflected and transmitted fields are derived and shown to be proportional to the probability of the occurrence of the slopes which will specularly reflect or transmit the incident wave into the observation direction. The bistatic scattering coefficients satisfy reciprocity but violate energy conservation. This is due to the neglect of the effects of multiple scattering and shadowing. The scattering coefficients are next modified to incorporate the shadowing effects. The sum of reflected and transmitted intensities are then shown to be always less than the incident intensity since only single scattering solutions are used. This will be used in the next section to derive the upper and lower bounds for the correct emissivity of a rough surface in the geometrical optics limit.

Stationary-Phase Method

The diffraction integrals are evaluated by the method of stationary-phase. The reflected fields are first calculated. From (14), the exponential phase factor is

$$\psi = \overline{k}_d \cdot \overline{r}' = k_{dx} x' + k_{dy} y' + k_{dz} f(x', y') \tag{68}$$

To determine the stationary-phase point, we set

$$\frac{\partial \psi}{\partial x'} = 0 = k_{dx} + k_{dz} \alpha_o \tag{69}$$

so that at stationary-phase point

$$\alpha_o = -\frac{k_{dx}}{k_{dz}} \tag{70}$$

Similarly, by differentiating the phase term ψ with respect to y', we get

$$\beta_o = -\frac{k_{dy}}{k_{dz}} \tag{71}$$

The slopes α_o and β_o are such that the incident and scattered wave direction form a specular reflection. This can be seen from the fact that from (12) we have

$$\hat{n}(\alpha_o, \beta_o) = (\overline{k}_s - \overline{k}_i)/|\overline{k}_d| \tag{72}$$

Replacing the surface slopes α and β by α_o and β_o, we obtain, using (14)

$$<|\overline{E}_s|^2> = \frac{k^2|E_o|^2}{16\pi^2 r^2}\left|(\overline{\overline{I}} - \hat{k}_s\hat{k}_s)\cdot \overline{F}(\alpha_o,\beta_o)\right|^2 <II^*> \qquad (73)$$

where

$$<II^*> = <\int_{A_o} d\overline{r}_\perp \int_{A_o} d\overline{r}'_\perp \, e^{i\overline{k}_{d\perp}\cdot(\overline{r}_\perp - \overline{r}'_\perp)} e^{ik_{dz}(f(\overline{r}_\perp) - f(\overline{r}'_\perp))} > \qquad (74)$$

The above integral can be solved by the method of asymptotics. For large k, contributions of the integral come from regions where (x',y') is close to (x,y). Expanding $f(x',y')$ about (x,y),

$$f(x',y') = f(x,y) + \alpha(x' - x) + \beta(y' - y) + \ldots \qquad (75)$$

and replacing the integration variables by

$$u = k(x - x') \qquad (76a)$$
$$v = k(y - y') \qquad (76b)$$

we obtain

$$<II^*> = \left\langle \frac{1}{k^2}A_o \int\int du\, dv \right.$$
$$\left. \times \exp\left[iu(q_x + \alpha q_z) + iv(q_y + \beta q_z) + 0(\frac{1}{k})\right]\right\rangle \qquad (77)$$

where

$$\overline{q} = \frac{\overline{k}_d}{k} \qquad (78)$$

Ignoring the $O(1/k)$ and higher order terms, we have

$$<II^*> = \frac{4\pi^2 A_o}{k^2} <\delta(q_x + \alpha q_z)\delta(q_y + \beta q_z)> \qquad (79)$$

Therefore

$$<\lim_{k\to\infty} II^*> = \frac{4\pi^2 A_o}{k^2}\int_{-\infty}^{\infty}\int_{-\infty}^{\infty} d\alpha\, d\beta\, \delta(q_x + \alpha q_z)\delta(q_y + \beta q_z)p(\alpha,\beta) \qquad (80)$$

where $p(\alpha,\beta)$ is the probability density function for the slopes at the surface. It follows that

$$<\lim_{k\to\infty} II^*> = \frac{4\pi^2 A_o}{k_{dz}^2}p\left(-\frac{k_{dx}}{k_{dz}}, -\frac{k_{dy}}{k_{dz}}\right) \qquad (81)$$

Kirchhoff Approach

For the Gaussian random rough surface (Problem 14)

$$p(\alpha, \beta) = \frac{1}{2\pi\sigma^2 |C''(0)|} \exp\left[-\frac{\alpha^2 + \beta^2}{2\sigma^2 |C''(0)|}\right] \tag{82}$$

where σ is the standard deviation of the height of rough surface and $C''(0)$ is the double derivative of the correlation function at $\rho = 0$. Thus, $\sigma|C''(0)|$ is the mean square surface slope s^2 and for the Gaussian correlation function of (45) with correlation length l

$$s^2 = \sigma^2 |C''(0)| = 2\frac{\sigma^2}{l^2} \tag{83}$$

Using (82) in (81) gives

$$<II^*> = \frac{2\pi A_o}{k_{dz}^2 \sigma^2 |C''(0)|} \exp\left[-\frac{k_{dx}^2 + k_{dy}^2}{2k_{dz}^2 \sigma^2 |C''(0)|}\right] \tag{84}$$

Another way to evaluate $<II^*>$ is to perform the ensemble average first, and then to approximate the integral. From (44) and (74)

$$<II^*> = \int_{-2L_x}^{2L_x} dx \int_{-2L_y}^{2L_y} dy\, (2L_x - |x|)(2L_y - |y|)$$
$$\times e^{i\overline{k}_{d\perp} \cdot \overline{r}_\perp} \exp\left[-k_{dz}^2 \sigma^2 (1 - C(\rho))\right] \tag{85}$$

Since $k_{dz}^2 \sigma^2 \gg 1$, most of the contribution comes from around the origin. Thus, expanding the integrand about the origin we have $1 - C(\rho) \approx \rho^2 |C''(0)|/2$, and substituting into (84) the integral can be evaluated readily by making use of the integral identity of (50). The final result for $<II^*>$ is the same as (84).

For an incident field with polarization b_i, the scattered intensity for polarization a_s is given by

$$<|E_s(\overline{r})|^2> = \frac{k^2 |E_o|^2}{16\pi^2 r^2} |\hat{a}_s \cdot \overline{F}_b(\alpha_o, \beta_o)|^2 <II^*> \tag{86}$$

where

$$\overline{F}_b(\alpha_o, \beta_o) = \overline{F}(\alpha_o, \beta_o)\big|_{\hat{e}_i = \hat{b}_i} \tag{87}$$

and using (15a), we find that

$$|\hat{a}_s \cdot \overline{F}_b(\alpha_o, \beta_o)|^2 = \frac{|\overline{k}_d|^4}{k^2 |\hat{k}_i \times \hat{k}_s|^4 k_{dz}^2} f_{ba} \tag{88}$$

with

$$f_{vv} = |(\hat{h}_s \cdot \hat{k}_i)(\hat{h}_i \cdot \hat{k}_s)R_h + (\hat{v}_s \cdot \hat{k}_i)(\hat{v}_i \cdot \hat{k}_s)R_v|^2 \qquad (89a)$$

$$f_{hv} = |(\hat{v}_s \cdot \hat{k}_i)(\hat{h}_i \cdot \hat{k}_s)R_h - (\hat{h}_s \cdot \hat{k}_i)(\hat{v}_i \cdot \hat{k}_s)R_v|^2 \qquad (89b)$$

$$f_{vh} = |(\hat{h}_s \cdot \hat{k}_i)(\hat{v}_i \cdot \hat{k}_s)R_h - (\hat{v}_s \cdot \hat{k}_i)(\hat{h}_i \cdot \hat{k}_s)R_v|^2 \qquad (89c)$$

$$f_{hh} = |(\hat{v}_s \cdot \hat{k}_i)(\hat{v}_i \cdot \hat{k}_s)R_h + (\hat{h}_s \cdot \hat{k}_i)(\hat{h}_i \cdot \hat{k}_s)R_v|^2 \qquad (89d)$$

and R_v and R_h are evaluated at

$$\hat{n} = \frac{k_{dx}/k_{dz}\hat{x} + k_{dy}/k_{dz}\hat{y} + \hat{z}}{\sqrt{k_{dx}^2/k_{dz}^2 + k_{dy}^2/k_{dz}^2 + 1}} \qquad (90)$$

Then, the bistatic scattering coefficients for the reflected intensities are, in view of (55) and (84)

$$\gamma_{ab}^r(\hat{k}_s, \hat{k}_i) = \frac{|\overline{k}_d|^4}{\cos\theta_i |\hat{k}_i \times \hat{k}_s|^4 k_{dz}^4} \frac{1}{2\sigma^2 |C''(0)|} \exp\left[-\frac{k_{dx}^2 + k_{dy}^2}{2k_{dz}^2 \sigma^2 |C''(0)|}\right] f_{ab} \qquad (91)$$

We note that from (89) the geometrical optics solution does not depend on whether the total or scattered field is used in the diffraction integral because $\overline{F}(\alpha, \beta)$ is evaluated at the stationary-phase point (Holzer and Sung, 1978).

In the backscattering direction $\hat{k}_s = -\hat{k}_i$. The backscattering cross sections are defined to be

$$\sigma_{ab}(\hat{k}_i) = \cos\theta_i \, \gamma_{ab}^r(-\hat{k}_i, \hat{k}_i) \qquad (92)$$

From (92), we obtain

$$\sigma_{hh}(\hat{v}_i) = \sigma_{vv}(\theta_i) = \frac{|R|^2}{\cos^4\theta_i 2\sigma^2 |C''(0)|} \exp\left[-\frac{\tan^2\theta_i}{2\sigma^2 |C''(0)|}\right] \qquad (93a)$$

$$\sigma_{vh}^r(\theta_i) = \sigma_{hv}^r(\theta_i) = 0 \qquad (93b)$$

where R is the reflection coefficient at normal incidence. We note that from (93b) that there is no depolarization in the backscattering direction.

The bistatic scattering coefficients for the transmitted waves can be derived in a similar manner. The stationary-phase method is used to evaluate the diffraction integral for the transmitted fields. The stationary phase points are given by

$$\alpha_o = -\frac{k_{tdx}}{k_{tdz}} \qquad (94a)$$

Kirchhoff Approach

$$\beta_o = -\frac{k_{tdy}}{k_{tdz}} \qquad (94b)$$

where α_o and β_o are the values that the slopes α and β assume at the stationary-phase point. We note that at the stationary-phase point

$$\hat{n} = \frac{k_{tdx}/k_{tdz}\hat{x} + k_{tdy}/k_{tdz}\hat{y} + \hat{z}}{\sqrt{k_{tdx}^2/k_{tdz}^2 + k_{tdy}^2/k_{tdz}^2 + 1}} \qquad (95)$$

and it can be shown that

$$\overline{k}_i - (\overline{k}_i \cdot \hat{n})\hat{n} = \overline{k}_t - (\overline{k}_t \cdot \hat{n})\hat{n} \qquad (96)$$

which is a statement of Snell's law that the tangential components of the wave vectors $\overline{k}_i$ and $\overline{k}_t$ must be equal. Thus, the slopes α_o and β_o are such that incident and transmitted wave directions form a specular transmission.

The transmitted field is obtained from (14) by replacing α and β by α_o and β_o and the associated integrals are evaluated in a manner similar to that of the scattered fields. The bistatic scattering coefficients for the transmitted intensities are, in view of (61)

$$\gamma_{ab}^t(\hat{k}_t, \hat{k}_i) = \frac{2k_1^2|\overline{k}_{td}|^2(\hat{n}\cdot\hat{k}_t)^2}{\cos\theta_i |\hat{k}_i \times \hat{k}_t|^4 k_{tdz}^4} \frac{\eta}{\eta_1} \frac{1}{\sigma^2|C''(0)|}$$

$$\times \exp\left[-\frac{k_{tdx}^2 + k_{tdy}^2}{2k_{tdz}^2 \sigma^2 |C''(0)|}\right] W_{ab} \qquad (97)$$

with

$$W_{vv} = \left|(\hat{h}_t \cdot \hat{k}_i)(\hat{h}_i \cdot \hat{k}_t)(1 + R_h) + (\hat{v}_t \cdot \hat{k}_i)(\hat{v}_i \cdot \hat{k}_t)\frac{\eta_1}{\eta}(1 + R_v)\right|^2 \qquad (98a)$$

$$W_{hv} = \left|-(\hat{v}_t \cdot \hat{k}_i)(\hat{h}_i \cdot \hat{k}_t)(1 + R_h) + (\hat{h}_t \cdot \hat{k}_i)(\hat{v}_i \cdot \hat{k}_t)\frac{\eta_1}{\eta}(1 + R_v)\right|^2 \qquad (98b)$$

$$W_{vh} = \left|(\hat{h}_t \cdot \hat{k}_i)(\hat{v}_i \cdot \hat{k}_t)(1 + R_h) - (\hat{v}_t \cdot \hat{k}_i)(\hat{h}_i \cdot \hat{k}_t)\cdot\frac{\eta_1}{\eta}(1 + R_v)\right|^2 \qquad (98c)$$

$$W_{hh} = \left|(\hat{v}_t \cdot \hat{k}_i)(\hat{v}_i \cdot \hat{k}_t)(1 + R_h) + (\hat{h}_t \cdot \hat{k}_i)(\hat{h}_i \cdot \hat{k}_t)\cdot\frac{\eta_1}{\eta}(1 + R_v)\right|^2 \qquad (98d)$$

The reflection coefficients R_h and R_v are to be taken at the stationary-phase point of (94) so that

$$R_h = \frac{k(\hat{n}\cdot\hat{k}_i) - k_1(\hat{n}\cdot\hat{k}_t)}{k(\hat{n}\cdot\hat{k}_i) + k_1(\hat{n}\cdot\hat{k}_t)} \qquad (99a)$$

$$R_v = \frac{k_1(\hat{n}\cdot\hat{k}_i) - k(\hat{n}\cdot\hat{k}_t)}{k_1(\hat{n}\cdot\hat{k}_i) + k(\hat{n}\cdot\hat{k}_t)} \qquad (99b)$$

Specular Surface Limit

In the limit $\sigma^2|C''(0)| \to 0$, which corresponds to the vanishing of the variance of the slope, a specular surface is obtained. In such a limit (Goodman, 1968)

$$\lim_{\sigma^2|C''(0)|\to 0} \frac{1}{2\sigma^2|C''(0)|} \exp\left[-\frac{k_{dx}^2 + k_{dy}^2}{2k_{dz}^2\sigma^2|C''(0)|}\right] = \pi\delta\left(\frac{k_{dx}}{k_{dz}}, \frac{k_{dy}}{k_{dz}}\right) \tag{100}$$

The δ functions can be expressed in terms of angular variables which imply that the scattering is nonzero only at $\theta_s = \theta_i$ and $\phi_s = \phi_i$. The bistatic scattering coefficients of (91) for the reflected intensities simplify to

$$\gamma_{ba}^r(\hat{k}_s, \hat{k}_i) = \frac{4\pi}{\sin\theta_i}|R_{bo}|^2 \delta(\theta_s - \theta_i)\delta(\phi_s - \phi_i)\delta_{ab} \tag{101}$$

where R_{vo} and R_{ho} are the Fresnel reflection coefficients of a flat surface. In a similar manner, the bistatic scattering coefficients for the transmitted intensities simplify to

$$\gamma_{ba}^t(\hat{k}_t, \hat{k}_i) = \frac{4\pi}{\sin\theta_{1i}}(1 - |R_{bo}|^2)\delta(\theta_t - \theta_{1i})\delta(\phi_t - \phi_i)\delta_{ba} \tag{102}$$

where

$$\theta_{1i} = \sin^{-1}\left(\frac{k}{k_1}\sin\theta_i\right) \tag{103}$$

Therefore, all the intensities are scattered into the specular reflection and transmission directions of a flat surface.

The bistatic scattering coefficients obtained in this section are single scattering solutions that neglect the multiple scattering and shadowing effects. In the present form, they satisfy the principle of reciprocity but violate the energy conservation. In the next section, the reciprocity and energy conservation relations will be investigated. Then the bistatic coefficients are modified to incorporate the shadowing effects and later used to study the emissivity of a rough surface in the geometrical optics limit.

Reciprocity and Conservation of Energy

We note that a reciprocity relation exists for the bistatic transmission and reflection coefficients obtained in (91) and (97). Consider two media, 1 and 2, with indices of refraction n_1 and n_2 and with wave numbers k_1 and k_2, respectively. The two media are separated by a rough interface. Then the bistatic transmission coefficient $\gamma_{ab}^{t_{21}}(\hat{k}_2, \hat{k}_1)$ signifies a wave incident from angle (θ_1, ϕ_1) with polarization b onto angle (θ_2, ϕ_2) in medium 2

Kirchhoff Approach

with polarization a. This is obtained from (97) by substituting $(\theta_i, \phi_i) = (\theta_1, \phi_1)$, $(\theta_t, \phi_t) = (\theta_2, \phi_2)$, $k_t = k_2$, and $k = k_1$. Similarly, $\gamma_{ba}^{t12}(\hat{k}_1, \hat{k}_2)$ is for an incident wave from medium 2 and can be obtained by substituting in (97), $(\theta_i, \phi_i) = (\theta_2, \phi_2)$, $(\theta_t, \phi_t) = (\theta_1, \phi_1)$, $k_t = k_1$ and $k = k_2$. The following reciprocity relation is seen to hold for the bistatic transmission coefficients (Problem 15).

$$n_1^2 \cos\theta_1 \, \gamma_{ab}^{t21}(\hat{k}_2, \hat{k}_1) = n_2^2 \cos\theta_2 \, \gamma_{ba}^{t12}(\hat{k}_1, \hat{k}_2) \tag{104}$$

For the bistatic reflection coefficients, we similarly obtain the following reciprocity relation

$$\cos\theta_s \, \gamma_{ab}^{rjj}(\hat{k}_i, \hat{k}_s) = \cos\theta_i \, \gamma_{ba}^{rjj}(\hat{k}_s, \hat{k}_i) \tag{105}$$

with $j = 1, 2$. In (105), $\gamma_{ab}^{r11}(\hat{k}_s, \hat{k}_i)$ and $\gamma_{ab}^{r22}(\hat{k}_s, \hat{k}_i)$ are the bistatic reflection coefficients in media 1 and 2 between an incident wave with polarization b and a scattered wave with polarization a.

In the definition for the bistatic scattering coefficients, we note that $S_o A_o \cos\theta_i$ is the power intercepted by the surface area normal to the direction of the incident wave. The total power reflected is $r^2 S_r$ integrated over the upper hemisphere. Similarly, the total transmitted power is $r^2 S_t$ integrated over the lower hemisphere. We define the reflectivity

$$r_b(\theta_i) = \frac{1}{4\pi} \sum_a \int_0^{\pi/2} d\theta_s \sin\theta_s \int_0^{2\pi} d\phi_s \, \gamma_{ab}^r(\hat{k}_s, \hat{k}_i) \tag{106a}$$

and the transmissivity

$$t_b(\theta_i) = \frac{1}{4\pi} \sum_a \int_0^{\pi/2} d\theta_t \sin\theta_t \int_0^{2\pi} d\phi_t \, \gamma_{ab}^t(\hat{k}_t, \hat{k}_i) \tag{106b}$$

where the summation a is over both the vertical and horizontal polarizations.

Making use of (89) through (91), we find the reflectivities for the vertical and horizontal polarizations to be

$$\begin{bmatrix} r_v(\theta_i) \\ r_h(\theta_i) \end{bmatrix} = \int_0^{\pi/2} d\theta_s \sin\theta_s \int_0^{2\pi} d\phi_s \, \frac{1}{2\pi\sigma^2 |C''(0)|}$$

$$\times \exp\left[-\frac{k_{dx}^2 + k_{dy}^2}{2k_{dz}^2 \sigma^2 |C''(0)|}\right] \frac{|\bar{k}_d|^4}{4\cos\theta_i |\hat{k}_i \times \hat{k}_s|^2 k_{dz}^4}$$

$$\cdot \begin{bmatrix} (\hat{h}_i \cdot \hat{k}_s)^2 |R_h|^2 + (\hat{v}_i \cdot \hat{k}_s)^2 |R_v|^2 \\ (\hat{v}_i \cdot \hat{k}_s)^2 |R_h|^2 + (\hat{h}_i \cdot \hat{k}_s)^2 |R_v|^2 \end{bmatrix} \tag{107}$$

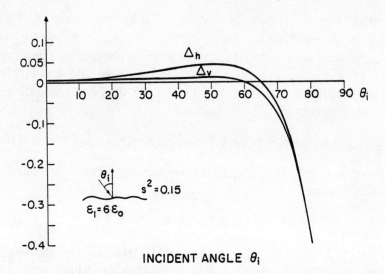

Fig. 2.18 Energy conservation relations for the geometrical optics solution.

where we made use of the fact that

$$|\hat{k}_i \times \hat{k}_s|^2 = (\hat{v}_i \cdot \hat{k}_s)^2 + (\hat{h}_i \cdot \hat{k}_s)^2 = (\hat{h}_s \cdot \hat{k}_i)^2 + (\hat{v}_s \cdot \hat{k}_i)^2 \qquad (108)$$

Similarly, making use of (97) and (98), the transmissivities for the vertical and horizontal polarizations are found to be

$$\begin{bmatrix} t_v(\theta_i) \\ t_h(\theta_i) \end{bmatrix} = \int_o^{\pi/2} d\theta_t \sin\theta_t \int_o^{2\pi} d\phi_t \frac{1}{2\pi\sigma^2 |C''(0)|}$$

$$\times \exp\left[-\frac{k_{tdx}^2 + k_{tdy}^2}{2k_{tdz}^2 \sigma^2 |C''(0)|}\right] \frac{k_1^2 |\overline{k}_{td}|^2 (\hat{n} \cdot \hat{k}_i)(\hat{n} \cdot \hat{k}_t)}{\cos\theta_i |\hat{k}_i \times \hat{k}_t|^2 k_{tdz}^4}$$

$$\cdot \begin{bmatrix} (\hat{h}_i \cdot \hat{k}_t)^2(1 - |R_h|^2) + (\hat{v}_i \cdot \hat{k}_t)^2(1 - |R_v|^2) \\ (\hat{v}_i \cdot \hat{k}_t)^2(1 - |R_h|^2) + (\hat{h}_i \cdot \hat{k}_t)^2(1 - |R_v|^2) \end{bmatrix} \quad (109)$$

where we made use of

$$|\hat{k}_i \times \hat{k}_t|^2 = (\hat{h}_t \cdot \hat{k}_i)^2 + (\hat{v}_t \cdot \hat{k}_i)^2 = (\hat{h}_i \cdot \hat{k}_t)^2 + (\hat{v}_i \cdot \hat{k}_t)^2 \qquad (110)$$

and

$$|1 + R_h|^2 = (1 - |R_h|^2)\frac{(\hat{n} \cdot \hat{k}_i)}{n_t(\hat{n} \cdot \hat{k}_t)} \qquad (111a)$$

$$\frac{|1 + R_v|^2}{n_t^2} = (1 - |R_v|^2)\frac{(\hat{n} \cdot \hat{k}_i)}{n_t(\hat{n} \cdot \hat{k}_t)} \qquad (111b)$$

with $n_t = k_t/k$.

Conservation of energy relations should also exist for the reflectivity and transmissivity functions. However, as pointed out before, because the shadowing and multiple scattering effects are ignored, conservation of energy is only approximately satisfied (Lynch and Wagner, 1968, 1970). To investigate the violation of the conservation of energy, we define

$$\Delta_a(\theta_i) = 1 - r_a(\theta_i) - t_a(\theta_i) \tag{112}$$

with $a = v, h$. In Figure 2.18, $\Delta_h(\theta_i)$ and $\Delta_v(\theta_i)$ are plotted as a function of incident angle. We see that energy loss becomes severe as incident angle increases. However, as incident angle is further increased, the trend reverses and the sum of reflected and transmitted energy becomes greater than unity. Near the normal angle of incidence there is little shadowing and most of the rough surface is illuminated. Therefore, since we only have the single scattering solution and have ignored multiple scattering, there is loss of energy. At higher angle of incidence, the multiple scattering solutions are still left out. However, the shadowing effect dominates and the single scattering solution overestimates scattered energy as incident angle is increased. We shall modify the bistatic scattering coefficients to include the effect of shadowing and show that in doing so the sum of reflected and transmitted energy is always less than unity.

Shadowing Effect

The bistatic scattering coefficients are now modified to account for the shadowing effect. The modified scattering coefficients satisfy the reciprocity but still not conserve energy since multiple scattering effects are neglected. The sum of reflected and transmitted energy is shown to be always less than unity. The modified bistatic scattering coefficients will be used to derive the upper and lower bound for the correct emissivity of the rough surface.

The bistatic reflection coefficients are first considered. The diffraction integral for the reflected field, (14), is modified with the addition of an illumination function $L(\hat{k}_i, \hat{k}_s, \bar{r}')$, as follows:

$$\overline{E}_s(\bar{r}) = \frac{ik\,e^{ikr}}{4\pi r} E_o(\overline{\overline{I}} - \hat{k}_s\hat{k}_s) \cdot \int_{A_o} d\bar{r}'_\perp \overline{F}(\alpha, \beta) L(\hat{k}_i, \hat{k}_s, \bar{r}') e^{i\bar{k}_d \cdot \bar{r}'} \tag{113}$$

where $L(\hat{k}_i, \hat{k}_s, \bar{r}') = 1$ if a ray having a direction $\hat{k}_i$ is not intersected by the surface and illuminates the point $\bar{r}'$ and if the line drawn from the point $\bar{r}'$ in the direction $\hat{k}_s$ does not strike the surface, and $L(\hat{k}_i, \hat{k}_s, \bar{r}') = 0$ otherwise (Sancer, 1969). The above integral is evaluated by the method of stationary-phase. The scattered intensity is, from (73) and (74),

$$<|\overline{E}_s|^2> = \frac{k^2|E_o|^2}{16\pi^2 r^2} \left|(\overline{\overline{I}} - \hat{k}_s\hat{k}_s) \cdot \overline{F}(\alpha_o, \beta_o)\right|^2 <II^*> \tag{114}$$

where $<II^*>$ is now modified to include shadowing

$$<II^*> = \left\langle \int_{A_o} d\bar{r}_\perp \int_{A_o} d\bar{r}'_\perp \, e^{i\bar{k}_{d\perp}\cdot(\bar{r}_\perp - \bar{r}'_\perp)} \right.$$
$$\left. \times L(\hat{k}_i, \hat{k}_s, \bar{r}_\perp) L(\hat{k}_i, \hat{k}_s, \bar{r}'_\perp) \, e^{ik_{dz}(f(\bar{r}_\perp) - f(\bar{r}'_\perp))} \right\rangle \quad (115)$$

The above integral is solved by the method of asymptotics as in (75) through (82). We also note that as $k \to \infty$,

$$\lim_{k \to \infty} L(\hat{k}_i, \hat{k}_s, \bar{r}_\perp) L(\hat{k}_i, \hat{k}_s, \bar{r}'_\perp)$$
$$= \lim_{k \to \infty} L(\hat{k}_i, \hat{k}_s, \bar{r}_\perp) L\left(\hat{k}_i, \hat{k}_s, x - \frac{u}{k}, y - \frac{v}{k}\right)$$
$$= L^2(\hat{k}_i, \hat{k}_s, \bar{r}_\perp) \quad (116)$$

Hence, we obtain the following expression analogous to (81).

$$<\lim_{k \to \infty} II^*> = \frac{A_o 4\pi^2}{k_{dz}^2} \int dL \, p(\alpha, \beta, L) \bigg|_{\alpha = -\frac{k_{dx}}{k_{dz}}, \beta = -\frac{k_{dy}}{k_{dz}}} L^2(\hat{k}_i, \hat{k}_s, \bar{r}_\perp) \quad (117)$$

where $p(\alpha, \beta, L)$ is the joint probability density function for α, β, and L. Since the process is homogeneous, the result above is independent of $\bar{r}$. Representing in terms of the conditional probability density

$$p(\alpha, \beta, L) = p(\alpha, \beta) p(L \mid \alpha, \beta) \quad (118)$$

where

$$p(L \mid \alpha, \beta) = P_L(\hat{k}_i, \hat{k}_s \mid \alpha, \beta) \delta(L - 1) + \left[1 - P_L(\hat{k}_i, \hat{k}_s \mid \alpha, \beta)\right] \delta(L) \quad (119)$$

with $P_L(\hat{k}_i, \hat{k}_s \mid \alpha, \beta)$ the probability that a point will be illuminated by rays having the directions $\hat{k}_i$ and $-\hat{k}_s$ given the values of the slope α and β at the point. Thus, using (82)

$$<II^*> = \frac{A_o 4\pi^2}{k_{dz}^2} \frac{1}{2\pi\sigma^2 |C''(0)|} \exp\left[-\frac{k_{dx}^2 + k_{dy}^2}{2k_{dz}^2 \sigma^2 |C''(0)|}\right]$$
$$\times P_L\left(\hat{k}_i, \hat{k}_s \mid -\frac{k_{dx}}{k_{dz}}, -\frac{k_{dy}}{k_{dz}}\right) \quad (120)$$

The bistatic reflection coefficients are modified to

$$\gamma_{ab}^{m r}(\hat{k}_s, \hat{k}_i) = \gamma_{ab}^r(\hat{k}_s, \hat{k}_i) P_L\left(\hat{k}_i, \hat{k}_s \mid -\frac{k_{dx}}{k_{dz}}, -\frac{k_{dy}}{k_{dz}}\right) \quad (121)$$

Kirchhoff Approach

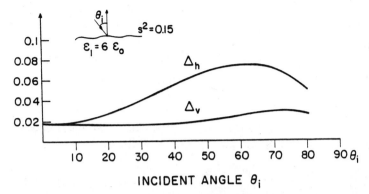

Fig. 2.19 Energy conservation for the modified geometrical optics solutions.

Shadowing effects have been studied by Beckmann (1965), Wagner (1967), Smith (1967), Sancer (1969), Brown (1984). We use the shadowing function derived by Smith and Sancer to modify the bistatic scattering coefficients.

$$P_L\left(\hat{k}_i, \hat{k}_s \mid -\frac{k_{dx}}{k_{dz}}, -\frac{k_{dy}}{k_{dz}}\right) = S(\theta_s, \theta_i) \tag{122}$$

where

$$S(\hat{k}_s, \hat{k}_i) = \begin{cases} \dfrac{1}{1+\Lambda(\mu_s)} & \phi_s = \phi_i + \pi,\ \theta_s \geq \theta_i \\ \dfrac{1}{\Lambda(\mu_i)+1} & \phi_s = \phi_i + \pi,\ \theta_i \geq \theta_s \\ \dfrac{1}{\Lambda(\mu_s)+\Lambda(\mu_i)+1} & \text{otherwise} \end{cases} \tag{123}$$

and

$$\mu = \cot\theta \tag{124}$$

$$\Lambda(\mu) = \frac{1}{2}\left[\sqrt{\frac{2}{\pi}}\frac{s}{\mu}e^{-\mu^2/2s^2} - \operatorname{erfc}\left(\frac{\mu}{\sqrt{2}s}\right)\right] \tag{125}$$

s^2 is the mean square surface slope [defined in (83)], and erfc is the complementary error function.

The bistatic transmission coefficients can be modified in a similar fashion. Thus,

$$\gamma_{ab}^{m_t}(\hat{k}_t, \hat{k}_i) = \gamma_{ab}^t(\hat{k}_t, \hat{k}_i) S_t(\theta_t, \theta_i) \tag{126}$$

where

$$S_t(\theta_t, \theta_i) = \frac{1}{1+\Lambda(\mu_t)+\Lambda(\mu_i)} \tag{127}$$

The modified reflectivity and transmissivity functions $r_b^m(\theta_i)$ and $t_b^m(\theta_i)$ are obtained from (106) by substituting in the modified bistatic scattering coefficients into the equation. The energy conservation is again studied by plotting Δ_v and Δ_h in Figure 2.19. We see that the sum of reflected and transmitted energy is always less than unity.

Emissivities

The emissivity of a rough surface can be calculated from the bistatic reflection coefficient. From (7), Section 4, Chapter 1, we have

$$e_a(\theta_i, \phi_i) = 1 - \frac{1}{4\pi} \sum_{b=v,h} \int_o^{\pi/2} d\theta_s \sin\theta_s \int_0^{2\pi} d\phi_s \, \gamma_{ba}^r(\theta_s, \phi_s; \theta_i, \phi_i) \quad (128)$$

The coherent and incoherent bistatic reflection coefficients derived from the Kirchhoff-approximated diffraction integrals in (59) can be used to calculate the emissivity of the rough surface. In terms of the coherent and incoherent reflectivity functions, the emissivity is given by

$$e_a(\theta_i) = 1 - r_{c_a}(\theta_i) - r_{i_a}(\theta_i) \quad (129)$$

where, for $\alpha = c$ or i, denoting coherent and incoherent, respectively

$$r_{\alpha_a}(\theta_i) = \frac{1}{4\pi} \sum_{b=v,h} \int_o^{\pi/2} d\theta_s \sin\theta_s \int_0^{2\pi} d\phi_s \, \gamma_{\alpha ba}^r(\theta_s, \phi_s; \theta_i, \phi_i) \quad (130)$$

In the above equation, the dependence on the azimuthal angle of incidence ϕ_i is dropped because the rough surface is isotropic and the results are independent of ϕ_i. After substituting in the explicit expressions for the bistatic scattering coefficients from (59) and carrying out the angular integration, we obtain (Tsang and Newton, 1982)

$$r_{c_a}(\theta_i) = |R_{ao}|^2 \exp\left[-4k^2\sigma^2 \cos^2\theta_i\right] \quad (131)$$

$$r_{i_a}(\theta_i) = \frac{k^2 l^2}{8\cos\theta_i} \int_0^{\pi/2} d\theta_s \sin\theta_s \exp\left[-\sigma^2 k^2 (\cos\theta_i + \cos\theta_s)^2\right]$$

$$\times \sum_{m=1}^{\infty} \frac{(k\sigma(\cos\theta_s + \cos\theta_i))^{2m}}{m!\,m} e^{-y_m}$$

$$\cdot \left\{ \left|(1+R_{ao}) - (1-R_{ao})\cos\theta_i\cos\theta_s\right|^2 \frac{I_1(x_m)}{x_m} \right.$$

$$+ \left|-(1+R_{ao})\cos\theta_s + (1-R_{ao})\cos\theta_i\right|^2$$

$$\left. \times \left(I_o(x_m) - \frac{I_1(x_m)}{x_m}\right) \right\} \quad (132)$$

where $a = v, h$, and

$$y_m = \frac{k^2 l^2 (\sin^2 \theta_i + \sin^2 \theta_s)}{4m} \tag{133}$$

$$x_m = \frac{k^2 l^2 \sin \theta_i \sin \theta_s}{2m} \tag{134}$$

and I_o and I_1 are the zeroth and first-order modified Bessel functions. We note that if the reflected field is used instead of the total field in the diffraction integral (Holzer and Sung, 1978), then, for $a = v$ or h

$$r_{i_a}(\theta_i) = |R_{ao}|^2 \frac{k^2 l^2}{8 \cos \theta_i} \int_o^{\pi/2} d\theta_s \sin \theta_s \exp\left[-\sigma^2 k^2 (\cos \theta_i + \cos \theta_s)^2\right]$$

$$\times \sum_{m=1}^{\infty} \frac{(k\sigma(\cos \theta_s + \cos \theta_i))^{2m}}{m!\, m} e^{-y_m}$$

$$\cdot \left\{ (1 + \cos \theta_i \cos \theta_s)^2 \frac{I_1(x_m)}{x_m} \right.$$

$$\left. + (\cos \theta_s + \cos \theta_i)^2 \left(I_o(x_m) - \frac{I_1(x_m)}{x_m}\right) \right\} \tag{135}$$

The difference between the incoherent reflectivities obtained using the total or the reflected field in the diffraction integral is due to the approximation made on the integrand $\overline{F}(\alpha, \beta)$. If the next-order term in the expansion of $\overline{F}(\alpha, \beta)$ about zero slopes is kept in (20), then the results obtained using the total and the reflected field can be shown to be the same. Note that in neglecting the shadowing effect, we are applying the tangent plane approximation even to the points on the rough surface that cannot be directly illuminated. The above model has some success in matching brightness temperature measurements from soils with rough surfaces (Tsang and Newton, 1982; Schmugge, 1983). The model with coherent reflectivity of (131) alone is discussed in Choudhury et al. (1979).

The emissivity may also be calculated in terms of the bistatic transmission coefficients from medium 1 to medium 0,

$$e_a(\theta_i, \phi_i) = \frac{1}{4\pi} \sum_{b=v,h} \int_o^{\pi/2} d\theta_1 \sin \theta_1$$

$$\times \int_o^{2\pi} d\phi_1\, \gamma_{ab}^{to_1}(\theta_i, \phi_i; \theta_1, \phi_1) \frac{n_1^2 \cos \theta_1}{n_o^2 \cos \theta_i} \tag{136}$$

The emissivity of a random rough surface in the high-frequency limit can be calculated using the bistatic scattering coefficients derived in the geometrical optics limit. The modified scattering coefficients, given by (121) and (126), incorporate the shadowing effect. A well-defined emissivity of a medium depends on (i) the satisfaction of reciprocity relations, and (ii) the satisfaction of conservation of energy by bistatic scattering coefficients. The modified scattering coefficients satisfy the reciprocity but violate the energy conservation since only the single-scattering solution is used. Thus, there is ambiguity and the results obtained using (128) and (136) are not the same. However, the sum of reflected and transmitted intensity is shown to be always less than unity and this fact can be made use of to derive the upper and lower limit of the correct emissivity.

The emissivity calculated using (128) represents the upper limit of the correct solution since the bistatic reflection coefficients are obtained using only the single-scattering solution. If the higher-order scattering effects are included, the net reflected intensity will be higher and the emissivity will always be lower. Thus, the upper bound solution for the emissivity is given by, in view of (128) and (106),

$$e_a^u(\theta_i) = 1 - r_a^m(\theta_i) \qquad (137)$$

where m denotes modified reflectivity with incorporation of shadowing effects according to (121). The emissivity calculated using (136) represents the lower limit of the correct solution. If the higher-order scattering effects at the rough boundary are included, the bistatic transmission coefficients will always increase. Consequently, more thermal emission from medium 1 will be transmitted and the emissivity will always increase. Using the reciprocity relations satisfied by the bistatic transmission coefficients, (104), the lower-bound solution for the emissivity is given by, in view of (136) and (106) and using modified transmission coefficients of (126),

$$e_a^l(\theta_i) = t_a^m(\theta_i) \qquad (138)$$

Therefore, the two solutions given by (137) and (138) represent the upper and lower limits of the correct solution, and the ambiguity is due to the violation of energy conservation.

6.2 Small Perturbation Method

The scattering of electromagnetic waves from a slightly rough surface can be studied using a perturbation method (Rice, 1963). It is assumed that the surface variations are much smaller than the incident wavelength and the slopes of the rough surface are relatively small. The small perturbation

Small Perturbation Method

method (SPM) makes use of the Rayleigh hypothesis to express the reflected and transmitted fields into upward- and downward-going waves, respectively. The field amplitudes are then determined from the boundary conditions and the divergence relations. The extended boundary condition (EBC) method may also be used with the perturbation method to solve for the scattered fields (Agarwal, 1977; Nieto-Vesperinas, 1982). In the EBC method, the surface currents on the rough surface are calculated first by applying the extinction theorem. The scattered fields can then be calculated from the diffraction integral by making use of the calculated surface fields. Both perturbation methods yield the same expansions for the scattered fields, because the expansions of the amplitudes of the scattered fields are unique within their circles of convergence (Maradudin, 1983). In this section we make use of the EBC method to formulate the problem. Even though the Rayleigh method is simpler in a sense that the scattered fields amplitudes are obtained directly, the EBC method is conceptually consistent with our previous sections on scattering from periodic surfaces. An investigation of the validity of the Rayleigh hypothesis can be found in Millar (1973).

Consider a plane wave in free space with electric field $\overline{E}_i = \hat{e}_i E_o \exp(i\overline{k}_i \cdot \overline{r})$ incident upon a slightly rough surface of a medium with permittivity ϵ_1. The rough surface is characterized by a random height distribution $z = f(\overline{r}_\perp)$ where $f(\overline{r}_\perp)$ is a random variable with zero mean, $<f(\overline{r}_\perp)> = 0$. Let f_{min} and f_{max} be the minimum and maximum values of the surface profile $f(\overline{r}_\perp)$. From the Huygen's principle and extinction theoreom, the total field $\overline{E}(\overline{r})$ in free space, and the transmitted field $\overline{E}_1(\overline{r})$ in the dielectric medium satisfies

$$\overline{E}_i(\overline{r}) + \int_{S'} dS' \left\{ i\omega\mu_o \overline{\overline{G}}(\overline{r}, \overline{r}') \cdot [\hat{n} \times \overline{H}(\overline{r}')] \right.$$
$$\left. + \nabla \times \overline{\overline{G}}(\overline{r}, \overline{r}') \cdot [\hat{n} \times \overline{E}(\overline{r}')] \right\}$$
$$= \begin{cases} \overline{E}(\overline{r}) & z > f(\overline{r}_\perp) \quad (1a) \\ 0 & z < f(\overline{r}_\perp) \quad (1b) \end{cases}$$

$$\int_{S'} dS' \left\{ i\omega\mu_1 \overline{\overline{G}}_1(\overline{r}, \overline{r}') \cdot [\hat{n}_d \times \overline{H}_1(\overline{r}')] + \nabla \times \overline{\overline{G}}_1(\overline{r}, \overline{r}') \cdot [\hat{n}_d \times \overline{E}_1(\overline{r}')] \right\}$$
$$= \begin{cases} 0 & z > f(\overline{r}_\perp) \quad (2a) \\ \overline{E}_1(\overline{r}) & z < f(\overline{r}_\perp) \quad (2b) \end{cases}$$

where S' denotes the rough surface in which the surface integration is to be carried out, $\hat{n}$ and $\hat{n}_d$ are the unit vectors normal to the rough surface and pointing into the free space and the dielectric medium, respectively, and $\overline{\overline{G}}(\bar{r},\bar{r}')$ and $\overline{\overline{G}}_1(\bar{r},\bar{r}')$ are the dyadic Green's functions for free space and the homogeneous dielectric of region 1, respectively, as given in (3), Section 6.1.

Since tangential fields are continuous, we can define surface field unknowns

$$dS'\eta\,\hat{n}\times\overline{H}(\bar{r}') = d\bar{r}'_\perp\,\overline{a}(\bar{r}'_\perp) = dS'\eta\,\hat{n}\times\overline{H}_1(\bar{r}') \tag{3a}$$

$$dS'\,\hat{n}\times\overline{E}(\bar{r}') = d\bar{r}'_\perp\,\overline{b}(\bar{r}'_\perp) = dS'\,\hat{n}\times\overline{E}_1(\bar{r}') \tag{3b}$$

Next we make use of the integral representation of dyadic Green's function as given in (15), Section 3.1, and substitute it into (1) and (2). Let f_{min} and f_{max} be, respectively, the minimum and maximum values of the surface profile $f(\bar{r}'_\perp)$. Evaluating (1b) for $z < f_{min}$ and (2a) for $z > f_{max}$ we obtain

$$\overline{E}_i(\bar{r}) = \frac{1}{8\pi^2}\int d\bar{k}_\perp\,e^{i\bar{k}_\perp\cdot\bar{r}_\perp}\,e^{-ik_z z}\frac{k}{k_z}\int d\bar{r}'_\perp\,e^{-i\bar{k}_\perp\cdot\bar{r}'_\perp}\,e^{ik_z f(\bar{r}'_\perp)}$$

$$\times\left\{\left[\hat{e}(-k_z)\hat{e}(-k_z) + \hat{h}(-k_z)\hat{h}(-k_z)\right]\cdot\overline{a}(\bar{r}'_\perp)\right.$$

$$\left.+ \left[-\hat{h}(-k_z)\hat{e}(-k_z) + \hat{e}(-k_z)\hat{h}(-k_z)\right]\cdot\overline{b}(\bar{r}'_\perp)\right\} \tag{4a}$$

$$0 = \frac{1}{8\pi^2}\int d\bar{k}_\perp\,e^{i\bar{k}_\perp\cdot\bar{r}_\perp}\,e^{ik_{1z}z}\frac{k_1}{k_{1z}}\int d\bar{r}'_\perp\,e^{-i\bar{k}_\perp\cdot\bar{r}'_\perp}\,e^{-ik_{1z}f(\bar{r}'_\perp)}$$

$$\times\left\{\frac{k}{k_1}\left[\hat{e}_1(k_{1z})\hat{e}_1(k_{1z}) + \hat{h}_1(k_{1z})\hat{h}_1(k_{1z})\right]\cdot\overline{a}(\bar{r}'_\perp)\right.$$

$$\left.+ \left[-\hat{h}_1(k_{1z})\hat{e}_1(k_{1z}) + \hat{e}_1(k_{1z})\hat{h}_1(k_{1z})\right]\cdot\overline{b}(\bar{r}'_\perp)\right\} \tag{4b}$$

The above equations are the extended boundary conditions, and can be used to solve for the surface fields along with the following equations, which are results of (3a) and (3b)

$$\hat{n}(\bar{r}'_\perp)\cdot\overline{a}(\bar{r}'_\perp) = 0 \tag{5a}$$

$$\hat{n}(\bar{r}'_\perp)\cdot\overline{b}(\bar{r}'_\perp) = 0 \tag{5b}$$

Using (12), Section 6.1, (5) can be rewritten as

$$a_z(\bar{r}'_\perp) = \left(\hat{x}\,\frac{\partial f(\bar{r}'_\perp)}{\partial x'} + \hat{y}\,\frac{\partial f(\bar{r}'_\perp)}{\partial y'}\right)\cdot\overline{a}_\perp(\bar{r}'_\perp) \tag{6}$$

Small Perturbation Method

$$b_z(\bar{r}'_\perp) = \left(\hat{x}\frac{\partial f(\bar{r}'_\perp)}{\partial x'} + \hat{y}\frac{\partial f(\bar{r}'_\perp)}{\partial y'}\right) \cdot \bar{b}_\perp(\bar{r}'_\perp) \qquad (7)$$

with a_z and b_z as the z components of $\bar{a}$ and $\bar{b}$, respectively.

Once the surface fields are obtained, then the scattered field in region 0 and the transmitted field in medium 1 can be derived by using (1a) and (2b). Thus, evaluating (1a) and (2b) for $z > f_{max}$ and $z < f_{min}$ respectively, we obtain

$$\bar{E}_s(\bar{r}) = -\frac{1}{8\pi^2}\int d\bar{k}_\perp\, e^{i\bar{k}_\perp\cdot\bar{r}_\perp}\, e^{ik_z z}\frac{k}{k_z}\int d\bar{r}'_\perp\, e^{-i\bar{k}_\perp\cdot\bar{r}'_\perp}e^{-ik_z f(\bar{r}'_\perp)}$$

$$\times \left\{\left[\hat{e}(k_z)\hat{e}(k_z) + \hat{h}(k_z)\hat{h}(k_z)\right]\cdot\bar{a}(\bar{r}'_\perp)\right.$$

$$\left. + \left[-\hat{h}(k_z)\hat{e}(k_z) + \hat{e}(k_z)\hat{h}(k_z)\right]\cdot\bar{b}(\bar{r}'_\perp)\right\} \qquad (8a)$$

$$\bar{E}_t(\bar{r}) = \frac{1}{8\pi^2}\int d\bar{k}_\perp\, e^{i\bar{k}_\perp\cdot\bar{r}_\perp}\, e^{-ik_{1z} z}\frac{k_1}{k_{1z}}\int d\bar{r}'_\perp\, e^{-i\bar{k}_\perp\cdot\bar{r}'_\perp}e^{ik_{1z} f(\bar{r}'_\perp)}$$

$$\times \left\{\frac{k}{k_1}\left[\hat{e}_1(-k_{1z})\hat{e}_1(-k_{1z}) + \hat{h}_1(-k_{1z})\hat{h}_1(-k_{1z})\right]\cdot\bar{a}(\bar{r}'_\perp)\right.$$

$$\left. + \left[-\hat{h}_1(-k_{1z})\hat{e}_1(-k_{1z}) + \hat{e}_1(-k_{1z})\hat{h}_1(-k_{1z})\right]\cdot\bar{b}(\bar{r}'_\perp)\right\} \qquad (8b)$$

Therefore, the objective is to solve for the surface fields using (4) and (5) and then to solve for the scattered fields using (8).

Equations (4), and (6) through (8) are exact. To solve for the surface fields, the perturbation method makes use of series expansions. Let

$$\bar{a}(\bar{r}'_\perp) = \sum_{m=0}^{\infty}\frac{\bar{a}^{(m)}(\bar{r}'_\perp)}{m!} \qquad (9a)$$

$$\bar{b}(\bar{r}'_\perp) = \sum_{m=0}^{\infty}\frac{\bar{b}^{(m)}(\bar{r}'_\perp)}{m!} \qquad (9b)$$

where $\bar{a}^m$ and $\bar{b}^m$ are, respectively, the mth-order solution of $\bar{a}$ and $\bar{b}$. We also have

$$e^{\pm ik_z f(\bar{r}'_\perp)} = \sum_{m=0}^{\infty}\frac{[\pm ik_z f(\bar{r}'_\perp)]^m}{m!} \qquad (10a)$$

$$e^{\pm ik_{1z} f(\bar{r}'_\perp)} = \sum_{m=0}^{\infty}\frac{[\pm ik_{1z} f(\bar{r}'_\perp)]^m}{m!} \qquad (10b)$$

In SPM, f and its derivatives are regarded as small parameters. The expansion of (9) and (10) are substituted into (4) to obtain the set of equations for the different-order solutions. From (7) and (9)

$$a_z^{(0)}(\bar{r}'_\perp) = b_z^{(0)}(\bar{r}'_\perp) = 0 \tag{11}$$

$$a_z^{(m)}(\bar{r}'_\perp) = m\left(\hat{x}\frac{\partial f(\bar{r}'_\perp)}{\partial x'} + \hat{y}\frac{\partial f(\bar{r}'_\perp)}{\partial y'}\right) \cdot \bar{a}_\perp^{(m-1)}(\bar{r}'_\perp) \tag{12a}$$

$$b_z^{(m)}(\bar{r}'_\perp) = m\left(\hat{x}\frac{\partial f(\bar{r}'_\perp)}{\partial x'} + \hat{y}\frac{\partial f(\bar{r}'_\perp)}{\partial y'}\right) \cdot \bar{b}_\perp^{(m-1)}(\bar{r}'_\perp) \tag{12b}$$

Thus, we are assuming

$$k_z f(\bar{r}'_\perp),\ k_{1z} f(\bar{r}'_\perp),\ \frac{\partial f}{\partial x'},\ \frac{\partial f}{\partial y'} \ll 1 \tag{13}$$

Substituting (9) and (10) into (4) and (6) through (7) and equating the same-order terms, we can calculate the surface fields to zeroth-order, first-order, etc. Then, from (8), the scattered fields can be obtained to different orders. In the following, we solve for the surface fields and scattered fields up to the second order. The zeroth-order solutions are just the reflected and transmitted fields of a flat surface. The first-order solution gives the lowest-order incoherent scattered intensities. However, the first-order solution does not give the depolarization effect in the backscattering direction. The second-order solution gives the lowest-order correction to the coherent reflection and transmission coefficients. Also, the depolarization of the backscattered power is manifested.

Zeroth-Order Solution

The zeroth-order solution can be obtained by keeping only the lowest-order terms in the expansion of (9). We first define the orthonormal system $(\hat{q}_i, \hat{p}_i, \hat{z}_i)$, which is given by

$$\hat{q}_i = \hat{x}\frac{k_{yi}}{k_{\rho i}} - \hat{y}\frac{k_{xi}}{k_{\rho i}} = \hat{e}(k_{zi}) \tag{14}$$

$\hat{z}_i = \hat{z}$ and $\hat{p}_i = \hat{z}_i \times \hat{q}_i = (\hat{x}k_{xi} + \hat{y}k_{yi})/k_{\rho i}$ where $k_{\rho i}^2 = k_{xi}^2 + k_{yi}^2$ and let

$$\bar{a}(\bar{r}'_\perp) = \hat{q}_i a_q(\bar{r}'_\perp) + \hat{p}_i a_p(\bar{r}'_\perp) + \hat{z}_i a_z(\bar{r}'_\perp) \tag{15a}$$

$$\bar{b}(\bar{r}'_\perp) = \hat{q}_i b_q(\bar{r}'_\perp) + \hat{p}_i b_p(\bar{r}'_\perp) + \hat{z}_i b_z(\bar{r}'_\perp) \tag{15b}$$

Small Perturbation Method

To solve for the zeroth-order solution, we note that

$$\overline{E}_i(\overline{r}) = \hat{e}_i E_o e^{i\overline{k}_{\perp i}\cdot\overline{r}_\perp - ik_{zi}z}$$

$$= \frac{\hat{e}_i}{4\pi^2} \int d\overline{k}_\perp e^{i\overline{k}_\perp\cdot\overline{r}_\perp - ik_z z} \int d\overline{r}'_\perp e^{i\overline{k}_{\perp i}\cdot\overline{r}'_\perp - i\overline{k}_\perp\cdot\overline{r}'_\perp} \quad (16)$$

Using (16) in (4a), we note that the $d\overline{r}'_\perp$ integration in (4a) for the zeroth order solution must produce a Dirac delta function of the form $\delta(\overline{k}_\perp - \overline{k}_{\perp i})$. Hence, in view of (11)

$$\hat{e}_i e^{i\overline{k}_{\perp i}\cdot\overline{r}'_\perp} = \frac{k}{2k_{zi}}\Big\{\Big[\hat{e}(-k_{zi})\hat{e}(-k_{zi}) + \hat{h}(-k_{zi})\hat{h}(-k_{zi})\Big]\cdot\overline{a}_\perp^{(0)}(\overline{r}'_\perp)$$

$$+ \Big[-\hat{h}(-k_{zi})\hat{e}(-k_{zi}) + \hat{e}(-k_{zi})\hat{h}(-k_{zi})\Big]\cdot\overline{b}_\perp^{(0)}(\overline{r}')\Big\} \quad (17a)$$

and from (4b), we have

$$\Big[\hat{e}_1(k_{1zi})\hat{e}_1(k_{1zi}) + \hat{h}_1(k_{1zi})\hat{h}_1(k_{1zi})\Big]\cdot\overline{a}_\perp^{(0)}(\overline{r}'_\perp)\frac{k}{k_1}$$

$$+ \Big[-\hat{h}_1(k_{1zi})\hat{e}_1(k_{1zi}) + \hat{e}_1(k_{1zi})\hat{h}_1(k_{1zi})\Big]\cdot\overline{b}_\perp^{(0)}(\overline{r}'_\perp) = 0 \quad (17b)$$

Using (11) and (15) in (17) and noting that the dot products of $\hat{p}_i$ and $\hat{q}_i$ with $\hat{e}$ and $\hat{h}$ can be calculated from (13), Section 3.1, we have, from (17a)

$$\hat{e}_i e^{i\overline{k}_{\perp i}\cdot\overline{r}'} = \frac{k}{2k_{zi}}\Big\{\hat{e}(-k_{zi})\Big(a_q^{(0)}(\overline{r}') + \frac{k_{zi}}{k}b_p^{(0)}(\overline{r}'_\perp)\Big)$$

$$+ \hat{h}\ (-k_{zi})\Big(\frac{k_{zi}}{k}a_p^{(0)}(\overline{r}'_\perp) - b_q^{(0)}(\overline{r}'_\perp)\Big)\Big\} \quad (18a)$$

Using (17b), we have

$$ka_q^{(0)}(\overline{r}') - k_{1zi}b_p^{(0)}(\overline{r}') = 0 \quad (18b)$$

$$\frac{kk_{1zi}}{k_1^2}a_p^{(0)}(\overline{r}') + b_q^{(0)}(\overline{r}') = 0 \quad (18c)$$

Since (18a) contains two scalar equations, (18) provides four equations for the four unknowns $a_p^{(0)}$, $a_q^{(0)}$, $b_p^{(0)}$, and $b_q^{(0)}$. Solving them and substituting back into $\overline{a}^{(0)}(\overline{r}'_\perp)$ and $\overline{b}^{(0)}(\overline{r}'_\perp)$ gives

$$\overline{a}^{(0)}(\overline{r}'_\perp) = \overline{a}^{(0)}(\overline{k}_{\perp i})\, e^{i\overline{k}_{\perp i}\cdot\overline{r}'_\perp} \quad (19a)$$

$$\overline{b}^{(0)}(\overline{r}'_\perp) = \overline{b}^{(0)}(\overline{k}_{\perp i})\, e^{i\overline{k}_{\perp i}\cdot\overline{r}'_\perp} \quad (19b)$$

where

$$\overline{a}_q^{(0)}(\overline{k}_{\perp i}) = [\hat{e}(-k_{zi}) \cdot \hat{e}_i]\frac{k_{zi}}{k}(1 - R_{ho}) \tag{20a}$$

$$\overline{a}_p^{(0)}(\overline{k}_{\perp i}) = [\hat{h}(-k_{zi}) \cdot \hat{e}_i](1 + R_{vo}) \tag{20b}$$

$$\overline{b}_q^{(0)}(\overline{k}_{\perp i}) = -[\hat{h}(-k_{zi}) \cdot \hat{e}_i]\frac{k_{zi}}{k}(1 - R_{vo}) \tag{20c}$$

$$\overline{b}_p^{(0)}(\overline{k}_{\perp i}) = [\hat{e}(-k_{zi}) \cdot \hat{e}_i](1 + R_{ho}) \tag{20d}$$

and R_{ho} and R_{vo} are the Fresnel reflection coefficients for the TE and TM waves

$$R_{ho} = \frac{k_{zi} - k_{1zi}}{k_{zi} + k_{1zi}} \tag{21a}$$

$$R_{vo} = \frac{\epsilon_1 k_{zi} - \epsilon_o k_{1zi}}{\epsilon_1 k_{zi} + \epsilon_o k_{1zi}} \tag{21b}$$

Using (19) in (8), the reflected and transmitted fields are given by

$$\overline{E}_s^{(0)} = \left\{ R_{ho}[\hat{e}(-k_{zi}) \cdot \hat{e}_i]\hat{e}(k_{zi}) + R_{vo}[\hat{h}(-k_{zi}) \cdot \hat{e}_i]\hat{h}(k_{zi}) \right\}$$

$$\times E_o\, e^{i\overline{k}_{i\perp}\cdot \overline{r}_\perp + ik_{zi} z} \tag{22}$$

$$\overline{E}_t^{(0)} = \left\{ (1 + R_{ho})[\hat{e}(-k_{zi}) \cdot \hat{e}_i]\hat{e}_1(-k_{1zi}) \right.$$

$$\left. + \frac{k}{k_1}(1 + R_{vo})[\hat{h}(-k_{zi}) \cdot \hat{e}_i]\hat{h}_1(-k_{1zi}) \right\} E_o\, e^{i\overline{k}_{i\perp}\cdot \overline{r}_\perp - ik_{1zi} z} \tag{23}$$

which are just the reflected and transmitted fields for a flat surface.

First-Order Solution

The first-order solution for the surface fields can be obtained by substituting (9) and (10) into (4), (6), (7), and (12) and equating first-order terms. From (12a) and (19a)

$$a_z^{(1)}(\overline{r}_\perp') = \left(\hat{x}\frac{\partial f(\overline{r}_\perp')}{\partial x'} + \hat{y}\frac{\partial f(\overline{r}_\perp')}{\partial y'} \right) \overline{a}_\perp^{(0)}(\overline{k}_{\perp i})\, e^{i\overline{k}_{\perp i}\cdot \overline{r}_\perp'} \tag{24}$$

To simplify (24), we introduce the Fourier transforms

$$F(\overline{k}_\perp) = \frac{1}{(2\pi)^2}\int d\overline{r}_\perp'\, f(\overline{r}_\perp')\, e^{-i\overline{k}_\perp \cdot \overline{r}_\perp'} \tag{25}$$

Small Perturbation Method

$$\overline{A}^{(1)}(\overline{k}_\perp) = \frac{1}{(2\pi)^2} \int d\overline{r}'_\perp \overline{a}^{(1)}(\overline{r}'_\perp) e^{-i\overline{k}_\perp \cdot \overline{r}'_\perp} \qquad (26a)$$

$$\overline{B}^{(1)}(\overline{k}_\perp) = \frac{1}{(2\pi)^2} \int d\overline{r}'_\perp \overline{b}^{(1)}(\overline{r}'_\perp) e^{-i\overline{k}_\perp \cdot \overline{r}'_\perp} \qquad (26b)$$

(Strictly speaking, the Fourier transforms do not exist for random functions and stochastic Fourier Stieltjes integral have to be defined (Tatarskii, 1971; Ishimaru, 1978). However, the final results for scattered intensities are not affected.) Multiply (24) by $\exp[-i\overline{k}_\perp \cdot \overline{r}'_\perp]/(2\pi)^2$ and integrate over $d\overline{r}'_\perp$. We obtain, by expressing $\partial f(\overline{r}'_\perp)/\partial x'$ and $\partial f(\overline{r}'_\perp)/\partial y'$ in terms of $F(\overline{k}_\perp)$,

$$A_z^{(1)}(\overline{k}_\perp) = \left\{ \frac{k_x k_{yi} - k_y k_{xi}}{k_{\rho i}} a_q^{(0)}(\overline{k}_{\perp i}) \right.$$

$$\left. + \left(\frac{k_x k_{xi} + k_y k_{yi}}{k_{\rho i}} - k_{\rho i} \right) a_p^{(0)}(\overline{k}_{\perp i}) \right\} i F(\overline{k}_\perp - \overline{k}_{\perp i}) \qquad (27a)$$

Similarly from (12b)

$$B_z^{(1)}(\overline{k}_\perp) = \left\{ \frac{k_x k_{yi} - k_y k_{xi}}{k_{\rho i}} b_q^{(0)}(\overline{k}_{\perp i}) \right.$$

$$\left. + \left(\frac{k_x k_{xi} + k_y k_{yi}}{k_{\rho i}} - k_{\rho i} \right) b_p^{(0)}(\overline{k}_{\perp i}) \right\} i F(\overline{k}_\perp - \overline{k}_{\perp i}) \qquad (27b)$$

Next we match both sides of equation (4a) to the first order. We note that

$$\left[\int d\overline{r}'_\perp e^{-i\overline{k}_\perp \cdot \overline{r}'_\perp} e^{ik_z f(\overline{r}'_\perp)} \overline{a}(\overline{r}'_\perp) \right]_{\text{first order}}$$

$$= \int d\overline{r}'_\perp e^{-i\overline{k}_\perp \cdot \overline{r}'_\perp} \left[ik_z f(\overline{r}'_\perp) \overline{a}^{(0)}(\overline{k}_{\perp i}) e^{i\overline{k}_{\perp i} \cdot \overline{r}'} + \overline{a}^{(1)}(\overline{r}') \right]$$

$$= (2\pi)^2 \left[ik_z F(\overline{k}_\perp - \overline{k}_{\perp i}) \overline{a}^{(0)}(\overline{k}_{\perp i}) + \overline{A}^{(1)}(\overline{k}_\perp) \right] \qquad (28)$$

Hence, the first-order equation from (4a) is

$$0 = \left[\hat{e}(-k_z)\hat{e}(-k_z) + \hat{h}(-k_z)\hat{h}(-k_z) \right]$$

$$\cdot \left[\overline{A}^{(1)}(\overline{k}_\perp) + ik_z \overline{a}^{(0)}(\overline{k}_{\perp i}) F(\overline{k}_\perp - \overline{k}_{\perp i}) \right]$$

$$+ \left[-\hat{h}(-k_z)\hat{e}(-k_z) + \hat{e}(-k_z)\hat{h}(-k_z) \right]$$

$$\cdot \left[\overline{B}^{(1)}(\overline{k}_\perp) + ik_z \overline{b}^{(0)}(\overline{k}_{\perp i}) F(\overline{k}_\perp - \overline{k}_{\perp i}) \right] \qquad (29a)$$

and from (4b), in a similar fashion, we obtain

$$0 = \frac{k}{k_1} \left[\hat{e}_1(k_{1z})\hat{e}_1(k_{1z}) + \hat{h}_1(k_{1z})\hat{h}_1(k_{1z}) \right]$$
$$\cdot \left[\overline{A}^{(1)}(\overline{k}_\perp) - ik_{1z}\overline{a}^{(0)}(\overline{k}_{\perp i})F(\overline{k}_\perp - \overline{k}_{\perp i}) \right]$$
$$+ \left[-\hat{h}_1(k_{1z})\hat{e}_1(k_{1z}) + \hat{e}_1(k_{1z})\hat{h}_1(k_{1z}) \right]$$
$$\cdot \left[\overline{B}^{(1)}(\overline{k}_\perp) - ik_{1z}\overline{b}^{(0)}(\overline{k}_{\perp i})F(\overline{k}_\perp - \overline{k}_{\perp i}) \right] \quad (29b)$$

Equations (29a) and (29b) are vector equations so that they comprise four scalar equations. Hence, (29) are four scalar equations for the four unknowns $A_q^{(1)}(\overline{k}_\perp)$, $A_p^{(1)}(\overline{k}_\perp)$, $B_q^{(1)}(\overline{k}_\perp)$, and $B_p^{(1)}(\overline{k}_\perp)$. After much algebraic manipulation, we obtain

$$\overline{A}_\perp^{(1)}(\overline{k}_\perp) = iF(\overline{k}_\perp - \overline{k}_{\perp i})\widetilde{\overline{A}}_\perp^{(0)}(\overline{k}_{\perp i}) \quad (30a)$$

$$\overline{B}_\perp^{(1)}(\overline{k}_\perp) = iF(\overline{k}_\perp - \overline{k}_{\perp i})\widetilde{\overline{B}}_\perp^{(0)}(\overline{k}_{\perp i}) \quad (30b)$$

where the explicit expansions for $\widetilde{A}_q^{(1)}$, $\widetilde{A}_p^{(1)}$, $\widetilde{B}_q^{(1)}$ and $\widetilde{B}_p^{(1)}$ are given in the Appendix. The solutions for $A_z^{(1)}(\overline{k}_\perp)$ and $B_z^{(1)}(\overline{k}_\perp)$ have already been given in (27).

The first-order scattered fields can now be obtained from (8a).

$$\overline{E}_s^{(1)} = -\frac{1}{2}\int d\overline{k}_\perp \, e^{i\overline{k}_\perp \cdot \overline{r}_\perp} e^{ik_z z} \frac{k}{k_z}$$
$$\times \left\{ \left[\hat{e}(k_z)\hat{e}(k_z) + \hat{h}(k_z)\hat{h}(k_z) \right] \right.$$
$$\cdot \left[\overline{A}^{(1)}(\overline{k}_\perp) - ik_z F(\overline{k}_\perp - \overline{k}_{\perp i})\overline{a}^{(0)}(\overline{k}_{\perp i}) \right]$$
$$+ \left[-\hat{h}(k_z)\hat{e}(k_z) + \hat{e}(k_z)\hat{h}(k_z) \right]$$
$$\left. \cdot \left[\overline{B}^{(1)}(\overline{k}_\perp) - ik_z F(\overline{k}_\perp - \overline{k}_{\perp i})\overline{b}^{(0)}(\overline{k}_{\perp i}) \right] \right\} \quad (31)$$

In view of (27) and (30) and the fact that

$$<\overline{F}(\overline{k}_\perp)> = \frac{1}{(2\pi)^2}\int d\overline{r}'_\perp \, e^{i\overline{k}_\perp \cdot \overline{r}'_\perp} <f(\overline{r}_\perp)> = 0 \quad (32)$$

therefore

$$<\overline{E}_s^{(1)}> = <\overline{E}_t^{(1)}> = 0 \quad (33)$$

Small Perturbation Method

Thus, the first-order solution does not modify the coherent reflection and transmission coefficients and we have to calculate the second-order solution to see the correction term for the coherent wave due to the rough surface.

The lowest-order incoherent coefficients can be derived from (31), by considering the vertically and horizontally polarized incident fields and calculating the vertically and horizontally polarized scattered fields. For an incident field with polarization $\hat{a}_i$, the scattered intensity with polarization $\hat{b}_s$ is given by

$$<|E_s^{(1)}|^2> = \int d\bar{k}_\perp f'_{ba} W(|\bar{k}_\perp - \bar{k}_{\perp i}|)$$
$$= \int d\Omega_s \, k^2 \cos\theta_s \, f'_{ba} W(|\bar{k}_\perp - \bar{k}_{\perp i}|) \quad (34)$$

where $W(|\bar{k}_\perp - \bar{k}_{\perp i}|)$ is the spectral density of the rough surface and is the Fourier transform of the correlation function. The quantities f'_{ba}, $b, a = v, h$ in (34) are listed in the Appendix.

The spectral density is

$$W(\bar{k}_\perp) = \frac{\sigma^2}{(2\pi)^2} \int d\bar{r}_\perp \, e^{i\bar{k}_\perp \cdot \bar{r}_\perp} C(\bar{r}_\perp) \quad (35)$$

and satisfies the relation

$$<F(\bar{k}'_\perp) F^*(\bar{k}_\perp)> = \delta(\bar{k}'_\perp - \bar{k}_\perp) W(|\bar{k}'_\perp|) \quad (36)$$

For a Gaussian correlation function of (45), Section 6.1, we have the spectral density given by

$$W(|\bar{k}_\perp - \bar{k}_{\perp i}|) = \frac{1}{4\pi} \sigma^2 l^2 \exp\left[-\frac{1}{4}(k_{dx}^2 + k_{dy}^2) l^2\right] \quad (37)$$

where

$$\bar{k}_{d\perp} = \bar{k}_\perp - \bar{k}_{\perp i} \quad (38)$$

and σ is the standard deviation of the surface height and l is the correlation length for $f(\bar{r}_\perp)$ in the transverse plane.

The bistatic scattering coefficients $\gamma_{ba}^r(\hat{k}_s, \hat{k}_i)$ are defined as the ratio of scattered power of polarization b_s per unit solid angle in direction $\hat{k}_s$ and the intercepted power of polarization a_i in direction $\hat{k}_i$ averaged over 4π radians. Therefore, in view of (34),

$$\gamma_{ba}^r(\hat{k}_s, \hat{k}_i) = 4\pi \frac{k^2 \cos\theta_s \, f'_{ba} W(|\bar{k}_\perp - \bar{k}_{\perp i}|)}{\cos\theta_i |E_o|^2} \quad (39)$$

where $\overline{k} = k\hat{k}_s$. Substituting the result of the Appendix and (37) into the above equation and rearranging terms, we obtain

$$\gamma_{ba}^r(\hat{k}_s, \hat{k}_i) = \frac{4k^4\sigma^2 l^2 \cos^2\theta_s \cos^2\theta_i}{\cos\theta_i} f_{ba} \exp\left[-\frac{1}{4}k_{d\rho}^2 l^2\right] \quad (40)$$

where

$$k_{d\rho}^2 = k^2\left[\sin^2\theta_s + \sin^2\theta_i - 2\sin\theta_s \sin\theta_i \cos(\phi_s - \phi_i)\right] \quad (41)$$

and

$$f_{hh} = \left|\frac{(k_1^2 - k^2)}{(k_z + k_{1z})(k_{zi} + k_{1zi})}\right|^2 \cos^2(\phi_s - \phi_i) \quad (42a)$$

$$f_{vh} = \left|\frac{(k_1^2 - k^2)kk_{1z}}{(k_1^2 k_z + k^2 k_{1z})(k_{zi} + k_{1zi})}\right|^2 \sin^2(\phi_s - \phi_i) \quad (42b)$$

$$f_{hv} = \left|\frac{(k_1^2 - k^2)kk_{1zi}}{(k_z + k_{1z})(k_1^2 k_{zi} + k^2 k_{1zi})}\right|^2 \sin^2(\phi_s - \phi_i) \quad (42c)$$

$$f_{vv} = \left|\frac{(k_1^2 - k^2)}{(k_1^2 k_z + k^2 k_{1z})(k_1^2 k_{zi} + k^2 k_{1zi})}\right.$$
$$\left. \times \left[k_1^2 k^2 \sin\theta_s \sin\theta_i - k^2 k_{1z} k_{1zi} \cos(\phi_s - \phi_i)\right]\right|^2 \quad (42d)$$

In the backscattering direction, $\hat{k}_s = -\hat{k}_i$. The backscattering cross sections per unit area are

$$\sigma_{hh}(\theta_i) = 4k^4\sigma^2 l^2 \cos^4\theta_i |R_{ho}|^2 \exp\left[-k^2 l^2 \sin^2\theta_i\right] \quad (43a)$$

$$\sigma_{vv}(\theta_i) = 4k^2\sigma^2 l^2 \cos^4\theta_i \left|\frac{(k_1^2 - k^2)(k_1^2 k^2 \sin^2\theta_i + k^2 k_{1z} k_{1zi})}{(k_1^2 k_{zi} + k^2 k_{1zi})^2}\right|^2$$
$$\times \exp\left[-k^2 l^2 \sin^2\theta_i\right] \quad (43b)$$

$$\sigma_{vh}(\theta_i) = \sigma_{hv}(\theta_i) = 0 \quad (43c)$$

Therefore, there is no depolarization in the backscattering direction.

Small Perturbation Method

The bistatic coefficients for the transmitted fields in medium 1 can be obtained from (8b). Following the same procedure, we obtain

$$\gamma_{ba}^t(\hat{k}_t, \hat{k}_i) = 4k_1^2 k^2 \sigma^2 l^2 \cos^2\theta_t \cos\theta_i \frac{\eta}{\eta_1} W_{ba} \exp\left[-\frac{1}{4}k_{td\rho}^2 l^2\right] \quad (44)$$

where

$$k_{td\rho}^2 = k_1^2 \sin\theta_t + k^2 \sin^2\theta_i - 2k_1 k \sin\theta_t \sin\theta_i \cos(\phi_t - \phi_i) \quad (45)$$

and

$$W_{hh} = \left|\frac{(k_1^2 - k^2)}{(k_z + k_{1z})(k_{zi} + k_{1zi})}\right|^2 \cos^2(\phi_t - \phi_i) \quad (46a)$$

$$W_{vh} = \left|\frac{(k_1^2 - k^2)k k_z}{(k_1^2 k_z + k^2 k_{1z})(k_{zi} + k_{1zi})}\right|^2 \sin^2(\phi_t - \phi_i) \quad (46b)$$

$$W_{hv} = \left|\frac{(k_1^2 - k^2)k k_{1zi}}{(k_z + k_{1z})(k_1^2 k_{zi} + k^2 k_{1zi})}\right|^2 \sin^2(\phi_t - \phi_i) \quad (46c)$$

$$W_{vv} = \left|\frac{(k_1^2 - k^2)k_1 k}{(k_1^2 k_z + k^2 k_{1z})(k_1^2 k_{zi} + k^2 k_{1zi})}\right.$$

$$\left. \times [k_1 k \sin\theta_t \sin\theta_i + k_z k_{1zi} \cos(\phi_t - \phi_i)]\right|^2 \quad (46d)$$

and $\overline{k}_1 = k_1 \hat{k}_t$.

The first-order solution gives the lowest-order incoherent scattered intensities. The bistatic reflection and transmission coefficients can be easily shown to satisfy the principle of reciprocity. However, in the first-order solution, there is no depolarization effect in the backscattering direction and the coherent reflection and transmission coefficients are not modified. Therefore, to calculate corrections to coherent reflection and transmission, we need to calculate the second-order solution.

Second-Order Solution

The second-order solution for the surface fields and the scattered fields can be calculated from (4), (7), and (8) by collecting and equating the second order terms. We consider both cases of horizontally and vertically polarized incidence. After much algebraic manipulations, $E_{hs}^{(2)}$ and $E_{vs}^{(2)}$ can be obtained. Taking the ensemble average gives the coherent reflected fields $< E_{hs}^{(2)} >$ and $< E_{vs}^{(2)} >$. The coherent field exists only in the specular derivation and does not contain any depolarization. For the case of horizontally polarized incidence

$$<E_{vs}^{(2)}> = 0 \qquad (47)$$

$$<E_{hs}^{(2)}> + <E_{hs}^{(0)}> = R_{01}\,\hat{e}(k_{zi})\,E_o\,e^{i\bar{k}_{i\perp}\cdot\bar{r}_\perp + ik_{zi}z} \qquad (48)$$

where R_{01} is the coherent reflection coefficient for horizontally polarized incident. The result for R_{01}, up to second order, is

$$R_{01} = R_{ho} + k_{zi}\frac{(k_1^2 - k^2)}{(k_{zi} + k_{1zi})^2}\sigma^2 l^2 \int_0^\infty k_\rho dk_\rho \exp\left[-\frac{1}{4}(k_\rho^2 + k_{\rho i}^2)l^2\right]$$

$$\times \left\{\left[-\frac{(k_1^2 - k^2)}{(k_z + k_{1z})}\frac{k_z k_{1z}}{k_\rho^2 + k_z k_{1z}} + k_{1zi}\right] I_o(x)\right.$$

$$\left. - \frac{k_1^2 - k^2}{k_z + k_{1z}}\frac{k_\rho^2}{k_\rho^2 + k_z k_{1z}}\left(I_o(x) - \frac{I_1(x)}{x}\right)\right\} \qquad (49)$$

where

$$x = \frac{1}{2}k_\rho k_{\rho i} l^2 \qquad (50)$$

and I_o and I_1 are the zeroth- and first-order modified Bessel functions.

The modified reflection coefficient for the vertical polarization can also be obtained by considering a vertically polarized incident field and calculating for the scattered fields. Following the same procedure, we obtain the reflection coefficient for vertically polarized incidence

$$S_{01} = R_{vo} - k_{zi}\frac{(k_1^2 - k^2)}{(k^2 k_{1zi} + k_1^2 k_{zi})^2}k^2 k_1^2 \sigma^2 l^2 \int_0^\infty k_\rho dk_\rho$$

$$\times \exp\left[-\frac{1}{4}(k_\rho^2 + k_{\rho i}^2)l^2\right]\left\{k_{1zi}\left(I_o(x) - 2\frac{k_\rho k_{\rho i}}{k_\rho^2 + k_z k_{1z}}I_1(x)\right)\right.$$

$$- \frac{k_1^2 - k^2}{k_z + k_{1z}}\frac{k_{1zi}^2}{k_1^2}I_o(x) + \frac{k_1^2 - k^2}{k_z + k_{1z}}\frac{k_{1zi}^2}{k_1^2}\frac{k_\rho^2}{k_\rho^2 + k_z k_{1z}}$$

$$\left.\times \left(I_o(x) - \frac{I_1(x)}{x}\right) - \frac{k_1^2 - k^2}{k_z + k_{1z}}\frac{k_\rho^2 k_{\rho i}^2}{k_\rho^2 + k_z k_{1z}}\frac{1}{k^2}I_o(x)\right\} \qquad (51)$$

The modified coherent transmission coefficients for the horizontal and vertical polarization can be calculated similarly to give

$$X_{01} = 1 + R_{ho} + k_{zi}\frac{(k_1^2 - k^2)}{(k_{zi} + k_{1zi})^2}\sigma^2 l^2 \int_0^\infty k_\rho dk_\rho$$

$$\times \exp\left[-\frac{1}{4}(k_\rho^2 + k_{\rho i}^2)l^2\right] \left\{ (k_z - k_{1z})\left(I_0(x) - \frac{k_\rho^2}{k_\rho^2 + k_z k_{1z}} \frac{I_1(x)}{x}\right) \right.$$

$$\left. - \frac{1}{2}(k_{zi} - k_{1zi})I_0(x) \right\} \tag{52}$$

and

$$Y_{01} = 1 + R_{vo} + k_{zi}\frac{(k_1^2 - k^2)}{(k_1^2 k_{zi} + k^2 k_{1zi})^2} k_1^2 \sigma^2 l^2 \int_0^\infty k_\rho dk_\rho$$

$$\times \exp\left[-\frac{1}{4}(k_\rho^2 + k_{\rho i}^2)l^2\right] \left\{ -\frac{1}{2}(k^2 k_{1zi} - k_1^2 k_{zi})I_0(x) \right.$$

$$- (k_1^2 - k^2)\frac{k_\rho^2 k_{\rho i}^2}{(k_1^2 k_z + k^2 k_{1z})}I_0(x) - \frac{k_1^2 - k^2}{k_z + k_{1z}}k_{zi}k_{1zi}I_0(x)$$

$$+ (k_1^2 - k^2)\frac{k_\rho^2 k_{zi}k_{1zi}}{(k_1^2 k_z + k^2 k_{1z})}\left(I_0(x) - \frac{I_1(x)}{x}\right)$$

$$\left. + \frac{k_z + k_{1z}}{k_1^2 k_z + k^2 k_{1z}}k_\rho k_{\rho i}(k^2 k_{1zi} - k_1^2 k_{zi})I_1(x) \right\} \tag{53}$$

The depolarization scattered intensity in the backscattering direction can be obtained by considering a horizontally polarized incident field and calculating the vertically polarized scattered intensity or vice versa. From reciprocity the two solutions σ_{vh} and σ_{hv} can be shown to be the same. Keeping only the terms that do not vanish in the backscattering direction and calculating the scattered intensity in the backscattering direction, the depolarized backscattering cross section per unit area can be calculated to be

$$\sigma_{hv}(\theta_i) = \sigma_{vh}(\theta_i) = 8\pi k^2 \cos^2\theta_i \left|\frac{2k(k_1^2 - k^2)^2 k_{1zi}k_{zi}}{(k_1^2 k_{zi} + k^2 k_{1zi})(k_{zi} + k_{1zi})}\right|^2$$

$$\times \int d\bar{k}'_\perp \left|\frac{(k_{xi}k'_y - k_{yi}k'_x)(k_{xi}k'_x + k_{yi}k'_y)}{k_{\rho i}^2(k^2 k'_z + k_1^2 k'_{1z})}\right|^2$$

$$\times W(|\bar{k}'_\perp - \bar{k}_{\perp i}|)W(|\bar{k}'_\perp + \bar{k}_{\perp i}|) \tag{54a}$$

Using the Gaussian correlation function, we get,

$$\sigma_{hv}(\theta_i) = \sigma_{vh}(\theta_i) = \frac{1}{2}k^6 \cos^4\theta_i \sigma^4 l^4$$

$$\times \left|\frac{(k_1^2 - k^2)^2 k_{1zi}}{(k_1^2 k_{zi} + k^2 k_{1zi})(k_{zi} + k_{1zi})}\right|^2 \exp\left[-\frac{1}{2}k_{\rho i}^2 l^2\right]$$

$$\times \int_0^\infty dk'_\rho \frac{k'^3_\rho}{|k^2 k'_{1z} + k_1^2 k'_z|^2} \exp\left[-\frac{1}{2}k'^2_\rho l^2\right] \quad (54b)$$

Over the years, extensive theoretical and experimental investigations have been performed on scattering and emission of random rough surfaces and applied to sea, planetary and soil surfaces. Most of the work utilized either the Kirchhoff approach or the small perturbation method (Beckmann and Spizzichino, 1963; Rice, 1963; Hagfors, 1964; Semenov, 1965, 1966; Kodis, 1966; Stogryn, 1967; Valenzuela, 1967, 1968; Beckmann, 1968; Ruck et al., 1970; Macdonald and Waite, 1971; Swift, 1974; Beckmann, 1975; Sung and Holzer, 1976; Sung and Ekerhardt, 1978; Bass and Fuks, 1979; Ulaby et al., 1981). Numerical simulations have also been done to test the validity of the methods (Axline and Fung, 1978; Chan and Fung, 1978). The KA and SPM have been superimposed to deal with composite rough surfaces (Fung and Chan, 1969; Peake et al., 1970; Burrows, 1973; Brown, 1978).

The small perturbation method is used to study the rough surface with the height small compared with a wavelength and the slope smaller than unity. The Kirchhoff approach has been used to study rough surfaces with large radius of curvature. In recent years, there has been considerable interest in the development of more general theories that can bridge these two limiting methods.

The full-wave approach (Bahar, 1978; Bahar and Barrick, 1983) has been used for composite surfaces that cannot be decomposed into small-scale perturbations and large-scale surfaces. It has also been used to study depolarization effects. The stochastic Fourier transform approach (Brown, 1982) makes use of the Fourier transform representation of the surface current and has been applied to the gently undulating and uniformly rough surfaces. The diagrammatic approach (Zipfel and DeSanto, 1972; DeSanto, 1974, 1983; DeSanto et al., 1980) makes use of the Feynman diagram and has been used to obtain the coherent intensity beyond the Kirchhoff approximation. Dyson and Bethe-Salpeter equations have also be derived analogous to that of random medium (Furutsu, 1983) by using an equivalent boundary condition on a flat plane lying above the rough surface. Studies have been reported on the combination of the Kirchhoff method for rough surface and the doubling method for volume scattering (Fung and Eom, 1981).

The Wiener-Hermite nonlinear functional representation (Wiener, 1958; Ogura, 1975) has been used in the formulation of rough surface scattering (Nakayama et al., 1981a,b). Also considered is the scattering by randomly placed bosses on a surface (Twersky, 1983). Recently, the first and second moments of the scattered field have been obtained based on the perturbation

Small Perturbation Method

method in a unified and consistent manner (Watson and Keller, 1983). Another approach is based on the extinction theorem (Nieto-Vesperinas, 1982), which is also the basis of the T-matrix method. The extinction theorem has been used extensively in the study of atomic-surface scattering problems (Marvin, 1980).

Appendix

Explicit Expressions for $\tilde{A}_q^{(1)}(\overline{k}_\perp)$, $\tilde{A}_p^{(1)}(\overline{k}_\perp)$, $\tilde{B}_q^{(1)}(\overline{k}_\perp)$, $\tilde{B}_p^{(1)}(\overline{k}_\perp)$, f'_{hh}, f'_{vh}, f'_{hv} and f'_{vv}.

$$\tilde{A}_q^{(1)}(\overline{k}_\perp) = \frac{(k_1^2 - k^2)}{(k_1^2 k_z + k^2 k_{1z})} k_\rho k_{\rho i} \left(\frac{k_x k_{yi} - k_y k_{xi}}{k_\rho k_{\rho i}} \right) a_p^{(0)}(\overline{k}_{\perp i})$$

$$+ \frac{k_\rho k_{\rho i}}{k} \left(\frac{k_x k_{xi} + k_y k_{yi}}{k_\rho k_{\rho i}} \right) b_p^{(0)}(\overline{k}_{\perp i})$$

$$+ \left(\frac{k k_1^2}{k_z k_{1z} + k_\rho^2} - \frac{k_z k_{1z} + k_\rho^2}{k} \right) \left(\frac{k_x k_{xi} + k_y k_{yi}}{k_\rho k_{\rho i}} \right)$$

$$\times \left(\frac{k_x k_{yi} - k_y k_{xi}}{k_\rho k_{\rho i}} \right) b_q^{(0)}(\overline{k}_{\perp i})$$

$$- \left[\frac{k k_1^2}{k_z k_{1z} + k_\rho^2} \left(\frac{k_x k_{yi} - k_y k_{xi}}{k_\rho k_{\rho i}} \right)^2 \right.$$

$$\left. + \frac{k_z k_{1z} + k_\rho^2}{k} \left(\frac{k_x k_{xi} + k_y k_{yi}}{k_\rho k_{\rho i}} \right)^2 \right] b_p^{(0)}(\overline{k}_{\perp i}) \quad \text{(A1)}$$

$$\tilde{A}_p^{(1)}(\overline{k}_\perp) = \frac{(k_1^2 - k^2)}{(k_1^2 k_z + k^2 k_{1z})} k_\rho k_{\rho i} \left(\frac{k_x k_{xi} + k_y k_{yi}}{k_\rho k_{\rho i}} \right) a_p^{(0)}(\overline{k}_{\perp i})$$

$$+ \frac{k_\rho k_{\rho i}}{k} \left(\frac{k_y k_{xi} - k_x k_{yi}}{k_\rho k_{\rho i}} \right) b_p^{(0)}(\overline{k}_{\perp i})$$

$$+ \left[\frac{k k_1^2}{k_z k_{1z} + k_\rho^2} \left(\frac{k_x k_{xi} + k_y k_{yi}}{k_\rho k_{\rho i}} \right)^2 \right.$$

$$\left. + \left(\frac{k_z k_{1z} + k_\rho^2}{k} \right) \left(\frac{k_x k_{yi} - k_y k_{xi}}{k_\rho k_{\rho i}} \right) \right] b_q^{(0)}(\overline{k}_{\perp i})$$

$$+ \left[\frac{k k_1^2}{k_z k_{1z} + k_\rho^2} - \frac{k_z k_{1z} + k_\rho^2}{k} \right] \left(\frac{k_x k_{xi} + k_y k_{yi}}{k_\rho k_{\rho i}} \right)$$

$$\times \left(\frac{k_y k_{xi} - k_x k_{yi}}{k_\rho k_{\rho i}} \right) b_p^{(0)}(\overline{k}_{\perp i}) \tag{A2}$$

$$\widetilde{B}_q^{(1)}(\overline{k}_\perp) = k\, a_p^{(0)}(\overline{k}_{\perp i}) - \frac{k}{k_z k_{1z} + k_\rho^2}$$

$$\times k_\rho k_{\rho i} \left(\frac{k_x k_{xi} + k_y k_{yi}}{k_\rho k_{\rho i}} \right) a_p^{(0)}(\overline{k}_{\perp i})$$

$$+ \left[\frac{(k_1^2 - k^2)}{(k_z + k_{1z})} - \frac{(k_1^2 - k^2)}{(k_1^2 k_z + k^2 k_{1z})} \right.$$

$$\left. \times k_\rho^2 \left(\frac{k_x k_{xi} + k_y k_{yi}}{k_\rho k_{\rho i}} \right)^2 \right] b_q^{(0)}(\overline{k}_{\perp i})$$

$$+ \frac{(k_1^2 - k^2)}{(k_1^2 k_z + k^2 k_{1z})} k_\rho^2 \left(\frac{k_x k_{xi} + k_y k_{yi}}{k_\rho k_{\rho i}} \right)$$

$$\times \left(\frac{k_x k_{yi} - k_y k_{xi}}{k_\rho k_{\rho i}} \right) b_p^{(0)}(\overline{k}_{\perp i}) \tag{A3}$$

$$\widetilde{B}_p^{(1)}(\overline{k}_\perp) = -k\, a_q^{(0)}(\overline{k}_{\perp i}) - \frac{k}{k_z k_{1z} + k_\rho^2}$$

$$\times k_\rho k_{\rho i} \left(\frac{k_y k_{xi} - k_x k_{yi}}{k_\rho k_{\rho i}} \right) a_p^{(0)}(\overline{k}_{\perp i})$$

$$- \frac{(k_1^2 - k^2)}{(k_1^2 k_z + k^2 k_{1z})} k_\rho^2 \left(\frac{k_x k_{xi} + k_y k_{yi}}{k_\rho k_{\rho i}} \right)$$

$$\times \left(\frac{k_y k_{xi} - k_x k_{yi}}{k_\rho k_{\rho i}} \right) b_q^{(0)}(\overline{k}_{\perp i})$$

$$+ \left[\frac{(k_1^2 - k^2)}{(k_z + k_{1z})} - \frac{(k_1^2 - k^2)}{(k_1^2 k_z + k^2 k_{1z})} \right.$$

$$\left. \times k_\rho^2 \left(\frac{k_y k_{xi} - k_x k_{yi}}{k_\rho k_{\rho i}} \right)^2 \right] b_p^{(0)}(\overline{k}_{\perp i}) \tag{A4}$$

$$f'_{hh} = \left| E_o (k_1^2 - k^2) \frac{2 k_{zi}}{(k_z + k_{1z})(k_{zi} + k_{1zi})} \left(\frac{k_x k_{xi} + k_y k_{yi}}{k_\rho k_{\rho i}} \right) \right|^2 \tag{A5}$$

$$f'_{vh} = \left| E_o(k_1^2 - k^2) \frac{2kk_{1z}k_{zi}}{(k_1^2 k_z + k^2 k_{1z})(k_{zi} + k_{1zi})} \left(\frac{k_x k_{yi} - k_y k_{xi}}{k_\rho k_{\rho i}} \right) \right|^2 \quad (A6)$$

$$f'_{hv} = \left| E_o(k_1^2 - k^2) \frac{2kk_{1zi}k_{zi}}{(k_z + k_{1z})(k_1^2 k_{zi} + k^2 k_{1zi})} \left(\frac{k_y k_{xi} - k_x k_{yi}}{k_\rho k_{\rho i}} \right) \right|^2 \quad (A7)$$

$$f'_{vv} = \left| E_o(k_1^2 - k^2) \frac{2k_1^2 k_{zi}}{(k_1^2 k_z + k^2 k_{1z})(k_1^2 k_{zi} + k^2 k_{1zi})} \right.$$
$$\left. \times \left[-k_\rho k_{\rho i} + \frac{k^2}{k_1^2} k_{1z} k_{1zi} \left(\frac{k_x k_{xi} + k_y k_{yi}}{k_\rho k_{\rho i}} \right) \right] \right|^2 \quad (A8)$$

PROBLEMS

2.1 1. Consider a half-space with permittivity $\epsilon_1 = 3.2\epsilon_o(1 + i0.01)$. Plot the vertically and horizontally polarized emissivity as a function of observation angle θ_i. Consider next the case $\epsilon_1 = 3.2\epsilon_o(1 + i0.1)$. Explain why the imaginary part of the permittivity has only a very small effect on the emissivity.

2. Repeat the above exercise for a two-layer medium with frequency = 10 GHz, $d = 30$ cm, $\epsilon_2 = 80\epsilon_o$. Explain why in this case the imaginary part of ϵ_1 plays a significant role.

2.2 Consider a two-layer medium with a linear profile for medium 1.

$$\epsilon_1(z) = \epsilon_1 \left[\frac{\epsilon_2}{\epsilon_1} + \frac{(z+d)}{d} \left(1 - \frac{\epsilon_2}{\epsilon_1} \right) \right]$$

Calculate the emissivity at nadir. The field solution can be expressed in terms of Bessel functions (Wait, 1962).

2.3 Consider a half-space medium with constant permittivity ϵ_1 and temperature profile

$$T_1(z) = \begin{cases} T_s \left[1 + \left(\frac{z}{d} \right) \left(1 - \frac{T}{T_s} \right) \right] & \text{for } 0 \geq z \geq -d \\ T & \text{for } z \leq -d. \end{cases}$$

Calculate the brightness temperature at nadir.

2.4 Consider a n-layer stratified medium. By using vector Green's theorem and the boundary conditions for the dyadic Green's function, derive the following symmetry relation in (11), Section 3.3:

$$\overline{\overline{G}}_{ol}(\bar{r}, \bar{r}') = \overline{\overline{G}}_{lo}^{t}(\bar{r}', \bar{r})$$

2.5 Consider the expression of brightness temperature at nadir for a two-layer medium with a constant temperature T.

$$T_B(d) = \left\{ 1 - \left| \frac{R_{o1} + R_{1t}e^{i2k_1 d}}{1 + R_{01}R_{1t}e^{i2k_1 d}} \right|^2 \right\} T$$

The brightness temperature is a function of the depth of the bottom interface d. Suppose the depth d is a random variable with probability density function

$$p(d) = \frac{1}{\sqrt{2\pi}\sigma} \exp\left[-\frac{d^2}{\sigma^2}\right]$$

Find the average brightness temperature defined by

$$<T_B> = \int_{-\infty}^{\infty} dd\, p(d)\, T_B(d)$$

You should expand the expression for $T_B(d)$ in a geometric series of $\exp(i2k_1 d)$. Calculate the result for $<T_B>$ when $k_1\sigma \gg 1$.

2.6 Consider the case where the periodic surface profile height is given by

$$f(x) = \begin{cases} h/2 & \text{for } 0 \leq x < P/2 \\ -h/2 & \text{for } P/2 \leq x < P \end{cases}$$

Calculate the $\overline{\overline{Q}}^{\pm}_{D_2}, \overline{\overline{Q}}^{\pm}_{N_2}, \overline{\overline{Q}}^{\pm}_{D_1}, \overline{\overline{Q}}^{\pm}_{N_1}$ matrix elements as expressed in (11) and (14), Section 5.2.

2.7 Use the extended boundary condition formalism to find the emissivity of a two layer medium with a sinusoidal rough surface for the first boundary and a flat surface for the second boundary. Restrict your attention to the use of horizontal polarization and observation angle in the direction $\phi = 0$.

2.8 Consider a Gaussian random process $f(x,y)$ with $<f>=0$ and

$$<f(x_1, y_2)\, f(x_2, y_2)> = \sigma^2 \exp\left[-\frac{(x_1 - x_2)^2 + (y_1 - y_2)^2}{L^2}\right]$$

Letting $\alpha = \partial f(x,y)/\partial x$, $\beta = \partial f(x,y)/\partial y$, calculate $<\alpha(x_1, y_1)\alpha(x_2, y_2)>$, $<\beta(x_1, y_1)\beta(x_2, y_2)>$ and $<\alpha(x_1, y_1)\beta(x_2, y_2)>$.

2.9 Let $\theta_i = 30\,\text{deg}$, $\phi_i = 0\,\text{deg}$, $\theta_s = 60\,\text{deg}$, $\phi_s = 60\,\text{deg}$, and $\epsilon_1 = 9\epsilon_o$. Calculate $\hat{k}_i$, $\hat{k}_s$, $\overline{F}(\alpha_o, \beta_o)$, $\hat{v}_i$, $\hat{h}_i$, $\hat{v}_s$ and $\hat{h}_s$ by using the formulas in Section 6.1.

2.10 The probability density function for N-dimensional Gaussian random vector $f^T = [f_1 \ f_2 \ \cdots \ f_N]$ is

$$p_{\underline{f}}(\underline{f}) = \frac{1}{(\sqrt{2\pi})^N |\Lambda_f|^{1/2}} \exp\left[-\frac{1}{2}(\underline{f} - \underline{m}_f)\Lambda_f^{-1}(\underline{f} - \underline{m}_f)\right]$$

where Λ_f is the covariance matrix. Consider the case $n = 2$ with

$$E(f_1^2) - [E(f_1)]^2 = \sigma_1^2$$

$$E(f_2^2) - [E(f_2)]^2 = \sigma_2^2$$

$$E[(f_1 - <f_1>)(f_2 - <f_2>)] = r\sigma_1\sigma_2$$

Derive an expression for $p_{\underline{f}}(f_1, f_2)$ and show that it is the same as (31), Section 6.1.

2.11 Let $f(x)$ be a stationary Gaussian random process with variance σ^2 and $<f(x)> = 0$. Show that

$$<\exp(i\alpha f(x))> = \exp\left[-\frac{\sigma^2 \alpha^2}{2}\right]$$

2.12 Let $f(x)$ be a stationary Gaussian random process with

$$<f(x)> = 0$$

$$<f(x_1)f(x_2)> = \sigma^2 \exp\left[-\frac{(x_1 - x_2)^2}{l^2}\right]$$

Calculate

$$\left\langle \exp\left[i\alpha\big(f(x_1) - f(x_2)\big)\right]\right\rangle$$

2.13 Consider the Kirchhoff approach to calculate the coherent and incoherent scattering coefficients. Suppose $\overline{F}(\alpha, \beta)$ in (20a), Section 6.1, is expanded to second order in slope.

$$\overline{F}(\alpha, \beta) = \overline{F}(0, 0) + \alpha \frac{\partial \overline{F}}{\partial \alpha}\bigg|_{\alpha,\beta=0} + \beta \frac{\partial \overline{F}}{\partial \beta}\bigg|_{\alpha,\beta=0}$$

$$+ \frac{\alpha^2}{2} \frac{\partial^2 \overline{F}}{\partial \alpha^2}\bigg|_{\alpha,\beta=0} + \frac{\beta^2}{2} \frac{\partial^2 \overline{F}}{\partial \beta^2}\bigg|_{\alpha,\beta=0}$$

$$+ \alpha\beta \frac{\partial^2 \overline{F}}{\partial \alpha \partial \beta}\bigg|_{\alpha,\beta=0}$$

Use the second-order expansion to calculate the incoherent scattering coefficients using a similar procedure as in (21) through (27), Section 6.1.

2.14 For the Gaussian random surface, show that

$$p(\alpha, \beta) = \frac{1}{2\pi\sigma^2 |C''(0)|} \exp\left[-\frac{\alpha^2 + \beta^2}{2\sigma^2 |C''(0)|}\right]$$

2.15 By using the bistatic scattering and transmission coefficients derived under the geometrical optics limit, show that the reciprocity relations in (104) and (105), Section 6.1, hold.

2.16 Consider the bistatic scattering coefficients for a Gaussian random rough surface $f(x, y)$ under the geometric optics approximation. Express your answer in terms of the slope probability density function $p(\alpha, \beta)$, where $\alpha = \partial f/\partial x$ and $\beta = \partial f/\partial y$.

2.17 Plot the backscattering coefficients for a random rough surface for $\sigma/L = 0.1$ as a function of incidence angle θ_i using the geometric optics approximation. Let $\epsilon_1 = 80\epsilon_o$.

2.18 Compare geometric optics backscattering coefficients of (93), Section 6.1, with those of the first-order small perturbation method in (40) through (43), Section 6.2, by letting $k\sigma$ be small in the geometric optics results and considering the correlation function to be $\exp(-\rho^2/l^2)$. Compare both cases of vertically polarized incidence and horizontally polarized incidence.

3

RADIATIVE TRANSFER THEORY – EXTINCTION MATRIX, EMISSION VECTOR, AND SCATTERING PHASE MATRICES

1. Introduction — 120

2. Stokes Parameters — 121
   - 2.1 Elliptical Polarization — 121
   - 2.2 Partial Polarization and Natural Light — 125
   - 2.3 Incoherent Addition of Stokes Parameters — 127
   - 2.4 Transformation of Stokes Parameters for a Rotation of the Axes — 128

3. Vector Radiative Transfer Equation: Constituents, Reciprocity, and Energy Conservation — 128
   - 3.1 Vector Radiative Transfer Equations — 129
   - 3.2 Scattering Function Matrix, Stokes Matrix, and Phase Matrix — 131
   - 3.3 Optical Theorem — 135
   - 3.4 Extinction Matrix for Nonspherical Particles — 138
   - 3.5 Emission Vector — 142
   - 3.6 Energy Conservation, Reciprocity and Reciprocal Relation between Active and Passive Remote Sensing — 148

4. Phase Matrix for Simple Objects — 155
   - 4.1 Rayleigh Phase Matrix — 155
   - 4.2 Eulerian Angles of Rotation — 158
   - 4.3 Phase Matrix for Small Ellipsoids with Prescribed Orientation Distribution — 160

4.4 Phase Matrix for Random Media　　　　　　　　　　162

5. Phase Matrix and Extinction Matrix for General Nonspherical Particles with Prescribed Orientation and Size Distribution: T-Matrix Approach　　　　　　　　　　168

    5.1 Vector Spherical Waves　　　　　　　　　　168
    5.2 Relation of T-Matrix to Scattering Amplitude Matrix, Extinction, and Scattering Cross Sections　　　　　　　　　　174
    5.3 Unitarity and Symmetry　　　　　　　　　　178
    5.4 Extended Boundary Condition Technique　　　　　　　　　　181
    5.5 Spheres　　　　　　　　　　188
    5.6 Spheroids　　　　　　　　　　191
    5.7 Rotation of T-Matrix　　　　　　　　　　193
    5.8 Extinction Coefficients and Phase Matrix for Axisymmetric Objects with Prescribed Orientation Distribution　　　　　　　　　　197
    5.9 Appendix　　　　　　　　　　199

6. Boundary Conditions for Radiative Transfer Equations　　　　　　　　　　200

    6.1 Planar Dielectric Interface　　　　　　　　　　200
    6.2 Rough Dielectric Interface　　　　　　　　　　203

Problems　　　　　　　　　　208

1 INTRODUCTION

The subject of radiative transfer is the analysis of radiation intensity in a medium that is able to absorb, emit, and scatter radiation. Radiative transfer theory was first initiated by Schuster in 1905 in an attempt to explain the appearance of absorption and emission lines in stellar spectra. Since then the subject of radiative transfer has been investigated principally by astrophysicists (Chandrasekhar, 1960), and in recent years by physicists for the problem of diffusion of neutrons (Davison, 1957; Case and Zweifel, 1967). Our interest in the radiative transfer theory lies in its application to the problem of remote sensing from scattering media.

In the active and passive remote sensing of low-loss areas such as snow and ice fields, the effects of scattering due to medium inhomogeneities play a dominant role. Two distinct theories are being used to deal with the problem of incorporating scattering effects: wave theory and the radiative transfer theory. In the wave theory, one starts out with Maxwell's equations, introduces the scattering and absorption characteristics of the medium, and tries to find solutions for the quantities of interest, such as brightness temperatures or backscattering cross sections. In principle, all the multiple scattering, diffraction, and interference effects can be included. However, the equations are generally complicated and approximations have to be made before numerical solutions can be calculated.

Radiative transfer theory, on the other hand, does not start with Maxwell's equations. It starts with the radiative transfer equations that govern the propagation of energy through the scattering medium. The development of the theory is heuristic and lacks the mathematical rigor of the wave theory. It is assumed in the radiative transfer theory that there is no correlation between fields and therefore, the addition of intensities is considered rather than the addition of fields. However, it has an advantage in that it is simple and, more importantly, includes multiple scattering effects. Though the transfer theory was developed on the basis of powers, it contains information about the correlation of fields (Ishimaru, 1975, 1978). The mutual coherence function is related to the Fourier transform of the specific intensity.

Scattering effects are accounted for by making use of two models: random fluctuations of permittivity (random medium approach) and discrete particles imbedded in a homogeneous medium (discrete scatterer approach). In this chapter, we shall focus on the development of vector radiative transfer equations including the polarization characteristics of electromagnetic propagation. The extinction matrix, phase matrix, and emission vector for these two types of scattering media will be derived. The solution of radiative transfer equations will be deferred to Chapter 4. Our objective is to study the scattering and propagation characteristics of the Stokes parameters in radiative transfer theory and to use the theory to calculate bistatic scattering cross sections and brightness temperatures.

2 STOKES PARAMETERS

In the radiative transfer theory, the polarization of the electromagnetic radiation is described by the four Stokes parameters (Chandrasekhar, 1960; Kraus, 1966). We will first discuss Stokes parameters for coherent, completely polarized waves and then the case of partially polarized and partially coherent radiation.

2.1 Elliptical Polarization

Consider a time harmonic elliptically polarized radiation field with time dependence $\exp(-i\omega t)$ propagating in the $\hat{k}$ direction, with complex electric field given by

$$\overline{E} = E_v \hat{v} + E_h \hat{h} \tag{1}$$

where $\hat{v}$ and $\hat{h}$ denote the two orthogonal polarizations. The four Stokes parameters are defined as

$$I = \frac{1}{\eta}\left[|E_v|^2 + |E_h|^2\right] \tag{2}$$

$$Q = \frac{1}{\eta}\left[|E_v|^2 - |E_h|^2\right] \tag{3}$$

$$U = \frac{2}{\eta} Re(E_v E_h^*) \tag{4}$$

$$V = \frac{2}{\eta} Im(E_v E_h^*) \tag{5}$$

where η is the wave impedance of the medium. It is also common to use the "modified Stokes parameters" given by

$$I_v = \frac{|E_v|^2}{\eta} \tag{6}$$

$$I_h = \frac{|E_h|^2}{\eta} \tag{7}$$

with U and V as defined in (4) and (5), respectively.

For an elliptically polarized wave, as in (1), let $E_v = |E_v|\exp(i\delta_v)$, $E_h = |E_h|\exp(i\delta_h)$, and $\delta = \delta_v - \delta_h$ be the phase difference between the two components. Elliptically polarized waves are then described by the three parameters, $|E_v|$, $|E_h|$, and δ. The Stokes parameters are

$$I_v = \frac{|E_v|^2}{\eta} \tag{8}$$

$$I_h = \frac{|E_h|^2}{\eta} \tag{9}$$

$$U = \frac{2}{\eta}|E_v||E_h|\cos\delta \tag{10}$$

$$V = \frac{2}{\eta}|E_v||E_h|\sin\delta. \tag{11}$$

Elliptical Polarization

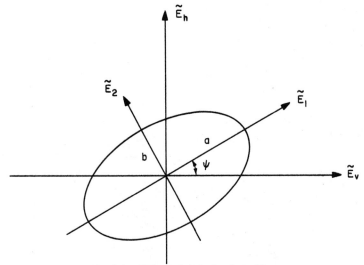

Fig. 3.1   Elliptical polarization ellipse.

From (8) through (11), it can be shown that

$$I^2 = Q^2 + U^2 + V^2 \tag{12}$$

so that only three Stokes parameters are required to describe the elliptically polarized waves.

However, the parameters $|E_v|$, $|E_h|$ and δ are not convenient in describing the polarization ellipse (Figure 3.1) with axes a and b. It is more convenient to use the quantities I, ψ and χ, where ψ is the angle between (E_v, E_h) with the axes of the ellipse (E_1, E_2), and χ obeys the relation

$$\tan \chi = \pm \frac{b}{a} \tag{13}$$

with + sign for left-hand polarization and − sign for right-hand polarization. Thus, χ describes the ratio of the axes of the polarization ellipse.

If we consider the electric fields in the time domain, then

$$\tilde{E}_v(t) = |E_v| \cos(\delta_v - \omega t) \tag{14}$$

$$\tilde{E}_h(t) = |E_h| \cos(\delta_h - \omega t) \tag{15}$$

By eliminating the t dependence from (12) and (13), we obtain the following equation of an ellipse (Figure 3.1)

$$\frac{(\tilde{E}_v)^2}{|E_v|^2} + \frac{(\tilde{E}_h)^2}{|E_h|^2} - \frac{2\tilde{E}_v \tilde{E}_h}{|E_v||E_h|} \cos \delta = \sin^2 \delta \tag{16}$$

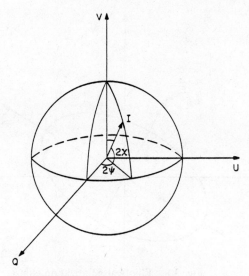

Fig. 3.2 Poincare sphere.

The polarization is left-handed if $\sin \delta > 0$ and right-handed if $\sin \delta < 0$. In the frame of major and minor axes,

$$\frac{\widetilde{E}_1^2}{a^2} + \frac{\widetilde{E}_2^2}{b^2} = 1 \tag{17}$$

Using (13) gives

$$\widetilde{E}_1(t) = E_o \cos \chi \sin \omega t \tag{18}$$

$$\widetilde{E}_2(t) = E_o \sin \chi \cos \omega t \tag{19}$$

with $E_o = |b/\sin \chi| = |a/\cos \chi|$. The corresponding complex amplitudes E_1 and E_2 are such that $\widetilde{E}_1(t) = Re(E_1 e^{-i\omega t})$, $\widetilde{E}_2(t) = Re(E_2 e^{-i\omega t})$. Hence, $E_1 = iE_o \cos \chi$ and $E_2 = E_o \sin \chi$. Since

$$E_v = E_1 \cos \psi - E_2 \sin \psi = iE_o \cos \chi \cos \psi - E_o \sin \chi \sin \psi \tag{20}$$

$$E_h = E_1 \sin \psi + E_2 \cos \psi = iE_o \cos \chi \sin \psi + E_o \sin \chi \cos \psi \tag{21}$$

applying the definition of Stokes parameters (2) through (7) to (20) and (21) gives

$$I = \frac{E_o^2}{\eta} \tag{22}$$

$$Q = I \cos 2\chi \cos 2\psi \tag{23}$$

$$U = I \cos 2\chi \sin 2\psi \tag{24}$$

$$V = I \sin 2\chi \tag{25}$$

From (22) through (25), it is obvious that (12) holds. Equations (22) through (25) may be compared with the cartesian coordinates of a point (r, θ, ϕ) on a sphere with radius $r = I$, $\theta = \pi/2 - 2\chi$, and $\phi = 2\psi$. The Cartesian coordinates are Q, U and V. This sphere is called the Poincare sphere. The north and south poles represent left-handed and right-handed circular polarization, respectively, and the equator represents linear polarization (Figure 3.2). The spherical surface generally represents an elliptically polarized wave.

2.2 Partial Polarization and Natural Light

In real life, quasi-monochromatic sources are very often encountered in which the field components E_v and E_h fluctuate with time. For example

$$E_v(t) = |E_v(t)| e^{i\delta_v(t)} \tag{1}$$

The wave is quasi-monochromatic if $|E_v(t)|$ and $\exp[i\delta_v(t)]$ are slowly varying compared with $\exp(-i\omega t)$. The Stokes parameters are now defined by a time-averaging procedure. Denoting time average by the brackets $<>_t$, we have

$$I_v = \frac{<|E_v|^2>_t}{\eta} \tag{2}$$

$$I_h = \frac{<|E_h|^2>_t}{\eta} \tag{3}$$

$$U = \frac{2}{\eta} Re <E_v E_h^*>_t \tag{4}$$

$$V = \frac{2}{\eta} Im <E_v E_h^*>_t \tag{5}$$

with

$$I = I_v + I_h \tag{6}$$

and

$$Q = I_v - I_h \tag{7}$$

Natural light is characterized by the fact that the intensity is the same in any direction perpendicular to the direction of propagation and that there is no correlation between E_v and E_h. Therefore, the necessary and sufficient conditions for light to be natural are

$$Q = U = V = 0 \tag{8}$$

Let the amplitudes $|E_v|$ and $|E_h|$, and phase difference δ remain constant for a fractional time interval t_1 having the values $|E_{v_1}|$, $|E_{h_1}|$, and δ_1. Similarly, let the values distinguished by the indices 2, 3, ... occur during fractional time intervals $t_2, t_3, \ldots$, and all the time intervals add up to T. Then

$$I_v = \frac{1}{\eta T} \sum_n t_n |E_{v_n}|^2 \qquad (9)$$

$$I_h = \frac{1}{\eta T} \sum_n t_n |E_{h_n}|^2 \qquad (10)$$

$$U = \frac{2}{\eta T} \sum_n t_n |E_{v_n}|^2 A_n \cos \delta_n \qquad (11)$$

$$V = \frac{2}{\eta T} \sum_n t_n |E_{v_n}|^2 A_n \sin \delta_n \qquad (12)$$

where

$$A_n = \frac{|E_{h_n}|}{|E_{v_n}|} \qquad (13)$$

denotes the ratio of $|E_h|$ and $|E_v|$ in the t_n interval. Then (Problem 2)

$$4 I_v I_h = \frac{4}{\eta^2 T^2} \left\{ \sum_n t_n^2 A_n^2 |E_{v_n}|^4 \right.$$
$$\left. + \sum_{\substack{n \\ (n,m)}} \sum_m t_n t_m |E_{v_n}|^2 |E_{v_m}|^2 (A_m^2 + A_n^2) \right\} \qquad (14)$$

$$U^2 + V^2 = \frac{4}{\eta^2 T^2} \left\{ \sum_n t_n^2 A_n^2 |E_{v_n}|^4 + 2 \sum_{\substack{n \\ (n,m)}} \sum_m t_n t_m \right.$$
$$\left. \times |E_{v_n}|^2 |E_{v_m}|^2 A_n A_m \cos(\delta_n - \delta_m) \right\} \qquad (15)$$

where (n, m) indicates that the summation is to be extended over all distinct pairs of intervals n and m. Subtracting (15) from (14) gives

$$4 I_v I_h - (U^2 + V^2) = \frac{4}{\eta^2 T^2} \sum_n \sum_{\substack{m \\ (n,m)}} t_n t_m |E_{v_n}|^2 |E_{v_m}|^2$$
$$\times \left[A_m^2 + A_n^2 - 2 A_n A_m \cos(\delta_n - \delta_m) \right] \qquad (16)$$

The right-hand side of (16) is always positive. Hence, generally

$$4I_v I_h \geq U^2 + V^2 \tag{17}$$

or

$$I^2 \geq Q^2 + U^2 + V^2 \tag{18}$$

The right-hand side of (16) is equal to zero if $\delta_n = \delta_m$ and $A_n = A_m$ for all n and m, which means that the amplitude ratio and phase difference of E_v and E_h stay constant. This is the case of elliptical polarization and $I^2 = Q^2 + U^2 + V^2$. Equation (18) implies that partially polarized light corresponds to a point inside the Poincare sphere of radius I.

2.3 Incoherent Addition of Stokes Parameters

When several independent waves are combined, the Stokes parameters of the mixture are the sum of the respective Stokes parameters of the separate waves. By independence, it is meant that the component streams of the mixture have no permanent phase relations among themselves. Consider an electric field that is the sum of fields from N different sources, so that

$$\overline{E} = \sum_{n=1}^{N} \overline{E}_n \tag{1}$$

and there is no correlation between $\overline{E}_m$ and $\overline{E}_n$ for $m \neq n$. We have

$$I_v = \sum_n I_{v_n} + \sum_n \sum_{\substack{m \\ (n,m)}} \frac{2}{\eta} Re < E_{v_n} E^*_{v_m} > \tag{2}$$

Since there is no correlation between E_{v_n} and E_{v_m}, the angular bracket in (2) vanishes. Hence,

$$I_v = \sum_n I_{v_n} \tag{3}$$

Similarly, $I_h = \sum_n I_{h_n}$, $U = \sum_n U_n$ and $V = \sum_n V_n$.

A good example of addition of Stokes parameters is the scattered waves from different particles in a volume. The phase of the scattered wave from a particle depends on the position of the particle. If the distribution of the particles is *sufficiently* random, then the scattered waves from the particles will be random in phase and the Stoke parameters of these *independent* or *incoherent* waves add. For the phase of the scattered waves from different particles to be random, the magnitude of randomness in particle positions must be comparable to or larger than a wavelength.

2.4 Transformation of Stokes Parameters for a Rotation of the Axes

Given the Stokes parameters for a pair of orthogonal directions, it may be desirable to calculate the Stokes parameters for a different pair of directions. We now consider the transformation of the Stokes parameters when the axes are rotated about the propagation direction. Suppose the axes are rotated through an angle ϕ so that $\hat{v}'$ makes an angle ϕ with $\hat{v}$ and an angle $\pi/2 - \phi$ with $\hat{h}$. Then

$$E_v' = E_v \cos \phi + E_h \sin \phi \tag{1}$$

$$E_h' = -E_v \sin \phi + E_h \cos \phi \tag{2}$$

By using (2) through (7), Section 2.1, the transformation of the Stokes parameters takes the following form

$$I_v' = I_v \cos^2 \phi + I_h \sin^2 \phi + \frac{U}{2} \sin(2\phi) \tag{3}$$

$$I_h' = I_v \sin^2 \phi + I_h \cos^2 \phi - \frac{U}{2} \sin(2\phi) \tag{4}$$

$$U' = -(I_v - I_h) \sin(2\phi) + U \cos(2\phi) \tag{5}$$

$$V' = V \tag{6}$$

By adding (3) and (4), we find that $I' = I$. Thus, I and V are invariant through a rotation of axes.

3 VECTOR RADIATIVE TRANSFER EQUATIONS: CONSTITUENTS, RECIPROCITY, AND ENERGY CONSERVATION

In this section, we derive the equation that governs the propagation of specific intensity in a medium containing random distribution of particles. The particles scatter and absorb the wave energy, and these characteristics should be included in a differential equation to be satisfied by the specific intensity. This equation is called the radiative transfer equation. There are three constituents of the radiative transfer equation. The extinction matrix describes the attenuation of specific intensity due to absorption and scattering. The phase matrix characterizes the coupling of intensities in two different

Vector Radiative Transfer Equations

directions due to scattering. The emission vector gives the thermal emission source of the specific intensity. Since the specific intensity is a four-element Stokes vector, the extinction and phase matrices are 4×4 matrices and the emission vector is a 4×1 column matrix. For spherical particles, the extinction matrix is diagonal and is a constant times the unit matrix. The emission vector has the first two elements equal and the last two elements equal to zero. For nonspherical particles, the extinction matrix is generally nondiagonal and the four elements of the emission vector are all nonzero. The resultant equations are called vector radiative transfer equations to distinguish them from the scalar transfer equation, which is a single differential equation dealing with intensity.

In this section, the constituents of the vector radiative transfer equations including all four Stokes parameters are derived. The phase matrix is derived by considering scattering by a single particle. The extinction matrix is obtained by coherent wave propagation theory using Foldy's approximation, and the emission vector is calculated by the fluctuation dissipation theorem. Since the constituents are derived by different means, it is important to ascertain that, when these constituents are combined to form the vector radiation transfer equations, reciprocity and energy conservation still hold. These issues will be studied by showing that within the framework of vector radiative transfer theory, the emissivities of vertically and horizontally polarized waves are related to bistatic scattering coefficients in the usual manner (Peake, 1959). In addition, the emissivity of the third and fourth Stokes parameters is also expressed in terms of bistatic scattering coefficients (Tsang, 1984a). Treatment of various aspects of radiative transfer theory can be found in texts by Chandrasekhar (1960), Sobolev (1963), Ozisik (1973) and in texts on neutron transport (Davison, 1957; Case and Zweifel, 1967).

3.1 Vector Radiative Transfer Equations

Let us consider a specific intensity $I(\bar{r}, \hat{s})$ incident upon a cylindrical elementary volume with unit cross section and length ds (Figure 3.3). The volume ds contains $n_o ds$ particles where n_o is the number of particles per unit volume. Each particle absorbs the power $\sigma_a I$, and scatters the power $\sigma_s I$. Therefore the decrease of the specific intensity $dI(\hat{r}, \hat{s})$ for the volume ds is

$$dI(\bar{r}, \hat{s}) = -ds(\sigma_a + \sigma_s)n_o I = -ds\sigma_t n_o I \qquad (1)$$

where σ_a and σ_s are, respectively, the scattering and absorption cross sections, and σ_t is the total cross section. At the same time, the specific

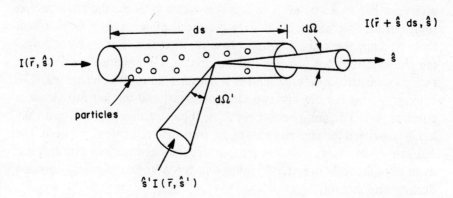

Fig. 3.3 Radiative energy transfer for specific intensity $I(\bar{r}, \hat{s})$ incident upon a cylindrical column of particles.

intensity is enhanced by the emission of particles $\kappa_\epsilon C_m T\, ds$ where

$$C_m = \frac{K}{\lambda_m^2} \tag{2}$$

λ_m is the wavelength in the medium and κ_ϵ is the emission coefficient, K is Boltzmann's constant, and T is the temperature. Also, the specific intensity increases because a portion of the specfic intensity $I(\bar{r}, \hat{s}')$ incident on the volume from other directions $\hat{s}'$ is scattered into the direction $\hat{s}$ and is added to the intensity $I(\bar{r}, \hat{s})$ (Figure 3.3).

To calculate this contribution, let us consider a specific intensity in the direction $\hat{s}'$ on a particle. The incident flux density through a small solid angle is $S_p = I(\bar{r}, \hat{s}')d\Omega'$, where S_p is the plane wave Poynting's vector. The power flux density S_r of the spherical wave scattered by a single particle in the direction $\hat{s}$ at distance R from the particle is then given by $(|f|^2/R^2)S_p$, where f is the scattering amplitude. Since the scattered intensity comes from all directions $\hat{s}'$, the specific intensity contribution in the direction $\hat{s}$ from $n_o\, ds$ particles is

$$n_o\, ds \int_{4\pi} d\Omega' |f|^2 I(\bar{r}, \hat{s}')$$

Combining all these contributions, we have

$$\frac{dI(\bar{r}, \hat{s})}{ds} = -n_o(\sigma_a + \sigma_s)I(\bar{r}, s') + \kappa_\epsilon C_m T + n_o \int_{4\pi} d\Omega'\, |f|^2 I(\bar{r}, \hat{s}') \tag{3}$$

Equation (3) can be generalized to the vector electromagnetic propagation. Using the property of incoherent addition of Stokes parameters, the vector radiative transfer equation for specific intensity is given by

$$\frac{d\bar{I}(\bar{r},\hat{s})}{ds} = -\bar{\bar{\kappa}}_e(\bar{r},\hat{s})\bar{I}(\bar{r},\hat{s}) - \kappa_{ag}(\bar{r},\hat{s})\bar{I}(\bar{r},\hat{s}) + \bar{J}_\epsilon$$
$$+ \int_{4\pi} d\Omega' \bar{\bar{P}}(\bar{r},\hat{s},\hat{s}') \cdot \bar{I}(\bar{r},\hat{s}') \qquad (4)$$

where $\bar{\bar{P}}(\bar{r},\hat{s},\hat{s}')$ is the phase matrix giving the contributions from direction $\hat{s}'$ into the direction $\hat{s}$. In (4), $\bar{\bar{\kappa}}_e$ is the extinction matrix for Stokes parameters due to the scatterers and $\bar{J}_\epsilon$ is the emission vector and κ_{ag} is absorption coefficient for the background medium which is assumed to be isotropic. Generally, extinction is a summation of absorption and scattering. For nonspherical particles, the extinction matrix is generally nondiagonal and the four elements of the emission vector are all nonzero. In active remote sensing, a wave is launched by a transmitter onto the medium. The thermal emission term is usually small when compared with the transmitter signal, and it can be neglected in active sensing. In the following sections, expressions for the phase matrix, extinction matrix, and the emission vector shall be derived. They can all be expressed in terms of the scattering function matrix.

3.2 Scattering Function Matrix, Stokes Matrix, and Phase Matrix

The Stokes matrix relates the Stokes parameters of the scattered wave to those of the incident wave whereas the scattering function matrix relates the scattered field to the incident field. For the case of incoherent addition of scattered waves, the phase matrix will just be the averaging of the Stokes matrices over orientation and size of the particles. Thus, we shall study the Stokes matrix of a single particle. Consider a plane wave

$$\bar{E} = (\hat{v}_i E_{v_i} + \hat{h}_i E_{h_i}) e^{i\bar{k}_i \cdot \bar{r}}$$
$$= \hat{e}_i E_o e^{i\bar{k}_i \cdot \bar{r}} \qquad (1)$$

impinging upon the particle. In spherical coordinates

$$\hat{k}_i = \sin\theta_i \cos\phi_i \hat{x} + \sin\theta_i \sin\phi_i \hat{y} + \cos\theta_i \hat{z} \qquad (2)$$

$$\hat{v}_i = \cos\theta_i \cos\phi_i \hat{x} + \cos\theta_i \sin\phi_i \hat{y} - \sin\theta_i \hat{z} \qquad (3)$$

$$\hat{h}_i = -\sin\phi_i \hat{x} + \cos\phi_i \hat{y} \tag{4}$$

In the direction $\hat{k}_s$, the far-field scattered wave $\overline{E}_s$ will be a spherical wave and is denoted by

$$\overline{E}_s = (E_{vs}\hat{v}_s + E_{hs}\hat{h}_s) \tag{5}$$

with

$$\hat{k}_s = \sin\theta_s \cos\phi_s \hat{x} + \sin\theta_s \sin\phi_s \hat{y} + \cos\theta_s \hat{z} \tag{6}$$

$$\hat{v}_s = \cos\theta_s \cos\phi_s \hat{x} + \cos\theta_s \sin\phi_s \hat{y} - \sin\theta_s \hat{z} \tag{7}$$

$$\hat{h}_s = -\sin\phi_s \hat{x} + \cos\phi_s \hat{y} \tag{8}$$

The scattered field will assume the form

$$\overline{E}_s = \frac{e^{ikr}}{r} \overline{\overline{F}}(\theta_s, \phi_s; \theta_i, \phi_i) \cdot \hat{e}_i E_o \tag{9}$$

where $\overline{\overline{F}}(\theta_s, \phi_s; \theta_i, \phi_i)$ is the scattering function matrix. Hence

$$\begin{bmatrix} E_{vs} \\ E_{hs} \end{bmatrix} = \frac{e^{ikr}}{r} \begin{bmatrix} f_{vv}(\theta_s, \phi_s; \theta_i, \phi_i) & f_{vh}(\theta_s, \phi_s; \theta_i, \phi_i) \\ f_{hv}(\theta_s, \phi_s; \theta_i, \phi_i) & f_{hh}(\theta_s, \phi_s; \theta_i, \phi_i) \end{bmatrix} \cdot \begin{bmatrix} E_{vi} \\ E_{hi} \end{bmatrix} \tag{10}$$

with

$$f_{ab}(\theta_s, \phi_s; \theta_i, \phi_i) = \hat{a}_s \cdot \overline{\overline{F}}(\theta_s, \phi_s; \theta_i, \phi_i) \cdot \hat{b}_i \tag{11}$$

and $a, b = v, h$. To relate the scattered Stokes parameters to the incident Stokes parameters, we apply the definitions in Section 2. Then

$$(\overline{I}_s)_{sp} = \frac{1}{r^2} \overline{\overline{L}}(\theta_s, \phi_s; \theta_i, \phi_i) \cdot \overline{I}_i \tag{12}$$

where $(\overline{I}_s)_{sp}$ and $\overline{I}_i$ are column matrices, and the subscript sp denotes spherical waves.

$$(\overline{I}_s)_{sp} = \begin{bmatrix} I_{v_s} \\ I_{h_s} \\ U_s \\ V_s \end{bmatrix}_{sp} \tag{13}$$

$$\overline{I}_i = \begin{bmatrix} I_{v_i} \\ I_{h_i} \\ U_i \\ V_i \end{bmatrix} \tag{14}$$

and $\overline{\overline{L}}(\theta_s, \phi_s; \theta_i, \phi_i)$ is the Stokes matrix.

$$\overline{\overline{L}}(\theta_s, \phi_s; \theta_i, \phi_i) = \begin{bmatrix} |f_{vv}|^2 & |f_{vh}|^2 \\ |f_{hv}|^2 & |f_{hh}|^2 \\ 2Re(f_{vv}f_{hv}^*) & 2Re(f_{vh}f_{hh}^*) \\ 2Im(f_{vv}f_{hv}^*) & 2Im(f_{vh}f_{hh}^*) \end{bmatrix}$$

$$\begin{matrix} Re(f_{vh}^*f_{vv}) & -Im(f_{vh}^*f_{vv}) \\ Re(f_{hh}^*f_{hv}) & -Im(f_{hv}f_{hh}^*) \\ Re(f_{vv}f_{hh}^* + f_{vh}f_{hv}^*) & -Im(f_{vv}f_{hh}^* - f_{vh}f_{hv}^*) \\ Im(f_{vv}f_{hh}^* + f_{vh}f_{hv}^*) & Re(f_{vv}f_{hh}^* - f_{vh}f_{hv}^*) \end{matrix} \Bigg]$$
(15)

Because of the incoherent addition of Stokes parameters, the phase matrix is equal to the average of the Stokes matrix over the distribution of particles in terms of size, shape, and orientation. For example, for ellipsoids with axes a, b, and c and orientation Eulerian angles α, β, γ, with respect to the principal frame, the phase matrix is

$$\overline{\overline{P}}(\theta, \phi; \theta', \phi') = n_o \int da \int db \int dc \int d\alpha \int d\beta \int d\gamma$$
$$\times p(a, b, c, \alpha, \beta, \gamma) \overline{\overline{L}}(\theta, \phi; \theta', \phi') \quad (16)$$

where $p(a, b, c, \alpha, \beta, \gamma)$ is the probability density function for the quantities a, b, c, α, β and γ. If the particles are all identical in shape and orientation, then,

$$\overline{\overline{P}}(\theta, \phi; \theta', \phi') = n_o \overline{\overline{L}}(\theta, \phi; \theta', \phi') \quad (17)$$

Reciprocal Relation for Scattering Function Matrix

In Chapter 1, we proved that for a reciprocal medium,

$$\overline{E}_a^s(\bar{r}_b) \cdot \hat{\beta} = \overline{E}_b^s(\bar{r}_a) \cdot \hat{\alpha} \quad (18)$$

where $\overline{E}_a^s(\bar{r}_b)$ is the field at the observation point $\bar{r}_b$ due to the dipole antenna source at $\bar{r}_a$ with dipole current in $\hat{\alpha}$ direction (Problem A, Figure 3.4) and $\overline{E}_b^s(\bar{r}_a)$ is the field at the observation point $\bar{r}_a$ due to dipole antenna at $\bar{r}_b$ and current in $\hat{\beta}$ direction (Problem B, Figure 3.4). Since both $\bar{r}_a$ and $\bar{r}_b$ are far away from the scatterer centered at the origin, the incident field on the scatterer due to sources at a and b are, respectively,

$$\overline{E}_a^i = \hat{\alpha}\left(-\frac{i\omega\mu Il}{4\pi r_a} e^{ikr_a} e^{-ik\hat{r}_a \cdot \bar{r}}\right) \quad (19)$$

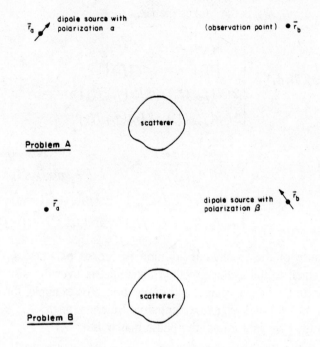

Fig. 3.4 Reciprocity for scattering function matrix $\overline{\overline{F}}$. Problem A: dipole at $\bar{r}_a$ with current in $\hat{\alpha}$ direction; observation at $\bar{r}_b$. Problem B: dipole at $\bar{r}_b$ with current in $\hat{\beta}$ direction; observation at $\bar{r}_a$.

and

$$\overline{E}_b^i = \hat{\beta}\left(-\frac{i\omega\mu Il}{4\pi r_b} e^{ikr_b} e^{-ik\hat{r}_b \cdot \bar{r}}\right) \qquad (20)$$

Hence, using the defining relation in (9) of the scattering function matrix, we have

$$\hat{\beta} \cdot \overline{E}_a^s(\bar{r}_b) = \frac{e^{ikr_b}}{r_b} \hat{\beta} \cdot \overline{\overline{F}}(\theta_b, \phi_b; \pi - \theta_a, \pi + \phi_a) \cdot \hat{\alpha}\left(-\frac{i\omega\mu Il}{4\pi r_a}\right) e^{ikr_a} \qquad (21)$$

and

$$\hat{\alpha} \cdot \overline{E}_b^s(\bar{r}_a) = \frac{e^{ikr_a}}{r_a} \hat{\alpha} \cdot \overline{\overline{F}}(\theta_a, \phi_a; \pi - \theta_b, \pi + \phi_b) \cdot \hat{\beta}\left(-\frac{i\omega\mu Il}{4\pi r_b}\right) e^{ikr_b} \qquad (22)$$

Equating (21) and (22) results in the following reciprocity relation for the scattering amplitude dyad

$$\hat{\beta} \cdot \overline{\overline{F}}(\theta_b, \phi_b; \pi - \theta_a, \pi + \phi_a) \cdot \hat{\alpha} = \hat{\alpha} \cdot \overline{\overline{F}}(\theta_a, \phi_a; \pi - \theta_b, \pi + \phi_b) \cdot \hat{\beta} \qquad (23)$$

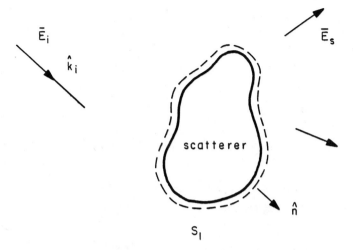

Fig. 3.5 Geometrical configuration for the derivation of optical theorem.

with $\alpha, \beta = v$ or h. Therefore,

$$\overline{\overline{F}}(\hat{k}_s; \hat{k}_i) = \overline{\overline{F}}^t(-\hat{k}_i, -\hat{k}_s) \tag{24}$$

3.3 Optical Theorem

The total cross section of a scatterer represents the total power loss from the incident wave due to the scattering and absorption of a wave by the scatterer. This loss is closely related to the behavior of the scattered field in the forward direction. This fundamental relation, called the optical theorem, relates the total cross section of the scatterer to the imaginary part of the forward amplitude. This theorem is used to calculate the total extinction cross section when the scattering amplitude is known.

Consider a plane wave incident on a scatterer of arbitrary shape and size (Figure 3.5). The field at any point in the lossless medium surrounding the object may be represented as the sum of the incident and scattered fields.

$$\overline{E} = \overline{E}_i + \overline{E}_s \tag{1a}$$

$$\overline{H} = \overline{H}_i + \overline{H}_s \tag{1b}$$

The time-average Poynting vector is

$$<\overline{S}> = \frac{1}{2} Re(\overline{E} \times \overline{H}^*) = <\overline{S}_i> + <\overline{S}_s> + <\overline{S}'> \tag{2}$$

where $<\overline{S}_i>$ and $<\overline{S}_s>$ are, respectively, Poynting vectors of incident and scattered waves and

$$<\overline{S}'> = \frac{1}{2}Re(\overline{E}_i \times \overline{H}_s^* + \overline{E}_s \times \overline{H}_i^*) \qquad (3)$$

Integrating the above Poynting vector over the surface S_1 (Figure 3.5)

$$\int_{S_1} <\overline{S}> \cdot \hat{n}\, ds = W_i + W_s + W' \qquad (4)$$

where W_i, W_s and W' are, integrations of $<\overline{S}_i> \cdot \hat{n}$, $<\overline{S}_s> \cdot \hat{n}$ and $<\overline{S}'> \cdot \hat{n}$ over S_1, respectively, and $\hat{n}$ is the outward normal to the surface S_1. Define

$$-W_a = \int_{S_1} <\overline{S}> \cdot \hat{n}\, ds \qquad (5)$$

The quantity W_a is the average power being absorbed by the scatterer, since the term on the right side of (5) represents the difference between the power flowing out of the volume enclosed by the surface S_1 and the power flowing into the volume.

Let the incident wave be linearly polarized.

$$\overline{E}_i = \hat{e}_i\, E_o\, e^{i\overline{k}_i \cdot \overline{r}} \qquad (6)$$

We have $\nabla \cdot <\overline{S}_i> = 0$ and $W_i = 0$ and we obtain from (4)

$$-W' = W_a + W_s \qquad (7)$$

which is the rate of extinction. Using (6) and the definition of W', we obtain

$$-W' = \frac{1}{2}Re\left\{E_0^* \int_{S_1} e^{-i\overline{k}_i \cdot \overline{r}} \right.$$
$$\left. \times \left[\hat{e}_i \cdot (\hat{n} \times \overline{H}_s) - \frac{1}{\eta}(\hat{k}_i \times \hat{e}_i) \cdot (\hat{n} \times \overline{E}_s)\right] ds\right\} \qquad (8)$$

Applying Huygen's principle to the scattered field and letting the surface be S_1, we obtain

$$\overline{E}_s(\overline{r}) = \int_{S_1} \left\{i\omega\mu\overline{\overline{G}}(\overline{r},\overline{r}') \cdot \left[\hat{n} \times \overline{H}_s(\overline{r}')\right] \right.$$
$$\left. + \nabla \times \overline{\overline{G}}(\overline{r},\overline{r}') \cdot \left[\hat{n} \times \overline{E}_s(\overline{r}')\right]\right\} ds' \qquad (9)$$

Optical Theorem

In the far field

$$\overline{\overline{G}}(\bar{r},\bar{r}') = (\hat{e}_s\hat{e}_s + \hat{h}_s\hat{h}_s)\frac{e^{ikr}}{4\pi r}e^{-i\bar{k}_s\cdot\bar{r}'} \qquad (10)$$

and

$$\overline{E}_s(\bar{r}) = \frac{e^{ikr}}{r}\overline{\overline{F}}(\hat{k}_s,\hat{k}_i)\cdot\hat{e}_i E_o \qquad (11)$$

where $(\hat{k}_s, \hat{h}_s, \hat{e}_s)$ is the orthonormal triad and $\overline{\overline{F}}(\hat{k}_s,\hat{k}_i)$ is the scattering amplitude in the scattered direction $\hat{k}_s$ for the incident wave in the direction $\hat{k}_i$. Substituting (10) and (11) in (9), we obtain

$$\overline{\overline{F}}(\hat{k}_s,\hat{k}_i)\cdot\hat{e}_i E_o = \frac{1}{4\pi}i\omega\mu\int_{S_1} e^{-i\bar{k}_s\cdot\bar{r}'}\Big\{(\hat{e}_s\hat{e}_s+\hat{h}_s\hat{h}_s)\cdot[\hat{n}\times\overline{H}_s(\bar{r}')]$$
$$+\frac{1}{\eta}\hat{k}_s\times(\hat{e}_s\hat{e}_s+\hat{h}_s\hat{h}_s)\cdot[\hat{n}\times\overline{E}_s(\bar{r}')]\Big\}ds' \qquad (12)$$

If we take the scattering amplitude in the forward direction in (12), i.e., $\hat{k}_s = \hat{k}_i$, and dot it with the unit vector $\hat{e}_i$, we obtain,

$$\hat{e}_i\cdot\overline{\overline{F}}(\hat{k}_i,\hat{k}_i)\cdot\hat{e}_i E_o = \frac{1}{4\pi}i\omega\mu\int_{S_1} e^{-i\bar{k}_i\cdot\bar{r}'}\Big\{\hat{e}_i\cdot[\hat{n}\times\overline{H}_s(\bar{r}')]$$
$$-\frac{1}{\eta}(\hat{k}_i\times\hat{e}_i)\cdot[\hat{n}\times\overline{E}_s(\bar{r}')]\Big\}ds' \qquad (13)$$

Comparing (13) with (8), we obtain

$$-W' = Im\left[\frac{2\pi}{\omega\mu}E_o^*\hat{e}_i\cdot\overline{\overline{F}}(\hat{k}_i,\hat{k}_i)\right]\cdot\hat{e}_i E_o \qquad (14)$$

Define the extinction cross section

$$\sigma_e = \frac{-W'}{\frac{1}{2\eta}|E_o|^2} \qquad (15)$$

In terms of σ_e and $\overline{\overline{F}}(\hat{k}_s,\hat{k}_i)$, the optical theorem reads

$$\sigma_e = \frac{4\pi}{k}Im\left[\hat{e}_i\cdot\overline{\overline{F}}(\hat{k}_i,\hat{k}_i)\cdot\hat{e}_i\right] \qquad (16)$$

Hence, the extinction cross sections for vertically polarized wave σ_{ev} and for horizontally polarized wave σ_{eh} are, respectively,

$$\sigma_{ev}(\hat{k}_i) = \frac{4\pi}{k}Im[f_{vv}(\hat{k}_i,\hat{k}_i)] \qquad (17a)$$

$$\sigma_{eh}(\hat{k}_i) = \frac{4\pi}{k} Im[f_{hh}(\hat{k}_i, \hat{k}_i)] \tag{17b}$$

Since the extinction cross section is in terms of the scattering amplitude in the forward direction, the optical theorem is also known as the extinction theorem or the forward scattering theorem. The Rayleigh scattering amplitude dyad for small spheres (Kong, 1975) with permittivity ϵ_s is

$$\overline{\overline{F}}(\hat{k}_i, \hat{k}_i) = \frac{3v_o k^2}{4\pi} \left[\frac{\epsilon_s - \epsilon}{\epsilon_s + 2\epsilon} \right] \tag{18}$$

where v_o is the volume of the sphere. Applying the optical theorem to (18) gives

$$\sigma_{ev}(\overline{k}_i) = \sigma_{eh}(\overline{k}_i) = \sigma_e = 3v_o k \, Im \left[\frac{\epsilon_s - \epsilon}{\epsilon_s + 2\epsilon} \right]$$

$$= v_o k \frac{\epsilon_{si}}{\epsilon} \left| \frac{3\epsilon}{\epsilon_s + 2\epsilon} \right|^2 \tag{19}$$

where ϵ_{si} is the imaginary part of ϵ_s. The extinction cross section given by (19) is equal to the absorption cross section and does not have the scattering cross section in it. Thus the optical theorem fails to give the correct result when applied to the scattering amplitude dyad (18) of small spheres. This is because (18) is not sufficiently accurate. It is derived from the Laplace equation rather than the full-wave equation. The accurate form of the scattering dyad, as shown in the section on Mie scattering, should include an additional imaginary term. We conclude that since the imaginary part of f_{jl} can be much smaller than the real part, the application of the optical theorem is meaningful only when both the real and imaginary parts of f_{jl}'s are calculated to sufficient accuracy.

The extinction coefficient in radiative transfer equation is customarily taken to be $n_o \sigma_e$, where n_o is the number of particles per unit volume. However, for a conglomeration of nonspherical particles, the extinction matrix is not derived by considering scattering by a single particle. It is obtained by coherent wave propagation theory under Foldy's approximation as discussed in the following section.

3.4 Extinction Matrix for Nonspherical Particles

For spherical particles the extinction matrix is diagonal

$$\overline{\overline{\kappa}}_e = \begin{bmatrix} \kappa_e & 0 & 0 & 0 \\ 0 & \kappa_e & 0 & 0 \\ 0 & 0 & \kappa_e & 0 \\ 0 & 0 & 0 & \kappa_e \end{bmatrix} \tag{1}$$

Extinction Matrix for Nonspherical Particles

For nonspherical particles, the extinction matrix is generally nondiagonal. The extinction coefficients can be identified with the attenuation of the coherent wave which can be calculated by using Foldy's approximation that will be discussed in Section 4.2, Chapter 6.

Let E_v and E_h be, respectively, the vertically and horizontally polarized components of the coherent wave. Then the following coupled equations hold for the coherent field along the propagation direction (θ, ϕ) (Oguchi, 1973; Ishimaru and Cheung, 1980; Ishimaru and Yeh, 1984; see also Section 4.2, Chapter 6). The direction is denoted by $\hat{s}$ with $\hat{s}(\theta, \phi) = \sin\theta\cos\phi\hat{x} + \sin\theta\sin\phi\hat{y} + \cos\theta\hat{z}$.

$$\frac{dE_v}{ds} = (ik + M_{vv})E_v + M_{vh}E_h \tag{2}$$

$$\frac{dE_h}{ds} = M_{hv}E_v + (ik + M_{hh})E_h \tag{3}$$

where s is the distance along the direction of propagation. Solving (2) and (3) yields two characteristic waves with defined polarization and attenuation rates. Thus, for propagation along any particular directions (θ, ϕ), there are only two attenuation rates. In (2) and (3)

$$M_{jl} = \frac{i2\pi n_o}{k} <f_{jl}(\theta, \phi; \theta, \phi)> \qquad j, l = v, h \tag{4}$$

where the angular bracket denotes average to be taken over the orientation and size distribution of the particles. For example, for ellipsoids with probability density function $p(a, b, c, \alpha, \beta, \gamma)$, the average is

$$<f_{jl}(\theta, \phi; \theta, \phi)> = \int_0^\infty da \int_0^\infty db \int_0^\infty dc \int_0^{2\pi} d\alpha$$
$$\times \int_0^\pi d\beta \int_0^{2\pi} d\gamma\, p(a, b, c; \alpha, \beta, \gamma)$$
$$\times f_{jl}(\theta, \phi; \theta, \phi; a, b, c; \alpha, \beta, \gamma) \tag{5}$$

Multiplying (2) by E_v^* and adding to its complex conjugate gives

$$\frac{d}{ds}\{|E_v|^2\} = 2Re(M_{vv})|E_v|^2 + 2Re(M_{vh}E_v^*E_h) \tag{6}$$

Using the definition of Stokes parameters I_v, I_h, U, and V, (6) takes the form

$$\frac{dI_v}{ds} = 2Re(M_{vv})I_v + Re(M_{vh})U + Im(M_{vh})V \tag{7}$$

Similar differential equations can be derived for dI_h/ds, dU/ds and dV/ds by making use of the (2) and (3)

$$\frac{dI_h}{ds} = 2Re(M_{hh})I_h + Re(M_{hv})U - Im(M_{hv})V \tag{8}$$

$$\frac{dU}{ds} = 2Re(M_{hv})I_v + 2Re(M_{vh})I_h + [Re(M_{vv}) + Re(M_{hh})]U \\ - [Im(M_{vv}) - Im(M_{hh})]V \tag{9}$$

$$\frac{dV}{ds} = -2Im(M_{hv})I_v + 2Im(M_{vh})I_h + [Im(M_{vv}) - Im(M_{hh})]U \\ + [Re(M_{vv}) + Re(M_{hh})]V \tag{10}$$

Identifying the extinction coefficients in radiative transfer theory as the attenuation rates in coherent wave propagation, we have the following general extinction matrix for nonspherical particles (Ishimaru and Cheung, 1980):

$$\bar{\bar{\kappa}}_e = \begin{bmatrix} -2ReM_{vv} & 0 & -Re(M_{vh}) & -Im(M_{vh}) \\ 0 & -2ReM_{hh} & -Re(M_{hv}) & Im(M_{hv}) \\ -2ReM_{hv} & -2ReM_{vh} & -(ReM_{vv}+ReM_{hh}) & (ImM_{vv}-ImM_{hh}) \\ 2ImM_{hv} & -2ImM_{vh} & -(ImM_{vv}-ImM_{hh}) & -(ReM_{vv}+ReM_{hh}) \end{bmatrix} \tag{11}$$

Equations (2) and (3) can be solved by seeking solutions of the form $\exp(iKs)$ where K is the effective propagation constant. There are generally two eigensolutions and two eigenvalues. The eigenvalues are given by

$$K_1 = k - \frac{i}{2}[M_{vv} + M_{hh} + r] \tag{12}$$

$$K_2 = k - \frac{i}{2}[M_{vv} + M_{hh} - r] \tag{13}$$

where

$$r = \left\{ (M_{vv} - M_{hh})^2 + 4M_{hv}M_{vh} \right\}^{1/2} \tag{14}$$

We shall take the sign of the radical such that in the case of small coupling terms M_{vh} and M_{hv}, r has the sign of $M_{vv} - M_{hh}$. With this convention, K_1 and K_2 will be close to vertical and horizontal polarizations, respectively, in the case of small cross polarization coupling. The eigenvectors are column matrices, $[1, b_1]$ and $[b_2, 1]$, respectively, with

$$b_1 = \frac{2M_{hv}}{M_{vv} - M_{hh} + r} \tag{15}$$

$$b_2 = \frac{2M_{vh}}{-M_{vv} + M_{hh} - r} \tag{16}$$

Thus, generally there are two characteristic waves in coherent wave propagation. In the case $M_{vh} = M_{hv} = 0$, K_1 and K_2 reduce to vertically and horizontally polarized waves, respectively.

To solve the set of equations (7) through (10) for coherent wave propagation of Stokes parameters, we seek solutions with dependence $\exp(-\beta s)$. Solutions of (7) through (10) give four eigenvalues and eigenvectors of the Stokes parameters (Tsang et al., 1984). The eigenvalues are denoted by $\beta_1, \beta_2, \beta_3$, and β_4.

$$\overline{\beta}(\hat{s}) = \begin{bmatrix} \beta_1(\theta,\phi) \\ \beta_2(\theta,\phi) \\ \beta_3(\theta,\phi) \\ \beta_4(\theta,\phi) \end{bmatrix} = -(ReM_{vv} + ReM_{hh}) + \frac{1}{2}\begin{bmatrix} -r - r* \\ -r + r* \\ r - r* \\ r + r* \end{bmatrix}$$

$$= \begin{bmatrix} 2ImK_1 \\ iK_2^* - iK_1 \\ iK_1^* - iK_2 \\ 2ImK_2 \end{bmatrix} \tag{17}$$

We shall represent the eigenvectors by a 4×4 eigenmatrix $\overline{\overline{E}}$ with column j as the eigenvector for β_j, $j = 1, 2, 3, 4$. The eigenmatrix is

$$\overline{\overline{E}}(\theta,\phi) = \begin{bmatrix} 1 & b_2^* & b_2 & |b_2|^2 \\ |b_1|^2 & b_1 & b_1^* & 1 \\ 2Re(b_1) & 1 + b_1 b_2^* & 1 + b_1^* b_2 & 2Re(b_2) \\ -2Im(b_1) & -i(1 - b_1 b_2^*) & i(1 - b_1^* b_2) & 2Im(b_2) \end{bmatrix} \tag{18}$$

We have inserted the (θ, ϕ) arguments for β_j and $\overline{\overline{E}}$ to indicate that the two quantities are generally angular dependent. The eigenvalues β_1 and

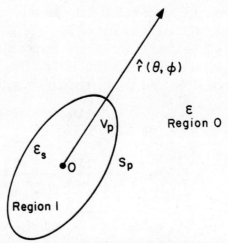

Fig. 3.6 Emission of a single nonspherical particle in the direction $\hat{r}(\theta,\phi)$.

β_4 and their eigenvectors are real whereas β_2 and β_3 and their associated eigenvectors are complex conjugates of each other. This ensures that the Stokes parameters are real quantities.

3.5 Emission Vector

In this section, we derive the emission vector for passive remote sensing of nonspherical particles (Tsang, 1984a). The fluctuation-dissipation theorem of Section 4, Chapter 2, is used to calculate the emission of a single nonspherical particle. The thermal emission of a conglomeration of particles is then introduced as a source term in the radiative transfer equation. Generally, all four Stokes parameters in the vector source term are nonzero and are proportional to the absorption coefficient in the backward direction.

Emission of a Single Nonspherical Particle

Consider a single nonspherical particle centered at the origin (Figure 3.6). The permittivity of the particle is $\epsilon_s = \epsilon'_s + i\epsilon''_s$. The particle occupies volume V_p which is enclosed by the surface S_p. To calculate the thermal emission in directions $\hat{r}(\theta,\phi)$ with

$$\hat{r}(\theta,\phi) = \sin\theta\cos\phi\hat{x} + \sin\theta\sin\phi\hat{y} + \cos\theta\hat{z} \tag{1}$$

we make use of the fluctuation-dissipation theorem. The thermal electromagnetic radiation of the particle can be attributed to an equivalent current source in the particle $\overline{J}(\overline{r},\omega)$, which has the following statistical covariance

Emission Vector

[(20), Chapter 2, Section 4.1].

$$<\overline{J}(\bar{r},\omega)\overline{J}^*(\bar{r}',\omega')> = \frac{4}{\pi}\omega\epsilon_s''\frac{h\nu}{e^{h\nu/KT}-1}\overline{\overline{I}}\delta(\omega-\omega')\delta(\bar{r}-\bar{r}') \quad (2)$$

where K is Boltzmann's constant, T is the temperature of the particle, ν is frequency, and h is Planck's constant. At microwave frequencies, we can further assume that $h\nu \ll KT$. The relation between quantities in the time domain and frequency domain is

$$\bar{j}(\bar{r},t) = \int_0^\infty d\omega\,\overline{J}(\bar{r},\omega)\,e^{-i\omega t} \quad (3)$$

We use regions 0 and 1, respectively, to denote the regions outside and inside the particle. The electric field in region 0 is

$$\overline{E}(\bar{r},\omega) = i\omega\mu\int_{V_p} d\bar{r}'\,\overline{\overline{G}}_{01}(\bar{r},\bar{r}';\omega)\cdot\overline{J}(\bar{r}',\omega) \quad (4)$$

where $\overline{\overline{G}}_{01}(\bar{r},\bar{r}';\omega)$ is the dyadic Green's function with unit dyad source at $\bar{r}'$. The radiation power flow P in solid angle $d\Omega$ in direction $\hat{r}$ for angular frequency $\Delta\omega$ about ω is

$$P = \lim_{r\to\infty} r^2 d\Omega \int_{\Delta\omega} d\omega \int_{\Delta\omega} d\omega' \frac{<\overline{E}(\bar{r},\omega)\cdot\overline{E}^*(\bar{r},\omega')>}{2\eta}e^{-i\omega t+i\omega' t} \quad (5)$$

with $\eta = \sqrt{\mu/\epsilon}$ as the characteristic impedance of region 0. Evaluation of the emission of all four Stokes parameters requires the calculation of $\hat{\alpha}\cdot<\overline{E}(\bar{r},\omega)\overline{E}^*(\bar{r}',\omega')>\cdot\hat{\beta}$ with $\alpha,\beta = v,h$ where $\hat{v}$ and $\hat{h}$ represent vertical and horizontal polarization vectors, respectively. Using the symmetry relation of dyadic Green's function

$$\overline{\overline{G}}_{01}(\bar{r},\bar{r}';\omega) = \overline{\overline{G}}_{10}^t(\bar{r}',\bar{r};\omega) \quad (6)$$

the locations of source and observation point can be interchanged. Hence, using (2), (4) and (6) in (5),

$$\lim_{r\to\infty}\frac{<\overline{E}(\bar{r},\omega)\overline{E}^*(\bar{r}',\omega')>r^2}{2\eta} = \lim_{r\to\infty}\frac{2\omega^3\mu^2}{\eta\pi}\epsilon_s''KT\delta(\omega-\omega')r^2$$

$$\times \int_{V_p} d\bar{r}'\,\overline{\overline{G}}_{10}^t(\bar{r}',\bar{r};\omega)\cdot\overline{\overline{G}}_{10}^*(\bar{r}',\bar{r};\omega) \quad (7)$$

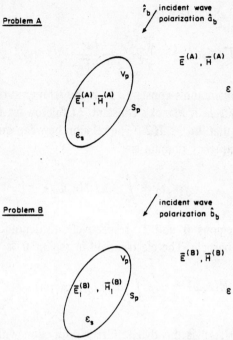

Fig. 3.7 Problem A: Plane wave incident on the nonspherical particle in the direction $\hat{r}_b = -\hat{r}$ and with incident polarization $\hat{a}_b$. Problem B: Plane wave incident on the particle in direction $\hat{r}_b$ and with incident polarization $\hat{b}_b$.

To calculate $\lim_{r \to \infty} \overline{\overline{G}}_{10}(\bar{r}', \bar{r}; \omega)$, consider problems (A) and (B) (Figure 3.7) with unit amplitude electromagnetic plane wave incident on the particle in the direction $\hat{r}_b$ which is in the opposite direction of $\hat{r}$.

$$\hat{r}_b = -\hat{r} = \hat{r}(\pi - \theta, \pi + \phi) \qquad (8)$$

The subscript b denotes backward. The polarizations of the incident wave in problems (A) and (B) are, respectively, $\hat{a}_b$ and $\hat{b}_b$. The field in region 0 is a sum of incident and scattered field, $\overline{E} = \overline{E}_i + \overline{E}_s$, and the field in the particle is denoted by $\overline{E}_1$. It follows that

$$\lim_{r \to \infty} \overline{\overline{G}}_{10}(\bar{r}', \bar{r}; \omega) \cdot \hat{a}_b = \frac{e^{ikr}}{4\pi r} \overline{E}_1^{(A)}(\bar{r}') \qquad (9)$$

A similar relation exists for $\lim_{r \to \infty} \overline{\overline{G}}_{10} \cdot \hat{b}_b$. Using these relations in (7) gives

$$\lim_{r\to\infty} \frac{\hat{a}_b \cdot \langle \overline{E}(\overline{r},\omega)\overline{E}^*(\overline{r}',\omega')\rangle \cdot \hat{b}_b}{2\eta} r^2 = \frac{KT}{2\pi\lambda^2}\delta(\omega-\omega')\eta\omega\epsilon_s''$$
$$\times \int_{V_p} d\overline{r}'\, \overline{E}_1^{(A)}(\overline{r}') \cdot \overline{E}_1^{(B)*}(\overline{r}') \quad (10)$$

where $\lambda = 2\pi/k$ is the wavelength. The integral in (10) is related to the absorption of electromagnetic radiation in the particle. Hence, emission in direction $\hat{r}$ is related to absorption in direction $\hat{r}_b$. To calculate the integral in (10), we note from Maxwell's equations that

$$\omega\epsilon_s'' \overline{E}_1^{(A)} \cdot \overline{E}_1^{(B)*} = -\frac{1}{2}\nabla \cdot \left(\overline{E}_1^{(A)} \times \overline{H}_1^{(B)*} + \overline{E}_1^{(B)*} \times \overline{H}_1^{(A)}\right) \quad (11)$$

By making use of (11), the integral in (10) can be converted to a surface integral over S_p. Making use of the continuity of tangential fields on S_p, we have

$$\int_{V_p} \omega\epsilon_s'' \overline{E}_1^{(A)}(\overline{r}') \cdot \overline{E}_1^{(B)*}(\overline{r}')\, d\overline{r}'$$
$$= -\frac{1}{2}\int_{S_p} d\overline{s}' \cdot \left(\overline{E}^{(A)}(\overline{r}') \times \overline{H}^{(B)*}(\overline{r}') + \overline{E}^{(B)*}(\overline{r}') \times \overline{H}^{(A)}(\overline{r}')\right)$$
$$= -\frac{1}{2}\int_{S_\infty} d\overline{s}' \cdot \left(\overline{E}^{(A)}(\overline{r}') \times \overline{H}^{(B)*}(\overline{r}') + \overline{E}^{(B)*}(\overline{r}') \times \overline{H}^{(A)}(\overline{r}')\right) \quad (12)$$

where S_∞ is a spherical surface at infinity. In arriving at the equality in (12), we have made use of

$$\nabla \cdot \left(\overline{E}^{(A)} \times \overline{H}^{(B)*} + \overline{E}^{(B)*} \times \overline{H}^{(A)}\right) = 0 \quad (13)$$

in region 0 and the divergence theorem.

To evaluate the surface integral at infinity in (12), we note that

$$\overline{E}_i^{(A)}(\overline{r}') = \hat{a}_b e^{ik\hat{r}_b \cdot \overline{r}'} \quad (14)$$

$$\overline{E}_i^{(B)*}(\overline{r}') = \hat{b}_b e^{-ik\hat{r}_b \cdot \overline{r}'} \quad (15)$$

On S_∞, the far-field solutions of the scattered fields can be made use of.

$$\overline{E}_s^{(A)}(\overline{r}') = \frac{e^{ikr'}}{r'}\sum_{\alpha'} \hat{\alpha}' f_{\alpha'a}(\hat{r}',\hat{r}_b) \quad (16)$$

$$\overline{E}_s^{(B)*}(\overline{r}') = \frac{e^{-ikr'}}{r'} \sum_{\beta'} \hat{\beta}' f_{\beta'b}^*(\hat{r}', \hat{r}_b) \tag{17}$$

where $f_{\alpha'\alpha}(\hat{r}', \hat{r}_b)$ with $\alpha', \alpha = v, h$ stands for scattering amplitude from direction $\hat{r}_b$ and polarization $\hat{\alpha}_b$ into direction $\hat{r}'$ and polarization $\hat{\alpha}'$. Using (14) through (17) in (12), we find that the product of incident field of (A) with incident field of (B) vanishes on integration. The integral of the product of scattered fields of (A) and (B) reduces to an angular integration of product of the scattering amplitudes. The integral of the product of incident field with scattered field can be calculated by the method of stationary phase (Ishimaru, 1978; Karam and Fung, 1982). The stationary-phase point is at $\theta' = \theta_b$, $\phi' = \phi_b$ with

$$\theta_b = \pi - \theta \tag{18}$$

$$\phi_b = \pi + \phi \tag{19}$$

which is the direction of the incident wave and is opposite to the direction of emission so that

$$\hat{r}_b = \sin\theta_b \cos\phi_b \hat{x} + \sin\theta_b \sin\phi_b \hat{y} + \cos\theta_b \hat{z} \tag{20}$$

The result is (Problem 30)

$$\eta \int_{V_p} d\overline{r}' \, \omega \epsilon_s'' \, \overline{E}_1^{(A)}(\overline{r}') \cdot \overline{E}_1^{(B)*}(\overline{r}')$$

$$= \frac{2\pi i}{k} \sum_{\beta'} \left[f_{\beta'b}^*(\hat{r}_b, \hat{r}_b)(\hat{\beta}_b' \cdot \hat{a}_b) - f_{\beta'a}(\hat{r}_b, \hat{r}_b)(\hat{\beta}_b' \cdot \hat{b}_b) \right]$$

$$- \sum_{\beta'} \int d\Omega' f_{\beta'a}(\hat{r}', \hat{r}_b) f_{\beta'b}^*(\hat{r}', \hat{r}_b) \tag{21}$$

Using (10) and (21) in (5), we find that the four elements of the Stokes emission power for a single particle are

$$P_j(\hat{r}) = \frac{\Delta\omega}{2\pi} d\Omega \, \frac{KT}{\lambda^2} \sigma_{aj}(\hat{r}_b) \tag{22}$$

where $j = 1, 2, 3$, and 4, and $\overline{\sigma}_a(\hat{r}_b)$ is the absorption cross section vector in direction $\hat{r}_b$.

$$\overline{\sigma}_a(\hat{r}) = \begin{bmatrix} \sigma_{a1}(\hat{r}) \\ \sigma_{a2}(\hat{r}) \\ \sigma_{a3}(\hat{r}) \\ \sigma_{a4}(\hat{r}) \end{bmatrix} \tag{23}$$

where

$$\sigma_{a1}(\hat{r}) = \frac{4\pi}{k} Im f_{vv}(\hat{r},\hat{r}) - \int d\Omega' \left[|f_{vv}(\hat{r}',\hat{r})|^2 + |f_{hv}(\hat{r}',\hat{r})|^2 \right] \quad (24)$$

$$\sigma_{a2}(\hat{r}) = \frac{4\pi}{k} Im f_{hh}(\hat{r},\hat{r}) - \int d\Omega' \left[|f_{vh}(\hat{r}',\hat{r})|^2 + |f_{hh}(\hat{r}',\hat{r})|^2 \right] \quad (25)$$

$$\sigma_{a3}(\hat{r}) = 2 Re \left\{ \frac{2\pi i}{k} \left[f_{vh}^*(\hat{r},\hat{r}) - f_{hv}(\hat{r},\hat{r}) \right] \right.$$
$$\left. - \int d\Omega' \left[f_{vv}(\hat{r}',\hat{r}) f_{vh}^*(\hat{r}',\hat{r}) + f_{hv}(\hat{r}',\hat{r}) f_{hh}^*(\hat{r}',\hat{r}) \right] \right\} \quad (26)$$

$$\sigma_{a4}(\hat{r}) = 2 Im \left\{ \frac{2\pi i}{k} \left[f_{vh}^*(\hat{r},\hat{r}) - f_{hv}(\hat{r},\hat{r}) \right] \right.$$
$$\left. - \int d\Omega' \left[f_{vv}(\hat{r}',\hat{r}) f_{vh}^*(\hat{r}',\hat{r}) + f_{hv}(\hat{r}',\hat{r}) f_{hh}^*(\hat{r}',\hat{r}) \right] \right\} \quad (27)$$

We recognize that $\sigma_{a1}(\hat{r})$ and $\sigma_{a2}(\hat{r})$ are, respectively, the absorption cross sections of vertically and horizontally polarized waves in direction $\hat{r}$.

Radiative Transfer Equations for Passive Remote Sensing of Nonspherical Particles

To derive the thermal emission source term in the vector radiative transfer equations, consider a cylindrical volume of cross-sectional area ΔA and length Δs containing n_o particles per unit volume. The Stokes emission power $\Delta \overline{P}$ in direction $\hat{s}$ is, in view of (5) and (22)

$$\Delta \overline{P}(\hat{s}) = n_o \Delta A \Delta s \frac{\Delta \omega}{2\pi} d\Omega \frac{KT}{\lambda^2} \overline{\sigma}_a(\hat{s}_b) \quad (28)$$

where

$$\hat{s}_b = -\hat{s} \quad (29)$$

Using the definition of specific intensity $\overline{I}(\hat{s})$ as defined in Chapter 1, we have

$$\Delta \overline{P}(\hat{s}) = \Delta \overline{I}(\hat{s}) d\Omega \frac{\Delta \omega}{2\pi} \Delta A \quad (30)$$

Comparing (28) and (30), we have $\Delta \overline{I}(\hat{s}) = n_o \Delta s \overline{\sigma}_a(\hat{s}_b) KT/\lambda^2$, so that the emission vector in the radiative transfer equations is $\overline{\kappa}_a(\hat{s}_b)CT$, where $\overline{\kappa}_a(\hat{s}_b)$ is the absorption coefficient vector

$$\overline{\kappa}_a(\hat{s}) = n_o \overline{\sigma}_a(\hat{s}) \quad (31)$$

with $\overline{\sigma}_a(\hat{s})$ given in (23) through (27), and

$$C = \frac{K}{\lambda^2} \tag{32}$$

The emission term can be inserted into the vector radiative transfer equations, which assume the following form

$$\frac{d\overline{I}(\overline{r},\hat{s})}{ds} = -\overline{\overline{\kappa}}_e(\hat{s}) \cdot \overline{I}(\overline{r},\hat{s}) + \overline{\kappa}_a(\hat{s}_b)CT(\overline{r}) \\ + \int d\Omega' \overline{\overline{P}}(\hat{s},\hat{s}') \cdot \overline{I}(\overline{r},\hat{s}') \tag{33}$$

where $\overline{\overline{\kappa}}_e(\hat{s})$ is the extinction matrix and $\overline{\overline{P}}(\hat{s},\hat{s}')$ is the phase matrix. They are defined in Sections 3.2 and 3.4, respectively. The elements of the absorption coefficient vector can be expressed in terms of the elements of the extinction matrix and the phase matrix by using (24) through (27), (31), and (15) through (17) (Section 3.2), and (11) (Section 3.4).

$$\kappa_{a1}(\hat{s}) = \kappa_{e11}(\hat{s}) - \int d\Omega'[P_{11}(\hat{s}',\hat{s}) + P_{21}(\hat{s}',\hat{s})] \tag{34}$$

$$\kappa_{a2}(\hat{s}) = \kappa_{e22}(\hat{s}) - \int d\Omega'[P_{12}(\hat{s}',\hat{s}) + P_{22}(\hat{s}',\hat{s})] \tag{35}$$

$$\kappa_{a3}(\hat{s}) = 2\kappa_{e13}(\hat{s}) + 2\kappa_{e23}(\hat{s}) - 2\int d\Omega'[P_{13}(\hat{s}',\hat{s}) + P_{23}(\hat{s}',\hat{s})] \tag{36}$$

$$\kappa_{a4}(\hat{s}) = -2\kappa_{e14}(\hat{s}) - 2\kappa_{e24}(\hat{s}) + 2\int d\Omega'[P_{14}(\hat{s}',\hat{s}) + P_{24}(\hat{s}',\hat{s})] \tag{37}$$

where κ_{aj} is the jth element of $\overline{\kappa}_a$, κ_{eij} and P_{ij} are, respectively, the ij element of $\overline{\overline{\kappa}}_e$ and $\overline{\overline{P}}$ with $i,j = 1,2,3,4$.

To take into account different particle sizes and orientations, the extinction and phase matrices and the absorption coefficient vector can be averaged over particle size and orientation distribution. The form of the radiative transfer equations will remain the same, except that all the constituents will be replaced by averaged quantities. For example, f_{ij} is replaced by $<f_{ij}>$ and $|f_{ij}|^2$ replaced by $<|f_{ij}|^2>$, etc. The extinction matrix, phase matrix, absorption vector, and temperature can also be inhomogeneous by allowing them to be functions of $\overline{r}$.

3.6 Energy Conservation, Reciprocity, and Reciprocal Relation between Active and Passive Remote Sensing

The constituents of the vector radiative transfer equations have been constructed in different manners. The extinction matrix is derived from Foldy's

Energy Conservation, Reciprocity

approximation in coherent wave propagation theory. The phase matrix is obtained by considering the scattering by a single particle. The emission vector has been derived from the fluctuation-dissipation approach in Section 3.5. Moreover, the third and fourth Stokes parameters have been introduced into these constituents. Thus, it is desirable to ascertain that when these constituents are combined heuristically to form the vector radiative transfer equations, energy conservation and reciprocity still hold. In this section, these two issues are studied by showing that within the framework of radiative transfer theory, the emissivities of vertically and horizontally polarized waves are related to the bistatic scattering coefficients in the manner as discussed in Section 4, Chapter 1. In addition, the emissions of the third and fourth Stokes parameters are also expressed in terms of bistatic scattering coefficients. The ability to express emissivity in terms of bistatic scattering coefficients is a demonstration of both reciprocity and energy conservation. In relating a wave incident on a medium embedded with nonspherical particles to the emission of the particles, the equality of absorption and emissivity is a result of reciprocity. The ability to express emissivity further as one minus scattering is a demonstration of energy conservation.

We first note that the reciprocity relation for scattering amplitude is, from (23), Section 3.2,

$$f_{jl}(\hat{s}, \hat{s}') = f_{lj}(\hat{s}'_b, \hat{s}_b) \tag{1}$$

The extinction and phase matrices of (15), Section 3.2, and (11), Section 3.4, do not exhibit a simple symmetry relation. To facilitate the derivation of the relation between active and passive sensing, we introduce a modified Stokes vector $\bar{J}$, modified extinction matrix $\bar{\bar{K}}(\hat{s})$, modified phase matrix $\bar{\bar{Q}}(\hat{s}, \hat{s}')$, and modified absorption vector $\bar{K}_a(\hat{s})$ as follows:

$$\bar{J} = \begin{bmatrix} J_1 \\ J_2 \\ J_3 \\ J_4 \end{bmatrix} = \begin{bmatrix} I_v \\ I_h \\ \frac{1}{\sqrt{2}} U \\ \frac{i}{\sqrt{2}} V \end{bmatrix} \tag{2}$$

$$\bar{\bar{K}}(\hat{s}) = \begin{bmatrix} \kappa_{e11} & 0 & \sqrt{2}\kappa_{e13} & -i\sqrt{2}\kappa_{e14} \\ 0 & \kappa_{e22} & \sqrt{2}\kappa_{e23} & -i\sqrt{2}\kappa_{e24} \\ \frac{1}{\sqrt{2}}\kappa_{e31} & \frac{1}{\sqrt{2}}\kappa_{e32} & \kappa_{e33} & -i\kappa_{e34} \\ \frac{i}{\sqrt{2}}\kappa_{e41} & \frac{i}{\sqrt{2}}\kappa_{e42} & i\kappa_{e34} & \kappa_{e44} \end{bmatrix}. \tag{3}$$

$$\overline{\overline{Q}}(\hat{s},\hat{s}') = \begin{bmatrix} P_{11} & P_{12} & \sqrt{2}P_{13} & -i\sqrt{2}P_{14} \\ P_{21} & P_{22} & \sqrt{2}P_{23} & -i\sqrt{2}P_{24} \\ \frac{1}{\sqrt{2}}P_{31} & \frac{1}{\sqrt{2}}P_{32} & P_{33} & -iP_{34} \\ \frac{i}{\sqrt{2}}P_{41} & \frac{i}{\sqrt{2}}P_{42} & iP_{43} & P_{44} \end{bmatrix} \quad (4)$$

$$\overline{K}_a(\hat{s}) = \begin{bmatrix} \kappa_{a1} \\ \kappa_{a2} \\ \frac{1}{\sqrt{2}}\kappa_{a3} \\ \frac{i}{\sqrt{2}}\kappa_{a4} \end{bmatrix} \quad (5)$$

In view of the reciprocity relation (1) of f_{ij} and expressions for phase and extinction matrices, we have the following simple symmetry relations for $\overline{\overline{K}}$ and $\overline{\overline{Q}}$

$$\overline{\overline{K}}(\hat{s}) = \overline{\overline{K}}^t(\hat{s}_b) \quad (6)$$

$$\overline{\overline{Q}}(\hat{s},\hat{s}') = \overline{\overline{Q}}^t(\hat{s}'_b,\hat{s}_b) \quad (7)$$

Using (5) and (34) through (37), Section 3.5, the relation between the elements of modified absorption vector $\overline{K}_a$ and $\overline{\overline{Q}}$ and $\overline{\overline{K}}$ is

$$K_{ai}(\hat{s}) = \sum_{j=1}^{2}\left[K_{ji}(\hat{s}) - \int d\Omega'\, Q_{ji}(\hat{s}',\hat{s})\right] \quad (8)$$

with $i = 1, 2, 3$ and 4. In (8), K_{ai} is the ith element of $\overline{K}_a$ and K_{ji} and Q_{ji} are, respectively, the ji elements of $\overline{\overline{K}}$ and $\overline{\overline{Q}}$. Note that the summation over j in (8) is carried out for $j = 1$ and 2.

Consider active remote sensing (Problem A) and passive remote sensing (Problem B) of a half space of nonspherical particles (Figure 3.8).

Problem A

In active remote sensing, a wave is launched in the direction $\hat{s}_{ob}$ at angle $(\pi - \theta_o, \pi + \phi_o)$. The governing equation is

$$\hat{s}\cdot\hat{z}\frac{d\overline{J}^{(A)}(z,\hat{s})}{dz} = -\overline{\overline{K}}(\hat{s})\cdot\overline{J}^{(A)}(z,\hat{s}) + \int d\Omega'\, \overline{\overline{Q}}(\hat{s},\hat{s}')\cdot\overline{J}^{(A)}(z,\hat{s}') \quad (9)$$

The boundary condition at $z = 0$ is

$$\overline{J}^{(A)}(z=0,\hat{s}) = \overline{J}_{inc}\,\delta(\hat{s}-\hat{s}_{ob}) \quad (10)$$

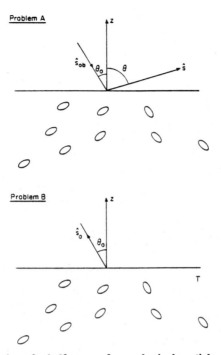

Fig. 3.8 Remote sensing of a half-space of nonspherical particles. Problem A: Active sensing with wave incident in direction $\hat{s}_{ob}$. Problem B: Passive sensing with emission in direction $\hat{s}_o$.

for $\hat{z} \cdot \hat{s} < 0$. In (10), $\delta(\hat{s} - \hat{s}_{ob})$ is the Dirac delta function of angular variables such that integration over a solid angle gives unity. The upward specific intensity at $z = 0$, $\overline{J}^A(z = 0, \hat{s})$ for $\hat{z} \cdot \hat{s} > 0$ gives scattered specific intensity. To classify the bistatic scattering coefficients, we distinguish the following four cases.

(i) Incident Wave Vertically Polarized

Let superscript t denote transpose.

$$\overline{J}_{inc}^t = (1,0,0,0) \tag{11}$$

The bistatic scattering coefficients are

$$\gamma_{j1}(\hat{s}, \hat{s}_{ob}) = \frac{4\pi \cos \theta}{\cos \theta_o} J_j^{(A)}(z = 0, \hat{s}) \tag{12}$$

with $j = 1, 2, 3$, and 4 and $\hat{z} \cdot \hat{s} > 0$.

(ii) Incident Wave Horizontally Polarized

$$\overline{J}^t_{inc} = (0, 1, 0, 0) \tag{13}$$

The corresponding bistatic scattering coefficients are

$$\gamma_{j2}(\hat{s}, \hat{s}_{ob}) = \frac{4\pi \cos\theta}{\cos\theta_o} J_j^{(A)}(z = 0, \hat{s}) \tag{14}$$

with $j = 1, 2, 3,$ and 4 and $\hat{z} \cdot \hat{s} > 0$.

(iii) Incident Wave Linearly Polarized with Equal Vertical and Horizontal Polarized Intensity

$$\overline{J}^t_{inc} = \left(\frac{1}{2}, \frac{1}{2}, \frac{1}{\sqrt{2}}, 0\right) \tag{15}$$

The associated bistatic scattering coefficients are

$$\gamma_{jp}(\hat{s}, \hat{s}_{ob}) = \frac{4\pi \cos\theta}{\cos\theta_o} J_j^A(z = 0, \hat{s}) \tag{16}$$

for $j = 1, 2, 3,$ and 4 and $\hat{z} \cdot \hat{s} > 0$.

(iv) Incident Wave Right-Hand Circularly Polarized

$$J^t_{inc} = \left(\frac{1}{2}, \frac{1}{2}, 0, -\frac{i}{\sqrt{2}}\right) \tag{17}$$

The bistatic scattering coefficients are

$$\gamma_{jR}(\hat{s}, \hat{s}_{ob}) = \frac{4\pi \cos\theta}{\cos\theta_o} J_j^A(z = 0, \hat{s}) \tag{18}$$

for $\hat{z} \cdot \hat{s} > 0$ and $j = 1, 2, 3,$ and 4.

Problem B

For passive remote sensing of the same medium of Problem A, we assume that the half-space medium has a uniform temperature T (Figure 3.8). The governing equation for the modified Stokes vector is

$$\hat{s} \cdot \hat{z} \frac{d\overline{J}^{(B)}}{dz}(z, \hat{s}) = -\overline{\overline{K}}(\hat{s}) \cdot \overline{J}^{(B)}(z, \hat{s}) + \overline{K}_a(\hat{s}_b)CT \\ + \int d\Omega' \overline{\overline{Q}}(\hat{s}, \hat{s}') \cdot \overline{J}^{(B)}(z, \hat{s}') \tag{19}$$

The boundary condition is, at $z = 0$

$$\overline{J}^{(B)}(z = 0, \hat{s}) = 0 \tag{20}$$

for $\hat{s} \cdot \hat{z} < 0$. The thermal emission into the upper region is given by $\overline{J}^{(B)}(z = 0, \hat{s})$ for $\hat{z} \cdot \hat{s} > 0$. The brightness temperature vector $\overline{T}_B(\hat{s})$ is defined as

$$\overline{T}_B(\hat{s}) = \begin{bmatrix} T_{B1} \\ T_{B2} \\ T_{B3} \\ T_{B4} \end{bmatrix} = \frac{1}{C} \overline{J}^{(B)}(z = 0, \hat{s}) \tag{21}$$

for $\hat{z} \cdot \hat{s} > 0$.

To derive the reciprocal relation between problems A and B, we let $\hat{s}$ be $\hat{s}_b$ in (19), take the transpose, and take the dot product of the equation with $\overline{J}^{(A)}(z, \hat{s})$. The equation is then subtracted from the dot product of (9) with $\overline{J}^{(B)t}(z, \hat{s}_b)$. The difference of the two equations is further integrated over $d\Omega$. We obtain, since $\hat{s}_b = -\hat{s}$,

$$\int d\Omega \, (\hat{s} \cdot \hat{z}) \frac{d}{dz} \{\overline{J}^{(B)t}(z, \hat{s}_b) \cdot \overline{J}^{(A)}(z, \hat{s})\}$$

$$= -\int d\Omega \, \overline{J}^{(B)t}(z, \hat{s}_b) \cdot \overline{\overline{K}}(\hat{s}) \cdot \overline{J}^{(A)}(z, \hat{s})$$

$$+ \int d\Omega \int d\Omega' \, \overline{J}^{(B)t}(z, \hat{s}_b) \cdot \overline{\overline{Q}}(\hat{s}, \hat{s}') \cdot \overline{J}^{(A)}(z, \hat{s}')$$

$$+ \int d\Omega \, \overline{J}^{(B)t}(z, \hat{s}_b) \cdot \overline{\overline{K}}^t(\hat{s}_b) \cdot \overline{J}^{(A)}(z, \hat{s})$$

$$- \int d\Omega \int d\Omega' \, \overline{J}^{(B)t}(z, \hat{s}'_b) \cdot \overline{\overline{Q}}^t(\hat{s}_b, \hat{s}'_b) \cdot \overline{J}^{(A)}(z, \hat{s})$$

$$- \int d\Omega \, CT \overline{K}_a^t(\hat{s}) \cdot \overline{J}^{(A)}(z, \hat{s}) \tag{22}$$

By invoking the symmetry relations of $\overline{\overline{K}}$ and $\overline{\overline{Q}}$ from (6) and (7), the sum of the first four terms on the right-hand side of (22) can be shown to vanish. We next integrate (22) from $-\infty$ to 0 and further apply the boundary conditions of (10) and (20). We get

$$(\hat{s}_{ob} \cdot \hat{z}) \overline{J}^{(B)t}(z = 0, \hat{s}_o) \cdot \overline{J}_{inc} = -CT \int_{-\infty}^{0} dz \int d\Omega \, \overline{K}_a^t(\hat{s}) \cdot \overline{J}^{(A)}(z, \hat{s}) \tag{23}$$

The right-hand side of (23) corresponds to the absorption of the Stokes vector by the particles in Problem A. It can be evaluated as follows. Integrate (9) over $d\Omega$ and sum the first two scalar equations of the vector equation.

$$\sum_{j=1}^{2} \int d\Omega \, (\hat{s} \cdot \hat{z}) \frac{dJ_j^{(A)}(z, \hat{s})}{dz} = -\int d\Omega \sum_{j=1}^{2} (\overline{\overline{K}}(\hat{s}) \cdot \overline{J}^{(A)}(z, \hat{s}))_j$$

$$+ \int d\Omega \int d\Omega' \sum_{j=1}^{2} \left(\overline{\overline{Q}}(\hat{s}, \hat{s}') \cdot \overline{J}^{(A)}(z, \hat{s}') \right)_j$$

$$= - \int d\Omega \, \overline{K}_a^t(\hat{s}) \cdot \overline{J}^{(A)}(z, \hat{s}) \qquad (24)$$

The second equality of (24) results from applying the relation (8) of $\overline{K}_a$ in terms of elements of $\overline{\overline{K}}$ and $\overline{\overline{Q}}$. Since $J_1^{(A)} + J_2^{(A)}$ is equal to the total intensity, the left-hand side of (24) is the divergence of energy flux. Hence, in the absence of absorption, the right-hand side of (24) is zero and flux is conserved. Next we integrate (24) over dz from $-\infty$ to zero and apply boundary condition of (10). Comparison with (23) gives

$$-\hat{s}_{ob} \cdot \hat{z} \overline{J}^{(B)t}(z=0, \hat{s}_o) \cdot \overline{J}_{inc} = - \sum_{j=1}^{2} (\hat{s}_{ob} \cdot \hat{z})(\overline{J}_{inc})_j CT$$

$$- \sum_{j=1}^{2} CT \int_{\hat{s} \cdot \hat{z} > 0} d\Omega \, (\hat{s} \cdot \hat{z}) \, J_j^{(A)}(z=0, \hat{s}) \qquad (25)$$

Equation (25) gives the relation between thermal emission of Problem B to scattering return in Problem A of active remote sensing. To obtain explicit expressions of brightness temperature (21) for the four Stokes parameters in terms of bistatic scattering coefficients, we apply the four cases of incidence *(i)* through *(iv)* to (25). We obtain

$$T_{B1}(\hat{s}_o) = T \left\{ 1 - \frac{1}{4\pi} \sum_{j=1}^{2} \int_{\hat{s} \cdot \hat{z} > 0} d\Omega \, \gamma_{j1}(\hat{s}, \hat{s}_{ob}) \right\} \qquad (26)$$

$$T_{B2}(\hat{s}_o) = T \left\{ 1 - \frac{1}{4\pi} \sum_{j=1}^{2} \int_{\hat{s} \cdot \hat{z} > 0} d\Omega \, \gamma_{j2}(\hat{s}, \hat{s}_{ob}) \right\} \qquad (27)$$

$$T_{B3}(\hat{s}_o) = \sqrt{2} T \left\{ 1 - \frac{1}{4\pi} \sum_{j=1}^{2} \int_{(\hat{s} \cdot \hat{z}) > 0} d\Omega \, \gamma_{jp}(\hat{s}, \hat{s}_{ob}) \right\}$$
$$- \frac{1}{\sqrt{2}} T_{B1}(\hat{s}_o) - \frac{1}{\sqrt{2}} T_{B2}(\hat{s}_o) \qquad (28)$$

$$T_{B4}(\hat{s}_o) = i\sqrt{2} T \left\{ 1 - \frac{1}{4\pi} \sum_{j=1}^{2} \int_{\hat{s} \cdot \hat{z} > 0} d\Omega \, \gamma_{jR}(\hat{s}, \hat{s}_{ob}) \right\}$$
$$- \frac{i}{\sqrt{2}} T_{B1}(\hat{s}_o) - \frac{i}{\sqrt{2}} T_{B2}(\hat{s}_o) \qquad (29)$$

Equations (25) and (26) are classical relations obtained by Peake (1959) and are derived here within the framework of vector radiative transfer equations. In addition, (28) and (29) express the brightness temperatures of the third and fourth Stokes parameters in terms of bistatic scattering coefficients.

4 PHASE MATRIX FOR SIMPLE OBJECTS

In this section, examples of the phase matrices for simple objects are given. We use Laplace equation to solve for the induced dipole moments in a sphere and in an ellipsoid due to an incident electric field. The radiation of the induced dipoles gives the scattered field of the object. Because of the usage of Laplace equation rather than the wave equation, the derived scattering function matrix is only valid in the low-frequency limit when the particle size is much smaller than the wavelength. The scattering function matrix for a random medium is also derived. The scattered field is calculated by using the Born approximation which is valid for small permittivity fluctuations.

The scattering amplitude function matrices are calculated in an approximate manner and the accuracy is insufficient for optical theorem to be applied. To derive the extinction cross sections, we calculate the scattering cross section by integrating the scattered intensity over scattered angles and the absorption cross section by considering the internal field of the scatterer. The extinction cross section is then obtained by adding the scattering and absorption cross sections.

To consider scattering by nonspherical particles with prescribed orientations, it is convenient to solve the problem in the coordinate system natural to the particle and then transform the results to the principal frame of the problem. Thus, we shall also discuss Eulerian angles of rotation which are used to relate the two sets of coordinate systems.

4.1 Rayleigh Phase Matrix

In this section, the phase matrix $\overline{\overline{P}}(\theta, \phi; \theta', \phi')$, the scattering coefficient κ_s, and the absorption coefficient κ_a will be derived for a homogeneous medium containing spherical Rayleigh scatterers. The background medium has a wavenumber k. Consider a plane wave with the electric field vector

$$\overline{E}_i(\overline{r}) = \hat{e}_i\, E_o\, e^{i\overline{k}_i \cdot \overline{r}} \tag{1}$$

incident on a sphere of radius a and permittivity $\epsilon_s = \epsilon'_s + i\epsilon''_s$, as shown in Figure 3.9. We solve for scattered fields in the Rayleigh limit when $ka \ll 1$.

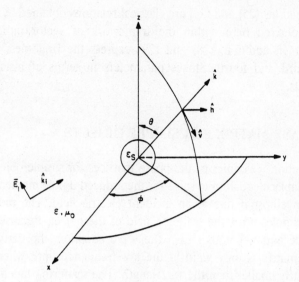

Fig. 3.9 Geometric configuration for scattering from a spherical particle in derivation of Rayleigh phase function.

The induced electric field $\overline{E}^{int}$ in the sphere is uniform and is in the same direction as the incident wave. The induced dipole moment is, from the solution of the Laplace equation,

$$\overline{p} = \frac{3(\epsilon_s - \epsilon)}{\epsilon_s + 2\epsilon} \epsilon v_o \hat{e}_i E_o \tag{2}$$

where $v_o = 4\pi a^3/3$ is the volume of the sphere. The scattered field $\overline{E}_s$ in the direction $\hat{k}_s$ can be attributed to the radiation of the dipole $\overline{p}$ and is (Problem 7)

$$\overline{E}_s = -\frac{\omega^2 \mu\, e^{ikr}}{4\pi r} \hat{k}_s \times (\hat{k}_s \times \overline{p}) \tag{3}$$

Substitute (2) in (3)

$$\overline{E}_s = \frac{k^2 e^{ikr}}{4\pi r} 3v_o y\, (\overline{\overline{I}} - \hat{k}_s \hat{k}_s) \cdot \hat{e}_i \overline{E}_o \tag{4}$$

where

$$y = \frac{\epsilon_s - \epsilon}{\epsilon_s + 2\epsilon} \tag{5}$$

Hence, the scattering function matrix is

$$\overline{\overline{F}}(\theta_s, \phi_s; \theta_i, \phi_i) = k^2 \frac{3v_o y}{4\pi} (\overline{\overline{I}} - \hat{k}_s \hat{k}_s) \cdot (\overline{\overline{I}} - \hat{k}_i \hat{k}_i) \tag{6}$$

Rayleigh Phase Matrix

From the scattering function matrix $\overline{\overline{F}}$, we can calculate the Stokes matrix $\overline{\overline{L}}$. For identical scatterers, the phase matrix $\overline{\overline{P}}(\hat{s}, \hat{s}') = n_o \overline{\overline{L}}(\hat{s}, \hat{s}')$. By using the orthonormal vectors $(\hat{v}_i, \hat{h}_i, \hat{k}_i)$ and $(\hat{v}_s, \hat{h}_s, \hat{k}_s)$ as given in (2) through (4) and (6) through (8), Section 3.2, $\overline{\overline{P}}(\theta, \phi; \theta', \phi')$ can be calculated readily. It is

$$\overline{\overline{P}}(\theta, \phi; \theta', \phi') = \begin{bmatrix} P_{11} & P_{12} & P_{13} & 0 \\ P_{21} & P_{22} & P_{23} & 0 \\ P_{31} & P_{32} & P_{33} & 0 \\ 0 & 0 & 0 & P_{44} \end{bmatrix} \qquad (7)$$

where

$$P_{11} = w[\sin^2\theta \sin^2\theta' + 2\sin\theta \sin\theta' \cos\theta \cos\theta' \cos(\phi - \phi')$$
$$+ \cos^2\theta \cos^2\theta' \cos^2(\phi - \phi')] \qquad (8)$$

$$P_{12} = w\cos^2\theta \sin^2(\phi - \phi') \qquad (9)$$

$$P_{13} = w[\cos\theta \sin\theta \sin\theta' \sin(\phi - \phi')$$
$$+ \cos^2\theta \cos\theta' \sin(\phi - \phi')\cos(\phi - \phi')] \qquad (10)$$

$$P_{21} = w\cos^2\theta' \sin^2(\phi - \phi') \qquad (11)$$

$$P_{22} = w\cos^2(\phi - \phi') \qquad (12)$$

$$P_{23} = -w\cos\theta' \sin(\phi - \phi')\cos(\phi - \phi') \qquad (13)$$

$$P_{31} = w[-2\sin\theta \sin\theta' \cos\theta' \sin(\phi - \phi')$$
$$- 2\cos\theta \cos^2\theta' \cos(\phi - \phi')\sin(\phi - \phi')] \qquad (14)$$

$$P_{32} = 2w\cos\theta \sin(\phi - \phi')\cos(\phi - \phi') \qquad (15)$$

$$P_{33} = w[\sin\theta \sin\theta' \cos(\phi - \phi')$$
$$+ \cos\theta \cos\theta' (\cos^2(\phi - \phi') - \sin^2(\phi - \phi'))] \qquad (16)$$

$$P_{44} = w[\sin\theta \sin\theta' \cos(\phi - \phi') + \cos\theta \cos\theta'] \qquad (17)$$

$$w = \frac{3}{8\pi}\kappa_s \qquad (18)$$

and κ_s is the scattering coefficient

$$\kappa_s = \frac{8\pi}{3} n_o k^4 a^6 |y|^2 = 2fk^4 a^3 |y|^2 \qquad (19)$$

In (19), $f = n_o v_o$ is the fractional volume occupied by the particles. The internal power absorption due to one single scatterer is $\int dv\, \omega\epsilon_s'' |\overline{E}^{int}(\overline{r})|^2/2$ which is equal to $v_o \omega \epsilon_s'' |3\epsilon/(\epsilon_s + 2\epsilon)|^2 |E_o|^2/2$. The absorption cross section σ_a, hence, is $v_o \omega \epsilon_s'' \eta |3\epsilon/(\epsilon_s + 2\epsilon)|^2$. The absorption coefficient due to the scatterers is $n_o \sigma_a$. Therefore, the absorption coefficient is

$$\kappa_a = n_o v_o \omega \epsilon_s'' \eta \left| \frac{3\epsilon}{\epsilon_s + 2\epsilon} \right|^2 = fk \frac{\epsilon_s''}{\epsilon} \left| \frac{3\epsilon}{\epsilon_s + 2\epsilon} \right|^2 \tag{20}$$

Extinction coefficient κ_e is the sum of κ_s and κ_a. The extinction matrix is diagonal with each element equal to κ_e. The emission vector is such that the first two elements are equal to $\kappa_a CT$ whereas the last two elements are zero.

4.2 Eulerian Angles of Rotation

For scattering by nonspherical particles with symmetry axes, it is often convenient to solve the scattering problem in the coordinate system natural to the scatterer, and then to transform the results to the principal coordinate system of the problem. To describe the transformation between the natural axes of the scatterer to that of the principal coordinate, the Eulerian angles of rotation can be used.

Let the set of natural axes of the scatterer be denoted by $\hat{x}_b, \hat{y}_b, \hat{z}_b$. We also use positive rotation to denote carrying right-handed screw in the positive direction along that axis. The Eulerian rotations are performed as follows (Figure 3.10):
1. A rotation $\alpha(0 \leq \alpha < 2\pi)$ about the z_b axis, bringing the frame of axis from initial position S_b into position $S'(x', y', z')$. The axis of this rotation is commonly called vertical.
2. A rotation $\beta(0 \leq \beta < \pi)$ about the y' axis of frame S' called line of nodes. Note that its position is in general different from the initial position of y_b axis of S_b. The resulting position of frame of axis is symbolized by $S''(x'', y'', z'')$.
3. A rotation $\gamma(0 \leq \gamma < 2\pi)$ about the z'' axis of S'' called the figure axis. The position of this axis depends on previous rotations α and β. The final rotation from S'' to S gives the principal axes x, y and z. Rotation will be interpreted as a rotation of the frame of references about the origin, the field points being supposedly fixed.

Consider a point P with coordinates x_b, y_b, z_b in the frame S_b, then its coordinates in S' frame (x', y', z') are related to coordinates x_b, y_b, z_b

Eulerian Angles of Rotation

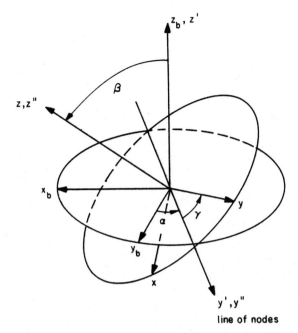

Fig. 3.10 Eulerian angles of rotation. Rotation from $(\hat{x}_b, \hat{y}_b, \hat{z}_b)$ to $(\hat{x}, \hat{y}, \hat{z})$. Rotation angles are α, β, γ with respect to $\hat{z}_b, \hat{y}'$ and $\hat{z}''$, successively.

as follows

$$\begin{bmatrix} x' \\ y' \\ z' \end{bmatrix} = \begin{bmatrix} \cos\alpha & \sin\alpha & 0 \\ -\sin\alpha & \cos\alpha & 0 \\ 0 & 0 & 1 \end{bmatrix} \cdot \begin{bmatrix} x_b \\ y_b \\ z_b \end{bmatrix} \qquad (1)$$

Rotation of β from S' to S'' gives

$$\begin{bmatrix} x'' \\ y'' \\ z'' \end{bmatrix} = \begin{bmatrix} \cos\beta & 0 & -\sin\beta \\ 0 & 1 & 0 \\ \sin\beta & 0 & \cos\beta \end{bmatrix} \cdot \begin{bmatrix} x' \\ y' \\ z' \end{bmatrix} \qquad (2)$$

and rotation of γ from S'' to S gives

$$\begin{bmatrix} x \\ y \\ z \end{bmatrix} = \begin{bmatrix} \cos\gamma & \sin\gamma & 0 \\ -\sin\gamma & \cos\gamma & 0 \\ 0 & 0 & 1 \end{bmatrix} \cdot \begin{bmatrix} x'' \\ y'' \\ z'' \end{bmatrix} \qquad (3)$$

Hence combining (1), (2) and (3) gives

$$\begin{bmatrix} x \\ y \\ z \end{bmatrix} = \overline{\overline{A}} \cdot \begin{bmatrix} x_b \\ y_b \\ z_b \end{bmatrix} \qquad (4)$$

with

$$a_{11} = \cos\gamma\cos\beta\cos\alpha - \sin\gamma\sin\alpha \qquad (5a)$$

$$a_{12} = \cos\gamma\cos\beta\sin\alpha + \sin\gamma\cos\alpha \qquad (5b)$$

$$a_{13} = -\cos\gamma\sin\beta \qquad (5c)$$

$$a_{21} = -\sin\gamma\cos\beta\cos\alpha - \cos\gamma\sin\alpha \qquad (5d)$$

$$a_{22} = -\sin\gamma\cos\beta\sin\alpha + \cos\gamma\cos\alpha \qquad (5e)$$

$$a_{23} = \sin\gamma\sin\beta \qquad (5f)$$

$$a_{31} = \sin\beta\cos\alpha \qquad (5g)$$

$$a_{32} = \sin\beta\sin\alpha \qquad (5h)$$

$$a_{33} = \cos\beta \qquad (5i)$$

where a_{ij} denotes the ij element of $\overline{\overline{A}}$. It can easily be shown that $\overline{\overline{A}}$ is unitary (Problem 11), i.e.,

$$\overline{\overline{A}}^t = \overline{\overline{A}}^{-1} \qquad (6)$$

The position vector of the point P is denoted by $\bar{r}$ and

$$\bar{r} = x\hat{x} + y\hat{y} + z\hat{z} = x_b\hat{x}_b + y_b\hat{y}_b + z_b\hat{z}_b \qquad (7)$$

Thus, we recognize that $\hat{x} \cdot \hat{x}_b = a_{11}$, $\hat{x} \cdot \hat{y}_b = a_{12}$, $\hat{x} \cdot \hat{z}_b = a_{13}$, etc. Hence, we have the following transformation between unit vectors in the two coordinate systems.

$$\hat{x}_b = a_{11}\hat{x} + a_{21}\hat{y} + a_{31}\hat{z} \qquad (8a)$$

$$\hat{y}_b = a_{12}\hat{x} + a_{22}\hat{y} + a_{32}\hat{z} \qquad (8b)$$

$$\hat{z}_b = a_{13}\hat{x} + a_{23}\hat{y} + a_{33}\hat{z} \qquad (8c)$$

$$\hat{x} = a_{11}\hat{x}_b + a_{12}\hat{y}_b + a_{13}\hat{z}_b \qquad (9a)$$

$$\hat{y} = a_{21}\hat{x}_b + a_{22}\hat{y}_b + a_{23}\hat{z}_b \qquad (9b)$$

$$\hat{z} = a_{31}\hat{x}_b + a_{32}\hat{y}_b + a_{33}\hat{z}_b \qquad (9c)$$

4.3 Phase Matrix for Small Ellipsoids with Prescribed Orientation Distribution

Consider a small ellipsoidal scatterer having natural axes $\hat{x}_b$, $\hat{y}_b$, and $\hat{z}_b$ with permittivity ϵ_s and the surface described by the equation

$$\frac{x_b^2}{a^2} + \frac{y_b^2}{b^2} + \frac{z_b^2}{c^2} = 1 \qquad (1)$$

Phase Matrix for Small Ellipsoids

The scattering function dyad $\overline{\overline{F}}$ is (Stratton, 1941; Problem 5)

$$\overline{\overline{F}} = \frac{k^2}{4\pi} v_o \frac{(\epsilon_s - \epsilon)}{\epsilon} \left\{ \frac{\hat{x}_b \hat{x}_b}{1 + v_d A_a} + \frac{\hat{y}_b \hat{y}_b}{1 + v_d A_b} + \frac{\hat{z}_b \hat{z}_b}{1 + v_d A_c} \right\} \quad (2)$$

where $v_o = 4\pi abc/3$ is the volume of the ellipsoid, and

$$v_d = \frac{abc}{2} \frac{(\epsilon_s - \epsilon)}{\epsilon} \quad (3)$$

$$A_a = \int_0^\infty \frac{ds}{(s + a^2) R_s} \quad (4)$$

$$A_b = \int_0^\infty \frac{ds}{(s + b^2) R_s} \quad (5)$$

$$A_c = \int_0^\infty \frac{ds}{(s + c^2) R_s} \quad (6)$$

$$R_s = [(s + a^2)(s + b^2)(s + c^2)]^{1/2} \quad (7)$$

By transformation of variable to $u = (s + a^2)(s + b^2)(s + c^2)$, it can be shown that a closed form solution exists for the sum of the three integrals in (4) through (6)

$$A_a + A_b + A_c = \frac{2}{abc} \quad (8)$$

For the case of spheroids ($a = b$), closed form solutions exist for the individual integrals, $A_a = A_b$ and A_c. For oblate spheroids ($c < a$), we have

$$A_c = \frac{2}{(a^2 - c^2)^{3/2}} \left[\frac{\sqrt{a^2 - c^2}}{c} - \tan^{-1} \frac{\sqrt{a^2 - c^2}}{c} \right] \quad (9)$$

and for the case of prolate spheroids ($c > a$), we have

$$A_c = -\frac{1}{c^3 e^3} \left[2e + \ln \frac{1-e}{1+e} \right] \quad (10)$$

where $e = \sqrt{1 - a^2/c^2}$ is the eccentricity. The other integrals for each type of spheroid can be calculated by using (8).

Using the scattering function matrix in (2), the Stokes matrix for one particle oriented with angles α, β, γ can be calculated readily. For example,

$$f_{vv}(\theta_s, \phi_s; \theta_i, \phi_i) = \frac{k^2}{4\pi} v_o \frac{(\epsilon_s - \epsilon)}{\epsilon} \left\{ \frac{(\hat{v}_s \cdot \hat{x}_b)(\hat{x}_b \cdot \hat{v}_i)}{1 + v_d A_a} \right.$$
$$\left. + \frac{(\hat{v}_s \cdot \hat{y}_b)(\hat{y}_b \cdot \hat{v}_i)}{1 + v_d A_b} + \frac{(\hat{v}_s \cdot \hat{z}_b)(\hat{z}_b \cdot \hat{v}_i)}{1 + v_d A_c} \right\} \quad (11a)$$

and
$$L_{11}(\theta_s, \phi_s; \theta_i, \phi_i) = |f_{vv}(\theta_s, \phi_s; \theta_i, \phi_i)|^2 \qquad (11b)$$

For randomly oriented ellipsoids, the phase matrix can be obtained by averaging over the Eulerian angles of rotation by a prescribed probability density function, $p(\alpha, \beta, \gamma)$. Hence,

$$\overline{\overline{P}}(\theta_s, \phi_s; \theta_i, \phi_i) = n_o \int_o^{2\pi} d\alpha \int_o^{\pi} d\beta \int_0^{2\pi} d\gamma$$
$$\times \overline{\overline{L}}(\theta_s, \phi_s; \theta_i, \phi_i; \alpha, \beta, \gamma) p(\alpha, \beta, \gamma) \qquad (12)$$

The absorption cross section for incident polarization $\hat{\beta}_i$ is

$$\sigma_{a\beta} = v_o k \frac{\epsilon_s''}{\epsilon} \left[\frac{(\hat{\beta}_i \cdot \hat{x}_b)^2}{|1 + v_d A_a|^2} + \frac{(\hat{\beta}_i \cdot \hat{y}_b)^2}{|1 + v_d A_b|^2} + \frac{(\hat{\beta}_i \cdot \hat{z}_b)^2}{|1 + v_d A_c|^2} \right] \qquad (13)$$

where $\beta = v$ or h, and ϵ_s'' is the imaginary part of ϵ_s. The absorption coefficient for $\hat{\beta}$ polarization is

$$\kappa_{a\beta}(\hat{k}_i) = n_o \int_o^{2\pi} d\alpha \int_o^{\pi} d\beta$$
$$\times \int_0^{2\pi} d\gamma \, p(\alpha, \beta, \gamma) \, \sigma_{a\beta}(\theta_i, \phi_i, \alpha, \beta, \gamma) \qquad (14)$$

The normalization of the probability density function is

$$\int_o^{2\pi} d\alpha \int_o^{\pi} d\beta \int_0^{2\pi} d\gamma \, p(\alpha, \beta, \gamma) = 1 \qquad (15)$$

The phase matrix, scattering, and absorption coefficients for oblate spheroids with prescribed orientation distribution that depends on β only is considered in Problem 9.

4.4 Phase Matrix for Random Media

A random medium is characterized by random fluctuation in permittivity. Consider a volume ΔV of the random medium (Figure 3.11). Inside the volume

$$\epsilon_1(\bar{r}) = \epsilon_{1m} + \epsilon_{1f}(\bar{r}) \qquad (1)$$

with ϵ_{1m} as the mean permittivity $<\epsilon_1(\bar{r})>$ so that $<\epsilon_{1f}(\bar{r})>= 0$. From Maxwell's equations,

$$\nabla \times \nabla \times \overline{E} - k_{1m}^2 \overline{E} = \omega^2 \mu \epsilon_{1f}(\bar{r}) \overline{E}(\bar{r}) \qquad (2)$$

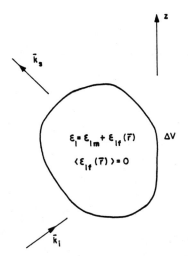

Fig. 3.11 Random medium of volume ΔV with permittivity $\epsilon_1(\bar{r}) = \epsilon_{1m} + \epsilon_{1f}(\bar{r})$.

where $k_{1m}^2 = \omega^2 \mu \epsilon_{1m}$.

The scattering phase matrix for the random medium is obtained by applying Born approximation with the far field solution (Tatarskii, 1961). We consider an incident plane wave on a volume ΔV of the random medium (Figure 3.11)

$$\bar{E}_i(\bar{r}) = (\hat{v}_i E_{vi} + \hat{h}_i E_{hi}) e^{i\bar{k}_i \cdot \bar{r}} \qquad (3)$$

Making use of the dyadic Green's function, we have the scattered field

$$\bar{E}_s(\bar{r}) = \omega^2 \mu \left[\bar{\bar{I}} + \frac{\nabla \nabla}{k_{1m}^2} \right] \cdot \int_{\Delta V} d\bar{r}' \frac{e^{ik_{1m}|\bar{r}-\bar{r}'|}}{4\pi |\bar{r}-\bar{r}'|} \epsilon_{1f}(\bar{r}') \bar{E}(\bar{r}') \qquad (4)$$

When $|\epsilon_{1f}(\bar{r})| \ll 1$, the Born approximation is to replace $\bar{E}(\bar{r})$ by $\bar{E}_i(\bar{r})$ inside the integral and making the far field approximation,

$$\bar{E}_s(\bar{r}) = (\hat{v}_s \hat{v}_s + \hat{h}_s \hat{h}_s) \cdot (\hat{v}_i E_{vi} + \hat{h}_i E_{hi}) \frac{W_1 e^{ik_{1m}r}}{r} \qquad (5)$$

with

$$W_1 = \frac{\omega^2 \mu}{4\pi} \int_{\Delta V} d\bar{r}' \epsilon_{1f}(\bar{r}') e^{ik_{1m}(\hat{k}_i - \hat{k}_s) \cdot \bar{r}'} \qquad (6)$$

Using (5) and (6), we can identify

$$f_{\alpha\beta}(\theta_s, \phi_s; \theta_i, \phi_i) = (\hat{\alpha}_s \cdot \hat{\beta}_i) W_1 \qquad (7)$$

with $\alpha, \beta = v, h$. Thus, phase matrix elements are calculated by using (7) and (15) through (17), Section 3.2, and are proportional to $<|W_1|^2>$ where

$$<|W_1|^2> = \frac{\omega^4 \mu^2}{16\pi^2} \int_{\Delta V} d\bar{r}'$$
$$\times \int_{\Delta V} d\bar{r}'' <\epsilon_{1f}(\bar{r}')\epsilon_{1f}^*(\bar{r}'')> e^{ik_{1m}(\hat{k}_i - \hat{k}_s)\cdot(\bar{r}' - \bar{r}'')} \quad (8)$$

The covariance function in the statistically homogeneous random medium depends on the distance of between two points as

$$<\epsilon_{1f}(\bar{r}')\epsilon_{1f}(\bar{r}'')> = \delta \epsilon_{1m}^2 b(\bar{r}' - \bar{r}') \quad (9)$$

where δ is the variance of the fluctuation and $b(\bar{r}' - \bar{r}'')$ is the normalized covariance function which is assumed to vanish as $|\bar{r}' - \bar{r}''| \to \infty$. We set

$$<|W|^2> = \frac{<|W_1|^2>}{\Delta V} = \frac{\delta k_{1m}^4}{16\pi^2} \int_{-\infty}^{\infty} d\bar{r}\, b(\bar{r})\, e^{ik_{1m}(\hat{k}_i - \hat{k}_s)\cdot\bar{r}} \quad (10)$$

From (17), Section 3.2, the phase matrix can be interpreted as Stokes matrix per unit volume. Hence,

$$\bar{\bar{P}}(\theta_s, \phi_s; \theta_i, \phi_i) = <|W(\theta_s, \phi_s; \theta_i, \phi_i)|^2>$$

$$\begin{bmatrix} (\hat{v}_s \cdot \hat{v}_i)^2 & (\hat{v}_s \cdot \hat{h}_i)^2 \\ (\hat{h}_s \cdot \hat{v}_i)^2 & (\hat{h}_s \cdot \hat{h}_i)^2 \\ 2(\hat{h}_s \cdot \hat{v}_i)(\hat{v}_s \cdot \hat{v}_i) & 2(\hat{v}_s \cdot \hat{h}_i)(\hat{h}_s \cdot \hat{h}_i) \\ 0 & 0 \end{bmatrix}$$

$$\begin{matrix} (\hat{v}_s \cdot \hat{h}_i)(\hat{v}_s \cdot \hat{v}_i) & 0 \\ (\hat{h}_s \cdot \hat{v}_i)(\hat{h}_s \cdot \hat{h}_i) & 0 \\ (\hat{v}_s \cdot \hat{v}_i)(\hat{h}_s \cdot \hat{h}_i) + (\hat{v}_s \cdot \hat{h}_i)(\hat{h}_s \cdot \hat{v}_i) & 0 \\ 0 & (\hat{v}_s \cdot \hat{v}_i)(\hat{h}_s \cdot \hat{h}_i) - (\hat{v}_s \cdot \hat{h}_i)(\hat{h}_s \cdot \hat{v}_i) \end{matrix} . \quad (11)$$

The scattering coefficients are

$$\kappa_{sv}(\theta, \phi) = \int_{4\pi} d\Omega' <|W(\theta', \phi'; \theta, \phi)|^2> [(\hat{v}' \cdot \hat{v})^2 + (\hat{h}' \cdot \hat{v})^2] \quad (12a)$$

$$\kappa_{sh}(\theta, \phi) = \int_{4\pi} d\Omega' <|W(\theta', \phi'; \theta, \phi)|^2> [(\hat{v}' \cdot \hat{h})^2 + (\hat{h}' \cdot \hat{h})^2] \quad (12b)$$

Thus, the scattering is proportional to the Fourier transform of the covariance function of permittivity fluctuation.

In the following, we consider several examples of correlation functions and the associated $<|W|^2>$. Assume for the random medium a correlation function that is Gaussian in the horizontal direction and exponential in the vertical direction.

$$<\epsilon_{1f}(\bar{r}')\epsilon_{1f}(\bar{r}'')> = \delta\epsilon_{1m}^2 \exp\left[-\frac{|z'-z''|}{l_z} - \frac{(|x'-x''|^2 + |y'-y''|^2)}{l_\rho^2}\right] \quad (13)$$

Corresponding to this correlation function, we have

$$<|W|^2> = \frac{\delta k_{1m}^4}{8\pi} \frac{l_\rho^2 l_z}{1 + k_{1m}^2 l_z^2(\cos\theta - \cos\theta')^2}$$
$$\times \exp\Big[-k_{1m}^2 l_\rho^2(\sin^2\theta + \sin^2\theta')/4$$
$$+ k_{1m}^2 l_\rho^2 \sin\theta\sin\theta'\cos(\phi - \phi')/2\Big] \quad (14)$$

Limit of a Laminar Structure $(l_\rho \to \infty)$

We characterize the scattering by the fluctuating permittivity with the correlation function

$$<\epsilon_{1f}(\bar{r}')\epsilon_{1f}^*(\bar{r}'')> = \delta\epsilon_{1m}^2 e^{-|z'-z''|/l_z} \quad (15)$$

Then we have, with (θ, ϕ) denoting the scattered directions $\hat{k}_s$ and (θ', ϕ') denoting the incident direction $\hat{k}_i$,

$$<|W|^2> = \frac{\delta k_{1m}^4}{2} \frac{l_z}{(1 + l_z^2|k_{iz} - k_{sz}|^2)} \delta(k_{ix} - k_{sx})\delta(k_{iy} - k_{sy}) \quad (16)$$

To convert to Dirac delta function of θ and ϕ, we make use of the following identity for transformation of the delta function

$$\delta(k_{ix} - k_{sx})\delta(k_{iy} - k_{sy}) = \frac{1}{|k_{1m}^2 \sin\theta'\cos\theta'|} \delta(\phi - \phi')$$
$$\times [\delta(\theta - \theta') + \delta(\theta' + \theta - \pi)] \quad (17)$$

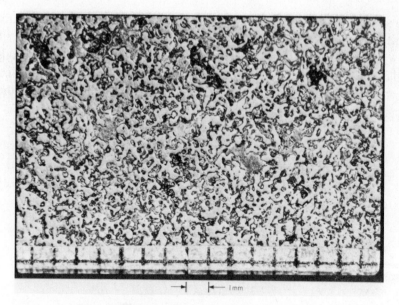

Fig. 3.12 Fine snow grain sample.

for θ and θ' between 0 and π. The above expression shows that there is coupling between the specular forward and backward propagating waves.

Limit of a Cylindrical Structure $(l_z \to \infty)$

For certain vegetation fields with cylindrical structures, we model the medium by the correlation function

$$<\epsilon_{1f}(\overline{r}')\epsilon_{1f}^*(\overline{r}'')> = \delta\epsilon_{1m}^2 \exp\left[-\frac{(x'-x'')^2 + (y'-y'')^2}{l_\rho^2}\right] \quad (18)$$

Thus, the function $<|W|^2>$ in the phase matrix is

$$<|W|^2> = \frac{\delta k_{1m}^3 l_\rho^2}{8} \delta(\cos\theta - \cos\theta')$$
$$\times \exp\left[-\frac{1}{2}k_{1m}^2 l_\rho^2 \sin^2\theta \left[1 - \cos(\phi - \phi')\right]\right] \quad (19)$$

The scattering pattern for this correlation function is cone-like.

The actual correlation functions of snow samples have been computed by directly applying the definition of correlation functions (Vallese and Kong, 1981). Consider the fine snow grain sample shown in Figure 3.12 and

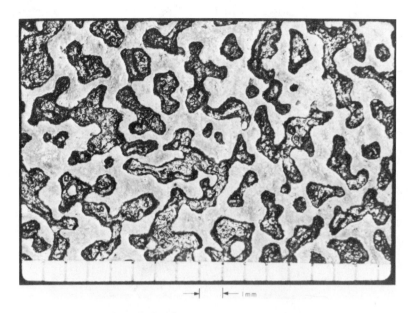

Fig. 3.13 Coarse snow grain sample.

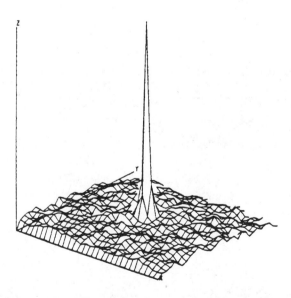

Fig. 3.14 Correlation function for fine snow grain sample, $l_\rho = l_z = 0.05$ mm.

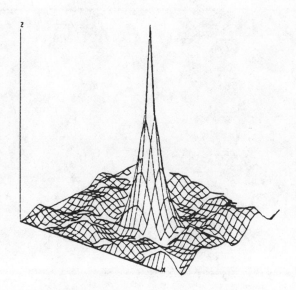

Fig. 3.15 Correlation function for coarse snow grain sample, $l_\rho = l_z = 0.3$ mm.

the coarse snow grain sample in Figure 3.13. The corresponding correlation functions are shown in Figures 3.14 and 3.15. The functions are seen to be exponential in form.

5 PHASE MATRIX AND EXTINCTION MATRIX FOR GENERAL NONSPHERICAL PARTICLES WITH PRESCRIBED ORIENTATION AND SIZE DISTRIBUTION: T-MATRIX APPROACH

5.1 Vector Spherical Waves

To represent the scattering by an arbitrarily-shaped particle, it is convenient to use the System Transfer Operator approach (also referred to as the T-matrix approach). The system transfer operator T is defined by

$$\overline{E}^s(\overline{r}) = T\overline{E}^E(\overline{r}) \tag{1}$$

where $\overline{E}^s(\overline{r})$ and $\overline{E}^E(\overline{r})$ are the scattered and exciting electric fields. Equation (1) implies that once T has been determined for each scatterer, the scattered field may be obtained from a knowledge of the exciting field

Vector Spherical Waves

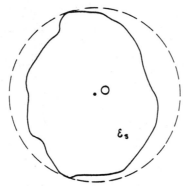

Fig. 3.16 Incident wave on a particle with circumscribing sphere.

and that T need be solved only once for each scatterer. The form of T is generally quite complex for nonspherical scatterers.

To represent the T-matrix, a circumscribing spherical surface (defined as the smallest sphere that encloses the particle) is used (Figure 3.16). Outside the circumscribing sphere, outgoing spherical waves can be used as a basis for the scattered field. The exciting fields can be expanded in regular spherical waves. Hence the T-matrix can be represented with spherical waves as basis functions.

In the spherical coordinate system, the scalar Helmholtz equation

$$\left(\nabla^2 + k^2\right)\psi = 0 \tag{2}$$

has the outgoing wave solution

$$\psi_{mn}(kr, \theta, \phi) = h_n(kr) P_n^m(\cos\theta) e^{im\phi} \tag{3}$$

with $n = 0, 1, 2, \ldots$, $m = 0, \pm 1, \ldots, \pm n$. In (3) h_n is a spherical Hankel function of first kind and P_n^m is the associated Legendre polynomial. We also define regular wave functions by

$$Rg\psi_{mn}(kr, \theta, \phi) = j_n(kr) P_n^m(\cos\theta) e^{im\phi} \tag{4}$$

where Rg stands for regular and j_n is the spherical Bessel function. The prefix *regular* is used to denote the fact that h_n is replaced by j_n which is finite at $r = 0$. The associated Legendre polynomial is defined as (Abramowitz and Stegun, 1965; Jackson, 1975)

$$P_n^m(x) = \frac{(-1)^m}{2^n n!}(1-x^2)^{m/2}\frac{d^{n+m}}{dx^{n+m}}(x^2-1)^n \tag{5}$$

for $m = 0, \pm 1, \ldots, \pm n$. Expression (5) holds for both positive and negative values of m. The relation between positive and negative values of m is

$$P_n^{-m}(x) = (-1)^m \frac{(n-m)!}{(n+m)!} P_n^m(x) \tag{6}$$

The spherical harmonic is defined as

$$Y_n^m(\theta, \phi) = P_n^m(\cos \theta) e^{im\phi} \tag{7}$$

with orthogonality relation

$$\int_0^\pi d\theta \sin\theta \int_0^{2\pi} d\phi \, Y_n^m(\theta, \phi) Y_{n'}^{-m'}(\theta, \phi) = (-1)^m \frac{4\pi}{2n+1} \delta_{mm'} \delta_{nn'} \tag{8}$$

The vector spherical harmonics are the three vector functions $\overline{V}_{mn}^{(\alpha)}(\theta, \phi)$, $\alpha = 1, 2,$ and 3. They are defined by

$$\overline{V}_{mn}^{(1)}(\theta, \phi) = \overline{P}_{mn}(\theta, \phi) = \hat{r} Y_n^m(\theta, \phi) \qquad (n = 0, 1, 2, \ldots) \tag{9}$$

$$\overline{V}_{mn}^{(2)}(\theta, \phi) = \overline{B}_{mn}(\theta, \phi) = r \nabla (Y_n^m(\theta, \phi))$$

$$= \left(\hat{\theta} \frac{dP_n^m(\cos\theta)}{d\theta} + \hat{\phi} \frac{im}{\sin\theta} P_n^m(\cos\theta) \right) e^{im\phi}$$

$$= \hat{r} \times \overline{C}_{mn}(\theta, \phi)$$

$$(n = 1, 2, 3, \ldots) \tag{10}$$

$$\overline{V}_{mn}^{(3)}(\theta, \phi) = \overline{C}_{mn}(\theta, \phi) = \nabla \times (\overline{r} Y_n^m(\theta, \phi))$$

$$= \left(\hat{\theta} \frac{im}{\sin\theta} P_n^m(\cos\theta) - \hat{\phi} \frac{dP_n^m(\cos\theta)}{d\theta} \right) e^{im\phi} \tag{11}$$

$$(n = 1, 2, 3, \ldots)$$

The orthogonality relation for vector spherical harmonics is (Morse and Feshbach, 1953)

$$\int_0^\pi d\theta \sin\theta \int_0^{2\pi} d\phi \, \overline{V}_{mn}^{(\alpha)}(\theta, \phi) \cdot \overline{V}_{-m'n'}^{(\beta)}(\theta, \phi) = \delta_{\alpha\beta} \delta_{mm'} \delta_{nn'} z_{\alpha mn} \tag{12}$$

where

$$z_{1mn} = (-1)^m \frac{4\pi}{2n+1} \tag{13}$$

Vector Spherical Waves

$$z_{2mn} = z_{3mn} = (-1)^m \frac{4\pi n(n+1)}{2n+1} \quad (14)$$

The three regular vector spherical waves $Rg\overline{L}_{mn}$, $Rg\overline{M}_{mn}$, and $Rg\overline{N}_{mn}$ are defined as ·

$$Rg\overline{L}_{mn}(kr,\theta,\phi) = \frac{\gamma'_{mn}}{k} \nabla(Rg\psi_{mn}(kr,\theta,\phi))$$

$$= \gamma'_{mn} \left\{ j'_n(kr)\overline{P}_{mn}(\theta,\phi) + \frac{j_n(kr)}{kr}\overline{B}_{mn}(\theta,\phi) \right\} \quad (15)$$

$$Rg\overline{M}_{mn}(kr,\theta,\phi) = \gamma_{mn} \nabla \times (\overline{r}\, Rg\psi_{mn}(kr,\theta,\phi))$$

$$= \gamma_{mn} j_n(kr)\overline{C}_{mn}(\theta,\phi) \quad (16)$$

$$Rg\overline{N}_{mn}(kr,\theta,\phi) = \frac{1}{k}\nabla \times Rg\overline{M}_{mn}(kr,\theta,\phi)$$

$$= \gamma_{mn}\left\{ \frac{n(n+1)j_n(kr)}{kr}\overline{P}_{mn}(\theta,\phi) \right.$$

$$\left. + \frac{(kr\, j_n(kr))'}{kr}\overline{B}_{mn}(\theta,\psi) \right\} \quad (17)$$

In (15) through (17)

$$\gamma'_{mn} = \sqrt{\frac{(2n+1)(n-m)!}{4\pi(n+m)!}} \quad (18)$$

$$\gamma_{mn} = \sqrt{\frac{(2n+1)(n-m)!}{4\pi n(n+1)(n+m)!}} \quad (19)$$

The vector spherical waves without the prefix Rg in front are the expressions in (15) through (17) with j_n replaced by h_n. We note that the index for $\overline{L}_{mn}$, is $n = 0, 1, 2, 3, \ldots$ whereas for $\overline{M}_{mn}$ and $\overline{N}_{mn}$, the index is $n = 1, 2, 3, \cdots$. The $\overline{L}$ functions satisfy the equation

$$\nabla(\nabla \cdot Rg\overline{L}_{mn}) + k^2 Rg\overline{L}_{mn} = 0 \quad (20)$$

whereas the $Rg\overline{M}$ and $Rg\overline{N}$ functions satisfy the vector wave equation

$$\nabla \times \nabla \times \overline{E} - k^2 \overline{E} = 0 \quad (21)$$

where $\overline{E}$ is $Rg\overline{M}_{mn}$ or $Rg\overline{N}_{mn}$.

The vector spherical wave functions can be expressed as integral representations of vector spherical harmonics (Problem 13)

$$Rg\overline{L}_{mn}(kr,\theta,\phi) = \frac{(-i)^{n-1}}{4\pi}\gamma'_{mn}\int_{4\pi}d\Omega'\,e^{ikr\,\hat{r}\cdot\hat{r}'}\overline{P}_{mn}(\theta',\phi') \quad (22)$$

$$Rg\overline{M}_{mn}(kr,\theta,\phi) = \frac{(-i)^{n}}{4\pi}\gamma_{mn}\int_{4\pi}d\Omega'\,e^{ikr\,\hat{r}\cdot\hat{r}'}\overline{C}_{mn}(\theta',\phi') \quad (23)$$

$$Rg\overline{N}_{mn}(kr,\theta,\phi) = \frac{(-i)^{n-1}}{4\pi}\gamma_{mn}\int_{4\pi}d\Omega'\,e^{ikr\,\hat{r}\cdot\hat{r}'}\overline{B}_{mn}(\theta',\phi') \quad (24)$$

With the aid of (22) through (24), it follows that

$$\overline{\overline{I}}e^{i\overline{p}\cdot\overline{r}} = \sum_{n,m}(-1)^{m}\frac{(2n+1)}{n(n+1)}i^{n}$$
$$\cdot\left\{-in(n+1)\frac{\overline{P}_{-mn}(\theta_{p},\phi_{p})}{\gamma'_{mn}}Rg\overline{L}_{mn}(pr,\theta,\phi)\right.$$
$$+\frac{\overline{C}_{-mn}(\theta_{p},\phi_{p})}{\gamma_{mn}}Rg\overline{M}_{mn}(pr,\theta,\phi)$$
$$\left.-\frac{i\overline{B}_{-mn}(\theta_{p},\phi_{p})}{\gamma_{mn}}Rg\overline{N}_{mn}(pr,\theta,\phi)\right\} \quad (25)$$

where (θ_p, ϕ_p) indicates the direction $\hat{p}$. A plane electromagnetic wave can be expressed in terms of spherical waves by taking the dot product of (25) with the incident electric field. Let the propagation direction of the incident wave be $\hat{k}_i = \sin\theta_i\cos\phi_i\hat{x} + \sin\theta_i\sin\phi_i\hat{y} + \cos\theta_i\hat{z}$.

$$\overline{E}_i = (E_{vi}\hat{v}_i + E_{hi}\hat{h}_i)e^{i\overline{k}_i\cdot\overline{r}}$$
$$= \sum_{n,m}(-1)^m\frac{1}{\gamma_{mn}}\frac{(2n+1)}{n(n+1)}i^n\left\{\left[E_{vi}[\hat{\theta}_i\cdot\overline{C}_{-mn}(\theta_i,\phi_i)]\right.\right.$$
$$\left.+ E_{hi}[\hat{\phi}_i\cdot\overline{C}_{-mn}(\theta_i,\phi_i)]\right]Rg\overline{M}_{mn}(kr,\theta,\phi)$$
$$+\left[E_{vi}[\hat{\theta}_i\cdot(-i\overline{B}_{-mn}(\theta_i,\phi_i))]\right.$$
$$\left.\left.+ E_{hi}[\hat{\phi}_i\cdot(-i\overline{B}_{-mn}(\theta_i,\phi_i))]\right]Rg\overline{N}_{mn}(kr,\theta,\phi)\right\} \quad (26)$$

Note that for $\overline{B}_{mn}$ and $\overline{C}_{mn}$ functions, the associated Legendre polynomial always appears in the context $P_n^m(\cos\theta)/\sin\theta$ which computationally is in

Vector Spherical Waves

the form zero divided by zero for $\theta = 0$. Hence, to facilitate the construction of $\overline{B}_{mn}$ and $\overline{C}_{mn}$ functions, we introduce t_n^m and s_n^m functions defined by

$$t_n^m(\cos\theta) = \sqrt{2\pi n(n+1)}\,\gamma_{mn}\, m\, \frac{P_n^m(\cos\theta)}{\sin\theta} \qquad (27)$$

$$s_n^m(\cos\theta) = \sqrt{2\pi n(n+1)}\,\gamma_{mn}\, \frac{dP_n^m(\cos\theta)}{d\theta} \qquad (28)$$

To compute t_n^m and s_n^m for $n \geq 1$, we use the following recurrence relations (Problem 14)

$$t_{m+1}^{m+1}(\cos\theta) = -\sin\theta \left(\frac{m+1}{m}\right)\left(\frac{2m+3}{2m+2}\right)^{1/2} t_m^m(\cos\theta) \qquad (29)$$

$$t_{m+1}^m(\cos\theta) = \sqrt{2m+3}\,\cos\theta\, t_m^m(\cos\theta) \qquad (30)$$

$$t_n^m(x) = \sqrt{\frac{2n+1}{n^2-m^2}} \Bigg[\sqrt{2n-1}\, x\, t_{n-1}^m(x)$$
$$- \left(\frac{(n-1)^2-m^2}{(2n-3)}\right)^{1/2} t_{n-2}^m(x)\Bigg] \qquad (31)$$

for $n \geq m+2$. The recurrence relation can be initialized by

$$t_1^1(\cos\theta) = -\sqrt{3}/2 \qquad (32)$$

Also,

$$t_n^0(\cos\theta) = 0 \qquad (33)$$

The computation of the $s_n^m(\cos\theta)$ function can be carried out by using the following relations.

(i) For $m > 0$:

$$s_m^m(\cos\theta) = \cos\theta\, t_m^m(\cos\theta) \qquad (34)$$

$$m\, s_n^m(\cos\theta) = n\cos\theta\, t_n^m(\cos\theta) - t_{n-1}^m(\cos\theta)\left(\frac{(2n+1)(n^2-m^2)}{2n-1}\right)^{1/2}$$
$$(n \geq m+1) \qquad (35)$$

(ii) For $m = 0$:

$$s_n^o(\cos\theta) = \frac{\cos\theta}{n-1}(4n^2-1)^{1/2}s_{n-1}^o(\cos\theta)$$

$$-\frac{n}{n-1}\left(\frac{2n+1}{2n-3}\right)^{1/2}s_{n-2}^o(\cos\theta) \qquad (36)$$

which is initialized by the relations

$$s_o^o(\cos\theta) = 0 \qquad (37)$$

$$s_1^o(\cos\theta) = -\sqrt{3/2}\sin\theta \qquad (38)$$

To calculate t_n^m and s_n^m for negative m, we can use the relations

$$t_n^{-m}(\cos\theta) = (-1)^{m+1}t_n^m(\cos\theta) \qquad (39)$$

$$s_n^{-m}(\cos\theta) = (-1)^m s_n^m(\cos\theta) \qquad (40)$$

Thus, to set up a computer code, we first decide on a maximum value of n equal to N_{max} and initialize the values by using (32) and (33). Next the values of t_m^m and t_{m+1}^m for $m = 1, 2, ..., N_{max}$ are calculated by (29) and (30). The values of t_n^m, for $n = m+2, m+3, ..., N_{max}$ are then computed by (31). The values for negative degrees m are obtained by using (39). The values of the functions s_n^m for $m > 0$ are obtained from the t_n^m values by (34) and (35). The function s_n^o is computed by using the recurrence relation of (36) which is initialized by (37) and (38). Negative degrees of s_n^m are calculated by using (40).

5.2 Relation of T-Matrix to Scattering Amplitude Matrix, Extinction, and Scattering Cross Sections

Consider a wave incident on an isolated scattering object with permittivity ϵ_s (Figure 3.16). The scattering object is centered at the origin. The center of the object is usually arbitrary and can be defined as the point of maximum symmetry. A scattering object is characterized by its permittivity ϵ_s, the location of its center and the shape of its boundary.

In the absence of other scattering objects, the incident field is equal to the exciting field. The incident field is expanded in terms of spherical waves. The source of the incident field is assumed to be outside the circumscribing sphere. Hence, for the region within the circumscribing sphere, the incident field can be expressed as

$$\overline{E}^E(\overline{r}) = \overline{E}^{inc}(\overline{r})$$
$$= \sum_{m,n}\left[a_{mn}^{E(M)}Rg\overline{M}_{mn}(kr,\theta,\phi) + a_{mn}^{E(N)}Rg\overline{N}_{mn}(kr,\theta,\phi)\right] \qquad (1)$$

n	m	combined index l
1	-1	1
1	0	2
1	1	3
2	-2	4
2	-1	5
2	0	6
2	1	7
2	2	8
3	-3	9
3	-2	10
3	-1	11
3	0	12
3	1	13
3	2	14
3	3	15
etc.		etc.

Table 3.1 Correspondence between l and (n, m).

Outside the circumscribing sphere, the scattered wave will be linear combinations of outgoing spherical waves.

$$\overline{E}^s(\overline{r}) = \sum_{m,n} \left[a_{mn}^{s(M)} \overline{M}_{mn}(kr, \theta, \phi) + a_{mn}^{s(N)} \overline{N}_{mn}(kr, \theta, \phi) \right] \quad (2)$$

The T-matrix is used to describe the linear relation between scattering coefficients a_{mn}^s and the exciting field coefficients a_{mn}^E.

$$a_{mn}^{s(M)} = \sum_{m'n'} \left[T_{mnm'n'}^{(11)} a_{m'n'}^{E(M)} + T_{mnm'n'}^{(12)} a_{m'n'}^{E(N)} \right] \quad (3)$$

$$a_{mn}^{s(N)} = \sum_{m'n'} \left[T_{mnm'n'}^{(21)} a_{m'n'}^{E(M)} + T_{mnm'n'}^{(22)} a_{m'n'}^{E(N)} \right] \quad (4)$$

To put the representation in (3) and (4) in a more compact form, the following notations can be used. Suppose the truncation of the infinite summation is at multipole $n = N_{max}$. Then the number of terms in $a_{mn}^{s(M)}$ is $3 + 5 + 7 + \ldots + (2N_{max} + 1) = N_{max}(N_{max} + 2)$. A combined index l can be used to represent the two indices n and m as follows

$$l = n(n+1) + m \quad (5)$$

The correspondence between l and (n, m) is shown in Table 3.1. Let

$$L_{max} = N_{max}(N_{max} + 2) \qquad (6)$$

Hence, letting $\overline{a}^{E(M)}$ and $\overline{a}^{E(N)}$ to denote the column matrices of dimension L_{max} representing the coefficients $a_l^{E(M)}$ and $a_l^{E(N)}$, respectively, and $\overline{a}^{s(M)}$ and $\overline{a}^{s(N)}$ to denote column matrices of coefficients $a_l^{s(M)}$ and $a_l^{s(N)}$, respectively, we have the matrix relation

$$\begin{bmatrix} \overline{a}^{s(M)} \\ \overline{a}^{s(N)} \end{bmatrix} = \begin{bmatrix} \overline{\overline{T}}^{(11)} & \overline{\overline{T}}^{(12)} \\ \overline{\overline{T}}^{(21)} & \overline{\overline{T}}^{(22)} \end{bmatrix} \begin{bmatrix} \overline{a}^{E(M)} \\ \overline{a}^{E(N)} \end{bmatrix} \qquad (7)$$

where $\overline{\overline{T}}^{(11)}$, $\overline{\overline{T}}^{(12)}$, $\overline{\overline{T}}^{(21)}$, and $\overline{\overline{T}}^{(22)}$ are matrices of dimension $L_{max} \times L_{max}$ representing the T-matrix coefficients $T_{ll'}^{(11)}$, $T_{ll'}^{(12)}$, $T_{ll'}^{(21)}$, and $T_{ll'}^{(22)}$, respectively. Further define

$$\overline{a}^s = \begin{bmatrix} \overline{a}^{s(M)} \\ \overline{a}^{s(N)} \end{bmatrix} \qquad (8)$$

$$\overline{a}^E = \begin{bmatrix} \overline{a}^{E(M)} \\ \overline{a}^{E(N)} \end{bmatrix} \qquad (9)$$

and

$$\overline{\overline{T}} = \begin{bmatrix} \overline{\overline{T}}^{(11)} & \overline{\overline{T}}^{(12)} \\ \overline{\overline{T}}^{(21)} & \overline{\overline{T}}^{(22)} \end{bmatrix} \qquad (10)$$

Hence, $\overline{a}^s$ and $\overline{a}^E$ are of dimension $2L_{max}$ and $\overline{\overline{T}}$ is of dimension $2L_{max} \times 2L_{max}$. Thus,

$$\overline{a}^s = \overline{\overline{T}} \overline{a}^E \qquad (11)$$

Using the T-matrix, the scattering amplitude dyad $\overline{\overline{F}}(\theta_s, \phi_s; \theta_i, \phi_i)$ can be calculated. For a plane wave incident in the direction $\hat{k}_i = (\theta_i, \phi_i)$,

$$\overline{E}^{inc} = \hat{e}_i E_o e^{i\overline{k}_i \cdot \overline{r}} \qquad (12)$$

the incident field coefficients are, from (26), Section 5.1,

$$a_{mn}^{E(M)} = (-1)^m \frac{1}{\gamma_{mn}} \frac{(2n+1)}{n(n+1)} i^n \, \overline{C}_{-mn}(\theta_i, \phi_i) \cdot \hat{e}_i E_o \qquad (13)$$

$$a_{mn}^{E(N)} = (-1)^m \frac{1}{\gamma_{mn}} \frac{(2n+1)}{n(n+1)} i^n \frac{\overline{B}_{-mn}(\theta_i, \phi_i)}{i} \cdot \hat{e}_i E_o \qquad (14)$$

Relation of T-Matrix

The asymptotic far-field solution for $kr \to \infty$ is, using large argument approximation for spherical Hankel functions,

$$\lim_{kr \to \infty} \overline{M}_{mn}(kr, \theta, \phi) = \gamma_{mn} \overline{C}_{mn}(\theta, \phi) i^{-n-1} \frac{1}{kr} e^{ikr} \tag{15}$$

$$\lim_{kr \to \infty} \overline{N}_{mn}(kr, \theta, \phi) = \gamma_{mn} \overline{B}_{mn}(\theta, \phi) i^{-n} \frac{1}{kr} e^{ikr} \tag{16}$$

Substituting (15) and (16) in (2) and making use of (3), (4), (13), (14), and the definition of the scattering amplitude dyad $\overline{\overline{F}}$, gives the following relation expressing $\overline{\overline{F}}$ in terms of the T-matrix elements

$$\overline{\overline{F}}(\theta, \phi,; \theta', \phi') = \frac{4\pi}{k} \sum_{n,m,n',m'} (-1)^{m'} i^{n'-n-1}$$

$$\times \left\{ \left[T^{(11)}_{mnm'n'} \gamma_{mn} \overline{C}_{mn}(\theta, \phi) + T^{(21)}_{mnm'n'} i \gamma_{mn} \overline{B}_{mn}(\theta, \phi) \right] \right.$$

$$\times \gamma_{-m'n'} \overline{C}_{-m'n'}(\theta', \phi')$$

$$+ \left[T^{(12)}_{mnm'n'} \gamma_{mn} \overline{C}_{mn}(\theta, \phi) + T^{(22)}_{mnm'n'} i \gamma_{mn} \overline{B}_{mn}(\theta, \phi) \right]$$

$$\left. \times \gamma_{-m'n'} \frac{\overline{B}_{-m'n'}(\theta', \phi')}{i} \right\} \tag{17}$$

Using the optical theorem, the extinction cross section for incident direction (θ_i, ϕ_i) can be calculated from (16), Section 3.3. The extinction matrix can be evaluated by using (4) and (11), Section 3.4. To find the scattering cross section for incident direction (θ_i, ϕ_i) and incident polarization β, we note that

$$\sigma_{s\beta}(\theta_i, \phi_i) = \int_{4\pi} d\Omega \left[|f_{v\beta}(\theta, \phi; \theta_i, \phi_i)|^2 + |f_{h\beta}(\theta, \phi; \theta_i, \phi_i)|^2 \right] \tag{18}$$

Substituting (17) in (18) and making use of orthogonality relations for vector spherical harmonics as given in (12), Section 5.1, we have (Problem 15)

$$\sigma_{s\beta}(\theta_i, \phi_i) = \frac{16\pi^2}{k^2} \sum_{n,m} \left\{ \left| \sum_{m'n'} i^{n'}(-1)^{m'} \gamma_{-m'n'} \right. \right.$$

$$\left[T^{(11)}_{mnm'n'} \overline{C}_{-m'n'}(\theta_i, \phi_i) \cdot \hat{\beta} + T^{(12)}_{mnm'n'} \frac{\overline{B}_{-m'n'}(\theta_i, \phi_i)}{i} \cdot \hat{\beta} \right] \Big|^2$$

$$+ \left| \sum_{m'n'} i^{n'}(-1)^{m'} \gamma_{-m'n'} \left[T^{(21)}_{mnm'n'} \overline{C}_{-m'n'}(\theta_i, \phi_i) \cdot \hat{\beta} \right. \right.$$

$$\left. \left. + T^{(22)}_{mnm'n'} \frac{\overline{B}_{-m'n'}(\theta_i, \phi_i)}{i} \cdot \hat{\beta} \right] \Big|^2 \right\} \tag{19}$$

with $\beta = v$ or h. Hence, given the T-matrix elements, the scattering function matrix elements $\overline{\overline{F}}$ can be computed according to (17). All the constituents of vector radiative transfer equations are in terms of $\overline{\overline{F}}$ elements, and thus can all be calculated.

5.3 Unitarity and Symmetry

In this section, we shall derive the unitarity and symmetry properties of the T-matrix. The unitarity property is a result of energy conservation for nonabsorptive scatterer and symmetry is a result of reciprocity.

A. Unitarity

The total field is a summation of incident and scattered fields. We use (1) and (2), Section 5.2, and also the fact that regular wave solution is a combination of outgoing waves (Hankel functions of first kind) and incoming waves (Hankel functions of second kind), i.e. $j_n = (h_n^{(1)} + h_n^{(2)})/2$. We use superscript (2) to denote that the spherical Hankel function of the second kind is used in the vector wave function. Defining the scattering matrix $\overline{\overline{S}}$ as

$$\overline{\overline{S}} = \overline{\overline{I}} + 2\overline{\overline{T}} = \overline{\overline{I}} + 2 \begin{bmatrix} \overline{\overline{T}}^{(11)} & \overline{\overline{T}}^{(12)} \\ \overline{\overline{T}}^{(21)} & \overline{\overline{T}}^{(22)} \end{bmatrix} \tag{1}$$

and using the expression for $\overline{E}^i$ and $\overline{E}^s$ in Section 5.2 gives the total field

$$\overline{E}(\overline{r}) = \frac{1}{2} \sum_{ll'} \left\{ (S_{ll'}^{(11)} a_{l'}^{E(M)} + S_{ll'}^{(12)} a_{l'}^{E(N)}) \overline{M}_l(kr, \theta, \phi) \right.$$
$$\left. + (S_{ll'}^{(21)} a_{l'}^{E(M)} + S_{ll'}^{(22)} a_{l'}^{E(N)}) \overline{N}_l(kr, \theta, \phi) \right\}$$
$$+ \frac{1}{2} \sum_l \left\{ a_l^{E(M)} \overline{M}_l^{(2)}(kr, \theta, \phi) + a_l^{E(N)} \overline{N}_l^{(2)}(kr, \theta, \phi) \right\} \tag{2}$$

In the far field, we make asymptotic approximations of the spherical vector wave functions of (15) and (16), Section 5.2. Thus,

$$\lim_{kr \to \infty} \overline{E}(\overline{r}) = \frac{e^{ikr}}{2kr} \overline{W}_1(\theta, \phi) + \frac{e^{-ikr}}{2kr} \overline{W}_2(\theta, \phi) \tag{3}$$

where

$$\overline{W}_1(\theta,\phi) = \sum_{mnm'n'} \left\{ (S^{(11)}_{mnm'n'} a^{(M)}_{m'n'} + S^{(12)}_{mnm'n'} a^{(N)}_{m'n'}) \overline{C}_{mn}(\theta,\phi) \right.$$
$$\left. + (S^{(21)}_{mnm'n'} a^{(M)}_{m'n'} + S^{(22)}_{mnm'n'} a^{(N)}_{m'n'}) i \overline{B}_{mn}(\theta,\phi) \right\} \gamma_{mn} i^{-n-1} \quad (4)$$

$$\overline{W}_2(\theta,\phi) = \sum_{mn} \left\{ a^{(M)}_{mn} \overline{C}_{mn}(\theta,\phi) + a^{(N)}_{mn} \frac{\overline{B}_{mn}(\theta,\phi)}{i} \right\} \gamma_{mn} i^{n+1} \quad (5)$$

and are perpendicular to $\hat{r}$. The complex Poynting vector is then

$$\overline{S} = \overline{E} \times \overline{H}^* = \frac{\hat{r}}{4\eta(k^2 r^2)} \left\{ |\overline{W}_1(\theta,\phi)|^2 - |\overline{W}_2(\theta,\phi)|^2 \right.$$
$$+ e^{-2ikr} [\overline{W}_2(\theta,\phi) \cdot \overline{W}_1^*(\theta,\phi)]$$
$$\left. - e^{2ikr} [\overline{W}_1(\theta,\phi) \cdot \overline{W}_2^*(\theta,\phi)] \right\} \quad (6)$$

The sum of the latter two terms in the curly brackets is purely imaginary. Hence, the time-average Poynting's vector is

$$<\overline{S}> = \frac{1}{2} Re(\overline{E} \times \overline{H}^*) = \frac{\hat{r}}{8\eta(kr)^2} \left\{ |\overline{W}_1(\theta,\phi)|^2 - |\overline{W}_2(\theta,\phi)|^2 \right\} \quad (7)$$

If the scattering object is lossless, i.e., the imaginary part of permittivity ϵ_s is zero, the scatterer is nonabsorptive and integration of (7) over a spherical surface at infinity should give zero. Hence,

$$\int_{4\pi} d\Omega \left\{ |\overline{W}_1(\theta,\phi)|^2 - |\overline{W}_2(\theta,\phi)|^2 \right\} = 0 \quad (8)$$

Since

$$\overline{C}^*_{mn}(\theta,\phi) = (-1)^m \frac{(n+m)!}{(n-m)!} \overline{C}_{-mn}(\theta,\phi) \quad (9)$$

$$\overline{B}^*_{mn}(\theta,\phi) = (-1)^m \frac{(n+m)!}{(n-m)!} \overline{B}_{-mn}(\theta,\phi) \quad (10)$$

Using (5), (9) and (10), and the orthogonality relation of vector spherical harmonics of (12), Section 5.1, we have

$$\int_{4\pi} d\Omega |\overline{W}_2(\theta,\phi)|^2 = \sum_{mn} \left\{ |a^{(M)}_{mn}|^2 + |a^{(N)}_{mn}|^2 \right\} = \bar{a}^+ \bar{a} \quad (11)$$

where $\bar{a}$ is the column matrix of dimension $2L_{max}$ containing the coefficients $a_{mn}^{(M)}$ and $a_{mn}^{(N)}$ and superscript $+$ denotes Hermitian conjugate. Similarly,

$$\int d\Omega |\overline{W}_1(\theta,\phi)|^2 = \sum_{mn} \left\{ \left| \sum_{m'n'} S_{mnm'n'}^{(11)} a_{m'n'}^{(M)} + S_{mnm'n'}^{(12)} a_{m'n'}^{(N)} \right|^2 \right.$$
$$\left. + \left| \sum_{m'n'} S_{mnm'n'}^{(21)} a_{m'n'}^{(M)} + S_{mnm'n'}^{(22)} a_{m'n'}^{(N)} \right|^2 \right\}$$
$$= \bar{a}^+ \overline{\overline{S}}^+ \overline{\overline{S}} \bar{a} \tag{12}$$

where the scattering matrix $\overline{\overline{S}}$ is defined in (1).

Equating (11) and (12) for lossless scatterer and noting that the equality is true for arbitrary incident wave coefficients $\bar{a}$, we have the unitarity condition for lossless particle

$$\overline{\overline{S}}^+ \overline{\overline{S}} = \overline{\overline{I}} \tag{13}$$

which, in terms of $\overline{\overline{T}}$ matrix, becomes

$$\overline{\overline{T}}^+ \overline{\overline{T}} = -\frac{1}{2}\left\{\overline{\overline{T}}^+ + \overline{\overline{T}}\right\} \tag{14}$$

B. Symmetry

From (24), Section 3.2, we have the symmetry relation for scattering amplitude matrix

$$\overline{\overline{F}}(\theta,\phi;\theta',\phi') = \overline{\overline{F}}^t(\pi-\theta',\pi+\phi';\pi-\theta,\pi+\phi) \tag{15}$$

Apply (15) to the relation in (17), Section 5.2, and use the properties

$$\overline{C}_{mn}(\pi-\theta,\pi+\phi) = (-1)^n \overline{C}_{mn}(\theta,\phi) \tag{16}$$

$$\overline{B}_{mn}(\pi-\theta,\pi+\phi) = (-1)^{n+1} \overline{B}_{mn}(\theta,\phi) \tag{17}$$

In the expression for $\overline{\overline{F}}^t(\pi-\theta',\pi+\phi';\pi-\theta,\pi+\phi)$, further interchange m and $-m'$, and n and n'. Then,

$$\overline{\overline{F}}^t(\pi-\theta',\pi+\phi';\pi-\theta,\pi+\phi)$$

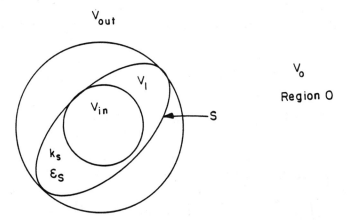

Fig. 3.17 Applying extended boundary condition to calculation of the T-matrix. Scatterer occupying V_1. Region outside circumscribing sphere is V_{out}. Region inside inscribing sphere is V_{in}.

$$= \frac{4\pi}{k} \sum_{mnm'n'} (-1)^m i^{n'-n-1} \gamma_{-m'n'} \gamma_{mn}$$

$$\cdot \left\{ T^{(11)}_{-m'n'(-m)n} \overline{C}_{mn}(\theta,\phi) \overline{C}_{-m'n'}(\theta',\phi') \right.$$

$$+ T^{(12)}_{-m'n'(-m)n} i\overline{B}_{mn}(\theta,\phi) \overline{C}_{-m'n'}(\theta',\phi')$$

$$- T^{(21)}_{-m'n'(-m)n} \overline{C}_{mn}(\theta,\phi) i\overline{B}_{-m'n'}(\theta',\phi')$$

$$\left. + T^{(22)}_{-m'n'(-m)n} \overline{B}_{mn}(\theta,\phi) \overline{B}_{-m'n'}(\theta',\phi') \right\} \quad (18)$$

Equating (18) to $\overline{\overline{F}}(\theta,\phi;\theta',\phi')$, and noting that the equality is true for arbitrary (θ,ϕ) and (θ',ϕ'), gives the following reciprocity relation for the T-matrix elements

$$T^{(ij)}_{-m'n'(-m)n} = (-1)^{m'+m} T^{(ji)}_{mnm'n'} \quad (19)$$

with $i,j = 1$ or 2.

5.4 Extended Boundary Condition Technique

The computation of the $\overline{\overline{T}}$ matrix through the extended boundary condition technique is introduced by Waterman (1965, 1968, 1971). Consider a scatterer occupying region V_1 enclosed by surface S and with wavenumber k_s and permittivity ϵ_s (Figure 3.17). The scatterer is centered at the origin. The region exterior to the scatterer is denoted as region 0. Then the electric fields $\overline{E}$ and $\overline{E}_1$ in regions 0 and 1, respectively, satisfy the following equations

$$\nabla \times \nabla \times \overline{E} - k^2 \overline{E} = 0 \qquad \text{for } \overline{r} \in V_o \qquad (1)$$

and

$$\nabla \times \nabla \times \overline{E}_1 - k_s^2 \overline{E}_1 = 0 \qquad \text{for } \overline{r} \in V_1 \qquad (2)$$

Let $\overline{\overline{G}}_o$ be the free space dyadic Green's function with wavenumber k. Making use of vector Green's theorem

$$\int dV \left\{ \overline{P} \cdot \nabla \times \nabla \times \overline{Q} - \overline{Q} \cdot \nabla \times \nabla \times \overline{P} \right\}$$
$$= \oint d\overline{S} \cdot \left\{ \overline{Q} \times \nabla \times \overline{P} - \overline{P} \times \nabla \times \overline{Q} \right\} \qquad (3)$$

performing the volume integration for V_o, and by inserting

$$\overline{P} = \overline{E} \qquad (4)$$

$$\overline{Q} = \overline{\overline{G}}_o(\overline{r}, \overline{r}') \cdot \overline{a} \qquad (5)$$

with $\overline{a}$ an arbitrary constant vector, we obtain

$$\begin{Bmatrix} \overline{E}(\overline{r}') \cdot \overline{a} & \text{if } \overline{r}' \in V_o \\ 0 & \text{if } \overline{r}' \in V_1 \end{Bmatrix} = \overline{E}^{inc}(\overline{r}') \cdot \overline{a}$$
$$+ \int_S dS\, \hat{n}(\overline{r}) \cdot \left\{ (\nabla \times \overline{E}(\overline{r})) \times (\overline{\overline{G}}_o(\overline{r}, \overline{r}') \cdot \overline{a}) \right.$$
$$\left. + \overline{E}(\overline{r}) \times \nabla \times (\overline{\overline{G}}_o(\overline{r}, \overline{r}') \cdot \overline{a}) \right\} \qquad (6)$$

where the integral is to be carried out over the surface of the particle and $\hat{n}$ is the outward normal to S.

Since $\overline{a}$ is arbitrary, we can cancel out the $\overline{a}$ on both sides of (6). We also recognize that for $\overline{r}' \in V_o$, the second term on the right-hand side of (6) is the scattered field. Therefore, for $\overline{r}' \in V_o$

$$\overline{E}(\overline{r}') = \overline{E}^{inc}(\overline{r}') + \overline{E}^s(\overline{r}') \qquad (7)$$

where

$$\overline{E}^s(\overline{r}') = \int_S dS \left\{ i\omega\mu\, \hat{n} \times \overline{H}(\overline{r}) \cdot \overline{\overline{G}}_o(\overline{r}, \overline{r}') + \hat{n} \times \overline{E}(\overline{r}) \cdot \nabla \times \overline{\overline{G}}_o(\overline{r}, \overline{r}') \right\} \qquad (8)$$

For $\bar{r}' \in V_1$

$$-\overline{E}^{inc}(\bar{r}') = \int_S dS \left\{ i\omega\mu \hat{n} \times \overline{H}(\bar{r}) \cdot \overline{\overline{G}}_o(\bar{r},\bar{r}') \right.$$
$$\left. + \hat{n} \times \overline{E}(\bar{r}) \cdot \nabla \times \overline{\overline{G}}_o(\bar{r},\bar{r}') \right\} \quad (9)$$

Equation (9) is known as the extended boundary condition, since it analytically extends $\bar{r}'$ to the region within the particle. In terms of spherical vector wave functions, the dyadic Green's function has the following expansion for $\bar{r}' \neq \bar{r}$ (Problem 16)

$$\overline{\overline{G}}_o(\bar{r},\bar{r}') = ik \sum_{n,m} (-1)^m \begin{cases} \overline{M}_{-mn}(kr,\theta,\phi) Rg\overline{M}_{mn}(kr',\theta',\phi') \\ + \overline{N}_{-mn}(kr,\theta,\phi) Rg\overline{N}_{mn}(kr',\theta',\phi') \\ \qquad \text{for } r > r' \\ Rg\overline{M}_{-mn}(kr,\theta,\phi) \overline{M}_{mn}(kr',\theta',\phi') \\ + Rg\overline{N}_{-mn}(kr,\theta,\phi) \overline{N}_{mn}(kr',\theta',\phi') \\ \qquad \text{for } r' > r \end{cases}$$
$$(10)$$

Let V_{in} be the region inside the inscribing sphere and V_{out} be the region outside the circumscribing sphere. The inscribing sphere is the largest sphere with its center at the origin contained within the scatterer. The circumscribing sphere is the smallest sphere with its center at the origin containing the scatterer.

The source for the incident field is assumed to be outside the circumscribing sphere. Hence, the incident field can be expanded in regular vector wave functions

$$\overline{E}^{inc}(\bar{r}') = \sum_{n,m} \left\{ a_{mn}^{(M)} Rg\overline{M}_{mn}(kr',\theta',\phi') + a_{mn}^{(N)} Rg\overline{N}_{mn}(kr',\theta',\phi') \right\}$$
$$(11)$$

Substitute (10) and (11) in (9). For $\bar{r}' \in V_{in}$, then $r' < r$ for all points $\bar{r}$ on S. Balancing the coefficients of $Rg\overline{M}_{mn}(kr',\theta',\phi')$ and $Rg\overline{N}_{mn}(kr',\theta',\phi')$ results in the equations for $a_{mn}^{(M)}$ and $a_{mn}^{(N)}$

$$\begin{bmatrix} a_{mn}^{(M)} \\ a_{mn}^{(N)} \end{bmatrix} = -ik(-1)^m \int_S dS\, \hat{n} \times i\omega\mu \overline{H}(\bar{r}) \cdot \begin{bmatrix} \overline{M}_{-mn}(kr,\theta,\phi) \\ \overline{N}_{-mn}(kr,\theta,\phi) \end{bmatrix}$$
$$- ik^2(-1)^m \int_S dS\, \hat{n} \times \overline{E}(\bar{r}) \cdot \begin{bmatrix} \overline{N}_{-mn}(kr,\theta,\phi) \\ \overline{M}_{-mn}(kr,\theta,\phi) \end{bmatrix} \quad (12)$$

For $\bar{r}' \in V_{out}$, i.e., when $\bar{r}'$ is outside the circumscribing sphere, we have $r < r'$ for all points $\bar{r}$ on S. Hence, substituting (10) in (8), results in the following expression of $\overline{E}^s(\bar{r})$ for $\bar{r}' \in V_{out}$

$$\overline{E}^s(\bar{r}') = \sum_{m,n} a_{mn}^{s(M)} \overline{M}_{mn}(kr', \theta', \phi') + a_{mn}^{s(N)} \overline{N}_{mn}(kr', \theta', \phi') \qquad (13)$$

where

$$\begin{bmatrix} a_{mn}^{s(M)} \\ a_{mn}^{s(N)} \end{bmatrix} = ik(-1)^m \int_S dS\, \hat{n} \times i\omega\mu\, \overline{H}(\bar{r}) \cdot \begin{bmatrix} Rg\overline{M}_{-mn}(kr, \theta, \phi) \\ Rg\overline{N}_{-mn}(kr, \theta, \phi) \end{bmatrix}$$

$$+ ik^2(-1)^m \int_S dS\, \hat{n} \times \overline{E}(\bar{r}) \cdot \begin{bmatrix} Rg\overline{N}_{-mn}(kr, \theta, \phi) \\ Rg\overline{M}_{-mn}(kr, \theta, \phi) \end{bmatrix} \qquad (14)$$

To solve the two equations in (12), we expand the surface fields in terms of the vector wave functions with wavenumber k_s. For $\bar{r}$ on S,

$$\begin{bmatrix} \hat{n} \times \overline{E}(\bar{r}) \\ \hat{n} \times \overline{H}(\bar{r}) \end{bmatrix} = \hat{n} \times \sum_{m'n'} \begin{bmatrix} c_{m'n'}^{(M)} \\ (k_s/i\omega\mu) d_{m'n'}^{(M)} \end{bmatrix} Re\overline{M}_{m'n'}(k_s r, \theta, \phi)$$

$$+ \hat{n} \times \sum_{m'n'} \begin{bmatrix} c_{m'n'}^{(N)} \\ (k_s/i\omega\mu) d_{m'n'}^{(N)} \end{bmatrix} Rg\overline{N}_{m'n'}(k_s r, \theta, \phi) \quad (15)$$

The tangential surface electric field has coefficients $c_{m'n'}$ and the tangential surface magnetic field has coefficients $d_{m'n'}$. We next apply the vector Green's theorem (3) to the interior region V_1 with $\overline{Q} = \overline{\overline{G}}_1(\bar{r}, \bar{r}') \cdot \bar{a}$, $\overline{P} = \overline{E}_1(\bar{r})$ where

$$\overline{\overline{G}}_1(\bar{r}, \bar{r}') = \left(\overline{\overline{I}} + \frac{1}{k_s^2} \nabla\nabla \right) \frac{e^{ik_s|\bar{r} - \bar{r}'|}}{4\pi |\bar{r} - \bar{r}'|} \qquad (16)$$

is the dyadic Green's function with the wave number of the scatterer. Use the extended boundary condition with $\bar{r}'$ in the exterior region V_o, and make use of the continuity of tangential surface fields and the expansion of surface fields as given in (15). It is readily shown that (Problem 17)

$$c_{m'n'}^{(M)} = d_{m'n'}^{(N)} \qquad (17)$$

and

$$c_{m'n'}^{(N)} = d_{m'n'}^{(M)} \qquad (18)$$

Extended Boundary Condition Technique

If we substitute (15), (17) and (18) in (12), we have,

$$a^{(M)}_{mn} = \sum_{m'n'} \left[P_{mnm'n'} c^{(M)}_{m'n'} + R_{mnm'n'} c^{(N)}_{m'n'} \right] \qquad (19)$$

$$a^{(N)}_{mn} = \sum_{m'n'} \left[S_{mnm'n'} c^{(M)}_{m'n'} + U_{mnm'n'} c^{(N)}_{m'n'} \right] \qquad (20)$$

where

$$P_{mnm'n'} = -ikk_s J^{(21)}_{mnm'n'} - ik^2 J^{(12)}_{mnm'n'} \qquad (21)$$

$$R_{mnm'n'} = -ikk_s J^{(11)}_{mnm'n'} - ik^2 J^{(22)}_{mnm'n'} \qquad (22)$$

$$S_{mnm'n'} = -ikk_s J^{(22)}_{mnm'n'} - ik^2 J^{(11)}_{mnm'n'} \qquad (23)$$

$$U_{mnm'n'} = -ikk_s J^{(12)}_{mnm'n'} - ik^2 J^{(21)}_{mnm'n'} \qquad (24)$$

$$\begin{bmatrix} J^{(11)}_{mnm'n'} \\ J^{(12)}_{mnm'n'} \\ J^{(21)}_{mnm'n'} \\ J^{(22)}_{mnm'n'} \end{bmatrix} = (-1)^m \int_S dS\, \hat{n}(\bar{r}) \cdot \begin{bmatrix} Rg\overline{M}_{m'n'}(k_s r, \theta, \phi) \times \overline{M}_{-mn}(kr, \theta, \phi) \\ Rg\overline{M}_{m'n'}(k_s r, \theta, \phi) \times \overline{N}_{-mn}(kr, \theta, \phi) \\ Rg\overline{N}_{m'n'}(k_s r, \theta, \phi) \times \overline{M}_{-mn}(kr, \theta, \phi) \\ Rg\overline{N}_{m'n'}(k_s r, \theta, \phi) \times \overline{N}_{-mn}(kr, \theta, \phi) \end{bmatrix} \qquad (25)$$

Equations (19) and (20) can be put in the following compact form by using the numbering system of Section 5.2. Let

$$\overline{\overline{Q}}^t = \begin{bmatrix} \overline{\overline{P}} & \overline{\overline{R}} \\ \overline{\overline{S}} & \overline{\overline{U}} \end{bmatrix} \qquad (26)$$

where $\overline{\overline{P}}, \overline{\overline{R}}, \overline{\overline{S}}, \overline{\overline{U}}$ are $L_{max} \times L_{max}$ matrices representing the coefficients $P_{mnm'n'}$, $R_{mnm'n'}$, $S_{mnm'n'}$, and $U_{mnm'n'}$, respectively. Also, let

$$\bar{a}^E = \begin{bmatrix} \bar{a}^{(M)} \\ \bar{a}^{(N)} \end{bmatrix} \qquad (27)$$

$$\bar{c} = \begin{bmatrix} \bar{c}^{(M)} \\ \bar{c}^{(N)} \end{bmatrix} \qquad (28)$$

then (19) and (20) become

$$\bar{a}^E = \overline{\overline{Q}}^t \bar{c} \qquad (29)$$

Solution of (29) gives

$$\bar{c} = (\overline{\overline{Q}}^t)^{-1} \bar{a}^E \qquad (30)$$

If the surface field expansion of (15) is substituted into the scattered field representation of (14), we obtain

$$\bar{a}^s = -Rg\overline{\overline{Q}}^t \bar{c} \tag{31}$$

where

$$\bar{a}^s = \begin{bmatrix} \bar{a}^{s(M)} \\ \bar{a}^{s(N)} \end{bmatrix} \tag{32}$$

$$Rg\overline{\overline{Q}}^t = \begin{bmatrix} Rg\overline{\overline{P}} & Rg\overline{\overline{R}} \\ Rg\overline{\overline{S}} & Rg\overline{\overline{U}} \end{bmatrix} \tag{33}$$

and $Rg\overline{\overline{P}}$, $Rg\overline{\overline{R}}$, $Rg\overline{\overline{S}}$ and $Rg\overline{\overline{U}}$ are the corresponding expressions in (21) through (24) with J elements replaced by RgJ elements. The RgJ elements are the corresponding expressions of (25) with vector wave function $\overline{M}_{-mn}$ and $\overline{N}_{-mn}$ replaced by $Rg\overline{M}_{-mn}$ and $Rg\overline{N}_{-mn}$, respectively. The definition of the $\overline{\overline{T}}$ matrix is

$$\bar{a}^s = \overline{\overline{T}} \bar{a}^E \tag{34}$$

Hence using (30), (31) and (34), we have the following relation between the $\overline{\overline{T}}$ matrix and $\overline{\overline{Q}}$ matrix elements

$$\overline{\overline{T}} = -Rg\overline{\overline{Q}}^t (\overline{\overline{Q}}^t)^{-1} \tag{35}$$

The surface integrals in (25) are usually calculated in spherical coordinates. For a surface defined by the equation

$$r = r(\theta, \phi) \tag{36}$$

we have

$$dS\,\hat{n}(\bar{r}) = r^2 \sin\theta\,\bar{\sigma}(\bar{r})\,d\theta\,d\phi \tag{37}$$

where

$$\bar{\sigma}(\bar{r}) = \hat{r} - \frac{r_\theta}{r}\hat{\theta} - \frac{r_\phi}{r\sin\theta}\hat{\phi} \tag{38}$$

In (38) r_θ and r_ϕ are the r function in (36) differentiated with respect to θ and ϕ respectively.

T-Matrix for Axisymmetric Objects

For axisymmetric objects, the equation for the surface (36) is independent of ϕ, so that $r_\phi = 0$. Substituting (37) and (38) in (25) the $d\phi$ integration gives zero unless $m' = m$. We obtain (Problem 18)

$$J^{(11)}_{mnm'n'} = 2\pi\delta_{mm'}(-1)^m \int_0^\pi d\theta \sin\theta\, r^2 \gamma_{mn'}\gamma_{-mn}(\bar{\sigma}\cdot\hat{r})$$

$$\times j_{n'}(k_s r) h_n(kr) \frac{(-im)}{\sin\theta}\left[P_{n'}^m(\cos\theta)\frac{dP_n^{-m}(\cos\theta)}{d\theta}\right.$$

$$\left. + \frac{dP_{n'}^m(\cos\theta)}{d\theta}P_n^{-m}(\cos\theta)\right] \quad (39a)$$

$$J^{(12)}_{mnm'n'} = 2\pi\delta_{m'm}(-1)^m \int_0^\pi d\theta \sin\theta\, r^2 \gamma_{mn'}\gamma_{-mn}\bar{\sigma}(\bar{r})$$

$$\cdot\left\{ \hat{r} j_{n'}(k_s r)\frac{(krh_n(kr))'}{kr}\left[\frac{m^2}{\sin^2\theta}P_{n'}^m(\cos\theta)P_n^{-m}(\cos\theta)\right.\right.$$

$$\left. + \frac{dP_{n'}^m(\cos\theta)}{d\theta}\frac{dP_n^{-m}(\cos\theta)}{d\theta}\right]$$

$$\left. -\hat{\theta} j_{n'}(k_s r)\frac{h_n(kr)}{kr}n(n+1)P_n^{-m}(\cos\theta)\frac{dP_{n'}^m(\cos\theta)}{d\theta}\right\} \quad (39b)$$

$$J^{(21)}_{mnm'n'} = 2\pi\delta_{mm'}(-1)^m \int_0^\pi d\theta \sin\theta\, r^2 \gamma_{mn'}\gamma_{-mn}\bar{\sigma}(\bar{r})$$

$$\cdot\left\{ -\hat{r} h_n(kr)\frac{(k_s r j_{n'}(k_s r))'}{k_s r}\left[\frac{dP_{n'}^m(\cos\theta)}{d\theta}\frac{dP_n^{-m}(\cos\theta)}{d\theta}\right.\right.$$

$$\left. + \frac{m^2}{\sin^2\theta}P_{n'}^m(\cos\theta)P_n^{-m}(\cos\theta)\right]$$

$$\left. +\hat{\theta}\frac{j_{n'}(k_s r)}{k_s r}h_n(kr)n'(n'+1)P_{n'}^m(\cos\theta)\frac{dP_n^{-m}(\cos\theta)}{d\theta}\right\} \quad (39c)$$

$$J^{(22)}_{mnm'n'} = 2\pi\delta_{mm'}(-1)^m \int_0^\pi d\theta \sin\theta\, r^2 \gamma_{mn'}\gamma_{-mn}\bar{\sigma}(\bar{r})$$

$$\cdot\left\{\hat{r}\frac{(k_s r j_{n'}(k_s r))'}{k_s r}\frac{(krh_n(kr))'}{kr}\frac{(-im)}{\sin\theta}\right.$$

$$\times\left[\frac{dP_{n'}^m(\cos\theta)}{d\theta}P_n^{-m}(\cos\theta) + P_{n'}^m(\cos\theta)\frac{dP_n^{-m}(\cos\theta)}{d\theta}\right]$$

$$+ \hat{\theta}\frac{im}{\sin\theta}P_{n'}^{m}(\cos\theta)P_{n}^{-m}(\cos\theta)$$

$$\times \left[n(n+1)\frac{(k_s r j_{n'}(k_s r))'}{k_s r}\frac{h_n(kr)}{kr}\right.$$

$$\left.+ n'(n'+1)\frac{j_{n'}(k_s r)}{k_s r}\frac{(krh_n(kr))'}{kr}\right]\Bigg\} \quad (39d)$$

where

$$\bar{\sigma}(\bar{r}) = \hat{r} - \frac{r_\theta}{r}\hat{\theta} \quad (40)$$

The expressions for $Rg\,J_{mnm'n'}^{(ij)}$, with $i,j = 1,2$, will be the corresponding expressions of (39) with $h_n(kr)$ replaced by $j_n(kr)$.

Since the expressions are proportional to $\delta_{mm'}$, there is no coupling between different m indices. Hence, instead of directly inverting the $\bar{\bar{Q}}$ matrix which is of dimension $2N_{max}(N_{max}+2)$, it is computationally more efficient to invert a block diagonal matrix formed by regrouping the elements according to the m index. The $\bar{\bar{Q}}$ matrix can be cast into block diagonal form with each block $\bar{\bar{Q}}^{(m)}$ labeled by the m index. The dimension of $\bar{\bar{Q}}^{(m)}$ is $2(N_{max} - |m| + 1)$ for $m \neq 0$ and $2N_{max}$ for $m = 0$. Since $m = 0, \pm 1, \pm 2, \ldots \pm N_{max}$, there are $2N_{max} + 1$ such blocks. The integrals in (39) are conveniently evaluated by means of Gauss–Legendre quadrature. The order of quadrature chosen must contain enough sample points to resolve the variation of r as a function of θ for objects of large aspect ratio, the variation of associated Legendre polynomial with respect to θ, and also the variation of Bessel function j_n and h_n with respect to θ, which can be rapid at high frequencies. The quantity N_{max} is chosen such that the T-matrix elements converge.

5.5 Spheres

For the case of spheres with radius a, we have $r = a$ and $\bar{\sigma} = \hat{r}$. Substituting these into (40), Section 5.4, the integrals can be evaluated readily. We have

$$J_{mnm'n'}^{(11)} = 0 \quad (1)$$

for all m, n, m', and n'. Similarly,

$$J_{mnm'n'}^{(22)} = 0 \quad (2)$$

$$J_{mnm'n'}^{(12)} = a^2 \delta_{mm'}\delta_{nn'}j_n(k_s a)\frac{[ka\,h_n(ka)]'}{ka} \quad (3)$$

$$J^{(21)}_{mnm'n'} = -a^2 \delta_{mm'} \delta_{nn'} h_n(ka) \frac{[k_s a\, j_n(k_s a)]'}{k_s a} \tag{4}$$

Substituting into the expressions for $\overline{\overline{P}}, \overline{\overline{R}}, \overline{\overline{S}}, \overline{\overline{U}}$, we obtain diagonal matrices for $\overline{\overline{P}}$ and $\overline{\overline{U}}$ while $\overline{\overline{R}}$ and $\overline{\overline{S}} = 0$. The corresponding expression for $Rg\,J$ elements are those of (1) through (4) with $h_n(ka)$ replaced by $j_n(ka)$. Hence, $Rg\,R_{mnm'n'} = Rg\,S_{mnm'n'} = 0$ and $Rg\,\overline{\overline{P}}$ and $Rg\,\overline{\overline{U}}$ are diagonal matrices. The $\overline{\overline{T}}$ matrix is also diagonal

$$\overline{\overline{T}} = \begin{bmatrix} \overline{\overline{T}}^{(11)} & 0 \\ 0 & \overline{\overline{T}}^{(22)} \end{bmatrix} \tag{5}$$

and

$$T^{(11)}_{mnm'n'} = \delta_{mm'} \delta_{nn'} T^{(M)}_n \tag{6}$$

$$T^{(22)}_{mnm'n'} = \delta_{mm'} \delta_{nn'} T^{(N)}_n \tag{7}$$

where

$$T^{(M)}_n = -\frac{j_n(k_s a)[ka\, j_n(ka)]' - j_n(ka)[k_s a\, j_n(k_s a)]'}{j_n(k_s a)[ka\, h_n(ka)]' - h_n(ka)[k_s a\, j_n(k_s a)]'} \tag{8}$$

$$T^{(N)}_n = -\frac{[k_s^2 a^2 j_n(k_s a)][ka\, j_n(ka)]' - [k^2 a^2 j_n(ka)][k_s a\, j_n(k_s a)]'}{[k_s^2 a^2 j_n(k_s a)][ka\, h_n(ka)]' - [k^2 a^2 h_n(ka)][k_s a\, j_n(k_s a)]'} \tag{9}$$

Extinction and Scattering Cross Section for Spheres

To calculate the extinction cross section of spheres, we use (5) through (9) in (17) and (19), Section 5.2. Thus, the extinction cross section for an incident wave with polarization $\beta = v$ or h, and direction (θ_i, ϕ_i)

$$\sigma_{e\beta}(\theta_i, \phi_i) = \frac{4\pi}{k^2} Im \Bigg\{ \sum_{mn} \frac{(-1)^m}{i} \frac{(2n+1)}{n(n+1)}$$
$$\times \Big[\hat{\beta} \cdot \overline{C}_{mn}(\theta_i, \phi_i) T^{(M)}_n \overline{C}_{-mn}(\theta_i, \phi_i) \cdot \hat{\beta}$$
$$+ \hat{\beta} \cdot \overline{B}_{mn}(\theta_i, \phi_i) T^{(N)}_n \overline{B}_{-mn}(\theta_i, \phi_i) \cdot \hat{\beta} \Big] \Bigg\} \tag{10}$$

We next use the results of the addition theorem of (5) and (6), Section 5.9, for vector spherical harmonics to get

$$\sum_m (-1)^m \hat{\beta} \cdot \overline{C}_{mn}(\theta, \phi) \overline{C}_{-mn}(\theta, \phi) \cdot \hat{\beta}$$
$$= \sum_m (-1)^m \hat{\beta} \cdot \overline{B}_{mn}(\theta, \phi) \overline{B}_{-mn}(\theta, \phi) \cdot \hat{\beta} = P'_n(1) = \frac{n(n+1)}{2} \tag{11}$$

Hence,
$$\sigma_{e\beta}(\theta_i, \phi_i) = -\frac{2\pi}{k^2} Re \sum_{n=1}^{\infty}(2n+1)[T_n^{(M)} + T_n^{(N)}] \quad (12)$$

and is independent of (θ_i, ϕ_i) and β. The scattering cross sections can be calculated by using (19), Section 5.2, and (5) through (9)

$$\sigma_{s\beta}(\theta_i, \phi_i) = \frac{4\pi}{k^2} \sum_{n,m} \frac{(n+m)!}{(n-m)!} \frac{2n+1}{n(n+1)} \Big\{ |T_n^{(M)}|^2 |\overline{C}_{-mn}(\theta_i, \phi_i) \cdot \hat{\beta}|^2$$
$$+ |T_n^{(N)}|^2 |\overline{B}_{-mn}(\theta_i, \phi_i) \cdot \hat{\beta}|^2 \Big\} \quad (13)$$

Using the conjugate relation in (9) and (10) in Section 5.3 of $\overline{C}_{mn}^*$ in terms of $\overline{C}_{-mn}$ and $\overline{B}_{mn}^*$ in terms of $\overline{B}_{-mn}$ and (11)

$$\sum_m \frac{(n+m)!}{(n-m)!} |\overline{C}_{-mn}(\theta_i, \phi_i) \cdot \hat{\beta}|^2 = \sum_m \frac{(n+m)!}{(n-m)!} |\overline{B}_{-mn}(\theta_i, \phi_i) \cdot \hat{\beta}|^2$$
$$= \frac{n(n+1)}{2} \quad (14)$$

Using (14) in (13) gives the result

$$\sigma_{s\beta}(\theta_i, \phi_i) = \frac{2\pi}{k^2} \sum_{n=1}^{\infty}(2n+1)\Big\{|T_n^{(M)}|^2 + |T_n^{(N)}|^2\Big\} \quad (15)$$

Small Spheres

For small dielectric spheres, $ka \ll 1$ and $k_s a \ll 1$, then $|h_n| \gg |j_n|$ in the expressions (8) and (9) for $T_n^{(M)}$ and $T_n^{(N)}$. The largest T-matrix element is the electric dipole term $T_1^{(N)}$, which is the term that needs to be retained in the low-frequency limit. However, as discussed in earlier sections, in order that the optical theorem be satisfied, it is important to keep the leading term of the imaginary part and the leading term of the real part of $T_1^{(N)}$. Using (9), it can be easily shown that for $ka \ll 1$ and $k_s a \ll 1$

$$T_1^{(N)} = T_{1r}^{(N)} + iT_{1i}^{(N)} \quad (16)$$

where $T_{1r}^{(N)}$ and $T_{1i}^{(N)}$ are both complex for lossy scatterers, and

$$T_{1i}^{(N)} = \frac{2}{3}(ka)^3 y \quad (17)$$

$$y = \frac{\epsilon_s - \epsilon}{\epsilon_s + 2\epsilon} \tag{18}$$

$$T_{1r} = -(T_{1i}^{(N)})^2 \tag{19}$$

Note that since $ka \ll 1$, we have $|T_{1r}^{(N)}| \ll |T_{1i}^{(N)}|$. The extinction cross section is, by using (12)

$$\begin{aligned}\sigma_e &= -\frac{6\pi}{k^2}\left[\operatorname{Re} T_{1r}^{(N)} - \operatorname{Im} T_{1i}^{(N)}\right] \\ &= \frac{4\pi}{k^2}(ka)^3\left[\operatorname{Im} y + \frac{2}{3}(ka)^3 \operatorname{Re} y^2\right]\end{aligned} \tag{20}$$

The scattering cross section is, by (15),

$$\sigma_s = \frac{6\pi}{k^2}|T_{1i}^{(N)}|^2 = \frac{8\pi}{3}(ka)^6 |y|^2 \tag{21}$$

The optical theorem is satisfied in (20). The extinction cross section includes both absorption and scattering. It is to be noted that the optical theorem is satisfied with (20) and (21) because the $T_{1r}^{(N)}$ term in (16) has been included in spite of the fact that it is much smaller than $T_{1i}^{(N)}$. Hence, in order to calculate extinction rates and cross sections, it is important to retain the leading term of the real part and the leading term of the imaginary part of the T-matrix elements.

5.6 Spheroids

For the case of spheroids, the surface is governed by the equation

$$\frac{x^2 + y^2}{a^2} + \frac{z^2}{c^2} = 1 \tag{1}$$

For $a > c$, the spheroid is oblate and for $a < c$, the spheroid is prolate. In spherical coordinates, (1) assumes the form

$$r = r(\theta) = \left[\frac{\sin^2 \theta}{a^2} + \frac{\cos^2 \theta}{c^2}\right]^{-1/2} \tag{2}$$

and from (40), Section 5.4,

$$\bar{\sigma} = \hat{r} + \hat{\theta} r^2 \sin\theta \cos\theta \left(\frac{1}{a^2} - \frac{1}{c^2}\right) \tag{3}$$

The T-matrix elements can be evaluated numerically in a straightforward manner. Certain symmetry relations also exist for the T-matrix elements of spheroids (Waterman, 1971). The integration over $\int_0^\pi d\theta \sin\theta$ can be broken into a sum of

$$\int_0^{\pi/2} d\theta \sin\theta + \int_0^{\pi/2} d\theta' \sin\theta'$$

with $\theta' = \pi - \theta$. Since $r(\pi - \theta) = r(\theta)$ for the case of spheroids, and making use of the property

$$P_n^m(\cos(\pi - \theta)) = (-1)^{n+m} P_n^m(\cos\theta) \tag{4}$$

it follows that

$$J_{mnm'n'}^{(11)} = J_{mnm'n'}^{(22)} = 0 \tag{5}$$

if $n + n' =$ even, and

$$J_{mnm'n'}^{(12)} = J_{mnm'n'}^{(21)} = 0 \tag{6}$$

if $n + n' =$ odd. Hence,

$$P_{mnm'n'} = U_{mnm'n'} = 0 \tag{7}$$

for $n + n' =$ odd, and

$$R_{mnm'n'} = S_{mnm'n'} = 0 \tag{8}$$

for $n + n' =$ even. By making use of the relation between the negative degree associated Legendre polynomial and that of the positive degree in (5), Section 5.1, it follows that

$$J_{(-m)n(-m')n'}^{(ij)} = \begin{cases} -J_{mnm'n'}^{(ij)} & \text{for } ij = 11 \text{ or } 22 \\ J_{mnm'n'}^{(ij)} & \text{for } ij = 12 \text{ or } 21 \end{cases} \tag{9}$$

Small Spheroids

For small spheroids, we shall only keep the dipole term $n = n' = 1$. The leading term of the real part and the leading term of the imaginary part of $J_{mnm'n'}^{(ij)}$ terms will be kept. The results are basically the same as that of Section 4.3, based on separation of variable approach for the Laplace equation, except for an additional term that ensures the optical theorem be satisfied. The only nonzero T-matrix elements are $T^{(22)}_{m1m'1}$ (Problem 31) with

$$T^{(22)}_{m1m'1} = \delta_{mm'} T_m \tag{10}$$

where

$$T_o = it_o - t_o^2 \tag{11}$$

$$T_1 = it_1 - t_1^2 \tag{12}$$

$$t_o = \frac{2}{9} \frac{k^3 a^2 c \left(\epsilon_s/\epsilon - 1\right)}{(1 + v_d A_c)} \tag{13}$$

$$t_1 = \frac{2}{9} \frac{k^3 a^2 c \left(\epsilon_s/\epsilon - 1\right)}{(1 + v_d A_a)} \tag{14}$$

and A_a and A_c are given in (8) through (10), Section 4.3, and v_d given in (3), Section 4.3.

Scattering of waves by single objects of sizes comparable or smaller than wavelength has been studied for many years. Books on classical techniques are Van de Hulst (1957), Bowman et al. (1969), Kerker (1969), Ruck et al. (1970), and Felsen and Marcuvitz (1973). Recent numerical techniques include the extended boundary condition method (Waterman, 1965, 1971; Barber and Yeh, 1975; Iskander, 1983), the integral equation method (Wu and Tsai, 1977; Glisson and Wilton, 1980; Holt, 1982) and the unimoment method (Mei, 1974; Chang and Mei, 1976; Morgan and Mei, 1979; Morgan, 1980; Hunka and Mei, 1981). Scattering by thin disks which are appropriate models for leaves has been considered in Weil and Chu (1976).

5.7 Rotation of T-Matrix

Very often the axes of symmetry of the scatterer do not coincide with the principal frame of reference. For example, in remote sensing of geophysical media, the z axis of the principal frame is chosen to coincide with the vertical axis. The axes of the scattering objects usually follow a prescribed orientation distribution. The T-matrix can first be calculated with respect to the natural frame of the particle and then related to the T-matrix in the principal frame. In this section the relation between the T-matrices of the two frames will be derived.

Let $\hat{x}_b, \hat{y}_b, \hat{z}_b$ be the natural axes of the scatterer. An Eulerian rotation of α, β, γ in succession will rotate the $(\hat{x}_b, \hat{y}_b, \hat{z}_b)$ axes to the $(\hat{x}, \hat{y}, \hat{z})$ axes of the principal frame. The relations between the $(\hat{x}_b, \hat{y}_b, \hat{z}_b)$ axes to the $(\hat{x}, \hat{y}, \hat{z})$ axes are as given in Section 4.2. In spherical coordinates, a point has coordinates (r, θ_b, ϕ_b) with respect to the natural frame and coordinates

(r, θ, ϕ) with respect to the principal frame. Rotation is interpreted as a rotation of the frame of references about the origin, the field points being supposedly fixed.

Consider an incident field $\overline{E}^{inc}(\overline{r})$ and a scattered field $\overline{E}^s(\overline{r})$ outside the circumscribing sphere. In the principal frame

$$\overline{E}^{inc}(\overline{r}) = \sum_{m,n} \left[a_{mn}^{(M)} Rg\overline{M}_{mn}(kr, \theta, \phi) + a_{mn}^{(N)} Rg\overline{N}_{mn}(kr, \theta, \phi) \right] \quad (1)$$

$$\overline{E}^s(\overline{r}) = \sum_{m,n} \left[a_{mn}^{s(M)} \overline{M}_{mn}(kr, \theta, \phi) + a_{mn}^{s(N)} \overline{N}_{mn}(kr, \theta, \phi) \right] \quad (2)$$

The relation between the incident field coefficients a_{mn} and the scattered field coefficient is through the T-matrix elements of the principal frame.

$$\overline{a}^s = \overline{\overline{T}}\,\overline{a} \quad (3)$$

In terms of the natural frame coordinates,

$$\overline{E}^{inc}(\overline{r}) = \sum_{m,n} \left[\hat{a}_{mn}^{(M)} Rg\overline{M}_{mn}(kr, \theta_b, \phi_b) + \hat{a}_{mn}^{(N)} Rg\overline{N}_{mn}(kr, \theta_b, \phi_b) \right] \quad (4)$$

$$\overline{E}^s(\overline{r}) = \sum_{m,n} \left[\hat{a}_{mn}^{s(M)} \overline{M}_{mn}(kr, \theta_b, \phi_b) + \hat{a}_{mn}^{s(N)} \overline{N}_{mn}(kr, \theta_b, \phi_b) \right] \quad (5)$$

where $\hat{a}_{mn}$ and $\hat{a}_{mn}^s$ are the coefficients for the incident field and the scattered field in the natural frame. The two sets of coefficients are related by the transition matrix $\hat{\overline{\overline{T}}}$ of the natural frame,

$$\hat{\overline{a}}^s = \hat{\overline{\overline{T}}}\,\hat{\overline{a}} \quad (6)$$

The calculation of the transition matrix is first performed in the natural frame in which we can utilize the symmetry properties of the scatterer. It is desirable to derive a relation that expresses the principal frame $\overline{\overline{T}}$ in terms of the natural frame $\hat{\overline{\overline{T}}}$. To do this requires a relation between the vector spherical wave functions for the two frames. This relation can be best expressed in terms of the representation of rotation group. From Edmonds (1957), we have (Problem 24)

$$\gamma_{mn} Y_n^m(\theta_b, \phi_b) = \sum_{m'} \gamma_{m'n} Y_n^{m'}(\theta, \phi) D_{m'm}^{(n)}(\alpha\beta\gamma) \quad (7)$$

where $D^{(n)}_{m'm}(\alpha\beta\gamma)$ is the representation of the rotation group with spherical harmonics as basis, and is

$$D^{(n)}_{m'm}(\alpha\beta\gamma) = e^{im'\gamma} d^{(n)}_{m'm}(\beta) e^{im\alpha} \tag{8}$$

with

$$d^{(j)}_{m'm}(\beta) = \left[\frac{(j+m')!(j-m')!}{(j+m)!(j-m)!}\right]^{1/2} \left(\cos\frac{\beta}{2}\right)^{m'+m}$$

$$\times \left(\sin\frac{\beta}{2}\right)^{m'-m} P^{(m'-m,m'+m)}_{j-m'}(\cos\beta) \tag{9}$$

where $P^{(a,b)}_n(x)$ is the Jacobi polynomial of order n and degree (a,b) (Abramowitz and Stegun, 1965). An alternative way of writing (9) is

$$d^{(j)}_{m'm}(\beta) = \sum_{(\sigma)} \frac{[(j+m)!(j-m)!(j+m')!(j-m')!]^{1/2}}{\sigma!(m+m'+\sigma)!(j-m-\sigma)!(j-m'-\sigma)!}$$

$$\times (-1)^{j-m'-\sigma} \left(\sin\frac{\beta}{2}\right)^{2j-2\sigma-m'-m} \left(\cos\frac{\beta}{2}\right)^{m'+m+2\sigma} \tag{10}$$

where the summation over σ is such that all the factorials in the summation are nonnegative. Thus, the allowable σ's are such that they must satisfy the following four conditions: (1) $\sigma \geq 0$; (2) $\sigma \geq -(m+m')$; (3) $\sigma \leq j-m$; and (4) $\sigma \leq j-m'$. Thus, for given j, m and m', there is only a finite number of terms in the summation (10), so that a computer code for $d^{(j)}_{m'm}(\beta)$ can be prepared readily.

The following are useful symmetry relations for the $d^{(j)}_{m'm}$ functions

$$d^{(j)}_{m'm}(-\beta) = d^{(j)}_{mm'}(\beta) = (-1)^{m'-m} d^{(j)}_{m'm}(\beta) \tag{11}$$

$D^{(j)}_{m'm}(\alpha\beta\gamma)$ is the representation of the rotation group element $D(\alpha,\beta,\gamma)$ the inverse of which is $D^{-1}(\alpha\beta\gamma)$. Hence

$$D^{-1}(\alpha,\beta,\gamma) = D(-\gamma,-\beta,-\alpha). \tag{12}$$

Since $D(\alpha\beta\gamma)$ is a unitary operator,

$$D^+(\alpha\beta\gamma) = D^{-1}(\alpha\beta\gamma). \tag{13}$$

Hence

$$D^{+(j)}_{m'm}(\alpha\beta\gamma) = D^{-1(j)}_{m'm}(\alpha\beta\gamma) = D^{*(j)}_{mm'}(\alpha\beta\gamma) = D^{(j)}_{m'm}(-\gamma,-\beta,-\alpha)$$

$$= e^{-i\alpha m'} e^{-i\gamma m} d^{(j)}_{m'm}(-\beta) = e^{-i\alpha m'} e^{-i\gamma m} d^{(j)}_{mm'}(\beta) \tag{14}$$

Two useful orthogonality relations are (Problem 23)

$$\sum_{m''} D^{-1(j)}_{m'm''}(\alpha\beta\gamma) D^{(j)}_{m''m}(\alpha\beta\gamma)$$

$$= \delta_{m'm} = e^{i(m-m')\alpha} \sum_{m''} d^{(j)}_{m''m'}(\beta) d^{(j)}_{m''m}(\beta) \qquad (15)$$

$$\frac{1}{8\pi^2} \int_0^{2\pi} d\alpha \int_0^{\pi} d\beta \sin\beta \int_0^{2\pi} d\gamma D^{(j_2)}_{m'_2 m_2}(\alpha\beta\gamma) D^{-1(j_1)}_{m_1 m'_1}(\alpha\beta\gamma)$$

$$= \frac{1}{2j_1+1} \delta_{m_1 m_2} \delta_{m'_1 m'_2} \delta_{j_1 j_2} \qquad (16)$$

Since $\nabla_b = \nabla$, $\bar{r}_b = \bar{r}$, and $r_b = r$, we can multiply by $\bar{r}$ and take the curl of both sides of (7) to give

$$\gamma_{mn} \overline{C}_{mn}(\theta_b, \phi_b) = \sum_{m'} \gamma_{m'n} \overline{C}_{m'n}(\theta, \phi) D^{(n)}_{m'm}(\alpha\beta\gamma) \qquad (17)$$

A similar relation exists for $\overline{B}_{mn}$. The inverse relations are also derived easily by using (15). Multiplying both sides of (17) by $h_n(kr)$ gives the transformation of vector spherical wave functions.

$$\overline{M}_{mn}(kr, \theta_b, \phi_b) = \sum_{m'} \overline{M}_{m'n}(kr, \theta, \phi) D^{(n)}_{m'm}(\alpha\beta\gamma) \qquad (18)$$

The corresponding inverse relation is

$$\overline{M}_{m''n}(kr, \theta, \phi) = \sum_m \overline{M}_{mn}(kr, \theta_b, \phi_b) D^{-1(n)}_{mm''}(\alpha\beta\gamma) \qquad (19)$$

Taking the curl of both sides of (18) and (19) gives similar relations for $\overline{N}_{mn}$ functions. The same relations hold for regular vector wave functions.

Substituting in (5) the relation between vector wave functions in the two frames and comparing the resulting equation with (2) give the relations between a^s_{mn} and $\hat{a}^s_{mn}$. The equations can be cast into a more compact form by defining a $\overline{\overline{D}}$ matrix of dimension $L_{max} \times L_{max}$ as

$$(\overline{\overline{D}})_{mnm_1 n_1} = \delta_{nn_1} D^{(n)}_{mm_1} \qquad (20)$$

Then (Problem 22),

$$(\overline{\overline{D}}^{-1})_{mnm_1 n_1} = \delta_{nn_1} D^{-1(n)}_{mm_1} \qquad (21)$$

Hence the relations between a_{mn}^s and $\hat{a}_{mn}^s$ are $\bar{a}^{s(M)} = \bar{\bar{D}}\,\hat{\bar{a}}^{s(M)}$, $\bar{a}^{s(N)} = \bar{\bar{D}}\,\hat{\bar{a}}^{s(N)}$ and the relations between $\hat{a}_{mn}$ and a_{mn} are $\bar{a}^{(M)} = \bar{\bar{D}}^{-1}\bar{a}^{(M)}$ and $\hat{\bar{a}}^{(N)} = \bar{\bar{D}}^{-1}\bar{a}^{(N)}$. Therefore, on substituting into (1) through (6), we have the transformation relation between $\bar{\bar{T}}$ and $\hat{\bar{\bar{T}}}$ as follows:

$$T^{(ij)}_{mnm'n'} = \sum_{m_2,m_3} D^{(n)}_{mm_2}(\alpha\beta\gamma)\,\hat{T}^{(ij)}_{m_2 n m_3 n'}\,D^{-1(n')}_{m_3 m'}(\alpha\beta\gamma) \qquad (22)$$

with $i,j = 1,2$. Thus if the $2L_{max} \times 2L_{max}$ matrices $\bar{\bar{D}}$ and $\bar{\bar{D}}^{-1}$ are defined by

$$\bar{\bar{D}} = \begin{bmatrix} \bar{\bar{D}} & 0 \\ 0 & \bar{\bar{D}} \end{bmatrix} \qquad (23)$$

$$\bar{\bar{D}}^{-1} = \begin{bmatrix} \bar{\bar{D}}^{-1} & 0 \\ 0 & \bar{\bar{D}}^{-1} \end{bmatrix} \qquad (24)$$

then we have the following relation

$$\bar{\bar{T}} = \bar{\bar{D}}\,\hat{\bar{\bar{T}}}\,\bar{\bar{D}}^{-1} \qquad (25)$$

5.8 Extinction Coefficients and Phase Matrix for Axisymmetric Objects with Prescribed Orientation Distribution

To calculate the extinction coefficients and phase matrix for randomly oriented axisymmetric objects, we make use of the relation between the scattering dyad $\bar{\bar{F}}$ and the $\bar{\bar{T}}$ matrix and the rotation matrix of Section 5.7. For an axisymmetric object, the T-matrix in the natural frame assumes the following form

$$\hat{T}_{mnm'n'} = \delta_{mm'}\hat{T}_{mnmn'} \qquad (1)$$

which has no coupling between azimuthal indices m. Thus, for an axisymmetric object with axes oriented at Eulerian angles $(\alpha\beta\gamma)$ with respect to the principal frame, the scattering function dyad is, from (17), Section 5.2,

$$\bar{\bar{F}}(\theta,\phi;\theta',\phi';\alpha\beta\gamma)$$
$$= \frac{4\pi}{k} \sum_{\substack{nmn'm' \\ m_2}} (-1)^{m'} i^{n'-n-1} D^{(n)}_{mm_2}(\alpha\beta\gamma)$$

$$\times \left\{ \left[\hat{T}^{(11)}_{m_2 n m_2 n'} \gamma_{mn} \overline{C}_{mn}(\theta,\phi) + \hat{T}^{(21)}_{m_2 n m_2 n'} i\gamma_{mn} \overline{B}_{mn}(\theta,\phi) \right] \right.$$

$$\gamma_{-m'n'} \overline{C}_{-m'n'}(\theta',\phi') D^{-1(n')}_{m_2 m'}(\alpha\beta\gamma)$$

$$- \left[\hat{T}^{(12)}_{m_2 n m_2 n'} \gamma_{mn} \overline{C}_{mn}(\theta,\phi) + \hat{T}^{(22)}_{m_2 n m_2 n'} i\gamma_{mn} \overline{B}_{mn}(\theta,\phi) \right]$$

$$\left. \gamma_{-m'n'} i\overline{B}_{-m'n'}(\theta',\phi') D^{-1(n')}_{m_2 m'}(\alpha\beta\gamma) \right\} \quad (2)$$

In view of the rotational symmetry of the object, we note from (2) that the scattering amplitude dyad $\overline{\overline{F}}$ is independent of the Eulerian angle α. The extinction coefficients and the phase matrix are calculated by an average over the orientation probability density function of the objects. For example,

$$<\overline{\overline{F}}(\theta,\phi,\theta',\phi')> = \int_0^{2\pi} d\gamma \int_0^{\pi} d\beta \sin\beta\, p(\beta,\gamma) \overline{\overline{F}}(\theta,\phi;\theta',\phi';\alpha\beta\gamma) \quad (3)$$

where $p(\beta,\gamma)$ is orientation probability density function satisfying the condition

$$\int_0^{2\pi} d\gamma \int_0^{\pi} d\beta \sin\beta\, p(\beta,\gamma) = 1 \quad (4)$$

The phase matrix is

$$\overline{\overline{P}}(\theta,\phi;\theta',\phi') = n_o \int_0^{2\pi} d\gamma \int_0^{\pi} d\beta \sin\beta\, p(\beta,\gamma) \overline{\overline{L}}(\theta,\phi;\theta',\phi';\alpha\beta\gamma) \quad (5)$$

where $\overline{\overline{L}}(\theta,\phi;\theta',\phi';\alpha\beta\gamma)$ is the Stokes matrix for one particle with orientation $(\alpha\beta\gamma)$ and is calculated from (2) and (15) (Section 3.2).

Special Case of Uniform γ Distribution

For the special case of uniform γ distribution, the probability density function is $p(\beta,\gamma) = p(\beta)/2\pi$ so that $\int_0^{\pi} d\beta \sin\beta\, p(\beta) = 1$. Then from (2)

$$<\overline{\overline{F}}(\theta,\phi;\theta,\phi;\alpha\beta\gamma)>$$

$$= \frac{4\pi}{k} \sum_{nmn'm_2} \int_0^{\pi} d\beta \sin\beta\, p(\beta) (-1)^m i^{n'-n-1} d^{(n)}_{mm_2}(\beta)$$

$$\left\{ \left[\hat{T}^{(11)}_{m_2 n m_2 n'} \gamma_{mn} \overline{C}_{mn}(\theta,\phi) + \hat{T}^{(21)}_{m_2 n m_2 n'} i\gamma_{mn} \overline{B}_{mn}(\theta,\phi) \right] \right.$$

$$\gamma_{-mn'} \overline{C}_{-mn'}(\theta,\phi) d^{(n')}_{mm_2}(\beta)$$

$$-\left[\hat{T}^{(12)}_{m_2nm_2n'}\gamma_{mn}\overline{C}_{mn}(\theta,\phi)+\hat{T}^{(22)}_{m_2nm_2n'}i\gamma_{mn}\overline{B}_{mn}(\theta,\phi)\right]$$

$$\gamma_{-mn'}i\overline{B}_{-mn'}(\theta,\phi)d^{(n')}_{mm_2}(\beta)\Big\} \qquad (6)$$

Since $\overline{C}_{mn}(\theta,\phi)$ and $\overline{B}_{mn}(\theta,\phi)$ are proportional to $\exp(im\phi)$ while $\overline{C}_{-mn'}(\theta,\phi)$ and $\overline{B}_{-mn'}(\theta,\phi)$ are proportional to $\exp(-im\phi)$, the ϕ dependence in (6) disappears. Therefore, the extinction matrix is a function of polarization β and angle θ and is not a function of ϕ. The medium is statistically azimuthal symmetric.

Special Case of Uniform γ and β Distribution

For the case of uniform γ and β distribution, $p(\beta) = 1/2$. From (16), Section 5.7,

$$\int_0^\pi d\beta \sin\beta\, d^{(n)}_{mm_2}(\beta) d^{(n')}_{mm_2}(\beta) = \frac{2}{2n+1}\delta_{nn'} \qquad (7)$$

The addition theorem in Section 5.9 (Appendix) gives

$$\sum_m (-1)^m \hat{\beta}\cdot\overline{C}_{mn}(\theta,\phi)\overline{B}_{-mn}(\theta,\phi)\cdot\hat{\beta}=0 \qquad (8)$$

Using (6) through (8) and (11) (Section 5.5), the summation over n' and m in (6) can be carried out and

$$<f_{\beta\beta}(\theta,\phi;\theta,\phi;\alpha\beta\gamma)>=\frac{1}{2ik}\sum_{nm_2}(\hat{T}^{(11)}_{m_2nm_2n}+\hat{T}^{(22)}_{m_2nm_2n}) \qquad (9)$$

and is independent of θ, ϕ and polarization β.

5.9 Appendix

The addition theorem for spherical harmonics is (Jackson, 1975)

$$P_\nu(\hat{r}\cdot\hat{r}')=\sum_\mu(-1)^\mu Y^\mu_\nu(\theta,\phi)Y^{-\mu}_\nu(\theta',\phi') \qquad (1)$$

where

$$\hat{r}=\sin\theta\cos\phi\hat{x}+\sin\theta\sin\phi\hat{y}+\cos\theta\hat{z} \qquad (2)$$

$$\hat{r}'=\sin\theta'\cos\phi'\hat{x}+\sin\theta'\sin\phi'\hat{y}+\cos\theta'\hat{z} \qquad (3)$$

so that

$$\hat{r}\cdot\hat{r}'=\cos\theta\cos\theta'+\sin\theta\sin\theta'\cos(\phi-\phi') \qquad (4)$$

Using (1), the addition theorem for vector spherical harmonics can be derived (Problem 27). It is

$$\hat{\phi}\frac{1}{\sin\theta}\frac{\partial^2 P_\nu}{\partial\phi\partial\theta'} + \hat{\theta}\frac{\partial^2 P_\nu}{\partial\theta\partial\theta'} = \sum_\mu (-1)^\mu \overline{B}_{\mu\nu}(\theta,\phi)[\hat{\theta}'\cdot\overline{B}_{-\mu\nu}(\theta',\phi')]$$

$$= -\sum_\mu (-1)^\mu \overline{B}_{\mu\nu}(\theta,\phi)[\hat{\phi}'\cdot\overline{C}_{-\mu\nu}(\theta',\phi')] \quad (5)$$

$$\hat{\phi}\frac{1}{\sin\theta\sin\theta'}\frac{\partial^2 P_\nu}{\partial\phi\partial\phi'} + \hat{\theta}\frac{1}{\sin\theta'}\frac{\partial^2 P_\nu}{\partial\theta\partial\phi'}$$

$$= \sum_\mu (-1)^\mu \overline{B}_{\mu\nu}(\theta,\phi)[\hat{\phi}'\cdot\overline{B}_{-\mu\nu}(\theta',\phi')]$$

$$= \sum_\mu (-1)^\mu \overline{B}_{\mu\nu}(\theta,\phi)[\hat{\theta}'\cdot\overline{C}_{-\mu\nu}(\theta',\phi')] \quad (6)$$

where the argument of the Legendre polynomials P_ν is $(\hat{r}\cdot\hat{r}')$. Similar relations can be obtained for $\sum_\mu (-1)^\mu \overline{C}_{\mu\nu}(\theta,\phi)\overline{C}_{-\mu\nu}(\theta,\phi)$ by taking the cross product of (5) and (6) with $\hat{r}$.

6 BOUNDARY CONDITIONS FOR RADIATIVE TRANSFER EQUATIONS

At a boundary separating two dielectric media, there are boundary conditions relating the Stokes parameters of the two media. We first consider the boundary conditions for a planar dielectric interface and then for a rough dielectric interface.

6.1 Planar Dielectric Interface

Consider a planar interface separating dielectric media 1 and 2 and a plane wave incident from medium 1 onto medium 2 along direction $\hat{k}_i$ (Figure 3.18). The incident wave generates a reflected wave in direction $\hat{k}_r$ and a transmitted wave in direction $\hat{k}_t$. We have three orthogonal systems $(\hat{v}_i, \hat{h}_i, \hat{k}_i)$, $(\hat{v}_r, \hat{h}_r, \hat{k}_r)$ and $(\hat{v}_t, \hat{h}_t, \hat{k}_t)$, with $\hat{h} = \hat{z}\times\hat{k}/|\hat{z}\times\hat{k}|$, and $\hat{v} = \hat{h}\times\hat{k}$. For the incident wave

$$\overline{E}_i = (\hat{v}_i E_{vi} + \hat{h}_i E_{hi})\, e^{ik_x x - ik_{1z} z} \quad (1a)$$

$$\overline{H}_i = \frac{1}{\eta_1}(\hat{v}_i E_{vi} - \hat{h}_i E_{hi})\, e^{ik_x x - ik_{1z} z} \quad (1b)$$

Planar Dielectric Interface

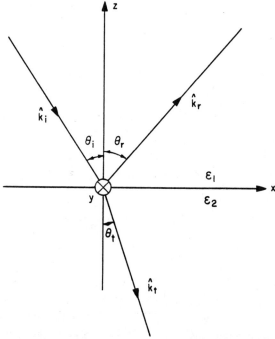

Fig. 3.18 Geometric configuration for planar dielectric interface: derivation of boundary conditions.

where $k_x = k_1 \sin \theta_i$, $k_{1z} = k_1 \cos \theta_i$. The reflected and transmitted waves are (Problem 21)

$$\overline{E}_r = (\hat{v}_r S_{12} E_{vi} + \hat{h}_r R_{12} E_{hi}) e^{ik_x x + ik_{1z} z} \tag{2a}$$

$$\overline{H}_r = \frac{1}{\eta_1}(\hat{h}_r S_{12} E_{vi} - \hat{v}_r R_{12} E_{hi}) e^{ik_x x + ik_{1z} z} \tag{2b}$$

$$\overline{E}_t = \left(\hat{v}_t \frac{\eta_2}{\eta_1} Y_{12} E_{vi} + \hat{h}_t X_{12} E_{hi}\right) e^{ik_x x - ik_{2z} z} \tag{3a}$$

$$\overline{H}_t = \left(\hat{h}_t \frac{1}{\eta_1} Y_{12} E_{vi} - \hat{v}_t \frac{1}{\eta_2} X_{12} E_{hi}\right) e^{ik_x x - ik_{2z} z} \tag{3b}$$

where R_{12} and S_{12} are reflection coefficients and X_{12} and Y_{12} are transmission coefficients for horizontally and vertically polarized waves, respectively.

$$R_{12} = \frac{k_{1z} - k_{2z}}{k_{1z} + k_{2z}} = X_{12} - 1 \tag{4a}$$

$$S_{12} = \frac{\epsilon_2 k_{1z} - \epsilon_1 k_{2z}}{\epsilon_2 k_{1z} + \epsilon_1 k_{2z}} = Y_{12} - 1 \tag{4b}$$

We also define reflectivity and transmissivity $r_{\beta 12}$ and $t_{\beta 12}$

$$r_{h12} = |R_{12}|^2 = 1 - t_{h12} \tag{5a}$$

$$r_{v12} = |S_{12}|^2 = 1 - t_{v12} \tag{5b}$$

From (1) we know that the plane wave Poynting vector for the incident wave is $S_{pi} = S_{vi} + S_{hi}$, where S_{vi} and S_{hi} signify the vertical and horizontal polarization components, respectively, with $S_{vi} = |E_{vi}|^2/\eta_1$, $S_{hi} = |E_{hi}|^2/\eta_1$. Similarly, for the reflected wave, $S_{vr} = r_{v12}S_{vi}$ and $S_{hr} = r_{h12}S_{hi}$. For the transmitted wave, from (3)

$$S_{vt} = \left|\frac{\eta_2}{\eta_1}Y_{12}\right|^2 \frac{\eta_1}{\eta_2} S_{vi} \tag{6a}$$

$$S_{ht} = |X_{12}|^2 \frac{\eta_1}{\eta_2} S_{hi} \tag{6b}$$

for θ_i less than the critical angle. Specific intensity is defined as power per unit area per unit solid angle and per unit frequency. Let $d\Omega$ denote differential solid angle. Using Snell's law, it follows that $d\Omega_i = d\Omega_r$, $\phi_t = \phi_i$ and

$$\sqrt{\epsilon_2}\sin\theta_t = \sqrt{\epsilon_1}\sin\theta_i \tag{7}$$

Differentiating (7) and multiplying the result with (7) give

$$\epsilon_2 \cos\theta_t \, d\Omega_t = \epsilon_1 \cos\theta_i \, d\Omega_i \tag{8}$$

Using (8), (6) takes the following form for the specific intensity relation between incident and transmitted waves

$$I_{\beta t} = \frac{\epsilon_2}{\epsilon_1} t_{\beta 12} I_{\beta i} \tag{9}$$

with $\beta = v, h$. Equation (9) relates the first two components of the Stokes parameters for the incident and transmitted intensities at a plane boundary. We note that there is a divergence of beam factor ϵ_2/ϵ_1 for the transmitted intensity.

The relation between the incident and reflected waves for specific intensities are $I_{vr} = r_{v12}I_{vi}$, and $I_{hr} = r_{h12}I_{hi}$. The final relations between the incident and reflected Stokes parameters and between the incident and transmitted Stokes parameters assume the following matrix form:

$$\bar{I}_r = \bar{\bar{T}}_{11}(\theta_i) \cdot \bar{I}_i \tag{10}$$

with

$$\overline{\overline{T}}_{11}(\theta_i) = \begin{bmatrix} r_{v12}(\theta_i) & 0 & 0 & 0 \\ 0 & r_{h12}(\theta_i) & 0 & 0 \\ 0 & 0 & Re(S_{12}R_{12}^*) & -Im(S_{12}R_{12}^*) \\ 0 & 0 & Im(S_{12}R_{12}^*) & Re(S_{12}R_{12}^*) \end{bmatrix} \quad (11)$$

and

$$\overline{I}_t = \overline{\overline{T}}_{12}(\theta_i) \cdot \overline{I}_i \quad (12)$$

with

$$\overline{\overline{T}}_{12}(\theta_i) = \frac{\epsilon_2}{\epsilon_1} \begin{bmatrix} t_{v12}(\theta_i) & 0 & 0 & 0 \\ 0 & t_{h12}(\theta_i) & 0 & 0 \\ 0 & 0 & \frac{\cos\theta_t}{\cos\theta_i} Re(Y_{12}X_{12}^*) & -\frac{\cos\theta_t}{\cos\theta_i} Im(Y_{12}X_{12}^*) \\ 0 & 0 & \frac{\cos\theta_t}{\cos\theta_i} Im(Y_{12}X_{12}^*) & \frac{\cos\theta_t}{\cos\theta_i} Re(Y_{12}X_{12}^*) \end{bmatrix} \quad (13)$$

Equation (13) applies when θ_i is less than the critical angle. For θ_i larger than the critical angle, $\overline{\overline{T}}_{12}(\theta_i) = 0$.

6.2 Rough Dielectric Interface

The boundary conditions satisfied by the specific intensities at rough dielectric interfaces are derived in this section. The scattered and the transmitted fields derived by a combination of Kirchhoff approximation and geometrical optics approach are used. Unlike the planar interface case where the coupling at the boundary is to only the specular reflection and transmission directions, the incident intensity is coupled to all of the reflected and transmitted directions. The shadowing effect is incorporated by modifying the coupling matrices. We note that since only the single scattering solution is used the reflected and transmitted intensities are always underestimated. Thus, if these boundary conditions are used with the active radiative transfer equations, then the scattering intensities obtained will be the lower limit since the values will always increase if higher order scattering effects at the rough surface are included. Similarly, using these boundary conditions with the

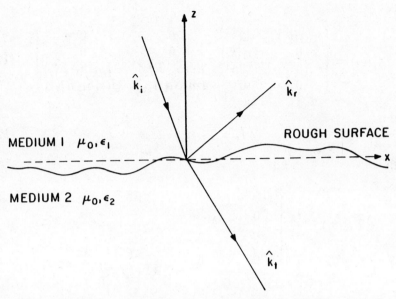

Fig. 3.19 Geometric configuration for rough dielectric interface: derivation of boundary conditions.

passive radiative transfer equations to solve for the thermal emission will give underestimated results. However, the upper and lower limits for the emissivities can be obtained by solving for the emissivities with the active and passive radiative transfer equations. With the active equations, we solve for the scattered intensity and calculate the emissivity by one minus the sum of all the reflected power. This gives the upper limit of the correct emissivity. If the higher-order scattering effects at the rough interface are included, the net reflected power will be higher and the emissivity will always be lower. With the passive radiative transfer equation, we can solve for the thermal emission for a uniform temperature case. Then the emissivity can be obtained by normalizing the calculated brightness temperatures by the physical temperature. This represents the lower limit of the correct emissivity. If the higher-order scattering effects at the rough interface are included, more thermal emission will always be transmitted and the emissivity always increases (Shin and Kong, 1982).

Consider a plane wave incident from medium 1 onto medium 2 along direction $\hat{k}_i$ upon a rough dielectric interface (Figure 3.19). The electric field of the incident wave is given by

$$\overline{E}_i = \hat{e}_i \, E_{ei} \, e^{i \overline{k}_i \cdot \overline{r}} \tag{1}$$

where $\overline{k}_i$ denotes the incident wave vector and $\hat{e}_i$ the polarization of the electric field vector. The rough surface is characterized by a random height distribution $z = f(\bar{r}_\perp)$ where $f(\bar{r}_\perp)$ is a Gaussian random variable with zero mean, $<f(\bar{r}_\perp)> = 0$. The incident field will generate the reflected and transmitted fields in media 1 and 2, respectively. The Kirchhoff-approximated diffraction integrals for the scattered fields are calculated as in Section 6.1, Chapter 2.

The stationary phase points for the reflected fields are given by

$$\alpha_o = -\frac{k_{1dx}}{k_{1dz}} \quad (2a)$$

$$\beta_o = -\frac{k_{1dy}}{k_{1dz}} \quad (2b)$$

and for the transmitted fields

$$\alpha'_o = -\frac{k_{2dx}}{k_{2dz}} \quad (3a)$$

$$\beta'_o = -\frac{k_{2dy}}{k_{2dz}} \quad (3b)$$

where

$$\bar{k}_{1d} = \bar{k}_i - \bar{k}_{1r} = \hat{x}k_{1dx} + \hat{y}k_{1dy} + \hat{z}k_{1dz} \quad (4a)$$

$$\bar{k}_{2d} = \bar{k}_i - \bar{k}_{2t} = \hat{x}k_{2dx} + \hat{y}k_{2dy} + \hat{z}k_{2dz} \quad (4b)$$

Thus, the boundary conditions for the specific intensities at a rough interface is given by (Problem 32)

$$\bar{I}_1(\hat{k}_s) = \int_0^{2\pi} d\phi_i \int_0^{\pi/2} d\theta_i \sin\theta_i \, \overline{\overline{R}}_{12}(\theta_s, \phi_s; \theta_i, \phi_i) \cdot \bar{I}_1(\hat{k}_i) \quad (5a)$$

$$\bar{I}_2(\hat{k}_t) = \int_0^{2\pi} d\phi_i \int_0^{\pi/2} d\theta_i \sin\theta_i \, \overline{\overline{T}}_{12}(\theta_t, \phi_t; \theta_i, \phi_i) \cdot \bar{I}_1(\hat{k}_i) \quad (5b)$$

where $\bar{I}_1$ and $\bar{I}_2$ are the column matrices for the specific intensities containing the four Stokes parameters

$$\bar{I}_\alpha = \begin{bmatrix} I_{v\alpha} \\ I_{h\alpha} \\ U_\alpha \\ V_\alpha \end{bmatrix} \quad \alpha = 1, 2 \quad (6)$$

The reflected and transmitted intensities at the directions $\hat{k}_s$ and $\hat{k}_t$ are given by integration of all the scattered intensities which are coupled to that direction from the incident intensities.

The explicit expressions for the coupling matrices $\overline{\overline{R}}_{12}$ and $\overline{\overline{T}}_{12}$ at the rough surface boundary are given by

$$\overline{\overline{R}}_{12}(\theta_s, \phi_s; \theta_i, \phi_i) = \frac{1}{\cos\theta_s} \frac{|\overline{k}_{1d}|^4}{4|\hat{k}_i \times \hat{k}_s|^4 k_{1dz}^4} \frac{1}{2\pi s^2}$$
$$\times \exp\left[-\frac{k_{1dx}^2 + k_{1dy}^2}{2k_{1dz}^2 s^2}\right] \overline{\overline{C}}_{12}^r(\theta_s, \phi_s; \theta_i, \phi_i) \quad (7a)$$

$$\overline{\overline{T}}_{12}(\theta_t, \phi_t; \theta_i, \phi_i) = \frac{1}{\cos\theta_t} \frac{k_2^2 |\overline{k}_{2d}|^2 (\hat{n} \cdot \hat{k}_t)^2}{|\hat{k}_i \times \hat{k}_t|^4 k_{2dz}^4} \frac{\eta_1}{\eta_2} \frac{1}{2\pi s^2}$$
$$\times \exp\left[-\frac{k_{2dx}^2 + k_{2dy}^2}{2k_{2dz}^2 s^2}\right] \overline{\overline{C}}_{12}^t(\theta_t, \phi_t; \theta_i, \phi_i) \quad (7b)$$

where s^2 is the mean square surface slope,

$$\overline{\overline{C}}_{12}^\alpha = \begin{bmatrix} <|f_{vv}^\alpha|^2> & <|f_{vh}^\alpha|^2> \\ <|f_{hv}^\alpha|^2> & <|f_{hh}^\alpha|^2> \\ 2Re<f_{vv}^\alpha f_{hv}^{\alpha*}> & 2Re<f_{vh}^\alpha f_{hh}^{\alpha*}> \\ 2Im<f_{vv}^\alpha f_{hv}^{\alpha*}> & 2Im<f_{vh}^\alpha f_{hh}^{\alpha*}> \\ Re<(f_{vh}^{\alpha*} f_{vv}^\alpha)> & -Im<(f_{vh}^{\alpha*} f_{vv}^\alpha)> \\ Re<(f_{hh}^{\alpha*} f_{hv}^\alpha)> & -Im<(f_{hv}^\alpha f_{hh}^{\alpha*})> \\ Re<(f_{vv}^\alpha f_{hh}^{\alpha*} + f_{vh}^\alpha f_{hv}^{\alpha*})> & -Im<(f_{vv}^\alpha f_{hh}^{\alpha*} - f_{vh}^\alpha f_{hv}^{\alpha*})> \\ Im<(f_{vv}^\alpha f_{hh}^{\alpha*} + f_{vh}^\alpha f_{hv}^{\alpha*})> & Re<(f_{vv}^\alpha f_{hh}^{\alpha*} - f_{vh}^\alpha f_{hv}^{\alpha*})> \end{bmatrix} \quad (8)$$

with $\alpha = r, t$,

$$f_{vv}^r = (\hat{h}_s \cdot \hat{k}_i)(\hat{h}_i \cdot \hat{k}_s) R_h + (\hat{v}_s \cdot \hat{k}_i)(\hat{v}_i \cdot \hat{k}_s) R_v \quad (9a)$$

$$f_{hv}^r = (\hat{v}_s \cdot \hat{k}_i)(\hat{h}_i \cdot \hat{k}_s) R_h - (\hat{h}_s \cdot \hat{k}_i)(\hat{v}_i \cdot \hat{k}_s) R_v \quad (9b)$$

$$f_{vh}^r = (\hat{h}_s \cdot \hat{k}_i)(\hat{v}_i \cdot \hat{k}_s) R_h - (\hat{v}_s \cdot \hat{k}_i)(\hat{h}_i \cdot \hat{k}_s) R_v \quad (9c)$$

$$f_{hh}^r = (\hat{v}_s \cdot \hat{k}_i)(\hat{v}_i \cdot \hat{k}_s) R_h + (\hat{h}_s \cdot \hat{k}_i)(\hat{h}_i \cdot \hat{k}_s) R_v \quad (9d)$$

and

$$f^t_{vv} = (\hat{h}_t \cdot \hat{k}_i)(\hat{h}_i \cdot \hat{k}_t)(1 + R'_h)$$
$$+ (\hat{v}_t \cdot \hat{k}_i)(\hat{v}_i \cdot \hat{k}_t)\frac{\eta_1}{\eta}(1 + R'_v) \qquad (10a)$$

$$f^t_{hv} = -(\hat{v}_t \cdot \hat{k}_i)(\hat{h}_i \cdot \hat{k}_t)(1 + R'_h)$$
$$+ (\hat{h}_t \cdot \hat{k}_i)(\hat{v}_i \cdot \hat{k}_t)\frac{\eta_1}{\eta}(1 + R'_v) \qquad (10b)$$

$$f^t_{vh} = (\hat{h}_t \cdot \hat{k}_i)(\hat{v}_i \cdot \hat{k}_t)(1 + R'_h)$$
$$- (\hat{v}_t \cdot \hat{k}_i)(\hat{h}_i \cdot \hat{k}_t)\frac{\eta_1}{\eta}(1 + R'_v) \qquad (10c)$$

$$f^t_{hh} = (\hat{v}_t \cdot \hat{k}_i)(\hat{v}_i \cdot \hat{k}_t)(1 + R'_h)$$
$$+ (\hat{h}_t \cdot \hat{k}_i)(\hat{h}_i \cdot \hat{k}_t)\frac{\eta_1}{\eta}(1 + R'_v) \qquad (10d)$$

R_v and R_h and R'_v and R'_h are the local reflection coefficients for the vertical and horizontal polarizations evaluated at the stationary phase points (α_o, β_o) and (α'_o, β'_o), respectively.

As mentioned in Chapter 2, the geometrical optics result used to derive the boundary conditions for a rough dielectric interface satisfies the principle of reciprocity but violates the principle of energy conservation. This is due to the neglect of the effects of multiple scattering and shadowing. These shadowing effects can be incorporated to modify the boundary conditions.

Following the same procedure as in (113) through (127), Section 6.1, Chapter 2, we obtain

$$\overline{\overline{R}}^m_{12}(\hat{k}_s; \hat{k}_i) = S(\hat{k}_s, \hat{k}_i)\overline{\overline{R}}_{12}(\hat{k}_s; \hat{k}_i) \qquad (11a)$$

$$\overline{\overline{T}}^m_{12}(\hat{k}_t; \hat{k}_i) = S(\hat{k}_t, \hat{k}_i)\overline{\overline{T}}_{12}(\hat{k}_t; \hat{k}_i) \qquad (11b)$$

where $S(\hat{k}_\alpha, \hat{k}_\beta)$ is the probability that a point will be illuminated by rays having the direction $\hat{k}_\beta$ and $-\hat{k}_\alpha$, given the value of the slope at the point. This has been discussed in Section 6.1, Chapter 2.

PROBLEMS

3.1 Calculate $I, Q, U, V, I_v, I_h, \chi,$ and ψ for the following elliptically polarized waves. Give their locations on the Poincare sphere.

1. $\overline{E} = (\hat{v} + \hat{h}) e^{i30 \text{ deg}}$
2. $\overline{E} = (\hat{v} + i\hat{h})$
3. $\overline{E} = (\hat{v} - i\hat{h})$
4. $\overline{E} = (\hat{v} + 3 e^{i30 \text{ deg}} \hat{h})$

3.2 By using (9) through (13), Section 2.2, show that (14) and (15), Section 2.2, are true. Note that (n, m) indicates that summation is to be extended over all distinct pairs of intervals n and m. That is, if we include $(n = 1, m = 2)$ then we have to exclude $(n = 2, m = 1)$.

3.3 Using the relation between the scattered and incident fields of (9), Section 3.2,

$$\overline{E}_s = \frac{e^{ikr}}{r} \overline{\overline{F}}(\theta_s, \phi_s; \theta_i, \phi_i) \cdot \hat{e}_i E_o \qquad (1)$$

and the definition of Stokes parameters, show that the Stokes matrix $\overline{\overline{L}}(\theta_s, \phi_s; \theta_i, \phi_i)$ is given by (15), Section 3.2.

3.4 Consider specific intensity in an infinite medium in the presence of the thermal emission source term. Assume the special case that $\overline{\kappa}_a(\hat{s}_b) = \overline{\kappa}_a(\hat{s})$. Show that the solution is $\overline{I} = (CT, CT, 0, 0)$.

3.5 Show that the scattering function dyad $\overline{\overline{F}}$ for small ellipsoids is given by (2), Section 4.3, by first calculating the induced dipole moment $\overline{p}$ from the Laplace equation (Stratton, 1941).

3.6 Consider the following definition of Eulerian angles of rotation (Karam and Fung, 1983). The natural frame and the principal frame are denoted respectively by $(\hat{x}_b, \hat{y}_b, \hat{z}_b)$ and $(\hat{x}, \hat{y}, \hat{z})$. The Eulerian angles are successive rotations $\alpha_1, \beta_1, \gamma_1$ about $\hat{z}, \hat{y},$ and $\hat{x}$ axes, respectively. Find the relations between this set of angles $(\alpha_1, \beta_1, \gamma_1)$ and the Eulerian angles in the text (α, β, γ).

3.7 Show that the expression of the radiation field of a dipole $\overline{p}$ located at the origin is as given by (3), Section 4.1,

$$\overline{E} = -\frac{\omega^2 \mu e^{ikr}}{4\pi r} \hat{k}_s \times (\hat{k}_s \times \overline{p}) \qquad (2)$$

Problems

where $\hat{k}_s$ is the observation direction.

3.8 Consider the case of Rayleigh phase functions. The scattering coefficient κ_s is found from

$$\kappa_{sv} = \int d\Omega'[P_{11}(\Omega,\Omega') + P_{21}(\Omega,\Omega')] \qquad (3)$$

$$\kappa_{sh} = \int d\Omega'[P_{12}(\Omega,\Omega') + P_{22}(\Omega,\Omega')] \qquad (4)$$

By making use of the two formulas above, show that

$$\kappa_s = 2fk^4a^3|y|^2 \qquad (5)$$

3.9 In this problem, we calculate the phase matrices, scattering coefficients, and absorption coefficients for small spheroids. Use the scattering amplitude matrix of small ellipsoids and let $a = b$. For such a case, the Stokes matrix is independent of angle α. Further assume that the probability density function $p(\alpha, \beta, \gamma)$ is uniform in γ. That is, $p(\alpha, \beta, \gamma) = p(\beta)$. Let $f = n_o v_o$ be the fractional volume occupied by the spheroids, and

$$a_r = \frac{(\epsilon_s/\epsilon) - 1}{1 + v_d A_c} \qquad (6)$$

$$a_\theta = \frac{(\epsilon_s/\epsilon) - 1}{1 + v_d A_a} \qquad (7)$$

$$b_r = kf\,Im(a_r) \qquad (8)$$

$$b_\theta = kf\,Im(a_\theta) \qquad (9)$$

Derive expressions for the phase matrix elements P_{11}, P_{12}, P_{21}, and P_{22}, the scattering coefficients $\kappa_{sv}(\theta')$ and $\kappa_{sh}(\theta')$, and the absorption coefficients $\kappa_{av}(\theta')$ and $\kappa_{ab}(\theta')$. Expressions for P_{11}, κ_{sv} and κ_{av} are (Tsang et al., 1981)

$$P_{11}(\theta, \phi; \theta', \phi') = \frac{k^4}{16\pi^2} v_o f \int_0^\pi d\beta\,\frac{\sin\beta}{2} p(\beta)$$

$$\times \left\{ \left[\sin\theta\sin\theta' + \cos\theta\cos\theta'\cos(\phi - \phi')\right]^2 |a_\theta|^2 \right.$$

$$\left. + \left[\frac{1}{8}\cos^2\theta\cos^2\theta'\sin^4\beta[1 + 2\cos^2(\phi - \phi')]\right.\right.$$

$$+ \frac{1}{2} \sin^2 \theta \cos^2 \theta' \sin^2 \beta \cos^2 \beta$$

$$+ 2 \sin \theta' \cos \theta' \sin \theta \cos \theta \sin^2 \beta \cos^2 \beta \cos(\phi - \phi')$$

$$+ \frac{1}{2} \sin^2 \theta' \cos^2 \theta \sin^2 \beta \cos^2 \beta$$

$$+ \sin^2 \theta \sin^2 \theta' \cos^4 \beta \Big] |a_r - a_\theta|^2$$

$$+ 2[\sin \theta \sin \theta' + \cos \theta \cos \theta' \cos(\phi - \phi')]$$

$$\cdot \Big[\frac{1}{2} \cos \theta \cos \theta' \sin^2 \beta \cos(\phi - \phi')$$

$$+ \sin \theta \sin \theta' \cos^2 \beta \Big] Re[a_\theta^*(a_r - a_\theta)] \Big\} \quad (10)$$

$$\kappa_{sv}(\theta', \phi') = \frac{k^4}{6\pi} v_o f \int_o^\pi d\beta \frac{\sin \beta}{2} p(\beta)$$

$$\times \left\{ |a_\theta|^2 + \left[\frac{1}{2} \cos^2 \theta' \sin^2 \beta + \sin^2 \theta' \cos^2 \beta \right] |a_r - a_\theta|^2 \right.$$

$$\left. + 2 \left[\sin^2 \theta' \cos^2 \beta + \frac{1}{2} \cos^2 \theta' \sin^2 \beta \right] Re[a_\theta^*(a_r - a_\theta)] \right\} (11)$$

$$\kappa_{av}(\theta', \phi') = \int_0^\pi d\beta \frac{\sin \beta}{2} p(\beta)$$

$$\times \left\{ b_\theta + \left[\frac{\cos^2 \theta' \sin^2 \beta}{2} + \sin^2 \theta' \cos^2 \beta \right] (b_r - b_\theta) \right\} \quad (12)$$

Normalization of probability is $\int_0^\pi d\beta \sin \beta \, p(\beta)/2 = 1$.

3.10 Consider a circularly polarized wave

$$\overline{E} = (\hat{v} + i\hat{h})$$

1. Find the Stokes parameters.

2. Find the Stokes parameters for a frame rotated 45 deg with respect to $(\hat{v}, \hat{h})$.

3.11 Show that the transformation matrix $\overline{\overline{A}}$ for Eulerian angles is unitary,

Problems

as given in (6), Section 4.2

$$\overline{\overline{A}}^t = \overline{\overline{A}}^{-1} \tag{13}$$

3.12 From the definition of integrals A_a, A_b and A_c in (4) through (7), Section 4.3,

1. Show that

$$A_a + A_b + A_c = \frac{2}{abc} \tag{14}$$

2. Show that for oblate spheroids A_c is given by (9), Section 4.3, and for prolate spheroids A_c is given by (10), Section 4.3.

3.13 Derive the integral representation of vector spherical waves in (22) through (24), Section 5.1.

3.14 By using the recurrence relations for associated Legendre polynomials (Stratton, 1941; Abramowitz and Stegun, 1965), derive the recurrence relations and properties of the $t_n^m(\cos\theta)$ and $s_n^m(\cos\theta)$ functions in (29) through (40), Section 5.1.

3.15 Derive the expression (19), Section 5.2, of the scattering cross section of a single nonspherical particle in terms of T-matrix elements.

3.16 Derive the vector spherical wave expansion of the dyadic Green's function in (10), Section 5.4 (Tai, 1971).

3.17 In the text, the expansions of surface fields $\hat{n} \times \overline{E}$ and $\hat{n} \times \overline{H}$ are made [(15), Section 5.4] for $\bar{r} \in S$. By applying vector Green's theorem with volume integration in the interior region of the scatterer and the extended boundary condition, show that (17) and (18), Section 5.4, are true

$$c_{mn}^{(M)} = d_{mn}^{(N)} \tag{15}$$

$$c_{mn}^{(N)} = d_{mn}^{(M)} \tag{16}$$

In applying vector Green's theorem, use Green's function with the wavenumber of the scatterer as given in (16), Section 5.4.

3.18 Derive the expression for $J_{mnm'n'}^{(ij)}$, $i, j = 1, 2$ for axisymmetric objects as given by (39), Section 5.4.

3.19 Show that for spheres, the T-matrix elements for the low-frequency limit $ka \ll 1$ are such that

$$T_n^{(M)} = T_{nr}^{(M)} + iT_{ni}^{(M)} \tag{17}$$

$$T_n^{(N)} = T_{nr}^{(N)} + iT_{ni}^{(N)} \tag{18}$$

where $T_{nr}^{(M)}, T_{ni}^{(M)}, T_{nr}^{(N)}, T_{ni}^{(N)}$ are all complex for lossy scatterers and

$$T_{nr}^{(M)} = -(T_{ni}^{(M)})^2 \tag{19}$$

$$T_{nr}^{(N)} = -(T_{ni}^{(N)})^2 \tag{20}$$

3.20 1. Calculate the absorption cross section σ_a and the scattering cross section σ_s for a Mie scatterer.

2. Calculate the extinction cross section σ_e from optical theorem.

3. Show that $\sigma_e = \sigma_a + \sigma_s$ exactly by using the explicit expressions for $T_n^{(M)}$ and $T_n^{(N)}$.

3.21 Verify the solutions of reflected and transmitted fields in (2) and (3), Section 6.1.

3.22 Show, by using orthogonality property of $\overline{\overline{D}}$ matrix elements, that

$$\overline{\overline{D}}\,\overline{\overline{D}}^{-1} = \overline{\overline{I}} \tag{21}$$

3.23 Derive the orthogonality relation of D matrix elements (15) and (16), Section 5.7, by making use of orthogonality relation of Jacobi polynomials (Abramowitz and Stegun, 1965).

3.24 Verify the transformation formula for spherical harmonics (7), Section 5.7, for the case $n = 1$ and $m = -1, 0, 1$ by writing out the explicit expressions from (9), Section 5.7, and checking against the result based on using the Eulerian angle transformation formula of Section 4.2.

3.25 Use the results of Section 5.8 to calculate $<\overline{\overline{F}}(\theta, \phi; \theta, \phi)>$ and the extinction matrix for oblate spheroids for the following cases: (i) arbitrary $p(\beta, \gamma)$; (ii) $p(\beta, \gamma) = p(\beta)/2\pi$; and (iii) $p(\beta, \gamma) = 1/4\pi$.

3.26 Prove the relation for the expansion of plane waves into vector spherical waves of (25), Section 5.1.

3.27 Derive the addition theorem for vector spherical harmonics as given

Problems

in the Appendix (Section 5.9).

3.28 In this problem we will prove the addition theorem for the Legendre polynomials (Jackson, 1975). Let P be a point on a sphere whose coordinates with respect to a fixed rectangular reference system are θ and ϕ. A second point Q has the coordinates θ_i and ϕ_i. The angle made by the axis OP with the axis OQ is γ. The Legendre polynomials at P with respect to new polar axis OQ are of the form $P_n(\cos\gamma)$ and our problem is to expand $P_n(\cos\gamma)$ in terms of the coordinates θ, ϕ and θ_i, ϕ_i.

The theorem on the expansion of an arbitrary function in spherical surface harmonics states that if $g(\theta, \phi)$ is an arbitrary function on the surface of a sphere which together with all its first and second derivatives is continuous, then $g(\theta, \phi)$ can be represented by an absolutely convergent series of surface harmonics.

$$g(\theta, \phi) = \sum_{n=0}^{\infty} \left[a_{no} P_n(\cos\theta) \right.$$
$$\left. + \sum_{m=1}^{\infty} (a_{nm} \cos m\phi + b_{nm} \sin m\phi) P_n^m(\cos\theta) \right] \quad (22)$$

Give expressions for a_{no}, a_{nm} and b_{nm} in terms of integrals of $g(\theta, \phi)$. Show that

$$\int_0^{2\pi} \int_0^{\pi} g(\theta, \phi) P_n(\cos\theta) \sin\theta d\theta d\phi = \frac{4\pi}{2n+1} [g(\theta, \phi)]_{\theta=0} \quad (23)$$

by noting $P_n(1) = 1$ and $P_n^m(1) = 0$.

Using the above expansion theorem to assume that $P_n(\cos\gamma)$ takes the following form

$$P_n(\cos\gamma) = \frac{c_o}{2} P_n(\cos\theta)$$
$$+ \sum_{m=1}^{n} (c_m \cos m\phi + d_m \sin m\phi) P_n^m(\cos\theta) \quad (24)$$

Multiplying both sides by $P_{n'}^{m'}(\cos\theta) \cos m'\phi$ and integrating over the unit sphere, calculate c_m and d_m. Combining the above results show that

$$P_n(\cos\gamma) = \sum_{m=0}^{n} (2 - \delta_m) \frac{(n-m)!}{(n+m)!} P_n^m(\cos\theta_i) P_n^m(\cos\theta)$$
$$\times \cos m(\phi - \phi_i) \quad (25)$$

where

$$\delta_m = \begin{cases} 1 & \text{if } m = 0 \\ 0 & \text{otherwise} \end{cases} \quad (26)$$

3.29 In both passive and active microwave remote sensing of random medium, the spectral density function of the spatial fluctuations of permittivity plays an important role. The spectral density function is defined as the three-dimensional Fourier transform of the normalized covariance function

$$\Phi(\bar{k}) = \frac{1}{8\pi^3} \int_{-\infty}^{\infty} d^3\bar{r}\, b(\bar{r})\, e^{i\bar{k}\cdot\bar{r}} \quad (27)$$

with the normalized covariance function $b(\bar{r})$ defined as

$$<\epsilon_{1f}(\bar{r}')\epsilon_{1f}^*(\bar{r}'')> = \delta\, \epsilon_{1m}^2\, b(\bar{r}' - \bar{r}'') \quad (28)$$

where δ is the variance of the fluctuations. In the case of passive remote sensing using the radiative transfer theory, the spectral density function appears in the phase matrix elements in terms of the function $<|W(\Omega,\Omega')|^2>$

$$<|W(\Omega,\Omega')|^2> = \frac{\pi k_{1m}^4 \delta}{2}\, \Phi(k_{1m}(\hat{k}_i - \hat{k}_s)) \quad (29)$$

where Ω, Ω' are the angles $(\theta,\phi), (\theta',\phi')$ corresponding to the directions of $\hat{k}_i$ and $\hat{k}_s$. Derive Φ for the following cases:
1. $b(\bar{r}) = e^{-|\bar{r}|/r_o}$
2. $b(\bar{r}) = e^{-|z|/l}$
3. $b(\bar{r}) = e^{-(x^2+y^2)/l_\rho^2}$
4. $b(\bar{r}) = e^{-|z|/l-(x^2+y^2)/l_\rho^2}$

3.30 In problems (A) and (B) of Figure 3.7, electric fields $\bar{E}^A$ and $\bar{E}^B$ on S_∞ are

$$\bar{E}^A(\bar{r}') = \hat{a}_b\, e^{ik\hat{r}_b\cdot\bar{r}'} + \frac{e^{ikr'}}{r'} \sum_{\alpha'} \hat{\alpha}'\, f_{\alpha'a}(\hat{r}', \hat{r}_b) \quad (30)$$

$$\bar{E}^B(\bar{r}') = \hat{b}_b\, e^{ik\hat{r}_b\cdot\bar{r}'} + \frac{e^{ikr'}}{r'} \sum_{\beta'} \hat{\beta}'\, f_{\beta'b}(\hat{r}', \hat{r}_b) \quad (31)$$

Problems

Use the above expressions and the method of stationary-phase to calculate

$$\int_{S_\infty} d\bar{S}' \left[\bar{E}^A(\bar{r}') \times \bar{H}^{B*}(\bar{r}') + \bar{E}^{B*}(\bar{r}') \times \bar{H}^A(\bar{r}') \right] \quad (32)$$

and verify the result of (21), Section 3.5.

3.31 In this problem, we will evaluate the T-matrix elements for small spheroids by using the integral representations of the $Rg\, J^{(ij)}_{mnm'n'}$ and $J^{(ij)}_{mnm'n'}$ elements as given by (39), Section 5.4. We shall keep only the dipole term $n = n' = 1$. The leading term of the real part and the leading term of the imaginary part of $J^{(ij)}_{mnm'n'}$ series will be retained.

The approximations on the spherical bessel function are $j_1(z) = z/3$ and $[z\, j_1(z)]'/z = 2/3$. From (5) and (6), Section 5.6

$$Rg\, J^{(11)}_{m1m'1} = Rg\, J^{(22)}_{m1m'1} = J^{(11)}_{m1m'1} = J^{(22)}_{m1m'1} = 0 \quad (33)$$

Put the approximations of spherical Bessel functions in the expressions for $Rg\, J^{(21)}_{m1m'1}$ and $Rg\, J^{(12)}_{m1m'1}$, and also make use of (9), Section 5.6,

$$Rg\, J^{(12)}_{1111} = Rg\, J^{(12)}_{-11(-1)1} = -\frac{k_s}{k} Rg\, J^{(21)}_{1111} = -\frac{k_s}{k} Rg\, J^{(21)}_{-11(-1)1} \quad (34)$$

where

$$Rg\, J^{(21)}_{1111} = \frac{k}{12} \int_0^\pi d\theta \sin\theta\, r^3$$

$$\times \left[-(1 + \cos^2\theta) + r^2 \sin^2\theta \cos^2\theta \left(\frac{1}{a^2} - \frac{1}{c^2} \right) \right] \quad (35)$$

If (2), Section 5.6, is substituted into (35), the integral can be readily evaluated and

$$Rg\, J^{(21)}_{1111} = -\frac{2ka^2 c}{9} \quad (36)$$

Hence, using (34) and (36)

$$Rg\, P_{(-1)1(-1)1} = Rg\, P_{1111} = 0 \quad (37)$$

$$Rg\, U_{(-1)1(-1)1} = Rg\, U_{1111} = Rg\, U_1 = -i\left(\frac{\epsilon_s}{\epsilon} - 1\right)\frac{2k^3 a^2 c}{9} \quad (38)$$

Similarly,
$$Rg\, P_{0101} = 0 \tag{39}$$

and
$$Rg\, U_{0101} = Rg\, U_o = -\frac{2}{9}ik^3 a^2 c \left(\frac{\epsilon_s}{\epsilon} - 1\right) \tag{40}$$

To calculate the $J^{(12)}_{m1m'1}$ and $J^{(21)}_{m1m'1}$ elements, we keep the leading term in the real part and imaginary part of $h_1(kr)$, so that $h_1(z) = -i/z^2 + z/3$ and $[zh_1(z)]'/z = i/z^3 + 2/3$. If we substitute into the expressions for $J^{(12)}_{m1m'1}$ and $J^{(21)}_{m1m'1}$, and then into $P_{m1m'1}$ and $U_{m1m'1}$, we have

$$P_{1111} = P_{-11(-1)(1)} = P_1$$
$$= -\frac{k_s}{8k} \int_0^\pi d\theta \sin\theta (\overline{\sigma}\cdot\hat{r})(1+\cos^2\theta) + Rg\, P_{1111} \tag{41}$$

and is not equal to zero, whereas

$$U_{1111} = U_{(-1)1(-1)1} = U_1$$
$$= \left(\frac{k_s^2}{4k^2} - \frac{1}{4}\right) \int_0^\pi d\theta \sin\theta \left[(\overline{\sigma}\cdot\hat{r})\frac{(1+\cos^2\theta)}{2}\right.$$
$$\left. + (\overline{\sigma}\cdot\hat{\theta})\sin\theta\cos\theta\right]$$
$$+ \frac{3}{8}\int_0^\pi d\theta \sin\theta(\overline{\sigma}\cdot\hat{r})(1+\cos^2\theta) + Rg\, U_1 \tag{42}$$

If we substitute expressions for $\overline{\sigma}$ and r, after some algebraic manipulation, we have

$$U_1 = Rg\, U_1 + 1 + v_d A_a \tag{43}$$

Similarly, $P_{0101} = P_o$ is not equal to zero and

$$U_{0101} = U_o = Rg\, U_o + 1 + v_d A_c \tag{44}$$

Thus the $\overline{\overline{Q}}^T$ matrix is a 6×6 diagonal matrix with diagonal elements P_1, P_o, P_1, U_1, U_o and U_1 while $Rg\, \overline{\overline{Q}}^T$ is a diagonal 6×6 matrix with diagonal elements $0, 0, 0, Rg\, U_1, Rg\, U_o, Rg\, U_1$. Hence, only nonzero T-matrix elements are $T^{(22)}_{m1m'1}$ as given by (10), Section 5.6, where

$$T_o = -\frac{Rg\, U_o}{U_o} \tag{45}$$

$$T_1 = -\frac{Rg\, U_1}{U_1} \tag{46}$$

Since $|Rg\, U_o| \ll |U_o|$ and $|Rg\, U_1| \ll |U_1|$, we can further approximate T_o and T_1 in (45) and (46). Show that (45) and (46) reduce to (11) and (12), Section 5.6, respectively.

3.32 Following the procedure for the Kirchhoff approach combined with the geometric optics approximation, derive the boundary conditions of (5) through (11), Section 6.2, for the Stokes parameters of a rough dielectric interface.

4

SOLUTIONS OF RADIATIVE TRANSFER EQUATIONS WITH APPLICATIONS TO REMOTE SENSING

1. Introduction   220

2. Iterative Method   220
    2.1 Thermal Emission in the Absence of Scattering   221
    2.2 Single Scattering for Isotropic Point Scatterers   224
    2.3 Passive Remote Sensing of a Layer of Spherical Particles   228
    2.4 Active Remote Sensing of a Half-Space of Spherical Particles   235
    2.5 Active Remote Sensing of a Layer of Nonspherical Particles   241
    2.6 Second-Order Scattering from Isotropic Point Scatterers   250
    2.7 Second-Order Solution for Small Spherical Particles   252

3. Discrete Ordinate–Eigenanalysis Method   258
    3.1 Radiative Transfer Equations of Passive Remote Sensing   258
    3.2 Closed-Form Solution for Laminar Structure   259
    3.3 Discrete Ordinate–Eigenanalysis Approach   263
    3.4 Thermal Emission of Three-Dimensional Random Medium   269
    3.5 Thermal Emission of a Layer of Spherical Scatterers Overlying a Homogeneous Dielectric Half-Space   272
    3.6 Active Remote Sensing of a Layer of Small Spherical Scatterers Overlying a Homogeneous Dielectric Half-Space   286

4. Method of Invariant Imbedding Applied to Problems

with Inhomogeneous Profiles 291

4.1 One-Dimensional Problem 291
4.2 Thermal Emission from an Inhomogeneous Slab – Three-Dimensional Problem 300
4.3 Thermal Emission of Three-Dimensional Random Medium 305
4.4 Thermal Emission of Layers of Spherical Scatterers in the Presence of Inhomogeneous Absorption and Temperature Profiles 305

Problems 312

1 INTRODUCTION

In this chapter, we discuss solutions for the radiative transfer equations, and illustrate the solutions for different kinds of scattering media. Three techniques are considered: the iterative method, the discrete ordinate–eigenanalysis method, and the method of invariant imbedding. The iterative approach is convenient for the case of small albedo when scattering is dominated by absorption. It also gives physical insight into the multiple scattering processes since there is a one-to-one correspondence between the iterative order and the order of multiple scattering. For cases of general albedo, the discrete ordinate–eigenanalysis method can be used for problems with homogeneous profiles. In this method, the continuum of propagation directions is first discretized into a finite number of directions by means of quadrature and the resulting system of equations is then solved by eigenanalysis. For cases where the media have inhomogeneous profiles, the method of invariant imbedding can be employed. The boundary value problem is reformulated into an initial value problem with slab thickness as the variable. The initial value problem is conveniently solved by stepping forward in slab thicknesses.

2 ITERATIVE METHOD

In the iterative method, scattering is treated as a small perturbation. The solutions of radiative transfer equations are decomposed into a series of perturbation orders with each order calculated by iteration of the previous order and with scattering as the iterative parameter.

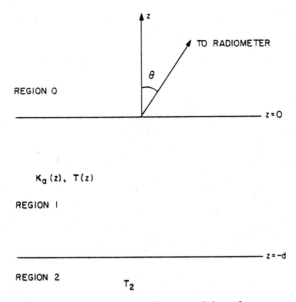

Fig. 4.1 Thermal emission from a layer of particles overlying a homogeneous dielectric half-space in the absence of scattering.

2.1 Thermal Emission in the Absence of Scattering

The zeroth-order solution is obtained by ignoring scattering. Consider the thermal emission by a layer of particles overlying a homogeneous dielectric half-space (Figure 4.1). When scattering is ignored, the radiative transfer equations assume the following form. For $0 < \theta < \pi/2$

$$\cos\theta \frac{d}{dz} I_\beta(\theta, z) = -\kappa_a(z) I_\beta(\theta, z) + \kappa_a(z) CT(z) \qquad (1)$$

$$-\cos\theta \frac{d}{dz} I_\beta(\pi - \theta, z) = -\kappa_a(z) I_\beta(\pi - \theta, z) + \kappa_a(z) CT(z) \qquad (2)$$

where $\kappa_a(z)$ and $T(z)$ are, respectively, the absorption and temperature profiles of region 1. In (1) and (2), we have assumed that there is no polarization dependence for the absorption coefficient $\kappa_a(z)$ and that the usage of two Stokes parameters is sufficient. The quantities $I_\beta(\theta, z)$ and $I_\beta(\pi - \theta, z)$ denote, respectively, upward- and downward-going specific intensities at angle θ and polarization β. The boundary conditions are

$$I_\beta(\pi - \theta, z = 0) = 0 \qquad (3)$$

$$I_\beta(\theta, z = -d) = r_\beta(\theta) I_\beta(\pi - \theta, z = -d) + [1 - r_\beta(\theta)] CT_2 \qquad (4)$$

where T_2 is the temperature of the lower half-space and $r_\beta(\theta)$ is the reflectivity for the boundary between regions 1 and 2. The specific intensities can be calculated by solving (1) through (4). The brightness temperature is then given by

$$T_{B\beta}(\theta) = \frac{1}{C} I_\beta(\theta, z=0) \tag{5}$$

The set of equations in (1) through (4) can be solved readily, and the solution for the brightness temperature is given by (Problem 1)

$$T_{B\beta}(\theta) = \int_{-d}^{0} dz' \, T(z') \kappa_a(z') \sec\theta \exp\left[-\int_{z'}^{0} dz'' \kappa_a(z'') \sec\theta\right]$$

$$+ r_\beta(\theta) \exp\left[-\int_{-d}^{0} dz' \kappa_a(z') \sec\theta\right] \int_{-d}^{0} dz' \kappa_a(z') T(z') \sec\theta$$

$$\times \exp\left[-\int_{-d}^{z'} dz'' \kappa_a(z'') \sec\theta\right]$$

$$+ [1 - r_\beta(\theta)] T_2 \exp\left[-\int_{-d}^{0} dz' \kappa_a(z') \sec\theta\right] \tag{6}$$

The first term in (6) corresponds to the upward emission of layer 1 and the second term corresponds to the downward emission of layer 1 that is reflected by the surface at $z = -d$ and further attenuated by layer 1 before reaching the radiometer. The third term is the upward emission of layer 2 and is attenuated by the intervening layer 1 in its upward path. The zeroth-order solution has been used extensively in the passive remote sensing of the atmosphere (Staelin, 1969; Grody, 1976; Barrett and Martin, 1981). The absorption coefficients include absorption of atmospheric gases, water vapor, water droplets, etc.

For small water droplets, absorption dominates over scattering, and the absorption cross section σ_a for a spherical droplet of radius a with permittivity ϵ_s is, from (20), Section 4.1 (Chapter 3),

$$\sigma_a = 3v_o k \, (Im \, y) \tag{7}$$

where

$$y = \frac{\epsilon_s - \epsilon}{\epsilon_s + 2\epsilon} \tag{8}$$

and v_o is the droplet size $4\pi a^3/3$. Let the drop-size distribution be governed by $n(a)$, which is the number of particles per unit volume per unit radius and has units m^{-4}. Then the number of particles per unit volume n_o is integration of $n(a)$ over da. The absorption coefficient is, on averaging over drop-size distribution, and using (7),

$$\kappa_a = \int_0^\infty da\, \sigma_a(a)\, n(a) = 3k\,(Im\, y)\, f \qquad (9)$$

where

$$f = \int_0^\infty da\, v_o\, n(a) \qquad (10)$$

is the fractional volume occupied by the droplets. Let ρ be the mass density of particles in units of $kg\, m^{-3}$, then

$$M = \rho f \qquad (11)$$

is the mass (kg) of particles per unit m^3 of volume of space in region 1. Hence, using (9) and (11)

$$\kappa_a = 3k\,(Im\, y)\,\frac{M}{\rho} \qquad (13)$$

Thus, in the Rayleigh absorption limit, even in the presence of particle size distribution, the absorption coefficient is linearly proportional to mass per unit volume M. This implies that the ability to invert the absorption profile $\kappa_a(z)$ in atmospheric sensing enables one to calculate the mass profile of cloud layers. This has important applications in meteorology. Single and multiple scattering effects in the passive remote sensing of cloud and rainfall have been considered by Zavody (1974), Tsang et al. (1977), and Wilheit et al. (1977).

For the case of nonspherical particles, it is necessary to use four Stokes parameters as discussed in Section 3.5, Chapter 3. This can be illustrated with a simple example. Consider the thermal emission of a slab of aligned small spheroids with axes of symmetry pointing in the $\hat{y}$ direction. Let the slab be of thickness d. At low frequency, scattering can be ignored. Hence, the radiative transfer equations become

$$\frac{d}{dz}\bar{I}(\hat{s}) = -\bar{\bar{\kappa}}(\hat{s})\cdot\bar{I}(\hat{s}) + \bar{\kappa}_a(\hat{s}_b)\, CT \qquad (13)$$

with

$$\bar{\kappa}_a(\hat{s}) = \begin{bmatrix} \kappa_{11}(\hat{s}) \\ \kappa_{22}(\hat{s}) \\ 2\kappa_{13}(\hat{s}) + 2\kappa_{23}(\hat{s}) \\ -2\kappa_{14}(\hat{s}) - 2\kappa_{24}(\hat{s}) \end{bmatrix} \qquad (14)$$

Then the thermal emission Stokes vector observed by a radiometer above the slab in the direction $\hat{s} = (\theta, \phi)$ is $\bar{I}_{out}(\theta, \phi)$ which is given by the expression

$$\bar{I}_{out}(\theta, \phi) = \left[\bar{\bar{I}} - \bar{\bar{E}}(\theta, \phi)\bar{\bar{D}}(-\bar{\beta}(\theta, \phi)d\sec\theta)\bar{\bar{E}}^{-1}(\theta, \phi)\right]\bar{I}_p \tag{15}$$

where

$$\bar{I}_p = \begin{bmatrix} CT \\ CT \\ 0 \\ 0 \end{bmatrix} \tag{16}$$

and $\bar{\bar{D}}(\bar{\beta}(\theta, \phi)d\sec\theta)$ is the diagonal matrix with the ii-th element equal to $\exp(+\beta_i(\theta, \phi)d\sec\theta)$ with $i = 1, 2, 3, 4$ and $\bar{\bar{E}}$ is the eigenmatrix given by (18), Section 3.4, Chapter 3.

If the conventional formulation of two Stokes parameters is used, then only the 2×2 matrix in the upper left-hand corner of the extinction matrix is used. The conventional solution $\bar{I}_{con}$ is thus, assuming that the third and fourth parameters are zero,

$$\bar{I}_{con} = \begin{bmatrix} CT(1 - e^{-\kappa_{11}d\sec\theta}) \\ CT(1 - e^{-\kappa_{22}d\sec\theta}) \\ 0 \\ 0 \end{bmatrix} \tag{17}$$

Thus the solutions $\bar{I}_{out}$ of (15) and $\bar{I}_{con}$ of (17) are generally different unless $M_{vh} = M_{hv} = 0$. For small spheroids with symmetry axes pointing in the $\hat{y}$ direction and observation in the x-z plane, it is readily shown that M_{vh} and M_{hv} are not equal to zero. Furthermore, cases can be constructed such that $\kappa_{11}d$, $\kappa_{22}d$, and $\beta_4 d$ are much larger than 1, while $\beta_1 d$ is much less than 1. Then the solution using four Stokes parameters are significantly different from that using two Stokes parameters (Problem 25).

2.2 Single Scattering for Isotropic Point Scatterers

The model of isotropic point scatterers assumes that the phase function is independent of angle and polarization. It is a convenient model to illustrate the multiple scattering processes in a scattering medium and facilitates comparison with the results of analytic wave theory. (This comparison will be made in Chapter 5.) In this section, we illustrate the formulation of integral equations from the radiative transfer equations and the boundary conditions.

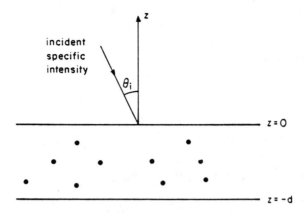

Fig. 4.2 Active remote sensing of a slab of point particles overlying a homogeneous half-space.

With slight generalizations, the method is used in subsequent sections on iterative solutions.

Consider specific intensity incident onto a slab of point scatterers overlying a homogeneous half-space (Figure 4.2). For point scatterers, the phase function is independent of angles. Let $\tau = \kappa_e z$ denote the optical thickness and $\mu = \cos\theta$. For $\mu > 0$,

$$\mu \frac{d}{d\tau} I(\tau, \mu, \phi) = - I(\tau, \mu, \phi)$$
$$+ \frac{\tilde{\omega}}{4\pi} \int_0^1 d\mu' \int_0^{2\pi} d\phi' [I(\tau, \mu', \phi') + I(\tau, -\mu', \phi')] \tag{1}$$

$$-\mu \frac{d}{d\tau} I(\tau, -\mu, \phi) = - I(\tau, -\mu, \phi)$$
$$+ \frac{\tilde{\omega}}{4\pi} \int_0^1 d\mu' \int_0^{2\pi} d\phi' [I(\tau, \mu', \phi') + I(\tau, -\mu', \phi')] \tag{2}$$

In (1) and (2), $I(\tau, \mu, \phi)$ and $I(\tau, -\mu, \phi)$ denote, respectively, upward- and downward-going specific intensities. The boundary conditions are

$$I(\tau = 0, -\mu, \phi) = \delta(\mu - \mu_i)\delta(\phi) \tag{3}$$

$$I(\tau = -\tau_d, \mu, \phi) = r(\mu) I(\tau = -\tau_d, -\mu, \phi) \tag{4}$$

In (1) through (4) $\tilde{\omega} = \kappa_s/\kappa_e$ is the albedo. In (4), $r(\mu)$ is the reflectivity at the boundary $z = -d$ at angle θ and $\tau_d = \kappa_e d$. We first cast the differential equations of (1) and (2) and boundary equations (3) and

(4) into integral equations. Multiply (1) by $\exp(\tau/\mu)/\mu$ and integrate over τ from $-\tau_d$ to τ.

$$e^{\tau/\mu} I(\tau, \mu, \phi) = e^{-\tau_d/\mu} I(\tau = -\tau_d, \mu, \phi)$$

$$+ \frac{\tilde{\omega}}{4\pi\mu} \int_0^1 d\mu' \int_0^{2\pi} d\phi' \int_{-\tau_d}^{\tau} d\tau' e^{\tau'/\mu} [I(\tau', \mu', \phi') + I(\tau', -\mu', \phi')]$$
(5)

Next multiply (2) by $e^{-\tau/\mu}/\mu$ and integrate from τ to 0.

$$e^{-\tau/\mu} I(\tau, -\mu, \phi) = I(\tau = 0, -\mu, \phi)$$

$$+ \frac{\tilde{\omega}}{4\pi\mu} \int_0^1 d\mu' \int_0^{2\pi} d\phi' \int_\tau^0 d\tau' e^{-\tau'/\mu} [I(\tau', \mu', \phi') + I(\tau', -\mu', \phi')]$$
(6)

Next the boundary conditions (3) and (4) are imposed on (5) and (6)

$$I(\tau, -\mu, \phi) = e^{\tau/\mu_i} \delta(\mu - \mu_i) \delta(\phi)$$

$$+ \frac{\tilde{\omega}}{4\pi\mu} \int_0^1 d\mu' \int_0^{2\pi} d\phi' \int_\tau^0 d\tau' e^{(\tau-\tau')/\mu}$$

$$\times [I(\tau', \mu', \phi') + I(\tau', -\mu', \phi')] \quad (7)$$

$$I(\tau, \mu, \phi) = e^{-(\tau+\tau_d)/\mu_i} r(\mu_i) e^{-\tau_d/\mu_i} \delta(\mu - \mu_i) \delta(\phi)$$

$$+ e^{-(\tau+\tau_d)/\mu} r(\mu) \frac{\tilde{\omega}}{4\pi\mu} \int_0^1 d\mu' \int_0^{2\pi} d\phi' \int_{-\tau_d}^0 d\tau'$$

$$\times e^{(-\tau_d-\tau')/\mu} [I(\tau', \mu', \phi') + I(\tau', -\mu', \phi')]$$

$$+ \frac{\tilde{\omega}}{4\pi\mu} \int_0^1 d\mu' \int_0^{2\pi} d\phi' \int_{-\tau_d}^\tau d\tau' e^{-(\tau-\tau')/\mu}$$

$$\times [I(\tau', \mu', \phi') + I(\tau', -\mu', \phi')] \quad (8)$$

Equations (7) and (8) form a set of coupled equations for $I(\tau, \mu, \phi)$ and $I(\tau, -\mu, \phi)$. In the iteration approach, the solutions are decomposed into a series of perturbation orders by treating $\tilde{\omega}$ as a small parameter.

$$I(\tau, \mu, \phi) = I^{(0)}(\tau, \mu, \phi) + I^{(1)}(\tau, \mu, \phi) + I^{(2)}(\tau, \mu, \phi) + \ldots \quad (9)$$

where $-1 < \mu < 1$. The perturbation orders also have one-to-one correspondence with multiple scattering processes, with $I^{(0)}$ representing reduced intensity of the original wave, $I^{(1)}$ representing single scattering solutions, $I^{(2)}$ representing double scattering solutions, etc. The scattered intensity in the upper half-space is $I(\tau = 0, \mu, \phi)$ and the bistatic scattering coefficient is

$$\gamma(\mu, \phi; \mu_i, \phi_i = 0) = 4\pi \frac{\cos\theta \, I(\tau = 0, \mu, \phi)}{\cos\theta_i} \tag{10}$$

Using (7) through (9), the zeroth-order solutions are

$$I^{(0)}(\tau, -\mu, \phi) = e^{\tau/\mu_i} \delta(\mu - \mu_i)\delta(\phi) \tag{11a}$$

$$I^{(0)}(\tau, \mu, \phi) = e^{-\tau/\mu_i} r(\mu_i) e^{-2\tau_d/\mu_i} \delta(\mu - \mu_i)\delta(\phi) \tag{11b}$$

which represent original waves that are attenuated as they propagate through the slab. To calculate the first-order bistatic scattering coefficient, we evaluate $I^{(1)}(\tau = 0, \mu, \phi)$ by using (8) and (11).

$$I^{(1)}(\tau = 0, \mu, \phi) = \frac{\tilde{\omega}}{4\pi\mu} \left\{ \frac{(1 - e^{-\tau_d(1/\mu + 1/\mu_i)})}{1/\mu + 1/\mu_i} \right.$$
$$\left. + r(\mu_i) \frac{(e^{-2\tau_d/\mu_i} - e^{-\tau_d/\mu - \tau_d/\mu_i})}{1/\mu - 1/\mu_i} \right\}$$
$$+ \frac{\tilde{\omega}}{4\pi\mu} r(\mu) e^{-\tau_d/\mu} \left\{ \frac{e^{-\tau_d/\mu} - e^{-\tau_d/\mu_i}}{1/\mu_i - 1/\mu} \right.$$
$$\left. + r(\mu_i) e^{-\tau_d/\mu_i} \frac{(1 - e^{-(\tau_d/\mu + \tau_d/\mu_i)})}{1/\mu + 1/\mu_i} \right\} \tag{12}$$

The first term in (12) represents scattering by a particle into the upper half-space (Figure 4.3a). The second term in (12) corresponds to single scattering of the reflected wave by a particle (Figure 4.3b). The third term represents scattering by a particle and the wave is further reflected by the boundary before going back into upper half-space (Figure 4.3c). The last term corresponds to single scattering of the reflected wave which is further reflected by the boundary (Figure 4.3d). We note that there are four scattering processes in single volume scattering in the presence of a reflective boundary. In the absence of the reflective boundary, single volume scattering will consist of only one term represented by Figure 4.3a. In backscattering directions, $\mu = \mu_i$ and $\phi = \pi$, we have, from (12)

$$I^{(1)}(\tau = 0, \mu = \mu_i, \phi = \pi) = \frac{\tilde{\omega}}{8\pi} \left(1 - e^{-2\tau_d/\mu_i}\right)$$
$$+ \frac{\tilde{\omega}}{2\pi\mu_i} r(\mu_i) \tau_d e^{-2\tau_d/\mu_i} + \frac{\tilde{\omega}}{8\pi} r^2(\mu_i) e^{-2\tau_d/\mu_i} \left(1 - e^{-2\tau_d/\mu_i}\right) \tag{13}$$

228   4. Solutions of Radiative Transfer Equations

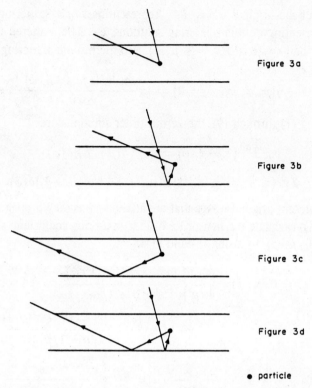

Fig. 4.3 Four scattering processes for single volume scattering in the presence of a reflective boundary.

In the absence of a reflective boundary, only the first term in (13) is retained and the result is identical to that of analytic wave theory (Chapter 5). However, when the reflective boundary is present, the scattering processes of Figures 4.3b through 4.3d lead to constructive interference and backscattering enhancement that are not accounted for in radiative transfer theory. Thus, the results in (13) are different from the corresponding results of analytic wave theory.

2.3 Passive Remote Sensing of a Layer of Spherical Particles

In this section, the solution for the brightness temperature of a layer of spherical particles is illustrated. For the case of thermal emission of spherical particles, because of azimuthal symmetry, the specific intensity is only a function of θ and the solutions of the radiative transfer equations are greatly simplified. The single scattering solution will be illustrated for a layer of cloud or rain droplets with drop-size distribution.

The radiative transfer equations are, for $0 \leq \theta \leq \pi$

$$\cos\theta \frac{d}{dz}\overline{I}(\theta, z) = -\kappa_e \overline{I}(\theta, z) + \kappa_a C \begin{bmatrix} T \\ T \\ 0 \\ 0 \end{bmatrix}$$

$$+ \int_0^\pi d\theta' \sin\theta' \int_0^{2\pi} d\phi' \overline{\overline{P}}(\theta, \phi; \theta', \phi') \cdot \overline{I}(\theta', z) \quad (1)$$

Because $\overline{I}$ is independent of ϕ', the $d\phi'$ integral in (1) can be readily carried out. It can be shown that

$$\int_0^{2\pi} d\phi' P_{ij}(\theta, \phi; \theta', \phi') = 0 \quad (2)$$

for $ij = 13, 14, 23, 24, 31, 32, 41$ and 42. In view of (2) and the absence of source term of the third and fourth Stokes parameters, we conclude that they are zero (Problem 3). Next define

$$[\alpha(\theta), \beta(\theta')] = \int_0^{2\pi} d\phi' P_{\alpha\beta}(\theta, \phi; \theta', \phi') \quad (3)$$

with $\alpha, \beta = v, h$ and $P_{vv} = P_{11}$, $P_{vh} = P_{12}$, $P_{hv} = P_{21}$ and $P_{hh} = P_{22}$. For the case of spherical particles, these can be evaluated by using the phase matrix elements for spherical particles as given in (15), Section 3.2, (17), Section 5.2, and (8) and (9), Section 5.5 of Chapter 3. They are (Problem 4)

$$[v(\theta), v(\theta')] = \frac{8\pi n_o}{k^2} \left| \sum_{n=1}^\infty \frac{T_n^{(N)}}{n(n+1)} s_n^o(\cos\theta) s_n^o(\cos\theta') \right|^2$$

$$+ \frac{16\pi n_o}{k^2} \sum_{m=1}^\infty \left| \sum_{n=m}^\infty [n(n+1)]^{-1} \right.$$

$$\times \left[T_n^{(M)} t_n^m(\cos\theta) t_n^m(\cos\theta') \right.$$

$$\left. \left. + T_n^{(N)} s_n^m(\cos\theta) s_n^m(\cos\theta') \right] \right|^2 \quad (4a)$$

$$[v(\theta), h(\theta')] = \frac{16\pi n_o}{k^2} \sum_{m=1}^\infty \left| \sum_{n=m}^\infty \frac{1}{n(n+1)} \right.$$

$$\times \left[T_n^{(M)} t_n^m(\cos\theta) s_n^m(\cos\theta') \right.$$

$$\left. \left. + T_n^{(N)} s_n^m(\cos\theta) t_n^m(\cos\theta') \right] \right|^2 \quad (4b)$$

$$[h(\theta), v(\theta')] = \frac{16\pi n_o}{k^2} \sum_{m=1}^{\infty} \left| \sum_{n=m}^{\infty} \frac{1}{n(n+1)} \right.$$

$$\times \left[T_n^{(M)} s_n^m(\cos\theta) t_n^m(\cos\theta') \right.$$

$$\left. \left. + T_n^{(N)} t_n^m(\cos\theta) s_n^m(\cos\theta') \right] \right|^2 \qquad (4c)$$

$$[h(\theta), h(\theta')] = \frac{8\pi n_o}{k^2} \left| \sum_{n=1}^{\infty} \frac{T_n^{(M)}}{n(n+1)} s_n^o(\cos\theta) s_n^o(\cos\theta') \right|^2$$

$$+ \frac{16\pi n_o}{k^2} \sum_{m=1}^{\infty} \left| \sum_{n=m}^{\infty} [n(n+1)]^{-1} \right.$$

$$\times \left[T_n^{(M)} s_n^m(\cos\theta) s_n^m(\cos\theta') \right.$$

$$\left. \left. + T_n^{(N)} t_n^m(\cos\theta) t_n^m(\cos\theta') \right] \right|^2 \qquad (4d)$$

The extinction, scattering, and absorption coefficients are, from Section 5.5 of Chapter 4,

$$\kappa_e = -\frac{2\pi n_o}{k^2} \operatorname{Re} \sum_{n=1}^{\infty} (2n+1) \{T_n^{(M)} + T_n^{(N)}\} \qquad (5)$$

$$\kappa_s = \frac{2\pi n_o}{k^2} \sum_{n=1}^{\infty} (2n+1) \{|T_n^{(M)}|^2 + |T_n^{(N)}|^2\} \qquad (6)$$

$$\kappa_a = \kappa_e - \kappa_s \qquad (7)$$

The radiative transfer equations now assume the following form, for $0 \leq \theta \leq \pi$ and $\beta = v, h$

$$\cos\theta \frac{d}{dz} I_\beta(\theta, z) = -\kappa_e I_\beta(\theta, z) + \kappa_a CT$$

$$+ \sum_{\alpha=v,h} \int_0^\pi d\theta' \sin\theta' \left(\beta(\theta), \alpha(\theta')\right) I_\alpha(\theta', z) \qquad (8)$$

Consider a layer of spherical particles bounded on both sides by free space (Figure 4.4). Then the boundary conditions are, for $0 \leq \theta \leq \pi/2$

$$I_\beta(\pi - \theta, z = 0) = 0 \qquad (9)$$

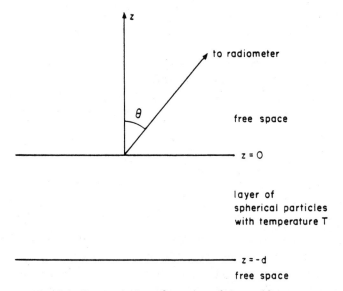

Fig. 4.4 Geometrical configuration of the problem.

$$I_\beta(\theta, z = -d) = 0 \tag{10}$$

To obtain iterative solution, the transfer equations and the boundary conditions can be cast into integral equations in manners similar to those in Section 2.2. The zeroth-order solutions are, for $0 < \vartheta < \pi/2$

$$I_\beta^{(0)}(\theta, z) = \frac{\kappa_a}{\kappa_e} CT \left(1 - e^{-\kappa_e \sec\theta(z+d)}\right) \tag{11}$$

$$I_\beta^{(0)}(\pi - \theta, z) = \frac{\kappa_a}{\kappa_e} CT \left(1 - e^{\kappa_e \sec\theta z}\right) \tag{12}$$

The brightness temperature for observation in region 0 is given by $T_{B\beta}(\theta) = I_\beta(\theta, z = 0)/C$. By carrying the iteration to first order, the solution is (Problem 5)

$$T_{B\beta}(\theta) = \frac{\kappa_a}{\kappa_e} T \Bigg\{ \left(1 - e^{-\kappa_e d \sec\theta}\right)(1 + \tilde{\omega})$$

$$- \sum_{\alpha=v,h} \int_0^{\pi/2} d\theta' \sin\theta' (\beta, \alpha') \sec\theta \, e^{-\kappa_e d \sec\theta}$$

$$\times \left(\frac{1 - e^{-\kappa_e d(\sec\theta' - \sec\theta)}}{\kappa_e (\sec\theta' - \sec\theta)} \right)$$

$$- \sum_{\alpha=v,h} \int_0^{\pi/2} d\theta' \sin\theta' (\beta, \alpha'(\pi - \theta')) \sec\theta$$

$$\times \left(\frac{1 - e^{-\kappa_e d(\sec \theta' + \sec \theta)}}{\kappa_e (\sec \theta + \sec \theta')} \right) \right\} \quad (13)$$

To illustrate the solution for cloud and rainfall droplets, a drop-size distribution must be introduced for the particles (Laws and Parsons, 1943; Marshall and Palmer, 1948). The following Γ-size distribution (Deirmendjian, 1969) can be used:

$$n(a) = K_1 a^P \exp(-K_2 a^Q) \quad (14)$$

where the constants P and Q are fitted empirically to the slopes of an experimentally obtained drop-size distribution. The unit of $n(a)$ is usually given in cm$^{-3}\mu^{-1}$ and has the meaning of number of particles per unit volume per unit radius. The extinction and absorption coefficients and the phase matrix elements are then obtained by replacing n_o in (4) through (7) by $n(a)$ and integration over a from 0 to ∞. The constants K_1 and K_2 can be expressed in terms of mode radius a_c (in microns) and specific water content M (gm/m^3).

The mode radius a_c is the drop-size at which $n(a)$ is at the maximum. By using (14) and $n'(a_c) = 0$, we have

$$K_2 = \frac{P}{Q a_c^Q} \quad (15)$$

The fractional volume f occupied by water droplets can be calculated by the formula

$$f = 10^{-12} \int_0^\infty da \, \frac{4\pi}{3} a^3 n(a) \quad (16)$$

where a is in microns.

The specific water content M is the mass of water per unit volume of space in the cloud or rain layer. Taking the specific density of water to be 1 gm/cm^3, we have

$$M = f \, 10^6 \quad (17)$$

Using (14), (16), and (17), we thus have

$$K_1 = \frac{3M Q K_2^R \times 10^6}{4\pi \Gamma(R)} \quad (18)$$

where Γ is gamma function and

$$R = \frac{P + 4}{Q} \quad (19)$$

Passive Remote Sensing of a Layer of Spherical Particles

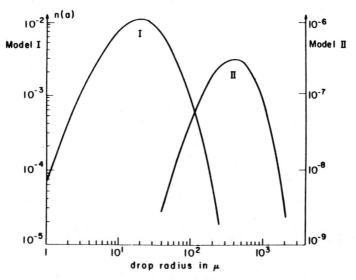

Fig. 4.5 Drop-size distribution for two models : 1. cloud, 2. rain.

In the following, we illustrate the results for the following two models:

1. Cloud (typical of stratus clouds):

$$a_c = 20\mu, \quad P = 5, \quad Q = 0.3, \quad M = 0.8 \text{ gm/m}^3$$

2. Rain (typical precipitation rate of 12 mm/h):

$$a_c = 400\mu, \quad P = 5, \quad Q = 0.5, \quad M = 0.5 \text{ gm/m}^3$$

For water with no salinity, the permittivity at 273 K is given by (Lane and Saxton, 1952; Saxton and Lane, 1952)

$$\epsilon_s = \left(5.5 + \frac{82.5}{1 - i3.59/\lambda}\right)\epsilon_o \qquad (20)$$

where λ is the wavelength in cm. We note that the fractional volumes for both cloud and rain are of the order 10^{-6} so that the concentrations of particles are sparse. The drop-size distributions for the two models are shown in Figure 4.5. The droplet size for rain is larger than that for cloud. However, there are fewer drops per unit volume in rain than in cloud. From the figures, we also note that the drop-size distribution is maximum at the mode radius.

In Figures 4.6a,b we plot the extinction and absorption coefficients for the two models. We note that within the frequency range considered,

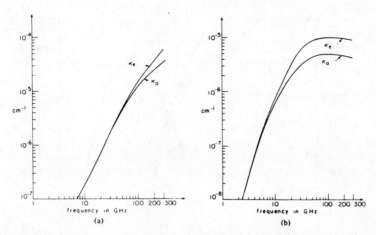

Fig. 4.6 Extinction and absorption coefficients (a) cloud model, (b) rain model.

κ_e and κ_a for cloud are monotonically increasing functions of frequency. As for rain, the curves exhibit both resonance and saturation. These are characteristics of Mie scattering except that the rapidly oscillating part of the Mie extinction coefficient has been averaged out by the integration over particle size distribution. Only a broad maximum is exhibited in Figure 4.6b. In the low-frequency range, the curves are rapidly increasing functions of frequency. Scattering is completely dominated by absorption and the albedo is practically zero. Thus at the low-frequency regime, the results of Section 2.1 where scattering is ignored are applicable. In Figures 4.7a,b we show the brightness temperature at nadir under the single scattering approximation. The temperature T is 273 K.

The numerical solution of the radiative transfer equation calculated by the discrete eigenanalysis technique (Section 3) is also shown for comparison. The cloud layer thickness is 2 km and the rain layer thickness is 1 km. We note from the results of Figure 4.6 that the cloud layer is optically thick $(\kappa_a d > 1)$ whereas the rain layer is optically thin $(\kappa_a d < 1)$ for frequencies larger than 100 GHz. Thus, scattering induces darkening for optically thick layers and brightening for optically thin layers. The no-scattering solution is obtained by setting $\tilde{\omega} = 0$ and is

$$T_{BO}(\theta) = T(1 - e^{-\kappa_a d \sec \theta}) \qquad (21)$$

In this section, we obtain analytic, closed-form solutions for the iterative solutions. The iterative solution can also be calculated by an entirely numerical procedure of discretizing the space coordinate z and the angular variable θ (Chang et al., 1976).

Active Remote Sensing of a Half-Space

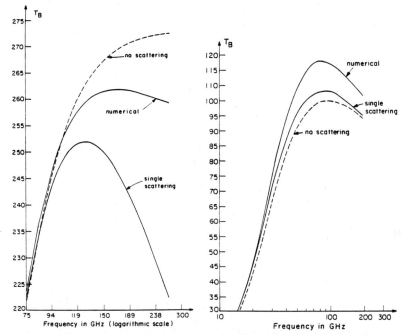

Fig. 4.7 Brightness temperatures of cloud model as function of frequency. Results obtained by different methods are shown for comparison. Shown in dotted lines is T_{BO}, the case where scattering is ignored. (a) Cloud layer thickness is 2 km. (b) Rain layer thickness is 1 km.

2.4 Active Remote Sensing of a Half-Space of Spherical Particles

The sharp increase of the scattering at the backscattering (retro-reflection) direction from some objects has been known for a long time. This is sometimes called the opposition effect or the glory effect. The glory appearing around the shadow of an airplane on a cloud below it, when viewed from the airplane, is an example of the peak in reflectivity. In astronomy, it is well known that the reflectivity of the moon increases sharply at full moon. In the case of volume scattering by a half-space of particles, the sharp peak in the backscattering direction may be caused by (1) the Mie scattering pattern and (2) the multiple scattering effect known as backscattering enhancement not accounted for in radiative transfer theory. The glory effect has also been observed in laboratory controlled experiments (Kuga and Ishimaru, 1984). The discussion on backscattering enhancement will be deferred to Chapter 5. In this section we will illustrate (1) by considering the bistatic scattering of a half-space of spherical particles (Figure 4.8).

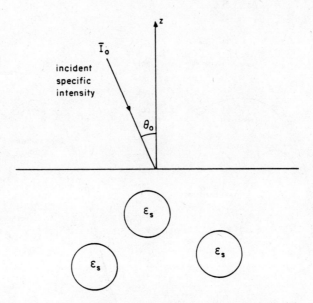

Fig. 4.8 Wave incident on a half-space of spherical particles.

The radiative transfer equation is, for $0 \leq \theta \leq \pi$

$$\cos\theta \frac{d}{dz}\overline{I}(\theta,\phi,z) = -\kappa_e \overline{I}(\theta,\phi,z)$$
$$+ \int_0^\pi d\theta' \sin\theta' \int_0^{2\pi} d\phi' \overline{\overline{P}}(\theta,\phi;\theta',\phi') \cdot \overline{I}(\theta',\phi',z) \quad (1)$$

where $\overline{I}$ is the column vector containing the four Stokes parameters. Consider specific intensity $\overline{I}_o$ incident onto the half-space of spherical particles from direction (θ_o, ϕ_o). Then the boundary condition is, for $0 < \theta < \pi/2$

$$\overline{I}(\pi - \theta, \phi, z = 0) = \overline{I}_o \delta(\cos\theta - \cos\theta_o) \delta(\phi - \phi_o) \quad (2)$$

Integral equations can be formulated using (1) and (2).

The zeroth-order solutions are, for $0 < \theta < \pi/2$

$$\overline{I}^{(0)}(\pi - \theta, \phi, z) = \overline{I}_o e^{\kappa_e z \sec\theta_o} \delta(\cos\theta - \cos\theta_o) \delta(\phi - \phi_o) \quad (3)$$

$$\overline{I}^{(0)}(\theta, \phi, z) = 0 \quad (4)$$

Thus, the zeroth-order solution for upward-going specific intensity is zero. The first-order upward-going specific intensity at $z = 0$ is

$$\overline{I}^{(1)}(\theta, \phi, z = 0) = \frac{\sec\theta}{(\sec\theta + \sec\theta_o)\kappa_e} \overline{\overline{P}}(\theta,\phi;\pi - \theta_o,\phi_o) \cdot \overline{I}_o \quad (5)$$

The first-order bistatic scattering coefficients can be calculated

$$\begin{bmatrix} \gamma_{vv}^{(1)} \\ \gamma_{vh}^{(1)} \\ \gamma_{hv}^{(1)} \\ \gamma_{hh}^{(1)} \end{bmatrix} (\theta, \phi; \theta_o, \phi_o) = \frac{4\pi \cos\theta}{\kappa_e(\cos\theta + \cos\theta_o)} \begin{bmatrix} P_{vv}(\theta, \phi; \pi - \theta_o, \phi_o) \\ P_{vh}(\theta, \phi; \pi - \theta_o, \phi_o) \\ P_{hv}(\theta, \phi; \pi - \theta_o, \phi_o) \\ P_{hh}(\theta, \phi; \pi - \theta_o, \phi_o) \end{bmatrix} \quad (6)$$

where
$$P_{ij}(\theta, \phi; \theta', \phi') = n_o |f_{ij}(\theta, \phi; \theta', \phi')|^2 \quad (7)$$

with $i, j = v, h$. For spherical particles, using the results of Section 5.5 of Chapter 3 and substituting into (17), Section 5.2 (Chapter 3),

$$\overline{\overline{F}}(\theta, \phi; \theta', \phi') = \frac{1}{ik} \sum_{n,m} (-1)^m \frac{(2n+1)}{n(n+1)} \left\{ T_n^{(M)} \overline{C}_{mn}(\theta, \phi) \overline{C}_{-mn}(\theta', \phi') \right.$$
$$\left. + T_n^{(N)} \overline{B}_{mn}(\theta, \phi) \overline{B}_{-mn}(\theta', \phi') \right\} \quad (8)$$

The summation over m in (8) can be carried out using the Appendix (Section 5.9) in Chapter 3. Thus, we have

$$\overline{\overline{F}}(\theta, \phi; \theta', \phi') = \frac{1}{ik} \sum_{n=1}^{\infty} \frac{(2n+1)}{n(n+1)}$$
$$\cdot \left\{ \hat{\theta}\hat{\theta}' \left[T_n^{(M)} \frac{1}{\sin\theta \sin\theta'} \frac{\partial^2 P_n}{\partial\phi\partial\phi'} + T_n^{(N)} \frac{\partial^2 P_n}{\partial\theta\partial\theta'} \right] \right.$$
$$+ \hat{\phi}\hat{\theta}' \left[-\frac{T_n^{(M)}}{\sin\theta'} \frac{\partial^2 P_n}{\partial\theta\partial\phi'} + \frac{T_n^{(N)}}{\sin\theta} \frac{\partial^2 P_n}{\partial\phi\partial\theta'} \right]$$
$$+ \hat{\theta}\hat{\phi}' \left[-\frac{T_n^{(M)}}{\sin\theta} \frac{\partial^2 P_n}{\partial\phi\partial\theta'} + \frac{T_n^{(N)}}{\sin\theta'} \frac{\partial^2 P_n}{\partial\theta\partial\phi'} \right]$$
$$\left. + \hat{\phi}\hat{\phi}' \left[T_n^{(M)} \frac{\partial^2 P_n}{\partial\theta\partial\theta'} + \frac{T_n^{(N)}}{\sin\theta \sin\theta'} \frac{\partial^2 P_n}{\partial\phi\partial\phi'} \right] \right\} \quad (9)$$

and the argument of the Legendre polynomial P_n in (9) is $[\cos\theta\cos\theta' + \sin\theta\sin\theta'\cos(\phi - \phi')]$.

Next we consider bistatic scattering in the plane of incidence by setting $\phi = \phi_o + \pi$ and $\phi' = \phi_o$. From (6) and (9), we have

$$\gamma_{vh}^{(1)}(\theta, \phi_o + \pi; \theta_o, \phi_o) = \gamma_{hv}^{(1)}(\theta, \phi_o + \pi; \theta_o, \phi_o) = 0 \quad (10)$$

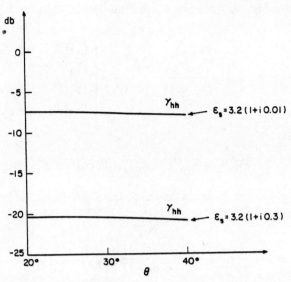

Fig. 4.9 Bistatic scattering coefficients γ_{vv} and γ_{hh} as a function of scattering angle θ for $\theta_i = 30\,\text{deg}$, $\phi = \pi + \phi_i$, fractional volume = 0.01, frequency = 10 GHz, $a = 0.1\,\text{cm}$, and for two ϵ_s: $\epsilon_s = 3.2(1+i0.01)$ and $\epsilon_s = 3.2(1+i0.3)$. In this figure, $ka = 0.209$.

Hence, the first-order bistatic scattering does not contain depolarization for scattering in the same plane. Depolarization is a higher-order effect for spherical particles. The like polarization scattering amplitudes are, from (9),

$$f_{vv}(\theta, \phi_o + \pi; \pi - \theta_o, \phi_o) = -\frac{1}{ik}\sum_{n=1}^{\infty}\frac{(2n+1)}{n(n+1)}$$
$$\times \left\{T_n^{(M)}\pi_n(\cos(\pi+\theta-\theta_o)) + T_n^{(N)}\tau_n(\cos(\pi+\theta-\theta_o))\right\} \quad (11a)$$

$$f_{hh}(\theta, \phi_o + \pi; \pi - \theta_o, \phi_o) = -\frac{1}{ik}\sum_{n=1}^{\infty}\frac{(2n+1)}{n(n+1)}$$
$$\times \left\{T_n^{(M)}\tau_n(\cos(\pi+\theta-\theta_o)) + T_n^{(N)}\pi_n(\cos(\pi+\theta-\theta_o))\right\} \quad (11b)$$

where
$$\pi_n(\mu) = \frac{dP_n(\mu)}{d\mu} \quad (12)$$

and
$$\tau_n(\mu) = \mu\pi_n(\mu) - (1-\mu^2)\frac{d\pi_n(\mu)}{d\mu} \quad (13)$$

Active Remote Sensing of a Half-Space

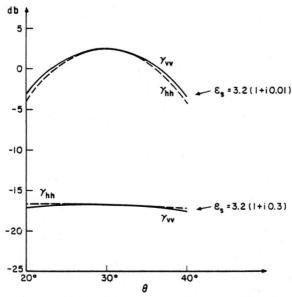

Fig. 4.10 Bistatic scattering coefficients γ_{vv} and γ_{hh} as a function of scattering angle θ for $\theta_i = 30 \deg$, $\phi = \pi + \phi_i$, fractional volume = 0.01, frequency = 50 GHz, $a = 1.0$ cm, and for two ϵ_s: $\epsilon_s = 3.2(1+i0.01)$ and $\epsilon_s = 3.2(1+i0.3)$. In this figure $ka = 10.47$.

are the π_n and τ_n functions used by Van de Hulst (1957). The a_n and b_n coefficients in his book are equal to $-T_n^{(N)*}$ and $-T_n^{(M)*}$, respectively. The relations of $S_1(\theta)$ and $S_2(\theta)$ in his book to the scattering amplitudes are $f_{vv}(\theta, \phi_o + \pi; \pi - \theta_o, \phi_o) = S_2^*(\pi + (\theta - \theta_o))/ik$ and $f_{hh}(\theta, \phi_o + \pi; \pi - \theta_o, \phi_o) = S_1^*(\pi + (\theta - \theta_o))/ik$. Extensive computer codes have been written to compute the extinction and scattering cross sections and the $S_1(\theta)$ and $S_2(\theta)$ functions. They are available commercially. Tabulations can also be found in Wichramasinghl (1973).

In Figure 4.9, we plot the bistatic scattering coefficients for small particles with $ka = 0.209$ and with two different permittivities ϵ_s. For small particles, the angular dependence of the bistatic scattering coefficient is small. Particles with a larger loss tangent have a larger albedo so that the bistatic scattering coefficients are smaller. In Figure 4.10, the bistatic scattering coefficients for larger particles ($ka = 10.47$) are shown. For the case $\epsilon_s = 3.2(1 + i0.01)$, there is a peak of the bistatic scattering coefficient in the backscattering direction $\theta = \theta_i$. This is a characteristic of the Mie scattering pattern for particles with small loss tangents and directly contributes to the *glory* effect. For more lossy particles, the peak in the backscattering direction disappears. This is evident from the bistatic

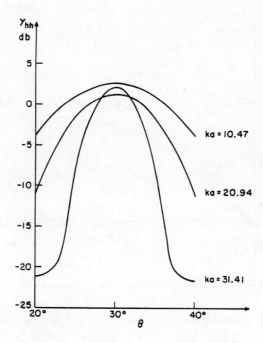

Fig. 4.11 Bistatic scattering coefficients γ_{hh} as a function of scattering angle θ for $\theta_i = 30\deg$, $\phi = \pi + \phi_i$, fractional volume $= 0.01$, frequency $= 50$ GHz, $\epsilon_s = 3.2(1 + i0.01)$, and $ka = 10.47, 20.94$, and 31.41.

scattering curve in Figure 4.10 for $\epsilon_s = 3.2(1 + i0.3)\epsilon_o$.

In Figure 4.11, the bistatic scattering coefficients for different ka values are illustrated. We note that the peak always occurs at the backscattering direction and the sharpness of the peak increases with particle size. For $ka = 31.41$, the peak value is $+2.13$ dB at $\theta = 30$ deg (backscattering) and is -21.71 dB at $\theta = 40$ deg, a difference of 23.84 dB. If the model of isotropic point scatterers is used, then from (12), Section 2.2, γ is equal to $\tilde{\omega}/2$ in the backscattering direction and does not predict the glory effect. For the curve $ka = 31.41$ in Figure 4.11, the corresponding values of κ_e and κ_s are, respectively, 0.564×10^{-2} cm^{-1} and 0.394×10^{-2} cm^{-1}, giving a value of 0.70 for $\tilde{\omega}$. Thus, γ is equal to -4.57 dB based on the isotropic scattering model. The model of isotropic scattering can also severely under-estimate the backscattering coefficient at wavelengths comparable to particle size. It is interesting to note that the results in Figure 4.11 have large absolute values for backscattering coefficients. Results of multiple scattering effects will be illustrated in Section 3 based on the discrete eigen-analysis technique (Shin and Kong, 1981; Cheung and Ishimaru, 1982).

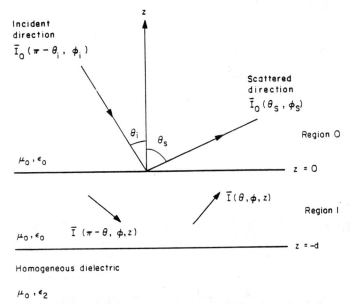

Fig. 4.12 Geometrical configuration for active remote sensing of a layer of nonspherical particles above a homogeneous dielectric half-space.

2.5 Active Remote Sensing of a Layer of Nonspherical Particles

The scattering of electromagnetic waves by nonspherical particles is important because (1) the particles in rain, ice crystals, fog, snow, leaves, etc., are nonspherical and (2) nonspherical particles can give strong depolarization return. It has been illustrated in Section 2.4 that for spherical particles, backscattering depolarization is a higher-order effect. Treatment of scattering by nonspherical particles must employ all four Stokes parameters (Sekera, 1966; Boerner et al., 1981) In this section, an iterative approach is applied to calculate the scattering cross section of a layer of randomly positioned and oriented nonspherical particles overlying a homogeneous dielectric half-space (Tsang et al., 1984). The orientation is described by a probability density function of the Eulerian angles of rotation (Chapter 3, Section 4.2).

Consider a collection of sparsely distributed nonspherical particles with permittivity ϵ_s embedded in region 1 with background permittivity ϵ_o above a half-space of homogeneous dielectric with permittivity ϵ_2 (Figure 4.12). The radiative transfer equations inside region 1 are

$$\cos\theta \frac{d}{dz}\bar{I}(\theta,\phi,z) = -\bar{\bar{\kappa}}_e(\theta,\phi) \cdot \bar{I}(\theta,\phi,z) + \bar{S}(\theta,\phi,z) \quad (1)$$

$$-\cos\theta \frac{d}{dz}\bar{I}(\pi-\theta,\phi,z) = -\bar{\bar{\kappa}}_e(\pi-\theta,\phi) \cdot \bar{I}(\pi-\theta,\phi,z) + \bar{W}(\theta,\phi,z) \quad (2)$$

where $\overline{I}(\theta, \phi, z)$ and $\overline{I}(\pi - \theta, \phi, z)$ represent, respectively, upward- and downward-going specific intensities and are 4×1 column matrices containing the four Stokes parameters. In (1) and (2), $\overline{S}(\theta, \phi, z)$ and $\overline{W}(\theta, \phi, z)$ are source terms representing scattering from other directions into the direction of propagation. They are given by

$$\overline{S}(\theta, \phi, z) = \int_0^{2\pi} d\phi' \int_0^{\pi/2} d\theta' \sin\theta' \left[\overline{\overline{P}}(\theta, \phi; \theta', \phi') \cdot \overline{I}(\theta', \phi', z) \right.$$
$$\left. + \overline{\overline{P}}(\theta, \phi; \pi - \theta', \phi') \cdot \overline{I}(\pi - \theta', \phi', z) \right] \quad (3)$$

$$\overline{W}(\theta, \phi, z) = \int_0^{2\pi} d\phi' \int_0^{\pi/2} d\theta' \sin\theta' \left[\overline{\overline{P}}(\pi - \theta, \phi; \theta', \phi') \cdot \overline{I}(\theta', \phi', z) \right.$$
$$\left. + \overline{\overline{P}}(\pi - \theta, \phi; \pi - \theta', \phi') \cdot \overline{I}(\pi - \theta', \phi', z) \right] \quad (4)$$

The extinction matrix $\overline{\overline{\kappa}}_e$ and the phase matrix for general nonspherical particles are described in Sections 3.2 and 3.4, Chapter 3.

Consider an incident wave with specific intensity $\overline{I}_o(\pi - \theta_i, \phi_i)$ impinging from region 0 upon the particles in the direction (θ_o, ϕ_o). Then the boundary conditions for the radiative transfer equations are as follows. At $z = 0$

$$\overline{I}(\pi - \theta, \phi, z = 0) = \overline{I}_o \delta(\cos\theta - \cos\theta_o) \delta(\phi - \phi_o) \quad (5)$$

and at $z = -d$

$$\overline{I}(\theta, \phi, z = -d) = \overline{\overline{R}}(\theta) \cdot \overline{I}(\pi - \theta, \phi, z = -d) \quad (6)$$

for $0 \leq \theta \leq \pi/2$. In (6), $\overline{\overline{R}}(\theta)$ is the reflectivity matrix for the interface separating region 1 and region 2 and is given by

$$\overline{\overline{R}}(\theta) = \begin{bmatrix} |R_v(\theta)|^2 & 0 & 0 & 0 \\ 0 & |R_h(\theta)|^2 & 0 & 0 \\ 0 & 0 & Re(R_v(\theta)R_h^*(\theta)) & -Im(R_v(\theta)R_h^*(\theta)) \\ 0 & 0 & Im(R_v(\theta)R_h^*(\theta)) & Re(R_v(\theta)R_h^*(\theta)) \end{bmatrix}$$
$$(7)$$

where R_v and R_h are, respectively, the Fresnel reflection coefficients for vertically and horizontally polarized waves.

To derive integral equations, we can regard $\overline{S}(\theta, \phi, z)$ and $\overline{W}(\theta, \phi, z)$ in (1) and (2) as source terms. The homogeneous solutions are then given by the eigensolutions of coherent propagation of Section 3.4 (Chapter 3).

The particular solution can be calculated by the method of variation of parameters. The arbitrary constants of the homogeneous solution are solved by imposing the boundary conditions of (5) and (6). The two coupled integral equations for the upward- and downward-going specific intensities are (Problem 6),

$$\bar{I}(\pi - \theta, \phi, z) = \bar{\bar{E}}(\pi - \theta, \phi)\bar{\bar{D}}(\beta(\pi - \theta, \phi)z \sec \theta)\bar{\bar{E}}^{-1}(\pi - \theta, \phi)\bar{I}_o$$
$$\times \delta(\cos \theta - \cos \theta_o)\delta(\phi - \phi_o)$$
$$+ \int_z^0 dz' \{\bar{\bar{E}}(\pi - \theta, \phi)\bar{\bar{D}}(\beta(\pi - \theta, \phi)(z - z') \sec \theta)$$
$$\cdot \bar{\bar{E}}^{-1}(\pi - \theta, \phi)\bar{W}(\theta, \phi, z')\} \quad (8)$$

$$\bar{I}(\theta, \phi, z) = \bar{\bar{E}}(\theta, \phi)\bar{\bar{D}}(-\beta(\theta, \phi) \sec \theta(z+d))\bar{\bar{E}}^{-1}(\theta, \phi)\bar{\bar{R}}(\theta)$$
$$\cdot \bar{\bar{E}}(\pi - \theta, \phi)\bar{\bar{D}}(-\beta(\pi - \theta, \phi)d \sec \theta)\bar{\bar{E}}^{-1}(\pi - \theta, \phi)\bar{I}_o$$
$$\times \bar{\delta}(\cos \theta - \cos \theta_o)\delta(\phi - \phi_o)$$
$$+ \bar{\bar{E}}(\theta, \phi)\bar{\bar{D}}(-\beta(\theta, \phi) \sec \theta(z+d))\int_{-d}^0 dz' \{\bar{\bar{E}}^{-1}(\theta, \phi)$$
$$\cdot \bar{\bar{R}}(\theta)\bar{\bar{E}}(\pi - \theta, \phi)\bar{\bar{D}}(-\beta(\pi - \theta, \phi) \sec \theta(z'+d))$$
$$\cdot \bar{\bar{E}}^{-1}(\pi - \theta, \phi)\bar{W}(\theta, \phi, z')\}$$
$$+ \int_{-d}^z dz' \bar{\bar{E}}(\theta, \phi)\bar{\bar{D}}(\beta(\theta, \phi) \sec \theta(z' - z))$$
$$\cdot \bar{\bar{E}}^{-1}(\theta, \phi)\bar{S}(\theta, \phi, z') \quad (9)$$

where $\bar{\bar{D}}(\beta(\theta, \phi)z \sec \theta)$ is a 4×4 diagonal matrix with the ii-th element equal to $\exp(\beta_i(\theta, \phi)z \sec \theta)$. The β_i are the eigenvalues of coherent propagation and $\bar{\bar{E}}$ is the associated eigenmatrix (Chapter 3, Section 3.4).

In applying the iteration method, we treat the first term on the right-hand side of (9) and (10) as the zeroth-order solution for the downward-going and upward-going specific intensities, respectively. The zeroth-order solution corresponds to coherent wave propagation of Stokes parameters. The first-order upward-going specific intensity at $z = 0$, is listed below (Problem 6). The first-order solution is also known as first-order multiple scattering

(Ishimaru, 1978).

$$I_l^{(1)}(\theta, \phi, z = 0) = \sum_{k,i} \sec\theta \left\{ \overline{\overline{E}}(\theta, \phi) \overline{\overline{D}}(-\beta(\theta, \phi) d \sec\theta) \right.$$

$$\left. \cdot \overline{\overline{E}}^{-1}(\theta, \phi) \overline{\overline{R}}(\theta) \overline{\overline{E}}(\pi - \theta, \phi) \right\}_{lk}$$

$$\times \left\{ \overline{\overline{E}}^{-1}(\pi - \theta, \phi) \overline{\overline{P}}(\pi - \theta, \phi; \theta_o, \phi_o) \overline{\overline{E}}(\theta_o, \phi_o) \right\}_{ki}$$

$$\times \frac{1 - e^{-\beta_k(\pi-\theta,\phi) d \sec\theta - \beta_i(\theta_o,\phi_o) d \sec\theta_o}}{\beta_k(\pi - \theta, \phi) \sec\theta + \beta_i(\theta_o, \phi_o) \sec\theta_o} \left\{ \overline{\overline{E}}^{-1}(\theta_o, \phi_o) \overline{\overline{R}}(\theta_o) \right.$$

$$\left. \cdot \overline{\overline{E}}(\pi - \theta_o, \phi_o) \overline{\overline{D}}(-\beta(\pi - \theta_o, \phi_o) d \sec\theta_o) \overline{\overline{E}}^{-1}(\pi - \theta_o, \phi_o) \overline{I}_o \right\}_i$$

$$+ \sum_{k,i} \sec\theta \left\{ \overline{\overline{E}}(\theta, \phi) \overline{\overline{D}}(-\beta(\theta, \phi) d \sec\theta) \overline{\overline{E}}^{-1}(\theta, \phi) \overline{\overline{R}}(\theta) \overline{\overline{E}}(\pi - \theta, \phi) \right\}_{lk}$$

$$\times \left\{ \overline{\overline{E}}^{-1}(\pi - \theta, \phi) \overline{\overline{P}}(\pi - \theta, \phi; \pi - \theta_o, \phi_o) \overline{\overline{E}}(\pi - \theta_o, \phi_o) \right\}_{ki}$$

$$\times \frac{e^{-\beta_k(\pi-\theta,\phi) d \sec\theta} - e^{-\beta_i(\pi-\theta_o,\phi_o) d \sec\theta_o}}{\beta_i(\pi - \theta_o, \phi_o) \sec\theta_o - \beta_k(\pi - \theta, \phi) \sec\theta} \left\{ \overline{\overline{E}}^{-1}(\pi - \theta_o, \phi_o) \overline{I}_o \right\}_i$$

$$+ \sec\theta \sum_{k,i} \overline{\overline{E}}_{lk}(\theta, \phi) \left\{ \overline{\overline{E}}^{-1}(\theta, \phi) \overline{\overline{P}}(\theta, \phi; \theta_o, \phi_o) \overline{\overline{E}}(\theta_o, \phi_o) \right\}_{ki}$$

$$\times \frac{e^{-\beta_k(\theta,\phi) d \sec\theta} - e^{-\beta_i(\theta_o,\phi_o) d \sec\theta_o}}{\beta_i(\theta_o, \phi_o) \sec\theta_o - \beta_k(\theta, \phi) \sec\theta} \left\{ \overline{\overline{E}}^{-1}(\theta_o, \phi_o) \overline{\overline{R}}(\theta_o) \right.$$

$$\left. \cdot \overline{\overline{E}}(\pi - \theta_o, \phi_o) \overline{\overline{D}}(-\beta(\pi - \theta_o, \phi_o) d \sec\theta_o) \overline{\overline{E}}^{-1}(\pi - \theta_o, \phi_o) \overline{I}_o \right\}_i$$

$$+ \sec\theta \sum_{k,i} \overline{\overline{E}}_{lk}(\theta, \phi) \left\{ \overline{\overline{E}}^{-1}(\theta, \phi) \overline{\overline{P}}(\theta, \phi; \pi - \theta_o, \phi_o) \overline{\overline{E}}(\pi - \theta_o, \phi_o) \right\}_{ki}$$

$$\times \frac{1 - e^{-\beta_k(\theta,\phi) d \sec\theta - \beta_i(\pi-\theta_o,\phi_o) d \sec\theta_o}}{\beta_k(\theta, \phi) \sec\theta + \beta_i(\pi - \theta_o, \phi_o) \sec\theta_o} \left\{ \overline{\overline{E}}^{-1}(\pi - \theta_o, \phi_o) \overline{I}_o \right\}_i \quad (10)$$

where the summation over indices k and i are from 1 to 4 and $()_{ij}$ denotes the ij-th element of the 4×4 matrix.

Next, we illustrate the numerical results of the backscattering cross section from randomly oriented oblate spheroids as functions of incident angle and frequency. The probability distribution is $p(\beta, \gamma)$ with $\int_0^{2\pi} d\gamma \int_0^\pi d\beta\, p(\beta, \gamma) = 1$ and $p(\beta, \gamma)$ is uniformly distributed between β_1

Active Remote Sensing of a Layer of Nonspherical Particles

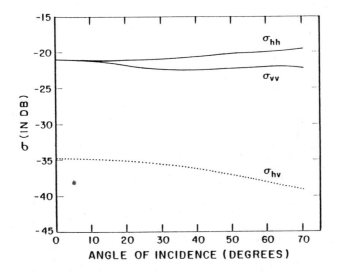

Fig. 4.13 Backscattering cross section as a function of incident angle at a frequency of 2 GHz. $d = 1$ meter, $\epsilon_2 = (10 + i)\epsilon_o$, $f = 0.004$, $\epsilon_s = (5 + i)\epsilon_o$, $a = 1$ cm, $c = 0.375$ cm, $\beta_1 = \pi/3$, $\beta_2 = \pi/2$, $\gamma_1 = 0$, and $\gamma_2 = 2\pi$.

and β_2, and γ_1 and γ_2

$$p(\beta, \gamma) = \begin{cases} \dfrac{\sin \beta}{(\cos \beta_1 - \cos \beta_2)(\gamma_2 - \gamma_1)} & \text{for } \beta_1 \leq \beta \leq \beta_2, \gamma_1 \leq \gamma \leq \gamma_2 \\ 0 & \text{otherwise} \end{cases} \quad (11)$$

We use f to denote the volume fraction of oblate spheroids, and use a and c to denote the major and minor axes of the spheroids, respectively.

The computations of the T-matrix elements in the natural frame are carried out in manners discussed in Sections 5.4 and 5.6, Chapter 3. In general, $N_{\max}$ will increase with the size of the aspect ratio of the spheroids. However, the amount of computation in calculating the T-matrix is small compared with averaging the extinction and phase matrix over orientation distribution. Hence, computation is much faster for aligned scatterers than for randomly oriented scatterers. Given the Eulerian angles, the rotation matrix can be calculated by using the formulas in Section 5.7, Chapter 3. The T-matrix elements for small spheroids can be computed by using the simpler formulas (10) through (14), Section 5.6, Chapter 3.

In Figures 4.13 and 4.14, the backscattering cross section are plotted as a function of incidence angle at 2 GHz and 5 GHz, respectively. The orientation distribution is such that the spheroids are more inclined in the vertical direction and uniformly distributed in the γ direction. Because of

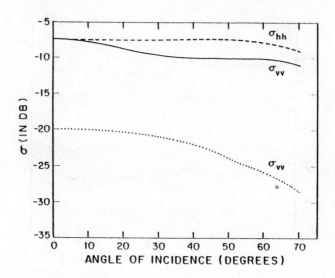

Fig. 4.14 Backscattering cross section as a function of incident angle at a frequency of 5GHz. $d = 1$ meter, $\epsilon_2 = (10 + i)\epsilon_o$, $f = 0.004$, $\epsilon_s = (5 + i)\epsilon_o$, $a = 1$ cm, $c = 0.375$ cm, $\beta_1 = \pi/3$, $\beta_2 = \pi/2$, $\gamma_1 = 0$, and $\gamma_2 = 2\pi$.

reciprocity, $\sigma_{vh} = \sigma_{hv}$, and because of azimuthal symmetry, $\sigma_{vv} = \sigma_{hh}$ at nadir. We note that the angular dependence of like polarizations is small with σ_{hh} increasing slightly with angle and σ_{vv} decreasing with angle. Backscattering cross sections for horizontally polarized waves are generally larger than vertically polarized waves because the electric field of the horizontally polarized waves can penetrate more effectively into the spheroids.

In Figure 4.15, we show the backscattering cross section as a function of frequency. The backscattering generally increases rapidly with frequency initially and then levels off at higher frequencies. At a frequency beyond 7 GHz, the albedo of the scattering medium becomes significant so that the first-order solution of the radiative transfer equations no longer gives satisfactory results. More accurate results can be obtained either by higher-order iteration or by solution of the radiative equations by numerical methods (Section 3). In the same figure, we also compare the results computed by the T-matrix method with that based on the low-frequency approximation of T-matrix elements of (10) through (14), Section 3.5.6. The two results differ for frequencies larger than 4 GHz.

In Figure 4.16, the backscattering cross section as a function of incident angle is illustrated for oblate spheroids that are more inclined in the horizontal direction. We note that the depolarization cross section increases

Active Remote Sensing of a Layer of Nonspherical Particles

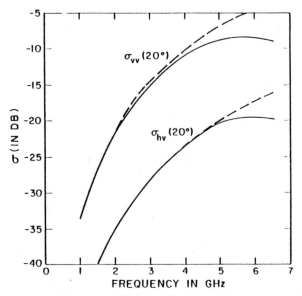

Fig. 4.15 Backscattering cross section as a function of frequency at an incident angle of 20 deg with $d = 1$ meter, $\epsilon_2 = (10 + i)\epsilon_o$, $f = 0.004$, $\epsilon_s = (5 + i)\epsilon_o$, $a = 1$ cm, $c = 0.375$ cm, $\beta_1 = \pi/3$, $\beta_2 = \pi/2$, $\gamma_1 = 0$, and $\gamma_2 = 2\pi$. Results are also compared with low-frequency approximation (dotted line) (Tsang at al., 1981).

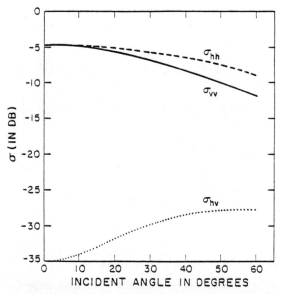

Fig. 4.16 Backscattering cross section as a function of incident angle at a frequency of 5 GHz. $d = 1$ meter, $\epsilon_2 = (10 + i)\epsilon_o$, $f = 0.004$, $\epsilon_s = (5 + i)\epsilon_o$, $a = 1$ cm, $c = 0.375$ cm, $\beta_1 = 0$, $\beta_2 = 30$ deg, $\gamma_1 = 0$, and $\gamma_2 = 2\pi$.

248   4. Solutions of Radiative Transfer Equations

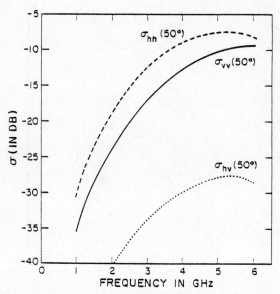

Fig. 4.17 Backscattering cross section as a function of frequency at an incident angle of 50 deg with $d = 1$ meter, $\epsilon_2 = (10+i)\epsilon_o$, $f = 0.004$, $\epsilon_s = (5+i)\epsilon_o$, $a = 1$ cm, $c = 0.375$ cm, $\beta_1 = 0$, $\beta_2 = \pi/6$, $\gamma_1 = 0$, and $\gamma_2 = 2\pi$.

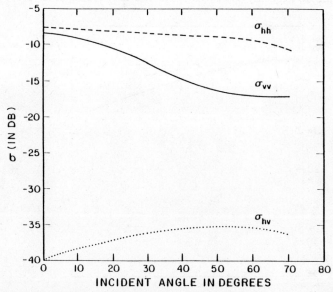

Fig. 4.18 Backscattering cross section as a function of incident angle at a frequency of 4 GHz. $d = 1$ meter. $\epsilon_2 = (10+i)\epsilon_o$, $f = 0.004$, $\epsilon_s = (5+i)\epsilon_o$, $a = 1$ cm, $c = 0.375$ cm, $\beta_1 = 25$ deg, $\beta_2 = 35$ deg, $\gamma_1 = 160$ deg, $\gamma_2 = 200$ deg, and $\phi_i = 0$.

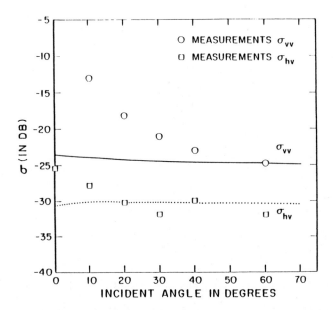

Fig. 4.19 Comparison of theoretical results with backscattering measurements from soybean at 1.1 GHz. The parameters used in the theoretical model are $\epsilon_2 = 4\epsilon_o$, $d = 1$ meter, $f = 0.003$, $\epsilon_s = (30.66 + i1.7)\epsilon_o$, $a = 1.5$ cm, $c = 0.02$ cm, $\beta_1 = 0$, $\beta_2 = \pi/2$, $\gamma_1 = 0$, and $\gamma_2 = 2\pi$.

with incident angle. In this case, σ_{hh} is also larger than σ_{vv}. The frequency dependence of backscattering cross sections for this orientation is shown in Figure 4.17. The trend is similar to that of vertically inclined spheroids.

In Figure 4.18, we consider the case when the symmetry axes of the spheroids are lying close to the direction $1/2\hat{x} + \sqrt{3}/2\hat{z}$. The distribution is no longer azimuthal symmetric. Hence, σ_{vv} is not equal to σ_{hh} at nadir. We note from the figure that σ_{hh} is significantly larger than σ_{vv}.

To illustrate the usefulness of the theory, a comparison is made between theoretical results and backscattering measurements from soybeans at 1.1 GHz in Figure 4.19 (Dobson et al., 1977). It is well known that scattering at a small angle of incidence is dominated by rough surface effects that have not been taken into account. From the figure, it can be seen that the theoretical results agree reasonably well with the experimental results at large angles of incidence.

Multiple scattering by aligned spheroids has been studied using the vector radiative transfer theory (Ishimaru et al., 1984). Scattering by randomly oriented circular disks has also been considered (Tsang et al., 1981; Karam and Fung, 1983).

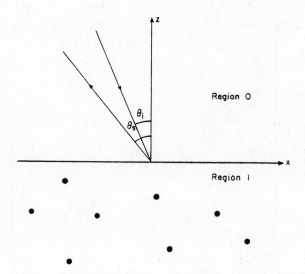

Fig. 4.20 Wave incident on a half-space of isotropic point scatterers.

2.6 Second Order Scattering from Isotropic Point Scatterers

The model of isotropic point scatterers is a convenient model to illustrate the multiple scattering processes in a scattering medium. In this section, we shall illustrate second-order scattering based on the iterative solution to the radiative transfer equations for a wave incident onto a half-space of point scatterers (Figure 4.20). The solution will be compared with the corresponding result from analytic wave theory in Chapter 5 where it will be shown that there is a difference of a factor of 2 between the two results in the direction of backscattering.

The differential radiative transfer equations will assume the form of (1) and (2), Section 2.2, and the boundary conditions are that of (3) and (4), Section 2.2, with $r(\mu)$ replaced by zero. The integral equations are, for $\mu > 0$

$$I(\tau, -\mu, \phi) = e^{\tau/\mu_i} \delta(\mu - \mu_i)\delta(\phi)$$
$$+ \frac{\tilde{\omega}}{4\pi\mu} \int_0^1 d\mu' \int_0^{2\pi} d\phi' \int_\tau^0 d\tau' \, e^{(\tau-\tau')/\mu}$$
$$\times [I(\tau', \mu', \phi') + I(\tau', -\mu', \phi')] \qquad (1)$$

$$I(\tau, \mu, \phi) = \frac{\tilde{\omega}}{4\pi\mu} \int_0^1 d\mu' \int_0^{2\pi} d\phi' \int_{-\infty}^\tau d\tau' \, e^{-(\tau-\tau')/\mu}$$
$$\times [I(\tau', \mu', \phi') + I(\tau', -\mu', \phi')] \qquad (2)$$

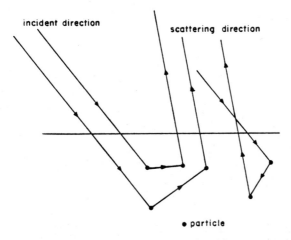

Fig. 4.21 Double scattering processes involving two particles.

The iterative solution will be carried to second order. The zeroth-order solutions are, for $\mu > 0$

$$I^{(0)}(\tau, -\mu, \phi) = \delta(\mu - \mu_i)\,\delta(\phi)\,e^{\tau/\mu_i} \tag{3a}$$

$$I^{(0)}(\tau, \mu, \phi) = 0 \tag{3b}$$

The first-order solutions are

$$I^{(1)}(\tau, \mu, \phi) = \frac{\tilde{\omega}}{4\pi\mu}\left(\frac{e^{\tau/\mu_i}}{1/\mu + 1/\mu_i}\right) \tag{4}$$

$$I^{(1)}(\tau, -\mu, \phi) = \frac{\tilde{\omega}}{4\pi\mu}\left(\frac{e^{\tau/\mu} - e^{\tau/\mu_i}}{1/\mu_i - 1/\mu}\right) \tag{5}$$

The second-order iterative solution can be calculated readily by substituting (4) and (5) into (1) and (2). The solution of the outgoing specific intensity at $\tau = 0$ is, to second order, (Problem 8)

$$I(\tau = 0, \mu, \phi) = \frac{\tilde{\omega}}{4\pi\mu}\left(\frac{1}{1/\mu + 1/\mu_i}\right)$$
$$+ \left(\frac{\tilde{\omega}}{4\pi}\right)^2 \frac{2\pi}{\mu} \int_0^1 \frac{d\mu'}{\mu'} \frac{\frac{1}{\mu} + \frac{1}{\mu_i} + \frac{2}{\mu'}}{\left(\frac{1}{\mu_i} + \frac{1}{\mu'}\right)\left(\frac{1}{\mu_i} + \frac{1}{\mu}\right)\left(\frac{1}{\mu'} + \frac{1}{\mu}\right)} \tag{6}$$

The first-order solution is proportional to the albedo $\tilde{\omega}$ and the second-order solution is proportional to $\tilde{\omega}^2$. In general, the nth order solution is

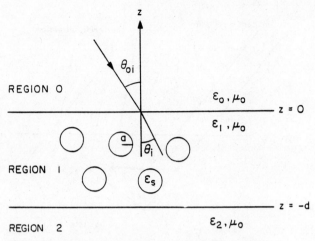

Fig. 4.22 Geometrical configuration of the problem.

proportional to $\tilde{\omega}^n$. Thus, the iterative solution converges quickly for small albedo when scattering is dominated by absorption. We also note that there is an integration over angles for the second-order solution because in the double scattering involving two particles, the direction of propagation from the first particle to the second particle can be arbitrary (Figure 4.21). Thus, the general n th order solution will involve an $(n-1)$ th-fold integration.

2.7 Second-Order Solution for Small Spherical Particles

When the effect of scattering is small, we can solve the radiative transfer equations through iterative approaches which give closed-form solutions. For the case of small spherical scatterers, the iterative solution assumes a much simpler form and the iterative solution has been carried out to second order. The radiative transfer equations and the boundary conditions are first cast into the integral equation form, using κ_s/κ_e (albedo) as small iteration parameters, which henceforth is known as κ_e iteration. The iterative procedure is applied to obtain zeroth-, first-, and second-order solutions. There is no depolarization of the backscattered intensities in the first-order solution for spherical scatterers. In the second-order solution, which accounts for the double scattering effect, we find depolarizations in the backscattered intensities. We also note that in active remote sensing, the results of scattering under the κ_e iteration are always smaller than the exact numerical solution. Thus, the κ_e iteration gives a lower bound of the scattering cross sections.

Consider a slab of homogeneous medium with permittivity $\epsilon_1 = \epsilon_1' + i\epsilon_1''$, containing small spherical particles with permittivity ϵ_s on top of a homogeneous medium with a permittivity $\epsilon_2 = \epsilon_2' + i\epsilon_2''$, as shown in Figure 4.22. For spherical particles the extinction matrix is κ_e times the unit dyad. The Rayleigh phase matrix derived in Section 4.1, Chapter 3, is used in this section. The specific intensity $\bar{I}(\theta, \phi, z)$ is a four-component vector that represents the four Stokes parameters.

To obtain the boundary conditions, first consider an incident wave with specific intensity $\bar{I}_{oi}(\pi - \theta_o, \phi_o)$ impinging from region 0, which is assumed to be free space, upon the scattering layer (Figure 4.22). The incident beam in region 0 assumes the form

$$\bar{I}_{oi}(\pi - \theta_o, \phi_o) = \bar{I}_{oi} \delta(\cos\theta_o - \cos\theta_{oi}) \delta(\phi_o - \phi_{oi}) \quad (1)$$

where the use of the Dirac delta function ensures that the incident power intercepted by the surface per unit area is

$$P_{oi} = \int_0^{\pi/2} d\theta_o \sin\theta_o \int_0^{2\pi} d\phi_o \cos\theta_o$$
$$\times \left\{ [\bar{I}_o(\pi - \theta_o, \phi_o)]_1 + [\bar{I}_o(\pi - \theta_o, \phi_o)]_2 \right\}$$
$$= (I_{oiv} + I_{oih}) \cos\theta_{oi} \quad (2)$$

The subscripts 1 and 2 in (3) denote the first and second components of the Stokes vector.

Assuming the incident beam given by (2) and using the boundary conditions [that are derived in Section 6.1, Chapter 3] for the Stokes parameters at planar dielectric interface, the boundary conditions that have to be satisfied by the solutions to the radiative transfer equations are, for $0 < \theta < \pi/2$, at $z = 0$

$$\bar{I}(\pi - \theta, \phi, z = 0) = \bar{\bar{T}}_{01}(\theta_o) \cdot \bar{I}_{oi}(\pi - \theta_o, \phi_o)$$
$$+ \bar{\bar{R}}_{10}(\theta) \cdot \bar{I}(\theta, \phi, z = 0) \quad (3a)$$

and at $z = -d$

$$\bar{I}(\theta, \phi, z = -d) = \bar{\bar{R}}_{12}(\theta) \cdot \bar{I}(\pi - \theta, \phi, z = -d) \quad (3b)$$

where we have broken up intensities in the scattering layer into upward-going intensities $\bar{I}(\theta, \phi, z)$ and downward-going intensities $\bar{I}(\pi - \theta, \phi, z)$. In the above equations, $\bar{\bar{T}}_{01}(\theta_o)$ represents the coupling from region 0 to region 1, $\bar{\bar{R}}_{10}(\theta)$ represents the coupling from upward-going intensities into downward-going intensities at the boundary of region 1 and region 0, and $\overline{\overline{R}}_{12}(\theta)$ represents the coupling from downward-going intensities into upward-going intensities at the boundary of region 1 and region 2. These coupling matrices are described in Section 6.1, Chapter 3.

Once the radiative transfer equations are solved subject to the boundary conditions, (3), the scattered intensity in the direction (θ_{os}, ϕ_{os}) in region 0 is

$$\overline{I}_o(\theta_{os}, \phi_{os}) = \overline{\overline{T}}_{10}(\theta_s) \cdot \overline{I}(\theta_s, \phi_s, z = 0) \tag{4}$$

where $\overline{\overline{T}}_{10}(\theta)$ represent the coupling from region 1 to region 0 and is obtained from $\overline{\overline{T}}_{01}(\theta_o)$ by interchanging the subscripts 0 and 1. The bistatic scattering coefficients $\gamma_{\beta\alpha}(\theta_{os}, \phi_{os}; \theta_{oi}, \phi_{oi})$ are defined as

$$\gamma_{\beta\alpha}(\theta_{os}, \phi_{os}; \theta_{oi}, \phi_{oi}) = 4\pi \frac{\cos\theta_{os}\, I_{os\beta}(\theta_{os}, \phi_{os})}{\cos\theta_{oi}\, I_{oi\alpha}} \tag{5}$$

where $\alpha, \beta = v$ or h. In the backscattering direction $\theta_{os} = \theta_{oi}$ and $\phi_{os} = \pi + \phi_{oi}$. The backscattering cross sections are defined to be

$$\sigma_{\beta\alpha}(\theta_{oi}) = \cos\theta_{oi}\, \gamma_{\beta\alpha}(\theta_{oi}, \pi + \phi_{oi}; \theta_{oi}, \phi_{oi}) \tag{6}$$

We cast the above radiative transfer equations into integral equation form using the albedo (κ_s/κ_e) as a small iterative parameter. The absorption coefficient is a sum of the absorption coefficient due to the scatterers κ_{ap} and that of the background medium κ_{ab}. The integral equations take the form

$$\overline{I}(\theta, \phi, z) = \sec\theta\, e^{-\kappa_e z \sec\theta} \int_{-d}^{z} dz'\, e^{\kappa_e z' \sec\theta}$$

$$\times \int_0^{\pi/2} d\theta' \sin\theta' \int_0^{2\pi} d\phi' \left\{ \overline{\overline{P}}(\theta, \phi; \theta', \phi') \cdot \overline{I}(\theta', \phi', z') \right.$$

$$\left. + \overline{\overline{P}}(\theta, \phi; \pi - \theta', \phi') \cdot \overline{I}(\pi - \theta', \phi', z') \right\}$$

$$+ \sec\theta\, \overline{A}(\theta, \phi)\, e^{-\kappa_e z \sec\theta} \tag{7}$$

$$\overline{I}(\pi - \theta, \phi, z) = \sec\theta\, e^{\kappa_e z \sec\theta} \int_z^0 dz'\, e^{\kappa_e z' \sec\theta}$$

$$\times \int_0^{\pi/2} d\theta' \sin\theta' \int_0^{2\pi} d\phi' \left\{ \overline{\overline{P}}(\pi - \theta, \phi; \theta', \phi') \cdot \overline{I}(\theta', \phi', z') \right.$$

$$\left. + \overline{\overline{P}}(\pi - \theta, \phi; \pi - \theta', \phi') \cdot \overline{I}(\pi - \theta', \phi', z') \right\}$$

$$+ \sec\theta\, \overline{B}(\theta, \phi)\, e^{\kappa_e z \sec\theta} \tag{8}$$

Second-Order Solution for Small Spherical Particles

where $\overline{A}(\theta, \phi)$ and $\overline{B}(\theta, \phi)$ are unknown coefficients and are determined from the boundary conditions. Refer to Problem 9 for expressions of $\overline{A}(\theta, \phi)$ and $\overline{B}(\theta, \phi)$.

The zeroth-order solutions are

$$\overline{I}^{(0)}(\theta, \phi, z) = e^{-\kappa_e z \sec \theta} \overline{\overline{F}}(\theta) \cdot \overline{\overline{R}}_{12}(\theta) \qquad (9)$$
$$\cdot \overline{\overline{T}}_{01}(\theta_o) \cdot \overline{I}_{oi}(\pi - \theta_o, \phi_o) e^{-2\kappa_e d \sec \theta}$$

$$\overline{I}^{(0)}(\pi - \theta, \phi, z) = e^{\kappa_e z \sec \theta} \overline{\overline{F}}(\theta) \cdot \overline{\overline{T}}_{01}(\theta_o) \cdot \overline{I}_{oi}(\pi - \theta_o, \phi_o) \qquad (10)$$

where

$$\overline{\overline{F}}(\theta) = \left[\overline{\overline{I}} - \overline{\overline{R}}_{10}(\theta) \cdot \overline{\overline{R}}_{12}(\theta) e^{-2\kappa_e d \sec \theta}\right]^{-1} \qquad (11)$$

Thus, the zeroth-order solution is entirely in the specular direction.

The first-order specific intensities are obtained in a straightforward manner. The first-order backscattering coefficients are

$$\sigma_{vv}^{(1)}(\theta_{oi}) = \frac{3}{2} \kappa_s \frac{\epsilon_o \cos^2 \theta_{oi}}{\epsilon_1 \cos^2 \theta_i} t_{v10}(\theta_i) t_{v01}(\theta_{oi}) [F_{11}(\theta_i)]^2$$
$$\times \left\{ (\sin^2 \theta_i - \cos^2 \theta_i)^2 2d \, r_{v12}(\theta_i) \, e^{-2\kappa_e d \sec \theta_i} \right.$$
$$+ \frac{1}{2\kappa_e \cos \theta_i} \left(1 - e^{-2\kappa_e d \sec \theta_i}\right)$$
$$\left. \times \left(1 + [r_{v12}(\theta_i)]^2 e^{-2\kappa_e d \sec \theta_i}\right) \right\} \qquad (12)$$

$$\sigma_{hh}^{(1)}(\theta_{oi}) = \frac{3}{2} \kappa_s \frac{\epsilon_o \cos^2 \theta_{oi}}{\epsilon_1' \cos^2 \theta_i} t_{h10}(\theta_i) t_{h01}(\theta_{oi}) [F_{22}(\theta_i)]^2$$
$$\times \left\{ 2d r_{h12}(\theta_i) e^{-2\kappa_e d \sec \theta_i} + \frac{1}{2\kappa_e \cos \theta_i} \right.$$
$$\left. \times \left(1 - e^{-2\kappa_e d \sec \theta_i}\right) \left(1 + [r_{h12}(\theta_i)]^2 e^{-2\kappa_e d \sec \theta_i}\right) \right\} \qquad (13)$$

$$\sigma_{hv}^{(1)}(\theta_{oi}) = \sigma_{vh}^{(1)}(\theta_{oi}) = 0 \qquad (14)$$

where

$$F_{11}(\theta) = \left[1 - r_{v10}(\theta) r_{v12}(\theta) e^{-2\kappa_e d \sec \theta}\right]^{-1} \qquad (15)$$

$$F_{22}(\theta) = \left[1 - r_{h10}(\theta) r_{h12}(\theta) e^{-2\kappa_e d \sec \theta}\right]^{-1} \qquad (16)$$

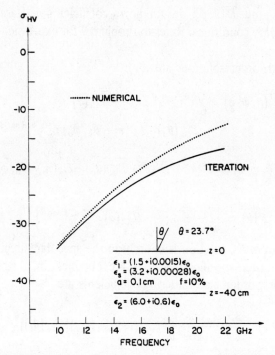

Fig. 4.23 Depolarization backscattering cross sections as a function of frequency for a 40 cm scattering layer.

We notice the absence of any depolarization effect in the backscattered power in the first-order solution since $\sigma_{hv}^{(1)}(\theta_{oi}) = \sigma_{vh}^{(1)}(\theta_{oi}) = 0$. This is expected because the first-order solution essentially accounts for the single scattering effect, and from symmetry arguments we can easily see that there is no depolarization of the backscattered power for the case of spherical scatterers. Therefore, to study the depolarization effect, we carry out the calculation to second-order. To obtain the second-order solution of backscattering cross sections requires only the calculation of the second-order upward-going intensities at $z = 0$, $I^{(2)}(\theta, \phi, z = 0)$. The final answer for the backscattering coefficient is complicated. The result for $\sigma_{hv}^{(2)}(\theta_i)$ is in the following form

$$\sigma_{hv}^{(2)}(\theta_{oi}) = 4 \frac{\epsilon_o \cos^2 \theta_{oi}}{\epsilon_1' \cos^2 \theta_i} \left(\frac{3}{8}\kappa_s\right)^2 t_{h10}(\theta_i) t_{v01}(\theta_{oi}) F_{22}(\theta_i) F_{11}(\theta_i)$$

$$\times \int_0^{\pi/2} d\theta' \sin\theta' \sec\theta' [Q_1(\theta_i, \theta') + Q_2(\theta_i, \theta') + Q_3(\theta_i, \theta')]$$
(17)

The explicit expressions for $Q_1(\theta_i, \theta'), Q_2(\theta_i, \theta')$, and $Q_3(\theta_i, \theta')$ are given in Shin and Kong (1981) (Problem 10). It can also be shown that $\sigma_{vh}^{(2)}(\theta_{oi}) = \sigma_{hv}^{(2)}(\theta_{oi})$, which is a statement of reciprocity.

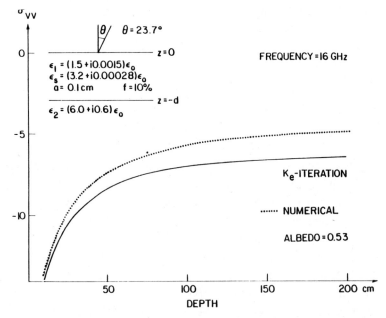

Fig. 4.24 Vertically polarized backscattering cross sections as a function of thickness of scattering layer at 16 GHz.

In Figure 4.23, the depolarization backscattering cross sections σ_{hv} are plotted as a function of frequency for the incident angle of 23.7 deg. The permittivity of the background medium ϵ_1 and the permittivity of the scatterers ϵ_s are $\epsilon_1 = (1.5 + i0.0015)\epsilon_o$ and $\epsilon_s = (3.2 + i0.00028)\epsilon_o$, which are typical values for snow and ice, respectively. The radius of the scatterers is 0.1 cm and the fractional volume occupied by the scatterers is 10%. The permittivity of the bottom layer is $\epsilon_2 = (6.0 + i0.6)\epsilon_o$, a typical value for the ground, and the thickness d of the scattering layer is 40 cm. The iterative solution is shown and also compared with the exact numerical solution calculated using the discrete-eigenanalysis method described in Section 3. We note that the iterative solution is always below that of the exact solution because multiple scattering of order higher than 2 has been neglected. At low frequencies where the albedo is small the iterative solutions agree well with the exact numerical solution. However, as the frequency is increased and the albedo becomes larger, the iterative solutions diverge from the numerical solution.

In Figure 4.24, we have plotted σ_{vv} as a function of thickness of the scattering layer at 16 GHz. For a small thickness, the iterative solutions agree well with the numerical solution, but as the thickness increases the iterative and numerical solutions are shown to diverge. Again the κ_e iterative solution is always lower than the numerical solution. We also note that the solution approaches a constant as the thickness increases to infinity.

3 DISCRETE ORDINATE-EIGENANALYSIS METHOD

The iterative method is convenient when the albedo is small so that the solution converges quickly. It also has simple closed-form solutions for the first two orders of solution. However, for general values of albedo, the integro-differential radiative transfer equations have to be solved numerically. The numerical solution includes all orders of multiple scattering within the framework of radiative transfer theory. The discrete ordinate-eigenanalysis method is applicable when the medium has homogeneous absorption and scattering profiles. In this method the continuum of propagation directions is discretized into a finite number of directions so that the integro-differential equations are converted into a system of ordinary differential equations with constant coefficients, the solution of which can be calculated by eigenanalysis. In Section 4, the method of invariant imbedding will be employed to treat problems with inhomogeneous profiles.

3.1 Radiative Transfer Equations for Passive Remote Sensing

The general radiative transfer equations for passive remote sensing are:

$$\cos\theta \frac{d}{dz}\overline{I}(\Omega, z) = \overline{J}_\epsilon - \overline{\overline{\kappa}}_{ab} \cdot \overline{I}(\Omega, z) - \overline{\overline{\kappa}}_e(\Omega) \cdot \overline{I}(\Omega, z)$$
$$+ \int d\Omega' \overline{\overline{P}}(\Omega, \Omega') \cdot \overline{I}(\Omega', z) \quad (1)$$

In this section, we shall consider only problems that have azimuthal symmetry about the z-axis. This symmetry includes both deterministic symmetry and statistical symmetry. In such a case, the third and fourth Stokes parameters are always zero, and the first two components are sufficient. Since both I_v and I_h are independent of the angle ϕ, the $d\phi'$ integration in (1) can be carried out analytically. The result is the following two equations

$$\cos\theta \frac{d}{dz}I_v(\theta, z) = \kappa_{av}(\theta)CT(z) - \kappa_{ev}(\theta)I_v(\theta, z)$$
$$+ \int_0^\pi d\theta' \sin\theta' [(v, v')I_v(\theta', z) + (v, h')I_h(\theta', z)] \quad (2)$$

$$\cos\theta \frac{d}{dz}I_h(\theta, z) = \kappa_{ah}(\theta)CT(z) - \kappa_{eh}(\theta)I_h(\theta, z)$$

$$+ \int_0^\pi d\theta' \sin\theta' [(h,v')I_v(\theta',z) + (h,h')I_h(\theta',z)] \quad (3)$$

where the extinction coefficients $\kappa_{ev}(\theta)$ and $\kappa_{eh}(\theta)$ now include both extinction by the scatterers and the absorption by the background medium, so that

$$\kappa_{e\beta}(\theta) = \kappa_{ab}(\theta) + \kappa_{e\beta}^{(s)}(\theta) \quad (4)$$

where $\beta = v, h$ and (s) superscript denotes scatterer. In (2) and (3) $[\alpha(\theta), \beta(\theta')]$ with $\alpha, \beta = v, h$ are as defined in (3), Section 2.3. In the following sections, we shall illustrate the solutions for scattering media with three different kinds of phase matrices: laminar structure, random media with three dimensional variations, and discrete spherical particles.

3.2 Closed-Form Solution for Laminar Structure

For media with laminar structure (15) through (17), Section 4.4, Chapter 3, scattering only couples two directions, the upward-going intensity $I(\theta, z)$ and its specular downward-going counterpart $I(\pi - \theta, z)$ at the same angle θ. The two polarizations are uncoupled. The result is a set of coupled equations for the upward and downward intensities that can be solved analytically.

Using (15) through (17), Section 4.4 (Chapter 3), we have $(h, v') = (v, h') = 0$. Let I_u denote upward-going specific intensity and I_d denote downward-going specific intensity. Denoting the scattering region as region 1, we have, for $0 < \theta < \pi/2$ (Problem 11)

$$\cos\theta \frac{d}{dz} I_u = -\kappa_e I_u + \kappa_a CT(z) + \frac{1}{2}\kappa_s(P_f I_u + P_b I_d) \quad (1)$$

$$\cos\theta \frac{d}{dz} I_d = \kappa_e I_d - \kappa_a CT(z) - \frac{1}{2}\kappa_s(P_f I_d + P_b I_u) \quad (2)$$

where the subscript u denotes upward; d, downward; f, forward; and b, backward. Extinction is sum of absorption and scattering $\kappa_e = \kappa_a + \kappa_s$. For horizontal polarization:

$$\kappa_s = \frac{\delta k_{1m}^2 l_z}{\cos\theta} \frac{1 + 2k_{1m}^2 l_z^2 \cos^2\theta}{1 + 4k_{1m}^2 l_z^2 \cos^2\theta} \quad (3)$$

$$P_b = \frac{1}{1 + 2k_{1m}^2 l_z^2 \cos^2\theta} \quad (4)$$

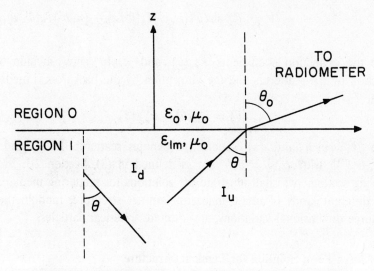

Fig. 4.25 Thermal emission of half-space of laminar structure.

For vertical polarization:

$$\kappa_s = \frac{\delta k_{1m}^2 l_z}{2 \cos \theta} \left[1 + \frac{\cos^2 2\theta}{1 + 4 k_{1m}^2 l_z^2 \cos^2 \theta} \right] \tag{5}$$

$$P_b = \frac{2 \cos^2 2\theta}{1 + 4 k_{1m}^2 l_z^2 \cos^2 \theta + \cos^2 2\theta} \tag{6}$$

and for both cases of polarization,

$$P_f = 2 - P_b \tag{7}$$

We illustrate the solution for the case of thermal emission of a half-space laminar medium (Figure 4.25). The following nonuniform temperature profile is assumed

$$T(z) = T_o + T_h e^{\gamma z} \tag{8}$$

A typical subsurface temperature profile of the Antarctica is shown in Figure 4.26. The temperature profile in the Amundsen-Scott Station (Lettau, 1971) in Antarctica is fitted with the exponentials as follows:
for December 31, (summer),

$$T_1(z) = 222 + 34 \, e^{0.81 z}$$

for August 31 (winter),

$$T_2(z) = 222 - 10 \, e^{0.37 z}$$

Closed-Form Solution for Laminar Structure

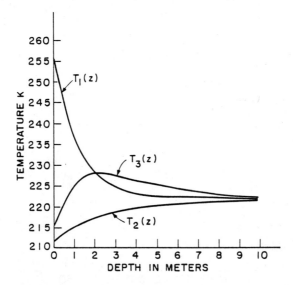

Fig. 4.26 Subsurface temperature distributions at the Amundsen-Scott Station in Antarctica (Lettau, 1971).

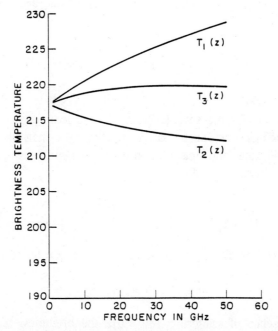

Fig. 4.27 Brightness temperature (without scattering) as a function of frequency for the three temperature distributions shown in Figure 4.26: $\epsilon'_{1m} = 1.8\epsilon_o$, $\epsilon''_{1m} = 0.00054\epsilon_o$.

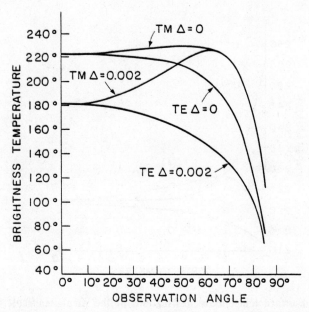

Fig. 4.28 Brightness temperature as a function of viewing angle for TE and TM waves at 20 GHz, $\epsilon'_{1m} = 1.8\epsilon_o$, $\epsilon''_{1m} = 0.00054\epsilon_o$, $l_z = 2$ mm, and $T = T_1(z)$.

for April 1 (autumn),

$$T_3(z) = 222 + 81\, e^{0.51z} - 88\, e^{0.66z}$$

where z is in meters.

The equations for the specific intensities are two coupled first-order differential equations then can be solved exactly. After matching boundary conditions, we find the following solution for the brightness temperature for the temperature distribution of (8) (Problem 11).

$$T_B(\theta_o) = \frac{2\kappa_a(1 - r_{01})}{(\alpha + \kappa_a) - r_{01}(\alpha - \kappa_a)} \left\{ T_o + \frac{\alpha}{\alpha + \gamma\cos\theta} T_h \right\} \qquad (9)$$

where r_{01} is the Fresnel reflectivity at the interface $z = 0$, and

$$\alpha = [\kappa_a(\kappa_a + \kappa_s P_b)]^{1/2} \qquad (10)$$

In the absence of scattering $(\delta = 0)$, the brightness temperatures for the three temperature distributions are illustrated in Figure 4.27. By correlating Figures 4.26 and 4.27, we note that high-frequency emission originates primarily from the surface while subsurface emission dominates the low-frequency brightness temperatures.

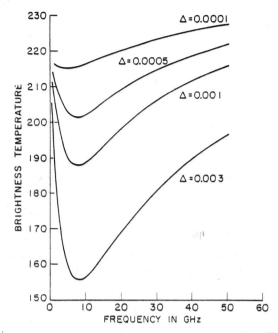

Fig. 4.29 Brightness temperature as a function of frequency for different scattering media for $T = T_1(z)$, $\epsilon'_{1m} = 1.8\epsilon_0$, $\epsilon''_{1m} = 0.00054\epsilon_0$, $l_z = 2\,\text{mm}$.

In Figure 4.28, the brightness temperature as a function of viewing angle is plotted for the case of $\delta = 0$ (no scattering) and $\delta = 0.002$ where δ is the variance of permittivity fluctuations. We note that the presence of scattering induces darkening, because scattering hinders the emission from reaching the radiometer.

In Figure 4.29, the brightness temperature is plotted as a function of frequency for different δ values. We note that scattering causes darkening in general. However, at high frequency, a significant amount of scattered energy is in the forward direction so that the emission can still reach the radiometer. This accounts for the minimum of brightness temperature at an intermediate frequency for the curve in Figure 4.29.

3.3 Discrete Ordinate–Eigenanalysis Approach

For general three-dimensional problems, the radiative transfer equations contain a continuum of directions that are coupled by scattering. In the discrete ordinate method, the directions are discretized into a finite number of directions by employing quadratures. A convenient quadrature is the Gauss-Legendre quadrature (Hildebrand, 1956; Abramowitz and Stegun,

1965).

Consider an integral

$$L = \int_{-1}^{1} d\mu \, f(\mu) \qquad (1)$$

over the interval -1 to 1. Then the integral can be approximated by

$$L = \sum_{j=-n}^{n} a_j f(\mu_j) \qquad (2)$$

where the summation j is carried over $j = \pm 1, \pm 2, \pm 3, \ldots \pm n$, μ_j are the $2n$ zeroes of the even-order Legendre polynomial $P_{2n}(\mu)$, and a_j are the Christoffel weighting functions. There are $2n$ μ_j values and $2n$ a_j values. They obey the relation

$$a_j = a_{-j} \qquad (3)$$

$$\mu_j = -\mu_{-j} \qquad (4)$$

for $j = \pm 1, \pm 2, \pm 3, \ldots \pm n$. Tabulation of μ_j values and a_j values can be found in mathematical handbooks (e.g., Abramowitz and Stegun, 1965). By letting $\mu' = \cos \theta'$, the integral in (2), Section 3.1, can be approximated by a quadrature formula as follows

$$\int_0^{\pi} d\theta' \sin \theta' [v(\theta), v(\theta')] I_v(\theta', z) = \sum_{j=-n}^{n} a_j [v(\theta), v_j] I_{v_j}(z) \qquad (5)$$

where

$$v_j = v(\theta = \cos^{-1} \mu_j) \qquad (6)$$

$$I_{v_j} = I_v(\theta = \cos^{-1} \mu_j) \qquad (7)$$

Thus, the continuum of directions $0 \leq \theta' \leq \pi$ is discretized into a finite $2n$ number of directions $\theta_j = \cos^{-1} \mu_j$, for $j = \pm 1, \pm 2, \ldots \pm n$. Since the zeroes of $P_{2n}(\mu)$ are evenly distributed about zero, for every upward direction $\theta_j = \cos^{-1} \mu_j$, there is a corresponding downward direction $\theta_{-j} = \pi - \theta_j = \cos^{-1} \mu_{-j}$. Hence there are n upward directions $\theta_j = \cos^{-1} \mu_j$ and n downward directions $\theta_{-j} = \pi - \theta_j$, with $j = 1, \ldots, n$.

We further allow the variable θ in (5) to assume the discretized values θ_j only. With this quadrature procedure, the following system of equations for passive remote sensing are obtained from (2) and (3), Section 3.1. For $i = \pm 1, \pm 2, \ldots, \pm n$

$$\mu_i \frac{d}{dz} I_{vi}(z) = \kappa_a CT - \kappa_{evi} I_{vi}(z)$$
$$+ \sum_{j=-n}^{n} a_j[(v_i, v_j) I_{vj}(z) + (v_i, h_j) I_{hj}(z)] \quad (8)$$

$$\mu_i \frac{d}{dz} I_{hi}(z) = \kappa_a CT - \kappa_{ehi} I_{hi}(z)$$
$$+ \sum_{j=-n}^{n} a_j[(h_i, v_j) I_{vj}(z) + (h_i, h_j) I_{hj}(z)] \quad (9)$$

where

$$I_{\beta i} = I_\beta(\mu_i, z) \quad (10)$$
$$\kappa_{e\beta i} = \kappa_{e\beta}(\mu_i) \quad (11)$$
$$(\alpha_i, \beta_j) = [\alpha(\mu_i), \beta(\mu_j)] \quad (12)$$

The choice of the order of quadrature n depends on the angular variation of the integrand in (5). If the variation is smooth, then a lower-order quadrature can be used. For Rayleigh scattering, an order of $n = 8$ is sufficient. For rapid variation of the integrand, a higher order of quadrature is required. Equations (8) and (9) constitute a system of $4n$ ordinary differential equations with constant coefficients.

To solve for the homogeneous solutions to (8) and (9), let

$$I_{\beta i}(z) = I_{\beta i} e^{\alpha z} \quad (13)$$

Substitute (13) in (8) and (9) in order to determine the $4n$ eigenvalues of α and the corresponding $4n$ eigenvectors. In matrix notation

$$\alpha \bar{\bar{\mu}} \cdot \bar{I}_u = -\bar{\bar{\kappa}}_e \cdot \bar{I}_u + \bar{\bar{F}} \cdot \bar{\bar{a}} \cdot \bar{I}_u + \bar{\bar{B}} \cdot \bar{\bar{a}} \cdot \bar{I}_d \quad (14)$$
$$-\alpha \bar{\bar{\mu}} \cdot \bar{I}_d = -\bar{\bar{\kappa}}_e \cdot \bar{I}_d + \bar{\bar{B}} \cdot \bar{\bar{a}} \cdot \bar{I}_u + \bar{\bar{F}} \cdot \bar{\bar{a}} \cdot \bar{I}_d \quad (15)$$

where $\bar{I}_u$ and $\bar{I}_d$ are two $2n \times 1$ matrices denoting upward- and downward-going specific intensities,

$$\bar{I}_u = \begin{bmatrix} I_{v1} \\ \vdots \\ I_{vn} \\ I_{h1} \\ \vdots \\ I_{hn} \end{bmatrix} \quad \bar{I}_d = \begin{bmatrix} I_{v-1} \\ \vdots \\ I_{v-n} \\ I_{h-1} \\ \vdots \\ I_{h-n} \end{bmatrix} \quad (16)$$

the matrices $\bar{\bar{\mu}}, \bar{\bar{\kappa}}_e$, and $\bar{\bar{a}}$ are $2n \times 2n$ diagonal matrices

$$\bar{\bar{\mu}} = \mathrm{diag}\,[\mu_1, \mu_2, \cdots, \mu_n, \mu_1, \mu_2, \cdots \mu_n] \tag{17}$$

$$\bar{\bar{\kappa}}_e = \mathrm{diag}\,[\kappa_{ev1}, \kappa_{ev2}, \cdots, \kappa_{evn}, \kappa_{eh1}, \kappa_{eh2}, \cdots, \kappa_{ehn}] \tag{18}$$

$$\bar{\bar{a}} = \mathrm{diag}\,[a_1, a_2, \cdots, a_n, a_1, a_2, \cdots a_n] \tag{19}$$

and the matrices $\bar{\bar{F}}$ and $\bar{\bar{B}}$ are forward and backward scattering phase matrices with dimensions of $2n \times 2n$.

$$\bar{\bar{F}} = \begin{bmatrix} (v_1, v_1) & \cdots & (v_1, v_n) & (v_1, h_1) & \cdots & (v_1, h_n) \\ \vdots & & \vdots & \vdots & & \vdots \\ (v_n, v_1) & \cdots & (v_n, v_n) & (v_n, h_1) & \cdots & (v_n, h_n) \\ (h_1, v_1) & \cdots & (h_1, v_n) & (h_1, h_1) & \cdots & (h_1, h_n) \\ \vdots & & \vdots & \vdots & & \vdots \\ (h_n, v_1) & \cdots & (h_n, v_n) & (h_n, h_1) & \cdots & (h_n, h_n) \end{bmatrix} \tag{20}$$

$$\bar{\bar{B}} = \begin{bmatrix} (v_1, v_{-1}) & \cdots & (v_1, v_{-n}) & (v_1, h_{-1}) & \cdots & (v_1, h_{-n}) \\ \vdots & & \vdots & \vdots & & \vdots \\ (v_n, v_{-1}) & \cdots & (v_n, v_{-n}) & (v_n, h_{-1}) & \cdots & (v_n, h_{-n}) \\ (h_1, v_{-1}) & \cdots & (h_1, v_{-n}) & (h_1, h_{-1}) & \cdots & (h_1, h_{-n}) \\ \vdots & & \vdots & \vdots & & \vdots \\ (h_n, v_{-1}) & \cdots & (h_n, v_{-n}) & (h_n, h_{-1}) & \cdots & (h_n, h_{-n}) \end{bmatrix} \tag{21}$$

From the symmetry relations for scattering phase functions, we have $(\alpha_i, \beta_j) = (\beta_{-j}, \alpha_{-i})$, with $\alpha, \beta = v, h$. Hence, both $\bar{\bar{F}}$ and $\bar{\bar{B}}$ are symmetric matrices.

The number of homogeneous equations can be reduced from $4n$ to $2n$ by defining

$$\bar{I}_+ = \bar{I}_u + \bar{I}_d \tag{22}$$

$$\bar{I}_- = \bar{I}_u - \bar{I}_d \tag{23}$$

Adding and subtracting (14) and (15) gives

$$\alpha \bar{\bar{\mu}} \cdot \bar{I}_- = \bar{\bar{A}} \cdot \bar{I}_+ \tag{24}$$

$$\alpha \bar{\bar{\mu}} \cdot \bar{I}_+ = \bar{\bar{W}} \cdot \bar{I}_- \tag{25}$$

where

$$\bar{\bar{A}} = -\bar{\bar{\kappa}}_e + \bar{\bar{F}} \cdot \bar{\bar{a}} + \bar{\bar{B}} \cdot \bar{\bar{a}} \tag{26}$$

$$\overline{\overline{W}} = -\overline{\overline{\kappa}}_e + \overline{\overline{F}} \cdot \overline{\overline{a}} - \overline{\overline{B}} \cdot \overline{\overline{a}} \tag{27}$$

Combining (24) and (25) gives the eigenvalue problem

$$(\overline{\overline{\mu}}^{-1} \cdot \overline{\overline{W}} \cdot \overline{\overline{\mu}}^{-1} \cdot \overline{\overline{A}} - \alpha^2)\overline{I}_+ = 0 \tag{28}$$

Thus, if α is an eigenvalue so is $-\alpha$. Equation (28) has $2n$ eigenvalues $\alpha_1^2, \alpha_2^2, \cdots, \alpha_{2n}^2$ and $2n$ eigenvectors $\overline{I}_{+1}, \overline{I}_{+2}, \cdots \overline{I}_{+2n}$. The solution is written in the following form

$$\overline{I}_+ = \sum_{l=1}^{2n} \left\{ P_l \overline{I}_{+l} e^{\alpha_l z} + P_{-l} \overline{I}_{+l} e^{-\alpha_l(z+d)} \right\} \tag{29}$$

The constants P_l and P_{-l} are to be determined from the boundary conditions. Also, $\overline{I}_-$ is determined from (24). Let superscript H denote homogeneous solution. Then,

$$\overline{I}_u^H = \frac{1}{2} \sum_{l=1}^{2n} \left\{ P_l (1 + \frac{1}{\alpha_l} \overline{\overline{\mu}}^{-1} \cdot \overline{\overline{A}}) \cdot \overline{I}_{+l} e^{\alpha_l z} \right.$$
$$\left. + P_{-l}(1 - \frac{1}{\alpha_l} \overline{\overline{\mu}}^{-1} \cdot \overline{\overline{A}}) \cdot \overline{I}_{+l} e^{-\alpha_l(z+d)} \right\} \tag{30}$$

$$\overline{I}_d^H = \frac{1}{2} \sum_{l=1}^{2n} \left\{ P_l (1 - \frac{1}{\alpha_l} \overline{\overline{\mu}}^{-1} \cdot \overline{\overline{A}}) \cdot \overline{I}_{+l} e^{\alpha_l z} \right.$$
$$\left. + P_{-l}(1 + \frac{1}{\alpha_l} \overline{\overline{\mu}}^{-1} \cdot \overline{\overline{A}}) \cdot \overline{I}_{+l} e^{-\alpha_l(z+d)} \right\} \tag{31}$$

The particular solution to (8) and (9) is (Problem 14), with superscript P denoting particular solution,

$$\overline{I}_u^P = \overline{I}_d^P = C\overline{T} \tag{32}$$

where $\overline{T}$ is a $2n \times 1$ column matrix with each element equal to the temperature T. The total solution is the sum of the homogeneous solution of (30) and (31) and the particular solution of (32).

The $4n$ unknown constants P_l and P_{-l} are to be determined from the boundary conditions. Suppose the scattering medium is bounded by dielectric interfaces at $z = 0$ and at $z = -d$ characterized by Fresnel reflectivity and transmissivity matrices (Figure 4.30). Then the boundary conditions are, in matrix form,

$$\bar{I}_d(z=0) = \bar{\bar{r}}_{10} \cdot \bar{I}_u(z=0) \tag{33}$$

$$\bar{I}_u(z=-d) = \bar{\bar{r}}_{12} \cdot \bar{I}_d(z=-d) + \bar{\bar{t}}_{12} \cdot C\bar{T}_2 \tag{34}$$

where

$$\bar{\bar{r}}_{10} = \text{diag}\,[r_{v10_1}, r_{v10_2}, \cdots, r_{v10_n}, r_{h10_1}, r_{h10_2}, \cdots, r_{h10_n}] \tag{35}$$

$$\bar{\bar{r}}_{12} = \text{diag}\,[r_{v12_1}, r_{v12_2}, \cdots, r_{v12_n}, r_{h12_1}, r_{h12_2}, \cdots, r_{h12_n}] \tag{36}$$

$$\bar{\bar{t}}_{12} = \text{diag}\,[t_{v12_1}, t_{v12_2}, \cdots, t_{v12_n}, t_{h12_1}, t_{h12_2}, \cdots, t_{h12_n}] \tag{37}$$

$$\bar{T}_2 = \begin{bmatrix} T_2 \\ T_2 \\ \vdots \\ T_2 \end{bmatrix} \tag{38}$$

T_2 is the temperature for the dielectric half-space below $z = -d$, $r_{\alpha 10}$ is the Fresnel reflectivity for the interface at $z = 0$ for α polarizations, and $r_{\alpha 12}$ and $t_{\alpha 12}$ are, respectively, Fresnel reflectivity and transmissivity for polarization α at the interface $z = -d$. Generalizations to other boundary conditions can be readily made. Equations (33) and (34) provide $4n$ equations for the $4n$ unknowns P_l and P_{-l}, $l = 1, 2, \ldots 2n$ and can be solved readily on a digital computer.

After the $4n$ unknown constants are determined, the brightness temperature $\bar{T}_B$ is given by

$$\bar{T}_B = \frac{1}{C} \bar{\bar{t}}_{10} \bar{I}_u(z=0) = \frac{1}{C} \bar{\bar{t}}_{10} \left\{ \bar{I}_u^H(z=0) + C\bar{T} \right\} \tag{39}$$

with

$$\bar{\bar{t}}_{10} = \bar{\bar{I}} - \bar{\bar{r}}_{10} \tag{40}$$

and $\bar{\bar{I}}$ as the $2n \times 2n$ unit matrix. Thus, $\bar{\bar{t}}_{10}$ is the transmissivity matrix for the n directions and the two polarizations at the boundary $z = 0$. In the following two sections, we illustrate the brightness temperature solutions for random media with three-dimensional variations and discrete spherical scatterers.

3.4 Thermal Emission of Three-Dimensional Random Medium

The phase matrix of a three-dimensional medium with correlation function

$$<\epsilon_{1f}(\bar{r}')\epsilon_{1f}(\bar{r}'')> = \delta\epsilon_{1m}^2 \exp\left[-\frac{|z'-z''|}{l_z} - \frac{(x'-x'')^2 + (y'-y'')^2}{l_\rho^2}\right] \quad (1)$$

has been derived in Section 4.4, Chapter 3. The coupling coefficients of passive remote sensing (v, v'), (v, h'), (h, v'), and (h, h') can be calculated by integration of corresponding phase matrix elements over $d\phi'$. They are (Tsang and Kong, 1976a)

$$(v, v') = Q(\theta, \theta') e^{-w} \left\{ \left[\sin^2\theta \sin^2\theta' + \frac{1}{2}\cos^2\theta \cos^2\theta'\right] I_0(w) \right.$$
$$\left. + 2\sin\theta \sin\theta' \cos\theta \cos\theta' I_1(w) + \frac{1}{2}\cos^2\theta \cos^2\theta' I_2(w) \right\} \quad (2)$$

$$(v, h') = Q(\theta, \theta') e^{-w} \frac{\cos^2\theta}{2} [I_0(w) - I_2(w)] \quad (3)$$

$$(h, v') = Q(\theta, \theta') e^{-w} \frac{\cos^2\theta'}{2} [I_0(w) - I_2(w)] \quad (4)$$

$$(h, h') = Q(\theta, \theta') e^{-w} \frac{1}{2} [I_0(w) + I_2(w)] \quad (5)$$

where

$$Q(\theta, \theta') = \frac{\delta k_{1m}^4 l_\rho^2}{4} \left[\frac{l_z}{1 + k_{1m}^2 l_z^2 (\cos\theta - \cos\theta')^2}\right]$$
$$\times \exp\left[-\frac{k_{1m}^2 l_\rho^2}{4}(\sin\theta - \sin\theta')^2\right] \quad (6)$$

$$w = \frac{k_{1m}^2 l_\rho^2}{2} \sin\theta \sin\theta' \quad (7)$$

The scattering coefficients are

$$\kappa_{sv}(\theta) = \int_0^\pi d\theta' \sin\theta' [(v', v) + (h', v)] \quad (8)$$

$$\kappa_{sh}(\theta) = \int_0^\pi d\theta' \sin\theta' [(v', h) + (h', h)] \quad (9)$$

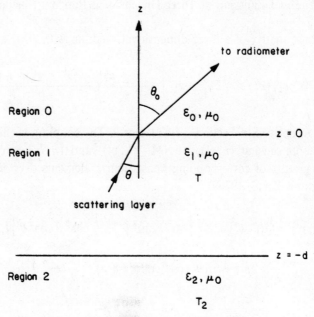

Fig. 4.30 Thermal emission of a scattering layer lying above a homogeneous dielectric half-space.

and the integration can be carried out using Gaussian-Legendre quadrature. The extinction coefficient is a summation of the absorption coefficient of the background medium and the scattering coefficient.

$$\kappa_{e\beta}(\theta) = \kappa_a + \kappa_{s\beta}(\theta) \tag{10}$$

where $\beta = v, h$ and $\kappa_a = 2k''_{1m}$.

In the following, we illustrate brightness temperature of a layer of random medium overlying a homogeneous dielectric half-space (Figure 4.30). In Figure 4.31, the vertically polarized brightness temperature for a half-space random medium as a function of viewing angle is plotted for different l_ρ values. We note that as l_ρ increases, the brightness temperature decreases because the albedo increases with l_ρ and scattering induces darkening for a half-space medium. We also note that for small l_ρ, scattering is diffuse while for $l_\rho = \infty$, results reduce to that of laminar structure where there is coupling only between the specularly related upward and downward directions. This is evident from Figure 4.31, as the brightness temperature for medium with the large l_ρ exhibits stronger angular dependence.

In Figure 4.32, the brightness temperature as a function of frequency is illustrated for a half-space random medium. We note that there is a broad minimum caused by resonant scattering analogous to the case of the laminar

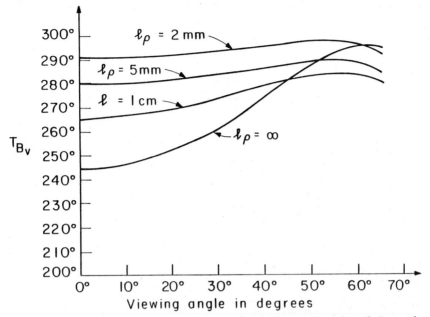

Fig. 4.31 Brightness temperature for vertical polarization as a function of observation angle for $\epsilon_{1m} = 1.8(1 + i0.0005)\epsilon_o$, frequency = 10 GHz, $l_z = 2$ mm, $T = 300$ K, and $\delta = 0.002$.

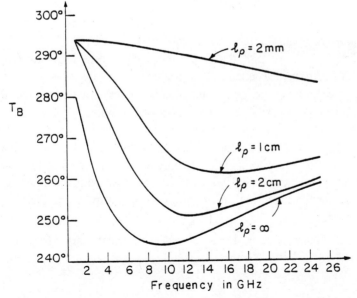

Fig. 4.32 Brightness temperature as a function of frequency for $\epsilon_{1m} = 1.8(1 + i0.0005)\epsilon_o$, $l_z = 2$ mm, $T = 300$ K, $\delta = 0.002$ and observation at nadir.

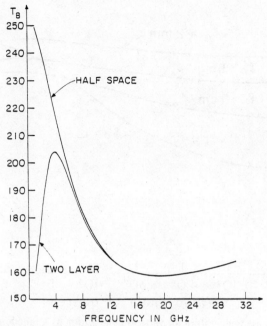

Fig. 4.33 Brightness temperature as a function of frequency for $\epsilon_{1m} = 3.2(1 + i0.0009)\epsilon_o$, $l_z = 1\,\text{mm}$, $l_\rho = 2\,\text{cm}$, $T = 273\,\text{K}$, $\delta = 0.02$ and a subsurface layer at the depth of 2 m with permittivity $\epsilon_2 = 77.2(1 + i0.17)\epsilon_o$.

structure of Section 3.2. The minimum cannot be uncovered by Rayleigh approximations or point scatterers (England, 1974, 1975).

In Figure 4.33, we plot the brightness temperature for a two-layer medium as a function of frequency. The presence of a subsurface boundary introduces a maximum at the low frequency side. At very low frequencies, the received emission at the radiometer originates primarily from the subsurface dielectric ϵ_2 which has a small emissivity. As frequency increases, brightness temperature increases since the medium ϵ_{1m}, which has a higher emissivity, begins to contribute to the received emission. As frequency further increases, scattering becomes dominant and causes a decrease in brightness temperature. This accounts for the maximum in brightness temperature in the figure.

3.5 Thermal Emission of a Layer of Spherical Scatterers Overlying a Homogeneous Dielectric Half-Space

In this section, we apply the discrete ordinate-eigenanalysis method to calculate the brightness temperature of a layer of spherical scatterers overlying a homogeneous dielectric half-space (Figure 4.30). The coupling coefficients $[v(\theta), v(\theta')]$, $[v(\theta), h(\theta')]$, $[h(\theta), v(\theta')]$ and $[h(\theta), h(\theta')]$, and the extinc-

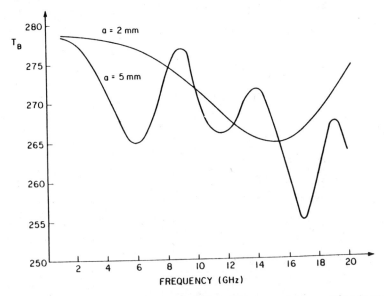

Fig. 4.34 Brightness temperatures as a function of frequency for $\epsilon_1 = 3(1+i0.0053)\epsilon_o$, $\epsilon_s = 8.3(1+i0.4)\epsilon_o$, $a = 5$ mm, $f = 0.03$, and $T = 300$ K. The results are compared for two different sphere sizes.

tion and absorption coefficients for spheres are listed in (4), Section 2.3. In general, the total absorption and extinction are the sum of the contributions of the background medium and the scatterers, i.e. $\kappa_a = \kappa_{ab} + \kappa_a^{(s)}$, and $\kappa_e = \kappa_{ab} + \kappa_e^{(s)}$ where κ_{ab} is absorption of background medium and superscript (s) denotes that of scatterers.

Half-Space Medium

In Figure 4.34, we plot the brightness temperature of a half-space of spherical scatterers. The parameters used correspond to a lunar regolith containing basaltic rock chips (England, 1975). The results are shown for scattering spheres of sizes $a = 5$ mm and $a = 2$ mm. From the figure, it can be seen that the resonance behavior of the smaller-sized particles occur at higher frequencies, while for the larger-sized particles, there are multiple resonant behavior in the frequency range plotted. In Figure 4.35, we compare the results of the computed brightness temperatures obtained with the Mie phase functions and Rayleigh phase functions. As the frequency increases, the Rayleigh result becomes inaccurate and Mie scattering must be used.

In Figure 4.36, the brightness temperatures of a half-space of medium containing nonabsorptive scatterers for two fractional volumes f are compared. As the concentration of scatterers increases, the scattering-induced darkening increases and the brightness temperature decreases. To illustrate

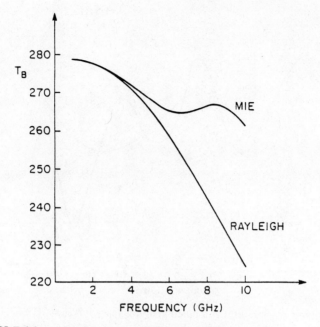

Fig. 4.35 Brightness temperatures as a function of frequency for $\epsilon_1 = 3(1+i0.0053)\epsilon_o$, $\epsilon_s = 8.3\epsilon_o$, $a = 5\,\text{mm}$, $f = 0.03$, and $T = 300\,\text{K}$. The results obtained with the Rayleigh- and the Mie-scattering phase functions are compared.

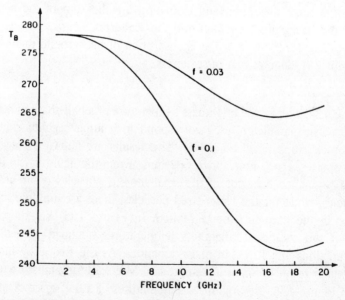

Fig. 4.36 Brightness temperatures as a function of frequency for $\epsilon_1 = 3(1+i0.0053)\epsilon_o$, $\epsilon_s = 8.3\epsilon_o$, $a = 2\,\text{mm}$, $T = 300\,\text{K}$ and for two different f values.

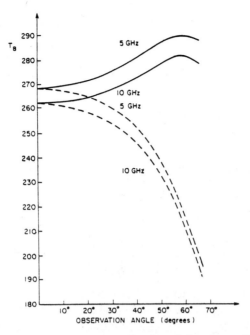

Fig. 4.37 Brightness temperatures as a function of the observation angle for $\epsilon_1 = 3(1 + i0.0053)\epsilon_o$, $\epsilon_s = 8.3\epsilon_o$, $a = 5$ mm, $f = 0.03$, and $T = 300$ K. Results for vertical polarization (solid line) and for horizontal polarizations are plotted at two different frequencies.

the angular dependence, the brightness temperatures for the vertical and horizontal polarizations at two different frequencies are plotted in Figure 4.37. We note that scattering also decreases the angular dependence of brightness temperatures.

In view of the fact that in practical cases the scatterers are not of the same size, the results are next extended to accommodate particle size distributions. We denote the number of particles per unit volume in the radius interval a and $a + da$ by $p(a)da$. In Figures 4.38 and 4.39, we plot the albedos and the brightness temperatures for the case of single scattering when $p(a)$ is uniform between $a = 2$ mm and $a = 5$ mm. Comparing with the case of uniform-sized particles, we find that the oscillations that are characteristic of Mie scattering have been smeared out and that the resonance locations are also shifted.

Experimental Data Matching with Snow Measurements

In Figures 4.40 through 4.44, we show results for interpretation of experimental data collected from a snow field (Shiue et al., 1978) using the discrete spherical scatterer model. A set of four microwave radiometers at frequencies

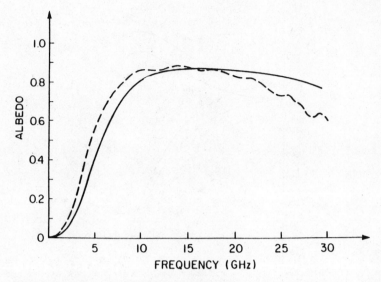

Fig. 4.38 The albedo as a function of frequency for $\epsilon_1 = 3(1+i0.0053)\epsilon_o$, $\epsilon_s = 8.3\epsilon_o$, and $f = 0.03$. The result for a uniform distribution of particle radius between 2 and 5 mm (solid line) is compared with that of single-sized spheres with radii of 5 mm (dashed line).

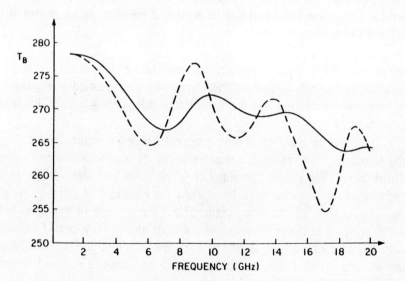

Fig. 4.39 Brightness temperatures as a function of frequency for $\epsilon_1 = 3(1+i0.0053)\epsilon_o$, $\epsilon_s = 8.3\epsilon_o$, $f = 0.03$, and $T = 300\,\text{K}$. The result for a uniform distribution of particle radii between 2 and 5 mm (solid line) is compared with that of single particles with radii of 5 mm (dashed line).

Thermal Emission of a Layer of Spherical Scatterers

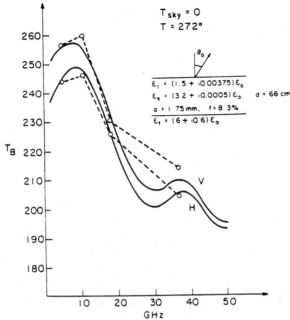

Fig. 4.40 Brightness temperature as a function of frequency. Comparison between theory and experiment.

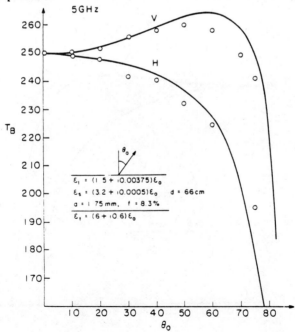

Fig. 4.41 Brightness temperature as a function of angle at 5 GHz. Comparison between theory and experiment.

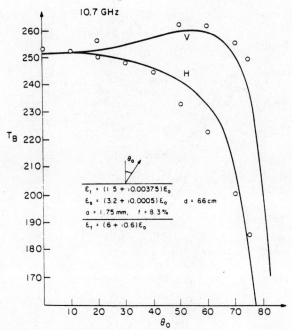

Fig. 4.42 Brightness temperature as a function of angle at 10.7 GHz. Comparison between theory and experiment.

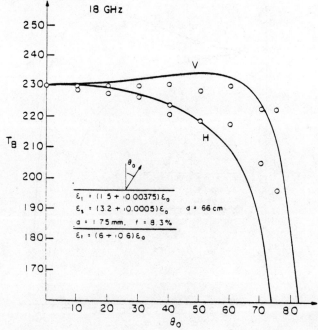

Fig. 4.43 Brightness temperature as a function of angle at 18 GHz. Comparison between theory and experiment.

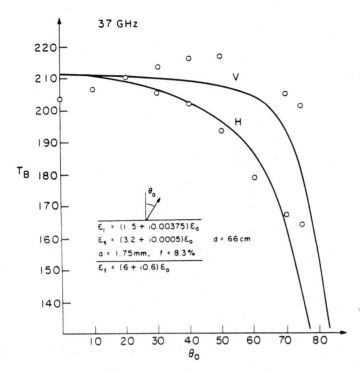

Fig. 4.44 Brightness temperature as a function of angle at 37 GHz. Comparison between theory and experiment.

5, 10.7, 18, and 37 GHz were used to measure the brightness temperatures of a snow field. In Figure 4.40, the brightness temperatures are plotted as a function of frequency for viewing angle of 33 deg, and matched with the result of a scattering layer containing 8.3% effective volume of scatterers with average diameter size of 3.5 mm.

Ground truth measurement for the depth gives $d = 66$ cm. The permittivity of the snow, the scatterer, and the ground are taken, respectively, to be $\epsilon_1 = (1.5 + i0.00375)\epsilon_o$, $\epsilon_s = (3.2 + i0.0005)\epsilon_o$, and $\epsilon_t = (6 + i0.6)\epsilon_o$. The angular dependence of the brightness temperatures at the four different frequencies are matched with the same theoretical model in Figures 4.41, 4.42, 4.43, and 4.44.

In applying the theoretical model to matching data collected from field measurements, both ϵ_1 and f are determined empirically by curve fitting of sufficient data points as functions of angle, frequency, depth, etc., over a specific area. The background dielectric constant ϵ_1 may be related to the actual physical volume occupied by ice particles constituting the snow. Snow is classified as a dense medium because the ice particles in snow occupy an

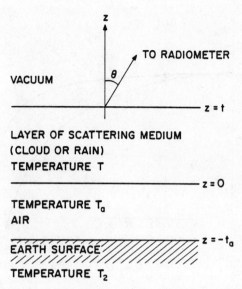

Fig. 4.45 Geometrical configuration of the problem of thermal emission of a layer of cloud or rainfall lying above the earth surface.

appreciable fractional volume. If the independent scattering model of radiative transfer theory is used together with the physical fractional volume of scatterers, the result will be an over-estimation of the scattering effect. This observation has been established both from the experimental standpoint of controlled laboratory experiments (Ishimaru and Kuga, 1982) and field measurements (Zwally, 1977; Kong et al., 1978), and from theoretical standpoint (Tsang and Kong, 1980a, 1982). The subject of scattering from dense medium will be treated in Chapter 6. Hence in matching the experimental data with radiative transfer theory in Figures 4.40 through 4.44, an empirical effective fractional volume of $f = 8.3\%$ is used to offset the over-estimation.

Thermal Emission of Cloud and Rainfall

The discrete-scatterer model has also been applied to investigate microwave thermal emission from a layer of cloud or rain (Tsang et al., 1977). The drop-size distribution and permittivity of water droplets in cloud and rainfall have been described in Section 2.3. We have used the discrete ordinate-eigenanalysis method to treat thermal emission of a layer of stratus cloud and rain (12 mm/hour) lying above the earth surface and also including the contributions of atmospheric gaseous absorption (Figure 4.45).

We first illustrate the brightness temperature for a single cloud or rain layer in free space in the absence of atmospheric gaseous absorption and the earth surface. The brightness temperature at 94 GHz as a function of layer thickness is shown in Figure 4.46. The observation angle is at nadir and the

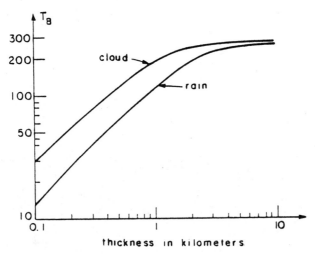

Fig. 4.46 Brightness temperatures as a function of layer thickness for cloud and rain models. Frequency is 94 GHz.

temperature T is at 273 K. The brightness temperature first increases as thickness increases and finally reaches a constant value corresponding to the half-space solution. To study the brightening and darkening effects due to scattering, we define $T_\Delta = T_B - T_{BO}$ where

$$T_{BO} = T(1 - e^{-\kappa_a t \sec\theta}) \qquad (3)$$

denotes the brightness temperature in the absence of scattering. A positive T_Δ means brightening due to scattering and a negative T_Δ means darkening.

In Figure 4.47, we plot T_Δ as a function of layer thickness. Scattering induces brightening for small optical thickness and induces darkening for large optical thickness. The brightening effect at nadir for small optical thickness is due to coupling of larger specific intensities from other directions into the nadir direction. Thus, scattering essentially increases the effective optical thickness at nadir.

We next investigate the brightness temperature for a cloud or rain layer over the earth surface. The boundary conditions for the radiative transfer equations are,

$$I_\beta(\pi - \theta, z = t) = 0 \qquad (4)$$

where $0 < \theta < \pi/2$ and β stands for v or h polarizations. This boundary condition in effect neglects the upper atmosphere, and also the cosmic background that contributes a frequency independent brightness temperature of approximately 3 K. The boundary condition at $z = 0$ is (Problem 16)

$$I_\beta(\theta, z = 0) = r_\beta(\theta) I_\beta(\pi - \theta, z = 0) e^{-2\kappa_{ag} t_a \sec\theta} + f_\beta(\theta) \qquad (5)$$

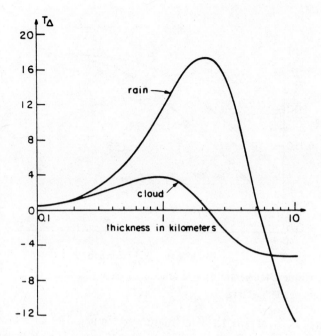

Fig. 4.47 Brightening and darkening effect T_Δ as a function of layer thickness for cloud and rain models. Frequency is 94 GHz.

where

$$f_\beta(\theta) = [1 - r_\beta(\theta)] \, CT_2 \, e^{-\kappa_{ag} t_a \sec \theta} + CT_a(1 - e^{-\kappa_{ag} t_a \sec \theta})$$
$$+ r_\beta(\theta) \, CT_a \, e^{-\kappa_{ag} t_a \sec \theta}(1 - e^{-\kappa_{ag} t_a \sec \theta}) \tag{6}$$

where $r_\beta(\theta)$ is the reflectivity for polarizations β for a specular surface at angle θ, and T_2 and T_a are the temperatures of the earth surface and the air layer, respectively. The air layer extends from $z = -t_a$ to $z = 0$ (Figure 4.45).

We note that the first term in (5) corresponds to the downward intensity of the cloud layer that is reflected by the earth surface and attenuated by the air layer in between. The second term in (5), $f_\beta(\theta)$, corresponds to the upward flux from the earth surface and the air layer. The first term in (6) corresponds to the upward emission from the earth surface attenuated in its uprising path by the intervening air layer. The second term corresponds to the upward emission by the air layer. The third term is the downward emission by the air layer that is reflected by the surface and also attenuated by the air layer. The discretized form of the boundary conditions of (5) and (6) using the finite number of angles based on the Gauss-Legendre quadrature are to be used in place of those of (33) and (34), Section 3.3. In

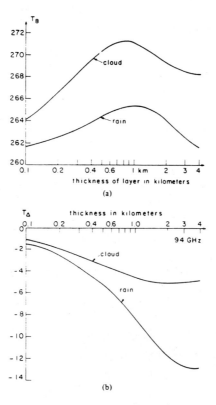

Fig. 4.48 Layer of cloud or rain on top of air layer and ocean surface. (a) Brightness temperature as a function of cloud or rain layer thickness. (b) Scattering effect T_Δ at 94 GHz for (a).

(5) and (6) κ_{ag} denotes the atmospheric gaseous absorption coefficient. In the cloud and rainfall layers, the absorption and extinction coefficients are to include κ_{ag}. Thus, $\kappa_a = \kappa_{ag} + \kappa_a^{(s)}$ and $\kappa_e = \kappa_{ag} + \kappa_e^{(s)}$ where $\kappa_a^{(s)}$ and $\kappa_e^{(s)}$ are absorption and extinction coefficients of the water droplets in cloud or rainfall.

The brightness temperature for a cloud or rain layer over ocean is illustrated in Figure 4.48. The permittivity of the ocean at this temperature is taken to be (Saxton and Lane, 1952)

$$\epsilon = \left(5.5 + \frac{65.5}{1 - i1.73/\lambda} + i\frac{80}{f}\right)\epsilon_o \qquad (7)$$

where f is frequency in GHz and λ is wavelength in cm. The ocean temperature is $T_2 = 293$ K. The cloud layer is assumed to be at a height of 6 km ($= t_a$) above the ocean surface. We examine the brightness temperature at two frequencies, 30 GHz and 94 GHz. For the purpose of

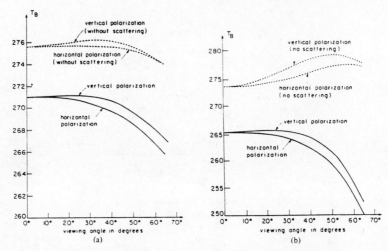

Fig. 4.49 Brightness temperature as a function of viewing angles. Shown in dotted lines are T_{BO} for vertical and horizontal polarizations. (a) Cloud layer of 1 km over ocean at 94 GHz. (b) Rain layer of 1 km over ocean at 94 GHz.

illustration, the atmospheric gaseous absorption coefficients κ_{ag} are taken to be 0.023 km^{-1} and 0.108 km^{-1} at 30 GHz and 94 GHz, respectively. The absorption coefficient κ_{ag} is assumed to be independent of height, although, in reality, κ_{ag} decreases with increasing altitude. The temperature of the air layer T_a is taken to be 293 K. In the absence of scattering, the brightness temperature of the two-layer model is given by (Problem 17)

$$T_{BO\beta}(\theta) = T\left\{1 - \exp\left[-(\kappa_a^{(s)} + \kappa_{ag})t \sec\theta\right]\right\}$$
$$+ T\, r_\beta(\theta) \exp\left[-2\kappa_{ag}t_a \sec\theta\right]\left\{1 - \exp\left[-(\kappa_a^{(s)} + \kappa_{ag})t \sec\theta\right]\right\}$$
$$\times \exp\left[-(\kappa_a^{(s)} + \kappa_{ag})t \sec\theta\right]$$
$$+ \frac{f_\beta(\theta)}{C} \exp\left[-(\kappa_a^{(s)} + \kappa_{ag})t \sec\theta\right] \quad (8)$$

We first discuss the brightness temperatures as observed from nadir and plotted as a function of thickness as shown in Figure 4.48a. The scattering effect is shown in Figure 4.48b by plotting the change in the brightness temperature compared with the no-scattering case $T_{\Delta\beta} = T_{B\beta} - T_{BO\beta}$. We note that for small thicknesses, scattering induces brightening but at the same time it also blocks the emission from the air layer and the ocean surface. These two effects tend to cancel and result in a relatively small brightening

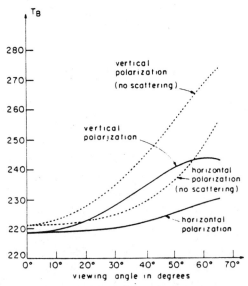

Fig. 4.50 Brightness temperature as a function of viewing angles. Shown in dotted lines are T_{BO} for vertical and horizontal polarizations. Rain layer of 1 km over ocean at 30 GHz.

or darkening effect. At 94 GHz, the atmospheric gaseous emission from the air layer is strong. Therefore, the scattering-induced blocking of emission is stronger than the scattering-induced brightening effect. Thus, we observe in Figure 4.48b, an overall darkening effect, even at small optical thickness. For large thicknesses, scattering causes darkening for both the cloud emission and the emission from the air and the ocean. These two effects reinforce each other and result in a larger darkening effect as compared with that shown in Figure 4.47.

In Figures 4.49a,b the brightness temperatures are plotted as a function of viewing angle at 94 GHz. The darkening effect is evident at larger angles of observation. In Figure 4.50, we plot the brightness temperatures as a function of viewing angle at 30 GHz. At 30 GHz, the atmospheric gaseous emission is lower so that we uncover the stronger polarization dependence due to the emission from the ocean surface. The scatter-induced darkening effect is stronger at larger angles of observation. It is also larger for the vertical polarizations because the emission from the ocean surface onto the cloud layer for this polarization is stronger.

In actual atmosphere, the atmospheric absorption and temperatures are not constants. They are functions of height. To treat scattering in the presence of inhomogeneous profiles of absorption and temperature, the method of invariant imbedding can be employed. The subject will be treated in Section 4 and illustrated for cloud, rainfall and atmospheric problems in Section 4.4.

3.6 Active Remote Sensing of a Layer of Small Spherical Scatterers Overlying a Homogeneous Dielectric Half-Space

In Section 2.7, the iteration approach was used to calculate the second-order scattering solution of a layer of small spherical scatterers overlying a homogeneous dielectric half-space (Figure 4.22). In this section, we use the discrete ordinate–eigenanalysis method to calculate the exact solution. All orders of multiple scattering within radiative transfer theory are included in this numerical solution.

For active remote sensing, unlike passive remote sensing, the azimuthal symmetry of the problem is lost even for spherical scatterers. This is due to the ϕ-dependence of the propagation direction of the incident wave and that of the incident polarization. A Fourier series expansion in the azimuthal direction is first used to eliminate the ϕ-dependence from the radiative transfer equations. Then, the set of equations without the ϕ-dependence is solved using the method of Gaussian quadrature, where the integrals in the radiative transfer equations are replaced by an appropriately weighted sum using the $2n$ zeroes of the even-order Legendre polynomial $P_{2n}(\cos\theta)$. The resulting system of first-order differential equations with constant coefficients are then solved by obtaining the eigenvalues and eigenvectors and by matching the boundary conditions.

Fourier Series Expansion in Azimuthal Direction

Starting with the radiative transfer equation, we first expand the intensities $\bar{I}(\theta, \phi, z)$ and the phase function matrix $\bar{\bar{P}}(\theta, \phi; \theta', \phi')$ into a cosine and sine series in the azimuthal direction. They are

$$\bar{I}(\theta, \phi, z) = \bar{I}^o(\theta, z) + \sum_{m=1}^{\infty} \left[\bar{I}^{mc}(\theta, z) \cos m(\phi - \phi_i) \right.$$
$$\left. + \bar{I}^{ms}(\theta, z) \sin m(\phi - \phi_i) \right] \tag{1}$$

$$\bar{\bar{P}}(\theta, \phi; \theta', \phi') = \bar{\bar{P}}^o(\theta, \theta') + \sum_{m=1}^{\infty} \left[\bar{\bar{P}}^{mc}(\theta, \theta') \cos m(\phi - \phi') \right.$$
$$\left. + \bar{\bar{P}}^{ms}(\theta, \theta') \sin m(\phi - \phi') \right] \tag{2}$$

with superscript m indicating the order of harmonics in the azimuthal direction, and superscripts c and s indicating, respectively, the cosine and sine dependences. We can substitute the Fourier series-expanded intensities and the phase function matrix into the radiative transfer equations and carry out the $d\phi'$ integration. Then, by collecting terms with the same cosine or sine dependence in the azimuthal direction we have (Problem 18), for $m = 0$

$$\cos\theta \frac{d}{dz}\overline{I}^o(\theta, z) = -\kappa_e(\theta)\overline{I}^o(\theta, z)$$

$$+ 2\pi \int_0^\pi d\theta' \sin\theta' \overline{\overline{P}}^o(\theta, \theta') \cdot \overline{I}^o(\theta', z) \quad (3)$$

and for $m \geq 1$

$$\cos\theta \frac{d}{dz}\overline{I}^{mc}(\theta,z) = -\kappa_e(\theta)\overline{I}^{mc}(\theta, z) + \pi \int_0^\pi d\theta' \sin\theta'$$

$$\times \left\{\overline{\overline{P}}^{mc}(\theta, \theta') \cdot \overline{I}^{mc}(\theta', z) - \overline{\overline{P}}^{ms}(\theta, \theta') \cdot \overline{I}^{ms}(\theta', z)\right\} \quad (4)$$

$$\cos\theta \frac{d}{dz}\overline{I}^{ms}(\theta,z) = -\kappa_e(\theta)\overline{I}^{ms}(\theta, z) + \pi \int_0^\pi d\theta' \sin\theta'$$

$$\times \left\{\overline{\overline{P}}^{ms}(\theta, \theta') \cdot \overline{I}^{mc}(\theta', z) + \overline{\overline{P}}^{mc}(\theta, \theta') \cdot \overline{I}^{ms}(\theta', z)\right\} \quad (5)$$

Solutions of Radiative Transfer Equations

The set of radiative transfer equations (3) through (5) without the ϕ-dependence can be solved using the method of Gaussian quadrature, where the integrals in the radiative transfer equations are replaced by weighted sum over $2n$ intervals between the $2n$ zeros of the even-order Legendre polynomial $P_{2n}(\theta)$. For each harmonic index m, the resulting system of first-order differential equations with constant coefficients can be solved by obtaining eigenvalues and eigenvectors. Thus, for the $m = 0$ mode, there are altogether $8n$ equations in (3) since there are $2n$ directions of propagation and four Stokes parameters for each direction of propagation. Hence, the eigen-problem is of dimension $8n$. For the $m \geq 1$ mode, there are $16n$ equations since there are $2n$ directions and four Stokes parameters for $\overline{I}^{mc}$ and four Stokes parameters for $\overline{I}^{ms}$. In practice, the solution is truncated at a maximum harmonic index $m = m_{max}$.

Boundary Conditions

To derive the boundary conditions, the incident intensities $\overline{I}_{oi}(\pi - \theta_o, \phi_o)$ are first expanded into the Fourier series

$$\overline{I}_{oi}(\pi - \theta_o, \phi_o) = \overline{I}_{oi}\,\delta(\cos\theta_o - \cos\theta_{oi})\,\delta(\phi_o - \phi_{oi})$$

$$= \overline{I}_{oi}\,\delta(\cos\theta_o - \cos\theta_{oi})$$

$$\times \left\{\frac{1}{2\pi} + \frac{1}{\pi}\sum_{m=1}^{\infty} \cos m(\phi_o - \phi_{oi})\right\} \quad (6)$$

Substituting the above expression along with the Fourier series-expanded intensities into the boundary conditions for a smooth interface and collecting terms with the same azimuthal dependence, we obtain, at $z = 0$

$$\overline{I}^o(\pi - \theta, z = 0) = \frac{1}{2\pi}\overline{\overline{T}}_{01}(\theta_o) \cdot \overline{I}_{oi}(\pi - \theta_o) + \overline{\overline{R}}_{10}(\theta) \cdot \overline{I}^o(\theta, z = 0) \quad (7)$$

and for $m \geq 1$

$$\overline{I}^{mc}(\pi - \theta, z = 0) = \frac{1}{\pi}\overline{\overline{T}}_{01}(\theta_o) \cdot \overline{I}_{oi}(\pi - \theta_o) + \overline{\overline{R}}_{10}(\theta) \cdot \overline{I}^{mc}(\theta, z = 0) \quad (8)$$

$$\overline{I}^{ms}(\pi - \theta, z = 0) = \overline{\overline{R}}_{10}(\theta) \cdot \overline{I}^{ms}(\theta, z = 0) \quad (9)$$

where

$$\overline{I}_{oi}(\pi - \theta_o) = \overline{I}_{oi}\delta(\cos\theta_o - \cos\theta_{oi}) \quad (10)$$

and at $z = -d$

$$\overline{I}^{\alpha}(\theta, z = -d) = \overline{\overline{R}}_{12}(\theta) \cdot \overline{I}^{\alpha}(\pi - \theta, z = -d) \quad (11)$$

for $\alpha = o, mc,$ or ms where $m \geq 1$. In (7) through (11), $\overline{\overline{R}}_{10}(\theta)$, $\overline{\overline{T}}_{01}(\theta_o)$ and $\overline{\overline{R}}_{12}(\theta)$ are as given in Section 6.1, Chapter 3.

Thus, the system of first-order ordinary differential equations for each mode index $m \leq m_{max}$ are solved by obtaining eigenvalues and eigenvectors and by matching the boundary conditions as in the case of passive remote sensing. Once the solutions to the radiative transfer equations without the ϕ-dependence are obtained, we can re-introduce their ϕ-dependence and calculate the scattered intensities in all directions, by recombining all the harmonic solutions to calculate the total solution according to (1).

The method above is generally applicable to scatterers of all sizes and shapes. In the following, the solution for small spherical Rayleigh scatterers is illustrated. The phase matrix is given in (7) through (17), Section 4.1, Chapter 3. Hence, $\overline{\overline{P}}^o(\theta, \theta')$, $\overline{\overline{P}}^{1c}(\theta, \theta')$, $\overline{\overline{P}}^{1s}(\theta, \theta')$, $\overline{\overline{P}}^{2c}(\theta, \theta')$ and $\overline{\overline{P}}^{2s}(\theta, \theta')$ can be readily calculated (Problem 19).

It is evident from the phase matrix that

$$\overline{\overline{P}}^{mc}(\theta, \theta') = \overline{\overline{P}}^{ms}(\theta, \theta') = 0 \quad (12)$$

for $m \geq 3$. Therefore, for $m \geq 3$, the radiative transfer equations for the Rayleigh phase function simplify to

$$\cos\theta\frac{d}{dz}\overline{I}^{mc}(\theta, z) = -\kappa_e\overline{I}^{mc}(\theta, z) \quad (13)$$

Active Remote Sensing of Small Spherical Scatterers

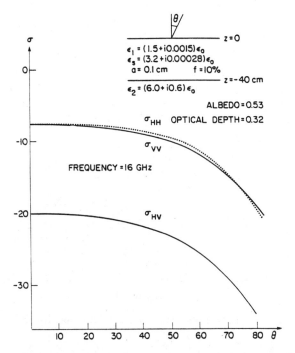

Fig. 4.51 Backscattering cross sections (like-polarization return and depolarization return) as a function of incidence angle.

$$\cos\theta \frac{d}{dz}\overline{I}^{ms}(\theta, z) = -\kappa_e \overline{I}^{ms}(\theta, z) \tag{14}$$

so that the solutions are $\overline{I}^{mc}(\theta, z) = \overline{A}^{mc} \exp(-\kappa_e z \sec\theta)$ and $\overline{I}^{ms}(\theta, z) = \overline{A}^{ms} \exp(-\kappa_e z \sec\theta)$ where $\overline{A}^{mc}$ and $\overline{A}^{ms}$ for $m \geq 3$ are constants to be determined by boundary conditions (Problem 20). The solutions of $\overline{I}^{mc}(\theta, z)$ and $\overline{I}^{ms}(\theta, z)$ for $m \leq 2$ are to be calculated by using eigenanalysis and boundary conditions.

In Figure 4.51, we plot σ_{vv}, σ_{hh} and σ_{hv} as a function of incident angle at 16 GHz to illustrate the angular dependence of the backscattering cross sections. We notice that σ_{hh} is larger than σ_{vv}. In Figure 4.52, vertically and horizontally polarized backscattering cross sections σ_{vv} and σ_{hh} are plotted as functions of the thickness of the scattering layer at a frequency of 16 GHz and at an incident angle of 30 deg. At a shallow depth, σ_{hh} is higher than σ_{vv}, and at a greater depth σ_{vv} is higher than σ_{hh}. We can attribute such phenomena to the bottom layer. When the thickness of the scattering layer is large and the effect of the bottom layer is small, σ_{vv} is greater than σ_{hh} because more intensities are being transmitted to be scattered by the particles in the vertical polarization. However, at the bottom interface, more intensities in the horizontal polarization are being

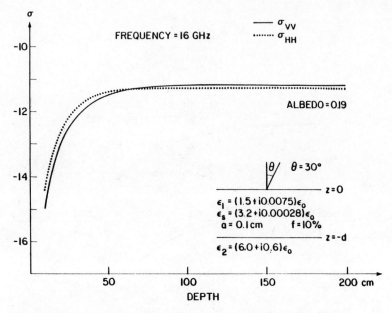

Fig. 4.52 Backscattering cross sections as a function of slab thickness d.

reflected to be scattered by the particles in the backward direction. Thus, for more shallow depths, we may have higher scattered intensities in the horizontal polarization.

In Figures 4.23 and 4.24, we have made comparisons of the results of the numerical approach with the second-order solution based on the iteration approach. It is to be noted that the second-order solution is always smaller than the numerical solution because higher-order scattering of orders larger than 2 has been neglected. Multiple scattering by larger spheres is considered in Cheung and Ishimaru (1982) and Ishimaru et al. (1982). Other theoretical treatment on radiative transfer theory as applied to remote sensing can be found in England (1974, 1975), Attema and Ulaby (1978), Blanchard and Rouse (1980), Fung and Chen (1981a,b), Tiuri (1982), and Mo et al. (1982,1984). Snow and ice measurements for active and passive remote sensing can be found in Rouse (1969), Waite and Macdonald (1969), Edgerton et al. (1971), Meir and Edgerton (1971), Johnson and Farmer (1971), Gloersen et al. (1973), Ketchum and Tooma (1973), Bryan and Larson (1975), Dunbar (1975), Meier (1975), Tooma et al. (1975), Kunzi et al. (1976), Elachi et al. (1976), Parashar et al. (1977), Shiue et al. (1978), Hofer and Shanda (1978), Hofer and Good (1979), Rango et al. (1979), Onstott et al. (1979), Campbell et al. (1980), Stiles and Ulaby (1980), Ulaby and Stiles (1980), and Matzler et al. (1982). Vegetation and soil effects can be found in MacDonald and Waite (1971), De Loor et al. (1974), Ulaby

(1975), Ulaby and Batlivala (1976), Bush and Ulaby (1976, 1978), Wang et al. (1984). Antenna effects on depolarization measurements are considered in Blanchard and Jean (1983). Principles of microwave remote sensing are considered in Moore (1966), Staelin (1969), Tomiyasu (1974), Long (1975), Flock (1979), Kritikos and Shiue (1979), and Ulaby et al. (1981). The transfer theory has also been used in light propagation through haze and cloud layers (Twomey et al., 1966; Kattawar et al., 1973).

4 METHOD OF INVARIANT IMBEDDING APPLIED TO PROBLEMS WITH INHOMOGENEOUS PROFILES

The method of solving the radiative transfer equations using the numerical quadrature eigenanalysis approach is limited to the case in which the scattering and absorption are homogeneous functions of depth. For the case of inhomogeneous profiles of scattering, absorption, and temperature, e.g., cloud and rainfall layers and snow layer in the Antarctica, the method of invariant imbedding can be employed to solve the radiative transfer equations. With the method of invariant imbedding (Bellman and Wing, 1975; Tsang and Kong, 1977), the boundary value problem of the radiative transfer equations is converted to an initial value problem starting at zero slab thickness. The equations so obtained incorporate the boundary conditions of the transfer equations. They are in the form of first-order ordinary differential equations and can be solved by standard methods of initial value problems by stepping forward in layer thickness. This method is also very convenient if brightness temperatures are to be determined as a function of slab thickness. We first consider the case of one-dimensional propagation when the fluxes are allowed to flow only in the forward and backward directions. Then the method is readily extended to the case of three-dimensional propagation.

4.1 One-Dimensional Problem

The problem of thermal emission of passive remote sensing is expressed in terms of reflection and transmission of an incident wave impinging on the inhomogeneous slab. Hence, the reflectivity and transmissivity functions are first studied.

A. *Reflection and Transmission by an Inhomogeneous Slab*

Consider a source intensity Q incident from medium 2 onto an non-emissive slab (Figure 4.53). The radiative transfer equations are given by

$$I_{r1}(z,s) = -\kappa_e(z) I_r(z,s) + f(z) I_r(z,s) + b(z) I_l(z,s) \qquad (1)$$

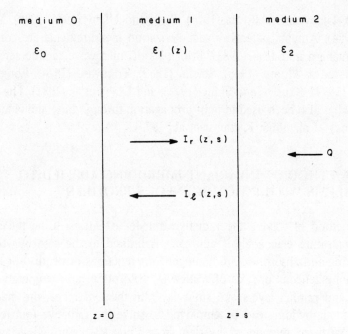

Fig. 4.53 Problem A: Reflection and transmission by an inhomogeneous slab.

$$-I_{l1}(z,s) = -\kappa_e(z)I_l(z,s) + b(z)I_r(z,s) + f(z)I_l(z,s) \qquad (2)$$

where $I_r(z,s)$ and $I_l(z,s)$ are specific intensities propagating to the right and to the left, respectively, $\kappa_e(z)$ is the extinction coefficient, $f(z)$ is the forward scattering coefficient and $b(z)$ is the backward scattering coefficient. The z argument in the extinction and scattering coefficients indicate that they are functions of position. In addition, the specific intensity is also a function of s, the thickness of the slab. We shall denote partial differentiation by a numerical index subscript denoting the variable position. Hence,

$$I_{r1}(z,s) = \frac{\partial I_r(z,s)}{\partial z} \qquad (3)$$

$$I_{r2}(z,s) = \frac{\partial I_r(z,s)}{\partial s} \qquad (4)$$

$$I_{r12}(z,s) = \frac{\partial^2 I_r(z,s)}{\partial z \partial s} \qquad (5)$$

This implies that

$$\frac{\partial}{\partial s}I_r(s,s) = I_{r1}(s,s) + I_{r2}(s,s) \qquad (6)$$

One-Dimensional Problem

The boundary condition for the radiative transfer equations are

$$I_r(0, s) = r_{01} I_l(0, s) \tag{7}$$

$$I_l(s, s) = r_{12} I_r(s, s) + t_{12} Q \tag{8}$$

where r_{01} is the reflectivity at the boundary at $z = 0$ and r_{12} and t_{12} are the reflectivity and transmissivity of the boundary at $z = s$, respectively.

In view of the linearity of (1) and (2), we can define a reflectivity function $R(s)$ and a transmissivity function $t(s)$ as follows

$$I_r(s, s) = R(s) t_{12} Q \tag{9}$$

$$I_l(0, s) = t(s) t_{12} Q \tag{10}$$

Both the reflectivity and transmissivity functions are functions of the thickness s. The aim is to form equations for $R(s)$ and $t(s)$ independent of the specific intensities $I_r(z, s)$ and $I_l(z, s)$. First we note that from (8) and (9)

$$I_l(s, s) = [1 + r_{12} R(s)] t_{12} Q \tag{11}$$

By substituting $z = s$ in the radiative transfer equations (1) and (2), $I_{r1}(s, s)$ and $I_{l1}(s, s)$ can be expressed in terms of $R(s)$ using (9) through (11). We have

$$I_{r1}(s, s) = [p(s) R(s) + b(s)] t_{12} Q \tag{12}$$

$$I_{l1}(s, s) = [\kappa_e(s) r_{12} R(s) - f(s) r_{12} R(s)$$
$$- b(s) R(s) + \kappa_e(s) - f(s)] t_{12} Q \tag{13}$$

where

$$p(s) = -\kappa_e(s) + f(s) + b(s) r_{12} \tag{14}$$

Next, we differentiate (9) and use (6)

$$\frac{dR(s)}{ds} t_{12} Q = I_{r1}(s, s) + I_{r2}(s, s) \tag{15}$$

To express I_{r2} and I_{l2} in terms of $I_{r1}(s, s)$ and $I_{l1}(s, s)$, we differentiate the radiative transfer equations and the associated boundary conditions, (1) and (2) and (7) and (8), to yield

$$I_{r21}(z, s) = -\kappa_e(z) I_{r2}(z, s) + f(z) I_{r2}(z, s) + b(z) I_{l2}(z, s) \tag{16}$$

$$-I_{l21}(z, s) = -\kappa_e(z) I_{l2}(z, s) + b(z) I_{r2}(z, s) + f(z) I_{l2}(z, s) \tag{17}$$

with boundary conditions

$$I_{r2}(0,s) = r_{01} I_{l2}(0,s) \tag{18}$$

$$I_{l2}(s,s) = r_{12} I_{r2}(s,s) + [r_{12} I_{r1}(s,s) - I_{l1}(s,s)] \tag{19}$$

On comparing (16) through (19) with the original transfer equations (1) and (2) and boundary conditions (7) and (8), we find that $I_{r2}(z,s)$ and $I_{l2}(z,s)$ satisfy the same equations and boundary conditions as $I_r(z,s)$ and $I_l(z,s)$ with a change of source term from $t_{12}Q$ to $r_{12} I_{r1}(s,s) - I_{l1}(s,s)$. Thus, by the principle of superposition and making use of the definition of the reflectivity function $R(s)$ and the transmissivity function $t(s)$, we have

$$I_{r2}(s,s) = R(s)[r_{12} I_{r1}(s,s) - I_{l1}(s,s)] \tag{20}$$

$$I_{l2}(0,s) = t(s)[r_{12} I_{r1}(s,s) - I_{l1}(s,s)] \tag{21}$$

In (12) and (13), $I_{r1}(s,s)$ and $I_{l1}(s,s)$ are in terms of $R(s)$ already. Hence, $I_{r2}(s,s)$ can be expressed in terms of $R(s)$ by using (20). The right-hand side of (15) can thus be expressed entirely in terms of $R(s)$ giving the following differential equation for $R(s)$

$$\frac{dR(s)}{ds} = b(s) + 2p(s)R(s) + q(s)R^2(s) \tag{22}$$

where

$$q(s) = (1 + r_{12}^2)b(s) - 2r_{12}(\kappa_e(s) - f(s)) \tag{23}$$

Similarly, differentiating (10) and using the (21) gives the differential equation for $t(s)$

$$\frac{dt(s)}{ds} = [p(s) + q(s)R(s)]t(s) \tag{24}$$

The initial conditions can be established by letting $s = 0$ in (9) and (10) and using (7) and (8)

$$R(s=0) = \frac{r_{01}}{1 - r_{01} r_{12}} \tag{25}$$

$$t(s=0) = \frac{1}{1 - r_{01} r_{12}} \tag{26}$$

The boundary value problem has been converted into an initial value problem which is readily solved with numerical method. Equations (22) and (25) form

One-Dimensional Problem

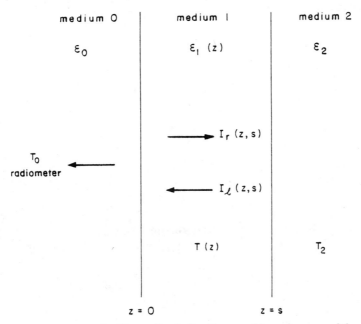

Fig. 4.54 Problem B: Thermal emission from an inhomogeneous slab.

an initial value problem for $R(s)$. Equations (24) and (26) form an initial value problem for $t(s)$ in terms of $R(s)$.

B. Thermal Emission from an Inhomogeneous Slab

We next consider the problem of emission (Figure 4.54). Medium 1 has a temperature distribution $T(z)$ and medium 2 has temperature T_2. The brightness temperature as a function of slab thickness is expressed in terms of the reflectivity and transmissivity function of Problem (A). The radiative transfer equations for Problem (B) are

$$I_{r1}(z,s) = -\kappa_e(z)I_r(z,s) + f(z)I_r(z,s)$$
$$+ b(z)I_l(z,s) + \kappa_a(z)CT(z) \qquad (27)$$

$$-I_{l1}(z,s) = -\kappa_e(z)I_l(z,s) + b(z)I_r(z,s)$$
$$+ f(z)I_l(z,s) + \kappa_a(z)CT(z) \qquad (28)$$

with boundary conditions

$$I_r(0,s) = r_{01} I_l(0,s) \qquad (29)$$

$$I_l(s,s) = r_{12} I_r(s,s) + t_{12} CT_2 \qquad (30)$$

We define the *reflected* temperature as

$$T_R(s) = \frac{1}{C} I_r(s, s) \tag{31}$$

The brightness temperature as observed by a radiometer in region 0 is

$$T_B(s) = \frac{1}{C} t_{01} I_l(0, s) \tag{32}$$

From (31)

$$\frac{dT_R(s)}{ds} = \frac{1}{C} [I_{r1}(s, s) + I_{r2}(s, s)] \tag{33}$$

The quantities $I_r(s, s)$ and $I_l(s, s)$ are expressed in terms of $T_R(s)$ through (30) and (31). Hence $I_{r1}(s, s)$ can be expressed in terms of $T_R(s)$ by letting $z = s$ in (27). To derive a relation for $I_{r2}(s, s)$ in terms of $T_R(s)$, we differentiate radiative transfer equations (27) and (28) and boundary conditions (29) and (30) with respect to s.

$$I_{r21}(z, s) = -\kappa_e(z) I_{r2}(z, s) + f(z) I_{r2}(z, s) + b(z) I_{l2}(z, s) \tag{34a}$$

$$-I_{l21}(z, s) = -\kappa_e(z) I_{l2}(z, s) + b(z) I_{r2}(z, s) + f(z) I_{l2}(z, s) \tag{34b}$$

$$I_{r2}(0, s) = r_{01} I_{l2}(0, s) \tag{35a}$$

$$I_{l2}(s, s) = r_{12} I_{r2}(s, s) + [r_{12} I_{r1}(s, s) - I_{l1}(s, s)] \tag{35b}$$

We note that the source terms disappear from the radiative transfer equations in (34). On comparison of (34) and (35) with Problem (A) of reflection and transmission, we note that they satisfy the same equations and boundary conditions with a change of source term from $t_{12}Q$ to $r_{12} I_{r1}(s, s) - I_{l1}(s, s)$ of Problem (B). Hence, using the definitions of reflectivity and transmissivity functions

$$I_{r2}(s, s) = R(s)[r_{12} I_{r1}(s, s) - I_{l1}(s, s)] \tag{36a}$$

$$I_{l2}(0, s) = t(s)[r_{12} I_{r1}(s, s) - I_{l1}(s, s)] \tag{36b}$$

The rest of the derivation is similar to the case of Problem (A) (Problem 21). We obtain

$$\frac{dT_R(s)}{ds} = [p(s) + R(s)q(s)]T_R(s) + [p(s)R(s) + b(s)]t_{12}T_2$$
$$+ \kappa_a(s)T(s)[1 + R(s)(1 + r_{12})] \tag{37}$$

One-Dimensional Problem

$$\frac{dT_B(s)}{ds} = t_{01}t(s)[T_R(s)q(s) + p(s)t_{12}T_2 + \kappa_a(s)T(s)(1+r_{12})] \quad (38)$$

with the initial conditions

$$T_R(s=0) = \frac{r_{01}t_{12}}{1-r_{01}r_{12}}T_2 \quad (39)$$

$$T_B(s=0) = \frac{t_{01}t_{12}}{1-r_{01}r_{12}}T_2 \quad (40)$$

Hence, the initial value problem for $T_R(s)$ and $T_B(s)$ is in terms of $R(s)$ and $t(s)$ of Problem (A) and must be solved in conjunction with the latter two quantities, which are governed by (22) and (24) through (26).

To illustrate the various effects due to nonuniform scattering, absorption, and temperature profiles, we use the laminar structure model of Section 3.2 which is now allowed to assume inhomogeneous profiles. In the theory developed, no restriction has been put on the profiles of $T(z)$, $\delta(z)$, $l(z)$, and $l_t(z)$ where $l_t(z)$ is defined as the loss tangent profile

$$l_t(z) = \epsilon''_{1m}(z)/\epsilon'_{1m} \quad (41)$$

We plot the numerical results corresponding to the case of a layer of ice over water. The parameters are taken to be $\epsilon'_{1m} = 3.2\epsilon_o$, $\epsilon_2 = 77.2(1+i0.17)\epsilon_o$, $T_2 = 273$ K, $l(z) = 1$ mm, and with the following profiles,

(a) Scattering profile:

$$\delta(z) = 0.01 + 0.01\exp(-0.005z)$$

$$T(z) = 273 \text{ K}$$

$$l_t(z) = 0.0009$$

(b) Scattering and temperature profiles:

$$\delta(z) = 0.01 + 0.01\exp(-0.005z)$$

$$T(z) = 273 - 33\exp(-0.01z)$$

$$l_t(z) = 0.0009$$

(c) Scattering and absorption profiles:

$$\delta(z) = 0.01 + 0.01\exp(-0.005z)$$

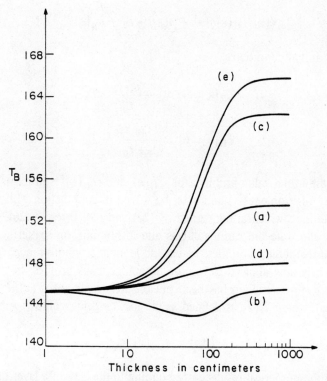

Fig. 4.55 Brightness temperatures as a function of layer thickness s for the case of laminar structure. Profiles (a), (b), (c), (d), and (e) of the case of ice over water are compared. Frequency is at 10 GHz.

$$T(z) = 273 \text{ K}$$

$$l_t(z) = 0.002 - 0.0011 \exp(-0.005z)$$

(d) Uniform profile:

$$\delta(z) = 0.02$$

$$T(z) = 273 \text{ K}$$

$$l_t(z) = 0.0009$$

(e) Scattering profile:

$$\delta(z) = 0.01 + 0.01 \exp(-0.03z)$$

$$T(z) = 273 \text{ K}$$

$$l_t(z) = 0.0009$$

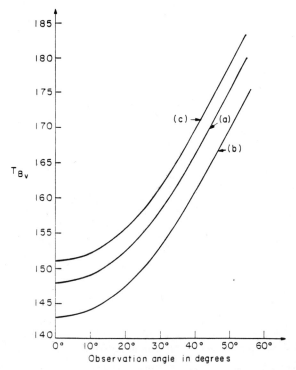

Fig. 4.56 Brightness temperatures for vertical polarization as a function of observation angle. The case is laminar ice over water. Thickness is 50 cm and frequency is at 10 GHz.

where z is distance in cm. In Figure 4.55, we plot the brightness temperature at 10 GHz as a function of layer thickness. We note that profiles (a), (c), (d), and (e) have very little difference in brightness temperatures at small layer thicknesses. For large layer thicknesses, profile (e) has the highest brightness temperature because there is least amount of scattering. For these four profiles, the brightness temperature increases as thickness increases because emission from ice is stronger than that from water. For profile (b), the brightness temperature first decreases as thickness increases because the temperature of ice in region 1 is lower than that of water.

In Figures 4.56 and 4.57, we show the angular and polarization dependences for the brightness temperature. We note that the Fresnel emissivities at the ice and water interface and at the air and ice interface are highly polarization- and angular-dependent. These two emissivities dominate the polarization and angular dependence of the brightness temperatures. Within the angular range considered, vertical polarization has a higher brightness temperature and is a monotonically increasing function of angles. The brightness temperature of horizontal polarization is a monotonically decreasing function

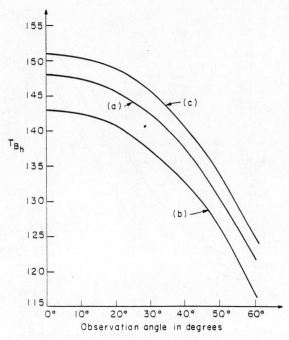

Fig. 4.57 Brightness temperatures for horizontal polarization as a function of observation angle. The case is laminar ice over water. Thickness is 50 cm and frequency is 10 GHz.

of angle.

4.2 Thermal Emission from an Inhomogeneous Slab – Three-Dimensional Problem

The invariant imbedding approach can be readily generalized to the case of three-dimensional problems. These include random media with three-dimensional variations and discrete scatterers. In three-dimensional problems, the radiative transfer equations can be cast into a system of first-order ordinary differential equations by making use of quadrature. Thus, instead of working with scalar quantities as in Section 4.1, we deal with matrix quantities. We assume that the radiative transfer equations and boundary conditions have been discretized with a finite number of $2n$ directions by using Gauss-Legendre quadrature $P_{2n}(\mu)$ as done in Section 3.3.

A. *Reflection and Transmission by an Inhomogeneous Slab*

Analogous to the one-dimensional medium case, the associated problem of reflection and transmission is to be solved in conjunction with the thermal emission problem. Consider a source intensity $\overline{Q}$ incident from medium 2 onto a non-emissive slab (Figure 4.53) where $\overline{Q}$ is a $2n \times 1$ vector containing the fluxes of the n directions with two polarizations for each direction.

In matrix notation, the discretized form of the radiative transfer equations are (Section 3.3)

$$\overline{\overline{\mu}} \cdot \overline{I}_{r1}(z,s) = -\overline{\overline{\kappa}}_e(z) \cdot \overline{I}_r(z,s) + \overline{\overline{F}}(z) \cdot \overline{\overline{a}} \cdot \overline{I}_r(z,s)$$
$$+ \overline{\overline{B}}(z) \cdot \overline{\overline{a}} \cdot \overline{I}_l(z,s) \qquad (1)$$

$$-\overline{\overline{\mu}} \cdot \overline{I}_{l1}(z,s) = -\overline{\overline{\kappa}}_e(z) \cdot \overline{I}_l(z,s) + \overline{\overline{B}}(z) \cdot \overline{\overline{a}} \cdot \overline{I}_r(z,s)$$
$$+ \overline{\overline{F}}(z) \cdot \overline{\overline{a}} \cdot \overline{I}_l(z,s) \qquad (2)$$

with boundary conditions

$$\overline{I}_r(0,s) = \overline{\overline{r}}_{01} \cdot \overline{I}_l(0,s) \qquad (3)$$

$$\overline{I}_l(s,s) = \overline{\overline{r}}_{12} \cdot \overline{I}_r(s,s) + \overline{\overline{t}}_{12} \cdot \overline{Q} \qquad (4)$$

for the problem of reflection and transmission by an inhomogeneous slab. The reflectivity function $\overline{\overline{R}}(s)$ and the transmissivity function $\overline{\overline{t}}(s)$ are defined as

$$\overline{I}_r(s,s) = \overline{\overline{R}}(s) \cdot \overline{\overline{a}} \cdot \overline{\overline{\mu}} \cdot \overline{\overline{t}}_{12} \cdot \overline{Q} \qquad (5)$$

$$\overline{I}_l(0,s) = \overline{\overline{t}}(s) \cdot \overline{\overline{a}} \cdot \overline{\overline{\mu}} \cdot \overline{\overline{t}}_{12} \cdot \overline{Q} \qquad (6)$$

The incident flux vector $\overline{Q}$ is taken to be

$$\overline{Q} = \begin{bmatrix} Q_{v_1} \\ Q_{v_2} \\ \vdots \\ Q_{v_n} \\ Q_{h_1} \\ \vdots \\ Q_{h_n} \end{bmatrix} \qquad (7)$$

In (1) through (7), all column matrices have dimension $2n$ and all the matrices have dimension $2n \times 2n$, as in Section 3.3.

The matrices $\overline{\overline{\mu}}$ and $\overline{\overline{a}}$ are, respectively, the discretized angle matrix and Christoffel matrix defined in (17) and (19), Section 3.3. The matrix $\overline{\overline{\kappa}}_e$

is the extinction matrix for the n directions for each of the two polarizations and is defined in (18), Section 3.3. The $\overline{\overline{F}}$ and $\overline{\overline{B}}$ matrices are forward and backscattering matrices defined in (20) and (21), Section 3.3, respectively, and denote the scattering coupling among the directions and the polarizations. The invariant imbedding method can accommodate for inhomogeneous profiles. Hence $\overline{\overline{\kappa}}_e(z)$, $\overline{\overline{F}}(z)$ and $\overline{\overline{B}}(z)$ are assumed to be functions of z.

With the aid of matrix notation, the derivation of the invariant imbedding equation is similar to that of the one-dimensional case. The differential equation for the reflectivity $\overline{\overline{R}}(s)$ and transmissivity function $\overline{\overline{t}}(s)$ are (Problem 22)

$$\frac{d}{ds}\overline{\overline{R}}(s) = \overline{\overline{\mu}}^{-1} \cdot \overline{\overline{B}}(s) \cdot \overline{\overline{\mu}}^{-1} - \overline{\overline{\mu}}^{-1} \cdot \overline{\overline{\kappa}}_e(s) \cdot \overline{\overline{R}}(s)$$

$$+ \overline{\overline{E}}(s) \cdot \overline{\overline{R}}(s) + \overline{\overline{R}}(s) \cdot \overline{\overline{X}}(s) \qquad (8)$$

$$\frac{d}{ds}\overline{\overline{t}}(s) = \overline{\overline{t}}(s) \cdot \overline{\overline{X}}(s) \qquad (9)$$

where

$$\overline{\overline{X}}(s) = -\overline{\overline{\kappa}}_e(s) \cdot \overline{\overline{\mu}}^{-1} + \overline{\overline{H}}(s) \cdot \overline{\overline{R}}(s) + \overline{\overline{E}}^t(s) + \overline{\overline{W}}_a(s) \cdot \overline{\overline{R}}(s) \qquad (10)$$

$$\overline{\overline{H}}(s) = -2\overline{\overline{a}} \cdot \overline{\overline{\kappa}}_e(s) \cdot \overline{\overline{r}}_{12} \qquad (11)$$

$$\overline{\overline{G}}(s) = \overline{\overline{F}}(s) + \overline{\overline{B}}(s) \cdot \overline{\overline{r}}_{12} \qquad (12)$$

$$\overline{\overline{E}}(s) = \overline{\overline{\mu}}^{-1} \cdot \overline{\overline{G}}(s) \cdot \overline{\overline{a}} \qquad (13)$$

$$\overline{\overline{W}}_a(s) = \overline{\overline{a}} \cdot \left[\overline{\overline{r}}_{12} \cdot \overline{\overline{F}}(s) \cdot \overline{\overline{a}} + \overline{\overline{F}}(s) \cdot \overline{\overline{r}}_{12} \cdot \overline{\overline{a}} \right.$$
$$\left. + \overline{\overline{r}}_{12} \cdot \overline{\overline{B}}(s) \cdot \overline{\overline{r}}_{12} \cdot \overline{\overline{a}} + \overline{\overline{B}}(s) \cdot \overline{\overline{a}} \right] \qquad (14)$$

and we have assumed $\overline{\overline{t}}_{12}, \overline{\overline{r}}_{12}, \overline{\overline{t}}_{01}, \overline{\overline{r}}_{01}$ and $\overline{\overline{\kappa}}_e$ to be diagonal matrices. The initial conditions are

$$\overline{\overline{R}}(s=0) = (\overline{\overline{a}} \cdot \overline{\overline{\mu}})^{-1} \cdot \overline{\overline{r}}_{01} \cdot (\overline{\overline{I}} - \overline{\overline{r}}_{12} \cdot \overline{\overline{r}}_{01})^{-1} \qquad (15)$$

$$\overline{\overline{t}}(s=0) = (\overline{\overline{a}} \cdot \overline{\overline{\mu}})^{-1} \cdot (\overline{\overline{I}} - \overline{\overline{r}}_{12} \cdot \overline{\overline{r}}_{01})^{-1} \qquad (16)$$

Equations (8) and (9) consist of $2(2n \times 2n) = 8n^2$ equations. They are initial value problems with thickness s as the independent variable and can be solved readily by stepping forward in thickness and using the initial conditions of (15) and (16). It can easily be shown from (8) through (16) that $\overline{\overline{R}}^t(s) = \overline{\overline{R}}(s)$, which is a statement of reciprocity (Problem 23).

B. *Thermal Emission from an Inhomogeneous Slab*

For the problem of thermal emission by the inhomogeneous slab (Figure 4.54), the radiative transfer equations become, for the case of a three-dimensional problem

$$\overline{\overline{\mu}} \cdot \overline{I}_{r1}(z,s) = -\overline{\overline{\kappa}}_e(z) \cdot \overline{I}_r(z,s) + \overline{\overline{F}}(z) \cdot \overline{\overline{a}} \cdot \overline{I}_r(z,s)$$
$$+ \overline{\overline{B}}(z) \cdot \overline{\overline{a}} \cdot \overline{I}_l(z,s) + C\overline{J}(z) \qquad (17)$$

$$-\overline{\overline{\mu}} \cdot \overline{I}_{l1}(z,s) = -\overline{\overline{\kappa}}_e(z) \cdot \overline{I}_l(z,s) + \overline{\overline{B}}(z) \cdot \overline{\overline{a}} \cdot \overline{I}_r(z,s)$$
$$+ \overline{\overline{F}}(z) \cdot \overline{\overline{a}} \cdot \overline{I}_l(z,s) + C\overline{J}(z) \qquad (18)$$

with the boundary conditions

$$\overline{I}_r(0,s) = \overline{\overline{r}}_{01} \cdot \overline{I}_l(0,s) \qquad (19)$$

$$\overline{I}_l(s,s) = \overline{\overline{r}}_{12} \cdot \overline{I}_r(s,s) + \overline{\overline{t}}_{12} \cdot C\overline{T}_2 \qquad (20)$$

The reflected temperature vector $\overline{T}_r(s)$ and the brightness temperature vector $\overline{T}_B(s)$ are defined as

$$\overline{T}_R(s) = \frac{1}{C}\overline{I}_r(s,s) \qquad (21)$$

$$\overline{T}_B(s) = \frac{1}{C}\overline{\overline{t}}_{01} \cdot \overline{I}_l(0,s) \qquad (22)$$

In (17) through (20), the emission vector is

$$\overline{J}(z) = \kappa_a(z)\, T(z)\, \overline{I} \qquad (23)$$

and

$$\overline{T}_2 = T_2\, \overline{I} \qquad (24)$$

where $\overline{I}$ is a column matrix with all $2n$ elements equal to unity.

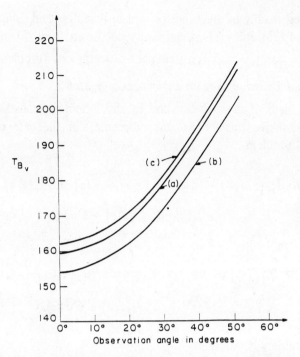

Fig. 4.58 Brightness temperatures for vertical polarization as a function of observation angle for profiles (a), (b), and (c). The case considered is ice over water with $l_\rho = 2$ cm. Thickness is 50 cm and frequency is 10 GHz. Other parameters are as in Section 4.1.

With the aid of matrix notation, the derivation of the invariant imbedding equation is similar to that for the one-dimensional case. The differential equations in terms of slab thickness s are (Problem 24)

$$\frac{d}{ds}\overline{T}_R(s) = -\overline{\overline{\mu}}^{-1} \cdot \overline{\overline{\kappa}}_e(s) \cdot \overline{T}_R(s) + \overline{\overline{E}}_s \cdot \overline{T}_R(s)$$

$$+ \overline{\overline{\mu}}^{-1} \cdot \overline{\overline{B}}(s) \cdot \overline{\overline{a}} \cdot \overline{\overline{t}}_{12} \cdot \overline{T}_2$$

$$+ \overline{\overline{\mu}}^{-1} \cdot \overline{J}(s) + \overline{\overline{R}}(s) \cdot \overline{U}(s) \tag{25}$$

$$\frac{d}{ds}\overline{T}_B(s) = \overline{\overline{t}}_{01} \cdot \overline{\overline{t}}(s) \cdot \overline{U}(s) \tag{26}$$

$$\overline{U}(s) = \overline{\overline{H}}(s) \cdot \overline{T}_R(s) + \overline{\overline{a}} \cdot (\overline{\overline{I}} + \overline{\overline{r}}_{12}) \cdot \overline{J}(s) + \overline{\overline{a}} \cdot \overline{\overline{G}}^t(s) \cdot \overline{\overline{a}} \cdot \overline{\overline{t}}_{12} \cdot \overline{T}_2$$

$$- \overline{\overline{a}} \cdot \overline{\overline{\kappa}}_e(s) \cdot \overline{\overline{t}}_{12} \cdot \overline{T}_2 + \overline{\overline{W}}_a(s) \cdot \overline{T}_R(s) \tag{27}$$

The initial conditions are

$$\overline{T}_R(s=0) = \overline{\overline{r}}_{01} \cdot (\overline{\overline{I}} - \overline{\overline{r}}_{12} \cdot \overline{\overline{r}}_{01})^{-1} \cdot \overline{\overline{t}}_{12} \cdot \overline{T}_2 \qquad (28)$$

$$\overline{T}_B(s=0) = \overline{\overline{t}}_{01} \cdot (\overline{\overline{I}} - \overline{\overline{r}}_{12} \cdot \overline{\overline{r}}_{01})^{-1} \cdot \overline{\overline{t}}_{12} \cdot \overline{T}_2 \qquad (29)$$

The quantities $\overline{\overline{H}}(s)$, $\overline{\overline{G}}(s)$, $\overline{\overline{E}}(s)$, and $\overline{\overline{W}}_a(s)$ are as defined in (11) through (14).

Equations (25) and (26) are $4n$ equations with the right-hand side dependent on $\overline{\overline{R}}(s)$ and $\overline{\overline{t}}(s)$ of Problem (A) and is to be solved in conjunction with Problem (A). Hence (8) and (9), and (25) and (26) consist of $8n^2 + 4n$ equations and can be solved by stepping forward in thickness s beginning with the initial conditions of (15) and (16) and (28) and (29). By making use of the fact that $\overline{\overline{R}}(s)$ is symmetric, the number of equations to be solved is reduced to $6n^2 + 5n$. All multiple scattering effects within the framework of radiative transfer theory are included. The procedure only involves matrix multiplication at each step. There is no eigenanalysis nor matrix inversions. The solutions for three-dimensional random medium and discrete scatterers are illustrated, respectively, in Sections 4.3 and 4.4.

4.3 Thermal Emission of Three-Dimensional Random Medium

In this section, we illustrate the results of Section 4.2 for random medium with three-dimensional variations. The coupling coefficients for passive remote sensing were derived in Section 3.4. The profiles (a), (b), and (c) of Section 4.1 are used. In Figures 4.58 and 4.59, we illustrate the angular and polarization variations of brightness temperatures for ice over water for $l_\rho = 2$ cm. Other parameters are as in Section 4.1. The brightness temperature of vertical polarization is higher than that of horizontal polarization. We note that for horizontal polarization, the brightness temperature first increases as the angle increases. This is in contrast to the case of a laminar structure (Figures 4.56 and 4.57). Three-dimensional variations allow coupling among specific intensities of different directions of propagation and polarizations. In Figures 4.58 and 4.59, scattering within the ice layer couples the two polarizations of the incoming emission from water, which is highly polarization and angular dependent.

4.4 Thermal Emission of Layers of Spherical Scatterers in the Presence of Inhomogeneous Absorption and Temperature Profiles

It is well known that atmospheric absorption and temperature profiles are inhomogeneous. Thus, in considering the scattering effects in passive remote

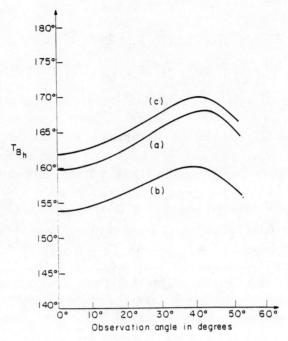

Fig. 4.59 Brightness temperatures for horizontal polarization as a function of observation angle for profiles (a), (b), and (c). The case considered is ice over water with $l_\rho = 2$ cm. Thickness is 50 cm and frequency is at 10 GHz. Other parameters are as in Section 4.1.

sensing of cloud and rainfall, the profile structure of the atmosphere must be taken into account. The formalism developed in Section 4.2 can be applied to the problem of layers of spherical scatterers in the presence of inhomogeneous profiles. In this section, we shall illustrate the results for a layered atmosphere consisting of a layer of ice particles and a layer of cloud or rainfall in the presence of background atmospheric gaseous absorption profile and temperature profiles (Figure 4.60).

The ice layer extends from $z = s_1$ to $z = s_2$. It is known that sometimes ice particles do exist in the upper atmosphere. The layer of cloud or rain extends from $z = s_3$ to $z = s_4$, and the earth surface is at $z = s$. The thickness s of the atmosphere is of the order of 100 km. We use $T(z)$ to describe the temperature profile of the atmosphere from $z = 0$ to $z = s$ and T_2 is the temperature of the earth surface. Let $\kappa_{ab}(z)$ denote the atmospheric background absorption profile from $z = 0$ to $z = s$ which includes absorption of atmospheric gases and water vapor but does not consider water droplets in cloud or rain nor the ice particles in the ice layer.

The water droplets in cloud or rain are described in Section 2.3 and

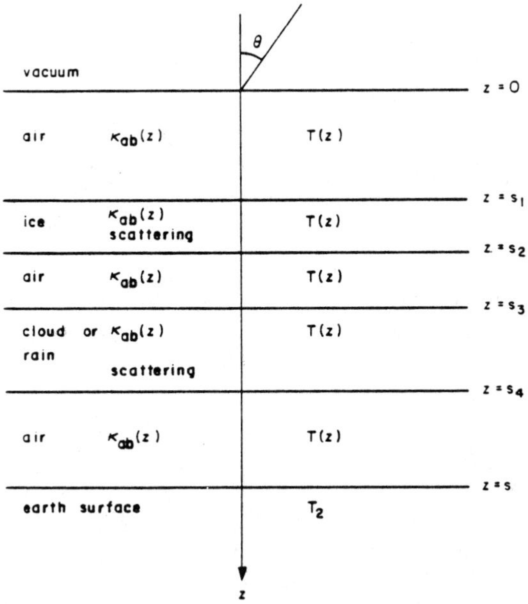

Thermal Emission of Layers of Spherical Scatterers

Fig. 4.60 Geometrical configuration of the problem. The radiometer is above $z = 0$. The earth surface is at $z = s$. There is an ice layer from $z = s_1$ to $z = s_2$ and a rain or cloud layer from $z = s_3$ to $z = s_4$. The atmosphere is described by temperature distribution $T(z)$ and a background absorption $\kappa_{ab}(z)$. There are also scattering and absorption due to the particles in the ice and cloud/rain layers.

are characterized by M (specific water content), P, Q, and mode radius a_c. The coupling coefficients (α, β') for Mie scattering are listed in Section 2.3. Let κ_{aw} and κ_{ew} denote the absorption and extinction coefficients of the water droplets in cloud or rain computed by using the Mie formula.

We shall characterize the ice particles by Γ drop-size distribution, and permittivity ϵ_{ice}. Let $(\kappa_a)_{ice}$ and $(\kappa_e)_{ice}$ denote the absorption and extinction coefficients of the ice particles computed by using the Mie formula. The coupling coefficients of scattering (α, β') are also computed according to the Mie formula of Section 2.3. Thus,

$$\kappa_a(z) = \begin{cases} 0 & \text{for } z \leq 0 \\ \kappa_{ab}(z) & \text{for } 0 \leq z < s_1 \\ \kappa_{ab}(z) + (\kappa_a)_{ice} & \text{for } s_1 \leq z < s_2 \\ \kappa_{ab}(z) & \text{for } s_2 \leq z < s_3 \\ \kappa_{ab}(z) + \kappa_{aw} & \text{for } s_3 \leq z < s_4 \\ \kappa_{ab}(z) & \text{for } s_4 \leq z < s \end{cases} \quad (1)$$

$$\kappa_e(z) = \begin{cases} 0 & \text{for } z < 0 \\ \kappa_{ab}(z) & \text{for } 0 \leq z < s_1 \\ \kappa_{ab}(z) + (\kappa_e)_{ice} & \text{for } s_1 \leq z < s_2 \\ \kappa_{ab}(z) & \text{for } s_2 \leq z < s_3 \\ \kappa_{ab}(z) + \kappa_{ew} & \text{for } s_3 \leq z < s_4 \\ \kappa_{ab}(z) & \text{for } s_4 \leq z < s \end{cases} \quad (2)$$

Scattering is present in the ice layer and the cloud or rainfall layer. The stepping forward procedure is, by using the invariant imbedding equations of (8), (9), (25), and (26), Section 4.2,

$$\bar{\bar{R}}(s + \Delta s) = \bar{\bar{R}}(s) + \frac{d\bar{\bar{R}}(s)}{ds}\Delta s \quad (3)$$

$$\bar{\bar{t}}(s + \Delta s) = \bar{\bar{t}}(s) + \frac{d\bar{\bar{t}}(s)}{ds}\Delta s \quad (4)$$

$$\bar{T}_R(s + \Delta s) = \bar{T}_R(s) + \frac{d\bar{T}_R(s)}{ds}\Delta s \quad (5)$$

$$\bar{T}_B(s + \Delta s) = \bar{T}_B(s) + \frac{d\bar{T}_B(s)}{ds}\Delta s \quad (6)$$

beginning with the initial conditions of zero thickness as given by (15), (16), (28), and (29), Section 4.2.

The stepping forward procedure is adaptive as the scattering functions $\bar{\bar{F}}(s)$, $\bar{\bar{B}}(s)$, $\bar{\bar{\kappa}}_e(s)$, $\bar{T}(s)$, and $\kappa_a(s)$ can be varied as s changes. The choice of the stepping interval Δs is also adaptive. We choose it to be of the order $0.01/\kappa_e(z)$ so that Δs can vary from 10 meters in the cloud or rain layers to 1 km in the upper atmosphere where $\kappa_{ab}(z)$ is very small. As the stepping procedure moves through the air layers (Fig 4.60) where there is only absorption and no scattering, we can use the simpler equations for the differential change of $d\bar{\bar{R}}(s)/ds$, $d\bar{\bar{t}}(s)/ds$, $d\bar{T}_R(s)/ds$, and $d\bar{T}_B(s)/ds$ by setting $\bar{\bar{F}}(s) = \bar{\bar{B}}(s) = 0$ in (8), (9), (25), and (26), Section 4.2.

We illustrate the results for an ice layer and a rain layer above a smooth ocean. The temperature of ocean is 26 deg C and the permittivity of smooth ocean is $(7.74 + i14)\epsilon_o$. In Figures 4.61 and 4.62, we plot a typical background absorption profile $\kappa_{ab}(z)$ and temperature profile $T(z)$. The background absorption includes gaseous absorption and absorption by water vapor but does not include absorption by water droplets in cloud or rainfall.

In Figure 4.63, we plot the vertical and horizontal brightness temperatures for (A) absence of cloud, rain, nor ice and (B) presence of a rain

Thermal Emission of Layers of Spherical Scatterers

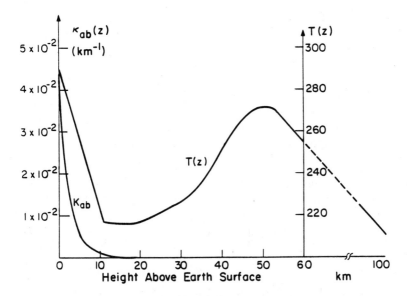

Fig. 4.61 Temperature profile and atmospheric background absorption profile at frequency = 94 GHz from earth surface to a height of 100 km. The absorption background profile includes the effects of gaseous absorption and absorption due to water vapor but does not include absorption of water droplets in cloud or rainfall. The IWC (integrated water content) is 0.5.

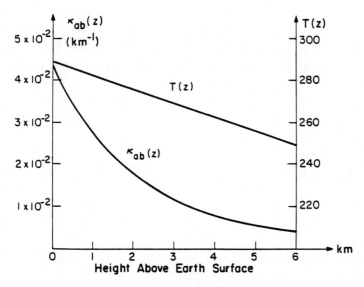

Fig. 4.62 Same as Figure 4.61, but with profile plotted from earth surface to a height of 6 km.

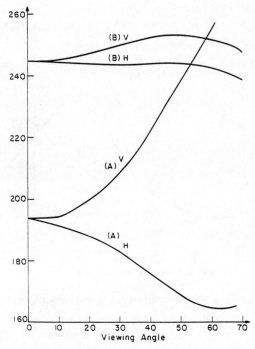

Fig. 4.63 Vertical V and horizontal H polarized brightness temperatures plotted as a function of viewing angles at frequency = 94 GHz. The atmospheric background absorption profile and temperature profile of Figure 4.61 is used. The temperature of earth is 26 deg C. The permittivity of earth is $(7.74 + i14)\epsilon_o$ which corresponds to smooth ocean of salinity 33.5. Earth surface is at $z = 100$ km. Curves: **(A)** No cloud or rain layer; **(B)** rain layer from $z = 97$ km to $z = 98$ km. Permittivity of water droplets in rain is $(6.147 + i7.277)\epsilon_o$. The rain obeys Γ drop size distribution with $M = 0.5$, $a_c = 400\mu$, $P = 5$, and $Q = 0.5$.

layer of thickness 1 km at a height of 2 km above the earth surface. In view of Figure 4.60, this corresponds to $s_3 = 97$ km and $s_4 = 98$ km. The permittivity of water in rain is taken to be $(6.147 + i7.277)\epsilon_o$. We note from the figure that Case (A) exhibits primarily the emission characteristics of ocean since the atmosphere is optically thin. In Case (B), the presence of the rain layer increases the total emission so that the brightness temperatures are above that of Case (A). Scattering also causes coupling among the different polarizations so that the angular and polarization dependences of the brightness temperatures are much weaker than those of case A.

In Figure 4.64, we consider the presence of an ice layer of (A) 0.5 km thick and (B) 1 km thick immediately above the rain layer. For the sake of simplicity, we have assumed that the ice particles have the same dropsize distribution as the water droplets in the rain layer. The permittivity of

Problems

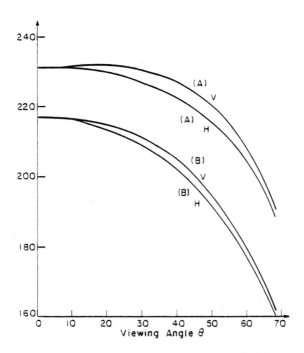

Fig. 4.64 Vertical V and horizontal H polarized brightness temperatures as a function of viewing angles at frequency = 94 GHz. The atmospheric background absorption profile and temperature profile of Figure 4.61 is used. The temperature of earth is 26 deg C. The permittivity of earth is $(7.74 + i14)\epsilon_o$ which corresponds to a smooth ocean of salinity 33.5. Earth surface is at $z = 100$ km. There is a rain layer from $z = 97$ km to $z = 98$ km. Permittivity of water droplets in rain is $(6.147 + i7.277)\epsilon_o$. The rain obeys Γ drop size distribution with $M = 0.5$, $a_c = 400\mu$, $P = 5$, and $Q = 0.5$. There is an ice layer above the rain layer extending from $z = s_1$ to $z = s_2$. The drop size distribution of ice spheres is taken to be identical to that of the rain layer. The permittivity of ice is $3.24(1+i5\times 10^{-4})\epsilon_o$. Curves: (A) Ice layer of 0.5 km thick with $s_1 = 96.5$ km and $s_2 = 97$ km; (B) ice layer of 1 km thick with $s_1 = 96$ km and $s_2 = 97$ km.

ice is taken to be $3.24(1 + i5 \times 10^{-4})\epsilon_o$. Thus, the ice layer has a very large albedo and is primarily a scattering medium. We note that ice causes darkening as ice particles scatter back the emission of the rain layer and the ocean. Case (B) has a larger ice layer thickness and a smaller brightness temperature. It is also interesting to observe that both vertical and horizontal brightness temperatures decrease with viewing angle. This is because there is more scattering at large viewing angles than at nadir.

PROBLEMS

4.1 Derive the solution of the brightness temperature [(6), Section 2.1] from (1) through (5), Section 2.1 for the case when scattering can be ignored.

4.2 Derive [(12), Section 2.2] of the first-order scattering solutions for a layer of point scatterers from the given integral equations (7) and (8), Section 2.2.

4.3 Show that (2), Section 2.3, is true and that the third and fourth Stokes parameters are zero for thermal emission of spherical particles.

4.4 1. Using (15) and (17), Section 5.2, and (8) and (9), Section 5.5, of Chapter 3, derive explicit expressions for the phase matrix elements $P_{ij}(\theta, \phi; \theta', \phi')$ with $i, j = 1, 2, 3, 4$ for spherical particles.

2. By performing integration over $d\phi'$ indicated in (3), Section 2.3, verify the expressions of $[\alpha(\theta), \beta(\theta)]$ in (4), Section 2.3.

4.5 Formulate the integral equations for the brightness temperature of a layer of spherical particles for Section 2.3. Then verify the single scattering solution of (13), Section 2.3.

4.6 Derive the integral equations (8) and (9), Section 2.5, and the first-order solution [(10), Section 2.5] for the active remote sensing of a layer of nonspherical particles.

4.7 Consider a layer of small oblate spheroids with major axis a and minor axis c. The spheroids are all aligned with the axes of symmetry in the direction $(\hat{x} + \hat{y} + \hat{z})/\sqrt{3}$. Derive closed-form expressions for the backscattering coefficients when the incident wave is in the x-y plane with $\phi_o = 0$.

4.8 Derive the second order solution [(8), Section 2.6] of active remote sensing of a half-space of point scatterers by using the integral equation [(1) and (2), Section 2.6].

4.9 In order to determine the coefficients $\overline{A}(\theta, \phi)$ and $\overline{B}(\theta, \phi)$ in (7) and (8), Section 2.7, we substitute $\overline{I}(\theta, \phi, z)$ and $\overline{I}(\pi - \theta, \phi, z)$ into the boundary conditions [(3), Section 2.7]. We then obtain two simultaneous equations for $\overline{A}(\theta, \phi)$ and $\overline{B}(\theta, \phi)$. Verify that the solutions of those two equations of $\overline{A}(\theta,\phi)$ and $\overline{B}(\theta,\phi)$ are

$$\begin{aligned}\overline{A}(\theta,\phi) =\;& \cos\theta\,\overline{\overline{F}}(\theta)\cdot\overline{\overline{R}}_{12}(\theta)\cdot\overline{\overline{T}}_{01}(\theta_o)\\
& \cdot \overline{I}_{oi}(\pi-\theta_o,\phi_o)\,e^{-2\kappa_e d\sec\theta}\\
& +\overline{\overline{F}}(\theta)\cdot\overline{\overline{R}}_{12}(\theta)e^{-2\kappa_e d\sec\theta}\\
& \times\int_{-d}^{0}dz'e^{-\kappa_e z'\sec\theta}\int_{0}^{\pi/2}d\theta'\sin\theta'\int_{0}^{2\pi}d\phi'\\
& \times\Big\{\overline{\overline{P}}(\pi-\theta,\phi;\theta',\phi')\cdot\overline{I}(\theta',\phi',z')\\
& \quad+\overline{\overline{P}}(\pi-\theta,\phi;\pi-\theta',\phi')\cdot\overline{I}(\pi-\theta',\phi',z')\Big\}\\
& +\overline{\overline{F}}(\theta)\cdot\overline{\overline{R}}_{12}(\theta)\cdot\overline{\overline{R}}_{10}(\theta)e^{-2\kappa_e d\sec\theta}\\
& \times\int_{-d}^{0}dz'e^{\kappa_e z'\sec\theta}\int_{0}^{\pi/2}d\theta'\sin\theta'\int_{0}^{2\pi}d\phi'\\
& \times\Big\{\overline{\overline{P}}(\theta,\phi;\theta',\phi')\cdot\overline{I}(\theta',\phi',z')\\
& \quad+\overline{\overline{P}}(\theta,\phi,;\pi-\theta',\phi')\cdot\overline{I}(\pi-\theta',\phi',z')\Big\}\quad(1)\end{aligned}$$

$$\begin{aligned}\overline{B}(\theta,\phi) =\;& \cos\theta\,\overline{\overline{F}}(\theta)\cdot\overline{\overline{T}}_{01}(\theta_o)\cdot\overline{I}_{oi}(\pi-\theta_o,\phi_o)\\
& +\overline{\overline{F}}(\theta)\cdot\overline{\overline{R}}_{12}(\theta)\int_{-d}^{0}dz'\,e^{-\kappa_e z'\sec\theta}\int_{r}^{\pi/2}d\theta'\sin\theta'\\
& \times\int_{0}^{2\pi}d\phi'\Big\{\overline{\overline{P}}(\theta,\phi;\theta',\phi')\cdot\overline{I}(\theta',\phi',z')\\
& \quad+\overline{\overline{P}}(\theta,\phi;\pi-\theta',\phi')\cdot\overline{I}(\pi-\theta',\phi',z')\Big\}\\
& +\overline{\overline{F}}(\theta)\cdot\overline{\overline{R}}_{10}(\theta)\cdot\overline{\overline{R}}_{12}(\theta)e^{-2\kappa_e d\sec\theta}\\
& \times\int_{-d}^{0}dz'\,e^{\kappa_e z'\sec\theta}\int_{0}^{\pi/2}d\theta'\sin\theta'\int_{0}^{2\pi}d\phi'\\
& \times\Big\{\overline{\overline{P}}(\pi-\theta,\phi;\theta',\phi')\cdot\overline{I}(\theta',\phi',z')\\
& \quad+\overline{\overline{P}}(\pi-\theta,\phi,;\pi-\theta',\phi')\cdot\overline{I}(\pi-\theta',\phi',z')\Big\}\quad(2)\end{aligned}$$

where $\overline{\overline{F}}(\theta)$ is given in (11), Section 2.7.

4.10 Derive the first-order and second-order backscattering coefficients for a half-space of small spherical scatterers (Shin and Kong, 1981) as given in (12), (13) and (17), Section 2.7.

4.11 Derive the coupling coefficients $[\alpha(\theta), \beta(\theta')]$ for laminar structures, the radiative transfer equations, and the closed-form solution [(9), Section 3.2] for a half-space medium of Section 3.2.

4.12 Extend the half-space laminar model of Section 3.2 for passive remote sensing to a multilayer model (Djermakoye and Kong, 1979).

4.13 Calculate the coupling coefficients $[\alpha(\theta), \beta(\theta')]$ for random medium with a cylindrical structure with l_ρ finite and $l_z \to \infty$. Solve the radiative transfer equations for a two layer model and calculate the brightness temperature (Chuang et al., 1980; Wang et al., 1984).

4.14 Show that the particular solution for the radiative transfer equations for thermal emission of a scattering layer with constant temperature T is equal to CT as given in (32), Section 3.3.

4.15 In the discrete ordinate–eigenanalysis approach of Section 3.3 to passive remote sensing, give an expression for the particular solution of the specific intensity when the temperature is an inhomogeneous profile $T_o + T_d \exp(\gamma z)$.

4.16 Show that the boundary conditions for the case of a cloud or rainfall layer lying above a layer of air on top of earth surface are given by (4) through (6), Section 3.5.

4.17 Show that in the absence of scattering, the brightness temperature of a cloud or rainfall layer lying above a layer of air on top of the earth surface is given by (8), Section 3.5.

4.18 Derive the differential equations (3) through (5), Section 3.6, for the specific intensities $I^{mc}(\theta, z)$ and $I^{ms}(\theta, z)$ for each harmonic index $m \geq 0$ for active remote sensing.

4.19 By using the phase matrix elements of spherical Rayleigh scatterers in (8) through (17), Section 4.1, Chapter 3, derive expressions for $\overline{\overline{P}}^o(\theta, \theta')$, $\overline{\overline{P}}^{1c}(\theta, \theta')$, $\overline{\overline{P}}^{1s}(\theta, \theta')$, $\overline{\overline{P}}^{2c}(\theta, \theta')$, and $\overline{\overline{P}}^{2s}(\theta, \theta')$ to be used in radiative transfer equations for active remote sensing. Also show that $\overline{\overline{P}}^{mc}(\theta, \theta') = \overline{\overline{P}}^{ms}(\theta, \theta') = 0$ for $m \geq 3$.

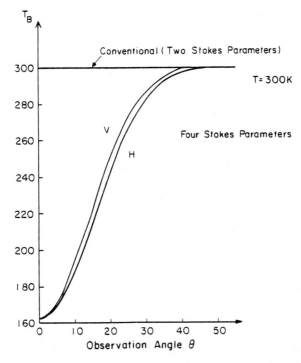

Fig. 4.65 Comparison between two Stokes parameter and four Stokes parameter results.

4.20 Derive closed-form expressions for the specific intensities $\bar{I}^{mc}(\theta, z)$ and $\bar{I}^{ms}(\theta, z)$ for $m \geq 3$ for active remote sensing of a layer of small spherical scatterers overlying a homogeneous dielectric half-space. Use (13) and (14), Section 3.6, as your starting point. Solve (13) and (14), Section 3.6, and match boundary conditions.

4.21 Consider the invariant imbedding approach to the solution of a one-dimensional problem in thermal emission. Derive equations (37) through (40), Section 4.1.

4.22 Consider the invariant imbedding approach to the analysis of the reflection and transmission of the three-dimensional problem. Derive (8) and (9), Section 4.2.

4.23 Show from (8) through (16), Section 4.2, that the reflectivity matrix $\overline{\overline{R}}(s)$ is symmetrical. That is $\overline{\overline{R}}^t(s) = \overline{\overline{R}}(s)$.

4.24 Consider the invariant imbedding approach to the solution of a three-dimensional problem in thermal emission. Derive (25) through (27), Section 4.2.

4.25 Consider thermal emission of a slab of small oblate spheroids with $a = 1\,\text{cm}, c = 0.05\,\text{cm}, f = 0.004$, with symmetry axis pointing in the $\hat{y}$ direction. Let the slab thickness be 100 meters and frequency be 1 GHz. Ignore scattering and compare numerically the solution based on four Stokes parameters [(15), Section 1.1] and two Stokes parameters [(17), Section 1.1]. Let $\epsilon_s = (30.68 + i1.55)\epsilon_o$ and $T = 300\,\text{K}$ and observation be in the x-z plane.

1. Show that $|T_1| \gg |T_o|$ where T_1 and T_o are the T-matrix elements for small spheroids as given in Section 5.6, Chapter 3.

3. Calculate $f_{\alpha\beta}(\theta, \phi; \theta, \phi)$, $\alpha, \beta = vv, vh, hv, hh$, the absorption vector and the extinction matrix.

4. Show that $\beta_4 d$, $\kappa_{11} d$, $\kappa_{22} d$ are much larger than 1 while $\beta_1 d$ is much smaller than 1.

5. Verify numerically that the conventional two Stokes parameter result gives brightness temperature to be 300 K while the four Stokes parameters result is completely different (Figure 4.65).

5

ANALYTIC WAVE THEORY FOR SCATTERING BY LAYERED RANDOM MEDIA

1. Introduction — 318

2. Scattering by Layered Random Media – Born Series — 319
   - 2.1 Active Remote Sensing of Half-Space Random Media — 319
   - 2.2 Depolarization Effects in the Active Remote Sensing of Half-Space Random Medium — 325
   - 2.3 Backscattering Enhancement: Volume Scattering — 328
   - 2.4 Active Remote Sensing of Layered Random Media — 330
   - 2.5 Backscattering Enhancement: Volume Scattering in the Presence of Reflective Boundary — 334

3. Analytic Wave Theory — 337
   - 3.1 Dyson's Equation for the Mean Field — 337
   - 3.2 Bilocal and Nonlinear Approximations — 342
   - 3.3 Bethe-Salpeter Equation — 347
   - 3.4 Ladder Approximation for Isotropic Point Scatterers and Relation to Radiative Transfer Theory — 351
   - 3.5 Backscattering Enhancement for Random Discrete Scatterers — 358
   - 3.6 Alternative Derivation of Radiative Transfer Equations from Ladder Approximation — 370
   - 3.7 Appendix — 374

4. Strong Permittivity Fluctuations — 375

4.1	Random Medium with Spherically Symmetric Correlation Function	376
4.2	Very Low Frequency Effective Permittivity	378
4.3	Effective Permittivity under the Bilocal Approximation	379
4.4	Backscattering Coefficients	381
4.5	Numerical Illustrations	384

5.	Modified Radiative Transfer Equations for Volume Scattering in the Presence of Reflective Boundaries	390
5.1	Introduction	390
5.2	Mean Green's Function and Mean Field	392
5.3	Modified Radiative Transfer Equations	399
5.4	Numerical Illustrations	403
5.5	Appendix	406

Problems   410

1   INTRODUCTION

The scattering of electromagnetic waves by random medium was first applied to the remote sensing of the Antarctic with the model of a half-space laminar structure by Gurvich et al. (1973). In active remote sensing, Stogryn (1974) studied the bistatic scattering cross sections in the low-frequency limit for a half-space random medium with a spherically symmetric correlation function. Initial success of data matching was also accomplished with the Born approximation. This method has been applied to obtain the bistatic scattering coefficients and the backscattering cross sections for a two-layer random medium (Zuniga and Kong, 1980) and, more generally, for a layered structure (Zuniga et al., 1979). The depolarization effect was subsequently studied (Zuniga et al., 1980) by carrying out the Born approximation to the second order to obtain the backscattering cross sections.

In order to assess the limitations of the radiative transfer theory, we will study analytic wave theory for scattering by layered random media in this chapter. First we will examine the Born series (Section 2) and then the Dyson's and Bethe-Salpeter equations (Section 3). The ladder-approximated Bethe-Salpeter equations are then used to derive the radiative transfer equations for a medium with no reflective boundary. This is accomplished by showing that the ladder approximated Bethe-Salpeter equation is equivalent to the Schwarzschild-Milne integral equation (Davison, 1958) for the case of isotropic point scatterers. However, the ladder approximation ignores cyclical diagrams and, like the radiative transfer theory, does not account for backscattering enhancement. This is illustrated in Sections 2.3, 2.5, and 3.5. The cyclical diagrams for point scatterers are resummed to obtain a cyclical transfer equation which is then solved numerically.

The variance of the permittivity fluctuations is usually large in geophysical media which usually consist of mixtures of constituents with very different dielectric properties. It should be noted that in solving Dyson's equation with the bilocal approximation, both the observation point and the source point are within the random medium and can coincide with each other in the domain of integration. To take care of the singular nature of the dyadic Green's functions, a strong permittivity fluctuation theory (SPF) that applies to both small and large variances of the permittivity functions is also developed in Section 4. It is also shown (Tsang et al., 1982a) that the effective permittivity at the very low-frequency limit is the same as the Polder and van Santern mixing formula (1946).

It was observed in experimental measurements (Blinn et al., 1972) that wave interference effects are significant in media with a subsurface reflective boundary. We take that into account by including the correlation between upward and downward waves that are specularly related. These are used to develop a set of modified radiative transfer equations (MRT) from the ladder-approximated Bethe-Salpeter equation in Section 5.

2 SCATTERING BY LAYERED RANDOM MEDIA – BORN SERIES

Of all the analytic wave approaches to the study of scattering by random medium, the Born approximation is the simplest. In performing the calculation, an integral equation is first formed for the electric field by using the unperturbed Green's function which is the Green's function in the absence of permittivity fluctuations. The integral equation is then solved by iteration with the iteration series known as the Born series. The first iterated term is the Born approximation. The n th-term corresponds to n th-order scattering. In this section, we will use the Born series to study scattering by layered random medium. The iteration series will be carried out to second order where depolarization effect is exhibited.

2.1 Active Remote Sensing of Half-Space Random Media

Consider a plane wave, $\overline{E}_{oi}$, incident upon a half-space random medium (Figure 5.1). The random medium is characterized by permittivity $\epsilon_1(\bar{r})$

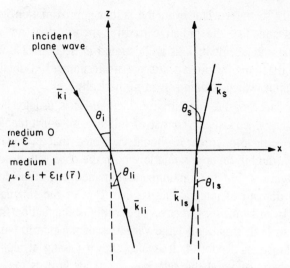

Fig. 5.1 Active remote sensing of half-space random media.

which is the sum of a mean part $<\epsilon_1>$ and a fluctuating part ϵ_{1f} the ensemble average of which vanishes. The medium above the random layer is homogeneous and all regions have permeability μ.

The electric fields in Regions 0 and 1, denoted by $\overline{E}$ and $\overline{E}_1$ respectively, satisfy vector wave equations as follows

$$\nabla \times \nabla \times \overline{E} - k^2 \overline{E} = 0 \qquad (1a)$$

$$\nabla \times \nabla \times \overline{E}_1 - k_{1m}^2 \overline{E}_1 = Q(\overline{r})\overline{E}_1 \qquad (1b)$$

where

$$Q(\overline{r}) = \omega^2 \mu \epsilon_{1f}(\overline{r}) \qquad (1c)$$

and

$$k_{1m}^2 = \omega^2 \mu <\epsilon_1(\overline{r})> \qquad (1d)$$

The solution to (1a) and (1b) may be written in terms of dyadic Green's functions for bounded media

$$\overline{E} = \overline{E}^{(0)} + \int_{V_1} d^3 r_1 \overline{\overline{G}}_{01}^{(0)}(\overline{r}, \overline{r}_1) Q(\overline{r}_1) \cdot \overline{E}_1(\overline{r}_1) \qquad (2a)$$

$$\overline{E}_1 = \overline{E}_1^{(0)} + \int_{V_1} d^3 r_1 \overline{\overline{G}}_{11}^{(0)}(\overline{r}, \overline{r}_1) Q(\overline{r}_1) \cdot \overline{E}_1(\overline{r}_1) \qquad (2b)$$

where $\overline{E}^{(0)}$ and $\overline{E}_1^{(0)}$ denote the unperturbed solutions of $\overline{E}$ and $\overline{E}_1$ when the permittivity fluctuation $\epsilon_{1f}(\overline{r})$ vanishes in (2a) and (2b), and

$\overline{\overline{G}}_{01}^{(0)}$ and $\overline{\overline{G}}_{11}^{(0)}$ are layered-media dyadic Green's functions in the absence of permittivity fluctuations as given in Chapter 2. We solve (2a) and (2b) iteratively. The first order solution is obtained by replacing $\overline{E}_1$ under the integral sign by $\overline{E}_1^{(0)}$. Hence $\overline{E} = \overline{E}^{(0)} + \overline{E}_s^{(1)}$, with

$$\overline{E}_s^{(1)}(\overline{r}) = \int_{V_1} d^3 r_1 \overline{\overline{G}}_{01}^{(0)}(\overline{r},\overline{r}_1) Q(\overline{r}_1) \cdot \overline{E}_1^{(0)}(\overline{r}_1) \qquad (3)$$

and V_1 is the lower half denoted as region 1. This approximation is often referred to as the first-order Born approximation, after Max Born who applied the technique in quantum mechanical potential scattering problems.

First-Order Scattered Intensity

Forming the absolute square of $\overline{E}_s^{(1)}$ and ensemble averaging we obtain the first-order scattered intensity in the form

$$<|\overline{E}_s^{(1)}(\overline{r})|^2> = \int_{V_1} d^3 r_1 \int_{V_1} d^3 r_2 \overline{\overline{G}}_{01}^{(0)}(\overline{r},\overline{r}_1) \cdot \overline{E}_1^{(0)}(\overline{r}_1)$$

$$\cdot \overline{\overline{G}}_{01}^{(0)*}(\overline{r},\overline{r}_2) \cdot \overline{E}_1^{(0)*}(\overline{r}_2) <Q(\overline{r}_1)Q^*(\overline{r}_2)> \qquad (4a)$$

where $<Q(\overline{r}_1)Q^*(\overline{r}_2)>$ is a two-point correlation function. For a statistically homogeneous medium, the correlation function depends only on the separation of the points $\overline{r}_1$ and $\overline{r}_2$. That is

$$<Q(\overline{r}_1)Q^*(\overline{r}_2)> = C(\overline{r}_1 - \overline{r}_2) \qquad (4b)$$

Thus, only fields scattered from fluctuations separated by distances less than or comparable to the correlation length will significantly contribute to scattered intensity. We note that the observation point $\overline{r}$ in equation (4a) is in the far field in region 0. Hence, the far-field approximation can be made for the Green's functions in (4a). The far-field dyadic Green's function $\overline{\overline{G}}_{01}^{(0)}(\overline{r},\overline{r}_1)$ can be calculated by using the saddle point method (Problem 1; (18) through (24), Section 3.2, Chapter 2). Let the observation point be at (r, θ_s, ϕ_s), so that

$$\overline{k}_{\perp s} = k_{xs}\hat{x} + k_{ys}\hat{y} \qquad (5)$$

with $k_{xs} = k \sin\theta_s \cos\phi_s$, and $k_{ys} = k \sin\theta_s \sin\phi_s$. Then, in the far field

$$\overline{\overline{G}}_{01}^{(0)}(\overline{r},\overline{r}_1) = \frac{e^{ikr}}{4\pi r} \overline{\overline{H}} \exp(-i\overline{k}_{1s} \cdot \overline{r}_1) \qquad (6)$$

where

$$\overline{\overline{H}} = X_{01s}\hat{e}(k_{zs})\hat{e}_1(k_{1zs}) + \frac{k}{k_{1m}}Y_{01s}\hat{h}(k_{zs})\hat{h}_1(k_{1zs}) \qquad (7)$$

$$\overline{k}_{1s} = \overline{k}_{\perp s} + \hat{z}k_{1zs} \qquad (8)$$

$$k_{zs} = (k^2 - k_{xs}^2 - k_{ys}^2)^{1/2} \qquad (9)$$

$$k_{1zs} = (k_{1m}^2 - k_{xs}^2 - k_{ys}^2)^{1/2} \qquad (10)$$

while X_{01} and Y_{01} are given in Section 3.2, Chapter 2, and $\hat{e}$ and $\hat{h}$ are defined in (13), (16), and (17), Section 3.1, Chapter 2. The subscript s in X_{01s} and Y_{01s} indicates that the k_z values are those of the scattered direction.

Let the incident field be in direction (θ_i, ϕ_i), so that

$$\overline{k}_{\perp i} = k_{xi}\hat{x} + k_{yi}\hat{y} \qquad (11)$$

with $k_{xi} = k\sin\theta_i \cos\phi_i$, and $k_{yi} = k\sin\theta_i \sin\phi_i$. In the following, we shall illustrate the case where the incident wave is horizontally polarized. The analysis for a vertically polarized incident wave will be similar. The incident field is

$$\overline{E}_i = E_o\,\hat{e}(-k_{zi})\,e^{-ik_{zi}z}\,e^{i\overline{k}_{\perp i}\cdot\overline{r}} \qquad (12)$$

where $k_{zi} = k\cos\theta_i$. The unperturbed field in region 1 is

$$\overline{E}_1^{(0)}(\overline{r}) = E_o\,X_{01i}\,\hat{e}_{1i}(k_{1zi})\,e^{i\overline{k}_{1i}\cdot\overline{r}} \qquad (13)$$

where $k_{1zi} = (k_{1m}^2 - k^2\sin^2\theta_i)^{1/2}$ and $\overline{k}_{1i} = \overline{k}_{\perp i} - \hat{z}k_{1zi}$ and the i in X_{01i} denotes that the k_z values are those corresponding to the incident direction. Substituting (6), (7), and (13) into (4a), and using (4b), we obtain

$$\langle|\overline{E}_s^{(1)}(\overline{r})|^2\rangle = \frac{|E_o|^2}{16\pi^2 r^2}\left|\left[X_{01s}\hat{e}(k_{zs})\hat{e}_1(k_{1zs})\right.\right.$$
$$\left.\left.+ \frac{k}{k_{1m}}Y_{01s}\hat{h}(k_{zs})\hat{h}_1(k_{1zs})\right]\cdot X_{01i}\hat{e}_{1i}(k_{1zi})\right|^2 B \quad (14)$$

where

$$B = \int_{V_1} d\overline{r}_a \int_{V_1} d\overline{r}_b\, C(\overline{r}_a - \overline{r}_b)\exp[i(\overline{k}_{1i} - \overline{k}_{1s})\cdot\overline{r}_a + i(\overline{k}_{1s}^* - \overline{k}_{1i}^*)\cdot\overline{r}_b] \quad (15)$$

The correlation function may be expressed as the Fourier transform of the spectral intensity Φ.

$$C(\bar{r}_1 - \bar{r}_2) = \delta k_{1m}^{\prime 4} \int_{-\infty}^{\infty} d\bar{\beta}\, \Phi(\bar{\beta})\, e^{-i\bar{\beta}\cdot(\bar{r}_1 - \bar{r}_2)} \qquad (16)$$

where δ is the variance of permittivity fluctuations as defined in Chapter 3 and k_{1m}' is the real part of k_{1m}. Substitution of (16) into (15) yields

$$B = 4\pi^2 A \delta k_{1m}^{\prime 4} \int_{-\infty}^{\infty} d\beta_z \frac{\Phi(k_{xi} - k_{xs}, k_{yi} - k_{ys}, \beta_z)}{(k_{1zi} + k_{1zs} + \beta_z)(k_{1zi}^* + k_{1zs}^* + \beta_z)} \qquad (17)$$

with A as the target area. Residue calculus can be used to perform the integration in (17). Since we are interested in low-absorption areas with $k_{1m}'' \ll k_{1m}'$, most of the contribution of the integral comes from the residue of the pole at $\beta_z = -k_{1zi}^* - k_{1zs}^*$. There may be other poles in the integrand of (17), but their residues are much smaller. Hence

$$B \simeq 4\pi^3 A \delta k_{1m}^{\prime 4} \frac{\Phi(k_{xi} - k_{xs}, k_{yi} - k_{ys}, -k_{1zi}^* - k_{1zs}^*)}{(k_{1zi}'' + k_{1zs}'')} \qquad (18)$$

We use superscripts $'$ and $''$ and to denote real and imaginary parts, respectively. The case of vertically polarized wave incidence can be studied in a similar manner (Problem 2).

The bistatic coefficients $\gamma_{\alpha\beta}(\hat{k}_s, \hat{k}_i)$ can be calculated. For the case $\phi_i = 0$, they are

$$\begin{bmatrix} \gamma_{vv} \\ \gamma_{hv} \\ \gamma_{vh} \\ \gamma_{hh} \end{bmatrix} = \frac{\delta\, k_{1m}^{\prime 4}\, \pi^2 \Phi}{\cos\theta_i (k_{1zi}'' + k_{1zs}'')}$$

$$\times \begin{bmatrix} |Y_{01s} Y_{01i}|^2 \dfrac{k^6 \sin^2\theta_s}{k_{1m}^8} \left| k_{xi} - \dfrac{k_{xs} k_{1zs} k_{1zi}}{(k_{xs}^2 + k_{ys}^2)} \right|^2 \\[6pt] \left| k_{1zi} \dfrac{k}{k_{1m}^2} X_{01s} Y_{01i} \right|^2 \sin^2\phi_s \\[6pt] \left| k_{1zs} \dfrac{k}{k_{1m}^2} Y_{01s} X_{01i} \right|^2 \sin^2\phi_s \\[6pt] |X_{01s} X_{01i}|^2 \cos^2\phi_s \end{bmatrix} \qquad (19)$$

where the argument of the function Φ is the same as in (18). We illustrate the numerical results for the following correlation function

$$C(\bar{r}) = \delta k_{1m}^{\prime 4} \exp(-r/r_o) \qquad (20)$$

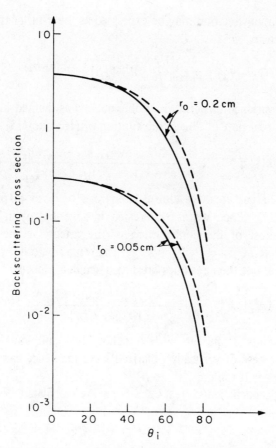

Fig. 5.2 Backscattering cross section as a function of incident angle. The frequency is 10 GHz. $\epsilon_{1m} = \epsilon_o(2.075 + i0.0025)$, $\delta = 0.27$. The solid lines are σ_{hh} and the dashed lines are σ_{vv}.

and the corresponding spectral density

$$\Phi(\overline{\beta}) = \frac{r_o^3}{\pi^2(1+\beta^2 r_o^2)^2} \tag{21}$$

In Figure 5.2, we plot, as a function of incident angle, the backscattering cross section $\sigma_{\beta\alpha} = \gamma_{\beta\alpha}(\theta_s = \theta_i, \phi_s = \pi + \phi_i)\cos\theta_i$. It is noted from (19) that $\sigma_{hv} = \sigma_{vh} = 0$ in the backscattering direction so that depolarization is absent. We see from Figure 5.2 that σ decreases as θ_i increases. In this case, σ_{vv} is slightly larger than σ_{hh} because the transmissivity of vertically polarized waves at the interface is larger.

2.2 Depolarization Effects in the Active Remote Sensing of Half-Space Random Medium

We note that the first-order scattering in Section 2.1 does not contain depolarization. Depolarization is a second-order effect in a random medium. The governing integral equations are (2a) and (2b), Section 2.1. The second-order scattered field $\overline{E}_s^{(2)}$ is obtained by approximating $\overline{E}_1(\bar{r})$ in (2a), Section 2.1, by first order solution

$$\overline{E}_1(\bar{r}) \simeq \overline{E}_1^{(0)}(\bar{r}) + \int_{V_1} d^3 r_1 \overline{\overline{G}}_{11}^{(0)}(\bar{r}, \bar{r}_1) \cdot Q(\bar{r}_1) \overline{E}_1^{(0)}(\bar{r}_1) \qquad (1)$$

Therefore, the second-order scattered field is given by

$$\overline{E}_s^{(2)} = \int_{V_1} d^3 r_1 \int_{V_1} d^3 r_2 \overline{\overline{G}}_{01}^{(0)}(\bar{r}, \bar{r}_1) \cdot \overline{\overline{G}}_{11}^{(0)}(\bar{r}_1, \bar{r}_2) \cdot \overline{E}_1^{(0)}(\bar{r}_2) Q(\bar{r}_1) Q(\bar{r}_2) \qquad (2)$$

Forming the absolute square of (2) yields the second-order scattered intensity

$$<|\overline{E}_s^{(2)}|^2>_{\mu\nu} = \int_{V_1} d^3 r_1 \cdots d^3 r_4 \left[\overline{\overline{G}}_{01}^{(0)}(\bar{r}, \bar{r}_1) \cdot \overline{\overline{G}}_{11}^{(0)}(\bar{r}_1, \bar{r}_2) \cdot \overline{E}_1^{(0)}(\bar{r}_2) \right]_{\mu\nu}$$

$$\cdot \left[\overline{\overline{G}}_{01}^{(0)}(\bar{r}, \bar{r}_3) \cdot \overline{\overline{G}}_{11}^{(0)}(\bar{r}_3, \bar{r}_4) \cdot \overline{E}_1^{(0)}(\bar{r}_4) \right]_{\mu\nu}^{*}$$

$$\times <Q(\bar{r}_1)Q(\bar{r}_2)Q^{*}(\bar{r}_3)Q^{*}(\bar{r}_4)> \qquad (3)$$

where the subscripts ν and μ denote the polarizations of the incident and scattered waves, respectively. Since we are concerned with depolarization effects, the case of $\mu \neq \nu$ will be considered.

The fourth moment of $Q(\bar{r})$ may be expanded in clusters where for Gaussian statistics, we have

$$<Q(1)Q(2)Q^{*}(3)Q^{*}(4)> = C(1-2)C(3-4) + C(1-3)C(2-4)$$

$$+ C(1-4)C(2-3) \qquad (4)$$

and $C(i-j)$ is the two-point correlation function for the random medium. The first term of (4) when substituted into (3) gives the square of the second-order mean field

$$|<\overline{E}_s^{(2)}>|^2 = \left| \int_V d^3 r_1 d^3 r_2 \overline{\overline{G}}_{01}^{(0)}(\bar{r}, \bar{r}_1) \cdot \overline{\overline{G}}_{11}^{(0)}(\bar{r}_1, \bar{r}_2) \right.$$

$$\left. \cdot \overline{E}_1^{(0)}(\bar{r}_2) C(\bar{r}_1 - \bar{r}_2) \right|^2 \qquad (5)$$

The mean scattered fields are specular in nature and therefore do not contribute in the backscattered direction except at normal incidence. In the case of normal incidence, it can be shown that the contribution of (5) is $O(\kappa_s^2/k_{1m}'^2)$ whereas the remaining terms in (4) lead to much larger contributions of $O(\kappa_s^2/k_{1m}''^2)$, where κ_s is the scattering coefficient. Hence, we include only the last two terms of (4) which give rise to depolarization effects.

The next step is to substitute for $\overline{\overline{G}}_{01}^{(0)}, \overline{\overline{G}}_{11}^{(0)}, \overline{E}_1^{(0)}$ and the Fourier transform of the correlation functions $C(\bar{r}_i - \bar{r}_j)$ in (3). The integrations can be carried out (Problem 3) and the cross-polarized backscattered intensity and cross sections, $\sigma_{hv} = \sigma_{vh}$ are obtained. For the case

$$C(\bar{r}) = \delta k_{1m}'^4 \exp\left[-\frac{|z|}{l_z} - \frac{x^2 + y^2}{l_\rho^2}\right] \quad (6)$$

the result is

$$\sigma_{vh}^{(2)}(\theta_i) = \sigma_{hv}^{(2)}(\theta_i) = \sigma_{d+}^{(2)} + \sigma_{d-}^{(2)} \quad (7)$$

where

$$\sigma_{d\pm}^{(2)} = \frac{4\pi^2 \epsilon_o \cos\theta_i}{\kappa_a^2 \epsilon_{1m}} t_v(\theta_{1i}) t_h(\theta_{1i})$$

$$\times \int_0^{\pi/2} d\theta \frac{\sin\theta \sec\theta}{\sec\theta_{1i} + \sec\theta} W_1(\theta, \theta_{1i}) W_3(\theta, \theta_{1i})$$

$$\times \left[W_2(\theta, \theta_{1i}) \sin^2\theta \left(\frac{1}{4} \cos^2\theta_{1i} \sin^2\theta + \cos^2\theta \sin^2\theta_{1i}\right) \right.$$

$$+ W_3(\theta, \theta_{1i}) \frac{\sec\theta_{1i}}{4(\sec\theta_{1i} + \sec\theta)} \left\{ |R_{10} + S_{10}\cos^2\theta|^2 \cos^2\theta_{1i} \right.$$

$$\left. \left. \pm |S_{10}|^2 \cos^2\theta \sin^2\theta \sin^2\theta_{1i} \right\} \right] \quad (8)$$

$$W_1(\theta, \theta_{1i}) = \left(\frac{\pi \delta k_{1m}'^4}{2}\right)^2 \exp\left[-\frac{k_{1m}'^2 l_\rho^2}{2}(\sin^2\theta + \sin^2\theta_{1i})\right] \quad (9)$$

$$W_2(\theta, \theta_{1i}) = \frac{l_z l_\rho^2}{4\pi^2[1 + k_{1m}'^2 l_z^2(\cos\theta - \cos\theta_{1i})^2]} \quad (10)$$

$$W_3(\theta, \theta_{1i}) = \frac{l_z l_\rho^2}{4\pi^2[1 + k_{1m}'^2 l_z^2(\cos\theta + \cos\theta_{1i})^2]} \quad (11)$$

Backscattering Enhancement: Volume Scattering

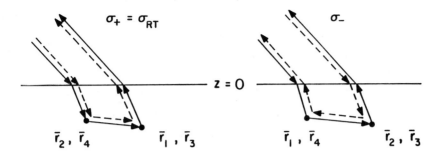

Fig. 5.3 Second order scattering in RT approach and wave approach: Solid line – Field; Dotted line – Conjugate field.

where R_{10} and S_{10} are Fresnel reflection coefficients as defined in (11), Section 3.2, Chapter 2, and are to be taken in the direction θ. The angle θ_{1i} is the incident direction in region 1 and obeys Snell's law $\theta_{1i} = \sin^{-1}[(\epsilon/\epsilon_{1m})^{1/2}\sin\theta_i]$. The direction θ is the intermediate scattering angle. Double scattering involves scattering from direction θ_{1i} into direction θ and then scattering from θ into the backscattering direction θ_{1i}. This explains the angular integration of angle θ in (8). The quantities t_v and t_h are the Fresnel transmissivities, so that $t_v = 1 - |S_{10}|^2$ and $t_h = 1 - |R_{10}|^2$. The quantity κ_a is the background absorption coefficient $2k''_{im}$.

We note that $\sigma_+^{(2)}$ is due to the second pair of correlations in (4) and $\sigma_-^{(2)}$ is from the third pair. The first pair has been shown to have negligible contribution in the backscattering direction. The problem of scattering from a half-space random medium has also been solved by using the radiative transfer (RT) approach (Tsang and Kong, 1978). On comparison with (8), the depolarization in RT is equal to $\sigma_{d+}^{(2)}$

$$\sigma_{d+}^{(2)} = (\sigma_d^{(2)})_{RT} \qquad (12)$$

Hence, the term $\sigma_{d-}^{(2)}$ is an additional contribution that is not accounted for in radiative transfer theory and is only present in analytic wave theory. The additional contributions are also present in the second order like-polarization return. Physically, the scattering processes that generate the contribution σ_+ and σ_- are illustrated in Figures 5.3a,b. The σ_- term directly contributes to the backscattering enhancement effect which will be discussed in the next section.

In Figure 5.4, we illustrate the depolarized backscattering cross section σ_{hv} as a function of incident angle from a half-space random medium. The Born approximation is used.

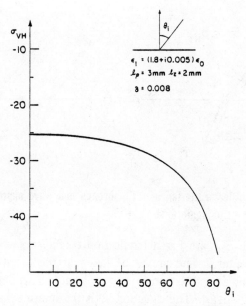

Fig. 5.4 Illustration of σ_{hv} calculated using the Born approximation for a half-space random medium.

2.3 Backscattering Enhancement: Volume Scattering

In the previous section, we note that the analytic wave theory results in terms that are not accounted for by the radiative transfer theory. The additional term (e.g., $\sigma^{(-)}$) is only significant in the neighborhood of the backscattering direction. Thus, when the bistatic scattering coefficient is plotted as a function of scattered angle for a fixed incident angle, the coefficient exhibits a sharp peak in the backscattering direction of ($\theta_s = \theta_i, \phi_s = \pi + \phi_i$). Such a phenomenon is known as backscattering enhancement and has been observed in laboratory-controlled experiments (Kuga and Ishimaru, 1984). The sharp increase of scattering in the backscattering direction has been known for a long time. The *glory effect* is usually attributed to the Mie scattering pattern as a result of single scattering by particles as discussed in the Section 2.4, Chapter 4.

In the case of volume scattering in the absence of a reflective boundary below the inhomogeneities, backscattering enhancement is a multiple scattering effect. As shown in Section 2.1, it is not present in first-order scattering. To illustrate backscattering enhancement, consider double scattering involving particles 1 and 2. The scattered electric E is a sum of two fields E_{12} and E_{21} (Figure 5.5)

$$E = E_{21} + E_{12} \tag{1}$$

Backscattering Enhancement: Volume Scattering

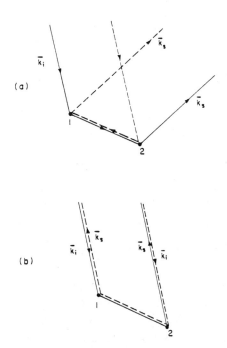

Fig. 5.5 Physical processes for double scattering involving two particles: (a) Scattered direction $\bar{k}_s$ is not in the backscattering direction. E_{21} is solid line and E_{12} is dotted line. (b) Scattered direction $\bar{k}_s$ is in backscattering direction, i.e., $\bar{k}_s = -\bar{k}_i$. E_{21} is solid line and E_{12} is dotted line. $E_{21} = E_{12}$ by reciprocity.

The field E_{21} is the scattered field that originates from the incident wave scattered by particle 1, traveling from particle 1 to particle 2, and finally being scattered by particle 2. The field E_{12} corresponds to the wave which is scattered by particle 2 before scattered by particle 1 (Figure 5.5). The averaged scattered intensity is

$$<|E|^2> = <|E_{21}|^2> + <|E_{12}|^2> + 2Re<E_{12}E_{21}^*> \qquad (2)$$

where the ensemble average is to be taken over particle positions. It is easily shown (Problem 4) from Figure 5.5a that the phase difference between E_{12} and E_{21} is $(\bar{k}_i + \bar{k}_s) \cdot \bar{r}_{12}$, where $\bar{r}_{12}$ is the vector joining particles 1 and 2. We consider the following two cases:

1. $\bar{k}_s$ is not in the backscattering direction, so that $\bar{k}_i + \bar{k}_s$ is not equal to zero (Figure 5.5a). In this case, the phase difference between E_{12} and E_{21} is not equal to zero and is given by $(\bar{k}_i + \bar{k}_s) \cdot \bar{r}_{12}$. Since the particle positions are random, the phase difference will also be random. On averaging over particle positions, $<E_{12}E_{21}^*>$ is equal to zero, and from (2)

$$<|E|^2> = <|E_{21}|^2> + <|E_{12}|^2> \qquad (3)$$

Thus, the intensities add. This is in agreement with the assumption of radiative transfer theory that scattered fields are uncorrelated so that the intensities add.

2. $\overline{k}_i + \overline{k}_s = 0$. This case corresponds to backscattering and is illustrated in Figure 5.5b. The phase difference vanishes. The two fields E_{12} and E_{21} are always in phase in spite of the randomness of particle positions. Thus, $<E_{12}E_{21}^*>$ is not equal to zero. From reciprocity, we actually have

$$E_{12} = E_{21} \qquad (4)$$

so that from (2)

$$<|E|^2> = 2<|E_{21}|^2> + 2<|E_{12}|^2> = 4<|E_{21}|^2> \qquad (5)$$

The result in (5) is twice that of (3). This phenomenon is not accounted for in the radiative transfer theory. We note that the result in (5) only occurs in the backscattering direction. For other directions, (3) is a good approximation. Thus, when the bistatic scattering coefficient is plotted as a function of scattered angle at a fixed incident angle, there is a sharp peak in the backscattering direction. Although we have only illustrated backscattering enhancement for second-order scattering, it can be shown that the phenomenon also exists in higher-order scattering by a similar argument.

2.4 Active Remote Sensing of Layered Random Media

In remote sensing applications, a realistic model might consist of partitioning the entire scattering region into subregions of random media, each with a characteristic correlation function. For example, in vegetation cover such as forest terrain, the subregions would be the leaf and trunk sections of trees, where the respective lateral and vertical correlation lengths are significantly different. A model of vertically stratified medium consisting of N random layers and $M-N$ $(M \geq N)$ homogeneous layers, with boundaries at $z = 0, -d_1, -d_2, \ldots, d_M$ (Figure 5.6) can be used

Here, $\Delta_m = 0$ or 1 according to whether the mth layer is homogeneous or random. The first-order scattered field is the superposition of the first-order scattered fields from each layer

$$\overline{E}_s^{(1)} = \sum_{m=1}^{M} \int_V d^3 r_1 \, \overline{\overline{G}}_{om}^{(0)}(\overline{r}, \overline{r}_1) \cdot Q_m(\overline{r}_1) \overline{E}_m^{(0)}(\overline{r}_1) \qquad (1)$$

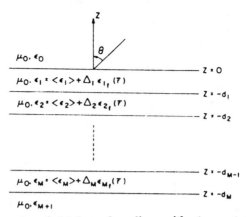

Fig. 5.6 Scattering geometry of M-layered medium, with $\Delta_m = 1$ or 0 ($m = 1, 2, \ldots, M$) according to whether the m-th layer is random or homogeneous.

Assume that the permittivity fluctuations in two different regions are uncorrelated

$$<Q_m(\bar{r}_1) Q_n^*(\bar{r}_2)> = \delta_{mn} C_m(\bar{r}_1 - \bar{r}_2) \tag{2}$$

Therefore, the scattered intensity takes the form

$$<|\overline{E}_s^{(1)}|^2> = \sum_{m=1}^{M} \int_{V_m} d^3 r_1 \int_{V_m} d^3 r_2 \left[\overline{\overline{G}}_{om}^{(0)}(\bar{r}, \bar{r}_1) \cdot \overline{E}_m^{(0)}(\bar{r}_1) \right]$$
$$\cdot \left[\overline{\overline{G}}_{om}^{(0)}(\bar{r}, \bar{r}_2) \cdot \overline{E}_m^{(0)}(\bar{r}_2) \right]^* C_m(\bar{r}_1 - \bar{r}_2) \tag{3}$$

The far-field dyadic Green's function with source point in the m-th layer has the form

$$\overline{\overline{G}}_{om}^{(0)}(\bar{r}, \bar{r}_1) = \frac{e^{ikr}}{4\pi r} \left[\overline{\overline{H}}_m e^{-i\bar{k}_m \cdot \bar{r}_1} + \overline{\overline{F}}_m e^{-i\bar{K}_m \cdot \bar{r}_1} \right] \tag{4}$$

where $\bar{k}_m = \bar{k}_\perp + \hat{z} k_{mz}$ and $\bar{K}_m = \bar{k}_\perp - \hat{z} k_{mz}$. Similarly, the unperturbed field in the m-th layer has the form

$$\overline{E}_m^{(0)} = \left[\overline{E}_{um}^t e^{ik_{mz} z} + \overline{E}_{dm}^t e^{-ik_{mz} z} \right] e^{i\bar{k}_\perp \cdot \bar{r}_\perp} \tag{5}$$

The coefficients in (4) and (5) are determined from the propagation matrix formalism (Chapter 2). Substituting (5) and (4) into (3), the first-order scattered intensity in region 0 can be calculated.

In the following, we illustrate the case of two adjacent random layers, $M = 2$. The backscattering cross section σ_{hh} for horizontally polarized waves is given by

$$\sigma_{hh} = \delta_1 k_1'^4 \pi^2 \frac{|X_{01i}|^4}{|D_{2i}|^4} \left[8d_1 \Phi_1(2\overline{k}_{\perp i}, 0) \left| R_{12i} + R_{23i} e^{i2k_{2zi}(d_2-d_1)} \right|^2 \right.$$

$$\times \left| 1 + R_{12i} R_{23i} e^{i2k_{2zi}(d_2-d_1)} \right|^2 e^{-4k_{1zi}''d_1}$$

$$+ \left\{ \left| R_{12i} + R_{23i} e^{i2k_{2zi}(d_2-d_1)} \right|^4 e^{-4k_{1zi}''d_1} \right.$$

$$\left. + \left| 1 + R_{12i} R_{23i} e^{i2k_{2zi}(d_2-d_1)} \right|^4 \right\}$$

$$\times \left\{ 1 - e^{-4k_{1zi}''d_1} \right\} \frac{\Phi_1(2\overline{k}_{\perp i}, 2k_{1zi})}{2k_{1zi}''} \Bigg]$$

$$+ \delta_2 k_2'^4 \pi^2 \frac{|X_{01i} X_{12i}|^4}{|D_{2i}|^4} \left[8(d_2-d_1) \Phi_2(2\overline{k}_{\perp i}, 0) \right.$$

$$\times |R_{23i}|^2 e^{-4k_{2zi}''(d_2-d_1)} e^{-4k_{1zi}''d_1}$$

$$+ \left\{ |R_{23i}|^4 e^{-4k_{2zi}''(d_2-d_1)} + 1 \right\}$$

$$\left. \times \left\{ 1 - e^{-4k_{2zi}''(d_2-d_1)} \right\} e^{-4k_{1zi}''d_1} \frac{\Phi_2(2\overline{k}_{\perp i}, 2k_{2zi})}{2k_{2zi}''} \right] \quad (6)$$

where R_{jl} and X_{jl}, $j,l = 0,1$, and 2 are Fresnel reflection and transmission coefficients as defined in Chapter 2, and

$$D_2 = 1 + R_{12} R_{23} e^{i2k_{2z}(d_2-d_1)}$$

$$+ R_{01} \left[R_{12} + R_{23} e^{i2k_{2z}(d_2-d_1)} \right] e^{i2k_{1z}d_1} \quad (7)$$

A similar expression for the backscattering cross section σ_{vv} of vertically polarized waves can also be obtained (Problem 5; Zuniga et al., 1979). We plot the typical behavior of the backscattering cross sections as a function of frequency for $\theta_{oi} = 30 \deg$ in Figure 5.7.

Note that the spectral variation of σ_{hh} exhibits two maxima due to resonant scattering within each random layer. This phenomenon of double resonance (or multiple resonances in the case of many random layers) may explain the spectral behavior observed in some backscattering data. The three-layer model is also useful in accounting for the diurnal change due to solar illumination observed in radar remote sensing of snow fields. In Figure

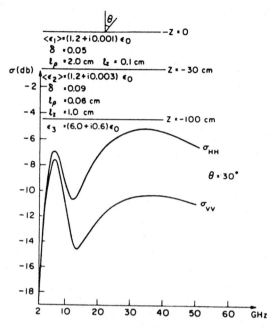

Fig. 5.7 σ_{hh} and σ_{vv} as a function of frequency at 30 deg for two adjacent random layers.

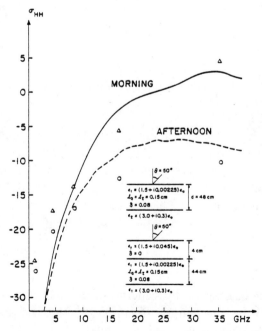

Fig. 5.8 σ_{hh} as a function of frequency at 50 deg for diurnal changes.

5.8, we show a match with experimental data obtained in a snow field which exhibits diurnal changes (Stiles and Ulaby, 1980b) for σ_{hh} plotted as a function of frequency at the viewing angle of 50 deg. A two-layer model is used to describe the morning data and a three-layer model for the afternoon data. In the three-layer model, the surface layer of 4 cm is assumed to represent wet snow.

Next we consider the special case of a two-layer random medium when the top interface is not reflective. In this case $\epsilon_1 = \epsilon_o + \epsilon_{1f}(\bar{r})$, $R_{01} = 0$, $X_{01} = 1$ and $D_{2i} = 1$. From (6), we have

$$\sigma_{hh} = \delta_1 k_1'^4 \pi^2 \left\{ 8d_1 \Phi_1(2\bar{k}_{\perp i}, 0) |R_{12i}|^2 e^{-4k_{1zi}''d_1} \right.$$

$$\left. + \left[|R_{12i}|^4 e^{-4k_{1zi}''d_1} + 1\right] \left[1 - e^{-4k_{1zi}''d_1}\right] \frac{\Phi_1(2\bar{k}_{\perp i}, 2k_{1zi})}{2k_{1zi}''} \right\} \quad (8)$$

Radiative transfer theory can also be applied to calculate the backscattering cross sections for a two-layer random medium (Problem 6). The result is the same as that of (8) except that the first term inside the large brackets is a factor of 4 instead of 8. This disagreement is due to the fact that backscattering enhancement is present in single-volume scattering in the presence of a reflective boundary below the inhomogeneities. This will be discussed in the next section.

2.5 Backscattering Enhancement: Volume Scattering in the Presence of Reflective Boundary

In this section, it will be shown that backscattering enhancement occurs in the single volume-scattering in the presence of a reflective boundary. The reflective surface can be smooth or rough. Consider a layer of inhomogeneities above a reflective boundary. The configuration is representative of a layer of vegetation lying above soil. Single scattering by a particle consists of four contributions (Figure 5.9). Hence, the electric field E is

$$E = E_p + E_{pb} + E_{bp} + E_{bpb} \quad (1)$$

where E_p is the field caused by scattering of the incident wave along $\bar{k}_i$ by the particle into the direction $\bar{k}_s$ (Figure 5.9a). The field E_{bp} is the wave that is scattered by the particle, and then reflected by the boundary into the direction $\bar{k}_s$ (Figure 5.9c). The field E_{pb} is the wave which first hits the boundary before hitting the particle (Figure 5.9c). Finally the field E_{bpb} is

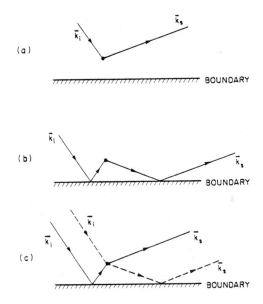

Fig. 5.9 Physical processes for single-volume scattering in the presence of reflection boundary. Scattered direction $\bar{k}_s$ is not in backscattering direction. (a) E_p; (b) E_{bpb}; (c) solid line, E_{pb}; dotted line, E_{bp}.

the field that hits the boundary, then the particle, and then the boundary (Figure 5.9b). The averaged intensity is

$$<|E|^2> = <|E_p|^2> + <|E_{bpb}|^2> + <|E_{bp}|^2> + <|E_{pb}|^2>$$
$$+ 2Re<E_p E_{bpb}^*> + 2Re<E_p E_{bp}^*>$$
$$+ 2Re<E_p E_{pb}^*> + 2Re<E_{bpb} E_{bp}^*>$$
$$+ 2Re<E_{bpb} E_{pb}^*> + 2Re<E_{bp} E_{pb}^*> \quad (2)$$

where the angular brackets $<\ >$ stand for averaging over particle positions. Consider the two cases:

1. Scattered direction not in backscattering direction. There are nonzero phase differences among the four field quantities as can be seen from Figure 5.9. The phase differences depend on the particle position. Hence, on averaging over particle positions that are assumed to be random, the cross terms in (2) vanish, and

$$<|E|^2> = <|E_p|^2> + <|E_{bpb}|^2> + <|E_{bp}|^2> + <|E_{pb}|^2> \quad (3)$$

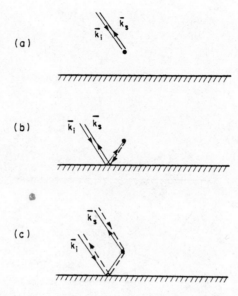

Fig. 5.10 Physical processes for single volume scattering in the presence of reflective boundary. Scattered direction $\bar{k}_s$ is in backscattering direction, $\bar{k}_s = -\bar{k}_i$. (a) E_p; (b) E_{bpb}; (c) Solid line, E_{pb}; dotted line, E_{bp}; $E_{bp} = E_{pb}$ by reciprocity.

The intensities of the four field quantities add to give the total intensity. This is in agreement with the assumption of radiative transfer theory.

2. Scattered direction along backscattering direction. The physical processes associated with the four field quantities are depicted in Figure 5.10. The phase difference between E_{bp} and E_{pb} vanishes. By reciprocity

$$E_{bp} = E_{pb} \qquad (4)$$

The nonzero phase differences still exist between E_p and the rest of the field quantities and between E_{bpb} and the rest of the field quantities. Hence, from (2)

$$<|E|^2> = <|E_p|^2> + <|E_{bpb}|^2> + 2<|E_{bp}|^2> + 2<|E_{pb}|^2> \qquad (5)$$

A comparison between (3) and (5) shows that there is backscattering enhancement which is thus a first-order effect for volume scattering in the presence of a reflective boundary. Backscattering enhancement still exists even if the reflective boundary is a rough surface. The physical processes for E_{pb} and E_{bp} are illustrated in Figure 5.11. By reciprocity, we still have $E_{bp} = E_{pb}$. Thus, although rough surfaces can convert a coherent wave into incoherent scattered energy, the backscattering enhancement phenomenon is preserved.

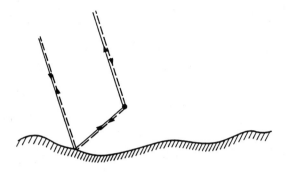

Fig. 5.11 Single-volume scattering in the presence of a rough surface. E_{bp} is still equal to E_{pb} by reciprocity. Solid line, E_{pb}; dotted line, E_{bp}.

3 ANALYTIC WAVE THEORY

In this section, we will use the diagrammatic approach (Frisch, 1968) to study propagation and scattering in random media. The diagrammatic approach leads to Dyson's equation for the mean field and the Bethe-Salpeter equation for the covariance of the field. Of the approximations to Dyson's equation; the bilocal and nonlinear approximations will be examined. We will also study the ladder approximation to the Bethe-Salpeter equation. It is shown that the ladder approximation is consistent with the radiative transfer theory for the case of half-space random media. However, the ladder approximation neglects the cyclical diagrams that directly contribute to backscattering enhancement. The case of uncorrelated point scatterers will be studied, whereas the problem of the correlated scatterer positions is deferred to Chapter 6.

Wave propagation and scattering by random medium have been studied for many years and has been applied to propagation in turbulent medium (Booker and Gordon, 1950; Bugnolo, 1960; Bremmer, 1964; Keller, 1964; Beran, 1968; Brown, 1971; Barabanenkov et al., 1971; Gurvich and Tatarskii, 1975; Ishimaru, 1978; Chow et al., 1981). In the case of a one-dimensional random medium, because of the simplicity of the equations, alternative approaches have been employed and the results have been shown to deviate significantly from that of the radiative transfer theory (Morrison et al., 1981; Papanicolaou and Keller, 1971; Kohler and Papanicolaou, 1973; Ogura, 1975).

3.1 Dyson's Equation for the Mean Field

Consider an unbounded random medium with permittivity $\epsilon(\bar{r}) = <\epsilon> + \epsilon_f(\bar{r})$. The dyadic Green's function $\overline{\overline{G}}(\bar{r}, \bar{r}_o)$ in the medium satisfies the vector wave equation

$$\nabla \times \nabla \times \overline{\overline{G}}(\bar{r},\bar{r}_o) - \omega^2\mu <\epsilon> \overline{\overline{G}}(\bar{r},\bar{r}_o) = Q(\bar{r})\overline{\overline{G}}(\bar{r},\bar{r}_o) + \overline{\overline{I}}\delta(\bar{r}-\bar{r}_o) \quad (1)$$

with $Q(\bar{r}) = \omega^2\mu\epsilon_f(\bar{r})$. The solution to (1) can be expressed as an integral equation by using the homogeneous dyadic Green's function $\overline{\overline{G}}^{(0)}(\bar{r},\bar{r}_o)$ which satisfies the wave equation with mean wavenumber $k_m = \omega\sqrt{\mu <\epsilon>}$.

$$\overline{\overline{G}}(\bar{r},\bar{r}_o) = \overline{\overline{G}}^{(0)}(\bar{r},\bar{r}_o) + \int d^3r_1 \overline{\overline{G}}^{(0)}(\bar{r},\bar{r}_1) \cdot Q(\bar{r}_1)\overline{\overline{G}}(\bar{r}_1,\bar{r}_o) \quad (2)$$

Equation (2) may be solved formally by iteration to generate the Neumann series for the dyadic Green's function

$$\overline{\overline{G}}(\bar{r},\bar{r}_o) = \overline{\overline{G}}^{(0)}(\bar{r},\bar{r}_o) + \int d^3r_1 \overline{\overline{G}}^{(0)}(\bar{r},\bar{r}_1) \cdot Q(\bar{r}_1)\overline{\overline{G}}^{(0)}(\bar{r}_1,\bar{r}_o)$$

$$+ \int d^3r_1 d^3r_2 \overline{\overline{G}}^{(0)}(\bar{r},\bar{r}_1) \cdot Q(\bar{r}_1)\overline{\overline{G}}^{(0)}(\bar{r}_1,\bar{r}_2)Q(\bar{r}_2) \cdot \overline{\overline{G}}^{(0)}(\bar{r}_2,\bar{r}_o)$$

$$+ \cdots \quad (3)$$

The first term in the perturbation series (3) is the zeroth-order approximation and represents the solution to the homogeneous problem in which there is no scattering. The next term is the first-order Born approximation and has the physical interpretation of single scattering of the incident wave.

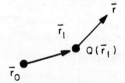

In general, the n th-order Born approximation represents the n-tuply scattered incident wave.

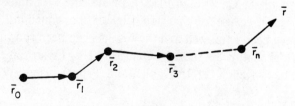

To discuss the statistics of the random fluctuations $\epsilon_f(\bar{r})$, a common assumption made is that $\epsilon_f(\bar{r})$ is a Gaussian stationary random process so that the first and second moments are sufficient to describe the statistics. Therefore, we have

$$<\epsilon_f(\bar{r})> = 0 \quad (4)$$

Dyson's Equation for the Mean Field

and in general, for integer n

$$<\epsilon_f(\bar{r}_1)\cdots\epsilon_f(\bar{r}_{2n+1})> = 0 \qquad (5a)$$

$$<\epsilon_f(\bar{r}_1)\cdots\epsilon_f(\bar{r}_{2n})> = \sum_{\substack{\text{distinct}\\\text{pairs}}} <\epsilon_f(\bar{r}_i)\epsilon_f(\bar{r}_j)>\cdots<\epsilon_f(\bar{r}_p)\epsilon_f(\bar{r}_s)> \qquad (5b)$$

where the summation extends over all possible distinct pairs of arguments to give $(2n-1)!!$ terms. For example, if $n=2$

$$<\epsilon_f(\bar{r}_1)\epsilon_f(\bar{r}_2)\epsilon_f(\bar{r}_3)\epsilon_f(\bar{r}_4)> = <\epsilon_f(\bar{r}_1)\epsilon_f(\bar{r}_2)><\epsilon_f(\bar{r}_3)\epsilon_f(\bar{r}_4)>$$
$$+ <\epsilon_f(\bar{r}_1)\epsilon_f(\bar{r}_3)><\epsilon_f(\bar{r}_2)\epsilon_f(\bar{r}_4)> \qquad (6)$$
$$+ <\epsilon_f(\bar{r}_1)\epsilon_f(\bar{r}_4)><\epsilon_f(\bar{r}_2)\epsilon_f(\bar{r}_3)>$$

In the case of non-Gaussian statistics, the moments of $\epsilon_f(\bar{r})$ may still be cluster-expanded but not exclusively in terms of the two-point correlation function (Frisch, 1968). Assuming Gaussian statistics and taking ensemble average of (3), the perturbation series for the mean dyadic Green's function is obtained

$$<\bar{\bar{G}}(\bar{r},\bar{r}_o)> = \bar{\bar{G}}^{(0)}(\bar{r},\bar{r}_o) + \int d^3r_1 d^3r_2 \bar{\bar{G}}^{(0)}(\bar{r},\bar{r}_1)\cdot\bar{\bar{G}}^{(0)}(\bar{r}_1,\bar{r}_2)$$
$$\cdot\bar{\bar{G}}^{(0)}(\bar{r}_2,\bar{r}_o) <Q(\bar{r}_1)Q(\bar{r}_2)>$$
$$+ \int d^3r_1 d^3r_2 d^3r_3 d^3r_4 \bar{\bar{G}}^{(0)}(\bar{r},\bar{r}_1)\cdot\bar{\bar{G}}^{(0)}(\bar{r}_1,\bar{r}_2)$$
$$\cdot\bar{\bar{G}}^{(0)}(\bar{r}_2,\bar{r}_3)\cdot\bar{\bar{G}}^{(0)}(\bar{r}_3,\bar{r}_4)\cdot\bar{\bar{G}}^{(0)}(\bar{r}_4,\bar{r}_o)$$
$$\times [<Q(\bar{r}_1)Q(\bar{r}_2)><Q(\bar{r}_3)Q(\bar{r}_4)>$$
$$+ <Q(\bar{r}_1)Q(\bar{r}_4)><Q(\bar{r}_2)Q(\bar{r}_3)>$$
$$+ <Q(\bar{r}_1)Q(\bar{r}_3)><Q(\bar{r}_2)Q(\bar{r}_4)>]$$
$$+\cdots \qquad (7)$$

If the iteration of (7) is continued, the expressions will become increasingly cumbersome to handle. Therefore, Feynman diagrams (Frisch, 1968) are introduced as follows:

• ≡ $Q(\bar{r}_1)$ ≡ vertex over which spatial integration is implied and contraction of dyadic indices

—— ≡ $\overline{\overline{G}}^{(0)}(\bar{r},\bar{r}_o)$

═══ ≡ $<\overline{\overline{G}}(\bar{r},\bar{r}_o)>$

⌢ ≡ $<Q(\bar{r}_1)Q(\bar{r}_2)>$

These rules set up a one-to-one correspondence between the analytical expressions in (7) and their graphical representations. Thus, (7) becomes

$$\equiv \;=\; \text{——} \;+\; \text{—⌢—} \;+\; \left[\text{—⌢—⌢—} \right.$$
$$+\; \text{—⌢⌢—} \;+\; \left. \text{—⌢⃝—} \right] \qquad (8)$$
$$+\; \cdots$$

where for example

$$\text{—⌢—} \equiv \int d^3r_1 d^3r_2 \overline{\overline{G}}^{(0)}(\bar{r},\bar{r}_1) \cdot \overline{\overline{G}}^{(0)}(\bar{r}_1,\bar{r}_2) \cdot \overline{\overline{G}}^{(0)}(\bar{r}_2,\bar{r}_o)$$
$$\times <Q(\bar{r}_1)Q(\bar{r}_2)>$$

A strongly connected diagram or irreducible diagram is defined as one in which the diagram cannot be divided without breaking the correlation connections. For example, the following diagrams are strongly connected:

All other diagrams are weakly connected or reducible. For example,

Dyson's Equation for the Mean Field

The mass operator, is defined as the sum of all strongly connected diagrams (minus end connectors)

$$\boxed{\times} \equiv \text{[diagram]} + \text{[diagram]} + \text{[diagram]} + \text{[diagram]} + \cdots \tag{9}$$

With the mass operator all the strongly connected diagrams of (8) are summed as

$$\text{—}\boxed{\times}\text{—} = \text{[diagram]} + \text{[diagram]} + \text{[diagram]} + \cdots \tag{10}$$

The weakly connected diagrams in (8) which contain two strongly connected elements may be summed as

$$\text{—}\boxed{\times}\text{—}\boxed{\times}\text{—} = \text{[diagram]} + \text{[diagram]} + \cdots \tag{11}$$

Continuing with this process, the Neumann series for the mean dyadic Green's function may be written in the equivalent form:

$$\begin{aligned}
\overline{\overline{}} &= \text{—} + \text{—}\boxed{\times}\text{—} + \text{—}\boxed{\times}\text{—}\boxed{\times}\text{—} \\
&\quad + \text{—}\boxed{\times}\text{—}\boxed{\times}\text{—}\boxed{\times}\text{—} + \cdots \\
&= \text{—} + \text{—}\boxed{\times}\left[\text{—} + \text{—}\boxed{\times}\text{—} \right. \\
&\quad \left. + \text{—}\boxed{\times}\text{—}\boxed{\times}\text{—} + \cdots \right]
\end{aligned} \tag{12}$$

or

$$= \; = \; - \; + \; -\!\!\otimes\!\!= \qquad (13)$$

Equation (13) is the diagrammatic representation of Dyson's equation. In analytical form, Dyson's equation becomes

$$<\overline{\overline{G}}(\bar{r},\bar{r}_o)> = \overline{\overline{G}}^{(0)}(\bar{r},\bar{r}_o) + \int d^3r_1 d^3r_2 \overline{\overline{G}}^{(0)}(\bar{r},\bar{r}_1) \cdot \\ \cdot \overline{\overline{Q}}(\bar{r}_1,\bar{r}_2) \cdot <\overline{\overline{G}}(\bar{r}_2,\bar{r}_o)> \qquad (14)$$

where $\overline{\overline{Q}}(\bar{r}_1,\bar{r}_2)$ is the dyadic mass operator. Dyson's equation is an exact equation for the mean dyadic Green's function. However, the mass operator as given in (9) is in the form of an infinite series and must be approximated to solve (14). In the next section, approximations to the mass operator will be examined.

3.2 Bilocal and Nonlinear Approximations

To solve Dyson's equation for the mean field, the mass operator is often approximated by retaining only the first term. This is called the bilocal or first-order smoothing approximation

$$\overline{\overline{Q}}(\bar{r}_1,\bar{r}_2) \equiv \; \otimes \; \approx \; \frown \; = C(\bar{r}_1 - \bar{r}_2)\overline{\overline{G}}^{(0)}(\bar{r}_1,\bar{r}_2) \qquad (1)$$

Under this approximation Dyson's equation becomes

$$= \; = \; - \; + \; -\!\!\frown\!\!= \qquad (2)$$

or

$$<\overline{\overline{G}}(\bar{r},\bar{r}_o)> = \overline{\overline{G}}^{(0)}(\bar{r},\bar{r}_o) + \int d^3r_1 d^3r_2 \overline{\overline{G}}^{(0)}(\bar{r},\bar{r}_1) \cdot \overline{\overline{G}}^{(0)}(\bar{r}_1,\bar{r}_2) \\ \cdot <\overline{\overline{G}}(\bar{r}_2,\bar{r}_o)> C(\bar{r}_1 - \bar{r}_2) \qquad (3)$$

To visualize the terms of the Neumann series that are included in the bilocal approximation to Dyson's equation, (2) can be iterated to obtain

$$= \; = \; - \; + \; \frown \; + \; \frown\frown \;$$
$$+ \; \ldots \tag{4}$$

Thus, the bilocal approximation accounts for those weakly connected diagrams that are formed by products of bilocal diagrams (i.e. $\frown$).

Bilocal Approximation Applied to Scalar Wave Equation

The bilocally approximated Dyson's equation for scalar wave propagation is given by (Tatarskii and Gertsenshtein, 1963)

$$<G(\bar{r},\bar{r}_o)> = G^{(0)}(\bar{r},\bar{r}_o) + \int d^3r_1 d^3r_2 G^{(0)}(\bar{r},\bar{r}_1)$$
$$\times G^{(0)}(\bar{r}_1,\bar{r}_2) <G(\bar{r}_2,\bar{r}_o)> C(\bar{r}_1-\bar{r}_2) \tag{5}$$

Operating on (5) with $(\nabla^2 + k_m^2)$ yields

$$(\nabla^2 + k_m^2) <G(\bar{r},\bar{r}_o)> = -\delta(\bar{r}-\bar{r}_o) - \int d^3r_1 G^{(0)}(\bar{r},\bar{r}_1)$$
$$\times <G(\bar{r}_1,\bar{r}_o)> C(\bar{r}-\bar{r}_1) \tag{6}$$

Equation (6) may be solved by using Fourier transforms (Problem 7). However, it is instructive to consider the case of small-scale fluctuations $k_m a \ll 1$ where a is the correlation length. In this case, $C(\bar{r}-\bar{r}_1)$ is sharply peaked at $\bar{r}=\bar{r}_1$, and (6) is approximated as

$$(\nabla^2 + k_m^2) <G(\bar{r},\bar{r}_o)> \simeq -\delta(\bar{r}-\bar{r}_o)$$
$$- <G(\bar{r},\bar{r}_o)> \int d^3r_1 G^{(0)}(\bar{r},\bar{r}_1) C(\bar{r}-\bar{r}_1) \tag{7a}$$

Changing integration variables to $\bar{R} = \bar{r} - \bar{r}_1$, (7a) can be rewritten as

$$\left[\nabla^2 + k_m^2 + \int d^3R \, G^{(0)}(\bar{R}) C(\bar{R})\right] <G(\bar{r},\bar{r}_o)> = -\delta(\bar{r}-\bar{r}_o) \tag{7b}$$

Thus, the randomness introduces an effective propagation constant, K, for the mean Green's function $<G(\bar{r},\bar{r}_o)>$ such that

$$K^2 = k_m^2 + \int d\bar{r} \, G^{(0)}(\bar{r}) C(\bar{r}) \tag{8}$$

Assuming a spherically symmetric correlation function of the form $C(\overline{R}) = \delta k_m'^4 \exp(-R/a)$ and taking $G^{(0)}(\overline{r}) = \exp(ik_m r)/4\pi R$ we find, on making use of $k_m a \ll 1$, that

$$K^2 = k_m^2 + \delta k_m'^4 a^2 (1 + 2ik_m a) \qquad (9)$$

The first term in (9) can be regarded as the zeroth-order solution and the second term in (9) as the first-order solution. Thus, for (9) to be valid, the first-order solution has to be smaller than the zeroth-order solution, giving the condition

$$\delta k_m^2 a^2 \ll 1 \qquad (10)$$

Since $k_m a \ll 1$, we may further simplify (9) to get

$$K \simeq k_m \left[1 + \frac{\delta k_m^2 a^2}{2} (1 + i2k_m a) \right] \qquad (11)$$

We see that the randomness has introduced an effective loss term that causes the coherent wave to decay exponentially as it propagates in the random medium. This decay occurs as a consequence of multiple scattering of the waves.

Bilocal Approximation Applied to Vector Wave Equation

In this case, the analog of (8) is (Tatarskii, 1964; Karal and Keller, 1964; Keller and Karal, 1966)

$$K^2 = k_m^2 + \int d\overline{r}\, \overline{\overline{G}}^{(0)}(\overline{r})\, C(\overline{r}) \qquad (12)$$

On integration, the second term in (12) is proportional to the unit dyad when $C(\overline{r})$ is spherically symmetric as will be assumed in this section.

The dyadic Green's function $\overline{\overline{G}}^{(0)}(\overline{r})$ is (Van Bladel, 1961)

$$\overline{\overline{G}}^{(0)}(\overline{r}) = PS\, \overline{\overline{G}}^{(0)}(\overline{r}) - \frac{\overline{\overline{I}}}{3k_m^2} \delta(\overline{r}) \quad \text{(sphere)} \qquad (13)$$

where PS stands for principal value with an infinitesimal volume of defined shape excluded about the point $\overline{r} = 0$. The expression (sphere) at the end of (13) indicates that a spherical infinitesimal volume has been chosen for the shape of the exclusion. The coefficient in front of the Dirac delta function of the second term depends on the shape of the exclusion volume. If shapes other than spherical are chosen, then the coefficient will be different from that of (13) (Problem 8).

For a spherical exclusion, (Problem 9)

$$PS\overline{\overline{G}}^{(0)}(\bar{r}) = PS\,G_1(\bar{r})\overline{\overline{I}} + PS\,G_2(\bar{r})\hat{r}\hat{r} \qquad (14)$$

where

$$G_1(\bar{r}) = (-1 + ik_m r + k_m^2 r^2)\frac{e^{ik_m r}}{4\pi k_m^2 r^3} \qquad (15)$$

$$G_2(\bar{r}) = (3 - 3ik_m r - k_m^2 r^2)\frac{e^{ik_m r}}{4\pi k_m^2 r^3} \qquad (16)$$

Using (14) through (16), we find that for spherically symmetric correlation functions

$$\int d\bar{r}\,C(\bar{r})\,PS\overline{\overline{G}}^{(0)}(\bar{r}) = \frac{2}{3}\overline{\overline{I}}\int_0^\infty dr\,C(r)\,r\,e^{ik_m r} \qquad (17)$$

Using (13) and (17) in (12) and using the exponential correlation function as in the scalar case, we find that in the low-frequency limit of $k_m a \ll 1$

$$K^2 = k_m^2 - \frac{\delta}{3}k_m^2 + \frac{2}{3}\delta k_m^4 a^2(1 + 2ik_m a) \qquad (18)$$

The result in (18) is valid if the zeroth-order solution [the first term in (18)] is much larger than the first-order solution [the rest of the terms in (18)]. Thus the validity conditions of (18) are

$$\delta \ll 1 \qquad (19)$$

and

$$\delta k_m^2 a^2 \ll 1 \qquad (20)$$

It is interesting to note, on comparing (10) with (19) and (20), that the vector analog has one more criterion than the scalar case. The weak permittivity fluctuation criterion of (19) is very restrictive in remote sensing applications because natural geophysical terrains are characterized by large permittivity fluctuations. Physically, in cases of large permittivity fluctuations, the internal field inside the inhomogeneity can be quite different from that of the background medium. The restriction in (19) can be eliminated by taking into account the singularity of the dyadic Green's function and will be done in Section 4.

Nonlinear Approximation

The terms in the mass operator can be rearranged to form a new renormalized mass operator (Problem 10). If only the first term of the new mass operator is retained, the nonlinear approximation to Dyson's equation is obtained.

$$\overline{\overline{=}} \;=\; \text{———} \;+\; \text{—⌒—} \qquad (21)$$

or

$$<\overline{\overline{G}}(\bar{r},\bar{r}_o)> = \overline{\overline{G}}^{(0)}(\bar{r},\bar{r}_o) + \int d^3r_1 d^3r_2 \overline{\overline{G}}^{(0)}(\bar{r},\bar{r}_1) \cdot <\overline{\overline{G}}(\bar{r}_1,\bar{r}_2)>$$
$$\cdot <\overline{\overline{G}}(\bar{r}_2,\bar{r}_o)> C(\bar{r}_1 - \bar{r}_2) \qquad (22)$$

The nonlinear Dyson's equation (22) may be solved by Fourier transforms. For simplicity, we shall consider the scalar analog of (22). The result is similar to (6) with $G^{(0)}$ replaced by $<G>$ in the integral

$$(\nabla^2 + k_m^2)<G(\bar{r},\bar{r}_o)> = -\delta(\bar{r} - \bar{r}_o) - \int d^3r_1 <G(\bar{r},\bar{r}_1)>$$
$$\times <G(\bar{r}_1,\bar{r}_o)> C(\bar{r} - \bar{r}_1) \qquad (23)$$

For the case $k_m a \ll 1$, the correlation function $C(\bar{r} - \bar{r}_1)$ is sharply peaked at $\bar{r} = \bar{r}_1$. Hence, the analog of (8) for the nonlinear approximation is

$$K^2 = k_m^2 + \int d\bar{r} <G(\bar{r})> C(\bar{r}) \qquad (24)$$

Assuming $<G(\bar{r})> = \exp(iKr)/4\pi r$ and $C(\bar{r}) = \delta k_m'^4 \exp(-r/a)$, we have

$$K^2 \simeq k_m^2 + \delta k_m'^4 a^2 (1 + 2iKa) \qquad (25)$$

Next we compare the result of bilocal approximation of (9) with that of nonlinear approximation of (25). Suppose k_m is real, then $Re(K^2)$ are the same for both approximations. The imaginary part of K^2 is related to the attenuation rate due to scattering. The two results of (9) and (25) for the imaginary part are different since one is $2\delta k_m^5 a^3$ while the other is $2\delta k_m'^4 K a^3$. Since $Re(K)$ is much greater than $Im(K)$, the attenuation rates based on the two approximations will be essentially the same if $Re(K)$ is approximately equal to k_m. This is usually the case for tenuous medium or for continuous random medium with small δ. Under these circumstances, the results of bilocal approximation and nonlinear approximation are essentially the same.

3.3 Bethe-Salpeter Equation

In the previous sections, Dyson's equation for the mean field is derived by resumming the averaged Neumann series for the field. In a similar fashion, an equation governing the second moment of the field can be derived by using the diagrammatic method.

We consider an infinite unbounded random medium with permittivity

$$\epsilon(\bar{r}) = <\epsilon> + \epsilon_f(\bar{r}) \qquad (1)$$

and $\epsilon_f(\bar{r})$ is assumed to be real. The solution to the scattering of electromagnetic waves in terms of dyadic Green's functions is given by (2), Section 3.1. The equation may be solved formally by iteration to yield the Neumann series solution for the random dyadic Green's function. If we perform the tensor product and ensemble average the resulting expression, the Neumann series solution for the covariance of the field is obtained. To facilitate the analysis, Feynman diagrams are introduced in conjunction with those of Section 3.1:

$$\boxed{\text{I}} \equiv <\bar{\bar{G}}(\bar{r},\bar{r}_o)\bar{\bar{G}}^*(\bar{r}',\bar{r}'_o)> \qquad (2)$$

$$\sim\!\!\sim\!\!\sim \equiv \bar{\bar{G}}(\bar{r},\bar{r}_o) \qquad (3)$$

$$\frown \text{ or } \big| \equiv <Q(\bar{r}_1)Q(\bar{r}_2)> \qquad (4)$$

Using diagrams, the equation for the Green's function can be represented as

$$\sim\!\!\sim\!\!\sim \;=\; \underline{\quad} \;+\; \underline{\quad}\!\bullet\!\!\sim\!\!\sim \qquad (5)$$

Iterating (5) yields

$$\sim\!\!\sim\!\!\sim \;=\; \underline{\quad} \;+\; \underline{\quad}\!\bullet\!\underline{\quad} \;+\; \underline{\quad}\!\bullet\!\underline{\quad}\!\bullet\!\underline{\quad}$$

$$+ \; \underline{\quad}\!\bullet\!\underline{\quad}\!\bullet\!\underline{\quad}\!\bullet\!\underline{\quad} \;+\; \cdots \qquad (6)$$

Two-level diagrams can be introduced where one level represents a vector space and the other level represents the complex conjugate vector space. For example,

$$\underset{\sim\sim\sim}{} \equiv \overline{\overline{G}}(\bar{r},\bar{r}_o)\overline{\overline{G}}^*(\bar{r}',\bar{r}'_o) \tag{7}$$

Upon forming the tensor product of the Neuman series (6), and ensemble averaging give

(8)

Weakly connected or reducible diagrams are those diagrams that can be derived without breaking a correlation connection.

Weakly Connected Examples:

All other diagrams are strongly connected or irreducible.

Strongly Connected Examples:

The intensity operator is the sum of all strongly connected diagrams (minus end connectors):

$$\boxed{X} \equiv \; | \; + \; \bowtie \; + \; \frown \; + \; \cdots \tag{9}$$

The sum of all strongly connected diagrams in (8) may be written in terms of the intensity operator as

Bethe-Salpeter Equation

The sum of all strongly and weakly connected diagrams in (8) containing only one element belonging to the intensity operator is given by:

[diagram]

The sum of all weakly connected diagrams in (8) containing any two elements of the intensity operator is given by

[diagram]

Continuing with this process, the series for the field correlation can be rewritten in the form

$$\quad (10)$$

or

$$\quad (11)$$

Equation (11) is the diagrammatic representation of the Bethe-Salpeter equation for the correlation of the field. In analytical form, the Bethe-Salpeter equation reads :

$$<G_{ij}(\bar{r},\bar{r}_o)G^*_{kl}(\bar{r}',\bar{r}'_o)> = <G_{ij}(\bar{r},\bar{r}_o)><G^*_{kl}(\bar{r}',\bar{r}'_o)>$$
$$+ \int d^3r_1 d^3r_2 d^3r'_1 d^3r'_2 <G_{im}(\bar{r},\bar{r}_1)><G^*_{kg}(\bar{r}',\bar{r}'_1)>$$
$$I_{mn,gr}(\bar{r}_1,\bar{r}_2;\bar{r}'_1,\bar{r}'_2) <G_{nj}(\bar{r}_2,\bar{r}_o)G^*_{rl}(\bar{r}'_2,\bar{r}'_o)>$$
$$\quad (12)$$

where $I_{mn,qr}$ is the intensity operator as given by (9).

Quite often the Bethe-Salpeter equation is written in terms of the field covariance rather than the field correlation. The field covariance is defined by the following diagram

$$\boxed{\tilde{I}} = \boxed{I} - \equiv \qquad (13)$$

Eliminating $\boxed{I}$ in (11) yields

$$\boxed{\tilde{I}} = \equiv \times \equiv + \equiv \times \boxed{\tilde{I}} \qquad (14)$$

The Bethe-Salpeter equation is an exact equation for the second moment of the field. However, the intensity operator as given by (9) is in the form of an infinite series and must be approximated in some fashion if useful solutions are to be obtained. A common approximation known as the ladder approximation is to retain only the first term of the series expansion

$$\boxed{\times} \simeq \big| = \delta(\bar{r}_1 - \bar{r}_2)\delta(\bar{r}'_1 - \bar{r}'_2)C(\bar{r}_1 - \bar{r}'_1) \qquad (15)$$

The Bethe-Salpeter equation then becomes

$$\boxed{\tilde{I}} = \equiv + \equiv \boxed{\tilde{I}} \qquad (16)$$

To see which terms of the Neumann series for the field covariance are included in the ladder approximation, we iterate (16) to obtain

$$\boxed{\tilde{I}} = \equiv + \equiv\equiv + \equiv\equiv\equiv \qquad (17)$$
$$+ \cdots$$

Physically, the ladder approximation accounts for multiple scattering of waves that propagate in the same direction and within a correlation length of each other (Figure 5.12). The ladder approximation together with the nonlinear approximation of the mean field satisfies energy conservation exactly (Problem 11) and should be used in conjunction with the nonlinear approximation. However it was shown in the previous section that for tenuous medium or for continuous random medium with small variance of permittivity fluctuations, we have $Re(K) \simeq k_m$ and the results of

Fig. 5.12 Physical picture of n th-order scattering in ladder approximation.

bilocal and nonlinear approximations are essentially the same. Under such circumstances, the bilocal approximation can be used in conjunction with the ladder approximation (Ishimaru, 1978; Fante, 1981). However for non-tenuous dense media to be treated in Chapter 6, such is generally not the case.

To study the fluctuation of intensities, the fourth-order moment of the field is to be calculated (Goodman, 1976; Tomiyasu, 1983). The fourth-order moment has been studied for propagation in turbulent medium (Gurvich and Tatarskii, 1975; Ishimaru, 1978; Dashen, 1979).

3.4 Ladder Approximation for Isotropic Point Scatterers and Relation to Radiative Transfer Theory

The model of isotropic point scatterers has been used in Chapter 4 to study the multiple scattering processes in radiative transfer theory. In this section, we examine the equation of analytic wave theory for point scatterers. The problem of finite-sized dielectric scatterers with correlated particle positions will be considered in Chapter 6. The field equations for the first two moments for point scatterers are analogous to those of random medium in Sections 3.1 through 3.3.

In this section, by using the model of isotropic point scatterers, the relation between radiative transfer theory and ladder approximation will be examined. It will be shown that the two are equivalent under the assumption that the rate of extinction of the coherent wave is much smaller than the wavenumber. The assumption is valid for problems in where multiple scattering effects are important.

Consider a wave $\psi_{inc}(\bar{r})$ incident on a single point scatterer located at $\bar{r}_\alpha$. Then the total field ψ is given by

$$\psi(\bar{r}) = \psi_{inc}(\bar{r}) + f \frac{e^{ik|\bar{r}-\bar{r}_\alpha|}}{|\bar{r}-\bar{r}_\alpha|} \psi_{inc}(\bar{r}_\alpha) \qquad (1)$$

where f is the scattering amplitude and k is the wavenumber of the background medium. Since $G^{(0)}(\bar{r}) = \exp(ikr)/4\pi r$ is the Green's function for the unperturbed problem, (1) can be written as

$$\psi(\bar{r}) = \psi_{inc}(\bar{r}) + \int d\bar{r}' \int d\bar{r}'' \, G^{(0)}(\bar{r},\bar{r}') T_\alpha(\bar{r}',\bar{r}'') \psi_{inc}(\bar{r}'') \qquad (2)$$

where

$$T_\alpha(\bar{r}',\bar{r}'') = 4\pi f \, \delta(\bar{r}' - \bar{r}_\alpha) \, \delta(\bar{r}'' - \bar{r}_\alpha) \qquad (3)$$

is the transition operator for the point scatterer at $\bar{r}_\alpha$. In diagrammatic notation, let

$$T_\alpha = \overset{\alpha}{\otimes} \qquad (4)$$

then (2) becomes

$$\psi = \psi_{inc} + \underline{\quad\quad}\overset{\alpha}{\otimes} \; \psi_{inc} \qquad (5)$$

where ——— stands for $G^{(0)}$.

For the case of the multiple scattering problem involving N particles located at $\bar{r}_1, \bar{r}_2, \ldots, \bar{r}_N$, the particle positions $\bar{r}_1, \bar{r}_2, \ldots, \bar{r}_N$ are taken to be random variables described by the joint probability density function $p_N(\bar{r}_1, \bar{r}_2, \ldots, \bar{r}_N)$. For the case of isotropic point scatterers, it is reasonable to assume that the particle positions are independent, so that

$$p_N(\bar{r}_1, \bar{r}_2, \ldots, \bar{r}_N) = \frac{1}{V^N} \qquad (6)$$

where V is the volume of the region of interest. Ensemble averaging can be taken over the field equation and a Dyson's equation and a Bethe-Salpeter equation can be obtained respectively for the first and second moments (Frisch, 1968). Retaining the first terms in the mass operator and the intensity operator leads, respectively, to analogous bilocal and ladder approximations. They are

$$\equiv \; = \; \underline{\quad\quad} + \underline{\quad\quad}\otimes\equiv \qquad (7)$$

for the bilocal-approximated Dyson's equation, and

$$\boxed{\text{I}} \; = \; \equiv \; + \; \boxed{\text{I}} \qquad (8)$$

for the ladder-approximated Dyson's equation, where the diagrammatic notations of Sections 3.1 and 3.3 are used. In (7) and (8) $\otimes$ stands for

Isotropic Point Scatterers

transition operator and a solid line joining two $\otimes$ denotes that the two $\otimes$ belong to the same scatterer. For every diagram involving s distinct scatterers $i_1, i_2, \ldots, i_s$, there is a factor n_o^s (n_o is the number of scatterers per unit volume) and an s-fold integration over particle positions $d\bar{r}_{i_1}, d\bar{r}_{i_2}, \ldots, d\bar{r}_{i_s}$. Thus, (7) in analytic form is

$$<G(\bar{r},\bar{r}_o)> = G^{(0)}(\bar{r},\bar{r}_o) + n_o \int d\bar{r}_i \int d\bar{r}_1 \int d\bar{r}_2 \, G^{(0)}(\bar{r},\bar{r}_1)$$
$$\times T_i(\bar{r}_1,\bar{r}_2) <G(\bar{r}_2,\bar{r}_o)> \qquad (9)$$

and (8) in analytic form is

$$<G(\bar{r},\bar{r}_o)G^*(\bar{r}',\bar{r}'_o)> = <G(\bar{r},\bar{r}_o)><G^*(\bar{r}',\bar{r}'_o)>$$
$$+ n_o \int d\bar{r}_i \int d\bar{r}_1 \int d\bar{r}_2 \int d\bar{r}_3 \int d\bar{r}_4$$
$$\times <G(\bar{r},\bar{r}_1)><G^*(\bar{r}',\bar{r}_2)> T_i(\bar{r}_1,\bar{r}_3)T_i^*(\bar{r}_2,\bar{r}_4)$$
$$\times <G(\bar{r}_3,\bar{r}_o)G^*(\bar{r}_4,\bar{r}'_o)> \qquad (10)$$

Using (3) in (9) and (10), we have

$$<G(\bar{r},\bar{r}_o)> = G^{(0)}(\bar{r},\bar{r}_o) + 4\pi n_o f \int d\bar{r}_i \, G^{(0)}(\bar{r},\bar{r}_i) <G(\bar{r}_i,\bar{r}_o)> \qquad (11)$$

and

$$<G(\bar{r},\bar{r}_o)G^*(\bar{r}',\bar{r}'_o)> = <G(\bar{r},\bar{r}_o)><G^*(\bar{r}',\bar{r}'_o)>$$
$$+ n_o(4\pi)^2 |f|^2 \int d\bar{r}_i <G(\bar{r},\bar{r}_i)><G^*(\bar{r}',\bar{r}_i)>$$
$$\times <G(\bar{r}_i,\bar{r}_o)G^*(\bar{r}_i,\bar{r}'_o)> \qquad (12)$$

The bilocal-approximated Dyson's equation in (11) can be solved readily. Operating on (11) by $(\nabla^2 + k^2)$, we have

$$(\nabla^2 + k^2 + 4\pi n_o f) <G(\bar{r},\bar{r}_o)> = -\delta(\bar{r} - \bar{r}_o) \qquad (13)$$

Hence, the effective propagation constant K obeys the equation

$$K^2 = k^2 + 4\pi n_o f \qquad (14)$$

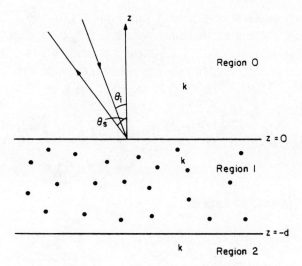

Fig. 5.13 Scattering of waves by a slab of isotropic point scatterers.

Generally, $Im\, f$ is not zero and $|4\pi n_o f| \ll k^2$. Hence,

$$K = k + \frac{2\pi n_o f}{k} \tag{15}$$

and the attenuation rate κ_e for the coherent wave is

$$\kappa_e = 2K'' = \frac{4\pi n_o}{k} Im\, f \tag{16}$$

Generally the extinction rate κ_e is much smaller than the wavenumber K'

$$K'' \ll K' \tag{17}$$

Next consider a plane wave incident upon a slab of point scatterers of thickness d in the direction (θ_i, ϕ_i) (Figure 5.13). Let k be the wavenumber of the upper half-space (region 0) and the lower half-space (region 2) and also the wavenumber of the background medium of the slab (region 1). It is assumed that the concentration of particles is low so that K' is approximately equal to k. The mean field in region 1 is

$$\psi_m(\bar{r}) = e^{i\overline{K}_i \cdot \bar{r}} \tag{18}$$

where

$$\overline{K}_i = k \sin\theta_i \cos\phi_i \hat{x} + k \sin\theta_i \sin\phi_i \hat{y} - K_{iz}\hat{z} \tag{19}$$

with

$$K_{iz} = (K^2 - k^2 \sin^2\theta_i)^{1/2} \simeq k\cos\theta_i + i\frac{K''}{\mu_i} \tag{20}$$

and $\mu_i = \cos\theta_i$. The approximate relation in (20) is a result of (17).

We shall calculate the bistatic scattering coefficient in region 0 by using ladder approximation. It is convenient to use Feynman diagrams because all the physical processes become transparent. Let $I_L(\bar{r})$ denote intensity in region 0, which is a summation of all the ladder terms. Diagrammatically

$$I_L(\bar{r}) = \quad\text{[diagram]} \quad + \quad\text{[diagram]}$$
$$+ \quad\text{[diagram]} \quad + \cdots \tag{21}$$

where

$\equiv\equiv\equiv$ = mean Green's function $<G_{11}(\bar{r},\bar{r}')>$ with both observation and source points in region 1

$\sim\sim$ = mean field ψ_m

$\approx\approx$ = mean Green's function $<G_{01}(\bar{r},\bar{r}')>$ with observation point in region 0 and source in region 1

The mean Green's function $<G_{11}>$ is

$$<G_{11}(\bar{r}_1,\bar{r}_2)> = \frac{e^{iK|\bar{r}_1-\bar{r}_2|}}{4\pi|\bar{r}_1-\bar{r}_2|}$$

$$= \frac{i}{8\pi^2}\int_{-\infty}^{\infty}d\bar{k}_\perp \frac{e^{i\bar{k}_\perp\cdot(\bar{r}_{1\perp}-\bar{r}_{2\perp})+iK_z|z_1-z_2|}}{K_z} \tag{22}$$

with $\bar{k}_\perp = k_x\hat{x}+k_y\hat{y}$ and $K_z = (K^2-k_x^2-k_y^2)^{1/2}$. Because the observation point is in far field in region 0, the mean Green's function $<G_{01}(\bar{r},\bar{r}')>$ is

$$<G_{01}(\bar{r},\bar{r}')> = \frac{e^{ikr}}{4\pi r}e^{-i\overline{K}_s\cdot\bar{r}'} \tag{23}$$

where $\overline{K}_s$ denotes scattered direction and is

$$\overline{K}_s = k\sin\theta_s\cos\phi_s\hat{x} + k\sin\theta_s\sin\phi_s\hat{y} + K_{zs}\hat{z} \tag{24}$$

where

$$K_{zs} \simeq k\cos\theta_s + i\frac{K''}{\mu_s} \tag{25}$$

with $\mu_s = \cos\theta_s$.

Next we define [L] as I_L with ends stripped. Hence diagrammatically,

$$I_L(\bar{r}) = \quad\text{[diagram]}\quad \text{[L]} \quad \text{[diagram]} \tag{26}$$

By summing the ladder series the operator equation for [L] is

$$\text{[L]} = \text{[diagram]} + \text{[diagram] [L]} \tag{27}$$

Equation (27) includes all the multiple scattering effects in the ladder terms. To solve the ladder intensity, let

$$L(\bar{r}_1, \bar{r}_2; \bar{r}_3, \bar{r}_4) = \begin{array}{c} {}^{1}\text{[L]}^{2} \\ {}_{3}\quad{}_{4} \end{array}$$

$$= F(\bar{r}_1, \bar{r}_2)\,\delta(\bar{r}_1 - \bar{r}_3)\,\delta(\bar{r}_2 - \bar{r}_4) \tag{28}$$
$$+ n_o(4\pi)^2|f|^2\,\delta(\bar{r}_1 - \bar{r}_2)\,\delta(\bar{r}_1 - \bar{r}_3)\,\delta(\bar{r}_2 - \bar{r}_4)$$

On substituting (28) into (27) and using the transition operator for point scatterers as given by (3), the F operator in (28) is found to obey the equation

$$F(\bar{r}_1, \bar{r}_2) = n_o|f|^2 \int_{V_1} d\bar{r}'\, \frac{e^{-\kappa_e|\bar{r}_1 - \bar{r}'|}}{|\bar{r}_1 - \bar{r}'|^2} F(\bar{r}', \bar{r}_2)$$
$$+ n_o|f|^4(4\pi)^2\, \frac{e^{-\kappa_e|\bar{r}_1 - \bar{r}_2|}}{|\bar{r}_1 - \bar{r}_2|^2} \tag{29}$$

Next define the two-dimensional Fourier transform by making use of the translational invariance property in the horizontal direction.

$$F(\bar{r}_1, \bar{r}_2) = \int_{-\infty}^{\infty} d\bar{k}_\perp\, e^{i\bar{k}_\perp \cdot (\bar{r}_{1\perp} - \bar{r}_{2\perp})}\, \tilde{F}(\bar{k}_\perp; z_1, z_2) \tag{30}$$

The bistatic scattering coefficient for ladder intensity is defined as

$$\gamma_L(\mu_s, \mu_i) = \frac{4\pi r^2}{A \cos\theta_i} I_L(\bar{r}) \tag{31}$$

where A is the area of the target under observation. Using (26), (28), and (30) in (31), it follows that

$$\gamma_L(\mu_s, \mu_i) = \frac{\tilde{\omega}}{\mu_i} \frac{1}{1/\mu_s + 1/\mu_i} \left(1 - e^{-\tau_d(1/\mu_s + 1/\mu_i)}\right)$$
$$+ \frac{\pi}{\mu_i} \int_{-d}^{0} dz' \int_{-d}^{0} dz'' \, \tilde{F}_o(z', z'') \, e^{\kappa_e(z'/\mu_s + z''/\mu_i)} \quad (32)$$

where $\tilde{F}_o(z', z'') = \tilde{F}(\bar{k}_\perp = 0; z', z'')$, $\tilde{\omega} = 4\pi n_o |f|^2/\kappa_e$ is the albedo and $\tau_d = \kappa_e d$ is the optical thickness of the slab. Thus the bistatic scattering coefficient depends only on $\tilde{F}(\bar{k}_\perp; z', z'')$ at $\bar{k}_\perp = 0$. Next, we make use of the plane wave representation of Green's function to calculate the absolute value of the Green's function squared. This will be in terms of a double integral of $\int d\bar{k}_{1\perp} \int d\bar{k}_{2\perp}$. Make the usual transformations of $\bar{k}_{A\perp} = (\bar{k}_{1\perp} + \bar{k}_{2\perp})/2$ and $\bar{k}_{D\perp} = \bar{k}_{1\perp} - \bar{k}_{2\perp}$. Hence

$$\frac{e^{-\kappa_e |\bar{r}_1 - \bar{r}_2|}}{|\bar{r}_1 - \bar{r}_2|^2} = \frac{1}{4\pi^2} \int_{-\infty}^{\infty} d\bar{k}_{A\perp} \int_{-\infty}^{\infty} d\bar{k}_{D\perp} \frac{e^{i\bar{k}_{D\perp} \cdot (\bar{r}_{1\perp} - \bar{r}_{2\perp})}}{K_{z(A+D)} K^*_{z(A-D)}}$$
$$\times e^{i(K_{z(A+D)} - K^*_{z(A-D)})|z_1 - z_2|} \quad (33)$$

where

$$K_{z(A+D)} = \left[K^2 - \left|\bar{k}_{A\perp} + \frac{\bar{k}_{D\perp}}{2}\right|^2\right]^{1/2} \quad (34)$$

$$K_{z(A-D)} = \left[K^2 - \left|\bar{k}_{A\perp} - \frac{\bar{k}_{D\perp}}{2}\right|^2\right]^{1/2} \quad (35)$$

Substitute (30) and (33) into (29), balance the terms on both sides of the resulting equation and finally set $\bar{k}_{D\perp}$ to zero to obtain an equation for the quantity $\tilde{F}_o(z_1, z_2)$. We have

$$\tilde{F}_o(z_1, z_2) = n_o |f|^2 \int_{-d}^{0} dz' \int_{-\infty}^{\infty} d\bar{k}_{A\perp} \frac{e^{-2K''_{zA}|z_1 - z'|}}{|K_{zA}|^2} \tilde{F}_o(z', z_2)$$
$$+ 4n_o^2 |f|^4 \int_{-\infty}^{\infty} d\bar{k}_{A\perp} \frac{e^{-2K''_{zA}|z_1 - z_2|}}{|K_{zA}|^2} \quad (36)$$

where $K_{zA} = (K^2 - k_{A\perp}^2)^{1/2}$ since $\bar{k}_{D\perp}$ has been set to zero to obtain equation for $\tilde{F}_o$. The integral over $d\bar{k}_{A\perp}$ in (36) can be expressed in terms of exponential integral E_1 (Appendix, Section 3.7).

Further define $\tau = \kappa_e z$ as optical distance and let

$$J_s(\tau) = \frac{1}{4n_o|f|^2} \int_{-d}^{0} dz_2 \, \tilde{F}_o(z, z_2) \, e^{\kappa_e z_2/\mu_i} \qquad (37)$$

Then it follows from (36) and (37) that

$$J_s(\tau) = \frac{\tilde{\omega}}{2} \int_{-\tau_d}^{0} d\tau' E_1(|\tau - \tau'|) J_s(\tau')$$

$$+ \frac{\tilde{\omega}}{2} \int_{-\tau_d}^{0} d\tau' E_1(|\tau - \tau'|) \, e^{\tau'/\mu_i} \qquad (38)$$

The ladder bistatic scattering coefficient is obtained by using (37) in (32).

$$\gamma_L(\mu_s, \mu_i) = \frac{\tilde{\omega}}{\mu_i} \frac{1}{1/\mu_s + 1/\mu_i} \left(1 - e^{-\tau_d(1/\mu_s + 1/\mu_i)}\right)$$

$$+ \frac{\tilde{\omega}}{\mu_i} \int_{-\tau_d}^{0} d\tau' \, e^{\tau'/\mu_s} \, J_s(\tau') \qquad (39)$$

Equation (38) is of the form of the Schwarzschild-Milne problem (Davison, 1958; Chandrasekhar, 1960; Ishimaru, 1978). Its equivalence to radiative transfer equations can be demonstrated as follows. Define upward-going diffuse specific intensity as

$$I(\tau, \mu) = \frac{\tilde{\omega}}{4\pi\mu} \int_{-\tau_d}^{\tau} d\tau' \, J_s(\tau') \, e^{(\tau'-\tau)/\mu} \qquad (40)$$

for $\mu > 0$ and downward-going diffuse specific intensity

$$I(\tau, -\mu) = \frac{\tilde{\omega}}{4\pi\mu} \int_{\tau}^{0} d\tau' \, J_s(\tau') \, e^{(\tau-\tau')/\mu} \qquad (41)$$

for $\mu > 0$. Differentiating (40) and (41) with respect to τ and using the integral equation for J_s as in (38) gives the radiative transfer equation for isotropic scatterers (Problem 24).

3.5 Backscattering Enhancement for Random Discrete Scatterers

Backscattering enhancement has been observed in laboratory-controlled optical experiments of scattering from discrete particles (Kuga and Ishimaru, 1984). The effect of backscattering enhancement is not contained in radiative transfer theory. It was shown in Section 3.4 that for volume scattering in the absence of a reflective boundary, radiative transfer theory is consistent with the ladder approximation. Thus, ladder approximation does not account for backscattering enhancement either. A comparison of Figures 5.5 and 5.12 shows that enhancement of backscattering is not exhibited because the cyclical diagrams that contribute significantly in the backscattering direction have been ignored (Barabanenkov, 1973; Watson, 1969; Zuniga et al., 1980; Tsang and Ishimaru, 1984). In Sections 2.2 and 2.3, backscattering enhancement was studied with the second-order Born approximation. In the case of turbulent medium, the enhancement effect has been studied by using the cumulative forward-scatter single-backscatter (CFSB) approximation, (deWolf, 1971, 1972; Ishimaru and Pinter, 1980; Kravtsov and Saichev, 1982; Yeh, 1983, Yang and Yeh, 1984), which is based on the assumption that essentially all scattered energy is in the forward direction. However, for the case of discrete scatterers a significant amount of scattered energy is not in the forward direction at all frequencies.

A second-order multiple scattering theory has been used to account for the enhancement in the backscattering direction (Tsang and Ishimaru, 1984). The theory predicts a sharp peak of angular width of the order $2K''/K'$ where K' and K'' are the real and imaginary parts, respectively, of the effective propagation constant. The second-order theory includes first-order scattering, the second-order ladder term, and the second-order cyclical term. However, the second-order solution is not adequate when the albedo and the optical thickness of the scattering medium become appreciable. In this section, by using the model of isotropic point scatterers, we investigate the multiple-scattering phenomenon by summing all the ladder and cyclical terms. As demonstrated in Section 3.4, the summation of all the ladder terms leads to the classical Schwarzschild-Milne integral equation with the exponential integral as kernel, the solution of which can be readily calculated. The summation of the cyclical terms leads to a two-variable cyclical-transfer integral equation with a more general kernel. The integral equation is then solved numerically, and the results thus include all the multiple scattering associated with the cyclical terms. Numerical results are illustrated as a function of scattering angle, albedo, and optical thickness. Results differ significantly from the second-order theory for appreciable optical thickness and albedo. The multiple-scattering solution also gives a sharper backscattering peak than the second-order solution. The ability to determine the angular distribution of the bistatic scattering coefficient around the backscattering direction has important applications in remote sensing because receivers generally have a finite angular width in their receiving pattern.

Consider a plane wave incident upon a slab of point scatterers as in Section 3.4 (Figure 5.13). Let $I_C(\bar{r})$ denote the intensity in region 0 that is a summation of all the cyclical terms. The total intensity is

$$I(\bar{r}) = I_L(\bar{r}) + I_C(\bar{r}) \tag{1}$$

and $I_L(\bar{r})$ is the ladder intensity as determined in Section 3.4. Diagrammatically

$$I_C(\bar{r}) = \quad [\text{diagrams}] \tag{2}$$

and the diagrammatic notations are as in Section 3.4. Next define the cyclical operator $\boxed{C}$ such that

$$I_C(\bar{r}) = \quad [\text{diagram with } C] \tag{3}$$

and the operator obeys the diagrammatic equation

$$\boxed{C} = \quad [\text{diagrams}] \tag{4}$$

Equations (3) and (4) include all the multiple-scattering effects in the cyclical terms. Our objective is to put (4) in a numerically tractable form and, by solving the equations numerically, to include all the multiple-scattering effects. For point scatterers, the transition operator for the particle located at $\bar{r}_j$ is

$$T_j(\bar{r}_1, \bar{r}_2) = 4\pi f \, \delta(\bar{r}_1 - \bar{r}_j) \, \delta(\bar{r}_2 - \bar{r}_j) \tag{5}$$

The bistatic scattering coefficient for the cyclical intensity is denoted by $\gamma_C(\mu_s, \mu_i)$ and is defined as

$$\gamma_C(\mu_s, \mu_i) = \frac{4\pi r^2}{A \cos \theta_i} I_C(\bar{r}) \tag{6}$$

where A is the area of the target under observation. Thus the total bistatic scattering coefficient is

$$\gamma(\mu_s,\mu_i) = \gamma_L(\mu_s,\mu_i) + \gamma_C(\mu_s,\mu_i) \qquad (7)$$

To study the cyclical equation of (4), let

$$C(\bar{r}_1,\bar{r}_2;\bar{r}_3,\bar{r}_4) = \begin{array}{c} {}_1{}^2 \\ \boxed{C} \\ {}_3{}^4 \end{array}$$

$$= A(\bar{r}_1,\bar{r}_2)\,\delta(\bar{r}_1-\bar{r}_4)\,\delta(\bar{r}_2-\bar{r}_3) \qquad (8)$$

Substituting (8) in (4) and using (5) gives the following integral equation for $A(\bar{r}_1,\bar{r}_2)$

$$A(\bar{r}_1,\bar{r}_2) = n_o^2(4\pi)^2|f|^4 \frac{e^{-\kappa_e|\bar{r}_1-\bar{r}_2|}}{|\bar{r}_1-\bar{r}_2|^2}$$

$$+ n_o^3(4\pi)^2|f|^6 \int_{V_1} d\bar{r}' \frac{e^{-\kappa_e|\bar{r}_1-\bar{r}'|}}{|\bar{r}_1-\bar{r}'|^2}\frac{e^{-\kappa_e|\bar{r}_2-\bar{r}'|}}{|\bar{r}_2-\bar{r}'|^2}$$

$$+ n_o^2|f|^4 \int_{V_1} d\bar{r}' \int_{V_1} d\bar{r}'' \frac{e^{-\kappa_e|\bar{r}_1-\bar{r}'|}}{|\bar{r}_1-\bar{r}'|^2}\frac{e^{-\kappa_e|\bar{r}_2-\bar{r}''|}}{|\bar{r}_2-\bar{r}''|^2} A(\bar{r}',\bar{r}'') \quad (9)$$

Because of translational invariance in the horizontal plane, let

$$A(\bar{r}_1,\bar{r}_2) = \int_{-\infty}^{\infty} d\bar{k}_{D\perp}\, e^{i\bar{K}_{D\perp}\cdot(\bar{r}_{1\perp}-\bar{r}_{2\perp})}\,\tilde{A}(\bar{k}_{D\perp};z_1,z_2) \qquad (10)$$

Substitute (10) in (9), and make use of the plane wave representation of Green's function in (33), Section 3.4. The product of the absolute square of Green's functions gives a quadruple integral $\int d\bar{k}_{1\perp}\int d\bar{k}_{2\perp}\int d\bar{k}_{3\perp}\int d\bar{k}_{4\perp}$. Next make use of the usual transformation of $\bar{k}_{A\perp} = (\bar{k}_{1\perp}+\bar{k}_{2\perp})/2$ and $\bar{k}_{D\perp} = (\bar{k}_{1\perp}-\bar{k}_{2\perp})$ for $\bar{k}_{1\perp}$ and $\bar{k}_{2\perp}$ as well as for $\bar{k}_{3\perp}$ and $\bar{k}_{4\perp}$. This gives equation (9) a kernel of the form

$$W(\tau,\alpha) = \frac{1}{2\pi}\int_{-\infty}^{\infty} d\bar{k}_{A\perp}\, \frac{e^{i(K_{z(A+D)}-K^*_{z(A-D)})\tau/\kappa_e}}{K_{z(A+D)}K^*_{z(A-D)}} \qquad (11)$$

where

$$K_{z(A+D)} = \left[K^2 - \left|\bar{k}_{A\perp} + \frac{\bar{k}_{D\perp}}{2}\right|^2\right]^{1/2} \qquad (12)$$

$$K_{z(A-D)} = \left[K^2 - \left|\overline{k}_{A\perp} - \frac{\overline{k}_{D\perp}}{2}\right|^2\right]^{1/2} \tag{13}$$

$$\alpha = \frac{|\overline{k}_{D\perp}|}{\kappa_e} \tag{14}$$

The function $W(\tau, \alpha)$ can be shown to be equal to (Appendix, Section 3.7)

$$W(\tau, \alpha) = \int_1^\infty dt \, \frac{e^{-\tau(\alpha^2 + t^2)^{1/2}}}{(\alpha^2 + t^2)^{1/2}} \tag{15}$$

Note that $W(\tau, \alpha = 0) = E_1(\tau)$. Further let

$$\tilde{A}(\overline{k}_{D\perp}; z_1, z_2) = \frac{\kappa_e^2}{\pi}\left[\Gamma(\tau_1, \tau_2; \alpha) + \frac{\tilde{\omega}^2}{2} W(|\tau_1 - \tau_2|; \alpha)\right] \tag{16}$$

Substituting (10) in (9) and making use of (11) and (16) gives the following cyclical transfer integral equation for Γ.

$$\Gamma(\tau_1, \tau_2; \alpha)$$

$$= \frac{\tilde{\omega}^3}{4} \int_{-\tau_d}^0 d\tau \, W(|\tau_1 - \tau|; \alpha) \, W(|\tau - \tau_2|; \alpha)$$

$$+ \frac{\tilde{\omega}^4}{8} \int_{-\tau_d}^0 d\tau \int_{-\tau_d}^0 d\tau' W(|\tau_1 - \tau|; \alpha) \, W(|\tau' - \tau_2|; \alpha) W(|\tau - \tau'|; \alpha)$$

$$+ \frac{\tilde{\omega}^2}{4} \int_{-\tau_d}^0 d\tau \int_{-\tau_d}^0 d\tau' W(|\tau_1 - \tau|; \alpha) \, W(|\tau' - \tau_2|; \alpha) \Gamma(\tau, \tau'; \alpha) \tag{17}$$

By using (3), (6), (8), (10), and (16), the cyclical bistatic coefficient can be expressed in terms of Γ as follows:

$$\gamma_C(\mu_s, \mu_i) = \frac{1}{\mu_i} \int_{-\tau_d}^0 d\tau_1 \int_{-\tau_d}^0 d\tau_2 \left[\frac{\tilde{\omega}^2}{2} W(|\tau_1 - \tau_2|; \alpha) + \Gamma(\tau_1, \tau_2; \alpha)\right]$$

$$\times \cos\left[\frac{k}{\kappa_e}(\mu_s - \mu_i)(\tau_1 - \tau_2)\right] e^{(1/\mu_s + 1/\mu_i)(\tau_1 + \tau_2)/2} \tag{18}$$

where

$$\alpha = \frac{|\overline{k}_{D\perp}|}{\kappa_e} = \frac{k}{\kappa_e}\left[(\sin\theta_i \cos\phi_i + \sin\theta_s \cos\phi_s)^2\right.$$

$$\left. + (\sin\theta_i \sin\phi_i + \sin\theta_s \sin\phi_s)^2\right]^{1/2} \tag{19}$$

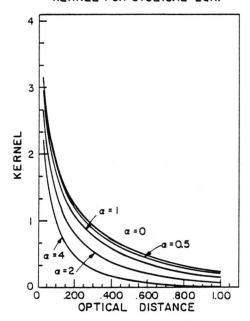

Fig. 5.14 The kernel $W(\tau, \alpha)$ as a function of τ for $\alpha = 0, 0.5, 1, 2$ and 4.

denotes the deviation from the backscattering direction.

It is interesting to note that the kernel of the ladder equation (38), Section 3.4, depends on $W(\tau, \alpha)$ with $\alpha = 0$ while the cyclical equation (17) has a kernel with α generally not equal to zero.

To calculate the cyclical bistatic coefficient requires a solution of the integral equation in (17). Once Γ is solved in (17), then the cyclical bistatic coefficient is calculated by using (18). Equation (17) is a two-variable integral equation with independent variables τ_1 and τ_2. Note that Γ in (17) is only coupled to the same α.

We next perform numerical calculation of the total bistatic scattering coefficient which is a sum of the ladder and the cyclical bistatic scattering coefficients according to (7). The task is to solve the integral equation (17) numerically and calculate the cyclical bistatic coefficient. The ladder bistatic coefficient can be obtained by solving the radiative transfer equation numerically via the method of Section 3.6, Chapter 4.

In Figure 5.14, we plot the kernel $W(\tau, \alpha)$ as a function of τ for different α's. The $W(\tau, \alpha)$ is calculated by numerically evaluating the integral in (15). The following small argument approximation can also be used for $\tau \ll 1$ and $\alpha\tau \ll 1$

$$W(\tau, \alpha) \simeq -\ln\tau - U - \ln\left(\frac{1+\sqrt{1+\alpha^2}}{2}\right) \quad (20)$$

where $U = 0.57721$ is Euler's constant. We note from Figure 5.14 that $W(\tau, \alpha)$ decreases exponentially with τ and also decreases with α. This implies that there is less scattering with increasing α consistent with the backscattering enhancement phenomenon. The kernel also has a logarithmic singularity at $\tau = 0$.

The integral equation of (17) is solved by the method of moments (Harrington, 1968) by discretizing τ_1 and τ_2 into intervals of width $\Delta\tau$. Pulse functions of width $\Delta\tau$ are used to expand the unknown variable $\Gamma(\tau_1, \tau_2; \alpha)$ and point matching is used to obtain the final set of matrix equations. To calculate the first two terms on the right-hand side of (17) and to evaluate the matrix elements for the third term, we perform a numerical integration over the patch. In view of the fast variation of W for all values of τ, a much smaller patch $\Delta\tau_s \ll \Delta\tau$ is chosen for this numerical integration procedure. In view of the logarithmic singularity of the kernel, the self-patch has a large contribution. Computation of the self-patch contribution is accomplished by using the small argument approximation of (20) and choosing a small enough $\Delta\tau_s$ such that $\alpha\Delta\tau_s \ll 1$. For example,

$$\int_{\tau_m - \Delta\tau_s}^{\tau_m + \Delta\tau_s} d\tau\, W(|\tau_m - \tau|) \simeq 2\Delta\tau_s P \quad (21)$$

where

$$P = -\ln\Delta\tau_s + 1 - U - \ln\left(\frac{1+\sqrt{1+\alpha^2}}{2}\right) \quad (22)$$

The symmetry property

$$\Gamma(\tau_1, \tau_2; \alpha) = \Gamma(\tau_2, \tau_1; \alpha) \quad (23)$$

is also used to reduce computation. Thus the number of basis functions for Γ is $N = (\tau_d/\Delta\tau)(\tau_d/\Delta\tau + 1)/2$. The two-variable integral equation is thus cast into a matrix equation of dimension N which is then solved numerically by standard techniques.

In the following we illustrate numerical results for $\kappa_e/k = 3.18 \times 10^{-3}$, $\theta_i = 0$, $\phi_i = 0$, $\phi_s = \pi$, θ_s from 0 deg to 1 deg and $\tau_d = 2$ and 4, and for albedos $\tilde{\omega}$ ranging from 0.4 to 0.95. For this range of angles, the values of α range from 0 to 5.49. For these values of α, it is sufficient to use $\Delta\tau = 0.25$ and $\Delta\tau_s = 0.025$. This corresponds to $N = 36$ for $\tau_d = 2$

and $N = 136$ for $\tau_d = 4$. The procedure is thus not suitable for large τ_d. However, it is found that for $\tilde{\omega} \leq 0.85$, a value of $\tau_d = 4$ approximates well the half-space solution.

For comparison, we also plot the single-scattering solution $\gamma^{(1)}$ and the result for the second-order multiple scattering theory $\gamma^{(2)}$. The second-order multiple scattering is calculated by adding the single-scattering solution, the second-order ladder and cyclical terms (Tsang and Ishimaru, 1984). Thus

$$\gamma^{(1)}(\mu_s, \mu_i) = \frac{\tilde{\omega}}{\mu_i} \frac{1 - e^{-\tau_d(1/\mu_s + 1/\mu_i)}}{1/\mu_s + 1/\mu_i} \qquad (24)$$

The second-order multiple scattering theory solution for a slab of isotropic point scatterers is

$$\gamma^{(2)}(\mu_s, \mu_i) = \gamma^{(1)}(\mu_s, \mu_i) + \gamma_L^{(2)}(\mu_s, \mu_i) + \gamma_C^{(2)}(\mu_s, \mu_i) \qquad (25)$$

where

$$\gamma_L^{(2)}(\mu_s, \mu_i) = \frac{\tilde{\omega}^2}{2\mu_i} \int_0^1 d\mu \Bigg\{ \frac{1}{1 + \mu/\mu_i} \Bigg[\frac{1 - e^{-\tau_d(1/\mu_s + 1/\mu_i)}}{1/\mu_s + 1/\mu_i} \\
- \frac{e^{-\tau_d(1/\mu + 1/\mu_i)} - e^{-\tau_d(1/\mu_s + 1/\mu_i)}}{1/\mu_s - 1/\mu} \Bigg] \\
+ \frac{1}{\mu/\mu_i - 1} \Bigg[\frac{1 - e^{-\tau_d(1/\mu_s + 1/\mu)}}{1/\mu_s + 1/\mu} - \frac{1 - e^{-\tau_d(1/\mu_s + 1/\mu_i)}}{1/\mu_s + 1/\mu_i} \Bigg] \Bigg\} \qquad (26)$$

is the second-order ladder term, and

$$\gamma_C^{(2)}(\mu_s, \mu_i) = \frac{\tilde{\omega}^2}{2\mu_i} \int_{-\tau_d}^0 d\tau_1 \int_{-\tau_d}^0 d\tau_2\, W(|\tau_1 - \tau_2|; \alpha) \qquad (27)$$

is the second-order cyclical term.

It is to be noted that in the backscattering direction of $\alpha = 0$ (Ishimaru, 1978)

$$\gamma_C = \gamma_L - \gamma^{(1)} \qquad (28)$$

In the following, equation (28) is verified by showing that the n th order ladder term $(n \geq 2)$ is equal to the n th order cyclical term at $\alpha = 0$. Diagrammatically, the n th order ladder term is

$$I_L^{(n)}(\bar{r}) = \qquad (29)$$

The nth order cyclical term is

$$I_C^{(n)}(\bar{r}) = \qquad\qquad\qquad\qquad \tag{30}$$

Hence using the expressions for $<G_{01}(\bar{r},\bar{r}')>$, $<G_{11}(\bar{r},\bar{r}')>$, $\psi_m(\bar{r})$, and the reciprocity property of $<G_{11}(\bar{r},\bar{r}')>$, it follows that

$$I_L^{(n)}(\bar{r}) = \frac{n_o^n (4\pi)^{2(n-1)} |f|^{2n}}{r^2} \int_{V_1} d\bar{r}_1 d\bar{r}_2 \cdots d\bar{r}_n$$

$$\times e^{2K_{iz}'' z_1} |G_{11}(\bar{r}_1,\bar{r}_2)|^2 \cdots |G_{11}(\bar{r}_{n-1},\bar{r}_n)|^2 e^{2K_{iz}'' z_n} \tag{31}$$

$$I_C^{(n)}(\bar{r}) = \frac{n_o^n (4\pi)^{2(n-1)} |f|^{2n}}{r^2} \int_{V_1} d\bar{r}_1 d\bar{r}_2 \cdots d\bar{r}_n \, e^{-i\overline{K}_s \cdot \bar{r}_1 + i\overline{K}_s^* \cdot \bar{r}_n}$$

$$\times |G_{11}(\bar{r}_1,\bar{r}_2)|^2 \cdots |G_{11}(\bar{r}_{n-1},\bar{r}_n)|^2 e^{i\overline{K}_i \cdot \bar{r}_n - i\overline{K}_i^* \cdot \bar{r}_1} \tag{32}$$

In the backscattering direction, $\alpha = 0$, we have $\overline{K}_{i\perp} + \overline{K}_{s\perp} = 0$ and $K_{iz} = K_{sz}$. Making these substitutions in (32) and comparing with (31) gives

$$I_C^{(n)}(\bar{r}) = I_L^{(n)}(\bar{r}) \tag{33}$$

for $n \geq 2$. Since there is no first-order cyclical term, it follows that

$$I_C(\bar{r}) = I_L(\bar{r}) - I_L^{(1)}(\bar{r}) \tag{34}$$

Thus condition (28) is true. Also, at $\alpha = 0$

$$\gamma = 2\gamma_C + \gamma^{(1)}$$
$$= 2\gamma_L - \gamma^{(1)} \tag{35}$$

The accuracy of the numerical solution of the cyclical transfer integral equation can be tested by checking that the condition (28) is satisfied at the backscattering direction of $\alpha = 0$. The numerical results given in Figures 5.15 to 5.19 satisfy (28) to within 1%.

In Figure 5.15, we plot γ, γ_L, $\gamma^{(1)}$, and $\gamma^{(2)}$ as a function of scattering angle θ_s for $\tilde{\omega} = 0.95$ and $\tau_d = 2$. We note that single scattering and ladder solutions have a much weaker angular dependence than the second-order multiple-scattering solution and the total multiple-scattering intensity γ. We also note that the second-order theory differs from the full

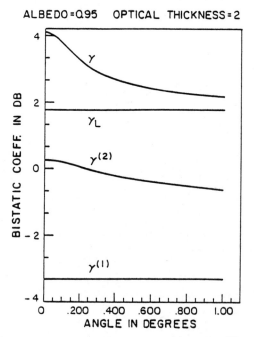

Fig. 5.15 Bistatic scattering coefficients γ, γ_L, $\gamma^{(1)}$, and $\gamma^{(2)}$ as a function of θ_s for $\tilde{\omega} = 0.95$ and $\tau_d = 2$.

multiple-scattering solution γ by close to 4 db. Hence the second-order scattering theory is not accurate for large albedo and appreciable optical thickness. In Figure 5.16, the various constituents of the results in Figure 5.15 are examined on a linear scale. We note that the full multiple-scattering cyclical term γ_C has a much narrower peak than the second-order cyclical term $\gamma_C^{(2)}$. This can be accounted for physically by the fact that higher order multiple scatterings are more sensitive to the path length difference created by angular deviation from the backscattering direction. The overall angular width of the peak in the backscattering direction is of the order of $\kappa_e/k = 3.18 \times 10^{-3} = 0.18$ deg.

In Figure 5.17, the results are illustrated for a smaller $\tilde{\omega} = 0.4$. Multiple scattering plays a lesser role. Compared to the case $\tilde{\omega} = 0.95$, the backscattering peak here is wider. The second-order solution is also closer to the full multiple scattering solution than that of $\tilde{\omega} = 0.95$.

In Figures 5.18 and 5.19, we plot, respectively, the bistatic scattering coefficients with $\tau_d = 2$ and $\tau_d = 4$ at $\tilde{\omega} = 0.85$. The case of $\tau_d = 4$ in Figure 5.19 approximates well the half-space solution. Comparison of the results in the two figures indicates that larger optical thickness gives a larger bistatic coefficient and a sharper backscattering peak.

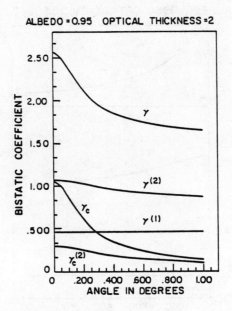

Fig. 5.16 Bistatic scattering coefficients γ, $\gamma^{(1)}$, $\gamma^{(2)}$, γ_C and $\gamma_C^{(2)}$ as a function of θ_s for $\tilde{\omega} = 0.95$ and $\tau_d = 2$.

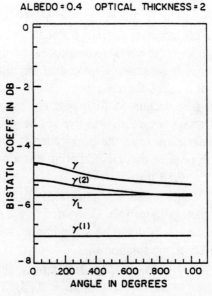

Fig. 5.17 Bistatic scattering coefficients γ, γ_L, $\gamma^{(1)}$, and $\gamma^{(2)}$ as a function of θ_s for $\tilde{\omega} = 0.4$ and $\tau_d = 2$.

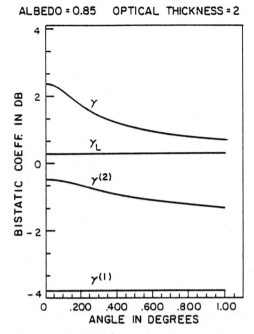

Fig. 5.18 Same as Figure 5.17 with $\tilde{\omega} = 0.85$ and $\tau_d = 2$.

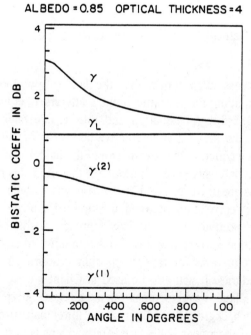

Fig. 5.19 Same as Figure 5.17 with $\tilde{\omega} = 0.85$ and $\tau_d = 4$.

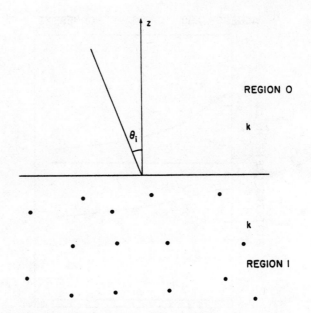

Fig. 5.20 Plane wave incident on a half-space of point scatterers.

The numerical results in Figures 5.15 to 5.19 contain 21 angles between 0 deg and 1 deg at an interval of 0.05 deg.

3.6 Alternative Derivation of Radiative Transfer Equation from Ladder Approximation

In Section 3.4, it has been shown that the ladder approximation of the Bethe-Salpeter equation for isotropic point scatterers is equivalent to the Schwarzschild-Milne integral equation and thus is equivalent to radiative transfer theory. In this section, an alternative derivation of radiative transfer equation is given by equating the covariances of the field to the Fourier transform of the specific intensity. The assumption is that the fields propagating in different directions are uncorrelated. For the case when a reflective boundary is present, there is correlation between the upward- and downward-going waves related by specular reflection. The approach in this section will be generalized in Section 5 to include this correlation effect to develop modified radiative transfer equations for random medium scattering in the presence of reflective boundaries. It will also be used in Chapter 6.

Consider a plane wave incident on a half-space of point scatterers (Figure 5.20). The background medium of the lower half-space $(z=0)$ is the same as that of the upper half-space. The incident field is

$$\psi_{inc}(\bar{r}) = \exp(ik_{ix}x - ik_{iz}z) \qquad (1)$$

where $k_{ix} = k\sin\theta_i$ and $k_{iz} = k\cos\theta_i$. The effective propagation constant for the lower half-space is given by (15), Section 3.4. The effective propagation constant is endowed with an imaginary part that causes the coherent wave to decay. Its magnitude, K is only slightly different from that of k so that the reflectivity for the coherent wave at the interface $z = 0$ is practically zero. Thus, the coherent field for the lower half-space is

$$<\psi(\bar{r})> = \exp(ik_{ix}x - iK_{iz}z) \qquad (2)$$

where

$$K_{iz} = (K^2 - k^2\sin^2\theta_i)^{1/2} \qquad (3)$$

The ladder-approximated Bethe-Salpeter equation [(12), Section 3.4], is an equation for the correlation of the field $<\psi(\bar{r})\psi^*(\bar{r}')>$. Define the incoherent field $\tilde{\psi}$ as

$$\tilde{\psi}(\bar{r}) = \psi(\bar{r}) - <\psi(\bar{r})> \qquad (4)$$

The equation for the covariance in region 1 is, under the ladder approximation

$$<\tilde{\psi}(\bar{r}_1)\tilde{\psi}^*(\bar{r}_2)> = n_o(4\pi)^2|f|^2 \int_{z_j<0} d\bar{r}_j <G(\bar{r}_1,\bar{r}_j)>$$

$$\times <G^*(\bar{r}_2,\bar{r}_j)> \left\{|<\psi(\bar{r}_j)>|^2 + <|\tilde{\psi}(\bar{r}_j)|^2>\right\} \quad (5)$$

and the equation holds for the lower half-space.

We next make a spectral decomposition of the incoherent field. Let $d\bar{k}_\perp = dk_x dk_y$ and

$$\tilde{\psi}(\bar{r}) = \int d\bar{k}_\perp\, e^{i\bar{k}_\perp\cdot\bar{r}_\perp} \left\{\tilde{\psi}_u(z,\bar{k}_\perp) e^{iK'_z z} + \tilde{\psi}_d(z,\bar{k}_\perp) e^{-iK'_z z}\right\} \quad (6)$$

where $\bar{k}_\perp = k_x\hat{x} + k_y\hat{y}$, $K_z = (K^2 - k_x^2 - k_y^2)^{1/2}$ and $K'_z = \mathrm{Re}\,K_z$. In (6) $\tilde{\psi}_u$ and $\tilde{\psi}_d$ are, respectively, upward-going and downward-going fields in the direction $\bar{k}_\perp$. There are two distance scales in the problem: the extinction distance $l_e = 1/2K''$ and the wavelength $= 2\pi/K'$. Generally, $l_e \gg 1/K'$. The spectral functions $\tilde{\psi}_u(z,\bar{k}_\perp)$ and $\tilde{\psi}_d(z,\bar{k}_\perp)$ are generally slowly varying on the wavelength scale.

For the half-space problem, we seek solutions of the form

$$<\tilde{\psi}_u(z_1,\bar{k}_{\perp 1})\tilde{\psi}_u^*(z_2,\bar{k}_{\perp 2})> = \delta(\bar{k}_{\perp 1} - \bar{k}_{\perp 2})J_u(z_1,z_2,\bar{k}_{\perp 1}) \quad (7)$$

$$<\tilde{\psi}_d(z_1,\overline{k}_{\perp 1})\tilde{\psi}_d^*(z_2,\overline{k}_{\perp 2})> = \delta(\overline{k}_{\perp 1} - \overline{k}_{\perp 2})J_d(z_1,z_2,\overline{k}_{\perp 2}) \qquad (8)$$

$$<\tilde{\psi}_u(z_1,\overline{k}_{\perp 1})\tilde{\psi}_d^*(z_2,\overline{k}_{\perp 2})> = 0 \qquad (9)$$

which assumes that incoherent fields propagating in two different directions are uncorrelated. The assumption is reasonable since incoherent fields are created as a result of scattering from particles that are randomly positioned. It is further assumed in (9) that upward- and downward-propagating incoherent waves with the same transverse wave-vector component are uncorrelated. When there is a reflective boundary present at $z < 0$, the assumption in (9) has to be modified (see Section 5).

Let

$$I_d^u(z,\overline{k}_\perp) = J_d^u(z,z,\overline{k}_\perp) \qquad (10)$$

represent the incoherent intensities of upward- and downward-going waves, respectively. The quantities I_u, I_d, J_u and J_d are slowly varying on the wavelength scale. Thus, for $|\overline{r}_1 - \overline{r}_2| \ll l_e$, we use (6) to calculate $<\tilde{\psi}(\overline{r}_1)\tilde{\psi}^*(\overline{r}_2)>$ and make use of (7) through (10) to get

$$<\tilde{\psi}(\overline{r}_1)\tilde{\psi}^*(\overline{r}_2)> = \int d\overline{k}_\perp e^{i\overline{k}_\perp \cdot (\overline{r}_{\perp 1} - \overline{r}_{\perp 2})}$$

$$\times \left\{ I_u\left(\frac{z_1+z_2}{2},\overline{k}_\perp\right) e^{iK'_z(z_1-z_2)} \right.$$

$$\left. + I_d\left(\frac{z_1+z_2}{2},\overline{k}_\perp\right) e^{-iK'_z(z_1-z_2)} \right\} \qquad (11)$$

Using the spectral representation of the mean Green's function, (2), for the mean field and (11), the right-hand side of (5) can be calculated (Problem 12). On equating that to (11), the factor $\int d\overline{k}_\perp \exp[i\overline{k}_\perp \cdot (\overline{r}_{\perp 1} - \overline{r}_{\perp 2})]$ can be canceled on both sides of the equations. This yields

$$I_u\left(\frac{z_1+z_2}{2},\overline{k}_\perp\right) e^{iK'_z(z_1-z_2)} + I_d\left(\frac{z_1+z_2}{2},\overline{k}_\perp\right) e^{-iK'_z(z_1-z_2)}$$

$$= \frac{\kappa_e \tilde{\omega}}{4\pi |K_z|^2} \int_{-\infty}^0 dz_j \, e^{iK_z|z_1-z_j| - iK_z^*|z_2-z_j|} \qquad (12)$$

$$\times \left\{ e^{2K''_{iz}z_j} + \int d\overline{k}_{\perp 1}[I_u(z_j,\overline{k}_{\perp 1}) + I_d(z_j,\overline{k}_{\perp 1})] \right\}$$

where

$$\kappa_e \tilde{\omega} = 4\pi n_o |f|^2 = \kappa_s \qquad (13)$$

with κ_s being the scattering rate and $\tilde{\omega}$ is the albedo. With no loss of generality, we let $z_1 > z_2$ in (12). The integral over dz_j in (12) can be broken into $\int_{-\infty}^{z_2} dz_j$, $\int_{z_2}^{z_1} dz_j$ and $\int_{z_1}^{0}$. Since $|z_1 - z_2| \ll l_e$, the integral $\int_{z_2}^{z_1} dz_j$ is much less than the other two. It can be neglected because it is of the order $O(1/K')$ whereas the other two are of the order $O(1/K'')$. Next we balance the phase dependences on the two sides of (12). On equating the $\exp[iK'_z(z_1 - z_2)]$ dependence, we have

$$I_u\left(\frac{z_1 + z_2}{2}, \overline{k}_\perp\right) = \frac{\kappa_e \tilde{\omega}}{4\pi |K_z|^2} e^{-K''_z(z_1+z_2)} \int_{-\infty}^{z_2} dz_j\, e^{2K''_z z_j}$$

$$\times \left\{ e^{2K''_{iz} z_j} + \int d\overline{k}_{\perp 1}[I_u(z_j, \overline{k}_{\perp 1}) + I_d(z_j, \overline{k}_{\perp 1})] \right\} \tag{14}$$

Balancing the phase dependence $\exp[-iK'_z(z_1 - z_2)]$ on both sides of (12) gives

$$I_d\left(\frac{z_1 + z_2}{2}, \overline{k}_\perp\right) = \frac{\kappa_e \tilde{\omega}}{4\pi |K_z|^2} e^{K''_z(z_1+z_2)} \int_{z_1}^{0} dz_j\, e^{-2K''_z z_j}$$

$$\times \left\{ e^{2K''_{iz} z_j} + \int d\overline{k}_{\perp 1}[I_u(z_j, \overline{k}_{\perp 1}) + I_d(z_j, \overline{k}_{\perp 1})] \right\} \tag{15}$$

Next let $z_1 = z_2 = z$ in (14) and (15) and consider only radiating waves in the spectral representation, by transforming $\overline{k}_\perp$ to the (θ, ϕ) direction. Let $k_x = K'\sin\theta\cos\phi$ and $k_y = K'\sin\theta\sin\phi$ so that

$$\int dk_\perp = \int_0^{\pi/2} d\theta \sin\theta \cos\theta \int_0^{2\pi} d\phi\, K'^2 \tag{16}$$

The attenuation rates in the z direction can be expressed as follows

$$K''_z = \frac{K''}{\mu} = \frac{\kappa_e}{2\mu} \tag{17}$$

$$K''_{iz} = \frac{K''}{\mu_i} = \frac{\kappa_e}{2\mu_i} \tag{18}$$

where $\mu = \cos\theta$ and $\mu_i = \cos\theta_i$. Further, let

$$I^u_d(z, \theta, \phi) = K'^2 \cos\theta\, I^u_d(z, \overline{k}_\perp) \tag{19}$$

By using (16) through (19) in (14) and (15), we obtain

$$I_u(z,\theta,\phi) = \frac{\tilde{\omega}}{4\pi\mu} \frac{e^{\kappa_e z/\mu_i}}{(1/\mu + 1/\mu_i)}$$

$$+ \frac{\tilde{\omega}\kappa_e}{4\pi\mu} \int_0^{\pi/2} d\theta_1 \sin\theta_1 \int_0^{2\pi} d\phi_1 \int_{-\infty}^z dz_j\, e^{-\frac{\kappa_e}{\mu}(z-z_j)}$$

$$\times [I_u(z_j,\theta_1,\phi_1) + I_d(z_j,\theta_1,\phi_1)] \qquad (20)$$

$$I_d(z,\theta,\phi) = \frac{\tilde{\omega}}{4\pi\mu} \left[\frac{e^{\kappa_e z/\mu} - e^{\kappa_e z/\mu_i}}{1/\mu_i - 1/\mu}\right]$$

$$+ \frac{\tilde{\omega}\kappa_e}{4\pi\mu} \int_0^{\pi/2} d\theta_1 \sin\theta_1 \int_0^{2\pi} d\phi_1 \int_z^0 dz_j\, e^{\frac{\kappa_e}{\mu}(z-z_j)}$$

$$\times [I_u(z_j,\theta_1,\phi_1) + I_d(z_j,\theta_1,\phi_1)] \qquad (21)$$

Equations (20) and (21) are identical to the integral form of the radiative transfer equations as given in (7) and (8) (Sections 2.2, Chapter 4) and taking the half-space limit (Problem 13). In view of the physical picture as represented by ladder approximation in Figure 5.12, it is clear that, in the absence of reflective boundaries, it is consistent with radiative transfer theory.

3.7 Appendix

The kernel $W(\tau,\alpha)$ is

$$W(\tau,\alpha) = \frac{1}{2\pi} \int_{-\infty}^\infty d\bar{k}_{A\perp} \frac{e^{i(K_{z(A+D)} - K^*_{z(A-D)})\tau/\kappa_e}}{K_{z(A+D)} K^*_{z(A-D)}} \qquad (1)$$

where $K_{z(A+D)}$, $K_{z(A-D)}$ and α are given respectively by (12), (13) and (14), Section 3.5. To evaluate the kernel $W(\tau,\alpha)$ as given by (1), we make use of the plane wave representation of Green's function. It follows that

$$W(\tau,\alpha) = \frac{1}{2\pi} \int_{-\infty}^\infty d\bar{r}_\perp\, e^{-i\bar{k}_{D\perp}\cdot\bar{r}_\perp} \frac{e^{-\kappa_e(r_\perp^2+z^2)^{1/2}}}{r_\perp^2 + z^2} \qquad (2)$$

Converting to polar coordinates and carrying out the angular integration gives

$$W(\tau,\alpha) = \int_0^\infty ds\, s\, J_0(\alpha s) \frac{\exp[-(s^2+\tau^2)^{1/2}]}{s^2+\tau^2} \qquad (3)$$

where J_o is the Bessel function of order zero. We next make use of the Sommerfeld integral identity which can be cast in the form

$$\frac{\exp[-\tau(\alpha^2+t^2)^{1/2}]}{(\alpha^2+t^2)^{1/2}} = \int_0^\infty ds\, s\, \frac{J_o(\alpha s)}{(s^2+\tau^2)^{1/2}} \exp[-(s^2+\tau^2)^{1/2}t] \quad (4)$$

Integration of (4) over t from 0 to 1 gives

$$W(\tau,\alpha) = \int_1^\infty dt\, \frac{\exp[-\tau(\alpha^2+t^2)^{1/2}]}{(\alpha^2+t^2)^{1/2}} \quad (5)$$

Note that setting $\overline{k}_{D\perp} = 0$ in (1) and (5) gives

$$W(\tau,\alpha=0) = E_1(\tau) \quad (6)$$

where $E_1(\tau)$ is exponential integral. Another useful formula for W is obtained by making an integral variable transformation of $t = 1/\mu$

$$W(\tau,\alpha) = \int_0^1 \frac{d\mu}{\mu} \frac{\exp[-(\tau/\mu)(1+\alpha^2\mu^2)^{1/2}]}{(1+\alpha^2\mu^2)^{1/2}} \quad (7)$$

The integration variable μ physically corresponds to $\cos\theta$ with θ being the angle of the direction of propagation.

4 STRONG PERMITTIVITY FLUCTUATIONS

It was noted in Section 3.2 that when the bilocal approximation is applied in a straightforward manner to vector electromagnetic wave equations, the results are only valid when the permittivity fluctuation variance δ is much less than 1. The results for weak permittivity fluctuations are listed in (18) through (20), Section 3.2. The small δ criterion severely restricts the usefulness of the random medium model in remote sensing problems. Geophysical media are often mixtures of substances with very different dielectric properties. For example, dry snow is a mixture of air and ice. Wet snow is a mixture of air, ice, and water. The approximate permittivities of air, ice, and water are, respectively, ϵ_o, $3.2\epsilon_o$ and $72\epsilon_o$ and are quite different from each other. Each constituent also occupies an appreciable fractional volume. In dry snow, the fractional volume of ice is between $10 \sim 40\%$ and that of air is between $60 \sim 90\%$. In wet snow, the fractional volume of water is between $0 \sim 10\%$. Thus, the variance δ of permittivity fluctuations is quite appreciable and is generally not much less than 1.

In the bilocal-approximated Dyson's equation, both the observation point and the source point of the dyadic Green's function are within the random medium and can coincide with each other within the domain of integration. Physically, this corresponds to the fact that the scatterers act as sources of radiation. By taking into account the singularity of the dyadic Green's function, a strong permittivity fluctuation theory of electromagnetic propagation can be derived (Ryzhov et al., 1965; Ryzhov and Tamoikin, 1970; Tamoikin, 1971) that is applicable to both small and large variances of the permittivity.

In this section, we apply the strong permittivity fluctuation theory (SPF) to study scattering from a layer of random medium with spherical correlation function. The first moment of the field is calculated by using the bilocal approximation, and the second moment of the field is calculated by using the distorted Born approximation. The correlation function is obtained by using simple physical arguments and is expressed in terms of the fractional volumes and the particle sizes of the constituents of the dielectric mixture. The low frequency assumption is made in the derivations. The low-frequency effective propagation constant has an imaginary part that is due to a combination of absorption and scattering. The scattering part is dependent on particle size and can be ignored in the very low-frequency limit. It is also shown that the derived result of the effective permittivity in the low frequency limit is identical to the Polder and van Santern mixing formula which has been successful in matching experimental measurements of dielectric constants of snow (Ambach and Denoth, 1980; Colbeck, 1980; Linlor et al., 1980). It is also identical to the result of the effective medium theory which has been successful in explaining dielectric constant measurement of rocks (Sen et al., 1981). In this section, we shall only deal with media with spherically symmetric correlation functions. Results for anisotropic correlation functions can be found in Tsang and Kong (1981).

4.1 Random Medium with Spherically Symmetric Correlation Function

Consider a medium with permittivity $\epsilon(\bar{r})$ which is a random function of position. The vector wave equation is

$$\nabla \times \nabla \times \overline{E} - k_o^2 \frac{\epsilon(\bar{r})}{\epsilon_o} \overline{E} = 0 \qquad (1)$$

where $k_o = \omega\sqrt{\mu_o\epsilon_o}$ is the free-space wavenumber. We rewrite (1) as follows

$$\nabla \times \nabla \times \overline{E} - k_o^2 \frac{\epsilon_g}{\epsilon_o} \overline{E} = k_o^2 \left(\frac{\epsilon(\bar{r}) - \epsilon_g}{\epsilon_o}\right) \overline{E} \qquad (2)$$

where the permittivity ϵ_g is chosen such that the effective permittivity calculated by the bilocal approximation differs from ϵ_g by a term that is dependent on particle size. The particle size-dependent term vanishes as $k_o \to 0$ so that ϵ_g is the effective permittivity at the very low frequency limit. The very low frequency limit is defined as the frequency range at which the scattering attenuation rate is very small compared with the absorption attenuation rate.

Let $\overline{\overline{G}}_g(\bar{r},\bar{r}')$ be the dyadic Green's function that satisfies a vector wave equation with wavenumber k_g with

$$k_g = \omega\sqrt{\mu_o \epsilon_g} \qquad (3)$$

Thus,

$$\nabla \times \nabla \times \overline{\overline{G}}_g(\bar{r},\bar{r}') - k_g^2 \overline{\overline{G}}_g(\bar{r},\bar{r}') = \overline{\overline{I}}\delta(\bar{r}-\bar{r}') \qquad (4)$$

Using (4) in (2), we have

$$\overline{E}(\bar{r}) = \overline{E}_o(\bar{r}) + k_o^2 \int d\bar{r}' \overline{\overline{G}}_g(\bar{r},\bar{r}') \frac{\epsilon(\bar{r}') - \epsilon_g}{\epsilon_o} \cdot \overline{E}(\bar{r}') \qquad (5)$$

where $\overline{E}_o(\bar{r})$ is the field that satisfies the homogeneous wave equation with wavenumber k_g. We note that in (5) both $\bar{r}$ the observation point and $\bar{r}'$ the source point of the dyadic Green's function $\overline{\overline{G}}_g(\bar{r},\bar{r}')$ are inside the random medium. The two points can coincide with each other within the domain of integration of $d\bar{r}'$. Thus, the singular nature of the dyadic Green's function at $\bar{r} = \bar{r}'$ must be taken into consideration. This can be attributed physically to the fact that the scatterers act as sources of radiation.

The singularity of the dyadic Green's function depends on the shape of the infinitesimal exclusion volume (Van Bladel, 1961; Problem 8). For permittivity fluctuations with spherical symmetric correlation functions, the correction to the effective permittivity calculated by the bilocal approximation will be small if a spherical exclusion volume is chosen for the singularity of $\overline{\overline{G}}_g$. Thus we decompose the dyadic Green's function $\overline{\overline{G}}_g(\bar{r},\bar{r}')$ as follows

$$\overline{\overline{G}}_g(\bar{r},\bar{r}') = PS\,\overline{\overline{G}}_g(\bar{r},\bar{r}') - \frac{\overline{\overline{I}}}{3k_g^2}\delta(\bar{r}-\bar{r}') \qquad (6)$$

where PS stands for principal value. For nonspherical correlation functions, other shapes of exclusion volumes are to be chosen (Tsang and Kong, 1981). Substituting (6) into (5) gives

$$\overline{F}(\bar{r}) = \overline{E}_o(\bar{r}) + k_o^2 \int d\bar{r}'\, PS\,\overline{\overline{G}}_g(\bar{r},\bar{r}')\xi(\bar{r}')\overline{F}(\bar{r}') \qquad (7)$$

where
$$\overline{F}(\bar{r}) = \frac{\epsilon(\bar{r}) + 2\epsilon_g}{3\epsilon_g} \overline{E}(\bar{r}) \tag{8}$$

$$\xi(\bar{r}) = 3\frac{\epsilon_g}{\epsilon_o}\left(\frac{\epsilon(\bar{r}) - \epsilon_g}{\epsilon(\bar{r}) + 2\epsilon_g}\right) \tag{9}$$

In (7) through (9), $\overline{F}(\bar{r})$ and $\overline{E}(\bar{r})$, respectively, play roles of external and internal fields. The diagrammatic method of random medium in Sections 3.1 and 3.2 can be applied to (7) using $\overline{F}(\bar{r})$ as the field quantity, $PS\overline{\overline{G}}_g(\bar{r},\bar{r}')$ as the propagator, and $k^2\xi(\bar{r})$ as the "scatterer." Analogous to (4), Section (3.1), we let

$$<\xi(\bar{r})>=0 \tag{10}$$

Then, applying the bilocal approximation to (7) gives an equation that is analogous to (5), Section 3.2,

$$<\overline{F}(\bar{r})>= \overline{E}_o(\bar{r}) + k_o^2 \int d\bar{r}'' \int d\bar{r}' PS\overline{\overline{G}}_g(\bar{r},\bar{r}') \cdot \overline{\overline{\xi}}_{eff}(\bar{r}',\bar{r}'') \cdot <\overline{F}(\bar{r}'')> \tag{11}$$

with
$$\overline{\overline{\xi}}_{eff}(\bar{r}',\bar{r}'') = k_o^2 PS\overline{\overline{G}}_g(\bar{r}',\bar{r}'')R_\xi(|\bar{r}'-\bar{r}''|) \tag{12}$$

where
$$R_\xi(|\bar{r}'-\bar{r}'|) =<\xi(\bar{r}')\xi(\bar{r}'')> \tag{13}$$

is the spherically symmetric correlation function of $\xi(\bar{r})$.

4.2 Very Low Frequency Effective Permittivity

At very low frequencies, the scattering attenuation rate can be neglected and the effective permittivity is given by ϵ_g which obeys the equation [from (9) and (10), Section 4.1]

$$\left\langle \frac{\epsilon(\bar{r}) - \epsilon_g}{\epsilon(\bar{r}) + 2\epsilon_g} \right\rangle = 0 \tag{1}$$

In this section, we shall show that ϵ_g, as given by (1), when applied to a dielectric mixture, is identical to that of the mixing formula of Polder and van Santern (1946).

Suppose there are n constituents in the mixture with ϵ_p and f_p ($p = 1,2,3,4,...,n$) as the permittivity and the fractional volume, respectively, of the pth constituent. Thus, $Pr[\epsilon(r) = \epsilon_p] = f_p$, with Pr standing for probability. Then

$$\sum_{p=1}^{n} f_p = 1 \tag{2}$$

and (1) takes the form

$$\sum_{p=1}^{n} \frac{\epsilon_p - \epsilon_g}{\epsilon_p + 2\epsilon_g} f_p = 0 \tag{3}$$

It follows from (2) and (3) that

$$\frac{1}{3\epsilon_g} = \sum_{p=1}^{n} \frac{f_p}{\epsilon_p + 2\epsilon_g} \tag{4}$$

Rearranging terms in (3) gives

$$\sum_{p=1}^{n} \frac{\epsilon_p - \epsilon_o}{\epsilon_p + 2\epsilon_g} f_p = (\epsilon_g - \epsilon_o) \sum_{p=1}^{n} \frac{f_p}{\epsilon_p + 2\epsilon_g} \tag{5}$$

where ϵ_o is the free space permittivity. Using (4) in (5), we obtain

$$\sum_{p=1}^{n} \frac{\epsilon_p - \epsilon_o}{\epsilon_p + 2\epsilon_g} f_p = \frac{\epsilon_g - \epsilon_o}{3\epsilon_g} \tag{6}$$

which is identical to Polder and van Santern's mixing formula [Polder and van Santern, 1946; (14)]. In Polder and van Santern's original derivation, the inhomogeneities are regarded as dipoles. This assumption is consistent with the low-frequency limit. The mixing formula in (3) is also identical to that of the effective medium theory (Kohler and Papanicolaou, 1981) and is also the starting point for the self-similar model for sedimentary rocks (Sen et al., 1981; Problem 16).

4.3 Effective Permittivity under the Bilocal Approximation

To calculate the effective permittivity under the bilocal approximation, we note from (8) and (9), Section 4.1, that

$$\xi(\bar{r})\overline{F}(\bar{r}) = \frac{\overline{D}(\bar{r})}{\epsilon_o} - \frac{\epsilon_g}{\epsilon_o}\overline{E}(\bar{r}) \tag{1}$$

where $\overline{D}(\bar{r}) = \epsilon(\bar{r})\overline{E}(\bar{r})$ is the electric displacement. Taking the ensemble average of (7), Section 4.1, and comparing with (11), Section 4.1, gives

$$<\xi(\bar{r})\overline{F}(\bar{r})> = \int d\bar{r}' \, \overline{\overline{\xi}}_{\mathit{eff}}(\bar{r} - \bar{r}') \cdot <\overline{F}(\bar{r}')> \tag{2}$$

From (8), Section 4.1

$$<\overline{F}(\overline{r})> = \frac{<\overline{D}(\overline{r})>}{3\epsilon_g} + \frac{2}{3}<\overline{E}(\overline{r})> \qquad (3)$$

Analogous to (2), we define $\overline{\overline{\epsilon}}_{eff}(\overline{r})$ such that

$$<\overline{D}(\overline{r})> = \int d\overline{r}' \, \overline{\overline{\epsilon}}_{eff}(\overline{r} - \overline{r}') <\overline{E}(\overline{r}')> \qquad (4)$$

Let $\overline{\overline{\epsilon}}_{eff}(\overline{k})$, $\overline{\overline{\xi}}_{eff}(\overline{k})$, $\overline{E}(\overline{k})$, $\overline{D}(\overline{k})$, and $\overline{F}(\overline{k})$ be, respectively, the three dimensional Fourier transforms of $\overline{\overline{\epsilon}}_{eff}(\overline{r})$, $\overline{\overline{\xi}}_{eff}(\overline{r})$, $\overline{E}(\overline{r})$, $\overline{D}(\overline{r})$, and $\overline{F}(\overline{r})$. For example

$$\overline{\overline{\epsilon}}_{eff}(\overline{k}) = \int d\overline{r} \, \overline{\overline{\epsilon}}_{eff}(\overline{r}) \, e^{-i\overline{k}\cdot\overline{r}} \qquad (5)$$

The $\overline{k}$ dependence of the effective permittivity $\overline{\overline{\epsilon}}_{eff}(\overline{k})$ represents spatial dispersion effects. Substituting (1) into (2) and taking Fourier transform of the resultant equation yield

$$\frac{<\overline{D}(\overline{k})>}{\epsilon_o} = \frac{\epsilon_g}{\epsilon_o} <\overline{E}(\overline{k})> + \overline{\overline{\xi}}_{eff}(\overline{k}) \cdot <\overline{F}(\overline{k})> \qquad (6)$$

Taking Fourier transform of (3) and then substituting into (6) give an expression of $<\overline{D}(\overline{k})>$ in terms of $<\overline{E}(\overline{k})>$. In view of the fact that $\overline{D}(\overline{k}) = \overline{\overline{\epsilon}}_{eff}(\overline{k}) \cdot <\overline{E}(\overline{k})>$, we have

$$\overline{\overline{\epsilon}}_{eff}(\overline{k}) = \epsilon_g \overline{\overline{I}} + \epsilon_o \left(\overline{\overline{I}} - \frac{\epsilon_o}{3\epsilon_g} \overline{\overline{\xi}}_{eff}(\overline{k}) \right)^{-1} \overline{\overline{\xi}}_{eff}(\overline{k}) \qquad (7)$$

Thus, the second term in (7) is the correction provided by the bilocal approximation to $\overline{\overline{\epsilon}}_{eff}(\overline{k})$. The validity of the bilocal approximation requires

$$|\overline{\overline{\xi}}_{eff}(\overline{k})| \ll 1 \qquad (8)$$

so that

$$\overline{\overline{\epsilon}}_{eff}(\overline{k}) \simeq \epsilon_g \overline{\overline{I}} + \epsilon_o \overline{\overline{\xi}}_{eff}(\overline{k}) \qquad (9)$$

For low frequencies, the spatial dispersion effects can be ignored so that the $\overline{k}$, argument in (9), can be replaced by zero. Thus

$$\overline{\overline{\epsilon}}_{eff} = \epsilon_g \overline{\overline{I}} + \epsilon_o \overline{\overline{\xi}}_{eff}^{(0)} \qquad (10)$$

where

$$\bar{\bar{\xi}}_{eff}^{(0)} = \int d\bar{r}\, \bar{\bar{\xi}}_{eff}(\bar{r}) \tag{11}$$

Using (12), Section 4.1, in (10) and (11) gives

$$\bar{\bar{\epsilon}}_{eff} = \epsilon_g \bar{\bar{I}} + k_o^2 \epsilon_o \int d\bar{r}\, PS\, \bar{\bar{G}}_g(\bar{r}) R_\xi(|\bar{r}|) \tag{12}$$

For spherically symmetric correlation functions, the integral in (12) is proportional to the unit dyad. The integral in (12) can be evaluated in a manner similar to (13) through (18), Section 3.2, by replacing k_m by k_g (Problem 17). Furthermore, in the low-frequency limit, $\exp(ik_g r) \simeq 1$. Hence,

$$\epsilon_{eff} = \epsilon_g + \epsilon_o \frac{2}{3} k_o^2 \int_0^\infty dr\, r\, R_\xi(r) + i\frac{2}{3} k_o^2 k_g \epsilon_o U \tag{13}$$

where

$$U = \int_0^\infty dr\, r^2\, R_\xi(r) \tag{14}$$

In (13), the real and imaginary parts of the second term are always much smaller than the real and imaginary parts, respectively, of the first term. Thus, the second term can be neglected. The third term, however, cannot be ignored because its imaginary part can be much larger than the imaginary part of the first term. Hence, (13) simplifies to

$$\epsilon_{eff} = \epsilon_g + i\frac{2}{3} k_o^2 k_g \epsilon_o U \tag{15}$$

We note from (15) that $Re(\epsilon_{eff}) \simeq Re(\epsilon_g)$. Both the absorption attenuation and scattering attenuation are included in the imaginary part of ϵ_{eff} whereas ϵ_g contains only the absorption effect in its imaginary part.

4.4 Backscattering Coefficients

Consider a plane wave incident on a half-space dielectric mixture with effective permittivity ϵ_{eff}, as derived in Section 4.3 (Figure 5.1). The field $\bar{F}(\bar{r})$ is decomposed into a coherent part $<\bar{F}(\bar{r})>$ and an incoherent part $\bar{\mathcal{F}}(\bar{r})$, $\bar{F}(\bar{r}) = <\bar{F}(\bar{r})> + \bar{\mathcal{F}}(\bar{r})$. The distorted Born approximation will be applied to calculate the incoherent scattered field. The distorted Born approximation corresponds physically to the single scattering of the coherent field. Diagrammatically, it corresponds to the first diagram on the right-hand side of (1), Section 3.6. It is also known as first-order multiple scattering

(Ishimaru, 1978). The scattered intensity $<|\overline{\mathcal{F}}(\bar{r})|^2>$ in the upper half space is, under the distorted Born approximation and using (7), Section 4.1

$$<|\overline{\mathcal{F}}(\bar{r})|^2> = k_o^4 \iint_A dx_1\, dy_1 \int_{-\infty}^0 dz_1 \iint_A dx_2\, dy_2 \int_{-\infty}^0 dz_2$$

$$\times C_\xi(|\bar{r}_1 - \bar{r}_2|)(<\overline{\overline{G}}_{01}(\bar{r},\bar{r}_1)> \cdot <\overline{F}(\bar{r}_1)>) \qquad (1)$$

$$\cdot (<\overline{\overline{G}}_{01}^*(\bar{r},\bar{r}_2)> \cdot <\overline{F}^*(\bar{r}_2)>)$$

where

$$C_\xi(|\bar{r}_1 - \bar{r}_2|) = <\xi(\bar{r}_1)\xi^*(\bar{r}_2)> \qquad (2)$$

A denotes the area of observation by the radar, and $<\overline{\overline{G}}_{01}(\bar{r},\bar{r}_1)>$ is the mean Green's function with $\bar{r}$ in the upper half-space and $\bar{r}_1$ in the lower half-space. For half-space scattering from a random medium, the presence of a boundary at $z=0$ can produce a boundary-layer effect (Rosenbaum, 1969) that has an analog in random discrete scatterer theory (Chapter 6). In view of the low frequency approximation, the influence of the boundary layer is negligible. As shown in Chapter 6, the coherent reflection and transmission coefficients for random scattering in the low-frequency limit are the classical Fresnel reflection and transmission coefficients. Thus, we shall make this assumption in calculating the half-space mean Green's function and the mean field in the lower half-space.

For incident horizontally polarized wave with amplitude E_o, the mean field in the random medium is

$$<\overline{F}(\bar{r})> = \hat{e}_{1i}\, E_o\, X_{01i}\, e^{i\overline{K}_i \cdot \bar{r}} \qquad (3)$$

For incident vertically polarized wave, the mean field is

$$<\overline{F}(\bar{r})> = \hat{h}_1(-K_{iz})\, E_o\, Y_{01i}\, \frac{k_o}{K}\, e^{i\overline{K}_i \cdot \bar{r}} \qquad (4)$$

where the subscript i is used to denote incident direction, $\overline{K}_i = \hat{x}k_{ix} + \hat{y}k_{iy} - \hat{z}K_{iz}$ and K is the effective propagation constant $K = \omega\sqrt{\mu_o \epsilon_{\text{eff}}}$. Since the observation point $\bar{r}$ in (3) is in the far field in region 0, hence

$$<\overline{\overline{G}}_{01}(\bar{r},\bar{r}')> = \frac{k_{sz}}{K_{sz}}\left[X_{10s}\hat{e}_s\hat{e}_{1s} + \frac{K}{k_o}Y_{10s}\hat{h}(k_{sz})\hat{h}_1(K_{sz})\right]$$

$$\times \frac{e^{ik_o r}}{4\pi r} \exp(-i\overline{K}_s \cdot \bar{r}') \qquad (5)$$

Backscattering Coefficients

where the subscript s denotes scattered direction. The expression in (5) is similar to the unperturbed Green's function $G_{01}^{(0)}(\bar{r},\bar{r}')$ of (6), Section 2.1, with the major difference being that the mean permittivity ϵ_{1m} and mean wavenumber k_{1m} are replaced respectively by the effective permittivity ϵ_{eff} and the effective wavenumber K. In (3) through (5), $\overline{K}_s = \hat{x}k_{sx} + \hat{y}k_{sy} + \hat{z}K_{sz}$, with $k_x = k\sin\theta\cos\phi$, $k_y = k\sin\theta\sin\phi$, $K_z = (K^2 - k_x^2 - k_y^2)^{1/2}$, $k_z = k\cos\theta$. In the incident direction, $(\theta,\phi) = (\theta_i,\phi_i)$; in the scattered direction $(\theta,\phi) = (\theta_s,\phi_s)$.

In (3) through (5), X_{01}, Y_{01}, X_{10}, and Y_{10} are Fresnel transmission coefficients as defined in (13), Section 3.2, Chapter 2, with k_{1z} and ϵ_1 replaced respectively by ϵ_{eff} and K_z. For example,

$$Y_{01} = \frac{2\epsilon_{eff}k_z}{\epsilon_{eff}k_z + \epsilon_o K_z}$$

In (3) and (5), $\hat{e}$ and $\hat{h}$ are, respectively, transverse electric (TE) and transverse magnetic (TM) polarization vectors in the upper region and $\hat{e}_1$ and $\hat{h}_1$ are the TE and TM polarization vectors in the lower half-space with medium 1 parameters replaced by the effective parameters of ϵ_{eff}, K, and K_z. Expressions for $\hat{e}$, $\hat{h}$, $\hat{e}_1$, and $\hat{h}_1$ are given in (13), Section 3.1, Chapter 2.

In the low-frequency limit, when the correlation lengths are short compared with the wavelength, $C_\xi(|r_1 - r_2|)$ is peaked at $\bar{r}_1 = \bar{r}_2$ and (1) can be approximated by

$$<|\bar{F}(\bar{r})^2|> = 4\pi k_o^4 W \iint_A dx_1\,dy_1 \int_{-\infty}^0 dz_1 \qquad (6)$$
$$\times \left|<\overline{\overline{G}}_{01}(\bar{r},\bar{r}_1)> \cdot <\overline{F}(\bar{r}_1)>\right|^2$$

where

$$W = \int_0^\infty dr\, r^2\, C_\xi(r) \qquad (7)$$

To calculate backscattering coefficients, we use the backscattering direction in the expression for $<\overline{\overline{G}}_{01}>$ with $k_{xs} = -k_{xi}$ and $k_{ys} = -k_{yi}$. Thus,

$$\sigma_{vv} = \frac{k_o^4 W k_{iz}^2}{4K_{iz}''|K_{iz}|^2} |Y_{10i}Y_{01i}|^2 \qquad (8)$$

$$\sigma_{hh} = \frac{k_o^4 W k_{iz}^2}{4K_{iz}''|K_{iz}|^2} |X_{10i}X_{01i}|^2 \qquad (9)$$

$$\sigma_{vh} = \sigma_{hv} = 0 \tag{10}$$

with $K''_{iz} = Im(K_{iz})$.

The distorted Born approximation is the first iteration of the Bethe-Salpeter equation and is valid for small albedo. The depolarization return for a random medium with a spherically symmetric correlation function is proportional to the square of the albedo. Calculation of the depolarization return requires a second iteration of the Bethe-Salpeter equation.

4.5 Numerical Illustrations

In this section, the results in Sections 4.2 through 4.4 are illustrated numerically using typical parameters encountered in microwave remote sensing of dry and wet snow. Dry snow is a mixture of two constituents, air and ice, with the fractional volume of ice between 10 and 40%. Wet snow is a mixture of three constituents, air, ice, and water, with the percentage of water between 0 and 10%. We use ϵ_b, ϵ_{s1}, and ϵ_{s2} to denote permittivities of air, ice, and water, respectively, and f_b, f_{s1} and f_{s2} to denote fractional volumes of air, ice, and water, respectively. The permittivity of ice, ϵ_{s1}, is chosen to be $3.2(1 + i0.001)\epsilon_o$ and the permittivity of water, ϵ_{s2}, obeys the relation

$$\frac{\epsilon_{s2}}{\epsilon_o} = 5.5 + \frac{82.5}{1 - i3.59/\lambda} \tag{1}$$

where λ is wavelength in centimeters.

A. Mixture of Two Constituents

The parameters are $\epsilon_b, f_b, \epsilon_s,$ and f_s with $f_b + f_s = 1$. The quantity ϵ_g is first determined by using (3), Section 4.2. We use the simple correlation functions (Problem 18)

$$R_\xi(r) = \begin{cases} <\xi^2> & \text{for } 0 \leq r \leq a \\ 0 & \text{otherwise} \end{cases} \tag{2}$$

$$C_\xi(r) = \begin{cases} <|\xi|^2> & \text{for } 0 \leq r \leq a \\ 0 & \text{otherwise} \end{cases} \tag{3}$$

We note from (9), Section 4.1, that $<\xi^2> = f_s\xi_s^2 + f_b\xi_b^2$ and $<|\xi|^2> = f_s|\xi_s|^2 + f_b|\xi_b|^2$ where $\xi_s = 3r_s\epsilon_g/\epsilon_o$, $\xi_b = 3r_b\epsilon_g/\epsilon_o$ and

$$r_s = \frac{\epsilon_s - \epsilon_g}{\epsilon_s + 2\epsilon_g} \tag{4}$$

$$r_b = \frac{\epsilon_b - \epsilon_g}{\epsilon_b + 2\epsilon_g} \tag{5}$$

Numerical Illustrations

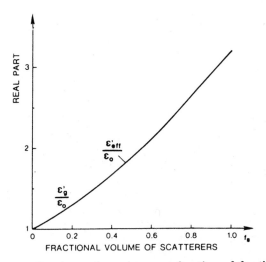

Fig. 5.21 Real part of ϵ_g/ϵ_o and $\epsilon_{eff}/\epsilon_o$ as a function of fractional volume of scatterers for $\epsilon_b = \epsilon_o$, $\epsilon_s = (3.2 + i0.0032)\epsilon_o$, frequency = 8.6 GHz, and $a = 0.5$ mm.

Using (13), Section 4.3, and (8) and (9), Section 4.4, gives

$$\epsilon_{eff} = \epsilon_g[1 + i2k_g^3 a^3(f_s r_s^2 + f_b r_b^2)] \tag{6}$$

$$\sigma_{hh} = 3|k_g|^4 a^3 \left[f_s|r_s|^2 + f_b|r_b|^2\right] \frac{k_{iz}^2}{|K_{iz}|^2} |X_{01i}X_{10i}|^2 \frac{1}{4K_{iz}''} \tag{7}$$

$$\sigma_{vv} = 3|k_g|^4 a^3 \left[f_s|r_s|^2 + f_b|r_b|^2\right] \frac{k_{iz}^2}{|K_{iz}|^2} |Y_{01i}Y_{10i}|^2 \frac{1}{4K_{iz}''} \tag{8}$$

In Figures 5.21 and 5.22, the real and imaginary parts of ϵ_g/ϵ_o and $\epsilon_{eff}/\epsilon_o$ are plotted as a function of the fractional volume of scatterers. The limit of $f_s = 1$ corresponds to the entire volume being occupied by scatterers. This is not attainable for particles of the same size and is only of theoretical interest. In that limit, as indicated in Figure 5.21, the real part of ϵ_g/ϵ_o and $\epsilon_{eff}/\epsilon_o$ is equal to 3.2. The imaginary part of $\epsilon_{eff}/\epsilon_o$ coincides with the imaginary part of ϵ_g/ϵ_o in the limit of $f_s = 1$, indicating that there is no attenuation due to scattering. In Figure 5.23, the backscattering coefficients are plotted as a function of incidence angle θ_i at a frequency of 8.6 GHz and for a half-space with a fractional volume of scatterers equal to 30%. The calculated effective permittivity is $(1.47 + i0.663 \times 10^{-3})\epsilon_o$. The results of the figure indicate that the polarization difference is quite small.

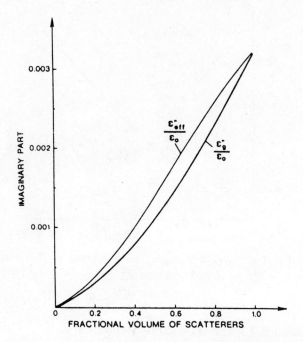

Fig. 5.22 Imaginary part of ϵ_g/ϵ_o and $\epsilon_{\text{eff}}/\epsilon_o$ as a function of fractional volume of scatterers for $\epsilon_b = \epsilon_o$, $\epsilon_s = (3.2 + i0.0032)\epsilon_o$, frequency = 8.6 GHz, and $a = 0.5$ mm.

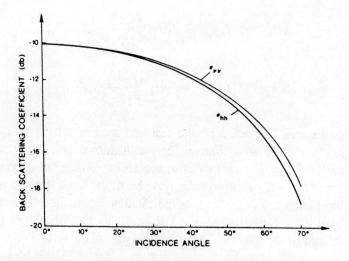

Fig. 5.23 Backscattering coefficients as function of incidence angle θ_i for $\epsilon_b = \epsilon_o$, $\epsilon_s = 3.2(1 + i0.001)\epsilon_o$, $f_s = 30\%$, $a = 0.5$ mm, and frequency = 8.6 GHz. Calculated value of ϵ_{eff} is $(1.47 + i0.663 \times 10^{-3})\epsilon_o$.

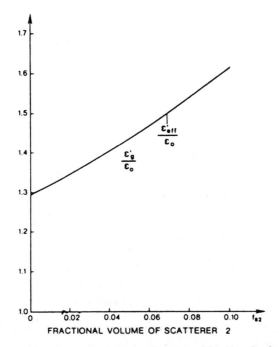

Fig. 5.24 Real part of ϵ_g/ϵ_o and $\epsilon_{eff}/\epsilon_o$ as function of fractional volume of scatterer 2 for $\epsilon_b = \epsilon_o$, $\epsilon_{s1} = 3.2(1 + i0.001)\epsilon_o$, $f_b = 80\%$, $a_1 = 0.5$ mm, $a_2 = 0.5$ mm, and frequency = 17 GHz.

B. Mixture of Three Constituents

For three constituents, we have a background medium with permittivity ϵ_b, two kinds of scatterers with permittivities ϵ_{s1} and ϵ_{s2}, and fractional volumes f_{s1} and f_{s2}. The scatterers are assumed to be spherical with radii a_1 and a_2. We have $f_b + f_{s1} + f_{s2} = 1$. The quantity ϵ_g is first calculated by using (3), Section 4.2. Assuming, without loss of generality, that $a_2 \geq a_1$, the correlation functions are (Problem 18)

$$R_\xi(r) = \begin{cases} <\xi^2> & \text{for } 0 \leq r \leq a_1 \\ \xi_{s2}^2 f_{s2}(1 + f_{s2}) & \text{for } a_1 < r \leq a_2 \\ 0 & \text{otherwise} \end{cases} \quad (9)$$

$$C_\xi(r) = \begin{cases} <|\xi|^2> & \text{for } 0 \leq r \leq a_1 \\ |\xi_{s2}|^2 f_{s2}(1 + f_{s2}) & \text{for } a_1 < r \leq a_2 \\ 0 & \text{otherwise} \end{cases} \quad (10)$$

In (15), Section 4.3, ϵ_{eff} is in terms of U and in (8) and (9), Section 4.4, σ_{vv} and σ_{hh} are in terms of W. The quantities U and W are

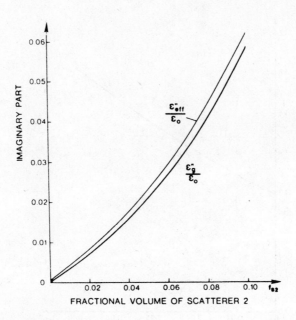

Fig. 5.25 Imaginary part of ϵ_g/ϵ_o and $\epsilon_{eff}/\epsilon_o$ as function of fractional volume of scatterer 2 for $\epsilon_b = \epsilon_o$, $\epsilon_{s1} = 3.2(1 + i0.001)\epsilon_o$, $f_b = 80\%$, $a_1 = 0.5$ mm, $a_2 = 0.5$ mm, and frequency = 17 GHz.

calculated from (14), Section 4.3, and (7), Section 4.4, respectively

$$U = \frac{3\epsilon_g^2}{\epsilon_o^2}\left\{(f_{s1}r_{s1}^2 + f_{s2}r_{s2}^2 + f_b r_b^2)a_1^3 + r_{s2}^2 f_{s2}(1 + f_{s2})(a_2^3 - a_1^3)\right\} \quad (11)$$

$$W = 3\frac{|\epsilon_g|^2}{\epsilon_o^2}\left\{(f_{s1}|r_{s1}|^2 + f_{s2}|r_{s2}|^2 + f_b|r_b|^2)a_1^3 + |r_{s2}|^2 f_{s2}(1 + f_{s2})(a_2^3 - a_1^3)\right\} \quad (12)$$

with $r_{s1} = (\epsilon_{s1} - \epsilon_g)/(\epsilon_{s1} + 2\epsilon_g)$, $r_{s2} = (\epsilon_{s2} - \epsilon_g)/(\epsilon_{s2} + 2\epsilon_g)$, and r_b as given by (5).

We use the theory for a mixture of three constituents to illustrate the application to wet snow. In Figures 5.24, 5.25, and 5.26, we plot, respectively, the real part of the effective dielectric constant, the imaginary part of the effective dielectric constant, and backscattering coefficients as a function of fractional volume of scatterer 2 from 0 to 10%. The permittivity of scatterer 2 is taken to be that of water and is given by (1). The fractional volume of

Numerical Illustrations

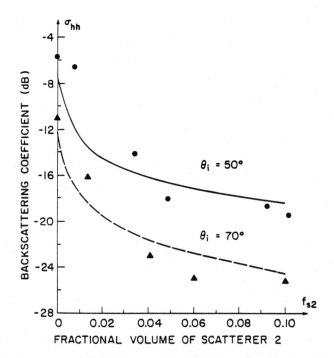

Fig. 5.26 Backscattering coefficients σ_{hh} at $\theta_i = 50\,\text{deg}$ and $\theta_i = 70\,\text{deg}$ as function of fractional volume of scatterer 2 for $\epsilon_b = \epsilon_o$, $\epsilon_{s1} = 3.2(1 + i0.001)\epsilon_o$, $f_b = 80\%$, $a_1 = 0.5$ mm, $a_2 = 0.5$ mm, and frequency = 17 GHz. Comparison with experimental data.

the background medium is fixed at 80%. Therefore, as f_{s2} increases from 0 to 10%, f_{s1} decreases from 20 to 10%. From Figure 5.26, we note that the backscattering coefficients decrease as the wetness of the snow is increased. the radii of both scatterers 1 and 2 are chosen to be 0.5 mm, which falls within the grain size of snow.

In Figure 5.26, the theoretical result is also used to match experimental data taken from wet snow (Stiles and Ulaby, 1980a). The frequency of operation for the data was 17 GHz and was for HH polarization. The data were plotted as a function of volumetric snow wetness, M_v, which we have equated with f_{s2}. The incident angles were 50 deg and 70 deg and the snow depth d was 30 cm with water equivalent W equal to 5 cm. Since ice has a smaller density than water, the quantity obtained by dividing the water equivalent W by snow depth d is generally about 5 to 20% smaller than the fractional volume of ice. Thus in Figure 5.26, we take the fractional volume of air to be 80%. The match is good in spite of the following effects, which have not been incorporated in the theory: (1) The data are taken over a two-layer medium whereas the theory is developed for the half-space case. However, it seems that the data shows relatively little effect of the bottom interface for frequencies between 8 and 17 GHz; (2) The temperature dependence of the permittivity of water has not been taken into account; (3) The frequency dependence of the loss tangent of ice has been neglected. It is noted that the above derivations are readily extended to treat the cases of more than two constituents and backscattering from layered media rather than the half-space medium. However, the construction of the correlation functions as illustrated in Problem 18 is crude and by no means exact. Better correlation functions can be derived from statistical analysis of the ground truth data or by more rigorous theoretical analysis (Stogryn, 1984).

5 MODIFIED RADIATIVE TRANSFER EQUATIONS FOR VOLUME SCATTERING IN THE PRESENCE OF REFLECTIVE BOUNDARIES

5.1 Introduction

In Section 3.5, the ladder-approximated Bethe-Salpeter equation is used to derive the radiative transfer equations for a half-space medium without any reflective boundary. This was accomplished by equating the covariance of the field to the Fourier transform of the specific intensity and assuming that field propagating in different directions are uncorrelated. For the case of a two-layer random medium (Figure 5.27), the presence of a reflective boundary will produce nonzero correlations between the upward- and downward-going waves that are specularly related. In this section, by incorporating this nonzero correlation, a set of modified radiative transfer equations are derived from the ladder-approximated Bethe-Salpeter equation. The interference effect due to the presence of reflective boundaries has also been observed in experimental data (Blinn et al., 1972; Carver, 1977).

Consider a slab of random medium separated by reflective boundaries from two homogeneous half-spaces (Figure 5.27). The permittivity of the random medium in Region 1 is

$$\epsilon_1(\bar{r}) = <\epsilon_1> + \epsilon_{1f}(\bar{r}) \tag{1}$$

The two homogeneous half-spaces have permittivities ϵ_o and ϵ_2, respectively for Region 0 and Region 2. Generally, ϵ_o, $<\epsilon_1>$ and ϵ_2 are quite different from each other. The electric field in Region 1 is decomposed into a mean field and an incoherent field

$$\overline{E}_1(\bar{r}) = \overline{E}_{1m}(\bar{r}) + \overline{\mathcal{E}}_1(\bar{r}) \tag{2}$$

Introduction

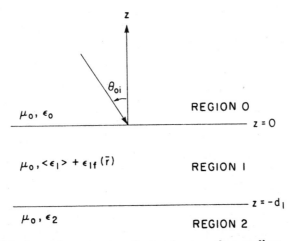

Fig. 5.27 Scattering geometry of a two-layer random medium.

The subscript m is used to denote the mean. Thus, $\overline{E}_{1m}(\bar{r}) \equiv <\overline{E}_1(\bar{r})>$. The mean wave number in Region 1 is $k_{1m} = \omega\sqrt{\mu_o <\epsilon_1>}$. Following Section 3.5 the incoherent field is next decomposed into a spectrum of upward- and downward-propagating plane waves represented by $\overline{\mathcal{E}}_u$ and $\overline{\mathcal{E}}_d$, respectively. We have

$$\overline{\mathcal{E}}_1(\bar{r}) = \int d\overline{\beta}_\perp \, e^{i\overline{\beta}_\perp \cdot \bar{r}_\perp} \left[\overline{\mathcal{E}}_u(z,\overline{\beta}_\perp) e^{i\beta'_{1mz}z} + \overline{\mathcal{E}}_d(z,\overline{\beta}_\perp) e^{-i\beta'_{1mz}z} \right] \quad (3)$$

where

$$\overline{\beta}_\perp = \beta_x \hat{x} + \beta_y \hat{y} \quad (4)$$

$$\beta_{1mz} = (k_{1m}^2 - \beta_x^2 - \beta_y^2)^{1/2} \quad (5)$$

and $\beta'_{1mz} = Re(\beta_{1mz})$. We use $\hat{e}$ and $\hat{h}$ to denote, respectively, TE and TM polarization vectors (Section 3.1). Decomposing $\overline{\mathcal{E}}_u$ and $\overline{\mathcal{E}}_d$ into TE and TM components, we have

$$\overline{\mathcal{E}}_u(z,\overline{\beta}_\perp) = \mathcal{E}_{hu}(z,\overline{\beta}_\perp)\hat{e}(\beta'_{1mz}) + \mathcal{E}_{vu}(z,\overline{\beta}_\perp)\hat{h}(\beta'_{1mz}) \quad (6)$$

$$\overline{\mathcal{E}}_d(z,\overline{\beta}_\perp) = \mathcal{E}_{hd}(z,\overline{\beta}_\perp)\hat{e}(-\beta'_{1mz}) + \mathcal{E}_{vd}(z,\overline{\beta}_\perp)\hat{h}(-\beta'_{1mz}) \quad (7)$$

where subscripts u and d stand for up and down, respectively, and subscripts h and v stand for horizontal and vertical polarization, respectively.

As in Section 3.5, there are two distance scales. The first one is a long-distance scale characterized by the quantities $l_v = [Im(k_v)]^{-1}$ and

$l_h = [Im(k_h)]^{-1}$ where k_v and k_h are the two effective propagation constants for v and h polarized waves, respectively. The long-distance scale is the extinction distance of the coherent wave and is much larger than the mean wavelength $\lambda_{1m} = 2\pi/k_{1m}$. It is also much larger than l_z with l_z being the correlation length of the permittivity fluctuations. Therefore, two points can be close on the l_v, l_h scale and yet far apart on a scale characterized by λ_{1m} or l_z. In deriving the modified radiative transfer (MRT) equations from the Bethe-Salpeter equation, we assume that for points z and z' close together on the l_v, l_h scale

$$<\mathcal{E}_{ju}(z,\overline{\alpha}_\perp)\mathcal{E}^*_{ku}(z',\overline{\beta}_\perp)> = \delta(\overline{\alpha}_\perp - \overline{\beta}_\perp)J_{jku}(z,z',\overline{\alpha}_\perp) \qquad (8)$$

$$<\mathcal{E}_{jd}(z,\overline{\alpha}_\perp)\mathcal{E}^*_{kd}(z',\overline{\beta}_\perp)> = \delta(\overline{\alpha}_\perp - \overline{\beta}_\perp)J_{jkd}(z,z',\overline{\alpha}_\perp) \qquad (9)$$

$$<\mathcal{E}_{ju}(z,\overline{\alpha}_\perp)\mathcal{E}^*_{kd}(z',\overline{\beta}_\perp)> = \delta(\overline{\alpha}_\perp - \overline{\beta}_\perp)J_{jkc1}(z,z',\overline{\alpha}_\perp) \qquad (10)$$

$$<\mathcal{E}_{jd}(z,\overline{\alpha}_\perp)\mathcal{E}^*_{ku}(z',\overline{\beta}_\perp)> = \delta(\overline{\alpha}_\perp - \overline{\beta}_\perp)J_{jkc2}(z,z',\overline{\alpha}_\perp) \qquad (11)$$

where $j,k = h$ or v. Thus, incoherent fields with different transverse directions of propagation are uncorrelated. However, for the same transverse directions, there exist correlations between upward- and downward-propagating incoherent fields due to the presence of reflecting dielectric interfaces at $z = 0$ and $z = -d_1$. These correlations are represented by J_{jkc1} and J_{jkc2} in (10) and (11). Furthermore, there exist cross correlations between TE and TM components giving rise to the third and fourth Stokes parameters for vector electromagnetic propagation.

In Section 5.2, we will study the mean (coherent) Green's function for the two-layer random medium using the nonlinear approximation (Rosenbaum, 1971; Dence and Spence, 1973) of Dyson's equations and in Section 5.3 the MRT equations will be derived. In Section 5.4, the MRT equations are solved by iteration and the solutions are compared with the results of the radiative transfer equations and those of Born approximation. However, because of the intrinsic limitation of the ladder approximation, the MRT does not account for backscattering enhancement as discussed in Sections 2.3 and 3.6. Derivations of the transport equation are customarily done with the ladder approximation (Barabanenkov and Finkelberg, 1968; Stott, 1968; Watson, 1969; Ishimaru, 1975, 1978; Furutsu, 1975; Besieris and Kohler, 1981; Fante, 1981).

5.2 Mean Green's Function and Mean Field

In this section the two-variable expansion procedure shall be used (Cole,

1968) to calculate the mean Green's function and mean field for the nonlinear-approximated Dyson's equation for a two-layer medium. We first illustrate the technique for a one-dimensional problem with $\epsilon_1(z) = <\epsilon_1> + \epsilon_{1f}(z)$ (Tsang and Kong, 1976). The results for a three-dimensional random medium will be listed. The coherent scalar wave for a half-space random medium has also been investigated with the bilocal approximation (Rosenbaum, 1969).

Dyson's equation under the nonlinear approximation is

$$\left(\frac{d^2}{dz^2} + k_{1m}^2\right) G_{11m}(z, z') = \delta k_{1m}^4 \int_{-d}^{0} dz_1\, G_{11m}(z, z_1)$$
$$\times e^{-|z-z_1|/l} G_{11m}(z_1, z') + \delta(z - z') \quad (1)$$

where we have taken $C(z - z_1) = \delta k_{1m}^4 e^{-|z-z_1|/l}$. Let $\delta \ll 1$. We solve (1) by using the two-variable expansion procedure. Define long-distance scales $\xi = \delta z, \xi' = \delta z'$, and $\xi_1 = \delta z_1$. From $\xi = \delta z$ it follows that

$$\frac{d^2}{dz^2} = \frac{\partial^2}{\partial z^2} + 2\delta \frac{\partial^2}{\partial z \partial \xi} + \delta^2 \frac{\partial^2}{\partial \xi^2} \quad (2)$$

The mean Green's function is expanded in perturbation series

$$G_{11m}(z, \xi; z', \xi') = G_{11mo}(z, \xi; z', \xi') + \delta G_{11m1}(z, \xi; z', \xi') + \cdots \quad (3)$$

Substituting (2) and (3) into (1), we find that on balancing terms to the zeroth-order in δ

$$\left(\frac{\partial^2}{\partial z^2} + k_{1m}^2\right) G_{11mo}(z, \xi; z', \xi') = \delta(z - z') \quad (4)$$

and to the first-order in δ

$$\left(\frac{\partial^2}{\partial z^2} + k_{1m}^2\right) G_{11m1}(z, \xi; z', \xi') = -2\frac{\partial^2}{\partial z \partial \xi} G_{11mo}(z, \xi; z', \xi')$$
$$+ k_{1m}^4 \int_{-d}^{0} dz_1\, G_{11mo}(z, z_1) e^{-|z-z_1|/l} G_{11mo}(z_1, z') \quad (5)$$

The solution to (4) can be written as

$$G_{11mo}(z, \xi; z', \xi') = \left[A(\xi)e^{ik_{1m}z} + B(\xi)e^{-ik_{1m}z}\right]$$
$$\times \left[U(\xi')e^{ik_{1m}z'} + W(\xi')e^{-ik_{1m}z'}\right] \quad (6a)$$

for $z > z'$ and

$$G_{11mo}(z,\xi;z',\xi') = \left[A(\xi')e^{ik_{1m}z'} + B(\xi')e^{-ik_{1m}z'}\right] \\ \times \left[U(\xi)e^{ik_{1m}z} + W(\xi)e^{-ik_{1m}z}\right] \tag{6b}$$

for $z < z'$. From (6), it is readily seen that the Green's function is symmetric.

Following the two-variable expansion procedure, (6) is substituted into (5) and secular terms are to be eliminated. The integrals in (5) can be calculated by noting that $A(\xi_1), B(\xi_1), U(\xi_1),$ and $W(\xi_1)$, as variables in the integrand of (5), are varying on the l_a and l_s scales. The l_a scale is the absorption scale defined by $l_a = 2\pi/\kappa_a$. The l_s scale is the scattering scale defined by $l_s = 2\pi/\kappa_s$. They vary much more slowly than $\exp(-|z - z_1|/l)$, which is 1 at $z = z_1$ and decays on the correlation length l scale. These terms can be taken out of the integral sign and substituted with their values at $\xi_1 = \xi$. We find that to eliminate the secular terms, the following equations must be satisfied

$$\frac{dA(\xi)}{d\xi} + g A(\xi) \left[B(\xi) U(\xi) \left(\frac{2 + i3k_{1m}l}{1 + i2k_{1m}l} \right) \right. \\ \left. + A(\xi) W(\xi) \left(\frac{1 - ik_{1m}l}{1 - i2k_{1m}l} \right) \right] = 0 \tag{7a}$$

$$\frac{dB(\xi)}{d\xi} - g B(\xi) \left[A(\xi) W(\xi) \left(\frac{2 - i3k_{1m}l}{1 - i2k_{1m}l} \right) \right. \\ \left. + B(\xi) U(\xi) \left(\frac{1 + ik_{1m}l}{1 + i2k_{1m}l} \right) \right] = 0 \tag{7b}$$

$$\frac{dU(\xi)}{d\xi} + g U(\xi) \left[B(\xi) U(\xi) \left(\frac{1 + ik_{1m}l}{1 + i2k_{1m}l} \right) \right. \\ \left. + A(\xi) W(\xi) \left(\frac{2 - i3k_{1m}l}{1 - i2k_{1m}l} \right) \right] = 0 \tag{7c}$$

$$\frac{dW(\xi)}{d\xi} - g W(\xi) \left[A(\xi) W(\xi) \left(\frac{1 - ik_{1m}l}{1 - i2k_{1m}l} \right) \right. \\ \left. + B(\xi) U(\xi) \left(\frac{2 + i3k_{1m}l}{1 + i2k_{1m}l} \right) \right] = 0 \tag{7d}$$

where

$$g = ik_{1m}^3 l \tag{8}$$

Combining (7a) with (7d), and (7b) with (7c), we obtain $d(A(\xi)W(\xi))/d\xi = 0$ and $d(B(\xi)U(\xi))/d\xi = 0$. Hence, $A(\xi)W(\xi) = L_1$ and $B(\xi)U(\xi) = L_2$, where L_1 and L_2 are constants independent of ξ. Substituting back into (7), we can solve for $A(\xi), B(\xi), U(\xi),$ and $W(\xi)$. Hence

$$A(\xi) = C_1 e^{-g(M_2+M_1)\xi} = \frac{L_1}{W(\xi)} \tag{9a}$$

$$B(\xi) = C_2 e^{g(N_1+N_2)\xi} = \frac{L_2}{U(\xi)} \tag{9b}$$

where

$$M_1 = L_1 \left(\frac{1 - ik_{1m}l}{1 - 2ik_{1m}l}\right) \tag{10a}$$

$$M_2 = L_2 \left(\frac{2 + i3k_{1m}l}{1 + i2k_{1m}l}\right) \tag{10b}$$

$$N_1 = L_1 \left(\frac{2 - i3k_{1m}l}{1 - i2k_{1m}l}\right) \tag{10c}$$

$$N_2 = L_2 \left(\frac{1 + ik_{1m}l}{1 + i2k_{1m}l}\right) \tag{10d}$$

Substituting (9) and (10) in (6) yields

$$G_{11mo}(z,z') = \left\{Ce^{-g(M_2+M_1)\delta z}e^{ik_{1m}z} + e^{g(N_1+N_2)\delta z}e^{-ik_{1m}z}\right\}$$
$$\times \left\{L_2 e^{-g(N_1+N_2)\delta z'}e^{ik_{1m}z'} + \frac{L_1}{C}e^{g(M_2+M_1)\delta z'}e^{-ik_{1m}z'}\right\} \tag{11}$$

for $z > z'$ and the expression for $z < z'$ can be obtained by symmetry. The quantity $C = C_1/C_2$ in (11) is a constant to be determined by the boundary conditions together with L_1 and L_2.

The Green's functions in (11) are continuous at $z = z'$. By (1) we must also have $[dG_{11mo}/dz]_{z=z'_+} - [dG_{11mo}/dz]_{z=z'_-} = 1$ which gives

$$L_1 - L_2 = 1/i2k_{1m} \tag{12}$$

We next impose the boundary condition of the continuity of G_{11mo} and dG_{11mo}/dz at $z = 0$ and at $z = -d$. The continuity at $z = 0$ gives

$$C = 1/R_{10} \tag{13}$$

where R_{10} is the reflection coefficient at $z = 0$, and

$$R_{10} = \frac{k_{1m} - k_o}{k_{1m} + k_o} = -R_{01} \qquad (14)$$

The boundary condition at $z = -d$ gives

$$\frac{CL_2}{L_1} e^{g[N_1 + N_2 + M_1 + M_2]\delta d} = R_{12} e^{i2k_{1m}d} \qquad (15)$$

where R_{12} is the reflection coefficient at $z = -d$

$$R_{12} = \frac{k_{1m} - k_2}{k_{1m} + k_2} \qquad (16)$$

The phase factors appearing in (11) may be written in terms of effective propagation constants η_1 and η_2 defined as:

$$\eta_1 \equiv \eta_1' + i\eta_1'' = k_{1m}[1 - k_{1m}^2 \delta l(N_1 + N_2)] \qquad (17a)$$

$$\eta_2 \equiv \eta_2' + i\eta_2'' = k_{1m}[1 - k_{1m}^2 \delta l(M_1 + M_2)] \qquad (17b)$$

Using (12) and (15), the unknowns L_1 and L_2 can be solved

$$L_1 = \frac{1}{i2k_{1m}D} \qquad (18a)$$

$$L_2 = \frac{R_{10} R_{12} e^{i(\eta_1 + \eta_2)d}}{i2k_{1m}D} \qquad (18b)$$

where

$$D = 1 + R_{01} R_{12} e^{i(\eta_1 + \eta_2)d} \qquad (19)$$

Combining (10), (17), and (18) gives the governing equations for the effective propagation constants η_1 and η_2

$$\eta_1 - \eta_2 = \frac{i}{2} \delta l k_{1m}^2 \qquad (20a)$$

$$\eta_1 + \eta_2 = 2k_{1m} + \frac{i}{2} k_{1m}^2 \frac{\delta l}{D} \left\{ \frac{3 - i4k_{1m}l}{1 - i2k_{1m}l} \right.$$

$$\left. + R_{10} R_{12} \left(\frac{3 + i4k_{1m}l}{1 + i2k_{1m}l} \right) e^{i(\eta_1 + \eta_2)d} \right\} \qquad (20b)$$

Mean Green's Function and Mean Field

Thus, η_1 and η_2 are first solved by (20). Then L_1 and L_2 are calculated by (18).

In terms of η_1 and η_2, the mean Green's function from (11) is as follows

$$G_{11mo}(z,z') = L_1(e^{i\eta_2 z} + R_{10} e^{-i\eta_1 z})(R_{12} e^{i(\eta_1+\eta_2)d+i\eta_1 z'} + e^{-i\eta_2 z'}) \quad (21)$$

for $z > z'$ and the expression for $z < z'$ is obtained by symmetry. The zeroth-order mean field can be readily calculated from the mean Green's function and is given by

$$E_{1m}(z) = E_u e^{i\eta_1 z} + E_d e^{-i\eta_2 z} \quad (22)$$

where the coefficients E_u and E_d follow by matching boundary conditions to zeroth-order at $z = 0$ and $z = -d$. The result is

$$E_{1m}(z) = i2k_{1m}L_1 X_{01}\left(e^{-i\eta_2 z} + R_{12} e^{i(\eta_1+\eta_2)d+i\eta_1 z}\right) \quad (23)$$

with $X_{01} = 1 + R_{01}$.

For the case of three-dimensional variations with $\epsilon_1(\bar{r}) = <\epsilon_1> + \epsilon_{1f}(\bar{r})$, the coherent dyadic Green's function satisfies Dyson's equation, which under the nonlinear approximation takes the form

$$\nabla \times \nabla \times \overline{\overline{G}}_{11m}(\bar{r},\bar{r}_o) - k_{1m}^2 \overline{\overline{G}}_{11m}(\bar{r},\bar{r}_o) = \overline{\overline{I}} \delta(\bar{r}-\bar{r}_o)$$
$$+ \int_{V_1} d^3 r_2 \overline{\overline{G}}_{11m}(\bar{r},\bar{r}_2) \cdot \overline{\overline{G}}_{11m}(\bar{r}_2,\bar{r}_o) C(\bar{r}-\bar{r}_2) \quad (24)$$

where $C(\bar{r}-\bar{r}_1)$ is the two-point correlation function, and the spatial integration extends over the random medium in region 1. The zeroth-order solution of the mean Green's function for a two-layer medium (Figure 5.27) is calculated by the two-variable expansion method (Problem 20; Zuniga and Kong, 1981) and is given by

$$\overline{\overline{G}}_{11m0}(\bar{r},\bar{r}_o) = \int d^2 \bar{k}_\perp \left\{\left[A_1 e^{i\eta_h z}\hat{e}(k_{1mz}) + A_2 e^{-i\eta_h z}\hat{e}(-k_{1mz})\right]\right.$$
$$\times \left[B_1 e^{i\eta_h z_o}\hat{e}(-k_{1mz}) + e^{-i\eta_h z_o}\hat{e}(k_{1mz})\right]$$
$$+ \left[C_1 e^{i\eta_v z}\hat{h}(k_{1mz}) + C_2 e^{-i\eta_v z}\hat{h}(-k_{1mz})\right]$$
$$\left.\times \left[D_1 e^{i\eta_v z_o}\hat{h}(-k_{1mz}) + e^{-i\eta_v z_o}\hat{h}(k_{1mz})\right]\right\} e^{i\bar{k}_\perp \cdot (\bar{r}_\perp - \bar{r}_{o\perp})} \quad (25)$$

for $z > z_o$. In (25) $k_{1mz} = (k_{1m}^2 - k_\perp^2)^{1/2}$. The mean Green's dyadic obeys the reciprocity symmetry relation $\overline{\overline{G}}_{11m}(\bar{r},\bar{r}_o) = \overline{\overline{G}}_{11m}^t(\bar{r}_o,\bar{r})$, where t stands for transpose of the dyad. Thus, the representation for $\overline{\overline{G}}_{11m0}(\bar{r},\bar{r}_o)$ with $z < z_o$ can be obtained from (25) by using the symmetry relation. In (25)

$$A_1 = \sigma/D_2(\bar{k}_\perp) \tag{26}$$

$$A_2 = \sigma R_{10}(\bar{k}_\perp)/D_2(\bar{k}_\perp) \tag{27}$$

$$B_1 = R_{12}(\bar{k}_\perp)\, e^{i2\eta_h d_1}, \tag{28}$$

$$C_1 = \sigma/F_2(\bar{k}_\perp) \tag{29}$$

$$C_2 = \sigma S_{10}(\bar{k}_\perp)/F_2(\bar{k}_\perp) \tag{30}$$

$$D_1 = S_{12}(\bar{k}_\perp)\, e^{i2\eta_v d_1} \tag{31}$$

where

$$\sigma = -[i8\pi^2 k_{1mz}]^{-1} \tag{32}$$

$$D_2(\bar{k}_\perp) = 1 + R_{01}(\bar{k}_\perp)R_{12}(\bar{k}_\perp)\, e^{i2\eta_h d_1} \tag{33}$$

$$F_2(\bar{k}_\perp) = 1 + S_{01}(\bar{k}_\perp)S_{12}(\bar{k}_\perp)\, e^{i2\eta_v d_1} \tag{34}$$

and R_{ij} and S_{ij} are the reflection coefficients for TE and TM polarizations with $i,j = 0,1,2$, with k_{1z} and ϵ_1 to be interpreted as k_{1mz} and ϵ_{1m}, respectively. The transmission coefficients are $X_{01} = 1 + R_{01}$ and $Y_{01} = 1 + S_{01}$.

It is to be noted in (25) that distinctive effective propagation constants were obtained for TE and TM polarizations and are denoted by η_h and η_v, respectively. Physically, this follows from the distinct permittivity correlations in the vertical and lateral directions. Consequently, the rates of scattering for TE and TM polarized waves are, in general, different and must be taken into account in the derivation of a transport equation for electromagnetic waves propagating in random media. The expressions for η_h and η_v are listed in the Appendix (Section 5.5). The effective propagation constants η_h and η_v are also functions of $\bar{k}_\perp$ indicating that they are angular dependent. The zeroth-order mean field propagating within the random layer due to an incident plane wave, $\overline{E}_{oi}$, of unit amplitude can be obtained from the mean

Green's function and takes the form (Problem 20)

$$\overline{E}_{1m0}(\bar{r}) = \left[E_{hui}\, e^{i\eta_{hi}z}\hat{e}(k_{1mzi}) + E_{hdi}\, e^{-i\eta_{hi}z}\hat{e}(-k_{1mzi}) \right] e^{i\bar{k}_{\perp i}\cdot\bar{r}_\perp}$$
$$+ \left[E_{vui}\, e^{i\eta_{vi}z}\hat{h}(k_{1mzi}) + E_{vdi}\, e^{-i\eta_{vi}z}\hat{h}(-k_{1mzi}) \right] e^{i\bar{k}_{\perp i}\cdot\bar{r}_\perp} \quad (35)$$

where the subscript i denotes that a quantity is to be evaluated at the incident wave direction. In (35)

$$E_{hui} = f_e(X_{01i}/D_{2i})\, R_{12i}\, e^{i2\eta_{hi}d_1} \tag{36a}$$

$$E_{hdi} = f_e(X_{01i}/D_{2i}) \tag{36b}$$

$$E_{vui} = f_m(k_o/k_{1m})(Y_{01i}/F_{2i})\, S_{12i}\, e^{i2\eta_{vi}d_1} \tag{36c}$$

$$E_{vdi} = f_m(k_o/k_{1m})(Y_{01i}/F_{2i}) \tag{36d}$$

where f_e and f_m denote the fraction of TE and TM components in the incident wave.

5.3 Modified Radiative Transfer Equations

By writing the field as a summation of coherent and incoherent fields as in (2), Section 5.1, the Bethe-Salpeter equation for the dyadic field covariance, in the ladder approximation is

$$<\overline{\mathcal{E}}_1(\bar{r})\overline{\mathcal{E}}_1^*(\bar{r}')> = \int_{V_1} d^3r_1 \int_{V_1} d^3r_2\, C(\bar{r}_1 - \bar{r}_2)$$
$$\times \left[\overline{\overline{G}}_{11m}(\bar{r},\bar{r}_1) \cdot \overline{E}_{1m}(\bar{r}_1) \cdot \overline{\overline{G}}_{11m}^*(\bar{r}',\bar{r}_2) \cdot \overline{E}_{1m}^*(\bar{r}_2) \right. \tag{1}$$
$$\left. + <\overline{\overline{G}}_{11m}(\bar{r},\bar{r}_1) \cdot \overline{\mathcal{E}}_1(\bar{r}_1) \cdot \overline{\overline{G}}_{11m}^*(\bar{r}',\bar{r}_2) \cdot \overline{\mathcal{E}}_1^*(\bar{r}_2)> \right]$$

We further assume that the correlation function is rotationally symmetric in the transverse x-y plane.

The incoherent field is decomposed into a spectrum of upward- and downward-going waves as in (3) through (7), Section 5.1. The correlation and cross-correlation of TE, TM, and upward- and downward-going waves are as in (8) through (11), Section 5.1. By substituting (7) through (11), Section 5.1, into (1), the procedure of the derivation of MRT equations is similar to that in Section 3.5 (Problem 21). There are two distance scales in the problem. The large distance scales correspond to the coherent wave decay distance and are $l_v = (\eta_v'')^{-1}$ and $l_h = (\eta_h'')^{-1}$. The short distance scales are wavelength $\lambda_{1m} = 2\pi/k_{1m}'$ and correlation lengths l_ρ and l_z. On substitution of (3) through (11), Section 5.1, into (1) and integrating, many terms can be discarded in view of the fact that $(l_v, l_h) \gg (\lambda_{1m}, l_\rho, l_z)$ and the correlations J_{jku}, J_{jkd}, J_{jkc1}, and J_{jkc2} $(j,k = v,h)$ are varying on the (l_v, l_h) scale. The phases of the terms in (1) are of the form $\exp[i\beta_{1mz}'(z-z')]$, $\exp[-i\beta_{1mz}'(z-z')]$, $\exp[i\beta_{1mz}'(z+z')]$ and $\exp[-i\beta_{1mz}'(z+z')]$. The phase factors $\exp[i\beta_{1mz}'(z-z')]$ and $\exp[-i\beta_{1mz}'(z-z')]$ correspond respectively to correlation of upward- and downward-going waves whereas the phase factors $\exp[i\beta_{1mz}'(z+z')]$ and $\exp[-i\beta_{1mz}'(z+z')]$ represent cross-correlation of upward and downward waves with the same transverse wave-vector. By balancing terms with the same phase factor, the modified radiative transfer equations are obtained (Zuniga and Kong, 1980). The final results are given below. Let $\overline{I}_u$ and $\overline{I}_d$ be 4×1 column matrices representing the four Stokes parameters of upward and downward waves, respectively. They are related to the correlations of (8) through (11), Section 5.1, as follows

$$\overline{I}_{\substack{u\\d}}(z, \overline{k}_\perp) = \begin{bmatrix} J_{hh\substack{u\\d}}(z,z,\overline{k}_\perp) \\ J_{vv\substack{u\\d}}(z,z,\overline{k}_\perp) \\ 2Re[J_{vh\substack{u\\d}}(z,z,\overline{k}_\perp)] \\ 2Im[J_{vh\substack{u\\d}}(z,z,\overline{k}_\perp)] \end{bmatrix} \qquad (2)$$

Since J_{jku} and J_{jkd} are correlations of incoherent waves $\overline{\mathcal{E}}$, the Stokes vectors $\overline{I}_u$ and $\overline{I}_d$ are those of the incoherent waves. The Stokes vectors of the coherent waves are denoted by $\overline{I}_{mu}$ and $\overline{I}_{md}$, respectively, for upward- and downward-going waves. Furthermore, there are constructive interferences in the directions that are specularly related to the incident direction giving Stokes vectors $\overline{I}_{mc1}$ and $\overline{I}_{mc2}$. The quantities $\overline{I}_{mu}$, $\overline{I}_{md}$, $\overline{I}_{mc1}$, and $\overline{I}_{mc2}$ are given in Appendix 5.5. The MRT equations in matrix form are

$$|\beta_{1mz}|^2 \frac{d}{dz}\overline{I}_u(z, \overline{\beta}_\perp) = -|\beta_{1mz}|^2 \overline{\overline{\eta}}(\overline{\beta}_\perp) \cdot \overline{I}_u(z, \overline{\beta}_\perp)$$
$$+ \overline{\overline{Q}}_{uu}(\overline{\beta}_\perp, \overline{k}_{\perp i}) \cdot \overline{I}_{mu}(z, \overline{k}_{\perp i})$$
$$+ \overline{\overline{Q}}_{ud}(\overline{\beta}_\perp, \overline{k}_{\perp i}) \cdot \overline{I}_{md}(z, \overline{k}_{\perp i})$$
$$+ \Delta_1 \overline{\overline{Q}}_{cl}(\overline{\beta}_\perp, \overline{k}_{\perp i}) \cdot \overline{I}_{mc1}(z, \overline{k}_{\perp i})$$

$$+ \int d^2\overline{k}_\perp \left[\overline{\overline{P}}_{uu}(\overline{\beta}_\perp, \overline{k}_\perp) \cdot \overline{I}_u(z, \overline{k}_\perp) \right.$$
$$\left. + \overline{\overline{P}}_{ud}(\overline{\beta}_\perp, \overline{k}_\perp) \cdot \overline{I}_d(z, \overline{k}_\perp) \right] \qquad (3)$$

$$-|\beta_{1mz}|^2 \frac{d}{dz}\overline{I}_d(z, \overline{\beta}_\perp) = -|\beta_{1mz}|^2 \overline{\overline{\eta}}(\overline{\beta}_\perp) \cdot \overline{I}_d(z, \overline{\beta}_\perp)$$
$$+ \overline{\overline{Q}}_{dd}(\overline{\beta}_\perp, \overline{k}_{\perp i}) \cdot \overline{I}_{md}(z, \overline{k}_{\perp i})$$
$$+ \overline{\overline{Q}}_{du}(\overline{\beta}_\perp, \overline{k}_{\perp i}) \cdot \overline{I}_{mu}(z, \overline{k}_{\perp i})$$
$$- \Delta_1 \overline{\overline{Q}}_{cl}(\overline{\beta}_\perp, \overline{k}_{\perp i}) \cdot \overline{I}_{mc2}(z, \overline{k}_{\perp i})$$
$$+ \int d^2\overline{k}_\perp \left[\overline{\overline{P}}_{dd}(\overline{\beta}_\perp, \overline{k}_\perp) \cdot \overline{I}_d(z, \overline{k}_\perp) \right.$$
$$\left. + \overline{\overline{P}}_{du}(\overline{\beta}_\perp, \overline{k}_\perp) \cdot \overline{I}_u(z, \overline{k}_\perp) \right] \qquad (4)$$

where $\Delta_1 = 1$ if $\overline{\beta}_\perp = \pm \overline{k}_{\perp i}$ and is zero otherwise. It is clear that these terms are of the constructive interference type only for $\overline{\beta}_\perp = \pm \overline{k}_{\perp i}$, when $\overline{\beta}_\perp$ coincides with the incident direction. The $\overline{\overline{Q}}$ and $\overline{\overline{P}}$ matrices appearing in (3) and (4) are scattering matrices for the mean and incoherent field intensities, respectively, and are defined in the Appendix (Section 5.5).

In (3) and (4) $\overline{\overline{\eta}}$ is the extinction matrix and is

$$\overline{\overline{\eta}}(\overline{\beta}_\perp) = \begin{bmatrix} 2\eta_h''(\overline{\beta}_\perp) & 0 & 0 & 0 \\ 0 & 2\eta_v''(\overline{\beta}_\perp) & 0 & 0 \\ 0 & 0 & (\eta_v'' + \eta_h'') & (\eta_v' - \eta_h') \\ 0 & 0 & -(\eta_v' - \eta_h') & (\eta_v'' + \eta_h'') \end{bmatrix} \qquad (5)$$

where superscript $'$ and $''$ denote real and imaginary parts respectively.

The boundary conditions at $z = 0$ and $z = -d_1$ for the MRT equations of (3) and (4) follow from the ladder-approximated Bethe-Salpeter equation. They are

$$\overline{I}_d(0, \overline{\beta}_\perp) = \overline{\overline{R}}_{10}(\overline{\beta}_\perp) \cdot \overline{I}_u(0, \overline{\beta}_\perp) \qquad (6a)$$

$$\overline{I}_u(-d_1, \overline{\beta}_\perp) = \overline{\overline{R}}_{12}(\overline{\beta}_\perp) \cdot \overline{I}_d(-d_1, \overline{\beta}_\perp) \qquad (6b)$$

where

$$\bar{\bar{R}}_{10}(\beta_\perp) = \begin{bmatrix} |R_{10}|^2 & 0 & 0 & 0 \\ 0 & |S_{10}|^2 & 0 & 0 \\ 0 & 0 & Re(R_{10}^* S_{10}) & -Im(R_{10}^* S_{10}) \\ 0 & 0 & Im(R_{10}^* S_{10}) & Re(R_{10}^* S_{10}) \end{bmatrix} \quad (7a)$$

$$\bar{\bar{R}}_{12}(\beta_\perp) = \begin{bmatrix} |R_{12}|^2 & 0 & 0 & 0 \\ 0 & |S_{12}|^2 & 0 & 0 \\ 0 & 0 & Re(R_{12}^* S_{12}) & -Im(R_{12}^* S_{12}) \\ 0 & 0 & Im(R_{12}^* S_{12}) & Re(R_{12}^* S_{12}) \end{bmatrix} \quad (7b)$$

The transmitted intensity $\bar{I}_{ou}$ from region 1 to region 0 is related to $\bar{I}_u$ of region 1 by

$$\bar{I}_{ou}(0, \bar{\beta}_\perp) = \bar{\bar{T}}_{10}(\bar{\beta}_\perp) \cdot \bar{I}_u(0, \bar{\beta}_\perp) \quad (8)$$

where

$$\bar{\bar{T}}_{10}(\beta_\perp) = \begin{bmatrix} |X_{10}|^2 & 0 & 0 & 0 \\ 0 & \left|\frac{\eta_o}{\eta_1} Y_{10}\right|^2 & 0 & 0 \\ 0 & 0 & Re(\frac{\eta_o}{\eta_1} Y_{10} X_{10}^*) & -Im(\frac{\eta_o}{\eta_1} Y_{10} X_{10}^*) \\ 0 & 0 & Im(\frac{\eta_o}{\eta_1} Y_{10} X_{10}^*) & Re(\frac{\eta_o}{\eta_1} Y_{10} X_{10}^*) \end{bmatrix} \quad (9)$$

and

$$\bar{I}_{ou}(0, \bar{\beta}_\perp) = \begin{bmatrix} <\mathcal{E}_{ohu}(\bar{\beta}_\perp, 0) \mathcal{E}_{ohu}^*(\bar{\beta}_\perp, 0)> \\ <\mathcal{E}_{ovu}(\bar{\beta}_\perp, 0) \mathcal{E}_{ovu}^*(\bar{\beta}_\perp, 0)> \\ 2Re <\mathcal{E}_{ovu}(\bar{\beta}_\perp, 0) \mathcal{E}_{ohu}^*(\bar{\beta}_\perp, 0)> \\ 2Im <\mathcal{E}_{ovu}(\bar{\beta}_\perp, 0) \mathcal{E}_{ohu}^*(\bar{\beta}_\perp, 0)> \end{bmatrix} \quad (10)$$

is the incoherent Stokes vector in Region 0. Equations (3) and (4), together with boundary conditions (6) and (8), form the complete set of modified radiative transfer equations for a two-layer random medium.

From the form of the MRT equations in (3), (4), (6), and (8), we note that they are entirely analogous to that of RT equations in Chapter 3. The significant difference is the existence of the $\bar{\bar{Q}}_{cl}$ matrix accounting for the correlations between specularly related upward and downward waves. The

MRT equations can also be converted to the standard form of radiative transfer equations by using solid angles and specific intensities (Problem 22).

5.4 Numerical Illustrations

As an application of the modified radiative transfer equations and to illustrate the significance of the $\overline{\overline{Q}}_{\mu\nu}$ matrix (particularly the term with factor Δ_1), we compute the backscattering cross section for a two-layer random medium by applying the first-order renormalization method. The method is equivalent to ignoring the scattering of incoherent intensity represented by the integral terms in (3) and (4). The resulting equations can be readily solved since the Stokes vectors $\overline{I}_{mu}$, $\overline{I}_{md}$, $\overline{I}_{mc1}$ and $\overline{I}_{mc2}$ corresponding to that of the coherent wave are known from Dyson's equation and are listed in the Appendix. The backscattering cross sections per unit area follow from the relation

$$\begin{pmatrix} \sigma_{hh} \\ \sigma_{vv} \end{pmatrix} = 4\pi k_o^2 \cos^2 \theta_{oi} \begin{pmatrix} I_{ouh} \\ I_{ouv} \end{pmatrix} \tag{1}$$

with $f_e = 1$ in I_{ouh} and $f_m = 1$ in I_{ouv}.

We consider a correlation function of the form of (6), Section 2.2. The backscattering cross sections per unit area for a two-layer random medium in the first-order renormalization (Problem 23) are:

$$\sigma_{hh} = \frac{\delta k_{1m}^{\prime 4} l_\rho^2 l_z}{4} \frac{|X_{10i}|^4}{|D_{2i}|^4} \frac{|k_{ozi}|^4}{|k_{1zi}|^4} e^{-k_o^2 l_\rho^2 \sin^2 \theta_{oi}}$$

$$\times \left\{ \frac{(1 - e^{-4\eta_{hi}^{\prime\prime} d_1})}{2\eta_{hi}^{\prime\prime}(1 + 4k_{1mzi}^{\prime 2} l_z^2)} (1 + |R_{12i}|^4 e^{-4\eta_{hi}^{\prime\prime} d_1}) \right.$$

$$\left. + 8d_1 |R_{12i}|^2 e^{-4\eta_{hi}^{\prime\prime} d_1} \right\} \tag{2a}$$

$$\sigma_{vv} = \frac{\delta k_{1m}^{\prime 4} l_\rho^2 l_z}{4} \frac{|Y_{10i}|^4}{|F_{2i}|^4} \frac{|k_{ozi}|^4}{|k_{1zi}|^4} e^{-k_o^2 l_\rho^2 \sin^2 \theta_{oi}}$$

$$\times \left\{ \frac{(1 - e^{-4\eta_{vi}^{\prime\prime} d_1})}{2\eta_{vi}^{\prime\prime}(1 + 4k_{1mzi}^{\prime 2} l_z^2)} (1 + |S_{12i}|^4 e^{-4\eta_{vi}^{\prime\prime} d_1}) \right.$$

$$\times \left| \frac{k_{1mzi}^2}{k_o^2} + \sin^2 \theta_{oi} \right|^2 + 8d_1 |S_{12i}|^2 e^{-4\eta_{vi}^{\prime\prime} d_1}$$

$$\left. \times \left| \frac{k_{1mzi}^2}{k_o^2} - \sin^2 \theta_{oi} \right|^2 \right\} \tag{2b}$$

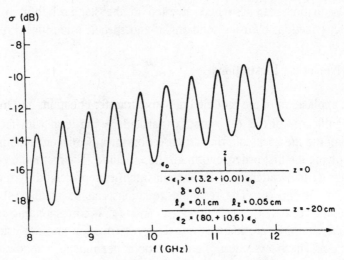

Fig. 5.28 $\sigma_{vv} = \sigma_{hh}$ from MRT theory as a function of frequency at nadir.

To illustrate the backscattering cross sections of (2) obtained from MRT theory, we plot in Figure 5.28 $\sigma_{vv} = \sigma_{hh}$ at nadir as a function of frequency for a 20 cm thick random layer. The coherent wave-like effects of MRT theory are apparent from the oscillatory behavior exhibited by the spectral dependence of σ_{hh} and σ_{vv}. However, for high-scattering media, the bottom boundary becomes shielded thus reducing the coherent behavior in backscattering. This is demonstrated in Figure 5.29 where $\sigma_{hh} = \sigma_{vv}$ is plotted at nadir as a function of frequency for the same parameters as Figure 5.28 but with increased δ.

A particularly significant coherent effect included in MRT theory arises from the Δ_1 terms in MRT equation. As discussed in Section 5.3, these constructive interference terms contribute only for $\overline{\beta}_\perp = \pm \overline{k}_{\perp i}$ (i.e., for forward or backward scattering). Physically, this is illustrated schematically in Figure 5.30 where we have indicated the propagation directions of the two scattered waves by dashed and solid lines for the cases of forward and backward scattering. Clearly, in each case the two waves interfere constructively at the far-field observation point, producing a significant contribution to the scattered intensity. Constructive interference terms of this type are not included in radiative transfer theories. Therefore, to gauge the error produced by omitting these constructive interference terms, we set Δ_1 equal to zero and re-derive the backscattering cross sections σ_{hh} and σ_{vv}. The results are similar to (2) except that the second terms within the curly brackets are reduced by a factor of 2. In the case of thin, low-loss layers, the additional contributions of the Δ_1 terms can be significant. However, for thick lossy layers, the error produced by omitting the Δ_1 terms is minimal.

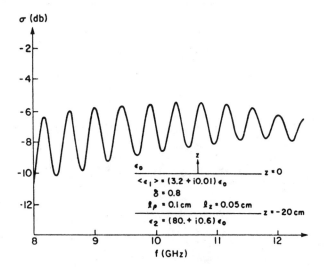

Fig. 5.29 $\sigma_{vv} = \sigma_{hh}$ from MRT theory as a function of frequency at nadir. Here the parameter values are identical to those of Figure 5.28 except $\delta = 0.8$.

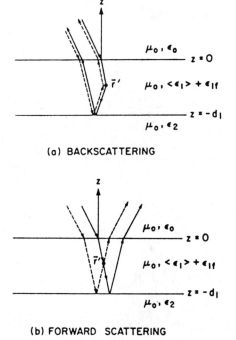

Fig. 5.30 Constructive interference path lengths for the mean field in the cases of backscattering and forward scattering.

The first- and second-order renormalization methods for the scalar theory have also been used to calculate scattering from half-space random medium (Fung and Fung 1977; Fung, 1979; Tan et al., 1980).

5.5 Appendix

The TE and TM effective propagation constants for a three-dimensional two-layer random medium in Section 5.2 are

$$\eta_h(\overline{k}_\perp) = k_{1mz} + \lambda^{TE}(\overline{k}_\perp) \tag{1}$$

$$\eta_v(\overline{k}_\perp) = k_{1mz} + \lambda^{TM}(\overline{k}_\perp) \tag{2}$$

where

$$\lambda^{TE}(\overline{k}_\perp) = \frac{\pi \delta k_{1m}'^4}{i k_{1mz}} \sum_n \int d\overline{k}_\perp' \, \mathrm{Res}\Phi(\overline{k}_\perp' - \overline{k}_\perp, \alpha_n^+)$$

$$\times \left\{ \frac{\cos^2(\phi - \phi')}{k_{1mz}' D_2(\overline{k}_\perp')} \left(\frac{(k_{1mz}' + \alpha_n^+)}{(k_{1mz}' + \alpha_n^+)^2 - k_{1mz}^2} \right. \right.$$

$$\left. - \frac{R_{10}(\overline{k}_\perp') R_{12}(\overline{k}_\perp') e^{i2\eta_h(\overline{k}_\perp')d_1}(k_{1mz}' - \alpha_n^+)}{(k_{1mz}' - \alpha_n^+)^2 - k_{1mz}^2} \right)$$

$$+ \frac{k_{1mz}' \sin^2(\phi - \phi')}{k_{1m}^2 F_2(\overline{k}_\perp')} \left(\frac{(k_{1mz}' + \alpha_n^+)}{(k_{1mz}' + \alpha_n^+)^2 - k_{1mz}^2} \right.$$

$$\left. \left. - \frac{S_{10}(\overline{k}_\perp') S_{12}(\overline{k}_\perp') e^{i2\eta_v(\overline{k}_\perp')d_1}(k_{1mz}' - \alpha_n^+)}{(k_{1mz}' - \alpha_n^+)^2 - k_{1mz}^2} \right) \right\}$$

$$+ \frac{1}{3} \frac{\delta k_{1m}^2}{k_{1mz}} \tag{3}$$

$$\lambda^{TM}(\overline{k}_\perp) = \frac{\pi \delta k_{1m}'^4}{i k_{1mz}} \sum_n \int d\overline{k}_\perp' \, \mathrm{Res}\Phi(\overline{k}_\perp' - \overline{k}_\perp, \alpha_n^+)$$

$$\times \left\{ \frac{k_{1mz}^2 \sin^2(\phi - \phi')}{k_{1m}^2 k_{1mz}' D_2(\overline{k}_\perp')} \left(\frac{(k_{1mz}' + \alpha_n^+)}{(k_{1mz}' + \alpha_n^+)^2 - k_{1mz}^2} \right. \right.$$

$$\left. - \frac{R_{10}(\overline{k}_\perp') R_{12}(\overline{k}_\perp') e^{i2\eta_h(\overline{k}_\perp')d_1}(k_{1mz}' - \alpha_n^+)}{(k_{1mz}' - \alpha_n^+)^2 - k_{1mz}^2} \right)$$

Appendix

$$+ \frac{1}{k'_{1mz}k^4_{1m}F_2(\overline{k}'_\perp)}$$

$$\times \left(\frac{(k'^2_{1mz}k^2_{1mz}\cos^2(\phi-\phi') + k^2_{1m}k'^2_\perp)(k'_{1mz}+\alpha^+_n)}{(k'_{1mz}+\alpha^+_n)^2 - k^2_{1mz}} \right.$$

$$+ \left. \frac{2k^2_{1mz}k'_{1mz}k_\perp k'_\perp \cos(\phi-\phi')}{(k'_{1mz}+\alpha^+_n)^2 - k^2_{1mz}} \right)$$

$$+ \frac{S_{10}(\overline{k}'_\perp)S_{12}(\overline{k}'_\perp)e^{i2\eta_v(\overline{k}'_\perp)d_1}}{k'_{1mz}k^4_{1m}F_2(\overline{k}'_\perp)}$$

$$\times \left(\frac{2k^2_{1mz}k'_{1mz}k_\perp k'_\perp \cos(\phi-\phi')}{(k'_{1mz}-\alpha^+_n) - k^2_{1mz}} \right.$$

$$\left. \left. - \frac{(k'_{1mz}-\alpha^+_n)(k'^2_{1mz}k^2_{1mz}\cos(\phi-\phi') + k^2_\perp k'^2_\perp)}{(k'_{1mz}-\alpha^+_n) - k^2_{1mz}} \right) \right\}$$

$$+ \frac{1}{3}\frac{\delta k^2_{1m}}{k_{1mz}} \tag{4}$$

where $(\phi - \phi')$ is the angle between $\overline{k}_\perp$ and $\overline{k}'_\perp$, δ is the variance of the permittivity fluctuations, and the summation $\sum_n$ extends over the poles, α^+_n of $\Phi(\overline{\alpha}_\perp, \alpha_z)$ in the upper complex α_z plane, and Φ is the Fourier transform of the two-point correlation function C as defined by (16), Section 2.1.

In Section 5.3, the scattering matrix $\overline{\overline{Q}}$ and coherent Stokes vector $\overline{I}_m$ are

$$\overline{\overline{Q}}_{uu}(\overline{\beta}_\perp, \overline{k}_{\perp i}) = \frac{\pi}{2}\delta k'^4_{1m}\Phi(\overline{k}_{\perp i} - \overline{\beta}_\perp, \beta_{1mz} - k_{1mzi})$$

$$\times \begin{bmatrix} \alpha_h[\hat{e}(\beta_{1mz})\cdot\hat{e}(k_{1mzi})]^2 & \alpha_h[\hat{e}(\beta_{1mz})\cdot\hat{h}(k_{1mzi})]^2 & 0 & 0 \\ \alpha_v[\hat{h}(\beta_{1mz})\cdot\hat{e}(k_{1mzi})]^2 & \alpha_v[\hat{h}(\beta_{1mz})\cdot\hat{h}(k_{1mzi})]^2 & 0 & 0 \\ 0 & 0 & 0 & 0 \\ 0 & 0 & 0 & 0 \end{bmatrix} \tag{5}$$

$$\overline{\overline{Q}}_{ud}(\overline{\beta}_\perp, \overline{k}_{\perp i}) = \text{same as } \overline{\overline{Q}}_{uu} \quad \text{except let } k_{1mzi} \to -k_{1mzi}, \tag{6}$$

$$\overline{\overline{Q}}_{dd}(\overline{\beta}_\perp, \overline{k}_{\perp i}) = \text{same as } \overline{\overline{Q}}_{uu} \quad \text{except let } \beta_{1mz} \to -\beta_{1mz} \text{ and}$$
$$k_{1mzi} \to -k_{1mzi}, \tag{7}$$

$$\overline{\overline{Q}}_{du}(\overline{\beta}_\perp, \overline{k}_{\perp i}) = \text{same as } \overline{\overline{Q}}_{uu} \text{ except let } \beta_{1mz} \to -\beta_{1mz} \tag{8}$$

$$\overline{\overline{Q}}_{c1}(\overline{\beta}_\perp, \overline{k}_{\perp i}) = \frac{\pi}{2} \delta k'^4_{1m} \Phi(\overline{k}_{\perp i} - \overline{\beta}_\perp, 0)$$

$$\times \begin{bmatrix} [\hat{e}(\beta_{1mz}) \cdot \hat{e}(k_{1mzi})] \\ \cdot [\hat{e}(-\beta_{1mz}) \cdot \hat{e}(-k_{1mzi})] & 0 & 0 & 0 \\ 0 & \begin{matrix}[\hat{h}(\beta_{1mz}) \cdot \hat{h}(k_{1mzi})] \\ \cdot [\hat{h}(-\beta_{1mz}) \cdot \hat{h}(-k_{1mzi})]\end{matrix} & 0 & 0 \\ 0 & 0 & 0 & 0 \\ 0 & 0 & 0 & 0 \end{bmatrix} \tag{9}$$

and

$$\overline{I}_{m^u_d}(z, \overline{\beta}_\perp) = \begin{bmatrix} |E_{h^u_d i}|^2 e^{2\eta''_{hi}(\overline{\beta}_\perp)z} \\ |E_{v^u_d i}|^2 e^{2\eta''_{vi}(\overline{\beta}_\perp)z} \\ 0 \\ 0 \end{bmatrix} \tag{10}$$

$$\overline{I}_{mc1}(z, \overline{k}_{\perp i}) = \begin{bmatrix} 2Re(E_{hui}E^*_{hdi}\gamma_{hi})e^{-2\eta''_{hi}z} \\ 2Re(E_{vui}E^*_{vdi}\gamma_{vi})e^{-2\eta''_{vi}z} \\ 0 \\ 0 \end{bmatrix} \tag{11}$$

$$\overline{I}_{mc2}(z, \overline{k}_{\perp i}) = \begin{bmatrix} 2Re(E_{hui}E^*_{hdi}\widetilde{\gamma}_{hi})e^{2\eta''_{hi}z} \\ 2Re(E_{vui}E^*_{vdi}\widetilde{\gamma}_{vi})e^{2\eta''_{vi}z} \\ 0 \\ 0 \end{bmatrix} \tag{12}$$

where

$$\alpha_h = \frac{1 - |R_{10}(\overline{\beta}_\perp)|^2 |R_{12}(\overline{\beta}_\perp)|^2 e^{-4\eta''_h(\overline{\beta}_\perp)d_1}}{|D_2(\overline{\beta}_\perp)|^2} \tag{13}$$

$$\alpha_v = \frac{1 - |S_{10}(\overline{\beta}_\perp)|^2 |S_{12}(\overline{\beta}_\perp)|^2 e^{-4\eta''_h(\overline{\beta}_\perp)d_1}}{|F_2(\overline{\beta}_\perp)|^2} \tag{14}$$

$$\gamma_{hi} = R^*_{12i} e^{-i2\eta^*_{hi}d_1} \frac{(1 + R_{01i}R_{12i}e^{i2\eta_{hi}d_1})}{|D_{2i}|^2} \tag{15}$$

$$\gamma_{vi} = S^*_{12i} e^{-i2\eta^*_{vi}d_1} \frac{(1 + S_{01i}S_{12i}e^{i2\eta_{vi}d_1})}{|F_{2i}|^2} \tag{16}$$

Appendix

$$\tilde{\gamma}_{hi} = -R_{10i}(D_{2i}^*/|D_{2i}|^2) \tag{17}$$

$$\tilde{\gamma}_{vi} = -S_{10i}(F_{2i}^*/|F_{2i}|^2) \tag{18}$$

We also have

$$\overline{\overline{P}}_{uu}(\overline{\beta}_\perp, \overline{k}_\perp) = \frac{\pi}{2}\delta\, k'^4_{1m}\, \Phi(\overline{k}_\perp - \overline{\beta}_\perp, \beta_{1mz} - k_{1mz}) \begin{bmatrix} \overline{\overline{A}} & \overline{\overline{B}} \\ \overline{\overline{C}} & \overline{\overline{D}} \end{bmatrix} \tag{19}$$

where

$$\overline{\overline{A}} = \begin{bmatrix} \alpha_h[\hat{e}(\beta_{1mz})\cdot\hat{e}(k_{1mz})]^2 & \alpha_h[\hat{e}(\beta_{1mz})\cdot\hat{h}(k_{1mz})]^2 \\ \alpha_v[\hat{h}(\beta_{1mz})\cdot\hat{e}(k_{1mz})]^2 & \alpha_v[\hat{h}(\beta_{1mz})\cdot\hat{h}(k_{1mz})]^2 \end{bmatrix} \tag{20}$$

$$\overline{\overline{B}} = \begin{bmatrix} \alpha_h[\hat{e}(\beta_{1mz})\cdot\hat{e}(k_{1mz})][\hat{e}(\beta_{1mz})\cdot\hat{h}(k_{1mz})] & 0 \\ \alpha_v[\hat{h}(\beta_{1mz})\cdot\hat{e}(k_{1mz})][\hat{h}(\beta_{1mz})\cdot\hat{h}(k_{1mz})] & 0 \end{bmatrix} \tag{21}$$

$$\overline{\overline{C}} = \begin{bmatrix} 2\alpha_{c1}[\hat{h}(\beta_{1mz})\cdot\hat{e}(k_{1mz})] & 2\alpha_{c1}[\hat{h}(\beta_{1mz})\cdot\hat{h}(k_{1mz})] \\ \times[\hat{e}(\beta_{1mz})\cdot\hat{e}(k_{1mz})] & \times[\hat{e}(\beta_{1mz})\cdot\hat{h}(k_{1mz})] \\ -2\alpha_{c2}[\hat{h}(\beta_{1mz})\cdot\hat{e}(k_{1mz})] & -2\alpha_{c2}[\hat{h}(\beta_{1mz})\cdot\hat{h}(k_{1mz})] \\ \times[\hat{e}(\beta_{1mz})\cdot\hat{e}(k_{1mz})] & \times[\hat{e}(\beta_{1mz})\cdot\hat{h}(k_{1mz})] \end{bmatrix} \tag{22}$$

$$\overline{\overline{D}} = \begin{bmatrix} D_{11} & D_{12} \\ D_{21} & D_{22} \end{bmatrix} \tag{23}$$

$$D_{11} = \alpha_{c1}\Big\{[\hat{h}(\beta_{1mz})\cdot\hat{h}(k_{1mz})][\hat{e}(\beta_{1mz})\cdot\hat{e}(k_{1mz})] \\ + [\hat{e}(\beta_{1mz})\cdot\hat{h}(k_{1mz})][\hat{h}(\beta_{1mz})\cdot\hat{e}(k_{1mz})]\Big\} \tag{24a}$$

$$D_{12} = \alpha_{c2}\Big\{[\hat{e}(\beta_{1mz})\cdot\hat{e}(k_{1mz})][\hat{h}(\beta_{1mz})\cdot\hat{h}(k_{1mz})] \\ - [\hat{e}(\beta_{1mz})\cdot\hat{h}(k_{1mz})][\hat{h}(\beta_{1mz})\cdot\hat{e}(k_{1mz})]\Big\} \tag{24b}$$

$$D_{21} = -\alpha_{c2}\Big\{[\hat{h}(\beta_{1mz})\cdot\hat{h}(k_{1mz})][\hat{e}(\beta_{1mz})\cdot\hat{e}(k_{1mz})] \\ + [\hat{e}(\beta_{1mz})\cdot\hat{h}(k_{1mz})][\hat{h}(\beta_{1mz})\cdot\hat{e}(k_{1mz})]\Big\} \tag{24c}$$

$$D_{22} = \alpha_{c1}\left\{[\hat{e}(\beta_{1mz}) \cdot \hat{e}(k_{1mz})][\hat{h}(\beta_{1mz}) \cdot \hat{h}(k_{1mz})]\right.$$
$$\left. - [\hat{e}(\beta_{1mz}) \cdot \hat{h}(k_{1mz})][\hat{h}(\beta_{1mz}) \cdot \hat{e}(k_{1mz})]\right\} \quad (24d)$$

and

$$\overline{\overline{P}}_{ud}(\overline{\beta}_\perp, \overline{k}_\perp) = \text{same as } \overline{\overline{P}}_{uu}(\overline{\beta}_\perp, \overline{k}_\perp) \quad (25)$$
$$\text{except let } k_{1mz} \to -k_{1mz},$$

$$\overline{\overline{P}}_{du}(\overline{\beta}_\perp, \overline{k}_\perp) = \text{same as } \overline{\overline{P}}_{uu}(\overline{\beta}_\perp, \overline{k}_\perp) \quad (26)$$
$$\text{except let } \beta_{1mz} \to -\beta_{1mz},$$

$$\overline{\overline{P}}_{dd}(\overline{\beta}_\perp, \overline{k}_\perp) = \text{same as } \overline{\overline{P}}_{uu}(\overline{\beta}_\perp, \overline{k}_\perp) \quad (27)$$
$$\text{except let } \beta_{1mz} \to -\beta_{1mz} \text{ and } k_{1mz} \to -k_{1mz}$$

and α_{c1} and α_{c2} are respectively the real and the imaginary parts of $(1 - R_{10} R_{12} S_{01}^* S_{12}^* e^{i2(\eta_h - \eta_v^*)d_1})/D_2 F_2^*$.

PROBLEMS

5.1 Use the saddle point method to derive the far-field approximation of the dyadic Green's function $\overline{\overline{G}}_{01}^{(0)}$ as given in (6), Section 2.1.

5.2 Calculate the scattered intensity for a half-space of random medium under the Born approximation. Consider the case of vertically polarized incidence. Your answer should be similar to (14), Section 2.1.

5.3 Derive the expression for the backscattering depolarization of a half-space random medium as given in (7) through (11), Section 2.2. Use the far-field approximation of $\overline{\overline{G}}_{01}^{(0)}$. Note that the far-field approximation cannot be used for $\overline{\overline{G}}_{11}^{(0)}$. You should use the integral representation of $\overline{\overline{G}}_{11}^{(0)}$ as given in Section 3, Chapter 2 (Zuniga et al., 1980).

5.4 Use Figure 5.5a to show that the phase difference between E_{12} and E_{21} is $(\overline{k}_i + \overline{k}_s) \cdot \overline{r}_{12}$.

5.5 Derive the backscattering coefficient σ_{vv} for a three-layer random medium under the Born approximation. Use (3), Section 2.4, as your

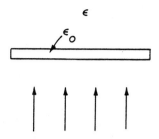

Fig. 5.31 External field applied to a disk-shaped cavity.

starting point with $M = 2$. Your answer should be similar to (6), Section 2.4.

5.6 Use radiative transfer theory to calculate the backscattering cross section of a layer of random medium above a homogeneous half-space. Let $\epsilon_1 = \epsilon_o + \epsilon_{1f}(\bar{r})$ so that the interface at $z = 0$ is not reflective. Use the technique in Chapter 4 to formulate an integral equation for the radiative transfer equations and the boundary conditions. Iterate the integral equations once to calculate the scattering cross section. Use κ_a iteration rather than κ_e iteration (Shin and Kong, 1981). Show that the answer is the same as (8), Section 2.4, except for the factor of 4 instead of 8 as indicated in Section 2.4.

5.7 Consider the bilocal approximation to the scalar wave equation as given by (6), Section 3.2. For a statistically homogeneous space, $< G(\bar{r}_1, \bar{r}_o) >$ is only a function of $|\bar{r}_1 - \bar{r}_o|$. Thus, the second term in (6), Section 3.2, is a convolution integral and it is possible to solve (6), Section 3.2, exactly by using the method of Fourier transformation. Calculate K by this method.

5.8 We shall study the behavior of the field inside a cavity in a polarizable medium with an externally applied electrostatic field. Assume that the applied field is along the $\hat{z}$ direction, i.e., $\overline{E}_a = \hat{z}E_o$ and the polarization vector is $\overline{P}$. Show that

1. If the shape of the cavity is a sphere, the field inside is $\overline{E} = \overline{E}_a + \overline{P}/3\epsilon_o$.

2. If the shape of the cavity is a disk (see Figure 5.31, a limiting case of a ellipsoid perpendicular to the applied field), the field inside is $\overline{E} = \overline{E}_a + \overline{P}/\epsilon_o$.

3. If the shape of the cavity is of a needle shape (a limiting case of

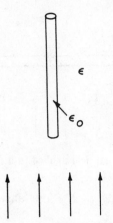

Fig. 5.32 External field applied to needle shaped cavity.

the two axes of an ellipsoid approaching zero), which is aligned in the same direction as the applied field, the field inside the cavity is (see Figure 5.32) $\overline{E} = \overline{E}_a$.

The field inside a cavity is related to the singular behavior of the dyadic Green's function when the observation point is in the source region. To overcome such a difficulty, a *principle value* integral is defined. Thus, the electric field when the observation point is in the source region is given as

$$\overline{E}(r) = i\omega\mu_o \, PS \int \overline{\overline{G}}(r,r') \cdot \overline{J}(r') \, dV' + \frac{\overline{\overline{L}} \cdot \overline{J}(r)}{i\omega\epsilon_o} \qquad (1)$$

where the principle value integral is obtained by performing the volume integration over the source region excluding a small cavity called *exclusion volume* with a certain shape and containing the observation point. The size of the cavity is allowed to shrink to zero eventually. Thus, the first term can be thought of as the contribution to the field at the observation point due to the source outside the exclusion volume. The latter term is the contribution to the field at the observation point due to the source inside the exclusion volume, which can be shown to be non-vanishing even when the exclusion volume shrinks to zero. In this definition, $\overline{E}(r)$ is unique whereas the value of each of the two terms on the right-hand side of the equation depends on the shape of the exclusion volume chosen. This is evident by finding the value of $\overline{\overline{L}}$ for different cavity shapes. Since the exclusion volume is vanishingly small, we can assume that $\overline{J}(r)$ is uniform inside the cavity. Show that

4. If the exclusion volume is chosen to be a sphere, $\overline{\overline{L}} = \overline{\overline{I}}/3$ where $\overline{\overline{I}}$ is the unit dyad.

5. If the exclusion volume is a disk, $\overline{\overline{L}} = \hat{z}\hat{z}$.

6. If the exclusion volume is a long cylinder, $\overline{\overline{L}} = \overline{\overline{I}}^t/2$, where $\overline{\overline{I}}^t$ is a unit dyad transverse to the axis of the cylinder. The correspondence of the above with the dielectric medium cavity problem can be noted if we assume $\overline{J}$ to consist of polarization current $\overline{J}_p$ only. In such a case, $\overline{J}_p = -i\omega \overline{P}$. The field inside the cavity is the principle value integral. It is given by $\overline{E}(r) = \overline{\overline{L}} \cdot \overline{P}/\epsilon_o$.

7. Explain the apparent disagreement for the long cylinder case.

5.9 Show that $PS\overline{\overline{G}}^{(0)}(\bar{r})$ is given by (14) through (16), Section 3.2, for a spherical exclusion volume.

5.10 In this problem, the terms of the mass operator will be arranged and a new mass operator will be obtained. To see this more clearly we begin with the Neumann series for the mean field as given by (8), Section 3.1.

Consider the diagram:

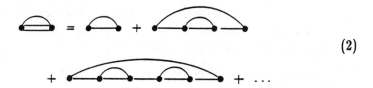

(2)

Equation (2) reproduces all strongly connected diagrams in the Neumann series for the mean field which have one outer correlation and both strongly and weakly connected diagrams inside. Next observe that all strong connected diagrams with two-crossed outer correlations may be summed by the following diagram:

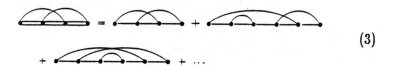

(3)

Show that all the strongly connected diagrams in the Neumann series with more than one outer correlation are of the crossed type. Show that by continuing the preceding process of resummation, all the strongly connected diagrams can be reproduced. By definition this is the mass operator:

$$\otimes = \frown + \overset{\frown}{\frown\frown} + \overset{\frown\frown}{\frown\frown\frown} + \cdots \qquad (4)$$

This is the same mass operator as that in (10), Section 3.1. If we retain only the first term of the new mass operator, we obtain the nonlinear approximation to Dyson's equation as given in (21), Section 3.2.

5.11 *The Principle of Energy Conservation*

Dyson's equation for the mean field and the Bethe-Salpeter equation for the covariance have been derived and are exact equations for the respective statistical quantities. This exact field problem constitutes an energy conservation system. However, the approximations imposed upon the Dyson and Bethe-Salpeter equations to the mass and intensity operators may be inconsistent with the required conservation of energy. Therefore, the question of an energetically self-consistent formulation for the approximated Dyson and Bethe-Salpeter equations is an important one. In this problem it is shown that the ladder approximation and the nonlinear approximation satisfy energy conservation exactly. The general relation between mass and intensity operators for energy conservation is also established.

Consider scalar wave propagation in a continuous random medium. The wave equation is

$$(\nabla^2 + k^2)\psi(\bar{r}) = -Q(\bar{r})\psi(\bar{r}) \qquad (5)$$

where $k^2 \equiv \omega^2 \mu <\epsilon>$ and $Q(\bar{r}) = \omega^2 \mu \epsilon_f(\bar{r})$. Show that

$$\nabla \cdot [\psi^*(\bar{r})\nabla\psi(\bar{r}) - \psi(\bar{r})\nabla\psi^*(\bar{r})] + (k^2 - k^{*2})\psi(\bar{r})\psi^*(\bar{r}) = 0 \qquad (6)$$

Equation (6) is an energy conservation law where the quantity in square brackets defines an energy flux vector, $\bar{S}_\psi$

$$\bar{S}_\psi \equiv \psi^*(\bar{r})\nabla\psi(\bar{r}) - \psi(\bar{r})\nabla\psi^*(\bar{r}) \qquad (7)$$

The Dyson and Bethe-Salpeter equations were found to be

$$-(\nabla^2 + k^2) <\psi(0)> = Q(0,1) <\psi(1)> \qquad (8)$$

Problems

$$-(\nabla^2 + k^2) <G(0,1)> = \delta(0-1) + Q(0,2) <G(2,1)> \quad (9)$$

$$<\psi^*(0')\psi(0)> = <\psi^*(0')><\psi(0)> + <G(0,1)><G^*(0',1')>$$
$$\times I(1,2;1',2') <\psi(2)\psi^*(2')> \quad (10)$$

where we now are using a shorthand notation: $0 \equiv \bar{r}_o, 1 \equiv \bar{r}_1, 2 \equiv \bar{r}_2$, etc., and the integration is implied over internal arguments. In (8) through (10), $Q(0,1)$ and $I(1,2;1',2')$ are the mass and intensity operators, respectively.

By using (8) through (10), show that

$$-\left\{<\nabla \cdot \bar{S}_\psi(0)> + (k^2 - k^{*2}) <\psi^*(0)\psi(0)>\right\}$$
$$= \lim_{0' \to 0} [(\nabla^2 + k^2) <\psi(0)\psi^*(0')>$$
$$- (\nabla'^2 + k^{*2}) <\psi(0)\psi^*(0')>]$$
$$= <\psi^*(0)> Q(0,1) <\psi(1)>$$
$$+ <G^*(0,1')> I(0,2;1',2') <\psi(2)\psi^*(2')>$$
$$+ Q(0,3) <G(3,1)><G^*(0,1')>$$
$$\times I(1,2;1',2') <\psi(2)\psi^*(2')>$$
$$- Q^*(0,1) <\psi^*(1)><\psi(0)>$$
$$- <G(0,1)> I(1,2;0,2') <\psi(2)\psi^*(2')>$$
$$- <G(0,1)> Q^*(0,3) <G^*(3,1')>$$
$$\times I(1,2;1',2') <\psi(2)\psi^*(2')> \quad (11)$$

Next apply the Bethe-Salpeter equation of (10) to the second and the fifth term on the right-hand side of (11) and show that

RHS of (11)
$$= <\psi^*(0)> Q(0,1) <\psi(1)>$$
$$+ <G^*(0,1')> I(0,2;1',2') <\psi(2)><\psi^*(2')>$$
$$+ <G^*(0,1')> I(0,2;1',2') <G(2,3)><G^*(2',3')>$$

$$\times I(3,4;3',4') <\psi(4)\psi^*(4')>$$

$$+ Q(0,3) <G(3,1)><G^*(0,1')>$$

$$\times I(1,2;1',2') <\psi(2)\psi^*(2')>$$

$$- Q^*(0,1) <\psi^*(1)><\psi(0)>$$

$$- <G(0,1)> I(1,2;0,2') <\psi(2)><\psi^*(2')>$$

$$- <G(0,1)> I(1,2;0,2') <G(2,3)><G^*(2',3')>$$

$$\times I(3,4;3',4') <\psi(4)\psi^*(4')>$$

$$- <G(0,1)> Q^*(0,3) <G^*(3,1')>$$

$$\times I(1,2;1',2') <\psi(2)\psi^*(2')> \qquad (12)$$

The ladder approximation of intensity operator and the nonlinear approximation to mass operator are, respectively,

$$I(1,2;1',2') = C(1-1')\delta(1-2)\delta(1'-2') \qquad (13)$$

$$Q(0,1) = <G(0,1)> C(0-1) \qquad (14)$$

By substituting (13) and (14) into (12), and making appropriate changes of dummy integration variables, show that the first term cancels the sixth term, the second term cancels the fifth term, the third term cancels the eighth term, and the fourth term cancels the seventh term. Hence the right-hand-side of (11) vanishes identically.

Next we examine the *integrated optical relation* by integrating over the 0 coordinate in (11) and (12). Since all the arguments of (12) are integration variables, dummy integration variables are to be interchanged. By keeping the arguments of the second and third terms unchanged and interchanging arguments for the rest of the terms in (12), show that

$$-\int d0 \left\{ <\nabla \cdot \overline{S}_\psi(0)> +(k^2 - k^{*2}) <\psi^*(0)\psi(0)> \right\}$$

$$= <\psi^*(2)> Q(2',2) <\psi(2)>$$

$$+ <G^*(0,1')> I(0,2;1',2') <\psi(2)><\psi^*(2')>$$

$$+ <G^*(0,1')> I(0,2;1',2') <G(2,3)><G^*(2',3')>$$

$$\times I(3,4;3',4')<\psi(4)\psi^*(4')>$$
$$+Q(2',2)<G(2,3)><G^*(2',3')>$$
$$\times I(3,4;3',4')<\psi(4)\psi^*(4')>$$
$$-Q^*(2,2')<\psi^*(2')><\psi(2)>$$
$$-<G(1',0)>I(0,2;1',2')<\psi(2)><\psi^*(2')>$$
$$-<G(1',0)>I(0,2;1',2')<G(2,3)><G^*(2',3')>$$
$$\times I(3,4;3',4')<\psi(4)\psi^*(4')>$$
$$-<G(2,3)>Q^*(2,2')<G^*(2',3')>$$
$$\times I(3,4;3',4')<\psi(4)\psi^*(4')> \quad (15)$$

where integration is performed over all repeated arguments in the right-hand side of (15). Show that the right-hand side of (15) vanishes identically if

$$Q(2',2)-Q^*(2,2')+[<G^*(0,1')>- \\ -<G(1',0)>]I(0,2;1',2')=0 \quad (16)$$

with integration over repeated arguments. Equation (16) is the general relation between mass and intensity operators for the integrated optical relation of (15) to vanish identically.

5.12 The spectral representation of the mean Green's function is

$$<G(\bar{r},\bar{r}_j)>=-\frac{i}{8\pi^2}\int d\bar{k}_\perp \frac{e^{i\bar{k}_\perp\cdot(\bar{r}_\perp-\bar{r}_{\perp j})+iK_z|z-z_j|}}{K_z} \quad (17)$$

Use (17), (2), Section 3.5, for the mean field and (11), Section 3.5, for the covariance of the incoherent field to show that the right-hand side of (5), Section 3.5, is equal to

$$\frac{\kappa_e\tilde{\omega}}{4\pi}\int_{-\infty}^0 dz_j \int d\bar{k}_\perp e^{i\bar{k}_\perp\cdot(\bar{r}_{\perp 1}-\bar{r}_{\perp 2})} \frac{e^{iK_z|z_1-z_j|-iK_z^*|z_2-z_j|}}{|K_z|^2}$$
$$\times\left[e^{2K_{iz}''z_j}+\int d\bar{k}_{\perp 1}\left\{I_u(z_j,\bar{k}_{\perp 1})+I_d(z_j,\bar{k}_{\perp 1})\right\}\right] \quad (18)$$

By equating (18) to (11), Section 3.5, show that (12), Section 3.5, is obtained.

5.13 Let the integral form of the radiative transfer equations in (7) and (8), Section 2.2, Chapter 4, assume the half-space limit without any reflective interface by letting $r(\mu) = 0$ and $\tau_d = -\infty$. For $\mu > 0$, let

$$I(\tau, -\mu, \phi) = e^{\tau/\mu_i}\delta(\mu - \mu_i)\delta(\phi) + I_d(\tau, \mu, \phi) \qquad (18)$$

$$I(\tau, \mu, \phi) = I_u(\tau, \mu, \phi) \qquad (19)$$

By substituting (18) and (19) into the RT equations, obtain equations for I_u and I_d and show that they are identical to (20) and (21), Section 3.5. Note that $\tau = \kappa_e z$.

5.14 Use (9), Section 3.6, in (8), Section 3.6, to derive the result in (10), Section 3.6. Solve the integral in (10), Section 3.6, exactly and show that it is equal to

$$L = \left(\frac{\tilde{\omega}}{4\pi}\right)^2 \frac{2\pi A}{r^2} \frac{1}{(1/\mu_s + 1/\mu_i)} [\mu_i \ln(1 + 1/\mu_i) + \mu_s \ln(1 + 1/\mu_s)] \qquad (20)$$

5.15 Derive (12), Section 3.6, from (11), Section 3.6. Next use transformation of variables and show that C is given by (14), Section 3.6.

5.16 The self-similar model is used to explain the dielectric response of rocks at very low frequency so that attenuation due to scattering can be ignored (Sen et al., 1981).

1. When an field $\overline{E}^{inc} = \hat{z}$ is incident on a spherical particle with permittivity ϵ_s, the potential outside the particle is

$$\Phi_{out} = -r\cos\theta + \frac{A}{r^2}\cos\theta \qquad (21)$$

The polarizability of the particle α is defined as $\bar{p} = \alpha \overline{E}^{int}$ where $\overline{E}^{int}$ is the field inside the particle. Show that

$$\alpha = 3v_o\epsilon_o \frac{\epsilon_s - \epsilon_o}{\epsilon_s + 2\epsilon_o} \qquad (22)$$

and

$$A = \frac{\epsilon_s - \epsilon_o}{2\epsilon_o + \epsilon_s} a^3 \qquad (23)$$

where $v_o = 4\pi a^3/3$ is the volume of the sphere. Thus, when A is calculated, the permittivity is given by

$$\frac{\epsilon_s}{\epsilon_o} = \frac{1 + 2A/a^3}{1 - A/a^3} \quad (24)$$

2. Consider a sphere of material of permittivity ϵ_m *(matrix)* coated with a material of permittivity ϵ_w *(water)*. The inner radius is b and the outer radius is a. An incident field $\overline{E}^{inc} = \hat{z}$ is incident on it. The potential outside is again given by the expression in (21) with a different A. Show that

$$A = a^3 \frac{(\epsilon_w - \epsilon_o)(\epsilon_m + 2\epsilon_w) + \eta(2\epsilon_w + \epsilon_o)(\epsilon_m - \epsilon_w)}{(\epsilon_w + 2\epsilon_o)(\epsilon_m + 2\epsilon_w) + \eta(2\epsilon_w - 2\epsilon_o)(\epsilon_m - \epsilon_w)} \quad (25)$$

Use the relation (24) in (25) to show that the overall permittivity of the coated sphere is

$$\epsilon_{cs} = \epsilon_w \left[\frac{\epsilon_m + 2\epsilon_w + 2\eta(\epsilon_m - \epsilon_w)}{\epsilon_m + 2\epsilon_w - \eta(\epsilon_m - \epsilon_w)} \right] \quad (26)$$

where $\eta = b^3/a^3$ is the volume fraction of the inner sphere.

3. Using the strong permittivity mixing formula of (3), Section 4.2, the effective permittivity of a mixture of n types coated spheres is

$$\sum_{p=1}^{n} f_p \frac{\epsilon_{cs}^{(p)} - \epsilon_g}{\epsilon_{cs}^{(p)} + 2\epsilon_g} = 0 \quad (27)$$

where f_p is the volume fraction of pth species and $\epsilon_{cs}^{(p)}$ is the corresponding coated sphere permittivity according to (26). Let η_p is the volume fraction of the inner sphere of the pth species. Consider the following simple model. All spheres in rocks have the same η. That is $\eta_p = \eta$ for all $p = 1, 2, \cdots, n$. Hence, $\epsilon_{cs}^{(p)} = \epsilon_{cs}$ for all p. From (27), $\epsilon_g = \epsilon_{cs}$. The rock is of porosity ϕ which is occupied by water. Hence $\eta = 1 - \phi$. Thus,

$$\epsilon_g = \epsilon_{cs} = \epsilon_w \left(\frac{\epsilon_m + 2\epsilon_w + 2(1 - \phi)(\epsilon_m - \epsilon_w)}{\epsilon_m + 2\epsilon_w - (1 - \phi)(\epsilon_m - \epsilon_w)} \right) \quad (28)$$

In the very low frequency limit, we have $\epsilon_m = \epsilon'_m + i\sigma_m/\omega$ and $\epsilon_w = \epsilon'_w + i\sigma_w/\omega$ where σ_m and σ_w are the conductivities, respectively, of rock matrix and water. Let $\epsilon_g = \epsilon'_g + i\sigma_g/\omega$ where σ_g is the effective conductivity. Show by using (28) that in the usual case when σ_w/ω dominates in the quasistatic limit

$$\sigma_g = \sigma_w \frac{2\phi}{3-\phi} \qquad (29)$$

Thus σ_g is proportional to ϕ for small ϕ. By using a differential approach, the self-similar model further shows that $\sigma_g \approx \phi^{3/2}$ for small ϕ (Sen et al., 1981). The effective permittivity of charged particles immersed in electrolytic solution and the associated dielectric enhancement effect have also been examined (Chew and Sen, 1982).

4. The Maxwell-Garnett mixing formula (1904) is

$$\frac{\epsilon - \epsilon_1}{\epsilon + 2\epsilon_1} = f_2 \frac{\epsilon_2 - \epsilon_1}{\epsilon_2 + 2\epsilon_1} \qquad (30)$$

where f_2 is the fractional volume of species 2. Fractional volume f_1 is equal to $1 - f_2$. The Maxwell-Garnett formula is identical to that of quasicrystalline approximation in the very low frequency limit (Chapter 6). Equation (30) is derived under the assumption that ϵ_1 is the permittivity of the background medium and ϵ_2 is that of the scatterer. In the case of rocks, assume that water is the background and the rock grains are the scatterers by letting $\epsilon_1 = \epsilon_w, \epsilon_2 = \epsilon_m$ and $f_2 = 1 - \phi$ in (30). Solve for ϵ and show that the same expression as (28) is obtained. The classical mixture formula can also be found in Bottcher (1952).

5.17 For a spherical exclusion volume $PS\overline{\overline{G}}_g(\bar{r})$ is given by (14), Section 3.2, with k_m replaced by k_g. Use the expression in (12), Section 4.3, to derive the effective permittivity in (13), Section 4.3.

5.18 Assume a mixture of two constituents. One constituent is background medium with permittivity ϵ_b and fractional volume f_b. The other constituent is scatterer with permittivity ϵ_s and fractional volume f_s. The scatterers are assumed to be spherical with radius a. We have $f_b + f_s = 1$. Next use simple physical arguments to construct correlation functions of $\xi(\bar{r})$ for the mixture. First look at the one-dimensional problem. Consider pulses of height ξ_s and width l (which corresponds to $2a$ in the three-dimensional case) distributed on a line. For the rest of the line, the amplitude is ξ_b (Figure 5.33a). The problem is to determine the correlation function. To simplify the problem, the line is partitioned into segments of length

Problems

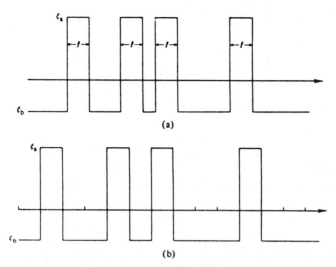

Fig. 5.33 (a) One-dimensional random process of $\xi(r)$ consisting of pulses of width l and height ξ_s. (b) Same as (a) except the line is partitioned into segments of length l.

l, with the amplitude in each partition assuming either the value ξ_s or ξ_b (Figure 5.33b). Let $\xi(k)$ be the value of amplitude of partition k. Then, $Pr(\xi(k) = \xi_s) = f_s$, $Pr(\xi(k) = \xi_b) = f_b$ and $<\xi(k)> = f_s\xi_s + f_b\xi_b = 0$. To calculate the correlation function, verify that

$$<\xi(k)\xi(k+m)> = f_s\xi_s^2 + f_b\xi_b^2 \quad \text{for } m = 0 \quad (31)$$

For $m \neq 0$,

$$\begin{aligned}<\xi(k)\xi(k+m)> &= Pr[\xi(k) = \xi_s, \xi(k+m) = \xi_s]\xi_s^2 \\ &+ Pr[\xi(k) = \xi_s, \xi(k+m) = \xi_b]\xi_s\xi_b \\ &+ Pr[\xi(k) = \xi_b, \xi(k+m) = \xi_s]\xi_b\xi_s \quad (32) \\ &+ Pr[\xi(k) = \xi_b, \xi(k+m) = \xi_b]\xi_b^2 \\ &= f_s^2\xi_s^2 + 2f_bf_s\xi_s\xi_b + f_b^2\xi_b^2 = 0\end{aligned}$$

Using the results for the three-dimensional continuum case gives (2) and (3), Section 4.5. Extend the method to treat a mixture of three constituents and derive (9) and (10), Section 4.5 (Tsang et al., 1982).

5.19 Dyson's equation for the mean scalar Green's function was derived in the text in the form

$$(\nabla^2 + k_m^2) <G(\bar{r},\bar{r}_o)> + \int d^3r_1 Q(\bar{r},\bar{r}_1) <G(\bar{r}_1,\bar{r}_o)> \qquad (33)$$
$$= -\delta(\bar{r} - \bar{r}_o)$$

where the mass operator, $Q(\bar{r},\bar{r}_1)$ is defined as the sum of all strongly connected diagrams.

1. Show that the mass operator is translationally invariant. In other words, show that $Q(\bar{r}_1,\bar{r}_2) = Q(\bar{r}_1 - \bar{r}_2)$.

2. Taking the source point at the origin (i.e., $\bar{r}_o = 0$), show that the solution to the above Dyson's equation may be written as

$$<G(\bar{r})> = \frac{1}{(2\pi)^3} \int d^3k \frac{e^{-i\bar{k}\cdot\bar{r}}}{k^2 - k_m^2 - \int d^3R\, Q(\bar{R}) e^{-i\bar{k}\cdot\bar{R}}} \qquad (34)$$

3. In the bilocal approximation, the mass operator is approximated as

$$Q(\bar{R}) \approx C(\bar{R}) \frac{e^{ik_m R}}{4\pi R} \qquad (35)$$

Taking the correlation function to be $C(R) = \delta k_m'^4 \exp(-R/a)$ show that in the limit $k_m a \ll 1$, the mean Green's function obtained in part (b) is a spherical wave, with an effective propagation constant given by

$$K^2 = k_m^2 + \frac{\delta k_m'^4 a^2}{(1 - ik_m a)^2} \qquad (36)$$

which is similar to the result obtained in the text using the Taylor expansion method.

5.20 Apply the two-variable expansion procedure to the nonlinear approximated Dyson's equation for a two-layer medium in (24), Section 5.2, and derive the results in (25) through (36), Section 5.2. Also derive the mean field given in (35), Section 5.2, from the mean Green's function.

5.21 Derive the MRT equations (3) and (4), Section 5.3, for a two-layer random medium by using (1), Section 5.3, as the starting point. The steps are similar to those in Section 3.5 and are outlined below (1), Section 5.3.

5.22 In this problem we convert the MRT equations of (3) and (4), Section 5.3, to the standard RT form by using solid-angle and specific intensities. First, make the substitutions $\beta_x = \beta_{1m} \sin\theta \cos\phi$,

Problems

$\beta_y = \beta_{1m} \sin\theta \sin\phi$, $\int d^2\beta_\perp = \int_0^{\pi/2} \sin\theta\, d\theta \int_0^{2\pi} d\phi\, \beta_{1m}^2 \cos\theta \equiv \int d\Omega\, \beta_{1m}^2 \cos\theta$ and

$$\overline{J}_u^u(z, \Omega) = (\beta_{1m}^2/\eta_1) \cos\theta \overline{I}_d^u(z, \overline{\beta}_\perp) \tag{37}$$

where $\eta_1 \equiv (\mu_o/\epsilon_{1m})^{1/2}$. Also, let $\overline{\overline{K}}(\Omega) = \cos\theta \overline{\overline{\eta}}(\overline{\beta}_\perp) - K_a \overline{\overline{I}}$, $K_a = 2\beta_{1m}'' \equiv 2Im(\beta_{1m})$, $\overline{\overline{Q}}_{\mu\nu}(\Omega, \Omega_i) = \overline{\overline{Q}}_{\mu\nu}(\overline{\beta}_\perp, \overline{k}_{\perp i})/\beta_{1m}^2 \cos\theta_i$, $\overline{\overline{Q}}_{cl}(\Omega, \Omega_i) = \overline{\overline{Q}}_{cl}(\overline{\beta}_\perp, \overline{k}_{\perp i})/\beta_{1m}^2 \cos\theta_i$, $\overline{\overline{P}}_{\mu\nu}(\Omega, \Omega') = \overline{\overline{P}}_{\mu\nu}(\overline{\beta}_\perp, \overline{k}_\perp')$.
Then, show that the MRT equations of (3) and (4), Section 5.3, become

$$\cos\theta \frac{d}{dz}\overline{J}_u(z, \Omega) = -K_a \overline{J}_u(z, \Omega) - \overline{\overline{K}} \cdot \overline{J}_u(z, \Omega)$$
$$+ \overline{\overline{Q}}_{ud}(\Omega, \Omega_i) \cdot \overline{J}_{md}(z, \Omega_i)$$
$$+ \Delta_1 \overline{\overline{Q}}_{cl}(\Omega, \Omega_i) \cdot \overline{J}_{mc1}(z, \Omega_i)$$
$$+ \int d\Omega' \left[\overline{\overline{P}}_{uu}(\Omega, \Omega') \cdot \overline{J}_u(z, \Omega') \right.$$
$$\left. + \overline{\overline{P}}_{ud}(\Omega, \Omega') \cdot \overline{J}_d(z, \Omega') \right] \tag{38}$$

$$-\cos\theta \frac{d}{dz}\overline{J}_d(z, \Omega) = -K_a \overline{J}_d(z, \Omega) - \overline{\overline{K}} \cdot \overline{J}_d(z, \Omega)$$
$$+ \overline{\overline{Q}}_{du}(\Omega, \Omega_i) \cdot \overline{J}_{mu}(z, \Omega_i)$$
$$+ \overline{\overline{Q}}_{dd}(\Omega, \Omega_i) \cdot \overline{J}_{md}(z, \Omega_i)$$
$$- \Delta_1 \overline{\overline{Q}}_{cl}(\Omega, \Omega_i) \cdot \overline{J}_{mc2}(z, \Omega_i)$$
$$+ \int d\Omega' \left[\overline{\overline{P}}_{du}(\Omega, \Omega') \cdot \overline{J}_u(z, \Omega') \right.$$
$$\left. + \overline{\overline{P}}_{dd}(\Omega, \Omega') \cdot \overline{J}_d(z, \Omega') \right] \tag{39}$$

with boundary conditions

$$\overline{J}_d(0, \Omega_1) = \overline{\overline{R}}_{10}(\Omega_1) \cdot \overline{J}_u(0, \Omega_1) \tag{40}$$

$$\overline{J}_u(-d_1, \Omega_1) = \overline{\overline{R}}_{12}(\Omega_1) \cdot \overline{J}_d(-d_1, \Omega_1) \tag{41}$$

$$\overline{J}_{ou}(0, \Omega_o) = \frac{\epsilon_o}{\epsilon_{1m}} \left[\frac{\cos\theta_o}{\cos\theta_1} \frac{\eta_1}{\eta_o} \overline{\overline{T}}_{10}(\Omega_1) \right] \cdot \overline{J}_u(0, \Omega_1) \tag{42}$$

5.23 In the first-order renormalization method, the scattering of the incoherent intensity is neglected in the Bethe-Salpeter equation. In this approximation, the MRT equations assume the form of (3) and (4), Section 5.3, without the integrals. Solve the result equations in conjunction with boundary conditions of (6) and (8), Section 5.3. Consider the backscattering direction $\overline{\beta}_\perp = -\overline{k}_{1i}$ in which case $\Delta_1 = 1$. Use the correlation function of (6), Section 2.2, and derive the backscattering cross sections σ_{vv} and σ_{hh} of (2), Section 5.4.

5.24 In this problem, it will be shown that the Schwarzschild-Milne integral equation is equivalent to the radiative transfer equation for isotropic point scatterers. The upward- and downward-going diffuse specific intensities are defined respectively by (40) and (41), Section 3.4. Show that the exponential integral is given by the following integral representation

$$E_1(\tau) = \int_0^1 \frac{d\mu}{\mu} e^{-\tau/\mu} \qquad (43)$$

or $\mu > 0$. Show from (40) and (41), Section 3.4, that

$$\frac{dI(\tau,\mu)}{d\tau} = -\frac{I(\tau,\mu)}{\mu} + \frac{\tilde{\omega}}{4\pi\mu} J_s(\tau) \qquad (44)$$

$$-\frac{dI(\tau,-\mu)}{d\tau} = -\frac{I(\tau,-\mu)}{\mu} + \frac{\tilde{\omega}}{4\pi\mu} J_s(\tau) \qquad (45)$$

Next add (40) and (41), Section 3.4, and integrate the sum over μ from 0 to 1 and show, by using (43), that

$$\int_0^1 d\mu [I(\tau,\mu) + I(\tau,-\mu)] = \frac{\tilde{\omega}}{4\pi} \int_{-\tau_d}^0 d\tau' J_s(\tau') E_1(|\tau'-\tau|) \qquad (46)$$

Using (38), Section 3.4, in (46), show that

$$\int_0^1 d\mu [I(\tau,\mu) + I(\tau,-\mu)] = \frac{J_s(\tau)}{2\pi} - \frac{\tilde{\omega}}{4\pi} \int_{-\tau_d}^0 d\tau' E_1(|\tau-\tau'|) e^{\tau'/\mu_i} \qquad (47)$$

Show that using (47) in (44) and (45) gives the radiative transfer equation for isotropic point scatterers.

6

SCATTERING BY RANDOM DISCRETE SCATTERERS

1.	Introduction	427
2.	Simple Model for Scattering from a Dense Medium	430
	2.1 Effective Permittivity	430
	2.2 Scattering Attenuation and Coherent Propagation Constant	432
	2.3 Coherent Reflection and Incoherent Scattering from a Half-Space of Scatterers	435
3.	Multiple Scattering Equations and Derivations	439
	3.1 Multiple Scattering Equations – Introduction	439
	3.2 Derivation of Multiple Scattering Equations Using N-Particle Green's Dyadic	440
	3.3 Derivation of Multiple Scattering Equations Using T-Matrix Formalism	446
	3.4 Summary	454
4.	Approximations of Multiple Scattering Equations	455
	4.1 Configurational Average of Multiple Scattering Equations	455
	4.2 Effective Field Approximation (Foldy's Approximation)	458
	4.3 Quasicrystalline Approximation	461
	4.4 Coherent Potential	464
	4.5 Low Frequency Solutions	470
	4.6 Energy Conservation and Second Moment	475
5.	Pair-Distribution Functions	479

5.1	Introduction	479
5.2	Percus-Yevick Equation	482
5.3	Structure Factor	485
5.4	Improvement of Percus-Yevick Approximation	486
5.5	Summary	489

6. Scattering of Electromagnetic Waves from a Half-Space of Dielectric Scatterers – Normal Incidence — 490

6.1	Introduction	490
6.2	Generalized Ewald-Oseen Extinction Theorem and the Lorentz-Lorenz Law	490
6.3	Effective Phase Velocity and Attenuation Rate in the Low-Frequency Limit	497
6.4	Dispersion Relations at Higher Frequencies	498

7. Scattering of Electromagnetic Waves from a Half-Space of Dielectric Scatterers – Oblique Incidence — 506

7.1	Introduction	506
7.2	Dispersion Relation and Coherent Reflected Wave	506
7.3	Vertically and Horizontally Polarized Incidence	511
7.4	Incoherent Scattered Intensity	515
7.5	Solution in Low-Frequency Limit	517
7.6	Results at Higher Frequency	520

8. Nonspherical Particles — 525

8.1	Introduction	525
8.2	Generalized Ewald-Oseen Extinction Theorem and Lorentz-Lorenz Law	525
8.3	Coherent Reflected Wave and Incoherent Scattered Field	529
8.4	Sparse Concentration of Particles	531
8.5	Low-Frequency Limit	533
8.6	Numerical Illustrations	537

9. Dispersion Relations Based on Coherent Potential — 542

9.1	Introduction	542
9.2	Dispersion Relation for Dielectric Spheres	543
9.3	Numerical Illustrations	546

10. Multiple Scattering of Second Moment	548
10.1 Introduction	548
10.2 First Moment for Half-Space Medium	550
10.3 Second Moment Radiative Transfer Equation and Cyclical Transfer Equation	555
Problems	563

1 INTRODUCTION

In Chapters 3 and 4, we have employed the radiative transfer theory to evaluate the scattering of waves by discrete scatterers. The phase matrix is constructed by assuming that the particles scatter independently. To obtain the phase matrix, the averaged Stokes matrix is multiplied by the number of particles per unit volume n_o. The extinction rate is also calculated in a similar manner. Hence, the extinction rate will be linearly proportional to the number of particles per unit volume and the fractional volume of particles $f = n_o v_o$, where v_o is the volume of a single particle (Figure 6.1). However, physical intuition indicates that the linear relation cannot be correct for arbitrary f. For example, at $f = 1$, when the entire volume is occupied by scatterers, the medium becomes a homogeneous medium. Hence, in the absence of absorption, scattering should be equal to zero in the limit $f = 1$ (Figure 6.1). Independent scattering is not valid for materials with an appreciable fractional volume of scatterers. This has been verified by controlled laboratory experiments (Ishimaru and Kuga, 1982). We shall call such materials dense medium. Foldy's approximation (Foldy, 1946), which is applicable to medium with a low concentration of particles is discussed in Section 4. However it gives an attenuation rate very similar to that of independent scattering and is not valid for dense media.

Many geological materials are dense media. They are often mixtures of substances, with each constituent occupying an appreciable fractional volume. The constituents usually have quite different refractive indices. For example, dry snow is a mixture of ice and air. The fractional volume of ice in dry snow is between 10 and 40%. Ice has a dielectric constant of 3.2 which is quite different from that of air. Wet snow has one more constituent, water, which has a dielectric constant of 80 and fractional volume between 0 to 10%. Rock is a mixture of rock grains and pores which may be filled with fluid or gas. The porosity of rocks is between 0 and 40%. Thus, scattering of waves by dense medium is important in geophysical remote sensing

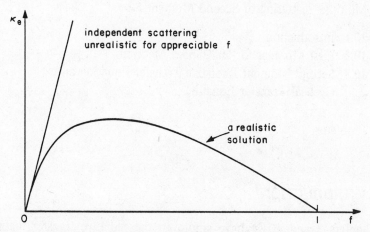

Fig. 6.1 Scattering attenuation as a function of fractional volume of particles: Independent scattering and a more realistic solution.

In this chapter, scattering of waves in dense medium will be studied. We shall begin with a simple model and then proceed with more rigorous multiple scattering theory.

When a coherent wave propagates through a scattering medium, it is attenuated by both absorption and scattering. The coherent wave propagation constant is denoted by K, where

$$K = K_r + iK_i \tag{1}$$

The attenuation K_i is a summation of the absorption K_{ia} and scattering K_{is}

$$K_i = K_{ia} + K_{is} \tag{2}$$

The scattering part of the attenuation is dependent on particle size. For cases where volume scattering is important

$$K_r \gg K_i \tag{3}$$

In a similar manner, the effective permittivity $\epsilon_{\it{eff}}$ is defined as

$$\epsilon_{\it{eff}} = \frac{K^2}{\omega^2 \mu} \tag{4}$$

Equation (3) gives the approximate relation

$$K^2 = (K_r + iK_{ia})^2 + 2iK_{is}K_r \tag{5}$$

Introduction

Class	Dielectric property of particles	Particle concentration	Particle positions	K_r	Relation of K_i with f
A	tenuous	sparse	independent	$\simeq k$	linear
B	non-tenuous	sparse	independent	$\simeq k$	linear
C	tenuous	dense	correlated	$\simeq k$	nonlinear
D	non-tenuous	dense	correlated	significantly different from k	nonlinear

Table 6.1 Classification of random discrete scatterers.

The effective permittivity includes absorption and scattering. The classical mixture formulas for effective permittivity applies when scattering attenuation can be ignored. Scattering is generally size dependent. For example, Rayleigh scattering gives a scattering cross section that is proportional to $k^4 a^6$. Hence at very low frequencies, scattering attenuation can be neglected. In such a limit the effective permittivity will be the same as the effective permittivity from classical mixture formula. Thus, the applicability of the mixture formula to a particular dense medium in the very low frequency limit will help in deciding the multiple scattering theory to be employed at higher frequencies.

Based on the preliminary discussion, random discrete scatterers can be classified as in Table 6.1. Particles are described as tenuous if their dielectric properties are only slightly different from the background medium. The characteristics of wave propagation in such media will be quite similar to that of continuous random medium which has been discussed in Chapter 5. When the concentration of particles is dense, then the particle positions are no longer independent and correlations of particle positions must be taken into account. The properties of K_r and K_i for the four classes of particles are described in Table 6.1 where k is used to denote the propagation constant of the background medium. Of the four classes of random discrete particles, class D coincides with most geophysical media and will be the main subject of study in this chapter.

Besides studying the effective propagation constants at microwave frequencies, the reflectivity of the coherent wave will also be studied. The classical Fresnel reflection coefficients are based on the fact that dipoles are induced in the particles at low frequencies (Born and Wolf, 1975). At fre-

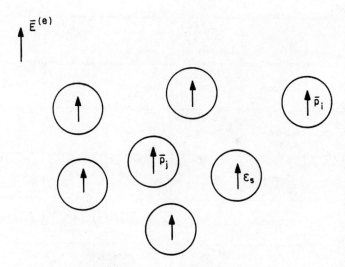

Fig. 6.2 Spheres with permittivity ϵ_s embedded in a background medium of permittivity ϵ (induced dipoles).

quencies where the particle size becomes appreciable, the Fresnel reflection formula has to be modified.

As discussed in Chapter 5, the total field can be decomposed into the coherent and incoherent components. The incoherent field will first be studied by making use of the distorted Born approximation which corresponds to single scattering of coherent wave. Theory of multiple scattering of the incoherent waves will be examined in Sections 4.6 and 10.

2 SIMPLE MODEL FOR SCATTERING FROM A DENSE MEDIUM

In this section, a simple physical model based on induced dipoles is used to calculate the effective permittivity, scattering attenuation, and bistatic scattering coefficients for electromagnetic scattering from a dense medium. The model gives physical insight into propagation and scattering in dense medium and the results are identical to the low-frequency results based on the quasicrystalline approximation to be treated in later sections. The final results for the scattering attenuation and bistatic scattering coefficients depend on the correlation of particle positions through the pair distribution function, which will be treated in Section 5.

2.1 Effective Permittivity

Consider a distribution of spheres with permittivity ϵ_s embedded in a background medium of permittivity ϵ (Figure 6.2). Under the excitation of an incident electromagnetic field, dipoles are induced in the spheres. The induced dipole moment $\overline{p}$ is $\overline{p} = \alpha \overline{E}^e$ where α is the molecular polarizability of the particle and $\overline{E}^e$ is the exciting field. The exciting field $\overline{E}^e$ is not the macroscopic field $\overline{E}$. It can be calculated by subtracting from $\overline{E}$ the field generated by a sphere of smooth polarization charge (Feynman et al., 1964). Thus, $\overline{E}^e = \overline{E} + \overline{P}/3\epsilon$ where $\overline{P}$ is polarization per unit volume. Since $\overline{P} = n_o \overline{p}$ where n_o is the number of particles per unit volume, $\overline{P}$ can be expressed in terms of $\overline{E}$

$$\overline{P} = \frac{n_o \alpha \overline{E}}{1 - (n_o \alpha / 3\epsilon)} \tag{1}$$

The displacement is $\overline{D} = \epsilon \overline{E} + \overline{P}$ and, since $\overline{D} = \epsilon_{eff} \overline{E}$, the effective permittivity ϵ_{eff} is

$$\epsilon_{eff} = \epsilon \left[\frac{1 + (2n_o \alpha / 3\epsilon)}{1 - (n_o \alpha / 3\epsilon)} \right] \tag{2}$$

The polarizability α of a sphere with permittivity ϵ_s in a background medium with permittivity ϵ is easily calculated (Problem 1). It is given by $\alpha = v_o 3 (\epsilon_s - \epsilon) \epsilon / (\epsilon_s + 2\epsilon)$ where $v_o = 4\pi a^3 / 3$ is the volume of one sphere. Substituting α into (2) and noting that $f = n_o v_o$ is the fractional volume occupied by the particles give the relation

$$\epsilon_{eff} = \epsilon \frac{1 + 2fy}{1 - fy} \tag{3}$$

where

$$y = \frac{\epsilon_s - \epsilon}{\epsilon_s + 2\epsilon} \tag{4}$$

Equation (3) can be regarded as a mixture formula since it gives the effective permittivity ϵ_{eff} in terms of the permittivities of the two species, ϵ_s and ϵ, and the respective fractional volumes f and $1 - f$. The effective permittivity ϵ_{eff} here does not include attenuation due to scattering which shall be calculated by considering the radiation of the induced dipoles. Equation (2) is known as the Clausius-Mossoti formula or the Lorentz-Lorenz formula. Equation (3) is known as the Maxwell-Garnett mixing formula (1904) and can be put in the following symmetrical form (Problem 2)

$$\frac{\epsilon_{eff} - \epsilon}{\epsilon_{eff} + 2\epsilon} = f \frac{\epsilon_s - \epsilon}{\epsilon_s + 2\epsilon} \tag{5}$$

which is also known as the Rayleigh mixing formula (Rayleigh, 1892).

2.2 Scattering Attenuation and Coherent Propagation Constant

The coherent propagation constant is calculated by including the attenuation due to scattering. The scattering power can be attributed to the radiation power of the induced dipoles of the previous section. The induced dipole model is valid in the low-frequency limit. It will be shown that the scattering attenuation rate depends on the correlation of particle positions. From Section 2.1, $\overline{E}^e = \overline{E}/(1 - fy)$. Therefore, the induced dipole, $\overline{p} = \alpha \overline{E}^e$, is

$$\overline{p} = \frac{3v_o \epsilon y}{1 - fy} \overline{E} \tag{1}$$

Since the displacement relation is $\overline{D} = \epsilon_{eff} \overline{E}$, the macroscopic field $\overline{E}$ satisfies the wave equation with wavenumber K with $K = \omega\sqrt{\mu \epsilon_{eff}}$. Since attenuation is small on the wavelength scale, we can approximate $\overline{E}$ by $\overline{E} = \hat{e} E_o \exp(i\overline{K}_r \cdot \overline{r})$ where $|\overline{K}_r| = K_r$ is the real part of K. Then the induced dipole moment $\overline{p}_i$ for the ith particle centered at $\overline{r}_i$ is (Figure 6.2)

$$\overline{p}_i = \frac{3v_o \epsilon y}{1 - fy} E_o e^{i\overline{K}_r \cdot \overline{r}_i} \hat{e} \tag{2}$$

The radiation field of the ith dipole is $\overline{E}_{si}(\overline{r})$, where

$$\overline{E}_{si}(\overline{r}) = \frac{\omega^2 \mu e^{ikR_i}}{4\pi R_i} (\hat{R}_i \times \overline{p}_i) \times \hat{R}_i \tag{3}$$

and $\overline{R}_i = \overline{r} - \overline{r}_i$ is the vector pointing from the ith dipole to the observation point. For $r \gg r_i$, (3) can be approximated by

$$\overline{E}_{si}(\overline{r}) = \overline{A} J_i \tag{4}$$

where $J_i = \exp(i\overline{\alpha} \cdot \overline{r}_i)$, $\overline{\alpha} = \overline{K}_r - k\hat{r}$ and $\overline{A}$ is a vector independent of i and is

$$\overline{A} = \frac{\omega^2 \mu e^{ikr}}{4\pi r} (\hat{r} \times \hat{e}) \times \hat{r} \frac{3v_o \epsilon y}{1 - fy} E_o \tag{5}$$

The total scattered field $\overline{E}_s(\overline{r})$ is the sum of the radiation fields of all the dipoles. Hence,

$$\overline{E}^s = \sum_{i=1}^{N} \overline{A} J_i \tag{6}$$

where N is the total number of particles. Thus, the radiation intensity I_s is

$$I_s = \frac{|\overline{A}|^2}{2\eta} \left\{ N + \sum_{i=1}^{N} \sum_{j>i} 2 \operatorname{Re}(J_i J_j^*) \right\} \tag{7}$$

Then, taking the configuration average of (7) gives

$$<I_s> = \frac{|A|^2}{2\eta}\{N + L\} \quad (8)$$

where

$$L = \sum_{i=1}^{N}\sum_{j>i} 2Re \int d\bar{r}_1 \ldots d\bar{r}_N\, p_N(\bar{r}_1, \ldots, \bar{r}_N)\, e^{i\bar{\alpha}\cdot(\bar{r}_i - \bar{r}_j)} \quad (9)$$

with $p_N(\bar{r}_1,\ldots,\bar{r}_N)$ being the N-particle probability density function for the particle centers $\bar{r}_1,\ldots,\bar{r}_N$. Since the particles are identical, we have

$$L = N(N-1)Re \int d\bar{r}_i d\bar{r}_j\, p_2(\bar{r}_i, \bar{r}_j)\, e^{i\bar{\alpha}\cdot(\bar{r}_i - \bar{r}_j)} \quad (10)$$

where p_2 is the two particle probability density function. Using Bayes' rule, $p_2(\bar{r}_i, \bar{r}_j) = p(\bar{r}_i|\bar{r}_j)\, p(\bar{r}_j)$ where $p(\bar{r}_i|\bar{r}_j)$ is the conditional probability of the ith particle at $\bar{r}_i$ given the jth particle at $\bar{r}_j$. Assume uniform particle distribution, so that $p(\bar{r}_j) = 1/V$ where V is the volume containing the particles. Further normalize the conditional probability by $p(\bar{r}_i|\bar{r}_j) = g_2(\bar{r}_i, \bar{r}_j)/V$ where $g_2(\bar{r}_i, \bar{r}_j)$ is the two-point distribution function. For radially symmetric problems $g_2(\bar{r}_i, \bar{r}_j) = g(\bar{r}_i - \bar{r}_j)$ and g is called the pair-distribution function (McQuarrie, 1976; Waseda, 1980). Then,

$$L = n_o N\, Re \int_V d\bar{r}\, g(\bar{r})\, e^{i\bar{\alpha}\cdot\bar{r}} \quad (11)$$

Because the particles are not interpenetrable, the pair-distribution function must satisfy the criterion

$$g(r) = 0 \qquad \text{for } r < b \quad (12)$$

where $b = 2a$ is the diameter of the sphere so that the volume integral in (13) is actually V less an exclusion volume S_e of radius b. Another criterion is that unless f is equal to maximum packing f_m, the two particles must be uncorrelated when their separation distance approaches infinity. Therefore

$$\lim_{r\to\infty} g(r) = 1 \qquad \text{for } f < f_m \quad (13)$$

Thus,

$$L = n_o N\, Re\left\{\int_{r>b} d\bar{r}[g(\bar{r}) - 1]\, e^{i\bar{\alpha}\cdot\bar{r}} + \int_{\substack{r>b \\ V}} d\bar{r}\, e^{i\bar{\alpha}\cdot\bar{r}}\right\} \quad (14)$$

Since $g(r) - 1 = 0$ for r greater than a few diameters of the sphere and at the low-frequency limit $\alpha b \ll 1$, the exponent for the first term in (14) can be set equal to zero.

The second term in (14) can be evaluated by using the divergence theorem

$$Re \int_{\substack{r>b \\ V}} d\bar{r} e^{i\bar{\alpha}\cdot\bar{r}} = Re \left[\frac{1}{\alpha^2}\int_{S_e} d\bar{S}\cdot\nabla\left(e^{i\bar{\alpha}\cdot\bar{r}}\right) - \frac{1}{\alpha^2}\int_{S_V} d\bar{S}\cdot\nabla\left(e^{i\bar{\alpha}\cdot\bar{r}}\right)\right]$$

$$= -\frac{b^2}{\alpha}2\pi \int_0^\pi d\theta \sin\theta \cos\theta \sin(\alpha b\cos\theta)$$

$$= -\frac{4\pi b^3}{3} \qquad (15)$$

In (15) S_e and S_V are, respectively, the spherical surface enclosing the exclusion volume and the surface enclosing the volume V. The S_V term corresponds to an *edge* effect giving rise to the extinction theorem (Born and Wolf, 1975) and is neglected in the calculation of radiated power. The S_e term can be evaluated readily since S_e is a spherical surface of radius b. Only the real part is needed. The third equality is a result of $\alpha b \ll 1$ in the low-frequency limit. Hence

$$<I_s> = N\frac{|\bar{A}|^2}{2\eta}\left[1 + 8f(3H_o - 1)\right] \qquad (16)$$

where

$$H_o = \int_1^\infty d\sigma\, \sigma^2[g(\sigma b) - 1] \qquad (17)$$

The total scattered power P_s is obtained by integrating $<I_s>$ over a 4π solid angle which can be carried out readily. Note that without loss of generality, we can let $\hat{e}$ to be in the $\hat{z}$ direction. The result is

$$P_s = \frac{E_o^2}{2\eta}Nk^4\frac{8\pi}{3}a^6\left|\frac{y}{1-fy}\right|^2(1 - 8f + 24fH_o) \qquad (18)$$

Now consider a cylindrical volume of area A and length l containing n_o particles per unit volume. The input power is $P_{in} = 1/2 E_o^2 A(\epsilon_{eff}/\mu)^{1/2}$ and the scattered power is given by (18) with $N = n_o Al$. The scattering attenuation rate denoted by $\kappa_s = 2K_{is}$ is given by $P_s/(lP_{in})$ and is

$$\kappa_s = \frac{2f}{K_r}k^5 a^3\left|\frac{y}{1-fy}\right|^2(1 - 8f + 24fH_o) \qquad (19)$$

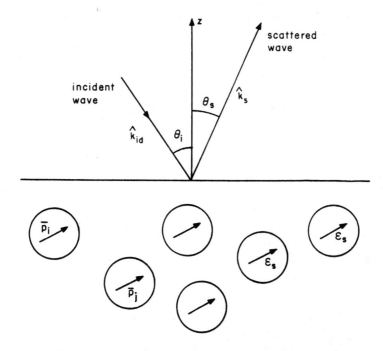

Fig. 6.3 Plane wave incident on a half-space of dielectric scatterers. Induced dipoles.

Using (5), Section 1, ϵ_{eff} of Section 2.1 and (19) gives

$$K^2 = k^2 \frac{(1+2fy)}{1-fy} + i2fk^5 a^3 \left|\frac{y}{1-fy}\right|^2 (1 - 8f + 24fH_o) \qquad (20)$$

Equation (20) gives the coherent propagation constant with scattering effects included. In view of (17), it depends on the pair-distribution function of particle positions. The result in (20) agrees with the quasicrystalline approximation in the low-frequency limit.

2.3 Coherent Reflection and Incoherent Scattering from a Half-Space of Scatterers

Consider a plane wave incident on a half-space of dielectric scatterers (Figure 6.3). We shall first consider vertically polarized incidence. The case of horizontally polarized incidence can be treated in a similar manner. The incident wave is in the direction (θ_i, ϕ_i) and the incident field is $\overline{E}_{inc} = \hat{\theta}_{id} E_o \exp(i\overline{k}_{id} \cdot \overline{r})$ with $\hat{k}_{id} = \sin\theta_i \cos\phi_i \hat{x} + \sin\theta_i \sin\phi_i \hat{y} - \cos\theta_i \hat{z}$, $\hat{\phi}_{id} = -\sin\phi_i \hat{x} + \cos\phi_i \hat{y}$, and $\hat{\theta}_{id} = \hat{\phi}_{id} \times \hat{k}_{id}$ forming the orthonormal triad. The subscript d is to denote the fact that the wave is propagating downward. In the lower region, the coherent effective propagation constant is K. The transmitted macroscopic field is $\overline{E}_t$ with propagation direction following Snell's law:

$$\overline{E}_t = \hat{\theta}_{td} T_{01}^{TM} \frac{k}{K} e^{i\overline{K}_d \cdot \overline{r}} \tag{1}$$

with K given by (20), Section 2.2, $\overline{K}_d = k \sin\theta_i \cos\phi_i \hat{x} + k \sin\theta_i \sin\phi_i \hat{y} - \sqrt{K^2 - k^2 \sin^2\theta_i}\,\hat{z}$, $\hat{\theta}_{td} = -\cos\theta_t \cos\phi_i \hat{x} - \cos\theta_t \sin\phi_i \hat{y} - \sin\theta_t \hat{z}$ and θ_t is the transmitted angle obeying Snell's law

$$\sin\theta_t = \frac{k}{K_r} \sin\theta_i \tag{2}$$

The Fresnel transmission coefficient for TM waves is T_{01}^{TM} with

$$T_{01}^{TM} = \frac{2K^2 k_{iz}}{K^2 k_{iz} + k^2 K_z} \tag{3}$$

where $k_{iz} = k \cos\theta_i$ and $K_z = \sqrt{K^2 - k^2 \sin^2\theta_i}$.

As in the previous section, dipoles are induced in the dielectric spheres and the dipole moment $\overline{p}_j$ of the jth particle, centered at $\overline{r}_j$ is, using (2), Section 2.2,

$$\overline{p}_j = \hat{\theta}_{td} \frac{3v_0 \epsilon y}{1 - fy} T_{01}^{TM} \frac{k}{K} e^{i\overline{K}_d \cdot \overline{r}} \tag{4}$$

The total radiation field of the dipoles is then, following Section 2.2

$$\overline{E}_s(\overline{r}) = \sum_{j=1}^{N} \overline{W} J_j \tag{5}$$

where $J_j = \exp[i(\overline{K}_d - \overline{k}_s) \cdot \overline{r}_j]$

$$\overline{W} = \frac{\omega^2 \mu e^{ikr}}{4\pi r}(\hat{k}_s \times \hat{\theta}_{td}) \times \hat{k}_s \frac{3v_0 \epsilon y}{1 - fy} T_{01}^{TM} \frac{k}{K} \tag{6}$$

and $\hat{k}_s = \sin\theta_s \cos\phi_s \hat{x} + \sin\theta_s \sin\phi_s \hat{y} + \cos\theta_s \hat{z}$ is the propagation direction of the scattered wave with $\overline{k}_s = k\hat{k}_s$.

Coherent Reflected Field

To calculate the coherent scattered field, the configuration average of (6) is taken

$$<\overline{E}_s(\overline{r})> = \overline{W} \sum_{j=1}^{N} <J_j> = \overline{W} N <J_j> \tag{7}$$

Since $p(\bar{r}_j) = 1/V$, we have, on evaluating $<J_j>$ by integrating over the lower half-space

$$<\overline{E}_s> = n_o \overline{W} \frac{4\pi^2 i}{(K_z + k_{iz})} \delta(k\sin\theta_s \cos\phi_s - k\sin\theta_i \cos\phi_i)$$
$$\times \delta(k\sin\theta_s \sin\phi_s - k\sin\theta_i \sin\phi_i) \quad (8)$$

Hence, the coherent reflected field is in the specular direction. For $\hat{k}_s$ in specular reflected direction

$$(\hat{k}_s \times \hat{\theta}_{td}) \times \hat{k}_s = -\cos(\theta_i + \theta_t)\hat{\theta}_i \quad (9)$$

Substituting (6) and (9) into (8) and making use of the relations $K^2/k^2 - 1 = 3fy/(1-fy)$, $K^2 - k^2 = (K_z + k_{iz})(K_z - k_{iz})$, $(K_z + k_{iz})\cos(\theta_i - \theta_t) = (k_{iz}K^2 + k^2 K_z)/(kK)$, and

$$\left(\frac{K_z - k_{iz}}{K_z + k_{iz}}\right) \frac{\cos(\theta_i + \theta_t)}{\cos(\theta_i - \theta_t)} = \frac{K^2 k_{iz} - k^2 K_z}{K^2 k_{iz} + k^2 K_z} \equiv R_{01}^{TM} \quad (10)$$

we get (Problem 4)

$$<\overline{E}_s> = -\hat{\theta}_i i 2\pi k_{iz} R_{01}^{TM} \frac{e^{ikr}}{r} \delta(k\sin\theta_s \cos\phi_s - k\sin\theta_i \cos\phi_i)$$
$$\times \delta(k\sin\theta_s \sin\phi_s - k\sin\theta_i \sin\phi_i) \quad (11)$$

Thus the coherent reflected wave is in the specular direction, containing no depolarization and obeys the Fresnel reflection formula for TM waves.

The case of horizontally polarized incident waves is treated in a similar manner. For an incident field of $\overline{E}_{inc} = \hat{\phi} \cdot E_o \exp(i\overline{k}_{id} \cdot \bar{r})$. The result for the coherent reflected field is

$$<\overline{E}_s(\bar{r})> = -\hat{\phi}_i i 2\pi k_{iz} R_{01}^{TE} \frac{e^{ikr}}{r} \delta(k\sin\theta_s \cos\phi_s - k\sin\theta_i \cos\phi_i)$$
$$\times \delta(k\sin\theta_s \sin\phi_s - k\sin\theta_i \sin\phi_i) \quad (12)$$

where

$$R_{01}^{TE} = \frac{k_{iz} - K_z}{k_{iz} + K_z} \quad (13)$$

is the Fresnel reflection coefficient for TE waves.

Incoherent Scattered Field

The incoherent scattered field $\overline{\mathcal{E}}_s$ is $\overline{\mathcal{E}}_s(\overline{r}) = \overline{E}_s(\overline{r}) - <\overline{E}_s(\overline{r})>$. Hence, the incoherent intensity is

$$<\overline{\mathcal{E}}_s(\overline{r}) \cdot \overline{\mathcal{E}}_s^*(\overline{r})> = |\overline{W}|^2 \left\{ \sum_{i=1}^{N} <|J_i|^2> \right.$$
$$\left. + \sum_{i=1}^{N} \sum_{\substack{j=1 \\ i \neq j}}^{N} <J_i J_j^*> - N^2 |<J_1>|^2 \right\} \quad (14)$$

Following a similar procedure as in the case of Section 2.2, we have

$$<\overline{\mathcal{E}}_s^* \cdot \overline{\mathcal{E}}_s> = |\overline{W}|^2 \left\{ \frac{n_o A_o}{2Im(K_z)} + n_o^2 \int d\overline{r}_i \int d\overline{r}_j \right.$$
$$\left. \times [g(\overline{r}_i - \overline{r}_j) - 1] e^{i\overline{K}_d \cdot \overline{r}_i - i\overline{K}_d^* \cdot \overline{r}_j} e^{-i\overline{k}_s \cdot \overline{r}_i + i\overline{k}_s \cdot \overline{r}_j} \right\} \quad (15)$$

where A_o is the area of the target area. The integral in (15) can be carried out readily. Hence.

$$<\overline{\mathcal{E}}_s^*(\overline{r}) \cdot \overline{\mathcal{E}}_s(\overline{r})> = |\overline{W}|^2 \frac{n_o A_o}{2Im(K_z)} \left\{ 1 + n_o \int d\overline{r} [g(\overline{r}) - 1] e^{i(Re\overline{K}_d - \overline{k}_s) \cdot \overline{r}} \right\} \quad (16)$$

where for vertically polarized incidence, $\overline{W}$ is given by (6) and for horizontally polarized incidence

$$\overline{W} = \frac{\omega^2 \mu e^{ikr}}{4\pi r} (\hat{k}_s \times \hat{\phi}_i) \times \hat{k}_s \frac{3v_o \epsilon y}{1 - fy} (1 + R_{01}^{TE}) \quad (17)$$

We can readily calculate the bistatic scattering coefficient and the backscattering coefficient. In the backscattering direction, $\theta_s = \theta_i$ and $\phi_s = \pi + \phi_i$, $\hat{k}_s \times (\hat{k}_s \times \hat{\phi}_i) = \hat{\phi}_s$, while $\hat{k}_s \times (\hat{k}_s \times \hat{\theta}_{td}) = -\cos(\theta_i - \theta_t)\hat{\theta}_s$.

In the low-frequency limit, the exponent in (16) can be replaced by zero. Hence, the backscattering coefficients are

$$\sigma_{vv}(\theta_i) = \sigma_{hh}(\theta_i) = \left| \frac{(K_z - k_{iz})k_{iz}}{n_o} \right|^2 \frac{n_o}{2\pi Im(K_z)}$$
$$\times \left\{ 1 + n_o \int_{-\infty}^{\infty} d\overline{r} [g(\overline{r}) - 1] \right\} \quad (18)$$

The results in (18) are identical to that of a more rigorous multiple scattering theory in the low-frequency limit to be treated in Section 7.5.

3 MULTIPLE SCATTERING EQUATIONS AND DERIVATIONS

3.1 Multiple Scattering Equations – Introduction

Consider a distribution of N particles in a volume V. The particles need not be identical in shape and size. However, interpenetration of particles is not allowed. Let $\overline{E}_{inc}$ be the incident field. The total field $\overline{E}$ is the sum of the incident and the scattered fields from all particles. Let $\overline{E}_{sj}$ denote the scattered field from the j th particle. Hence,

$$\overline{E} = \overline{E}_{inc} + \sum_{j=1}^{N} \overline{E}_{sj} \qquad (1)$$

Next, a field $\overline{E}_j^E$ (called the j th particle exciting field) is defined such that the relation between the scattered field $\overline{E}_{sj}$ and $\overline{E}_j^E$ is that of the single particle j in the absence of all other particles. The relation is expressed in terms of the transition operator T_j of Sections 5.1 and 5.2, Chapter 3,

$$\overline{E}_{sj} = T_j \overline{E}_j^E \qquad (2)$$

Note that $\overline{E}_j^E$ is a function of all the particles. Hence $\overline{E}_{sj}$ is also a function of all the particles. However, T_j depends on the particle j only. Next, the exciting field for j th particle is assumed to be equal to the total field less its own scattered field. Thus,

$$\overline{E}_j^E = \overline{E} - \overline{E}_{sj} \qquad (3)$$

Hence combining (1) through (3), we have

$$\overline{E}_j^E = \overline{E}_{inc} + \sum_{\substack{l=1 \\ l \neq j}}^{N} T_l \overline{E}_l^E \qquad (4)$$

with $j = 1, \ldots, N$. The single-particle transition operator T_l is assumed to be known. Hence equation (4) consists of N equations for the N unknowns $\overline{E}_1^E, \overline{E}_2^E, \ldots, \overline{E}_N^E$ and, in principle, can be solved.

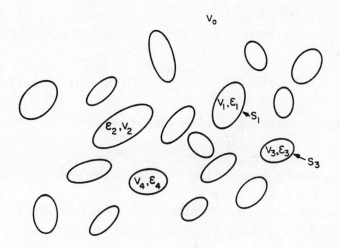

Fig. 6.4 Particles $1, 2, \ldots, N$ occupying regions $V_1, V_2, \ldots, V_N$ and bounded by surfaces $S_1, S_2, \ldots, S_N$, respectively.

The derivation of the multiple scattering equation (4) has been done in a heuristic manner. In the following two sections, two derivations of (4) will be given. The first derivation utilizes the N-particle Green's dyadic and the second derivation is based on the vector spherical wave expansion of the T-matrix formalism.

The early contributions to multiple scattering theory for discrete scatterers were made by Foldy (1945), Lax (1951, 1952), Waterman and Truell (1961), Twersky (1962, 1964, 1967), Fikioris and Waterman (1964), Mathur and Yeh (1964), Vezzetti and Keller (1967), and Frisch (1968).

3.2 Derivation of Multiple Scattering Equations Using N-Particle Green's Dyadic

Consider N particles occupying regions $V_1, V_2, V_3, \ldots, V_N$ bounded, respectively, by surfaces $S_1, S_2, \ldots, S_N$. The jth particle has permittivity ϵ_j, permeability μ, and wavenumber k_j. The particles need not be identical in shape and size (Figure 6.4). The background region is denoted by V_o. The N-particle dyadic Green's function $\overline{\overline{G}}(\bar{r}, \bar{r}')$ is defined as follows:

$$\overline{\overline{G}}(\bar{r}, \bar{r}') = \overline{\overline{G}}_{ij}(\bar{r}, \bar{r}') \tag{1}$$

The convention of the dyadic Green's function is (Tai, 1971) that the first subscript denotes the region of the observation point and the second subscript denotes the region of the source.

Derivation of Multiple Scattering Equations

Using the definition (1), we shall show that the following integral equation

$$\overline{\overline{G}}(\bar{r},\bar{r}') = \overline{\overline{G}}_o(\bar{r},\bar{r}') + \sum_{l=1}^{N} \int_{V_l} d\bar{r}'' \overline{\overline{G}}_o(\bar{r},\bar{r}'')(k_l^2 - k^2)\overline{\overline{G}}(\bar{r}'',\bar{r}') \quad (2)$$

holds for all $\bar{r}$ and $\bar{r}'$. In (2), $\overline{\overline{G}}_o(\bar{r},\bar{r}')$ is the free-space dyadic Green's function. Using operator notation, (2) can be put in a compact form. We have also used the convention that when two dyads are placed next to each other, their dot product is assumed.

Consider the following two cases:

1. Case A: Source point $\bar{r}'$ is in region V_o. The governing equations and the boundary conditions are

$$\nabla \times \nabla \times \overline{\overline{G}}_{oo}(\bar{r},\bar{r}') - k^2 \overline{\overline{G}}_{oo}(\bar{r},\bar{r}') = \overline{\overline{I}}\delta(\bar{r}-\bar{r}') \quad (3)$$

$$\nabla \times \nabla \times \overline{\overline{G}}_{jo}(\bar{r},\bar{r}') - k_j^2 \overline{\overline{G}}_{jo}(\bar{r},\bar{r}') = 0 \quad (4)$$

for $j = 1,2,\cdots,N$. For $\bar{r}$ on surface S_j

$$\hat{n} \times \overline{\overline{G}}_{oo}(\bar{r},\bar{r}') = \hat{n} \times \overline{\overline{G}}_{jo}(\bar{r},\bar{r}') \quad (5)$$

$$\hat{n} \times \nabla \times \overline{\overline{G}}_{oo}(\bar{r},\bar{r}') = \hat{n} \times \nabla \times \overline{\overline{G}}_{jo}(\bar{r},\bar{r}') \quad (6)$$

Apply the vector Green's theorem to the vectors $\overline{\overline{G}}_o(\bar{r},\bar{r}'')\cdot\bar{a}$ and $\overline{\overline{G}}_{oo}(\bar{r},\bar{r}')\cdot\bar{b}$, where $\bar{a}$ and $\bar{b}$ are constant vectors, by integrating over volume V_o and making use of vector wave equation (3), to obtain

$$-\sum_{l=1}^{N}\int d\overline{S}_l(\bar{r}) \cdot \left\{ \left[\overline{\overline{G}}_{oo}(\bar{r},\bar{r}')\cdot\bar{b}\right] \times \nabla \times \left[\overline{\overline{G}}_o(\bar{r},\bar{r}'')\cdot\bar{a}\right] \right.$$

$$\left. - \left[\overline{\overline{G}}_o(\bar{r},\bar{r}'')\cdot\bar{a}\right] \times \nabla \times \left[\overline{\overline{G}}_{oo}(\bar{r},\bar{r}')\cdot\bar{b}\right] \right\}$$

$$= \begin{cases} \bar{a}\cdot\overline{\overline{G}}_o(\bar{r}'',\bar{r}')\cdot\bar{b} - \bar{a}\cdot\overline{\overline{G}}_{oo}(\bar{r}'',\bar{r}')\cdot\bar{b} & \text{for } \bar{r}'' \text{ in region } 0 \\ \bar{a}\cdot\overline{\overline{G}}_o(\bar{r}'',\bar{r}')\cdot\bar{b} & \text{for } \bar{r}'' \text{ not in region } 0 \end{cases}$$
(7)

Note that $\bar{r}'$ has to be in region 0 while $\bar{r}''$ can be anywhere. In deriving (7), we have used the symmetry property of the Green's dyadic. Next apply vector Green's theorem to the two vectors $\overline{\overline{G}}_{lo}(\bar{r},\bar{r}')\cdot\bar{b}$ and $\overline{\overline{G}}_o(\bar{r},\bar{r}'')\cdot\bar{a}$

by integrating over volume V_l. Making use of vector wave equation (4) and summing over $l = 1, 2, \cdots, N$ we obtain

$$-\sum_{l=1}^{N}\int d\overline{S}_l(\bar{r}) \cdot \left\{ \left[\overline{\overline{G}}_{lo}(\bar{r},\bar{r}') \cdot \bar{b}\right] \times \nabla \times \left[\overline{\overline{G}}_o(\bar{r},\bar{r}'') \cdot \bar{a}\right] \right.$$

$$\left. - \left[\overline{\overline{G}}_o(\bar{r},\bar{r}'') \cdot \bar{a}\right] \times \nabla \times \left[\overline{\overline{G}}_{lo}(\bar{r},\bar{r}') \cdot \bar{b}\right] \right\} \quad (8)$$

$$= \sum_{l=1}^{N}\int dV_l(\bar{r}) \left\{ \left[\overline{\overline{G}}_o(\bar{r},\bar{r}'') \cdot \bar{a}\right] \cdot (k^2 - k_l^2) \left[\overline{\overline{G}}_{lo}(\bar{r},\bar{r}') \cdot \bar{b}\right] \right\}$$

$$+ \begin{cases} 0 & \text{for } \bar{r}'' \text{ in region } 0 \\ \bar{a} \cdot \overline{\overline{G}}_{lo}(\bar{r}'',\bar{r}') \cdot \bar{b} & \text{for } \bar{r}'' \text{ in region } l \end{cases}$$

In view of the boundary conditions (5) and (6), equations (7) and (8) are equal. Hence, equating (7) and (8) and canceling constant vectors $\bar{a}$ and $\bar{b}$ on both sides of the equation give

$$\overline{\overline{G}}_o(\bar{r}'',\bar{r}') + \sum_{l=1}^{N}\int dV_l(\bar{r})\overline{\overline{G}}_o(\bar{r}'',\bar{r})(k_l^2 - k^2)\overline{\overline{G}}_{lo}(\bar{r},\bar{r}')$$

$$= \begin{cases} \overline{\overline{G}}_{oo}(\bar{r}'',\bar{r}') & \bar{r}'' \text{ in region } 0 \\ \overline{\overline{G}}_{lo}(\bar{r}'',\bar{r}') & \bar{r}'' \text{ in region } l \end{cases} \quad (9)$$

2. Case B: Source point $\bar{r}'$ is in region V_j, $j \neq 0$. The governing equations and the boundary conditions are

$$\nabla \times \nabla \times \overline{\overline{G}}_{jj}(\bar{r},\bar{r}') - k_j^2 \overline{\overline{G}}_{jj}(\bar{r},\bar{r}') = \overline{\overline{I}}\delta(\bar{r}-\bar{r}') \quad (10)$$

$$\nabla \times \nabla \times \overline{\overline{G}}_{lj}(\bar{r},\bar{r}') - k_l^2 \overline{\overline{G}}_{lj}(\bar{r},\bar{r}') = 0 \quad (11)$$

for $l \neq j, l = 0, 1, 2, \ldots, N$. The boundary conditions for $\bar{r}$ on S_l, $l = 1, \ldots, N$, are such that $\hat{n} \times \overline{\overline{G}}_{oj}(\bar{r},\bar{r}') = \hat{n} \times \overline{\overline{G}}_{lj}(\bar{r},\bar{r}')$ and $\hat{n} \times \nabla \times \overline{\overline{G}}_{oj}(\bar{r},\bar{r}') = \hat{n} \times \nabla \times \overline{\overline{G}}_{lj}(\bar{r},\bar{r}')$. Similar derivations as in Case A can be applied to this set of Green's functions (Problem 7) to give

$$\overline{\overline{G}}_o(\bar{r}'',\bar{r}') + \sum_{l=1}^{N}\int dV_l(\bar{r})\overline{\overline{G}}_o(\bar{r}'',\bar{r})(k_l^2 - k^2)\overline{\overline{G}}_{lj}(\bar{r},\bar{r}')$$

$$= \begin{cases} \overline{\overline{G}}_{oj}(\bar{r}'',\bar{r}') & \text{for } \bar{r}'' \text{ in region } 0 \\ \overline{\overline{G}}_{mj}(\bar{r}'',\bar{r}') & \text{for } \bar{r}'' \text{ in region } m \text{ and } m \neq j \\ \overline{\overline{G}}_{jj}(\bar{r}'',\bar{r}') & \text{for } \bar{r}'' \text{ in region } j \end{cases} \quad (12)$$

Derivation of Multiple Scattering Equations

Comparing equations (9) and (12), it readily follows that the two equations can be combined into a single equation (2) which holds for all locations of $\bar{r}$ and $\bar{r}'$.

Operator Notations

The operator notation is next introduced to put the scattering equations in a more compact form. We let $\bar{\bar{G}}_{op}$ be the N-particle dyadic Green's operator and $\bar{\bar{G}}_o$ be the free-space dyadic Green's operator. Dirac's notations will be used. In coordinate representation

$$\bar{\bar{G}}(\bar{r},\bar{r}') = <\bar{r}|\bar{\bar{G}}_{op}|\bar{r}'> \tag{13}$$

Thus, the N-particle dyadic Green's function is equal to taking the N-particle dyadic Green's operator in the coordinate representation. The free-space Green's function is

$$\bar{\bar{G}}_o(\bar{r},\bar{r}') = <\bar{r}|\bar{\bar{G}}_{oop}|\bar{r}'> \tag{14}$$

Define a potential operator of the jth particle, U_{jop} as follows. The jth particle is centered at $\bar{r}_j$ and occupies region r_j. Then the potential function for jth particle is

$$U_j(\bar{r}-\bar{r}_j) = \begin{cases} 0 & \text{for } \bar{r} \notin V_j \\ k_j^2 - k^2 & \text{for } \bar{r} \in V_j \end{cases} \tag{15}$$

The potential operator U_{jop} is such that

$$<\bar{r}|U_{jop}|\bar{r}'> = <\bar{r}|U_j(\bar{r}_{op}-\bar{r}_j)|\bar{r}'> = U_j(\bar{r}-\bar{r}_j)<\bar{r}|\bar{r}'> \tag{16}$$

where $\bar{r}_{op}$ is the position operator and $<\bar{r}|\bar{r}'> = \delta(\bar{r}-\bar{r}')$. The free-space dyadic Green's operator can be expressed in terms of momentum operator $\bar{p}_{op}$ (Problem 8)

$$\bar{\bar{G}}_{oop} = -\left(\bar{p}_{op}\bar{p}_{op} - p_{op}^2\bar{\bar{I}}_{op} + k^2\bar{\bar{I}}_{op}\right)^{-1} \tag{17}$$

The inner product of a momentum eigenstate and a position eigenstate is $<\bar{r}|\bar{p}> = e^{i\bar{p}\cdot\bar{r}}$. The orthogonality relation for momentum eigenstates is $<\bar{p}|\bar{p}'> = (2\pi)^3\delta(\bar{p}-\bar{p}')$. The unit dyad operator can be represented in coordinate and momentum representation as

$$\bar{\bar{I}}_{op} = \int d\bar{r}|\bar{r}>\bar{\bar{I}}<\bar{r}| = \int \frac{d\bar{p}}{(2\pi)^3}|\bar{p}>\bar{\bar{I}}<\bar{p}| \tag{18}$$

In operator notation, (2) can be written in the compact form (Problem 9)

$$\bar{\bar{G}}_{op} = \bar{\bar{G}}_{oop} + \bar{\bar{G}}_{oop} \sum_{j=1}^{N} U_{jop} \bar{\bar{G}}_{op} \qquad (19)$$

Equation (2) for the dyadic Green's function is the operator equation (19) in coordinate representation. We shall suppress the *op* (operator) subscript since it can be inferred from the context whether a quantity is a function or an operator.

Transition Operator

Consider the case of a single scatterer centered at the origin. From (19) the single-particle Green's operator $\bar{\bar{G}}_s$ obeys the equation

$$\bar{\bar{G}}_s = \bar{\bar{G}}_o + \bar{\bar{G}}_o \bar{\bar{U}} \bar{\bar{G}}_s \qquad (20)$$

Following Frisch (1968), we use the following diagrammatic notations

$$\bar{\bar{G}}_o = \text{———} \qquad (21)$$

$$\bar{\bar{U}} = \bullet \qquad (22)$$

Hence (20) assumes the following diagrammatic form

$$\bar{\bar{G}}_s = \text{———} + \text{———}\bullet\, \bar{\bar{G}}_s \qquad (23)$$

The formal perturbation expansion of (23) is

$$\bar{\bar{G}}_s = \text{———} + \text{——}\bullet\text{——} + \text{——}\bullet\text{——}\bullet\text{——} \\ + \text{——}\bullet\text{——}\bullet\text{——}\bullet\text{——} + \cdots \qquad (24)$$

The dyad transition operator $\bar{\bar{T}}$ is defined to be

$$\bar{\bar{T}} = \bullet + \bullet\text{——}\bullet + \bullet\text{——}\bullet\text{——}\bullet + \cdots \qquad (25)$$

so that

$$\bar{\bar{G}}_s = \bar{\bar{G}}_o + \bar{\bar{G}}_o \bar{\bar{T}} \bar{\bar{G}}_o \qquad (26)$$

From (25), it follows that the transition operator $\bar{\bar{T}}$ satisfies the equation

$$\bar{\bar{T}} = \bar{\bar{U}} + \bar{\bar{U}} \bar{\bar{G}}_o \bar{\bar{T}} \qquad (27)$$

known as the Lippmann–Schwinger equation (Newton, 1966). The operator $\overline{\overline{T}}$ can be written formally as

$$\overline{\overline{T}} = \left(\overline{\overline{I}} - \overline{\overline{U}}\,\overline{\overline{G}}_o\right)^{-1} \overline{\overline{U}} \tag{28}$$

For a jth particle centered at $\bar{r}_j$, then the transition operator is $\overline{\overline{T}}_j$ with

$$\overline{\overline{T}}_j = \left(\overline{\overline{I}} - \overline{\overline{U}}_j \overline{\overline{G}}_o\right)^{-1} \overline{\overline{U}}_j \tag{29}$$

Multiple Scattering Equations

The problem of N-particle scattering is governed by (19). Let

$$\overline{\overline{U}}_j = U_j \overline{\overline{I}} = U(\bar{r} - \bar{r}_j)\overline{\overline{I}} \tag{30}$$

and define the jth-particle exciting operator $\overline{\overline{G}}_j$ as

$$\overline{\overline{G}}_j = \overline{\overline{G}}_o + \overline{\overline{G}}_o \sum_{\substack{l=1 \\ l \neq j}}^{N} \overline{\overline{U}}_l \overline{\overline{G}} \tag{31}$$

then, from (19) and (31), we have

$$\overline{\overline{G}} = \overline{\overline{G}}_j + \overline{\overline{G}}_o \overline{\overline{U}}_j \overline{\overline{G}} \tag{32}$$

Multiplying both sides of (32) by $\overline{\overline{U}}_j$ and making use of (29) gives

$$\overline{\overline{U}}_j \overline{\overline{G}} = \overline{\overline{T}}_j \overline{\overline{G}}_j \tag{33}$$

Substituting (33) in (31) and (19) gives, respectively,

$$\overline{\overline{G}}_j = \overline{\overline{G}}_o + \overline{\overline{G}}_o \sum_{\substack{l=1 \\ l \neq j}}^{N} \overline{\overline{T}}_l \overline{\overline{G}}_l \tag{34}$$

and

$$\overline{\overline{G}} = \overline{\overline{G}}_o + \overline{\overline{G}}_o \sum_{j=1}^{N} \overline{\overline{T}}_j \overline{\overline{G}}_j \tag{35}$$

Equations (34) and (35) are the multiple scattering equations postulated in Section 3.2 and rigorously derived here.

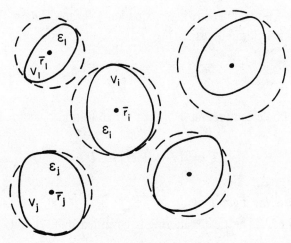

Fig. 6.5 Particles $1, 2, \ldots, N$ occupying regions $V_1, V_2, \ldots, V_N$ and bounded by surfaces $S_1, S_2, \ldots, S_N$, respectively. They are enclosed by non-overlapping circumscribing spheres.

We can apply the source state $|\bar{\xi}>$ to the right-hand side of (34) and (35). With $\overline{\overline{G}}|\bar{\xi}>$ being the total field state, $|\overline{E}>$, $\overline{\overline{G}}_o|\bar{\xi}>=|\overline{E}_{inc}>$ being the incident field state, and $|\overline{E}_j>=\overline{\overline{G}}_j|\bar{\xi}>$ being the exciting field of the j th particle,

$$|\overline{E}>=|\overline{E}_{inc}> +\overline{\overline{G}}_o \sum_{j=1}^{N} \overline{\overline{T}}_j |\overline{E}_j> \tag{36}$$

$$|\overline{E}_j>=|\overline{E}_{inc}> +\overline{\overline{G}}_o \sum_{\substack{l=1 \\ l \neq j}}^{N} \overline{\overline{T}}_l |\overline{E}_l> \tag{37}$$

3.3 Derivation of Multiple Scattering Equations Using T-Matrix Formalism

The multiple scattering equations can also be derived by extending the T-matrix formalism to an arbitrary number of particles. The formalism, discussed in Chapter 3, is based on the extended boundary condition and utilizes vector spherical wave functions as basis. The derivation below follows that of Peterson and Strom (1973). The tools used in the expansion include the spherical wave transformations associated with a translation of the origin.

Consider N scatterers bounded by surfaces $S_1, S_2, \ldots, S_N$ occupying regions $V_1, V_2, \ldots, V_N$. The scatterers are centered at $\bar{r}_1, \bar{r}_2, \ldots, \bar{r}_N$. It is also assumed that the scatterers are enclosed by circumscribing spheres that do not overlap each other (Figure 6.5). We consider a coordinate system with origin 0 outside the particles. Let the background region be denoted by V_o. The i th scatterer has permittivity equal to ϵ_i, wavenumber k_i, and permeability μ. Consider an incident field expressed in terms of spherical wave functions

$$\overline{E}^{inc}(\bar{r}) = \sum_{m,n} \left[a_{mn}^{(M)} Rg\overline{M}_{mn}(kr,\theta,\phi) + a_{mn}^{(N)} Rg\overline{N}_{mn}(kr,\theta,\phi) \right] \quad (1)$$

Applying the extended boundary condition technique gives the following equation for the total field

$$\begin{cases} \overline{E}(\bar{r}') & \text{if } \bar{r}' \in V_o \\ 0 & \text{if } \bar{r}' \notin V_o \end{cases} = \sum_{j=1}^{N} \int_{S_j} dS \left\{ \overline{\overline{G}}_o(\bar{r}',\bar{r}) \cdot i\omega\mu \hat{n}(\bar{r}) \times \overline{H}(\bar{r}) \right.$$

$$\left. + \nabla' \times \overline{\overline{G}}_o(\bar{r}',\bar{r}) \cdot \hat{n}(\bar{r}) \times \overline{E}(\bar{r}) \right\} + \overline{E}^{inc}(\bar{r}') \quad (2)$$

By using the boundary conditions and the translational invariance of the free-space dyadic Green's function, equation (2) becomes

$$\begin{cases} \overline{E}(\bar{r}') & \text{if } \bar{r}' \in V_o \\ 0 & \text{if } \bar{r}' \notin V_o \end{cases}$$

$$= \sum_{j=1}^{N} \int_{S_j} dS \left\{ \overline{\overline{G}}_o(\overline{r'r_j}, \overline{rr_j}) \cdot i\omega\mu \hat{n}(\bar{r}) \times \overline{H}_j(\bar{r}) \right. \quad (3)$$

$$\left. + \nabla_{\overline{r'r_j}} \times \overline{\overline{G}}_o(\overline{r'r_j},\overline{rr_j}) \cdot \hat{n}(\bar{r}) \times \overline{E}_j(\bar{r}) \right\} + \overline{E}_{inc}(\bar{r}')$$

where $\overline{E}_j$ and $\overline{H}_j$ denote the fields in the j th region and $\overline{rr_j}$ denotes the vector pointing from $\bar{r}_j$ to $\bar{r}$.

Let $\bar{r}'$ be in the V_l region and within the inscribing sphere of the l th particle. The inscribing sphere for the l th particle is the smallest sphere centered at $\bar{r}_l$ and contained inside the l th particle. Then
(a) $|\overline{r'r_l}| < |\overline{rr_l}|$ for $\bar{r}$ on S_l.
(b) $|\overline{r'r_j}| > |\overline{rr_j}|$ for $\bar{r}$ on S_j and $j \neq l$.
(c) $|\overline{r'r_l}| < |\overline{r_l r_j}|$ for $l \neq j$.
We can use the appropriate expansion of the dyadic Green's function $\overline{\overline{G}}_o$ in (10), Section 5.4 (Chapter 3), based on the conditions (a) and (b) and the boundary conditions of continuous tangential fields in (3). Hence

$$0 = ik \sum_{n,m} \int_{S_l} dS(-1)^m \left[Rg\overline{M}_{mn}(k\overline{r'r_l})\overline{M}_{-mn}(k\overline{rr_l}) \right.$$

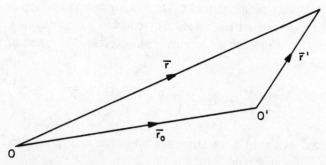

Fig. 6.6 Translational addition theorem for vector spherical waves. Three vectors $\bar{r}$, $\bar{r}'$, and $\bar{r}_o$ form a triangle.

$$+ Rg\overline{N}_{mn}(k\overline{r'r_l})\overline{N}_{-mn}(k\overline{rr_l}) \Big] \cdot i\omega\mu\hat{n}(\bar{r}) \times \overline{H}_l(\bar{r})$$

$$+ ik^2 \sum_{n,m} \int_{S_l} dS(-1)^m \Big[Rg\overline{N}_{mn}(k\overline{r'r_l})\overline{M}_{-mn}(k\overline{rr_l})$$

$$+ Rg\overline{M}_{mn}(k\overline{r'r_l})\overline{N}_{-mn}(k\overline{rr_l}) \Big] \cdot \hat{n}(\bar{r}) \times \overline{E}_l(\bar{r})$$

$$+ ik \sum_{\substack{j=1 \\ j \neq l}}^{N} \sum_{n,m} \int_{S_j} dS(-1)^m \Big[\overline{M}_{mn}(k\overline{r'r_j}) Rg\overline{M}_{-mn}(k\overline{rr_j})$$

$$+ \overline{N}_{mn}(k\overline{r'r_j}) Rg\overline{N}_{-mn}(k\overline{rr_j}) \Big] \cdot i\omega\mu\hat{n}(\bar{r}) \times \overline{H}_j(\bar{r})$$

$$+ ik^2 \sum_{j \neq l} \sum_{n,m} \int_{S_j} dS(-1)^m \Big[\overline{N}_{mn}(k\overline{r'r_j}) Rg\overline{M}_{-mn}(k\overline{rr_j})$$

$$+ \overline{M}_{mn}(k\overline{r'r_j}) Rg\overline{N}_{-mn}(k\overline{rr_j}) \Big] \cdot \hat{n}(\bar{r}) \times \overline{E}_j(\bar{r})$$

$$+ \sum_{m,n} [a_{mn}^{(M)} Rg\overline{M}_{mn}(k\bar{r}') + a_{mn}^{(N)} Rg\overline{N}_{mn}(k\bar{r}')] \qquad (4)$$

for $l = 1, 2, \ldots, N$. The translational addition theorem for vector spherical waves is next introduced in (4).

Translation Addition Theorem for Vector Spherical Waves

Consider three vectors $\bar{r}$, $\bar{r}'$ and $\bar{r}_o$ forming a triangle (Figure 6.6) such that

$$\bar{r} = \bar{r}' + \bar{r}_o \qquad (5)$$

Derivation of Multiple Scattering Equations

Then the translation addition theorem is (Cruzan, 1961, 1962)

$$Rg\overline{M}_{mn}(k\bar{r}) = \sum_{\nu\mu} \left\{ RgA_{\mu\nu mn}(k\bar{r}_o) Rg\overline{M}_{\mu\nu}(k\bar{r}') + RgB_{\mu\nu mn}(k\bar{r}_o) Rg\overline{N}_{\mu\nu}(k\bar{r}') \right\} \quad (6)$$

$$Rg\overline{N}_{mn}(k\bar{r}) = \sum_{\nu\mu} \left\{ RgB_{\mu\nu mn}(k\bar{r}_o) Rg\overline{M}_{\mu\nu}(k\bar{r}') + RgA_{\mu\nu mn}(k\bar{r}_o) Rg\overline{N}_{\mu\nu}(k\bar{r}') \right\} \quad (7)$$

and for $r_o > r'$

$$\overline{M}_{mn}(k\bar{r}) = \sum_{\nu\mu} \left\{ A_{\mu\nu mn}(k\bar{r}_o) Rg\overline{M}_{\mu\nu}(k\bar{r}') + B_{\mu\nu mn}(k\bar{r}_o) Rg\overline{N}_{\mu\nu}(k\bar{r}') \right\} \quad (8)$$

$$\overline{N}_{mn}(k\bar{r}) = \sum_{\nu\mu} \left\{ B_{\mu\nu mn}(k\bar{r}_o) Rg\overline{M}_{\mu\nu}(k\bar{r}') + A_{\mu\nu mn}(k\bar{r}_o) Rg\overline{N}_{\mu\nu}(k\bar{r}') \right\} \quad (9)$$

where

$$A_{\mu\nu mn}(k\bar{r}_o) = \frac{\gamma_{mn}}{\gamma_{\mu\nu}} (-1)^\mu \sum_p a(m,n \mid -\mu,\nu \mid p) a(n,\nu,p) \\ \times h_p(kr_o) Y_p^{m-\mu}(\theta_o, \phi_o) \quad (10)$$

$$B_{\mu\nu mn}(k\bar{r}_o) = \frac{\gamma_{mn}}{\gamma_{\mu\nu}} (-1)^{\mu+1} \sum_p a(m,n \mid -\mu,\nu \mid p, p-1) b(n,\nu,p) \\ \times h_p(kr_o) Y_p^{m-\mu}(\theta_o, \phi_o) \quad (11)$$

and γ_{mn} is defined in (19), Section 5.1, Chapter 3. The expressions $RgA_{\mu\nu mn}(k\bar{r}_o)$ and $RgB_{\mu\nu mn}(k\bar{r}_o)$ are respectively those in (10) and (11) with $h_p(k\bar{r}_o)$ replaced by $j_p(k\bar{r}_o)$. In (10) and (11)

$$a(m,n \mid \mu,\nu \mid p) = (-1)^{m+\mu}(2p+1) \\ \times \left[\frac{(n+m)!(\nu+\mu)!(p-m-\mu)!}{(n-m)!(\nu-\mu)!(p+m+\mu)!} \right]^{1/2} \\ \begin{pmatrix} n & \nu & p \\ m & \mu & -(m+\mu) \end{pmatrix} \begin{pmatrix} n & \nu & p \\ 0 & 0 & 0 \end{pmatrix} \quad (12)$$

$$a(m,n \mid \mu,\nu \mid p,q) = (-1)^{m+\mu}(2p+1)$$

$$\times \left[\frac{(n+m)!(\nu+\mu)!(p-m-\mu)!}{(n-m)!(\nu-\mu)!(p+m+\mu)!}\right]^{1/2}$$

$$\begin{pmatrix} n & \nu & p \\ m & \mu & -(m+\mu) \end{pmatrix} \begin{pmatrix} n & \nu & q \\ 0 & 0 & 0 \end{pmatrix} \qquad (13)$$

$$a(n,\nu,p) = \frac{i^{\nu-n+p}}{2\nu(\nu+1)}\Big[2\nu(\nu+1)(2\nu+1)$$

$$+ (\nu+1)(n+\nu-p)(n+p-\nu+1) \qquad (14)$$

$$- \nu(n+\nu+p+2)(\nu+p-n+1)\Big]$$

$$b(n,\nu,p) = -\frac{(2\nu+1)}{2\nu(\nu+1)} i^{\nu+p-n}\Big[(n+\nu+p+1)(\nu+p-n)$$

$$\times (n+p-\nu)(n+\nu-p+1)\Big]^{1/2} \qquad (15)$$

and

$$\begin{pmatrix} j_1 & j_2 & j_3 \\ m_1 & m_2 & -(m_1+m_2) \end{pmatrix} \qquad (16)$$

are the Wigner $3j$ symbols (Edmonds. 1957). The Wigner $3j$ symbols can be calculated by a computer code. Rotenberg (1959) also contains a tabulation of some of the values. Nonzero values of Wigner $3j$ symbols require that in (16), $|m_i| \leq j_i$, $i = 1, 2, 3$ and $|j_1 - j_2| \leq j_3 \leq j_1 + j_2$.

Since $\overline{r'r_j} = \overline{r'r_l} + \overline{r_lr_j}$ and $\overline{r'} = \overline{r'r_l} + \overline{r_l}$, we can use the translation addition theorem to expand $\overline{M}_{mn}(k\overline{r'r_j})$, $\overline{N}_{mn}(k\overline{r'r_j})$, $Rg\overline{M}_{mn}(k\overline{r'})$ and $Rg\overline{N}_{mn}(k\overline{r'})$ in (4) in terms of vector spherical waves $Rg\overline{M}_{mn}(k\overline{r'r_l})$ and $Rg\overline{N}_{mn}(k\overline{r'r_l})$ with a common center at $\overline{r}_l$. This is possible because of the conditions (a), (b), and (c). Next, the coefficients of $Rg\overline{M}_{mn}(k\overline{r'r_l})$ and $Rg\overline{N}_{mn}(k\overline{r'r_l})$ in the resulting equations are balanced to obtain two sets of equations listed in Problem 10.

Following Section 5.4, Chapter 3, we then expand the tangential surface fields in the following manner. For $\overline{r} \in S_j$,

$$\hat{n}(\overline{r}) \times \overline{E}_j(\overline{r}) = \hat{n} \times \sum_{n'm'} \Big[c_{m'n'}^{(M)(j)} Rg\overline{M}_{m'n'}(k_j\overline{rr_j})$$

$$+ c_{m'n'}^{(N)(j)} Rg\overline{N}_{m'n'}(k_j\overline{rr_j})\Big] \qquad (17)$$

Derivation of Multiple Scattering Equations

$$\hat{n}(\bar{r}) \times i\omega\mu \overline{H}_j(\bar{r}) = k_j \hat{n} \times \sum_{n'm'} \left[c_{m'n'}^{(N)(j)} Rg\overline{M}_{m'n'}(k_j\overline{rr_j}) \right.$$
$$\left. + c_{m'n'}^{(M)(j)} Rg\overline{N}_{m'n'}(k_j\overline{rr_j}) \right] \quad (18)$$

where the superscript j signifies the jth particle and k_j is the wavenumber of the jth-particle. Substituting (17) and (18) into the two equations in Problem 10, manipulations similar to Section 5.4, Chapter 3, can be done. Matrix elements $P_{\mu\nu m'n'}^{(l)}$, $R_{\mu\nu m'n'}^{(l)}$, $S_{\mu\nu m'n'}^{(l)}$ and $U_{\mu\nu m'n'}^{(l)}$ as defined in (21) through (24), Section 5.4 (Chapter 3), are reproduced with superscript (l) denoting the lth particle. Two sets of equations are obtained and can be put in a compact form by using the index and matrix notation of Section 5.4, Chapter 3. For example $\overline{\overline{P}}$ is a $L_{max} \times L_{max}$ matrix representation of the elements $P_{\mu\nu mn}$ and index (μ, ν) is combined into a single index l as in Table 3.1 of Chapter 3. We also use $\overline{\overline{A}}$ and $\overline{\overline{B}}$ to respectively represent $A_{mn\mu\nu}$ and $B_{mn\mu\nu}$. Thus

$$Rg\overline{\overline{A}}(k\bar{r}_l)\bar{a}^{(M)} + Rg\overline{\overline{B}}(k\bar{r}_l)\bar{a}^{(N)} = \overline{\overline{P}}^{(l)} \bar{c}^{(M)(l)} + \overline{\overline{R}}^{(l)} \bar{c}^{(N)(l)}$$
$$+ \sum_{\substack{j=1 \\ j \neq l}}^{N} \overline{\overline{A}}(k\overline{r_lr_j}) \left[Rg\overline{\overline{P}}^{(j)} \bar{c}^{(M)(j)} + Rg\overline{\overline{R}}^{(j)} \bar{c}^{(N)(j)} \right]$$
$$+ \sum_{\substack{j=1 \\ j \neq l}}^{N} \overline{\overline{B}}(k\overline{r_lr_j}) \left[Rg\overline{\overline{S}}^{(j)} \bar{c}^{(M)(j)} + Rg\overline{\overline{U}}^{(j)} \bar{c}^{(N)(j)} \right] \quad (19)$$

$$Rg\overline{\overline{B}}(k\bar{r}_l)\bar{a}^{(M)} + Rg\overline{\overline{A}}(k\bar{r}_l)\bar{a}^{(N)} = \overline{\overline{S}}^{(l)} \bar{c}^{(M)(l)} + \overline{\overline{U}}^{(l)} \bar{c}^{(N)(l)}$$
$$+ \sum_{\substack{j=1 \\ j \neq l}}^{N} \overline{\overline{B}}(k\overline{r_lr_j}) \left[Rg\overline{\overline{P}}^{(j)} \bar{c}^{(M)(j)} + Rg\overline{\overline{R}}^{(j)} \bar{c}^{(N)(j)} \right]$$
$$+ \sum_{\substack{j=1 \\ j \neq l}}^{N} \overline{\overline{A}}(k\overline{r_lr_j}) \left[Rg\overline{\overline{S}}^{(j)} \bar{c}^{(M)(j)} + Rg\overline{\overline{U}}^{(j)} \bar{c}^{(N)(j)} \right] \quad (20)$$

for $l = 1, 2, \ldots, N$ where all matrices are of dimensions $L_{max} \times L_{max}$ and column vectors are of dimension $L_{max} \times 1$. Following Section 5.4, Chapter 3, we further let

$$\bar{a}_{inc} = \begin{bmatrix} \bar{a}^{(M)} \\ \bar{a}^{(N)} \end{bmatrix} \quad (21)$$

$$\bar{c} = \begin{bmatrix} \bar{c}^{(M)} \\ \bar{c}^{(N)} \end{bmatrix} \qquad (22)$$

$$\bar{\bar{Q}}^t = \begin{bmatrix} \bar{\bar{P}} & \bar{\bar{R}} \\ \bar{\bar{S}} & \bar{\bar{U}} \end{bmatrix} \qquad (23)$$

and

$$\bar{\bar{\sigma}}(k\bar{r}) = \begin{bmatrix} \bar{\bar{A}}(k\bar{r}) & \bar{\bar{B}}(k\bar{r}) \\ \bar{\bar{B}}(k\bar{r}) & \bar{\bar{A}}(k\bar{r}) \end{bmatrix} \qquad (24)$$

so that the matrices are of dimension $2L_{max} \times 2L_{max}$ and column vectors are of dimension $2L_{max} \times 1$. Then equations (19) and (20) can be represented by a single matrix equation

$$\bar{\bar{Q}}^{(l)t}\bar{c}^{(l)} + \sum_{\substack{j=1 \\ j \neq l}}^{N} \bar{\bar{\sigma}}(k\bar{r}_l\bar{r}_j) Rg\bar{\bar{Q}}^{(j)t}\bar{c}^{(j)} = Rg\bar{\bar{\sigma}}(k\bar{r}_l)\bar{a}_{inc} \qquad (25)$$

for $l = 1, 2, \ldots N$.

Next define the exciting field coefficients $\bar{w}^{(l)}$ of the l th particle as

$$\bar{w}^{(l)} = \bar{\bar{Q}}^{(l)t}\bar{c}^{(l)} \qquad (26)$$

and note that the T-matrix of the l th particle is

$$\bar{\bar{T}}^{(l)} = -Rg\bar{\bar{Q}}^{(l)t}\left(\bar{\bar{Q}}^{(l)t}\right)^{-1} \qquad (27)$$

The multiple scattering equations of (25) can be expressed in terms of the T-matrices of the particles.

$$\bar{w}^{(l)} = \sum_{\substack{j=1 \\ j \neq l}}^{N} \bar{\bar{\sigma}}(k\bar{r}_l\bar{r}_j)\bar{\bar{T}}^{(j)}\bar{w}^{(j)} + Rg\bar{\bar{\sigma}}(k\bar{r}_l)\bar{a}_{inc} \qquad (28)$$

for $l = 1, 2, \ldots, N$.

The physical interpretation of (28) is that the field exciting the l th particle is the sum of the incident field and the scattered fields from all particles except from itself. The $\bar{\bar{\sigma}}$ matrices are transformation matrices so that all the terms have a common center at $\bar{r}_l$. The scattered field from the j th particle is given by $\bar{\bar{T}}^{(j)}\bar{w}^{(j)}$. Thus, the multiple scattering equations corresponding to (4), Section 3.1, have been derived. Equation (28) contains N equations for the N unknowns $\bar{w}^{(1)}, \bar{w}^{(2)}, \ldots, \bar{w}^{(N)}$ and in principle can be solved by simultaneous solutions of equations given the particle locations $\bar{r}_j$ and particle $\bar{\bar{T}}_j$ matrices.

After the exciting field $\bar{w}^{(l)}$ is solved, the scattered field can be calculated in the following manner. Use equation (3) and let $\bar{r}'$ lie in region V_o. Hence, from (3)

$$\bar{E}(\bar{r}') = \bar{E}^{inc}(\bar{r}') + \bar{E}^S(\bar{r}') \qquad (29)$$

where

$$\bar{E}^S(\bar{r}') = \sum_{j=1}^{N} \int_{S_j} dS \left[\bar{\bar{G}}_o(\bar{r}'r_j, \bar{r}r_j) \cdot i\omega\mu\, \hat{n}(\bar{r}) \times \bar{H}_j(\bar{r}) \right. \qquad (30)$$
$$\left. + \nabla_{\bar{r}'r_j} \times \bar{\bar{G}}_o(\bar{r}'r_j, \bar{r}r_j) \cdot \hat{n}(\bar{r}) \times \bar{E}_j(\bar{r}) \right]$$

for $\bar{r}' \in V_o$. To calculate (30), we further let $\bar{r}'$ lie outside all the circumscribing spheres of the particles so that $|\bar{r}'r_j| > |\bar{r}r_j|$ for $\bar{r}$ lying on S_j. Using the appropriate expansion of the dyadic Green's function and the expansion of the surface fields, it follows that (Problem 11) for $\bar{r}$ lying outside all the circumscribing spheres of the particles,

$$\bar{E}^S(\bar{r}) = \sum_{j=1}^{N} \left\{ \sum_{mn} a_{mn}^{S(M)(j)} \bar{M}_{mn}(k\bar{r}r_j) + a_{mn}^{S(N)(j)} \bar{N}_{mn}(k\bar{r}r_j) \right\} \qquad (31)$$

Let $\bar{a}^{S(j)}$ be a $2L_{max} \times 1$ column matrix defined by

$$\bar{a}^{S(j)} = \begin{bmatrix} \bar{a}^{S(M)(j)} \\ \bar{a}^{S(N)(j)} \end{bmatrix} \qquad (32)$$

and $\bar{a}^{S(M)(j)}$ and $\bar{a}^{S(N)(j)}$ are $L_{max} \times 1$ matrices representing $a_{mn}^{S(M)(j)}$ and $a_{mn}^{S(N)(j)}$. The scattered field coefficients $\bar{a}^{S(j)}$ satisfy the relation (Problem 11)

$$\bar{a}^{S(j)} = \bar{\bar{T}}^{(j)} \bar{w}^{(j)} \qquad (33)$$

Hence, once the exciting field coefficients $\bar{w}^{(j)}$ are solved from (28), the coefficients of the scattered field are calculated by (33). The final scattered field is then calculated via (31).

For the case of plane wave incidence, we can further simplify (28). Consider an incident wave with electric field

$$\bar{E}^{inc}(\bar{r}) = \bar{A}\, e^{i\bar{k}_i \cdot \bar{r}} \qquad (34)$$

with $\hat{k}_i = \sin\theta_i \cos\phi_i \hat{x} + \sin\theta_i \sin\phi_i \hat{y} + \cos\theta_i \hat{z}$. Then $\overline{E}^{inc}(\bar{r})$ can be written in the alternate form

$$\overline{E}^{inc}(\bar{r}) = e^{i\bar{k}_i \cdot \bar{r}_l} \overline{A} e^{i\bar{k}_i \cdot (\bar{r}-\bar{r}_l)}$$

$$= e^{i\bar{k}_i \cdot \bar{r}_l} \sum_{n,m} \left[a^{(M)}_{mn} Rg\overline{M}_{mn}(k\overline{rr_l}) + a^{(N)}_{mn} Rg\overline{N}_{mn}(k\overline{rr_l}) \right] \quad (35)$$

The second equality in (35) is a result of comparing the expressions (1) and (34). Note that the coefficients $a^{(M)}_{mn}$ and $a^{(N)}_{mn}$ in (35) are the same as those in (1). Apply the translation addition theorem to $Rg\overline{M}_{mn}(k\overline{rr_l})$ and $Rg\overline{N}_{mn}(k\overline{rr_l})$ with the vectors $\overline{rr_l}$, $\bar{r}$, and $-\bar{r}_l$ as the three vectors of the triangle. Equating with (1) and balancing coefficients of $Rg\overline{M}_{mn}(k\bar{r})$ and $Rg\overline{N}_{mn}(k\bar{r})$ give

$$a^{(M)}_{mn} = e^{i\bar{k}_i \cdot \bar{r}_l} \sum_{\mu\nu} \left[RgA_{mn\mu\nu}(-k\bar{r}_l) a^{(M)}_{\mu\nu} + RgB_{mn\mu\nu}(-k\bar{r}_l) a^{(N)}_{\mu\nu} \right] \quad (36)$$

$$a^{(N)}_{mn} = e^{i\bar{k}_i \cdot \bar{r}_l} \sum_{\mu\nu} \left[RgB_{mn\mu\nu}(-k\bar{r}_l) a^{(M)}_{\mu\nu} + RgA_{mn\mu\nu}(-k\bar{r}_l) a^{(N)}_{\mu\nu} \right] \quad (37)$$

In matrix notation, (36) and (37) become

$$\bar{a}_{inc} = e^{i\bar{k}_i \cdot \bar{r}_l} Rg\overline{\overline{\sigma}}(-k\bar{r}_l) \bar{a}_{inc} \quad (38)$$

for any $\bar{r}_l$. Using the property (Problem 12)

$$Rg\overline{\overline{\sigma}}(k\bar{r}) Rg\overline{\overline{\sigma}}(-k\bar{r}) = \overline{\overline{I}} \quad (39)$$

and substituting (38) and (39) into (28) give the following simplified form of the multiple scattering equations for plane wave incidence

$$\overline{w}^{(l)} = \sum_{\substack{j=1 \\ j \neq l}}^{N} \overline{\overline{\sigma}}(k\overline{r_l r_j}) \overline{\overline{T}}^{(j)} \overline{w}^{(j)} + e^{i\bar{k}_i \cdot \bar{r}_l} \bar{a}_{inc} \quad (40)$$

for $l = 1, 2, \ldots, N$. In Sections 6, 7 and 8, we will use (40) as the starting point to consider multiple scattering by a random distribution of particles.

3.4 Summary

In Section 3, the multiple scattering equations are derived via two methods as given in Sections 3.2 and 3.3. The results in Section 3.2 are in operator form without choosing any particular representation or coordinate system. The results in Section 3.3, however, are based on utilization of the vector spherical waves as basis and using the T-matrix formalism and extended boundary conditions. These lead to the condition of nonoverlapping circumscribing spheres for the particles. This condition is particularly restrictive for nonspherical particles.

4 APPROXIMATIONS OF MULTIPLE SCATTERING EQUATIONS

4.1 Configurational Average of Multiple Scattering Equations

To calculate the coherent and incoherent fields, ensemble averages are to be taken of the multiple scattering equations in Section 3. The concept of configurational averaging will be used. The probability density function of finding the first particle at $\bar{r}_1$, the second particle at $\bar{r}_2, \ldots$ and the N th particle at $\bar{r}_N$ is designated by $p(\bar{r}_1, \bar{r}_2, \ldots, \bar{r}_N)$. Using Bayes' rule it can be expressed in terms of conditional probabilities

$$p(\bar{r}_1, \ldots, \bar{r}_N) = p(\bar{r}_i)\, p(\bar{r}_1, \bar{r}_2, \ldots \bar{r}_i', \ldots, \bar{r}_N \mid \bar{r}_i)$$

$$= p(\bar{r}_i)\, p(\bar{r}_j \mid \bar{r}_i)\, p(\bar{r}_1, \bar{r}_2, \ldots, \bar{r}_i', \ldots \bar{r}_j' \ldots, \bar{r}_N \mid \bar{r}_i, \bar{r}_j) \quad (1)$$

where the $\prime$ superscript indicates that the term is absent. In (1), $p(\bar{r}_i)$ is the probability density function of finding particle i at $\bar{r}_i$ and $p(\bar{r}_j \mid \bar{r}_i)$ is the conditional probability of finding the j th particle at $\bar{r}_j$ given the i th particle at $\bar{r}_i$.

Following Lax (1951), the states of the particle $s_1, s_2, \ldots, s_N$ can also be introduced. The state of the particle can indicate its size and, in the case of the nonspherical particle, its orientation. The probability of the set of N particles located in the volume element $d\bar{r}_1, d\bar{r}_2, \ldots, d\bar{r}_N$ with their states in the region $ds_1, ds_2, \ldots, ds_N$ is given by $p(\bar{r}_1, \bar{r}_2, \ldots, \bar{r}_N; s_1, s_2, \ldots, s_N)\, d\bar{r}_1 d\bar{r}_2 \ldots d\bar{r}_N ds_1 ds_2 \ldots ds_N$, a quantity whose integral is normalized to unity.

If the particles are uniformly distributed, then the positions of all the particles are equally probable within volume V, and hence the single particle probability density function is $p(\bar{r}_i) = 1/V$. The multiple scattering equations in terms of operators are given in (34) and (35), Section 3.2. We note that while $\bar{\bar{G}}$ and $\bar{\bar{G}}_l$, $l = 1, 2, \ldots, N$ are functions of all the particles, $\bar{\bar{T}}_l$ is only a function of the l th particle.

Taking configurational average of (35), Section 3.2, gives

$$E(\overline{\overline{G}}) = \overline{\overline{G}}_o + \overline{\overline{G}}_o \sum_{j=1}^{N} E(\overline{\overline{T}}_j \overline{\overline{G}}_j) \qquad (2)$$

By using Bayes' rule

$$E(\overline{\overline{T}}_j \overline{\overline{G}}_j) = \int d\bar{r}_1 \ldots d\bar{r}_N\, ds_1 \ldots ds_N\, p(\bar{r}_1 \ldots \bar{r}_N; s_1 \ldots s_N) \overline{\overline{T}}_j \overline{\overline{G}}_j$$

$$= \int d\bar{r}_j\, ds_j\, \overline{\overline{T}}_j\, p(\bar{r}_j; s_j) \int d\bar{r}_1 \ldots d\bar{r}'_j \ldots d\bar{r}_N\, ds_1 \ldots ds'_j \ldots ds_N$$

$$\times p(\bar{r}_1, \ldots, \bar{r}'_j, \ldots, \bar{r}_N; s_1 \ldots s'_j \ldots s_N \mid \bar{r}_j, s_j) \overline{\overline{G}}_j$$

$$= E\left(\overline{\overline{T}}_j E_j(\overline{\overline{G}}_j)\right) \qquad (3)$$

where E is the expectation value and E_j is the conditional average given the position and state of the jth particle. After the average as indicated in (3) is taken, every term in the summation of (2) will be identical. Hence,

$$E(\overline{\overline{G}}) = \overline{\overline{G}}_o + N\overline{\overline{G}}_o \dot{E}\left(\overline{\overline{T}}_j E_j(\overline{\overline{G}}_j)\right) \qquad (4)$$

Similarly, taking the configurational average of (34), Section 3.2, gives

$$E_l(\overline{\overline{G}}_l) = \overline{\overline{G}}_o + \overline{\overline{G}}_o(N-1) E_l\left(\overline{\overline{T}}_j E_{jl}(\overline{\overline{G}}_j)\right) \qquad (5)$$

with $j \neq l$, where E_{jl} is the conditional average given the properties and the positions of particles j and l. Thus,

$$E_l\left(\overline{\overline{T}}_j E_{jl}(\overline{\overline{G}}_j)\right) = \int d\bar{r}_j\, ds_j\, \overline{\overline{T}}_j\, p(\bar{r}_j; s_j \mid \bar{r}_l; s_l) E_{jl}(\overline{\overline{G}}_j) \qquad (6)$$

with $j \neq l$.

The equations (4) and (5) indicate that the total average is given in terms of the conditional average with one particle fixed, and the conditional average with one particle fixed is given in terms of the conditional average with two particles fixed. In a similar manner, we can express the conditional average with n particles fixed in terms of $n+1$ particles fixed. A hierarchy of equations is thus generated. Truncation at various stages leads to different approximations. In the following sections, the results of the first two truncations will be given. We note that for statistical homogeneous medium, the average Green's operator $<\overline{\overline{G}}>$ is diagonal in the momentum representation

$$<\bar{p}|E(\overline{\overline{G}})|\bar{p}'> \equiv \overline{\overline{G}}(\bar{p}) <\bar{p}|\bar{p}'> \qquad (7)$$

The free-space dyadic Green's operator in momentum representation is

$$<\bar{p}|\overline{\overline{G}}_o|\bar{p}'> = \overline{\overline{G}}_o(\bar{p}) <\bar{p}|\bar{p}'> \qquad (8)$$

where

$$\overline{\overline{G}}_o(\bar{p}) = -\left(\bar{p}\bar{p} - p^2\overline{\overline{I}} + k^2\overline{\overline{I}}\right)^{-1}$$

$$= -\frac{1}{k^2}\hat{p}\hat{p} + \frac{1}{p^2 - k^2}(\hat{\theta}_p\hat{\theta}_p + \hat{\phi}_p\hat{\phi}_p) \qquad (9)$$

where $p = |\bar{p}|$. The unit vectors $(\hat{p}, \hat{\theta}_p, \hat{\phi}_p)$ of the spherical coordinate system for direction (θ_p, ϕ_p) have been employed.

By the translational theorem of operators, $\overline{\overline{U}}_j$ is related to $\overline{\overline{U}}$ centered at the origin by the operator equation (Merzbacher, 1970)

$$\overline{\overline{U}}_j = e^{-i\bar{p}\cdot\bar{r}_j}\overline{\overline{U}}e^{i\bar{p}\cdot\bar{r}_j} \qquad (10)$$

In view of the commutativity of $\overline{\overline{G}}_o$ with $\bar{p}$, we have, by using (29), Section 3.2, (Problem 13)

$$\overline{\overline{T}}_j = e^{-i\bar{p}\cdot\bar{r}_j}\overline{\overline{T}}e^{i\bar{p}\cdot\bar{r}_j} \qquad (11)$$

Also let $\overline{\overline{U}}_p(\bar{p},\bar{p}') = <\bar{p}|\overline{\overline{U}}|\bar{p}'>$ and $\overline{\overline{T}}_p(\bar{p},\bar{p}') = <\bar{p}|\overline{\overline{T}}|\bar{p}'>$ to denote respectively the momentum representation of $\overline{\overline{U}}$ and $\overline{\overline{T}}$. The dispersion relation of the coherent wave is obtained by setting the inverse of $\overline{\overline{G}}(\bar{p})$ equal to zero. From (25), Section 3.2, it follows that

$$\overline{\overline{T}} = \overline{\overline{U}}\,\overline{\overline{G}}_s\overline{\overline{G}}_o^{-1} \qquad (12)$$

Hence, in momentum representation,

$$\overline{\overline{T}}_p(\bar{p}_1,\bar{p}_2) = \int d\bar{r}_2 \int d\bar{r}_1\, e^{-i\bar{p}_1\cdot\bar{r}_1}\overline{\overline{U}}(\bar{r}_1)\overline{\overline{G}}_s(\bar{r}_1,\bar{r}_2)e^{i\bar{p}_2\cdot\bar{r}_2}\overline{\overline{G}}_o^{-1}(\bar{p}_2) \qquad (13)$$

Hence if the dyadic Green's function $\overline{\overline{G}}_s$ for a single particle is known, then the $\overline{\overline{T}}$ operator in momentum representation can be calculated via

(13). This has been performed for the case of a spherical particle (Problem 14; Tsang and Kong, 1980).

Relation Between Scattering Dyad $\overline{\overline{F}}$ and Transition Operator in Momentum Representation

To relate the scattering dyad $\overline{\overline{F}}$ to the transition operator, we note that

$$|\overline{E}_s> = \overline{\overline{G}}_o \overline{\overline{T}} |\overline{E}_{inc}> \qquad (14)$$

Suppose that the incident field state is given by a momentum eigenstate with $<\overline{r}|\overline{E}_{inc}> = \hat{e}_i e^{i\overline{k}_i \cdot \overline{r}}$. Then from (14)

$$\overline{E}_s(\overline{r}) \equiv <\overline{r}|\overline{E}_s> = \int d\overline{r}'\, \overline{\overline{G}}_o(\overline{r},\overline{r}') \int \frac{d\overline{p}'}{(2\pi)^3} e^{i\overline{p}'\cdot \overline{r}'} \overline{\overline{T}}_p(\overline{p}',\overline{k}_i) \cdot \hat{e}_i \qquad (15)$$

In the far-field limit, the integral in (15) can be evaluated readily. Let $\hat{k}_s$ represent the observation direction. Then

$$\overline{E}_s(\overline{r}) = (\hat{\theta}_s\hat{\theta}_s + \hat{\phi}_s\hat{\phi}_s) \cdot \frac{e^{ikr}}{4\pi r} \overline{\overline{T}}_p(k\hat{k}_s, k\hat{k}_i) \cdot \hat{e}_i \qquad (16)$$

The far-field scattering dyad $\overline{\overline{F}}(\hat{k}_s, \hat{k}_i)$ is defined in (9), Section 3.2, Chapter 3. Hence, comparing with (16) gives

$$\overline{\overline{F}}(\hat{k}_s, \hat{k}_i) \cdot \hat{e}_i = \frac{1}{4\pi}(\hat{\theta}_s\hat{\theta}_s + \hat{\phi}_s\hat{\phi}_s) \cdot \overline{\overline{T}}_p(k\hat{k}_s, k\hat{k}_i) \cdot \hat{e}_i \qquad (17)$$

Thus, we note that the scattering amplitude is related to the $\overline{\overline{T}}$ operator when the two arguments of $\overline{\overline{T}}$ in the momentum representation have magnitude equal to the wavenumber k of the background medium. However, $\overline{\overline{T}}_p$ can have arguments of arbitrary magnitude and thus is more general than the far-field scattering dyad $\overline{\overline{F}}$ (Newton, 1966).

4.2 Effective Field Approximation (Foldy's Approximation)

In the effective field approximation (EFA), truncation is carried out at the first equation of the hierarchy. It is assumed that

$$E_j(\overline{\overline{G}}_j) \simeq E(\overline{\overline{G}}) \qquad (1)$$

The approximation is valid for a sparse concentration of particles. Hence, (4), Section 4.1, becomes, on using the approximation in (1)

$$E(\overline{\overline{G}}) = \overline{\overline{G}}_o + N\overline{\overline{G}}_o E(\overline{\overline{T}}_j) E(\overline{\overline{G}}) \qquad (2)$$

Effective Field Approximation (Foldy's Approximation)

The averaged $\overline{\overline{T}}_j$ in momentum representation is, using (3) and (5), Section 4.1,

$$<\bar{p}|E(\overline{\overline{T}}_j)|\bar{p}'> = \int d\bar{r}_j\, ds_j\, p(\bar{r}_j, s_j) \overline{\overline{T}}_j \tag{3}$$

Since the particles are sparsely distributed, it is reasonable to assume that the state of the particle is independent of its position. Hence, $p(\bar{r}_j, s_j) = p(\bar{r}_j)p(s_j)$ and using (11), Section 4.1,

$$<\bar{p}|E(\overline{\overline{T}}_j)|\bar{p}'> = \frac{1}{V} <T_p(\bar{p},\bar{p})>_{st} <\bar{p}|\bar{p}'> \tag{4}$$

where

$$<T_p(\bar{p},\bar{p})>_{st} = \int ds\, T_p(\bar{p},\bar{p})\, p(s) \tag{5}$$

is the average of $T_p(\bar{p},\bar{p})$ over its states. Substituting (4) in the momentum representation of (2) and using (8) and (9), Section 4.1, give

$$\overline{\overline{G}}(\bar{p}) = \overline{\overline{G}}_o(\bar{p}) + n_o \overline{\overline{G}}_o(\bar{p}) <\overline{\overline{T}}_p(\bar{p},\bar{p})>_{st} \overline{\overline{G}}(\bar{p}) \tag{6}$$

By setting the inverse of $\overline{\overline{G}}(\bar{p})$ equal to zero, the following dispersion relation is obtained.

$$\det\left[\overline{\overline{G}}_o^{-1}(\bar{p}) - n_o <\overline{\overline{T}}_p(\bar{p},\bar{p})>_{st}\right] = 0 \tag{7}$$

The value of $\bar{p}$ that satisfies (7) will be the value of the effective propagation vector $\overline{K}$. In (7) 'det' stands for determinant. Using (9), Section 4.1, the contents of the square brackets in (7) can be written as

$$-k^2 \hat{p}\hat{p} + (p^2 - k^2)(\hat{\theta}_p\hat{\theta}_p + \hat{\phi}_p\hat{\phi}_p) - n_o <\overline{\overline{T}}_p(\bar{p},\bar{p})>_{st} \tag{8}$$

It is convenient to use $(\hat{p}, \hat{\theta}_p, \hat{\phi}_p)$ as basis for the dyads in (6). We note that the effective field approximation is valid for a sparse concentration of particles so that the coherent effective propagation constant is close to k. We look for the solution of p that is close to k in equation (7). Hence the arguments of $\overline{\overline{T}}_p$ in (8) can be replaced by $k\hat{p}$. The first term in (8) is much larger than the other terms. Thus, the determinant of (8) is approximately equal to $-k^2$ times the determinant of the 2×2 matrix formed by $\hat{\theta}_p$ and $\hat{\phi}_p$ components. Hence, the dispersion relation is

$$0 \simeq -k^2 \left\{ \begin{vmatrix} p^2 - k^2 - n_o \hat{\theta}_p \cdot <\overline{\overline{T}}_p(k\hat{p}, k\hat{p})>_{st} \cdot \hat{\theta}_p \\ -n_o \hat{\phi}_p \cdot <\overline{\overline{T}}_p(k\hat{p}, k\hat{p})>_{st} \cdot \hat{\theta}_p \end{vmatrix} \right.$$

$$\left. \begin{matrix} -n_o \hat{\theta}_p \cdot <\overline{\overline{T}}_p(k\hat{p}, k\hat{p})>_{st} \cdot \hat{\phi}_p \\ p^2 - k^2 - n_o \hat{\phi}_p \cdot <\overline{\overline{T}}_p(k\hat{p}, k\hat{p})>_{st} \cdot \hat{\phi}_p \end{matrix} \right\} \tag{9}$$

Using the relation between $\overline{\overline{F}}$ and $\overline{\overline{T}}$ in (17), Section 4.1, the dispersion relation can be expressed in terms of the $\overline{\overline{F}}$ dyad for propagation in the direction $\hat{k}_i$. Note that $\hat{\theta}$ corresponds to $\hat{v}$ and $\hat{\phi}$ corresponds to $\hat{h}$.

$$\begin{vmatrix} K^2 - k^2 - 4\pi n_o <f_{vv}(\theta_i,\phi_i;\theta_i,\phi_i)>_{st} & -4\pi n_o <f_{vh}(\theta_i,\phi_i;\theta_i,\phi_i)>_{st} \\ -4\pi n_o <f_{hv}(\theta_i,\phi_i;\theta_i,\phi_i)>_{st} & K^2 - k^2 - 4\pi n_o <f_{hh}(\theta_i,\phi_i;\theta_i,\phi_i)>_{st} \end{vmatrix} = 0 \quad (10)$$

with $\hat{v}_i$ and $\hat{h}_i$ (vertical and horizontal polarizations) as the basis for (10). In (10), $<f_{\alpha\beta}>_{st}$ denotes averaging over the state (e.g., orientation and size). The eigenvalues for K in (10) are the effective propagation constants and the corresponding eigenvectors give the characteristic wave polarizations associated with the propagation constants.

Following Ishimaru and Cheung (1980), let

$$M_{\alpha\beta}(\theta_i,\phi_i) = <f_{\alpha\beta}(\theta_i,\phi_i;\theta_i,\phi_i)>_{st} \frac{i2\pi n_o}{k} \quad (11)$$

Solving (10) gives values of K^2 and since K is close to k, the two effective coherent propagation constants are K_1 and K_2 with

$$K_{\frac{1}{2}} = k - \frac{i}{2}\left[M_{vv} + M_{hh} \pm \{(M_{vv} - M_{hh})^2 + 4M_{vh}M_{hv}\}^{1/2}\right] \quad (12)$$

and the associated eigenvectors are, for $M_{vh}, M_{hv} \neq 0$,

$$\frac{<E_v>}{<E_h>} = \frac{2M_{vh}}{-M_{vv} + M_{hh} \pm \{(M_{vv} - M_{hh})^2 + 4M_{vh}M_{hv}\}^{1/2}} \quad (13)$$

where $<E_v>/<E_h>$ is the ratio of vertical to horizontal polarization in the characteristic wave. Note that there are only two propagation constants and characteristic waves for a specified direction of propagation. For the case $<f_{vh}>_{st} = <f_{hv}>_{st} = 0$, when there is no coupling between the two polarizations, then the eigensolutions are

$$K_1 = k + \frac{2\pi n_o}{k} <f_{vv}>_{st} \quad (14)$$

with the characteristic wave vertically polarized, while

$$K_2 = k + \frac{2\pi n_o}{k} <f_{hh}>_{st} \quad (15)$$

with the characteristic wave horizontally polarized.

The results of (12) through (15) can be expressed in terms of coupled differential equations for the coherent field. First, we note that since K is close to k, the matrix equation of (10) can be approximated as

$$\begin{bmatrix} iK - ik - M_{vv} & -M_{vh} \\ -M_{hv} & iK - ik - M_{hh} \end{bmatrix} \begin{bmatrix} <E_v> \\ <E_h> \end{bmatrix} = 0 \qquad (16)$$

Equation (16) can be expressed in the form of coupled differential equations for the coherent field along the propagation direction $(\theta_i, \phi_i) = \hat{s}$,

$$\frac{d<E_v>}{ds} = (ik + M_{vv})<E_v> + M_{vh}<E_h> \qquad (17a)$$

$$\frac{d<E_h>}{ds} = M_{hv}<E_v> + (ik + M_{hh})<E_h> \qquad (17b)$$

with s as the distance along the direction of propagation. The coupled equations in (17) have been used to derive the extinction matrix for radiative transfer equations in Section 3.4, Chapter 3. The Foldy's approximation has also been studied (Ishimaru, 1978; Brown, 1980) and has been applied to scattering from vegetation (Lang, 1981; Lang and Sidhu, 1983). They have also been used extensively to study wave propagation in rain (Oguchi, 1973, 1983; Chu, 1974; Crane, 1974; Cox, 1981). In the next section, it shall be shown that Foldy's approximation is a special case of the quasicrystalline approximation.

4.3 Quasicrystalline Approximation

The quasicrystalline approximation (QCA) is a higher-order approximation than the effective field approximation. Truncation is made at the second stage of the hierarchy of equations. To impose the quasicrystalline approximation, it is convenient to use the $\overline{\overline{Q}}$ operator defined by $\overline{\overline{Q}}_j = \overline{\overline{T}}_j \overline{\overline{G}}_j \overline{\overline{G}}_o^{-1}$ and recast the multiple scattering equations of (34) and (35), Section 3.2, into the following form

$$\overline{\overline{G}} = \overline{\overline{G}}_o + \overline{\overline{G}}_o \sum_{j=1}^{N} \overline{\overline{Q}}_j \overline{\overline{G}}_o \qquad (1)$$

$$\overline{\overline{Q}}_l = \overline{\overline{T}}_l + \overline{\overline{T}}_l \overline{\overline{G}}_o \sum_{\substack{j=1 \\ j \neq l}}^{N} \overline{\overline{Q}}_j \qquad (2)$$

The configurational average of the equations becomes

$$E(\overline{\overline{G}}) = \overline{\overline{G}}_o + N\overline{\overline{G}}_o E\left(E_j(\overline{\overline{Q}}_j)\right)\overline{\overline{G}}_o \qquad (3)$$

$$E_l(\overline{\overline{Q}}_l) = \overline{\overline{T}}_l + (N-1)\overline{\overline{T}}_l\overline{\overline{G}}_o E_l\left(E_{lj}(\overline{\overline{Q}}_j)\right) \qquad (4)$$

with $j \neq l$. Under the quasicrystalline approximation, it is assumed that

$$E_{lj}(\overline{\overline{Q}}_j) \simeq E_j(\overline{\overline{Q}}_j) \qquad (5)$$

so that (4) becomes

$$E_l(\overline{\overline{Q}}_l) = \overline{\overline{T}}_l + (N-1)\overline{\overline{T}}_l\overline{\overline{G}}_o E_l\left(E_j(\overline{\overline{Q}}_j)\right) \qquad (6)$$

Equation (6) can also be re-expressed in terms of the potential operator $\overline{\overline{U}}_l$.

$$E_l(\overline{\overline{Q}}_l) = \overline{\overline{U}}_l + \overline{\overline{U}}_l\overline{\overline{G}}_o E_l(\overline{\overline{Q}}_l) + (N-1)\overline{\overline{U}}_l\overline{\overline{G}}_o E_l\left(E_j(\overline{\overline{Q}}_j)\right) \qquad (7)$$

After the conditional average is taken, $E_l(\overline{\overline{Q}}_l)$ is only a function of the l th particle.

We shall consider cases of identical particles. Taking (7) in momentum representation and using the translational property (10), Section 4.1, of $\overline{\overline{U}}_j$ give

$$<\bar{p}|E_l(\overline{\overline{Q}}_l)|\bar{p}'> = e^{-i(\bar{p}-\bar{p}')\cdot\bar{r}_l}\overline{\overline{U}}_p(\bar{p},\bar{p}')$$

$$+ \int \frac{d\bar{p}''}{(2\pi)^3}\overline{\overline{U}}_p(\bar{p},\bar{p}'')\, e^{-i(\bar{p}-\bar{p}'')\cdot\bar{r}_l}\overline{\overline{G}}_o(\bar{p}'') <\bar{p}''|E_l(\overline{\overline{Q}}_l)|\bar{p}'>$$

$$+ N\int d\bar{r}_j\, p(\bar{r}_j\mid\bar{r}_l) \int \frac{d\bar{p}''}{(2\pi)^3}\, e^{-i(\bar{p}-\bar{p}'')\cdot\bar{r}_l}\overline{\overline{U}}_p(\bar{p},\bar{p}'')$$

$$\cdot \overline{\overline{G}}_o(\bar{p}'') <\bar{p}''|E_j(\overline{\overline{Q}}_j)|\bar{p}'> \qquad (8)$$

The conditional probability $p(\bar{r}_j\mid\bar{r}_l)$ can be expressed in terms of pair distribution function. We further assume that the medium is statistically homogeneous so that $p(\bar{r}_j\mid\bar{r}_l) = g(\bar{r}_j - \bar{r}_l)/V$. Since the pair distribution function asymptotically approaches 1, define the Fourier transform as follows. Let

$$q(\bar{r}) = n_o[g(\bar{r}) - 1] + \delta(\bar{r}) \qquad (9)$$

and
$$q_p(\bar{p}) = \int d\bar{r}\, e^{-i\bar{p}\cdot\bar{r}} q(\bar{r}) \tag{10}$$

Next assume a trial solution in (8)

$$<\bar{p}|E_l(\bar{\bar{Q}}_l)|\bar{p}'> = e^{-i(\bar{p}-\bar{p}')\cdot\bar{r}_l}\, \bar{\bar{Q}}_p(\bar{p},\bar{p}') \tag{11}$$

and let
$$\bar{\bar{C}}_p(\bar{p},\bar{p}_1) = \bar{\bar{Q}}_p(\bar{p},\bar{p}_1)\left[\bar{\bar{I}} + n_o\bar{\bar{G}}_o(\bar{p}_1)\bar{\bar{Q}}_p(\bar{p}_1,\bar{p}_1)\right]^{-1} \tag{12}$$

Substituting (9) through (12) into (8) gives the following integral equation

$$\bar{\bar{C}}_p(\bar{p}_1,\bar{p}_2) = \bar{\bar{U}}_p(\bar{p}_1,\bar{p}_2)$$
$$+ \int \frac{d\bar{p}_3}{(2\pi)^3}\bar{\bar{U}}_p(\bar{p}_1,\bar{p}_3)\bar{\bar{G}}_o(\bar{p}_3)q_p(\bar{p}_3-\bar{p}_2)\bar{\bar{C}}_p(\bar{p}_3,\bar{p}_2) \tag{13}$$

By taking the average of (3), it follows that (Problem 16) the averaged Green's operator obeys (7), Section 4.1, with

$$\bar{\bar{G}}(\bar{p}) = \left\{\bar{\bar{G}}_o^{-1}(\bar{p}) - n_o\bar{\bar{C}}_p(\bar{p},\bar{p})\right\}^{-1} \tag{14}$$

Hence, to calculate the dispersion relation, the unknown $\bar{\bar{C}}_p(\bar{p},\bar{p}')$ is solved through (13) and then the effective propagation constant are provided by solving the equation

$$\det\left[\bar{\bar{G}}_o^{-1}(\bar{p}) - n_o\bar{\bar{C}}_p(\bar{p},\bar{p})\right] = 0 \tag{15}$$

By using the Lippmann-Schwinger equation of (27), Section 3.2, the quantity $\bar{\bar{C}}_p(\bar{p}_1,\bar{p}_2)$ can be put in an integral equation in terms of the transition operator $\bar{\bar{T}}_p(\bar{p}_1,\bar{p}_2)$ (Problem 16)

$$\bar{\bar{C}}_p(\bar{p}_1,\bar{p}_2) = \bar{\bar{T}}_p(\bar{p}_1,\bar{p}_2)$$
$$+ n_o\int d\bar{p}_3\bar{\bar{T}}_p(\bar{p}_1,\bar{p}_3)\bar{\bar{G}}_o(\bar{p}_3)H(\bar{p}_3-\bar{p}_2)\bar{\bar{C}}_p(\bar{p}_3,\bar{p}_2) \tag{16}$$

where
$$H(\bar{p}) = \frac{1}{(2\pi)^3}\int_{-\infty}^{\infty} d\bar{r}\, h(\bar{r})\, e^{-i\bar{p}\cdot\bar{r}} \tag{17}$$

and $h(\bar{r}) = g(\bar{r}) - 1$.

If we define the operator $\overline{\overline{C}}_j$ that is a function of $\bar{r}_j$ such that

$$<\bar{p}_1|\overline{\overline{C}}_j|\bar{p}_2> = e^{-i(\bar{p}_1-\bar{p}_2)\cdot \bar{r}_j} \overline{\overline{C}}_p(\bar{p}_1,\bar{p}_2) \tag{18}$$

Equation (16) can then be put in operator form

$$\overline{\overline{C}}_j = \overline{\overline{T}}_j + n_o \int d\bar{r}_l\, h(\bar{r}_l - \bar{r}_j) \overline{\overline{T}}_j \overline{\overline{G}}_o \overline{\overline{C}}_l \tag{19}$$

The operator equation that is equivalent to (14) is then

$$E(\overline{\overline{G}}) = \overline{\overline{G}}_o + n_o \overline{\overline{G}}_o \int d\bar{r}_j\, \overline{\overline{C}}_j E(\overline{\overline{G}}) \tag{20}$$

Comparing with (14), Section 3.1, Chapter 5, it can be concluded that $n_o \int d\bar{r}_j \overline{\overline{C}}_j$ is the mass operator for discrete scatterers under the quasicrystalline approximation.

EFA is a Special Case of QCA

For a sparse concentration of scatterers, it is reasonable to assume that the particle positions are independent of each other. The conditional probability $p(\bar{r}_j \mid \bar{r}_l)$ can be approximated by $p(\bar{r}_j \mid \bar{r}_l) = p(\bar{r}_j) = 1/V$ so that the pair correlation function becomes $g(\bar{r}) = 1$. Hence it follows from (9) and (10) that $q_p(\bar{p}) = 1$. Substituting into (13) and comparing with (27), Section 3.2, shows that

$$\overline{\overline{C}}_p(\bar{p}_1,\bar{p}_2) = \overline{\overline{T}}_p(\bar{p}_1,\bar{p}_2) \tag{21}$$

in this limit. Hence the dispersion relation of QCA of (15) reduces to that of EFA [(7), Section 4.2].

4.4 Coherent Potential

The idea of coherent potential (CP) originates from the observation that in the derivation of the multiple scattering equations in Section 3, the background medium dyadic Green's operator is used. The potential operator $\overline{\overline{U}}$ is proportional to $k_s^2 - k^2$ and is a measure of the difference in permittivity from the background medium. However, it seems that as the concentration of particles increases, the coherent wave will be propagating in an effective medium K, and the scattering potential is a result of the difference in wavenumber from K rather than k. The idea of coherent potential is to introduce the Green's operator with wavenumber K (Soven, 1966, 1967) and is analogous to the nonlinear approximation in random medium theory (Section 3.2, Chapter 5).

Coherent Potential

Consider the N-particle scattering equation of (19), Section 3.2,

$$\left(\overline{\overline{G}}_o^{-1} - \sum_{j=1}^{N} \overline{\overline{U}}_j\right)\overline{\overline{G}} = \overline{\overline{I}} \tag{1}$$

Suppose the basic equation (1) is rewritten by adding and subtracting an operator $n_o\overline{\overline{w}}(\bar{p})$ on the left-hand side of the equation. Then,

$$\left[\overline{\overline{G}}_o^{-1} - n_o\overline{\overline{w}}(\bar{p}) - \sum_{j=1}^{N}\left(\overline{\overline{U}}_j - \frac{\overline{\overline{w}}(\bar{p})}{V}\right)\right]\overline{\overline{G}} = \overline{\overline{I}} \tag{2}$$

where $\overline{\overline{w}}(\bar{p})$ is the coherent potential operator which is constant in space but may be a function of the momentum operator $\bar{p}$. From (2), we note that the background Green's operator and the potential operator are modified. Let

$$\overline{\overline{G}}_c^{-1} = \overline{\overline{G}}_o^{-1} - n_o\overline{\overline{w}}(\bar{p}) \tag{3}$$

be the coherent Green's operator and

$$\overline{\overline{\tilde{U}}}_j = \overline{\overline{U}}_j - \frac{\overline{\overline{w}}(\bar{p})}{V} \tag{4}$$

as the modified potential operator for each scatterer. Then, using (3) and (4) in (2)

$$\overline{\overline{G}} = \overline{\overline{G}}_c + \overline{\overline{G}}_c \sum_{j=1}^{N} \overline{\overline{\tilde{U}}}_j \overline{\overline{G}} \tag{5}$$

is the N-particle scattering equation. Equation (5) is completely analogous to the original N-particle scattering equation (19), Section 3.2, with $\overline{\overline{G}}_o$ replaced by $\overline{\overline{G}}_c$ and $\overline{\overline{U}}_j$ replaced by $\overline{\overline{\tilde{U}}}_j$. The process of taking configurational averages and truncating the hierarchy of equations at various stages as performed in the previous two sections can be repeated, giving new dispersion relations. However, these new dispersion relations depend on the choice of the coherent potential operator $\overline{\overline{w}}(\bar{p})$, which has not yet been determined. By treating $\overline{\overline{w}}(\bar{p})$ as an adjustable parameter, it may be used to improve the results. The *consistent* choice for $\overline{\overline{w}}(\bar{p})$ or the coherent potential choice is choosing $\overline{\overline{w}}(\bar{p})$ such that

$$<\overline{\overline{G}}> = \overline{\overline{G}}_c \tag{6}$$

Hence, the final result of the averaged Green's operator is equal to the original coherent Green's operator. The coherent potential shall be introduced into the effective field approximation and the quasicrystalline approximation.

Effective Field Approximation–Coherent Potential

When the effective field approximation is applied to the N-particle scattering equation (5) in a manner completely analogous to Section 4.2, with $\overline{\overline{G}}_c$ playing the role of $\overline{\overline{G}}_o$ and $\overline{\overline{U}}$ playing the role of $\overline{\overline{U}}$, the result for the averaged Green's operator in momentum representation is

$$\overline{\overline{G}}(\bar{p}) = \left[\overline{\overline{G}}_c^{\,-1}(\bar{p}) - n_o \overline{\overline{T}}_p(\bar{p},\bar{p}) \right]^{-1} \tag{7}$$

with the modified transition operator $\overline{\overline{T}}$ obeying the equation $\overline{\overline{T}} = \overline{\overline{U}} + \overline{\overline{U}}\,\overline{\overline{G}}_c\,\overline{\overline{T}}$. The coherent potential choice of $\overline{\overline{w}}(\bar{p})$ is such that (6) is satisfied, which, in view of (7), imposes the condition

$$\overline{\overline{T}}_p(\bar{p},\bar{p}) = 0 \tag{8}$$

That is, the momentum representation of $\overline{\overline{T}}$ with both arguments equal to $\bar{p}$ is zero.

By using (4) and $\overline{\overline{T}} = \overline{\overline{U}} + \overline{\overline{T}}\,\overline{\overline{G}}_c\,\overline{\overline{U}}$ it follows that

$$\overline{\overline{T}}\,\overline{\overline{G}}_c\,\overline{\overline{\tilde{G}}}^{\,-1} + \frac{\overline{\overline{w}}_p}{V} = \left(\overline{\overline{I}} + \overline{\overline{T}}\,\overline{\overline{G}}_c \right) \overline{\overline{U}} \tag{9}$$

where

$$\overline{\overline{\tilde{G}}} = \left(\overline{\overline{G}}_c^{\,-1} + \frac{\overline{\overline{w}}_p}{V} \right)^{-1} \tag{10}$$

Define $\overline{\overline{\tilde{t}}}$ that satisfies the Lippman-Schwinger equation

$$\overline{\overline{\tilde{t}}} = \overline{\overline{U}} + \overline{\overline{U}}\,\overline{\overline{\tilde{G}}}\,\overline{\overline{\tilde{t}}} \tag{11}$$

Eliminating $\overline{\overline{U}}$ from (9) and (11) and using (10), we have, after some algebra

$$\overline{\overline{T}} = \overline{\overline{G}}_c^{\,-1}\,\overline{\overline{\tilde{G}}}\,\overline{\overline{\tilde{t}}}\,\overline{\overline{\tilde{G}}}\,\overline{\overline{G}}_c^{\,-1} - \frac{\overline{\overline{w}}_p}{V}\,\overline{\overline{\tilde{G}}}\,\overline{\overline{G}}_c^{\,-1} \tag{12}$$

We take (12) in momentum representation and define $<\bar{p}|\overline{\overline{G}}_c|\bar{p}_1> = \overline{\overline{G}}_c(\bar{p}) <\bar{p}|\bar{p}_1>$ and $<\bar{p}|\overline{\overline{\tilde{G}}}|\bar{p}_1> = \overline{\overline{\tilde{G}}}(\bar{p}) <\bar{p}|\bar{p}_1>$. To normalize the resulting equation, we multiply by $(1/V)\,e^{-i(\bar{p}-\bar{p}_1)\cdot\bar{r}_j}$ and integrate over

$d\vec{r}_j$, noting that $\int d\vec{r}_j (1/V) e^{-i(\vec{p}-\vec{p}_1)\cdot\vec{r}_j} = <\vec{p}|\vec{p}_1>/V$ and $\int d\vec{r}_j = V$. Thus (12) becomes

$$\bar{\bar{\bar{T}}}_p(\bar{p},\bar{p}) = \left[\bar{\bar{G}}_c^{-1}(\bar{p})\bar{\bar{\bar{G}}}(\bar{p})\bar{\bar{\bar{t}}}_p(\bar{p},\bar{p}) - \bar{\bar{w}}(\bar{p})\right]\bar{\bar{\bar{G}}}(\bar{p})\bar{\bar{G}}_c^{-1}(\bar{p}) \qquad (13)$$

Thus, the coherent potential condition of $\bar{\bar{\bar{T}}}_p(\bar{p},\bar{p}) = 0$ implies that $\bar{\bar{w}}(\bar{p}) = \bar{\bar{G}}_c^{-1}(\bar{p})\bar{\bar{\bar{G}}}(\bar{p})\bar{\bar{\bar{t}}}_p(\bar{p},\bar{p})$. Furthermore by taking $V \to \infty$ in (10), and noting that $\bar{\bar{w}}_p\bar{\bar{G}}_c(\bar{p})$ remains finite, we have $\bar{\bar{\bar{G}}}(\bar{p}) \to \bar{\bar{G}}_c(\bar{p})$. Thus the choice of coherent potential is

$$\bar{\bar{w}}(\bar{p}) = \bar{\bar{\bar{t}}}_p(\bar{p},\bar{p}) \qquad (14)$$

Hence, the final results for the averaged Green's operator are (3) and (14). The quantity $\bar{\bar{\bar{t}}}$ satisfies (11), the momentum representation which becomes (on letting $V \to \infty$)

$$\bar{\bar{\bar{t}}}_p(\bar{p}_1,\bar{p}_2) = \bar{\bar{U}}_p(\bar{p}_1,\bar{p}_2) + \int \frac{d\bar{p}_3}{(2\pi)^3} \bar{\bar{U}}_p(\bar{p}_1,\bar{p}_3)\bar{\bar{G}}_c(\bar{p}_3)\bar{\bar{\bar{t}}}_p(\bar{p}_3,\bar{p}_2) \qquad (15)$$

The dispersion relation for EFA-CP is

$$\det\left[\bar{\bar{G}}_o^{-1}(\bar{p}) - n_o\bar{\bar{\bar{t}}}_p(\bar{p},\bar{p})\right] = 0 \qquad (16)$$

Since $\bar{\bar{G}}_c$ depends on $\bar{\bar{\bar{t}}}_p$ via (3) and (14), the integral equation (15) is nonlinear.

Quasicrystalline Approximation-Coherent Potential

The quasicrystalline approximation can be applied to the N-particle scattering equation (5). Manipulations are performed as in Section 4.3 with $\bar{\bar{G}}_c$ playing the role of $\bar{\bar{G}}_o$ and $\bar{\bar{U}}$ playing the role of $\bar{\bar{U}}$. The result for the averaged Green's operator is

$$\bar{\bar{G}}(\bar{p}) = \left[\bar{\bar{G}}_c^{-1}(\bar{p}) - n_o\bar{\bar{\tilde{C}}}_p(\bar{p},\bar{p})\right]^{-1} \qquad (17)$$

where $\bar{\bar{\tilde{C}}}_p$ satisfies (13), Section 4.3, with $\bar{\bar{U}}$ replaced by $\bar{\bar{U}}$ and $\bar{\bar{G}}_o$ replaced by $\bar{\bar{G}}_c$. Using the definition of $\bar{\bar{U}}$ of (4) in the equation for $\bar{\bar{\tilde{C}}}_p$ yields the result

$$\bar{\bar{W}}_p(\bar{p}_1,\bar{p}_2) = \bar{\bar{U}}_p(\bar{p}_1,\bar{p}_2)$$

$$+ \int \frac{d\bar{p}_3}{(2\pi)^3} \bar{\bar{U}}_p(\bar{p}_1,\bar{p}_3)\bar{\bar{G}}_c(\bar{p}_3)q_p(\bar{p}_3-\bar{p}_2)\bar{\bar{\tilde{C}}}_p(\bar{p}_3,\bar{p}_2) \qquad (18)$$

where

$$\overline{\overline{W}}_p(\bar{p},\bar{p}_1) = \left[\overline{\overline{I}} + \frac{\overline{\overline{w}}_p(\bar{p})}{V}\overline{\overline{G}}_c(\bar{p})q_p(\bar{p}-\bar{p}_1)\right] \cdot \overline{\overline{C}}_p(\bar{p},\bar{p}_1)$$
$$+ \frac{\overline{\overline{w}}_p(\bar{p})}{V} <\bar{p}|\bar{p}_1> \tag{19}$$

Next, use (19) to express $\overline{\overline{C}}_p$ in terms of $\overline{\overline{W}}_p$ and substitute into (18). We obtain

$$\overline{\overline{C}}_p(\bar{p},\bar{p}_1) = \overline{\overline{U}}_p(\bar{p},\bar{p}_1) + \int \frac{d\bar{p}_2}{(2\pi)^3}\overline{\overline{U}}_p(\bar{p},\bar{p}_2)\overline{\overline{G}}_c(\bar{p}_2)q_p(\bar{p}_2-\bar{p}_1)$$
$$\cdot \left[\overline{\overline{I}} + \frac{\overline{\overline{w}}(\bar{p}_2)}{V}\overline{\overline{G}}_c(\bar{p}_2)q_p(\bar{p}_2-\bar{p}_1)\right]^{-1}\overline{\overline{C}}_p(\bar{p}_2,\bar{p}_1) \tag{20}$$

where

$$\overline{\overline{C}}_p(\bar{p},\bar{p}_1) = \overline{\overline{W}}_p(\bar{p},\bar{p}_1)\left[\overline{\overline{I}} + \frac{1}{V}\overline{\overline{G}}_c(\bar{p}_1)q_p(0)\overline{\overline{w}}_p(\bar{p}_1)\right] \tag{21}$$

To normalize properly (19), we use the technique as in EFA-CP by multiplying both equations with $(1/V)e^{-i(\bar{p}-\bar{p}_1)\cdot \bar{r}_j}$ and integrating over $d\bar{r}_j$. Hence, (19) becomes

$$\overline{\overline{W}}_p(\bar{p},\bar{p}) = \left[\overline{\overline{I}} + \frac{\overline{\overline{w}}_p(\bar{p})}{V}\overline{\overline{G}}_c(\bar{p})q_p(0)\right]\overline{\overline{C}}_p(\bar{p},\bar{p}) + \overline{\overline{w}}_p(\bar{p}) \tag{22}$$

Taking the limit of $V \to \infty$, we note that $q_p(0)/V \to 0$. The coherent potential condition of (6) implies that $\overline{\overline{C}}_p(\bar{p},\bar{p}) = 0$ so that, in view of (21) and (22), it becomes

$$\overline{\overline{w}}_p(\bar{p}) = \overline{\overline{C}}_p(\bar{p},\bar{p}) \tag{23}$$

with $\overline{\overline{C}}_p$ satisfying the integral equation of (20). In view of the vanishing of q_p/V, equation (20) becomes

$$\overline{\overline{C}}_p(\bar{p}_1,\bar{p}_2) = \overline{\overline{U}}_p(\bar{p}_1,\bar{p}_2)$$
$$+ \int \frac{d\bar{p}_3}{(2\pi)^3}\overline{\overline{U}}_p(\bar{p}_1,\bar{p}_3)\overline{\overline{G}}_c(\bar{p}_3)q_p(\bar{p}_3-\bar{p}_2)\overline{\overline{C}}_p(\bar{p}_3,\bar{p}_2) \tag{24}$$

The dispersion relation is

$$\det\left[\overline{\overline{G}}_o^{-1}(\bar{p}) - n_o\overline{\overline{C}}_p(\bar{p},\bar{p})\right] = 0 \tag{25}$$

Low Frequency Solutions

In view of the fact that $\overline{\overline{G}}_c$ depends on $\overline{\overline{C}}_p$ via (3) and (23), the integral equation (24) is nonlinear.

If a modified transition operator $\overline{\overline{t}}$ is defined such that it satisfies the Lippmann-Schwinger equation with propagator $\overline{\overline{G}}_c$, i.e.

$$\overline{\overline{t}}_p(\bar{p}_1,\bar{p}_2) = \overline{\overline{U}}_p(\bar{p}_1,\bar{p}_2) + \int \frac{d\bar{p}_3}{(2\pi)^3} \overline{\overline{U}}_p(\bar{p}_1,\bar{p}_3)\overline{\overline{G}}_c(\bar{p}_3)\overline{\overline{t}}_p(\bar{p}_3,\bar{p}_2) \qquad (26)$$

then $\overline{\overline{C}}(\bar{p}_1,\bar{p}_2)$ can also be expressed in an integral equation in terms of the operator $\overline{\overline{t}}$ (Problem 16)

$$\overline{\overline{C}}_p(\bar{p}_1,\bar{p}_2) = \overline{\overline{t}}_p(\bar{p}_1,\bar{p}_2)$$
$$+ n_o \int d\bar{p}_3 \overline{\overline{t}}_p(\bar{p}_1,\bar{p}_3)\overline{\overline{G}}_c(\bar{p}_3)H(\bar{p}_3-\bar{p}_2)\overline{\overline{C}}_p(\bar{p}_3,\bar{p}_2) \qquad (27)$$

where $H(\bar{p})$ is defined in (17), Section 4.3.

Following the case of QCA, we define $\overline{\overline{t}}_j$ and $\overline{\overline{C}}_j$ operators to be functions of $\bar{r}_j$ such that

$$<\bar{p}_1|\overline{\overline{t}}_j|\bar{p}_2> = e^{-i(\bar{p}_1-\bar{p}_2)\cdot\bar{r}_j} \overline{\overline{t}}_p(\bar{p}_1,\bar{p}_2) \qquad (28)$$

$$<\bar{p}_1|\overline{\overline{C}}_j|\bar{p}_2> = e^{-i(\bar{p}_1-\bar{p}_2)\cdot\bar{r}_j} \overline{\overline{C}}_p(\bar{p}_1,\bar{p}_2) \qquad (29)$$

Then the operator form of (27), (26) and (24) respectively are

$$\overline{\overline{C}}_j = \overline{\overline{t}}_j + n_o \int d\bar{r}_l \overline{\overline{t}}_j \overline{\overline{G}}_c h(\bar{r}_l-\bar{r}_j)\overline{\overline{C}}_l \qquad (30)$$

$$\overline{\overline{t}}_j = \overline{\overline{U}}_j + \overline{\overline{U}}_j \overline{\overline{G}}_c \overline{\overline{t}}_j \qquad (31)$$

$$\overline{\overline{C}}_j = \overline{\overline{U}}_j + \int d\bar{r}_l \overline{\overline{U}}_j \overline{\overline{G}}_c q(\bar{r}_l-\bar{r}_j)\overline{\overline{C}}_l \qquad (32)$$

It also follows from (23) and (25) that the operator equation for the mean Green's operator is

$$E(\overline{\overline{G}}) = \overline{\overline{G}}_c = \overline{\overline{G}}_o + n_o \overline{\overline{G}}_o \int d\bar{r}_j \overline{\overline{C}}_j E(\overline{\overline{G}}) \qquad (33)$$

Hence the mass operator under QCA-CP is $n_o \int d\bar{r}_j \overline{\overline{C}}_j$. The result equations in (30), (31) and (33) are similar to that of Korringa and Mills (1972).

4.5 Low Frequency Solutions

In this section, the effective propagation constants under EFA, QCA, EFA-CP and QCA-CP will be calculated for spherical particles in the low-frequency limit (Tsang and Kong, 1980a). Both the real part of the effective propagation constant K_r and the imaginary part of the effective propagation constant K_i will be studied. When volume scattering is important, $K_i \ll K_r$. In solving the integral equations of the various approximations for such cases, the leading term of the real part and the leading term of the imaginary part will be kept while the other terms can be safely discarded.

It was seen in Section 4.4 that the introduction of the coherent potential into the effective field approximation and the quasicrystalline approximation results in a nonlinear equation for the mean Green's function. This is analogous to the relation between nonlinear approximation and the bilocal approximation of random medium treated in Section 3.2, Chapter 5, where it was found that the difference between the results of the two approximations are insignificant if the real part of the effective propagation constant K_r is approximately equal to k_m. For the case of class D random discrete scatterers (Table 6.1), the permittivity of the particles is significantly different from that of the background medium. When the particles are densely distributed, the real part of the effective wavenumber K_r is appreciably different from k. For example, in snow, the dielectric constant of ice crystals is 3.2 and for typical concentration of ice crystals, K_r/k is of the order of 1.4. Because of the difference between K_r and k, the introduction of any nonlinear approximation can significantly change the effective propagation constant. Thus the results of EFA-CP differ significantly from EFA and the results of QCA-CP differ significantly from QCA.

EFA

The Lippmann-Schwinger equations for the transition operator is

$$\overline{\overline{T}} = \overline{\overline{U}} + \overline{\overline{U}}\,\overline{\overline{G}}_o\,\overline{\overline{T}} \tag{1}$$

In space representation, $<\bar{r}|\overline{\overline{U}}|\bar{r}'> = U(\bar{r})\,\overline{\overline{I}}\,<\bar{r}|\bar{r}'>$ where

$$U(\bar{r}) = \begin{cases} k_s^2 - k^2 & r \leq a \\ 0 & r > a \end{cases} \tag{2}$$

and a is the radius of the particle of permittivity ϵ_s. To solve (1), a mixed representation $<\bar{r}|\overline{\overline{T}}|\bar{p}> \equiv \overline{\overline{T}}_m(\bar{r},\bar{p})$ can be used. Applying to (1) gives

$$\overline{\overline{T}}_m(\bar{r},\bar{p}) = U(\bar{r})\,\overline{\overline{I}}\,e^{i\bar{p}\cdot\bar{r}}$$
$$+ U(\bar{r})\,\overline{\overline{I}}\int d\bar{r}'\,\overline{\overline{G}}_o(\bar{r},\bar{r}')\,\overline{\overline{T}}_m(\bar{r}',\bar{p}) \tag{3}$$

Low Frequency Solutions

Since the particles are spherical, the singularity of the dyadic Green's function will be separated out in the following manner

$$\overline{\overline{G}}_o(\bar{r},\bar{r}') = PS\overline{\overline{G}}_o(\bar{r},\bar{r}') - \frac{\overline{\overline{I}}}{3k^2}\delta(\bar{r}-\bar{r}') \tag{4}$$

as in (13), Section 3.2, Chapter 5. In (4), PS stands for principal value with a spherical exclusion volume at $\bar{r}=\bar{r}'$. For nonspherical particles, a nonspherical exclusion volume is to be chosen giving a different coefficient for the dirac delta function in the second term of (4). At low frequencies, $\bar{p}$ can be set equal to zero in (3). We further let $\overline{\overline{T}}_m(\bar{r}) = \overline{\overline{T}}_m(\bar{r},\bar{p}=0)$. Hence

$$\overline{\overline{T}}_m(\bar{r}) = U(\bar{r})\overline{\overline{I}} - \frac{U(\bar{r})}{3k^2}\overline{\overline{T}}_m(\bar{r})$$

$$+ U(\bar{r})\int d\bar{r}'\ PS\overline{\overline{G}}_o(\bar{r},\bar{r}')\overline{\overline{T}}_m(\bar{r}') \tag{5}$$

To solve (5), let $\overline{\overline{T}}_m(\bar{r}) = T_m(\bar{r})\overline{\overline{I}}$ with

$$T_m(\bar{r}) = \begin{cases} T_m & r \le a \\ 0 & r > a \end{cases} \tag{6}$$

Substitute (6) into (5) and set $\bar{r}=0$. The integral in (5) is of the same form as (17), Section 3.2, Chapter 5, and can be evaluated readily. The integral equation in (5) reduces to the form

$$T_m = (k_s^2 - k^2) - \frac{(k_s^2 - k^2)}{3k^2}T_m$$

$$+ T_m(k_s^2 - k^2)\frac{2}{3}\int_0^a dr\ r\ e^{ikr} \tag{7}$$

For small particles, ka is much less than 1 and the real part of the third term in (7) is much smaller than the real part of the first and the second terms, and can be neglected. However, the imaginary part contributes to the leading term of the imaginary part of T_m and cannot be neglected. Hence,

$$T_m = (k_s^2 - k^2) - \frac{(k_s^2 - k^2)}{3k^2}T_m + T_m(k_s^2 - k^2)\frac{i2ka^3}{9} \tag{8}$$

Solving (8) gives

$$T_m = 3k^2 y\left[1 + \frac{2}{3}ik^3 a^3 y\right] \tag{9}$$

where

$$y = \frac{\epsilon_s - \epsilon}{\epsilon_s + 2\epsilon} \tag{10}$$

Using the results of (6) and (9), the momentum representation of $\overline{\overline{T}}$ can be calculated from the mixed representation. In the low frequency limit,

$$\overline{\overline{T}}_p(\bar{p}_1, \bar{p}_2) = T_m v_o \overline{\overline{I}} \tag{11}$$

where $v_o = 4\pi a^3/3$ is the volume of the particle. The dispersion relation is given by (7), Section 4.2. Hence, under EFA

$$K^2 = k^2 + 3k^2 fy \left[1 + i\frac{2}{3}k^3 a^3 y\right] \tag{12}$$

with $f = n_o v_o$ denoting the fractional volume occupied by the particles.

QCA

Integral equation in (16), Section 4.3 is to be solved with $\overline{\overline{T}}_p(\bar{p}_1, \bar{p}_2)$ as given by (11). Hence,

$$\overline{\overline{C}}_p(\bar{p}_1, \bar{p}_2) = T_m v_o \overline{\overline{I}}$$

$$+ fT_m \int_{-\infty}^{\infty} d\bar{p}_3 \, \overline{\overline{G}}_o(\bar{p}_3) H(\bar{p}_3 - \bar{p}_2) \overline{\overline{C}}_p(\bar{p}_3, \bar{p}_2) \tag{13}$$

In the low frequency limit, $\bar{p}_2$ can be set to zero in (13). Further, let

$$\overline{\overline{C}}_p(\bar{p}_1, \bar{p}_2) = c \overline{\overline{I}} \tag{14}$$

Substituting (14) into (13), the integral in (13) can be put in the space domain. Further, make use of (17), Section 4.3, and the decomposition of the dyadic Green's function as in (4) to obtain

$$c = T_m v_o + \frac{fT_m}{3k^2} c$$

$$+ cfT_m \int_{-\infty}^{\infty} d\bar{r} \, PS \, \overline{\overline{G}}_o(\bar{r}) \left[g(\bar{r}) - 1\right] \tag{15}$$

The integral in (15) can be calculated as in (17), Section 3.2, Chapter 5. The real part of the third term in (15) is small compared to that of the first two terms. The imaginary part contributes directly to the leading term of the imaginary part of c and has to be retained. Thus,

$$c = T_m v_o + fT_m c \left\{ \frac{1}{3k^2} + \frac{2}{3}ik \int_0^{\infty} dr \, r^2 \left[g(r) - 1\right] \right\} \tag{16}$$

The integral in (16) is of the order $O(a^3)$. Solving for c in (16), using the expression for T_m as given in (9), and noting that $Re(T_m) \gg Im(T_m)$ and $Re(c) \gg Im(c)$, gives

$$c = \frac{3v_o k^2 y}{1-fy}\left\{1 + i\frac{2}{3}\frac{(ka)^3 y}{1-fy}\right.$$

$$\left.\times\left[1 + 4\pi n_o \int_0^\infty dr\, r^2 [g(r) - 1]\right]\right\} \quad (17)$$

The dispersion relation is (15), Section 4.3. Hence under QCA,

$$K^2 = k^2 + n_o c \quad (18)$$

The results in (17) and (18) can be shown to agree with that of the simple model as given in (20) of Section 2.2. Note that $g(r) = 0$ for $r < b$.

EFA-CP

Under EFA-CP, the equation for the modified transition operator $\bar{\bar{\hat{t}}}$ is, from (15), Section 4.4,

$$\bar{\bar{\hat{t}}} = \bar{\bar{U}} + \bar{\bar{U}}\bar{\bar{G}}_c \bar{\bar{\hat{t}}} \quad (19)$$

Comparing (19) with the transition operator $\bar{\bar{T}}$ as given by (1), it can be concluded that $\bar{\bar{\hat{t}}}$ will only be appreciably different from $\bar{\bar{T}}$ if K is substantially different from k. To solve (19), we assume that in the low frequency limit $\bar{\bar{G}}_c$ assumes the same expression as $\bar{\bar{G}}_o$ with k replaced by K. For example, $\bar{\bar{G}}_c$ is the expression of (9), Section 4.1, with k replaced by K. Hence,

$$\bar{\bar{G}}_c(\bar{r},\bar{r}') = PS\,\bar{\bar{G}}_c(\bar{r},\bar{r}') - \frac{\bar{\bar{I}}}{3K^2}\delta(\bar{r} - \bar{r}') \quad (20)$$

The solution of the integral equation in (19) can be obtained in a manner analogous to that of (2) through (8). Thus,

$$\bar{\bar{\hat{t}}}_p(\bar{p}_1, \bar{p}_2) = \hat{t}_m v_o \bar{\bar{I}} \quad (21)$$

where $\hat{t}_m$ satisfies the following equation analogous to (8)

$$\hat{t}_m = (k_s^2 - k^2) - \frac{(k_s^2 - k^2)\hat{t}_m}{3K^2} + \hat{t}_m (k_s^2 - k^2)\frac{i2Ka^3}{9} \quad (22)$$

The $(k_s^2 - k^2)$ factor in (22) arises from the potential $\overline{\overline{U}}$ which, as indicated in (19), is unchanged. Other k's in (8) are replaced by K to give (22). Solving (22) gives

$$\hat{t}_m = \frac{z}{1 + z/(3K^2)} \left[1 + i\frac{2}{9} \frac{Ka^3 z}{1 + z/(3K^2)} \right] \qquad (23)$$

On applying (16), Section 4.4, the dispersion relation is, under EFA-CP,

$$K^2 = k^2 + f\hat{t}_m \qquad (24)$$

which is a nonlinear equation for K^2.

QCA-CP

The governing integral equation is (26) and (27), Section 4.4, with $\overline{\overline{t}}_p(\overline{p}_1, \overline{p}_2)$ as given by (21). The procedure of solution of (27), Section 4.4, is similar to the case of QCA. Hence, in the low frequency limit

$$\overline{\overline{C}}_p(\overline{p}_1, \overline{p}_2) = \hat{c}\,\overline{\overline{I}} \qquad (25)$$

and $\hat{c}$ obeys the following equation that is analogous to (16)

$$\hat{c} = v_o \hat{t}_m + f\hat{t}_m \hat{c} \left[\frac{1}{3K^2} + \frac{2}{3}iK \int_0^\infty dr\, r^2 \left[g(r) - 1 \right] \right] \qquad (26)$$

Use the expression of $\hat{t}_m$ as given by (23), solve $\hat{c}$ from (26) and retain only the leading term of the real part and the leading term of the imaginary part of $\hat{c}$ to get

$$\hat{c} = \frac{v_o z}{1 + z(1 - f)/(3K^2)} \left\{ 1 + i\frac{2}{9}Ka^3 \frac{z}{1 + z(1 - f)/(3K^2)} \right.$$

$$\left. \times \left[1 + 4\pi n_o \int_0^\infty dr\, r^2 \left[g(r) - 1 \right] \right] \right\} \qquad (27)$$

On applying (25), Section 4.4, the dispersion relation under QCA-CP is

$$K^2 = k^2 + n_o \hat{c} \qquad (28)$$

which is a nonlinear equation for K^2. Comparing the results of QCA of (17) and (18) and QCA-CP of (27) and (28), it can be concluded that the two results differ appreciably if K_r deviates significantly from k.

In the very low frequency limit, the scattering attenuation term that is dependent on particle size in (27) can be neglected, and the mixture formula for $\epsilon_{eff} = K^2/\omega\mu_o$ is, on using the first term in (27),

$$\epsilon_{eff} = \epsilon + \frac{3f(\epsilon_s - \epsilon)\epsilon_{eff}}{3\epsilon_{eff} + (\epsilon_s - \epsilon)(1 - f)} \qquad (29)$$

With simple algebra, it can be shown (Problem 33) that (29) is the same as

$$\epsilon_{eff} = \frac{\epsilon(1 - f) + f\epsilon_s + f\dfrac{(\epsilon - \epsilon_s)\epsilon_s}{3\epsilon_{eff} - \epsilon + \epsilon_s}}{1 + f\dfrac{\epsilon - \epsilon_s}{3\epsilon_{eff} + \epsilon_s - \epsilon}} \qquad (30)$$

The very low frequency result in (30) has also been derived by Kohler and Papanicolaou [1981, (8.23)].

Numerical illustrations of EFA-CP and QCA-CP will be given in Section 9. In this section, the results of the various approximations on the multiple scattering equations are given. The final results are all into the momentum representation and are based on the operator formalism of Section 3.2. The equations of the effective field approximation and quasicrystalline approximation can also be solved in the space domain by using the T-matrix formalism based on spherical waves as given in Section 3.3. This latter approach will be taken in Sections 6, 7, and 8. However, it is the operator formalism in the momentum domain that facilitates the introduction of the coherent potential. The coherent potential method has been applied in solid state physics (Soven, 1966, 1967, 1969; Velicky et al., 1968; Faulkner, 1970; Gyorffry, 1970; Korringa and Mills, 1972; Lax, 1973; Roth, 1974; Schwartz and Bansil, 1974; Faulkner, 1979; Ziman, 1979; Singh, 1981), to electromagnetic wave propagation (Tsang and Kong, 1980a; Kohler and Papanicolaou, 1981) and to elastic wave propagation (Devaney, 1980; Schwartz and Plona, 1984; Schwartz and Johnson, 1984).

4.6 Energy Conservation and Second Moment

In Chapter 5, it was shown in Problem 5.11 that in the case of random medium, energy conservation is satisfied exactly if the nonlinear approximation of the mean field is used in conjunction with the ladder approximation of the second moment of the field. It was also mentioned that the difference between the results of the nonlinear approximation and the bilocal approximation is small when the real part of the effective propagation constant is approximately equal to that of the background medium or the mean wavenumber. Thus it is possible to use the bilocal approximation with the ladder approximation for turbulent medium and continuous random medium with small variance of permittivity fluctuations to derive the radiative transfer equations (Ishimaru, 1978; Fante, 1981). In Sections 3.4 and 3.6 of Chapter 5, the effective field approximation and ladder approximation are used to derive the radiative transfer equations for sparsely distributed isotropic scatterers. It was not required to introduce the EFA-CP approximation. However, in this chapter, we are concerned with densely distributed non-tenuous particles where there is an appreciable difference between K_r and k so that the results of QCA and QCA-CP can be significantly different. Hence it is important to examine the issue of energy conservation for dense medium.

The scalar wave equation shall be used in this section. Extensions can readily be made to the vector wave equation by using dyadic notation. The condition of energy conservation for QCA-CP and the approximation for the intensity operator is examined. The condition also suggests that a modified ladder approximation should be used for the second moment of the field.

The scalar wave equation for particles with wavenumber k_s in a background medium of wavenumber k is

$$-(\nabla^2 + k^2)\psi(\bar{r}) = \sum_{j=1}^{N} U(\bar{r} - \bar{r}_j)\psi(\bar{r}) \quad (1)$$

The potential $U_j(\bar{r}) = U(\bar{r} - \bar{r}_j)$ is equal to $k_s^2 - k^2$ inside the particle j and zero outside particle j. It shall be assumed that both k_s and k are real. Following Problem 5.11, an energy flux vector $\bar{S}_\psi(\bar{r})$ can be defined

$$\bar{S}_\psi \equiv \psi^*(\bar{r})\nabla\psi(\bar{r}) - \psi(\bar{r})\nabla\psi^*(\bar{r}) \quad (2)$$

Since k and k_s are both real, it follows that $\nabla \cdot \bar{S}_\psi = 0$. Energy conservation is provided by the integrated optical relation

$$\int d\bar{r}\, E\left[\nabla \cdot \bar{S}_\psi(\bar{r})\right] = 0 \quad (3)$$

The Dyson's equation for the mean field and the mean Green's function are respectively

$$-(\nabla^2 + k^2)E\left[\psi(\bar{r})\right] = \int d\bar{r}'\, M(\bar{r},\bar{r}')E\left[\psi(\bar{r}')\right] \quad (4)$$

$$-(\nabla^2 + k^2)E\left[G(\bar{r},\bar{r}')\right] = \delta(\bar{r}-\bar{r}') + \int d\bar{r}''\, M(\bar{r},\bar{r}'')E\left[G(\bar{r}'',\bar{r}')\right] \quad (5)$$

where $M(\bar{r},\bar{r}') \equiv <\bar{r}|M|\bar{r}'>$ is the coordinate representation of the mass operator M. In this section, the infinite space Green's function is $G_o(\bar{r},\bar{r}') = \exp(ik|\bar{r}-\bar{r}'|)/4\pi|\bar{r}-\bar{r}'|$. The second moment of the field obeys the equation

$$E[\psi(\bar{r})\psi^*(\bar{r}')] = E[\psi(\bar{r})]E[\psi^*(\bar{r}')]$$
$$+ \int d\bar{r}_1 \int d\bar{r}_2 \int d\bar{r}'_1 \int d\bar{r}'_2 \, E[G(\bar{r},\bar{r}_1)]\, E[G^*(\bar{r}',\bar{r}'_1)]$$
$$\cdot I(\bar{r}_1,\bar{r}_2;\bar{r}'_1,\bar{r}'_2)\, E[\psi(\bar{r}_2)\psi^*(\bar{r}'_2)] \qquad (6)$$

where I is the intensity operator. In Problem 5.11, it was shown that the integrated optical relation (3) is satisfied if the mass operator and the intensity operator satisfies the relation

$$M(\bar{r}'_2,\bar{r}_2) - M^*(\bar{r}_2,\bar{r}'_2) + \int d\bar{r} \int d\bar{r}' \left\{ E[G^*(\bar{r},\bar{r}')] \right.$$
$$\left. - E[G(\bar{r}',\bar{r})] \right\} I(\bar{r},\bar{r}_2;\bar{r}',\bar{r}'_2) = 0 \qquad (7)$$

The governing equations for QCA-CP are given in Section 4.4. The mass operator under QCA-CP is, from (32) and (33), Section 4.4,

$$M = n_o \int d\bar{r}_j \, \hat{C}_j \qquad (8)$$

where $\hat{C}_j$ satisfies the equation

$$\hat{C}_j = U_j + \int d\bar{r}_l \, U_j \, G_c \, q(\bar{r}_l - \bar{r}_j) \, \hat{C}_l \qquad (9)$$

and

$$q(\bar{r}) = n_o h(\bar{r}) + \delta(\bar{r}) \qquad (10)$$

Next denote the Hermitian adjoint of an operator by superscript $+$ so that for an operator A, $<\bar{r}|A^+|\bar{r}'> \equiv A^*(\bar{r}',\bar{r})$. Since $U(\bar{r})$ is real, hence $U_j^+ = U_j$. From (8) and (9), it follows that

$$M - M^+ = n_o \int d\bar{r}_j \int d\bar{r}_l \, q(\bar{r}_l,\bar{r}_j) \left[U_j \, G_c \, \hat{C}_l - \hat{C}_l^+ \, G_c^+ \, U_j \right] \qquad (11)$$

In (11), substitute the first U_j by using the Hermitian adjoint of (9) and the second U_j by using (9). Hence,

$$M - M^+ = n_o \int d\bar{r}_j \int d\bar{r}_l \, q(\bar{r}_l,\bar{r}_j)$$

$$\times \left\{ \left[\hat{C}_j^+ - \int d\bar{r}_m\, q(\bar{r}_m, \bar{r}_j)\, \hat{C}_m^+ G_c^+ U_j \right] G_c \hat{C}_l \right.$$

$$\left. - \hat{C}_l^+ G_c^+ \left[\hat{C}_j - \int d\bar{r}_m\, q(\bar{r}_m, \bar{r}_j) U_j G_c \hat{C}_m \right] \right\} \quad (12)$$

By interchanging dummy integration variables, the second and fourth terms in (12) cancel each other. We further make use of the symmetry property of $q(\bar{r}_l, \bar{r}_j)$. Hence

$$M - M^+ = n_o \int d\bar{r}_j \int d\bar{r}_l\, q(\bar{r}_l, \bar{r}_j)\, \hat{C}_j^+ (G_c - G_c^+) \hat{C}_l \quad (13)$$

Taking (13) in coordinate representation gives

$$M(\bar{r}_2', \bar{r}_2) - M^*(\bar{r}_2, \bar{r}_2') = n_o \int d\bar{r}_j \int d\bar{r}_l \int d\bar{r} \int d\bar{r}'\, q(\bar{r}_l, \bar{r}_j)\, \hat{C}_j^*(\bar{r}', \bar{r}_2')$$

$$\cdot [G_c^*(\bar{r}, \bar{r}') - G_c(\bar{r}', \bar{r})] \hat{C}_l(\bar{r}, \bar{r}_2) \quad (14)$$

Comparing (14) with the energy conservation condition of (7) and using the fact that $E(G) = G_c$ suggests that the approximation for the intensity operator that is energetically consistent with QCA-CP is

$$I(\bar{r}, \bar{r}_2; \bar{r}', \bar{r}_2') = n_o \int d\bar{r}_j \int d\bar{r}_l\, q(\bar{r}_l, \bar{r}_j)\, \hat{C}_j^*(\bar{r}', \bar{r}_2')\, \hat{C}_l(\bar{r}, \bar{r}_2) \quad (15)$$

Using (10) in (15) gives

$$I(\bar{r}, \bar{r}_2; \bar{r}', \bar{r}_2') = n_o \int d\bar{r}_j\, \hat{C}_j^*(\bar{r}', \bar{r}_2')\, C_j(\bar{r}, \bar{r}_2)$$

$$+ n_o^2 \int d\bar{r}_j \int d\bar{r}_l\, [g(\bar{r}_l, \bar{r}_j) - 1]\, \hat{C}_j^*(\bar{r}', \bar{r}_2')\, \hat{C}_l(\bar{r}, \bar{r}_2) \quad (16)$$

The first term in (16) corresponds to the ladder approximation of scattering of the field and field conjugate from the same particle j. The second term includes scattering of field from particle l and scattering of field conjugate from particle j. The weighting factor $g(\bar{r}_l, \bar{r}_j) - 1$ states that the major contribution arises from particles l and j that are within a few diameters of each other. Equation (16) shall be labelled as the modified ladder approximation.

The second moment for discrete scatterers has been studied by Frisch (1968) by using the diagrammatic technique. The exact intensity operator

Pair-Distribution Functions

is a summation of infinite number of terms as given in equation (6.37) of Frisch (1968). The first two terms of that infinite summation are

$$I(\bar{r},\bar{r}_2;\bar{r}',\bar{r}'_2) = n_o \int d\bar{r}_j\, T_j^*(\bar{r}',\bar{r}'_2)\, T_j(\bar{r},\bar{r}_2)$$
$$+ n_o^2 \int d\bar{r}_j \int d\bar{r}_l\, [g(\bar{r}_l,\bar{r}_j) - 1]\, T_j^*(\bar{r}',\bar{r}'_2)\, T_l(\bar{r},\bar{r}_2) \quad (17)$$

where T_j is the transition operator for particle j. Comparing (16) and (17), we note that the two results are similar in nature. It can be seen from (30) and (31), Section 4.4, that $\hat{C}_j$ includes the transition operator of particle j together with short range effect from particles in the neighborhood of j.

In Section 10, radiative transfer equations will be derived based on the modified ladder approximation. It shall be shown that if QCA is used for the first moment, the derived radiative transfer equations satisfy energy conservation in the special case when $K_r \simeq k$. It is further shown that when QCA-CP is used for the first moment, then the result radiative transfer equations satisfy energy conservation for tenuous as well as non-tenuous dense media.

5 PAIR-DISTRIBUTION FUNCTIONS

5.1 Introduction

In applying the quasicrystalline approximation and the quasicrystalline approximation with coherent potential, the pair-distribution function of particle positions must be specified. In Section 4.3, we have shown that in the special case of independent particle position, $g(\bar{r}) = 1$, the dispersion relation for the quasicrystalline approximation reduces to that of effective field approximation. Another approximation to the pair-distribution function is the hole-correction (HC) approximation, given by $g(r) = 0$ for $r < b$ and $g(r) = 1$ for $r \geq b$ where b is the diameter of the circumscribing sphere of the particle. For the case of spherical particles with radius a, b is equal to $2a$. The hole-correction approximation takes into account the fact that the particles cannot interpenetrate each other and further assumes uniform distribution outside the hole.

Neither the independent position approximation nor the hole-correction approximation is correct when the fractional volume of scatterers, f, is appreciable. It is easier to visualize this for the case of one-dimensional scatterers. The hole-correction approximation is illustrated in Figure 6.7A.

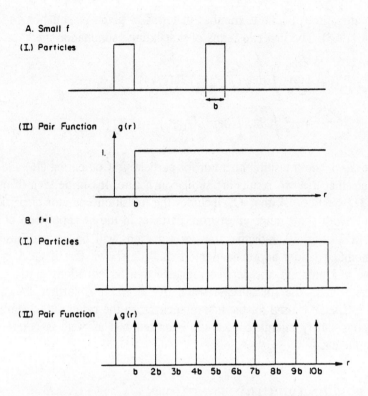

Fig. 6.7 Pair-distribution function for one-dimensional particles: (A) Small f; particles and pair function, (B) $f = 1$; particles and pair function.

Next we imagine that f is equal to unity so that the entire volume V is occupied by scatterers. In such a case, the centers of these one-dimensional particles will be separated by integral multiples of b from each other. The pair-distribution function $g(r)$ will be zero for $r \neq mb$ where m is any nonzero integer. It consists of delta functions at the position of r equal to an integral multiple of b (Figure 6.7B). Thus, the hole-correction approximation is very poor in such a limit. When f is not equal to 1 but appreciable, the pair-distribution function will be an interpolation between A and B in Figure 6.7. We also note that as the two particle separation r approaches infinity, the positions of the particles should be independent of each other, hence $\lim_{r \to \infty} g(r) = 1$ for f not equal to maximum concentration.

The study of pair-distribution functions is a subject of interest in statistical mechanics (Green, 1969; Finney, 1970; McQuarrie, 1976; Waseda, 1980). Based on the form of the pair-distribution function, substances are classified into three different types: (1) gas; (2) liquid and amorphous solid; and (3) crystalline solid. The three forms of the pair-distribution function are

Introduction

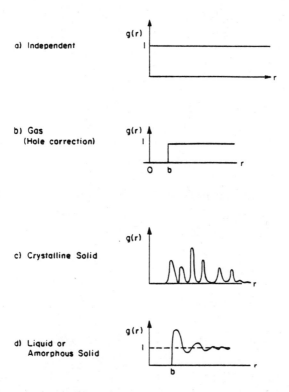

Fig. 6.8 Pair functions for (a) independent particle position, (b) gas, (c) crystalline solid, and (d) liquid and amorphous solid.

illustrated in Figure 6.8. The case of gas with particles sparsely distributed is considered to be a system of extreme disorder, so that the hole-correction approximation or the independent position approximation is a good description of the pair-distribution function. In the opposite extreme, the case of crystalline solid with relative positions of the particles fixed is a case of extreme order. The pair-distribution function exhibits sharp peaks (Figure 6.8c). The case of a liquid and amorphous solid is a system of partial order and is an interpolation between the two extreme cases of gas and crystalline solid (Figure 6.8d). Extensive experimental and theoretical investigations have been carried out for the pair-distribution function of liquid and amorphous solid (Waseda, 1980). Experimental techniques are largely based on X-ray, electron or neutron diffraction studies measuring the bistatic scattering of the structure. To a first-order approximation, the bistatic scattering intensity is proportional to the structure factor which is related to the Fourier transform of the pair-distribution functions. The pair-distribution function has also been studied by simulations (Hawley et al., 1967). In the following sections, we shall list some of the various results of pair-distribution functions that have been used in the study of multiple scattering of waves in dense media.

5.2 Percus-Yevick Equation

When $g(\bar{r})$ is equal to 1, the particle positions are independent of each other. Thus a total influence h can be defined as

$$h(\bar{r}_{12}) = g(\bar{r}_{12}) - 1 \tag{1}$$

that describes the total influence of particle 1 on particle 2. The total influence is decomposed into a sum of direct and indirect correlation functions. The direct correlation function is denoted by $c(\bar{r})$

$$h(\bar{r}_{12}) = c(\bar{r}_{12}) + \text{indirect} \tag{2}$$

The direct correlation function is such that it satisfies the following integral equation

$$h(\bar{r}_{12}) = c(\bar{r}_{12}) + n_o \int d\bar{r}_3 c(\bar{r}_{13}) h(\bar{r}_{32}) \tag{3}$$

which is known as the Ornstein-Zernike equation and can be regarded as the defining equation for the direct correlation function $c(\bar{r})$. The physical interpretation of (3) is that the indirect influence of particle 1 on particle 2 is a result of particle 1 acting directly on a particle at $\bar{r}_3$, which in turn exerts total influence on particle 2. The indirect influence is averaged over particle positions $\bar{r}_3$ and weighted by the number of particles per unit volume n_o as indicated in (3).

The Ornstein-Zernike equation consists of two unknowns $c(\bar{r})$ and $h(\bar{r})$ in one equation. To solve for the unknowns, an approximation is to be made on the relation between $c(\bar{r})$ and $h(\bar{r})$ reducing (3) to one equation and one unknown. The Percus-Yevick approximation can be introduced in the following heuristic manner.

The potential energy between two particles is governed by $u(\bar{r})$ where r is their separation. In the presence of other particles, there is an equivalent potential energy between the two particles $w(\bar{r})$ such that

$$g(\bar{r}) = e^{-\beta w(\bar{r})} \tag{4}$$

where β is the Boltzmann factor $1/KT$. Let

$$g(\bar{r}) = e^{-\beta u(\bar{r})} y(\bar{r}) \tag{5}$$

so that
$$y(\bar{r}) = e^{-\beta(w(\bar{r})-u(\bar{r}))} \tag{6}$$

is a measure of the indirect influence of other particles on the pair function. When $y = 1$, we have $w(\bar{r}) = u(\bar{r})$ and there is no indirect influence. The Percus-Yevick approximation consists of equating $h - c$ to $y - 1$

$$h(\bar{r}) - c(\bar{r}) = y(\bar{r}) - 1 \tag{7}$$

Substituting (7) into (3) and using (5) gives

$$y(\bar{r}_{12}) = 1 + n_o \int d\bar{r}_3 f(\bar{r}_{13}) y(\bar{r}_{13}) h(\bar{r}_{32}) \tag{8}$$

where
$$f(\bar{r}) = e^{-\beta u(\bar{r})} - 1 \tag{9}$$

In view of (1) and (5),
$$h(\bar{r}) = e^{-\beta u(\bar{r})} y(\bar{r}) - 1 \tag{10}$$

Substituting (10) into (8) gives an integral equation for the single unknown $y(\bar{r})$. The equation is known as the Percus-Yevick equation (Percus and Yevick, 1958).

The Percus-Yevick equation has a closed-form solution for the case of hard-sphere potential (Wertheim, 1963, 1964). The hard-sphere potential is described by

$$u(\bar{r}) = \begin{cases} \infty & \text{for } r < b \\ 0 & \text{for } r \geq b \end{cases} \tag{11}$$

which says that in the absence of other particles, the potential energy between the two particles is infinite when they overlap each other (thus disallowing interpenetration) and is zero otherwise.

Substituting (11) into (8) and using (9) and (10) result in the following integral equation

$$y(\bar{r}) = 1 + n_o \int_{\substack{r'<b \\ |\bar{r}-\bar{r}'|<b}} d\bar{r}' y(\bar{r}') - n_o \int_{\substack{r'<b \\ |\bar{r}-\bar{r}'|>b}} d\bar{r}' y(\bar{r}')[y(\bar{r}-\bar{r}') - 1] \tag{12}$$

In view of (7) and (10)

$$c(\bar{r}) = f(\bar{r}) y(\bar{r}) = \begin{cases} -y(\bar{r}) & \text{for } r < b \\ 0 & \text{for } r > b \end{cases} \tag{13}$$

A closed-form solution can be obtained for $c(\bar{r})$ via the integral equation (12) (Thiele, 1963; Wertheim, 1963). Let the distance be normalized by b

$$x = \frac{r}{t} \tag{14}$$

The solution for $c(\bar{r})$ is

$$-c(x) = \alpha + \beta x + \delta x^3 \tag{15}$$

where

$$\alpha = \frac{(1+2f)^2}{(1-f)^4} \tag{16}$$

$$\beta = -6f\frac{(1+f/2)^2}{(1-f)^4} \tag{17}$$

$$\delta = \frac{f(1+2f)^2}{2(1-f)^4} \tag{18}$$

$$f = \frac{n_o \pi b^3}{6} \tag{19}$$

For spheres, $b = 2a$ and f given in (19) is the fractional volume of spherical scatterers. For the case of nonspherical particles, $b = 2a$ describes the diameter for the exclusion sphere around the particle and f in (19) is the exclusion volume f_e.

The Percus-Yevick pair-distribution function is expressed in terms of the inverse Laplace transform. The final result assumes the following form

$$g(x) = \sum_{n=1}^{\infty} g_n(x) \tag{20}$$

where

$$g_n(x) = \frac{(-1)^{n+1}}{2\pi i f x} \int_{\delta'-i\infty}^{\delta'+i\infty} dt\, t\, e^{t(x-n)} \left(\frac{L(t)}{S(t)}\right)^n \tag{21}$$

$$L(t) = 12f[(1+f/2)t + (1+2f)] \tag{22}$$

$$S(t) = (1-f)^2 t^3 + 6f(1-f)t^2 + 18f^2 t - 12f(1+2f) \tag{23}$$

The function $S(t)$ is a cubic polynomial with three roots. The quantity δ' is chosen such that the path of integration in (21) is a line to the right of all the three roots of $S(t)$. By Cauchy's theorem, $g_n(x) = 0$ for $x < n$ and $g_n(x)$ for $x \geq n$ is calculated by summing the residues of the integrand

Structure Factor

at the three roots $t_o, t_1,$ and t_2 of the cubic polynomial $S(t)$. In view of (20), $g(x) = 0$ for $x < 1$.

5.3 Structure Factor

X-ray diffraction study of liquid structure utilizes the fact that the bistatic scattering properties of the liquid are proportional to the structure factor. In Section 7, by using the distorted Born approximation to scattering from a half-space of particles, the incoherent intensity is also found to be proportional to the structure factor. The structure factor is a quantity directly related to the Fourier transform of the pair-distribution function.

Define the Fourier transform of the total correlation function as

$$H(\bar{p}) = \frac{1}{(2\pi)^3} \int_{-\infty}^{\infty} d\bar{r}\, e^{-i\bar{p}\cdot\bar{r}} h(\bar{r}) \tag{1}$$

Let $C(\bar{p})$ denote the Fourier transform of the direct correlation function $c(\bar{r})$. The integral in the Ornstein-Zernike equation of (3), Section 5.2, is a convolution integral. Thus, in the Fourier transform domain, the integral in the Ornstein-Zernike equation is proportional to the product of $C(\bar{p})$ and $H(\bar{p})$. Solving the equation gives

$$H(\bar{p}) = \frac{C(\bar{p})}{1 - n_o(2\pi)^3 C(\bar{p})} \tag{2}$$

The structure factor $S(\bar{p})$ is defined as

$$S(\bar{p}) = 1 + n_o(2\pi)^3 H(\bar{p}) \tag{3}$$

We have seen in Section 2 that the integral of $g(\bar{r}) - 1$ enters into the final result of scattering attenuation and bistatic scattering. The integral can be expressed in terms of $H(\bar{p})$ at $\bar{p} = 0$. From (1)

$$\int_{-\infty}^{\infty} d\bar{r}\, [g(\bar{r}) - 1] = (2\pi)^3 H(\bar{p} = 0) \tag{4}$$

The Quantity $H(\bar{p})$ for Percus-Yevick Hard-Sphere Pair Function

The closed-form solution for the direct correlation function of Percus-Yevick hard spheres is given in (15) through (19), Section 5.2. The Fourier transform $C(\bar{p})$ can be calculated readily and $H(\bar{p})$ is obtained from (2) and $H(\bar{p})$ is only a function of $pb, b,$ and f (fractional exclusion volume f_e). In this case $C(\bar{p})$ is

$$C(\bar{p}) = C_{PY}(pb, b, f) \tag{5}$$

with (Problem 18)

$$(2\pi)^3 n_o C_{PY}(pb, b, f)$$
$$= 24f \left\{ \frac{(\alpha+\beta+\delta)}{u^2} \cos u - \frac{(\alpha+2\beta+4\delta)}{u^3} \sin u \right.$$
$$\left. - \frac{2(\beta+6\delta)}{u^4} \cos u + \frac{2\beta}{u^4} + \frac{24\delta}{u^5} \sin u + \frac{24\delta}{u^6}(\cos u - 1) \right\} \quad (6)$$

where $u = pb$ and α, β, δ are given in (16) through (18), Section 5.2, respectively. The closed-form solution of $H(\bar{p} = 0)$ is obtained by letting $p = 0$ in (2), and in (5) and (6). We get (Problem 19)

$$n_o(2\pi)^3 H_{PY}(\bar{p} = 0) = n_o \int_{-\infty}^{\infty} d\bar{r}\,[g_{PY}(\bar{r}) - 1] = -1 + \frac{(1-f)^4}{(1+2f)^2} \quad (7)$$

5.4 Improvement of Percus-Yevick Approximation

The pair-distribution function of a classical fluid is related to the pressure both by the virial equation of the canonical ensemble and by the equation for the compressibility of the grand canonical ensemble (McQuarrie, 1976). The two values of the pressure given by the two equations should be the same if the exact pair-distribution function is used and generally do not agree if approximate pair-distribution functions are used. This defect exists (Rowlinson, 1965) for the Percus-Yevick approximation.

Equation of State from Pressure Equation

The equation of state from the pressure equation of the canonical ensemble is (McQuarrie, 1976)

$$\frac{p}{KT} = n_o - \frac{n_o^2}{6KT} \int_0^\infty dr\, r\, u'(r)\, g(r) 4\pi r^2 \quad (1)$$

where p is pressure, K is Boltzmann's constant, and T is temperature. For the case of hard spheres, (1) becomes

$$\frac{p}{n_o KT} = 1 + \frac{2\pi b^3}{3} n_o g(r = b_+) \quad (2)$$

In view of the hard sphere potential and relations among y, c, and g, we have $g(r) = y(r)$ for $r > b$ and $c(r) = -y(r)$ for $r < b$. For the

Percus-Yevick equation, y, y' and y'' are continuous at $r = b$ (Wertheim, 1963). Hence,

$$g(r = b_+) = -c(x = 1) \tag{3}$$

Using (3) in (2) and the results for $c(r)$ in Section 5.2 give the following equation of state

$$\frac{p}{n_o KT} = \frac{1 + 2f + 3f^2}{(1-f)^2} \tag{4}$$

Expanding (4) in power series of f gives

$$\frac{p}{n_o KT} = 1 + 4f + 10f^2 + 16f^3 + \ldots \tag{5}$$

Equation of State from Compressibility Equation

The equation of state from the compressibility equation of grand canonical ensemble is (McQuarrie, 1976)

$$\frac{1}{KT}\frac{\partial p}{\partial n_o} = \left\{1 + n_o \int d\bar{r}\, [g(\bar{r}) - 1]\right\}^{-1} \tag{6}$$

For the Percus-Yevick pair-distribution function, the integral in (6) can be evaluated and given in (7), Section 5.3. Since $n_o = 6f/(\pi b^3)$, the differential equation in (6) becomes

$$\frac{\pi b^3}{6KT}\frac{\partial p}{\partial f} = \frac{(1 + 2f)^2}{(1 - f)^4} \tag{7}$$

Integrating (7) with respect to f and noting that $p = 0$ for $f = 0$ give the following result for the state equation.

$$\frac{p}{n_o KT} = \frac{1 + f + f^2}{(1 - f)^3} \tag{8}$$

Expanding (8) in power series of f gives

$$\frac{p}{n_o KT} = 1 + 4f + 10f^2 + 19f^3 + \ldots \tag{9}$$

Comparing (4) and (8) shows that the two state equations are different. However a comparison of (5) and (9) reveals that the difference is on the order of f^3. Hence, the difference between the two state equations will be small for $0 \leq f \leq 0.4$.

Improvement of the Percus-Yevick Result

The pair-distribution function can also be calculated by Monte Carlo simulations. Extensive computer experiments have been performed for hard spheres (Alder and Hecht, 1969; Barker and Henderson, 1971). A comparison between the computer results and the Percus-Yevick results shows that the latter suffers from three defects: (1) $g_{PY}(r/b, f)$ is too small near $r = b$; (2) oscillation at large r has the consequence that the main maximum of structure factor is too high; and (3) $g(r)$ oscillates slightly out of phase with the computer result of g. Verlet and Weis (1972) propose to correct the result with the following new pair-distribution function which we shall denote by g_{VW}

$$g_{VW}(r/b, f) = \begin{cases} 0 & \text{for } r < b \\ g_{PY}(r/b_w, f_w) + \delta g_1(r) & \text{for } r \geq b \end{cases} \quad (10)$$

where

$$f_w = f - f^2/16 \quad (11)$$

$$b_w = [f_w/f]^{1/3} b \quad (12)$$

$$\delta g_1(r) = \frac{A}{r} e^{-\mu(r-b)} \cos \mu(r - b) \quad (13)$$

with $\delta g_1(r)$ being a short-range term as indicated in (13). The parameters A and μ are given below.

$$\frac{A}{b} = \frac{3}{4} f_w^2 \frac{(1 - 0.7117 f_w - 0.114 f_w^2)}{(1 - f_w)^4} \quad (14)$$

$$\mu b = 24 \frac{A}{b} \frac{(1 - f_w)^2}{f_w(1 + f_w/2)} \quad (15)$$

The pair-distribution function g_{VW} obtained in this manner differs from the computer simulation result by at most 0.03 (Verlet and Weis, 1972). The appendix of McQuarrie (1976) contains a listing of the program that computes the g_{VW} function for $b \leq r \leq 5b$. A computer code for the Percus-Yevick pair function can also be prepared. Throop and Bearman (1965) have a tabulation of some of the values. The $H(\bar{p})$ function for the Verlet-Weis pair-distribution function can be calculated in a manner similar to that of Percus-Yevick pair function (Problem 20; Tsang and Kong, 1983). In Figure 6.9, we compare the Percus-Yevick and Verlet-Weis pair-distribution functions. For $0 \leq f \leq 0.4$, the difference is small with the only significant difference being at $r/b = 1$. Comparison of the structure

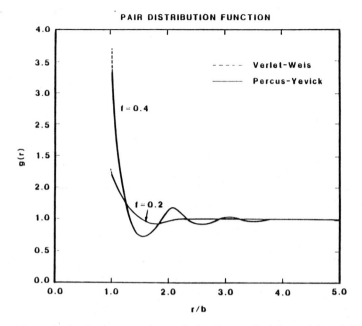

Fig. 6.9 Pair-distribution functions of the Percus-Yevick model and Verlet-Weis model.

factors of the Percus-Yevick and Verlet-Weis distributions also show that the difference between them is insignificant for $0 \leq f \leq 0.4$.

5.5 Summary

In this section we have studied the pair distribution functions, notably the Percus-Yevick and the Verlet-Weis results. It is also shown that the two results are practically the same when the fractional volume is between 0 and 0.4. The Percus-Yevick equation has only been considered for radial symmetric hard-sphere potential and for identical particles. For media in remote sensing, the particle sizes generally obey a distribution law. The pair-distribution function of these more complicated cases should be subjects of future study. It should also be pointed out that in employing results of statistical mechanics of liquids, we have assumed that the distribution of particles in geophysical media behaves very much like that of liquid molecules. Such an assumption, though reasonable, needs to be verified. For example, it will be interesting to investigate directly the pair-distribution function of ice particles in snow experimentally and theoretically and to make a comparison with the Percus-Yevick pair function.

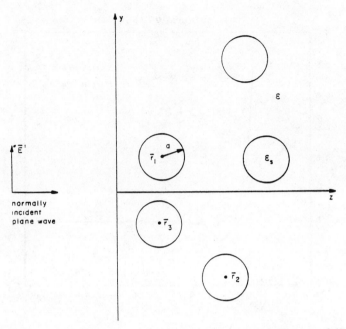

Fig. 6.10 Plane electromagnetic wave normally incident on a half-space of dielectric scatterers.

6 SCATTERING OF ELECTROMAGNETIC WAVES FROM A HALF-SPACE OF DIELECTRIC SCATTERERS – NORMAL INCIDENCE

6.1 Introduction

In this section, we consider a plane electromagnetic wave normally incident onto a half-space of identical dielectric spherical scatterers (Figure 6.10). The quasicrystalline approximation and T-matrix formalism will be used. Two sets of equations are obtained, viz., the generalized Ewald-Oseen extinction theorem and the generalized Lorentz-Lorenz law (Born and Wolf, 1975). The generalized Lorentz-Lorenz law is a set of homogeneous equations. Setting the determinant of the coefficient matrix equal to zero gives the dispersion relation for the coherent effective propagation constant K.

6.2 Generalized Ewald-Oseen Extinction Theorem and the Lorentz-Lorenz Law

The quasicrystalline approximation can be applied to the multiple scattering equations of the T-matrix formalism (Section 3.3), in a manner similar to that in Section 4. We start with the matrix equation of (40), Section 3.3, and

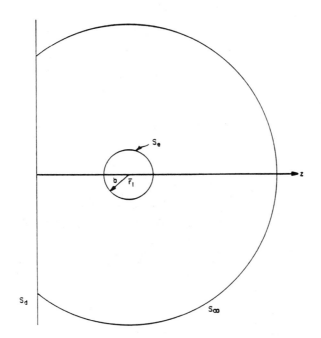

Fig. 6.11 Volume integration bounded by surface $S_d, S_e,$ and S_∞.

take conditional average by holding the l th particle fixed. The incident wave vector is $\bar{k}_i = \hat{z}k$ Note that, on conditional ensemble averaging, $E_l(\overline{w}^{(l)})$ will only be a function of $\bar{r}_l$. Let

$$E_l(\overline{w}^{(l)}) = \overline{w}(\bar{r}_l) \tag{1}$$

The first term on the right-hand side of (3.3.40) will become

$$\sum_{\substack{j=1 \\ j \neq l}}^{N} \int d\bar{r}_j\, p(\bar{r}_j \mid \bar{r}_l)\, \overline{\overline{\sigma}}(k\overline{r_l r_j})\, \overline{\overline{T}}_j\, E_{jl}(\overline{w}_j)$$

The quasicrystalline approximation consists of equating $E_{jl}(\overline{w}_j)$ to $E_j(\overline{w}_j)$. Using the definition of the pair-distribution function and the fact that the particles are identical, we obtain the equation (Problem 21)

$$\overline{w}(\bar{r}_1) = n_o \int_{z_2>0} d\bar{r}_2\, g(\bar{r}_2 - \bar{r}_1)\, \overline{\overline{\sigma}}(k\overline{r_1 r_2})\, \overline{\overline{T}}\, \overline{w}(\bar{r}_2) + e^{ikz_1}\, \bar{a}_{inc} \tag{2}$$

where $\overline{\overline{T}}$ is the T matrix for spheres and is given in Section 5.5, Chapter 3.

The integration volume in (2) is the half-space $z_2 > 0$, excluding a spherical volume of radius b centered about $\bar{r}_1$ (Figure 6.11). This is due to the fact that particles do not interpenetrate each other so that $p(\bar{r}_2 \mid \bar{r}_1) = 0$ for $|\bar{r}_2 - \bar{r}_1| < b$. For $z_1 \geq b$, the excluded volume is a complete sphere of radius b centered at $\bar{r}_1$. For $b > z_1 \geq 0$, a truncated sphere is excluded size of which depends on z_1. This complication indicates the necessity of a separate treatment of the region close to the boundary giving rise to a boundary layer effect.

To solve (2), assume the following trial solution

$$\overline{w}(\bar{r}_1) = \bar{a}_E \, e^{+iKz_1} \tag{3}$$

The representation in (3) is assumed to be valid for all $z_1 \geq 0$, even in the boundary layer region where a more complicated dependence on z_1 should be expected. In the subsequent treatment of evaluating the integral in (2), the truncated shape of the excluded sphere for $b > z_1 \geq 0$ is also neglected. Equation (3) can be considered as a first approximation to an iteration procedure for calculating the solution (Fikioris and Waterman, 1964).

The incident field is $\overline{E}^{inc} = \hat{y}\, e^{ikz}$ and expanding in vector spherical waves

$$\overline{E}^{inc} = \sum_{n=1}^{\infty} \frac{1}{2} \frac{i^n(2n+1)}{n(n+1)} [4\pi(2n+1)]^{1/2}$$

$$\times \left\{ Rg\overline{M}_{1n}(kr,\theta,\phi) - Rg\overline{M}_{-1n}(kr,\theta,\phi) \right.$$

$$\left. + Rg\overline{N}_{1n}(kr,\theta,\phi) + Rg\overline{N}_{-1n}(kr,\theta,\phi) \right\} \tag{4}$$

The incident field coefficients can then be obtained by using (4) and (1) and (21), Section 3.3. In view of the form of the incident wave in (4) and the symmetry of the particle positions, the exciting field coefficients $a_{mn}^{(E)(M)}$ and $a_{mn}^{(E)(N)}$ will contain only the $m = \pm 1$ spherical harmonic modes and equal to zero if $m \neq \pm 1$. Using this property, substitute (3) into (2) and use $\bar{\bar{\sigma}}$ as given by (10), (11), and (24), Section 3.3. The key integral to be evaluated is

$$I = \int_{z_2 > 0} d\bar{r}_2 \, g(\bar{r}_2 - \bar{r}_1) \, e^{iKz_2} \, h_p(k\,|\overline{r_1 r_2}|) P_p(\cos\theta_{r_1 r_2}) \tag{5}$$

Note that because the integration $\int d\bar{r}_2$ contains the component $\int_0^{2\pi} d\phi_2$, the integral will be zero unless the harmonic indices m of $\bar{\bar{\sigma}}$ and $\overline{w}$ in (2) are equal. Thus, the integral in (5) only contains Legendre polynomials rather than associated Legendre polynomials.

The integral in (5) can be decomposed in the following manner.

$$I = I_1 + I_2 \tag{6}$$

where

$$I_1 = e^{iKz_1} \int_{\substack{|\bar{r}_2-\bar{r}_1|>b \\ z_2 \geq 0}} d\bar{r}_2 \, e^{iK(z_2-z_1)} h_p(k|\overline{r_1 r_2}|) P_p(\cos\theta_{r_1 r_2}) \tag{7}$$

and

$$I_2 = e^{iKz_1} \int_{\substack{|\bar{r}_2-\bar{r}_1|>b \\ z_2 \geq 0}} d\bar{r}_2 \, [g(\bar{r}_2-\bar{r}_1)-1] e^{iK(z_2-z_1)} \\ \times h_p(k|\overline{r_1 r_2}|) P_p(\cos\theta_{r_1 r_2}) \tag{8}$$

Because of translational invariance in the horizontal plane x-y, there is no loss of generality in assuming $\bar{r}_1$ to be on the z axis. To evaluate integral I_1 in (7) note that, with $y_p = h_p(kr)P_p(\cos\theta)$

$$e^{iKz} y_p = \frac{1}{K^2-k^2}[e^{iKz}\nabla^2(y_p) - y_p \nabla^2 e^{iKz}] \tag{9}$$

Let $\bar{r}_2 - \bar{r}_1 = \bar{r}$ in (7). Use (9) and Green's theorem to transform the volume integral into three surface integrals over S_d, S_e and S_∞ (Figure 6.11). The integral over S_∞ can be shown to vanish.

Since complete spherical exclusion has been assumed (ignoring boundary layer effects), the surface integral over S_e can be evaluated by making use of the fact that the integral of the product of Legendre polynomial with exponential (Abramowitz and Stegun, 1965) is equal to a spherical Bessel function. Hence

$$\int_{S_e} dS \left[e^{iKz} \frac{\partial y_p}{\partial r} - y_p \frac{\partial e^{iKz}}{\partial r} \right]_{r=b} = -4\pi i^n (K^2-k^2) L_p(k, K \mid b) \tag{10}$$

where

$$L_p(k, K \mid b) = -\frac{b^2}{(K^2-k^2)} \left[k h'_p(kb) j_p(Kb) - K h_p(kb) j'_p(Kb) \right] \tag{11}$$

For the surface integral over S_d, the following Kasterin's representation (Problem 22) can be used.

$$y_p = h_p(kr) P_p(\cos\theta) = (-i)^p P_p \left(\frac{1}{ik} \frac{\partial}{\partial z} \right) h_o(kr) \tag{12}$$

Thus

$$\int_{S_d} dS \left[e^{iKz} \frac{\partial y_p}{\partial z} - y_p \frac{\partial e^{iKz}}{\partial z} \right]_{z=-z_1}$$

$$= 2\pi e^{-iKz_1}(-i)^{p+1} \int_0^\infty \rho\, d\rho$$

$$\times \left\{ \left[\frac{\partial}{\partial z} - iK \right] P_p \left(\frac{1}{ik} \frac{\partial}{\partial z} \right) \frac{e^{ik\sqrt{\rho^2+z^2}}}{\sqrt{\rho^2+z^2}} \right\}_{z=-z_1}$$

$$= -2\pi\, i^{p+1} \frac{(k+K)}{k} e^{-iKz_1 + ikz_1} \qquad (13)$$

Substituting (10) and (13) into (7) gives

$$I_1 = \frac{2\pi i(-i)^p}{(K-k)} \frac{e^{ikz_1}}{k^2} + 4\pi(-i)^p L_p(k, K \mid b)\, e^{iKz_1} \qquad (14)$$

To evaluate I_2 of (8), we note that the volume of integration in (8) consists of the half-space $z_2 > 0$ less a sphere of radius b centered at the point $\bar{r}_1$. We have seen from Figure 6.9 that for $0 \leq f \leq 0.4$, $g(\bar{r}_2 - \bar{r}_1) - 1$ is practically zero for $|\bar{r}_2 - \bar{r}_1|$ larger than a few b's. If the point $\bar{r}_1$ is at least several diameters deep in the scattering medium (thus ignoring boundary layer effects), the volume of integration in (8) can be extended to infinite space. Thus, letting $\bar{r} = \bar{r}_2 - \bar{r}_1$ in (8) and making the above assumption give

$$I_2 = 4\pi(-i)^p\, e^{iKz_1} M_p(k, K \mid b) \qquad (15)$$

where

$$M_p(k, K \mid b) = \int_b^\infty dr\, r^2 [g(r) - 1] h_p(kr) j_p(Kr) \qquad (16)$$

Substituting (14) and (15) into (2) gives two types of terms in (2). One type of terms has a $\exp(ikz_1)$ dependence corresponding to waves travelling with the propagation constant of the incident wave. The other type of terms has a $\exp(iKz_1)$ dependence corresponding to waves travelling with the propagation constant of the effective medium. The terms with propagation constant k should balance each other giving the generalized Ewald-Oseen extinction theorem (E-O Theorem). The medium generates a wave that *extinguishes* the original incident wave. Balancing the terms with propagation constant K gives the generalized Lorentz-Lorenz law (L-L law).

On examining the generalized Ewald-Oseen extinction theorem, and using the relations (Problem 23)

$$a(-\mu, n \mid \mu, \nu \mid p) = \frac{(n-\mu)!(\nu+\mu)!}{(n+\mu)!(\nu-\mu)!} a(\mu, n \mid -\mu, \nu \mid p) \tag{17}$$

and

$$a(-\mu, n \mid \mu, \nu \mid p, p-1)$$
$$= -\frac{(n-\mu)!(\nu+\mu)!}{(n+\mu)!(\nu-\mu)!} a(\mu, n \mid -\mu, \nu \mid p, p-1) \tag{18}$$

it follows that $a_{-1\nu}^{(E)(M)} = -a_{1\nu}^{(E)(M)}$ and $a_{-1\nu}^{(E)(N)} = a_{1\nu}^{(E)(N)}$. The generalized Ewald-Oseen extinction theorem assumes the form of the following two equations

$$\sqrt{4\pi(2\nu+1)} \frac{1}{2} \frac{i^\nu (2\nu+1)}{\nu(\nu+1)} \delta_{\mu 1}$$

$$+ n_o \sum_{n,p} (-1)^{p+\mu} \frac{\gamma_{1n}}{\gamma_{1\nu}} \frac{(2\pi i) i^p}{(K-k)k^2}$$

$$\times \left\{ T_n^{(M)} a_{1n}^{(E)(M)} a(\mu, n \mid -\mu, \nu \mid p) a(n, \nu, p) \right.$$

$$\left. - T_n^{(N)} a_{1n}^{(E)(N)} a(\mu, n \mid -\mu, \nu \mid p, p-1) b(n, \nu, p) \right\} = 0 \tag{19}$$

$$\sqrt{4\pi(2\nu+1)} \frac{1}{2} \frac{i^\nu (2\nu+1)}{\nu(\nu+1)} \delta_{\mu 1}$$

$$+ n_o \sum_{n,p} (-1)^{p+\mu} \frac{\gamma_{1n}}{\gamma_{1\nu}} \frac{(2\pi i) i^p}{(K-k)k^2}$$

$$\times \left\{ -T_n^{(M)} a_{1n}^{(E)(M)} a(\mu, n \mid -\mu, \nu \mid p, p-1) b(n, \nu, p) \right.$$

$$\left. + T_n^{(N)} a_{1n}^{(E)(N)} a(\mu, n \mid -\mu, \nu \mid p) a(n, \nu, p) \right\} = 0 \tag{20}$$

Next, make use of the following summation identities (Problem 24)

$$-\sum_p (-i)^p a(1, n \mid -1, \nu \mid p, p-1) b(n, \nu, p)$$

$$= \sum_p (-i)^p a(1, n \mid -1, \nu \mid p) a(n, \nu, p) = -i^{\nu-n} \frac{(2\nu+1) n(n+1)}{2\nu(\nu+1)}$$

$$\tag{21}$$

and define $X_{1n}^{(M)}$ and $X_{1n}^{(N)}$, respectively, in terms of $a_{1n}^{(E)(M)}$ and $a_{1n}^{(E)(N)}$ by

$$a_{1n}^{(E)} = X_{1n} \frac{1}{2} \frac{i^n(2n+1)}{n(n+1)} \sqrt{4\pi(2n+1)} \tag{22}$$

In (22), the superscripts (M) or (N) has been suppressed for $a_{1n}^{(E)}$ and X_{1n}. Then, the two equations (19) and (20) of the generalized Ewald-Oseen extinction theorem can be simplified to a single equation (Problem 24)

$$K - k = -\frac{\pi i n_o}{k^2} \sum_n \left(T_n^{(M)} X_{1n}^{(M)} + T_n^{(N)} X_{1n}^{(N)} \right) (2n+1) \tag{23}$$

Balancing terms of dependence $\exp(iKz_1)$ (with propagation constant K) in (2) gives the generalized Lorentz-Lorenz law below.

$$X_{1\nu}^{(M)} = -2\pi n_o \sum_{n,p} (2n+1) \left[L_p(k, K|b) + M_p(k, K|b) \right]$$

$$\times \left\{ T_n^{(M)} X_{1n}^{(M)} a(1, n| -1, \nu|p) A(n, \nu, p) \right.$$

$$\left. + T_n^{(N)} X_{1n}^{(N)} a(1, n| -1, \nu|p, p-1) B(n, \nu, p) \right\} \tag{24}$$

$$X_{1\nu}^{(N)} = -2\pi n_o \sum_{n,p} (2n+1) \left[L_p(k, K|b) + M_p(k, K|b) \right]$$

$$\times \left\{ T_n^{(M)} X_{1n}^{(M)} a(1, n| -1, \nu|p, p-1) B(n, \nu, p) \right.$$

$$\left. + T_n^{(N)} X_{1n}^{(N)} a(1, n| -1, \nu|p) A(n, \nu, p) \right\} \tag{25}$$

where

$$A(n, \nu, p) = \frac{1}{n(n+1)(2\nu+1)} \left[2\nu(\nu+1)(2\nu+1) \right.$$

$$+ (\nu+1)(n+\nu-p)(n+p-\nu+1) \tag{26}$$

$$\left. - \nu(n+\nu+p+2)(\nu+p-n+1) \right]$$

$$B(n, \nu, p) = \frac{1}{n(n+1)} \left[(n+\nu+p+1)(\nu+p-n) \right.$$

$$\left. \times (n+p-\nu)(n+\nu-p+1) \right]^{1/2} \tag{27}$$

Equations (24) and (25) form a system of simultaneous homogeneous equations for the unknown amplitudes $X_{1\nu}^{(M)}$ and $X_{1\nu}^{(N)}$. For a nontrivial solution, the determinant of the coefficients must vanish yielding an equation for the effective wavenumber K in terms of k, the scattering coefficients $T_n^{(M)}$ and $T_n^{(N)}$, and the fractional volume of the scatterers. This is the dispersion relation for the composite medium. Hence, the procedure is to first find the singular solution K of the generalized Lorentz-Lorenz law leaving an arbitrary constant to be next determined by the single scalar equation (23) of the generalized Ewald-Oseen extinction theorem.

6.3 Effective Phase Velocity and Attenuation Rate in the Low-Frequency Limit

A closed-form solution can be obtained for the effective propagation constant K in the low-frequency Rayleigh limit. In such a case, only the electric dipole term $T_1^{(N)}$ contributes in (24) and (25), Section 6.2. This gives an equation involving the unknown amplitude $X_{11}^{(N)}$ only. Setting the coefficient equal to zero gives the dispersion relation

$$1 + 6\pi n_o T_1^{(N)} \sum_p a(1,1 \mid -1,1 \mid p)$$

$$\times A(1,1,p)(L_p(k,K \mid b) + M_p(k,K \mid b)) = 0 \quad (1)$$

As discussed in (16) through (21), Section 5.5, Chapter 3, in the low-frequency limit, $T_1^{(N)} = T_{1r}^{(N)} + iT_{1i}^{(N)}$ and $T_{1r}^{(N)} = -T_{1i}^{(N)^2}$. In calculating the effective propagation constant K, it is to be noted that $K_i = Im(K)$ is a combination of absorption and scattering effects. We note that $K_r = Re(K)$ is of order $O(1)$ and K_i is of order $O(f)$ for absorption and is of order $O(fk^3a^3)$ for scattering. Since $|T_{1i}^{(N)}| \ll 1$, therefore $|T_{1r}^{(N)}| \ll |T_{1i}^{(N)}|$. However, $T_{1r}^{(N)}$ cannot be neglected because it contributes to the attenuation rate K_i. Thus, in performing the low-frequency approximation of terms in (1), we shall (i) keep only the leading term of the real part and the leading term of the imaginary part, (ii) ignore real part of a term if it is less than $O(k^3a^3)$ of the corresponding imaginary part, and (iii) ignore imaginary part of a term if it is less than $O(k^3a^3)$ of the corresponding real part.

Thus the low-frequency approximations for $T_1^{(N)}$ is given by (16) through (19), Section 5.5, Chapter 3, and the low-frequency limit of L_p is

$$L_p(k,K \mid b) = -\frac{i}{k(K^2 - k^2)}(K/k)^p - \delta_{po}\frac{b^3}{3} \quad (2)$$

Only the $p=0$ term in L_p has a nonzero second term in (2) because for $p \geq 1$, the real part is of an order smaller than $O(k^3 a^3)$ of the imaginary part. Similarly, $M_p(k, K \mid b) = \delta_{po} M_o$ with

$$M_o = \int_b^\infty dr\, r^2 [g(r) - 1] \tag{3}$$

Only the real part of $M_p(k, K \mid b)$ for $p = 0$ contributes. Other real and imaginary terms are smaller than the corresponding real and imaginary terms of $L_p(k, K \mid b)$. Thus, in the Rayleigh limit, knowledge of an integral of $g(r) - 1$ is required rather than a detailed behavior of the pair-distribution function $g(r)$ as a function of r.

Substituting the low-frequency approximation in (1) gives

$$1 + \frac{2\pi n_o T_1^{(N)}}{k} \frac{i}{(K^2 - k^2)} [2 + (K/k)^2] + 4\pi n_o T_1^{(N)} \left(\frac{b^3}{3} - M_o\right) = 0 \tag{4}$$

An expression for effective propagation K can be obtained from (4). The solution can be further simplified by making use of the property $K_i \ll K_r$. The final result is

$$K^2 = k^2 + \frac{3fk^2 y}{1 - fy}\left[1 + i\frac{2}{3}k^3 a^3 \frac{y}{1-fy}\left\{1 + 4\pi n_o \int_0^\infty dr\, r^2 [g(r) - 1]\right\}\right] \tag{5}$$

where $f = 4n_o \pi a^3 / 3$ is the fractional volume of scatterers and y is given in (4), Section 2.1. The result in (5) is the same as the result obtained through the simple model in (3), Section 2.1, and (19), Section 2.2.

Using the Percus-Yevick pair-distribution function result of (7), Section 5.3, in (5), the result for the effective propagation constant is

$$K^2 = k^2 + \frac{3fk^2 y}{1 - fy}\left[1 + i\frac{2}{3}k^3 a^3 y \frac{(1-f)^4}{(1-fy)(1+2f)^2}\right] \tag{6}$$

In Figures 6.12 and 6.13, we show, respectively, the normalized phase velocity k/K_r and the effective loss tangent $2K_i/K_r$ as a function of fractional volume of scatterers f using the result of (6) for $ka = 0.1$. We note that the phase velocity decreases with f. The loss tangent first increases with f, then saturates and decreases as f further increases.

6.4 Dispersion Relations at Higher Frequencies

For larger values of ka, (24) and (25), Section 6.2, can be solved numerically. In this section, we shall show the numerical results of the effective

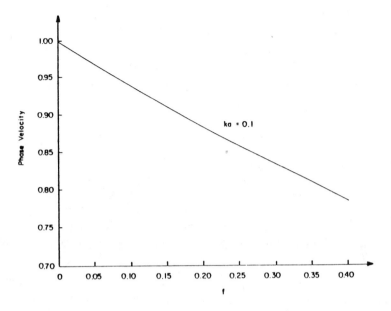

Fig. 6.12 Normalized phase velocity k/K_r in the Rayleigh limit as a function of fractional volume of scatterers f for $ka = 0.1$, and $\epsilon_s = 3.24\epsilon_0$.

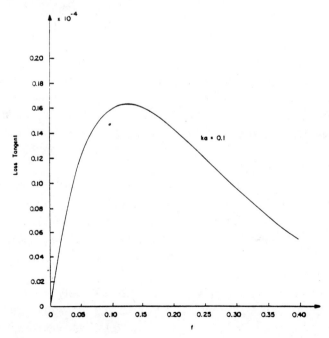

Fig. 6.13 Effective loss tangent $2K_i/K_r$ as a function of fractional volume of scatterers f for $ka = 0.1$ and $\epsilon_s = 3.24\epsilon_0$.

propagation constant K for ka ranging from 0.05 to 2.5 and f ranging from 0 to 0.4 which represent typical parameters in the microwave remote sensing of snow. For these values of ka and f, the determinant of the coefficient matrix was computed numerically by retaining a maximum of eight simultaneous homogeneous complex equations. The Wigner $3j$ symbols are generated by a computer code and checked against tabulated values (Rotenberg, 1959). The elements of $M_p(k, K \mid b)$ for $p = 0, 1, \ldots, 8$ were computed by numerically evaluating the integral in (16), Section 6.2, for r between b and $4b$. For f between 0 and 0.4, the value of $g(r) - 1$ is practically zero for r larger than $4b$.

For a given value of ka and f, the roots of the determinant were searched for in the complex K plane $(K_r + iK_i)$ using Mueller's method (Ralston, 1965). There are two good initial guesses: (i) We first note that for media with sparse concentration of particles, the exciting field will be approximately the same as the incident field, so that $X_{1n}^{(M)}$ and $X_{1n}^{(N)}$ both are approximately equal to 1. Replacing them by 1 in the generalized Ewald-Oseen extinction theorem of (23), Section 6.2, gives the result of an effective propagation constant $K^{(F)}$ under Foldy's approximation

$$K^{(F)} = k - \frac{\pi i n_o}{k^2} \sum_{n=1}^{\infty} (2n+1) \left(T_n^{(M)} + T_n^{(N)} \right) \qquad (1)$$

(ii) The second initial guess is provided by the low-frequency solution in (5), Section 6.3. These two guesses could be used systematically to obtain quick convergence of roots at increasingly higher values of ka. In the following, we shall illustrate results for the normalized phase velocity v_p defined as k/K_r, and the effective loss tangent (LT) defined as $2K_i/K_r$. The Percus-Yevick pair-distribution function is used unless otherwise specified (Tsang et al., 1982b; Tsang and Kong, 1982).

In Figures 6.14 and 6.15, we plot, respectively, v_p and LT versus ka for two different f values and for real ϵ_s. We note that the phase velocity first decreases with increasing frequency and then oscillates as frequency increases further. The oscillation is a characteristic of resonant scattering at higher frequencies. The loss tangent first increases rapidly with frequency and then saturates at high frequencies.

In Figures 6.16 and 6.17, we plot v_p and LT versus ka for two different f values and for complex ϵ_s representing absorptive scatterers. On comparing the results of Figures 6.15 and 6.17, we note that at low frequencies, absorption dominates over scattering as the loss tangent in Figure 6.17 is much larger than that in Figure 6.15. At higher frequencies, the loss tangents are comparable in magnitude in the two cases.

Dispersion Relations at Higher Frequencies

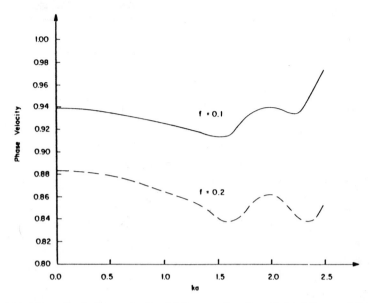

Fig. 6.14 Normalized phase velocity k/K_r as a function of ka for $\epsilon_s = 3.24\epsilon_o$ and $f = 0.1$ and 0.2.

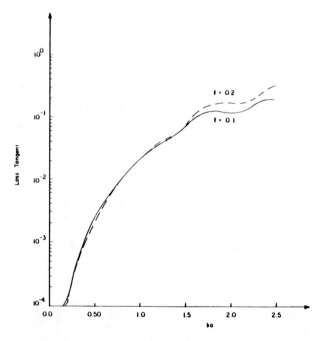

Fig. 6.15 Effective loss tangent $2K_i/K_r$ as a function of ka for $\epsilon_s = 3.24\epsilon_o$ and $f = 0.1$ and 0.2.

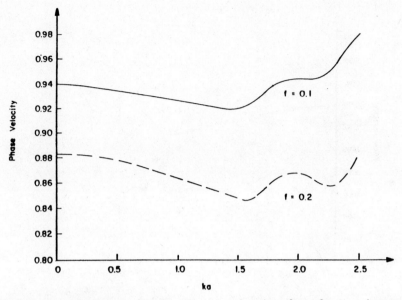

Fig. 6.16 Normalized phase velocity k/K_r as a function of ka for $\epsilon_s = (3.2375 + i0.18)\epsilon_o$ and $f = 0.1$ and 0.2.

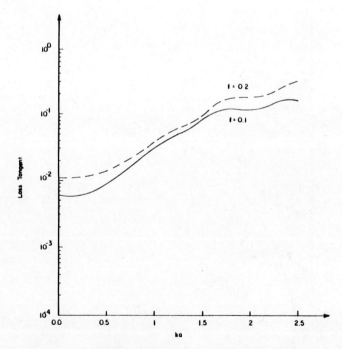

Fig. 6.17 Effective loss tangent $2K_i/K_r$ as a function of ka for $\epsilon_s = (3.2375 + i0.18)\epsilon_o$ and $f = 0.1$ and 0.2.

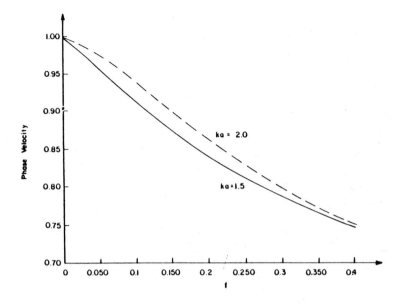

Fig. 6.18 Normalized phase velocity k/K_r as a function of f for $\epsilon_s = 3.24\epsilon_o$ and $ka = 1.5$ and 2.0.

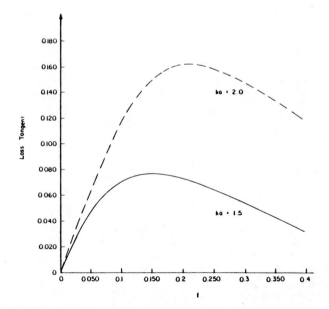

Fig. 6.19 Effective loss tangent $2K_i/K_r$ as a function of f for $\epsilon_s = 3.24\epsilon_o$ and $ka = 1.5$ and 2.0.

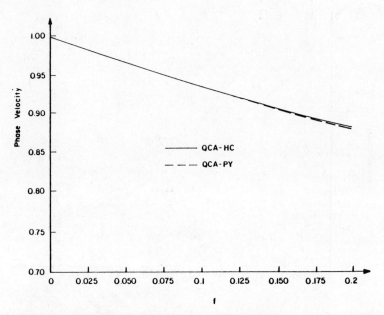

Fig. 6.20 Normalized phase velocity k/K_r as a function of f for $\epsilon_s = 3.24\epsilon_o$ and $ka = 0.5$. The QCA-PY and QCA-HC results are compared.

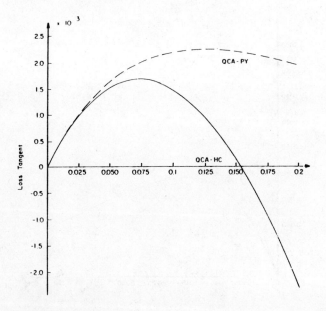

Fig. 6.21 Effective loss tangent $2K_i/K_r$ as a function of f for $\epsilon_s = 3.24\epsilon_o$ and $ka = 0.5$. The QCA-PY and QCA-HC results are compared.

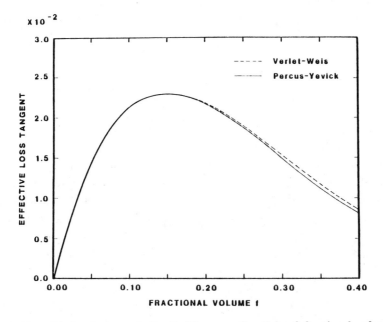

Fig. 6.22 Effective loss tangent $2K_i/K_r$ as a function of fractional volume of scatterers for $\epsilon_s = 3.24\epsilon_o$ and $ka = 1$. The results of Percus-Yevick model (PY) and of Verlet-Weis model (VW) are compared.

In Figures 6.18 and 6.19, we show, respectively, v_p and LT versus f for two different ka values. We note that the phase velocity decreases as f increases since the phase velocity for the scatterer is lower. As a function of f, LT first rises to a peak and then decreases as f further increases.

In Figures 6.20 and 6.21, we compare the results based on the Percus-Yevick (PY) pair function and the hole-correction (HC) pair function for $ka = 0.5$. We note that the HC result for v_p agrees very well with the PY result. However, for the loss tangent, the two results agree only for small f (up to 0.04). Furthermore, the HC result, after increasing to a peak, decreases rapidly as f further increases. For f larger than 0.15 (approximately), it decreases to negative values which are physically unacceptable because they indicate negative scattering and growth of the coherent wave.

In Figure 6.22, we compare the effective loss tangents computed by using the Percus-Yevick and the Verlet-Weis pair-distribution function. The difference is insignificant. Twersky (1977, 1978, 1979, 1983) has also applied his multiply-scattered amplitude formalism with quasicrystalline approximation to study propagation in small-spaced lossy-scatterers.

7 SCATTERING OF ELECTROMAGNETIC WAVES FROM A HALF-SPACE OF DIELECTRIC SCATTERERS – OBLIQUE INCIDENCE

7.1 Introduction

In this section, we consider a plane electromagnetic wave obliquely incident on a half-space of densely packed dielectric scatterers embedded in free-space. Both vertically and horizontally polarized incident waves are considered. The coherent reflectivity and the bistatic scattering coefficients will be calculated. The T-matrix, together with the quasi-crystalline approximation, is used to generate a set of homogeneous equations (the generalized Lorentz-Lorenz law), and a set of inhomogeneous equations (the generalized Ewald-Oseen extinction theorem). The coherent reflected wave is calculated by solving these two sets of equations. The incoherent scattered wave is calculated by applying the distorted Born approximation. For the case of spherical scatterers, it is also shown that the effective propagation constant is independent of direction and polarization. The advantage of this approach is that in the low-frequency limit, the scattered wave produced by the scatterers contains a reflected coherent part that obeys Snell's law, gives the Fresnel reflection coefficients for TE and TM waves, and reproduces the Brewster angle effect and the Clausius-Mosotti relation. In addition to these classical results, closed-form expressions are also derived for the bistatic scattering coefficients. At higher frequencies, the coherent reflected wave and the bistatic scattering coefficients are calculated numerically. The bistatic scattering coefficients, under the distorted Born approximation, are proportional to the structure factor (Waseda, 1980). The theory is also applied to match backscattering data from dry snow at microwave frequencies. The concept of Ewald-Oseen extinction theorem and Lorentz-Lorenz law is also applicable to random medium (Problem 34).

7.2 Dispersion Relation and Coherent Reflected Wave

Consider a plane wave incident on a half-space of N identical dielectric scatterers with radius a and permittivity ϵ_s and centered at $\bar{r}_1, \bar{r}_2, \ldots \bar{r}_N$ (Figure 6.23). The background medium has permittivity ϵ. The same medium also occupies the region $z \geq 0$. The incident wave can be expressed as in (1), Section 3.1. The incident wave is propagating downward and we let $\bar{k}_{id}$ to denote the downward propagating incident wave vector

$$\bar{k}_{id} = k(\sin\theta_i \cos\phi_i \hat{x} + \sin\theta_i \sin\phi_i \hat{y} - \cos\theta_i \hat{z})$$
$$= k_{ix}\hat{x} + k_{iy}\hat{y} - k_{iz}\hat{z} \qquad (1)$$

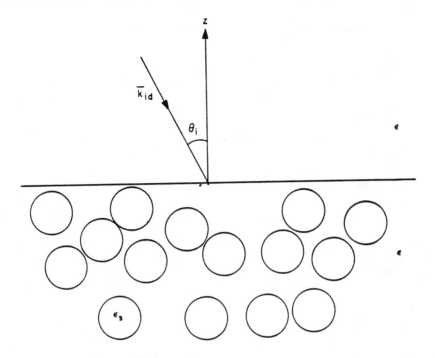

Fig. 6.23 Plane electromagnetic wave obliquely incident on a half-space of densely packed dielectric scatterers.

Thus the incident direction is $(\theta, \phi) = (\pi - \theta_i, \phi_i)$.

For the case of vertically polarized incident plane wave, the coefficients $\overline{a}^{inc}$ are, using results of (25), Section 5.1, Chapter 3,

$$a_{mn}^{inc(M)} = (-1)^m \frac{1}{\gamma_{mn}} \frac{(2n+1)}{n(n+1)} i^n \hat{\theta}_{id} \cdot \overline{C}_{-mn}(\pi - \theta_i, \phi_i) \quad (2)$$

$$a_{mn}^{inc(N)} = (-1)^m \frac{1}{\gamma_{mn}} \frac{(2n+1)}{n(n+1)} i^{n-1} \hat{\theta}_{id} \cdot \overline{B}_{-mn}(\pi - \theta_i, \phi_i) \quad (3)$$

where $\hat{\theta}_{id}$ is the vertically polarized vector for the downward-propagating wave $\overline{k}_{id}$,

$$\hat{\theta}_{id} = -\hat{x} \cos\theta_i \cos\phi_i - \hat{y} \cos\theta_i \sin\phi_i - \hat{z} \sin\theta_i \quad (4)$$

For an incident horizontally polarized wave, the coefficients $a_{mn}^{inc(M)}$ and $a_{mn}^{inc(N)}$ are obtained by taking the dot product in (2) and (3) with $\hat{\phi}_i$ instead of $\hat{\theta}_{id}$. Using quasicrystalline approximation to the T-matrix multiple scattering equation, the equation for the conditional-averaged exciting field coefficients is similar to that in (2), Section 6.2.

$$\overline{w}_E(\overline{r}_1) = n_o \int_{z_2<0} d\overline{r}_2\, g(\overline{r}_2 - \overline{r}_1)\, \overline{\overline{\sigma}}(k\overline{r}_1\overline{r}_2)\, \overline{\overline{T}}\, \overline{w}_E(\overline{r}_2) + e^{i\overline{k}_{id}\cdot\overline{r}_1}\, \overline{a}^{inc} \quad (5)$$

To solve (5), let

$$\overline{w}_E(\overline{r}_1) = \overline{a}_E\, e^{i\overline{K}_d\cdot\overline{r}_1} \quad (6)$$

where $\overline{K}_d$ denotes the downward-going effective wavevector

$$\overline{K}_d = K_x \hat{x} + K_y \hat{y} - K_z \hat{z} \quad (7)$$

and the effective wavenumber is $K = (K_x^2 + K_y^2 + K_z^2)^{1/2}$. Substituting (6) into the integral in (5), manipulations can be performed in a manner similar to those of the case of normal incidence in Section 6. Boundary layer effects are again ignored. The integral in (5) gives rise to two waves with dependences

$$e^{i\overline{K}_d\cdot\overline{r}_1} \quad (8)$$

and

$$e^{iK_x z_1 + iK_y y_1 - i(k^2 - K_x^2 - K_y^2)^{1/2} z_1} \quad (9)$$

The first wave given by (8) propagates with $\overline{K}_d$ and will balance the $\overline{w}_E(\overline{r}_1)$ term on the left-hand side of (5), giving rise to the generalized Lorentz-Lorenz law. The second wave given by (9) should extinguish the incident wave term in (5) in accordance with the generalized Ewald-Oseen extinction theorem (Born and Wolf, 1975) which means that it has the same propagation vector as $\overline{k}_{id}$. Thus equating the wavevector in (9) with $\overline{k}_{id}$ gives the following phase matching conditions

$$K_x = k_{ix} \quad (10)$$

$$K_y = k_{iy} \quad (11)$$

Using (10) and (11) in (7) gives Snell's law

$$K \sin \theta_t = k \sin \theta_i \quad (12)$$

with

$$\overline{K}_d = K \sin \theta_t \cos \phi_i \hat{x} + K \sin \theta_t \sin \phi_i \hat{y} - K \cos \theta_t \hat{z} \quad (13)$$

The effective propagation constant K is to be determined by the generalized Lorentz-Lorenz law. After K is determined, θ_t is calculated by Snell's law as given in (12). The transmitted angle θ_t is slightly complex since the effective propagation constant K is complex due to a combination of scattering and absorption.

Balancing the term with wave dependence in (8) with the left-hand side of (5) gives the the generalized Lorentz-Lorenz law as follows (Problem 25):

$$\overline{a}_E = n_o \overline{\overline{U}}^{(K)} \overline{\overline{T}} \overline{a}_E \qquad (14)$$

where

$$\overline{\overline{U}}^{(K)} = \begin{bmatrix} \overline{\overline{X}}^{(K)} & \overline{\overline{Z}}^{(K)} \\ \overline{\overline{Z}}^{(K)} & \overline{\overline{X}}^{(K)} \end{bmatrix} \qquad (15)$$

and $\overline{\overline{X}}^{(K)}$ and $\overline{\overline{Z}}^{(K)}$ are $L_{max} \times L_{max}$ matrices with

$$X^{(K)}_{mn\mu\nu} = \sum_p \frac{\gamma_{\mu\nu}}{\gamma_{mn}} (-1)^m a(\mu,\nu \mid -m, n \mid p) a(\nu, n, p) S^{(K)}_{p(\mu-m)} \qquad (16)$$

$$Z^{(K)}_{mn\mu\nu} = \sum_p \frac{\gamma_{\mu\nu}}{\gamma_{mn}} (-1)^{m+1} a(\mu,\nu \mid -m, n \mid p, p-1) b(\nu, n, p) S^{(K)}_{p(\mu-m)} \qquad (17)$$

$$S^{(K)}_{p(\mu-m)} = S_p (-1)^p 4\pi i^p P_p^{\mu-m} (\cos(\pi - \theta_t)) e^{i(\mu-m)\phi_i} \qquad (18)$$

$$S_p = \int_b^\infty dr\, r^2 \left[g(r) - 1\right] j_p(Kr) h_p(kr)$$

$$- \frac{b^2}{K^2 - k^2} \left[k j_p(Kb) h'_p(kb) - K j'_p(Kb) h_p(kb)\right] \qquad (19)$$

The indices (m,n) and (μ,ν) are in accordance with the numbering system in Section 5.2 (Chapter 3) and Section 3.3. The generalized Lorentz-Lorenz law of (14) is a homogeneous system of equations for the coefficients $\overline{a}_E$. Setting the determinant equal to zero gives the dispersion relation for the effective propagation constant K. The dispersion relation is generally directional and polarization dependent for non-spherical scatterers. We shall show that for the case of spherical scatterers, the dispersion relation is independent of polarization and direction of propagation.

Balancing the terms with the wave dependence given in (9) to those of the incident wave in (5) gives the generalized Ewald-Oseen extinction theorem. The results can be simplified by making use of the following two relations

$$\sum_p (-i)^p a(\nu, n, p) a(\mu, \nu \mid -m, n \mid p) P_p^{\mu-m} (\cos\theta) e^{i(\mu-m)\phi}$$

$$= i^{\nu-n} (-1)^{\nu+n} \frac{(2n+1)}{n(n+1)} \overline{B}_{\mu\nu}(\theta,\phi) \cdot \overline{B}_{-mn}(\theta,\phi) \qquad (20)$$

$$\sum_p (-i)^p b(\nu,n,p) a(\mu,\nu \mid -m,n \mid p,p-1) P_p^{\mu-m}(\cos\theta) e^{i(\mu-m)\phi}$$

$$= i^{\nu-n+1}(-1)^{\nu+n+1}\frac{(2n+1)}{n(n+1)}\overline{B}_{\mu\nu}(\theta,\phi)\cdot\overline{C}_{-mn}(\theta,\phi) \quad (21)$$

They can be derived (Problem 26) from the orthogonality property of vector spherical harmonics, the completeness relation of spherical harmonics, and the following relation which express a product of spherical harmonics and vector spherical harmonics in terms of vector spherical harmonics (Problem 27)

$$Y_p^{\mu-m}(\theta,\phi)\overline{B}_{-\mu\nu}(\theta,\phi)$$

$$= \sum_l \frac{i^{l-\nu-p}(-1)^{p+\mu}}{2p+1} a(-\mu,\nu \mid m,l \mid p) a(\nu,l,p)\overline{B}_{-ml}(\theta,\phi)$$

$$+ \sum_l \frac{i^{l-\nu-p-1}(-1)^{p+\mu}}{2p+1}$$

$$\times a(-\mu,\nu \mid m,l \mid p,p-1) b(\nu,l,p)\overline{C}_{-ml}(\theta,\phi) \quad (22)$$

A relation for $Y_p^{\mu-m}(\theta,\phi)\overline{C}_{-\mu\nu}(\theta,\phi)$ similar to (22) can be obtained by taking the cross product of (22) with $\hat{r}$. Using the results of (20) and (21), the generalized Ewald-Oseen extinction theorem assumes the following simplified form

$$n_o \overline{\overline{U}}^{(k)} \overline{\overline{T}} \overline{a}_E = -\overline{a}^{inc} \quad (23)$$

where

$$\overline{\overline{U}}^{(k)} = \begin{bmatrix} \overline{\overline{X}}^{(k)} & \overline{\overline{Z}}^{(k)} \\ \overline{\overline{Z}}^{(k)} & \overline{\overline{X}}^{(k)} \end{bmatrix} \quad (24)$$

with $\overline{\overline{X}}^{(k)}$ and $\overline{\overline{Z}}^{(k)}$ being $L_{max} \times L_{max}$ matrices, and

$$X_{mn\mu\nu}^{(k)} = \frac{\gamma_{\mu\nu}}{\gamma_{mn}}(-1)^{m+n+\nu} 2\pi i^{\nu-n+1}\cdot\frac{2n+1}{n(n+1)}\frac{1}{(K_z-k_{iz})k_{iz}k}$$

$$\cdot\overline{B}_{\mu\nu}(\pi-\theta_i,\phi_i)\cdot\overline{B}_{-mn}(\pi-\theta_i,\phi_i) \quad (25)$$

$$Z_{mn\mu\nu}^{(k)} = \frac{\gamma_{\mu\nu}}{\gamma_{mn}}(-1)^{m+n+\nu+1} i^{\nu-n}\frac{2n+1}{n(n+1)}\frac{2\pi}{(K_z-k_{iz})k_{iz}k}$$

$$\cdot\overline{B}_{\mu\nu}(\pi-\theta_i,\phi_i)\cdot\overline{C}_{-mn}(\pi-\theta_i,\phi_i) \quad (26)$$

The generalized Lorentz-Lorenz law of (14) is a system of homogeneous equations whereas the generalized Ewald-Oseen extinction theorem (23) is a system of inhomogeneous equations of the same dimension as (14). In the following sections, we shall show that for the case of spherical scatterers, the set of inhomogeneous equations in the E-O theorem (23) can be reduced to a single scalar equation. Hence, the procedure is first to find a singular solution of the generalized Lorentz-Lorenz law together with the dispersion relation for K leaving an arbitrary constant to be next determined by that single scalar equation that is equivalent to (23).

After the conditional average of the exciting field has been evaluated in the manner outlined above, the coherent reflected wave can be calculated by taking ensemble average of the scattered field, which obeys equations (31) and (32), Section 3.3. The ensemble average of the coherent reflected field is

$$<\overline{E}^s(\bar{r})> = n_o \sum_{mn} \int d\bar{r}_j \left[\overline{M}_{mn}(k\overline{rr}_j) T_n^{(M)} a_{mn}^{E(M)} e^{i\overline{K}_d \cdot \bar{r}_j} \right.$$
$$\left. + \overline{N}_{mn}(k\overline{rr}_j) T_n^{(N)} a_{mn}^{E(N)} e^{i\overline{K}_d \cdot \bar{r}_j} \right] \quad (27)$$

7.3 Vertically and Horizontally Polarized Incidence

In this section, we shall calculate the coherent reflected wave by solving the Lorentz-Lorenz law and Ewald-Oseen extinction theorem for the case of vertically and horizontally polarized incidence.

(A) Vertically Polarized Incidence

For the case of vertically polarized incidence, the incident field coefficients are given in (2) and (3), Section 7.2. To solve the generalized Lorentz-Lorenz law, assume the following trial solution

$$\bar{a}_E = \begin{bmatrix} \bar{a}^{E(M)} \\ \bar{a}^{E(N)} \end{bmatrix} \quad (1)$$

where $\bar{a}^{E(M)}$ and $\bar{a}^{E(N)}$ are column matrices of dimension L_{max} and

$$\bar{a}_{mn}^{E(M)} = (-1)^m \frac{1}{\gamma_{mn}} \frac{(2n+1)}{n(n+1)} i^n Y_n^{(M)} \left[\hat{\theta}_{td} \cdot \overline{C}_{-mn}(\pi - \theta_t, \phi_i) \right] \quad (2)$$

$$\bar{a}_{mn}^{E(N)} = (-1)^m \frac{1}{\gamma_{mn}} \frac{(2n+1)}{n(n+1)} i^{n-1} Y_n^{(N)} \left[\hat{\theta}_{td} \cdot \overline{B}_{-mn}(\pi - \theta_t, \phi_i) \right] \quad (3)$$

Note that in (2) and (3), we have assumed that the coefficients $Y_n^{(M)}$ and $Y_n^{(N)}$ are independent of the m index. This turns out to be true for the case of spherical scatterers. Substitute (2) and (3) in the L-L law of (14), Section 7.2, and noting that the T-matrix elements of spheres are diagonal and independent of index m, the summation over index μ can be carried out by using the following relations.

$$a(\nu, n, p) \sum_\mu a(\mu, \nu \mid -m, n \mid p) a(-\mu, \nu \mid m, l \mid p)$$

$$= -\delta_{ln}(2p+1) i^{n-\nu+p} a(1, \nu \mid -1, n \mid p) \quad (4)$$

$$\sum_\mu a(\mu, \nu \mid -m, n \mid p) a(-\mu, \nu \mid m, l \mid p, p-1) = 0 \quad (5)$$

$$b(\nu, n, p) \sum_\mu a(\mu, \nu \mid -m, n \mid p, p-1) a(-\mu, \nu \mid m, l \mid p, p-1)$$

$$= -\delta_{ln} i^{n+p-\nu} a(1, \nu \mid -1, n \mid p, p-1)(2p+1) \quad (6)$$

which can be derived by using the orthogonality relation of Wigner $3j$ symbols (Problem 28). We arrive at the following equations for the Lorentz-Lorenz law.

$$Y_n^{(M)} = 4\pi n_o \sum_{\nu p} \frac{(2\nu+1)}{\nu(\nu+1)} \frac{n(n+1)}{(2n+1)} i^{\nu-n-p} S_p$$

$$\times \left\{ -T_\nu^{(M)} Y_\nu^{(M)} a(1, \nu \mid -1, n \mid p) a(\nu, n, p) \right.$$

$$\left. + T_\nu^{(N)} Y_\nu^{(N)} a(1, \nu \mid -1, n \mid p, p-1) b(\nu, n, p) \right\} \quad (7)$$

$$Y_n^{(N)} = 4\pi n_o \sum_{\nu p} \frac{(2\nu+1)}{\nu(\nu+1)} \frac{n(n+1)}{(2n+1)} i^{\nu-n-p} S_p$$

$$\times \left\{ T_\nu^{(M)} Y_\nu^{(M)} a(1, \nu \mid -1, n \mid p, p-1) b(\nu, n, p) \right.$$

$$\left. - T_\nu^{(N)} Y_\nu^{(N)} a(1, \nu \mid -1, n \mid p) a(\nu, n, p) \right\} \quad (8)$$

The homogeneous system of the equations in (7) and (8) is identical to that of the case of normal incidence of (24) and (25), Section 6.2. Thus the dispersion relation is directionally independent.

Vertically and Horizontally Polarized Incidence

The generalized Ewald-Oseen extinction theorem can also be greatly simplified for spherical scatterers. On substituting (1) through (3) into the E-O theorem of (23), Section 7.2, and making use of the addition theorem for vector spherical harmonics (Chapter 3, Section 5.9) with $\theta = \pi - \theta_i$, $\phi = \phi_i$, $\theta' = \pi - \theta_t$ and $\phi' = \phi_i$, it follows that E-O theorem reduces to a single inhomogeneous scalar equation (Problem 29)

$$-\frac{(K_z - k_{iz})k_{iz}k}{2\pi}$$

$$= -in_o \sum_\nu T_\nu^{(M)} \frac{2\nu+1}{\nu(\nu+1)} Y_\nu^{(M)} \frac{P_\nu^1(\cos(\theta_i - \theta_t))}{\sin(\theta_i - \theta_t)}$$

$$+ in_o \sum_\nu T_\nu^{(N)} \frac{2\nu+1}{\nu(\nu+1)} Y_\nu^{(N)} \left\{ \cot(\theta_i - \theta_t) P_\nu^1(\cos(\theta_i - \theta_t)) \right.$$

$$\left. + \nu(\nu+1) P_\nu(\cos(\theta_i - \theta_t)) \right\} \quad (9)$$

By using (7) through (9), the conditional average of the exciting field can be solved. Equations (7) and (8) are homogeneous linear equations for the coefficients $Y_n^{(M)}$ and $Y_n^{(N)}$. For a nontrivial solution to exist, the determinant must vanish giving the dispersion equation from which K is determined. The solution for K is the same as that for normal incidence as in Section 6.2. There is also an eigenvector associated with the solution of K with one arbitrary constant multiplying the eigenvector. That arbitrary constant is then determined by using the single inhomogeneous equation of (9). The values of $Y_n^{(M)}$ and $Y_n^{(N)}$ are then uniquely determined. The conditional-averaged exciting field is then given by (2) and (3).

To calculate the coherent reflected field, substitute (2) and (3) in (27), Section 7.2. The $d\bar{r}_j$ integration can be carried out readily. Further simplification can be made by using the addition theorem for vector spherical harmonics (Chapter 3, Section 5.9) with $\theta = \theta_i$, $\phi = \phi_i$, $\theta' = \pi - \theta_t$ and $\phi' = \phi_i$. The final result for the coherent reflected field is

$$<\bar{E}^s(\bar{r})> = \hat{\theta}_i 2\pi n_o e^{i\bar{k}_i \cdot \bar{r}} \frac{i}{kk_{iz}} \frac{1}{(k_{iz} + K_z)}$$

$$\times \sum_n (-1)^n \frac{(2n+1)}{n(n+1)} \left\{ Y_n^{(M)} T_n^{(M)} \frac{P_n^1(\cos(\theta_i + \theta_t))}{\sin(\theta_i + \theta_t)} \right.$$

$$\left. + Y_n^{(N)} T_n^{(N)} \left[\cot(\theta_i + \theta_t) P_n^1(\cos(\theta_i + \theta_t)) \right. \right.$$

$$+ n(n+1)P_n(\cos(\theta_i + \theta_t))]\} \qquad (10)$$

where $\bar{k}_i = k(\sin\theta_i \cos\phi_i \hat{x} + \sin\theta_i \sin\phi_i \hat{y} + \cos\theta_i \hat{z})$ denotes the upward propagating direction that is specularly related to the downward incident wavevector $\bar{k}_{id}$. Thus, the coherent reflected wave is in the specular direction and contains no depolarization.

(B) Horizontally Polarized Incidence

A similar derivation can be made for the case of a horizontally polarized incident wave. To solve the Lorentz-Lorenz law, we let $\bar{a}^E$ assume the expressions as in (2) and (3) with $\bar{\theta}_{td}$ replaced by $\bar{\phi}_i$. Substituting into the Lorentz-Lorenz law of (14), Section 7.2, we obtain equations that are identical to (7) and (8). This means that the dispersion relation for the horizontally polarized wave is the same as that for the case of the vertically polarized incident wave. Thus, the dispersion relation is polarization independent besides being angular independent. Substituting into the Ewald-Oseen extinction theorem of (23), Section 7.2, and performing algebraic simplification as in the vertically polarized case, the E-O theorem reduces to the following single scalar equation

$$-\frac{(K_z - k_{iz})k_{iz}k}{2\pi}$$

$$= in_o \sum_\nu \frac{2\nu+1}{\nu(\nu+1)} Y_\nu^{(M)} T_\nu^{(M)} \left\{ \cot(\theta_i - \theta_t) P_\nu^1(\cos(\theta_i - \theta_t)) \right.$$

$$\left. + \nu(\nu+1) P_\nu(\cos(\theta_i - \theta_t)) \right\}$$

$$- in_o \sum_\nu \frac{2\nu+1}{\nu(\nu+1)} Y_\nu^{(N)} T_\nu^{(N)} \frac{P_\nu^1(\cos(\theta_i - \theta_t))}{\sin(\theta_i - \theta_t)} \qquad (11)$$

The procedure to calculate the coherent exciting field is to first solve for the propagation constant K through (7) and (8) with its associated eigenvector leaving the arbitrary constant to be calculated by the single inhomogeneous equation of (11). The quantities $Y_\nu^{(M)}$ and $Y_\nu^{(N)}$ will be uniquely determined by these two steps. The coherent reflected wave is next calculated by using (27), Section 7.2. The result is

$$<\bar{E}^s(\bar{r})> = \hat{\phi}_i \, i \frac{n_o 2\pi \, e^{i\bar{k}_i \cdot \bar{r}}}{(k_{iz} + K_z)k k_{iz}} \sum_n \frac{(2n+1)}{n(n+1)}(-1)^n$$

$$\times \left\{ T_n^{(M)} Y_n^{(M)} \left[\cot(\theta_i + \theta_t) P_n^1(\cos(\theta_i + \theta_t)) \right. \right.$$

$$+ n(n+1)P_n\left(\cos(\theta_i + \theta_t)\right)\Big]$$

$$+ T_n^{(N)} Y_n^{(N)} \frac{P_n^1\left(\cos(\theta_i + \theta_t)\right)}{\sin(\theta_i + \theta_t)}\Bigg\} \quad (12)$$

The coherent field is in the specular direction and is horizontally polarized. There is no depolarization in the coherent reflected field.

7.4 Incoherent Scattered Intensity

The coherent scattered field is in the specular direction. To calculate the backscattering coefficients, we need to study the incoherent intensity. To first-order approximation, the incoherent intensity can be calculated by the distorted Born approximation which takes into account single scattering of the coherent field. The incoherent scattered intensity is

$$<\overline{\mathcal{E}}^s(\bar{r}) \cdot \overline{\mathcal{E}}^{s*}(\bar{r})> = <\overline{E}^s(\bar{r}) \cdot \overline{E}^{s*}(\bar{r})>$$
$$- <\overline{E}^s(\bar{r})> \cdot <\overline{E}^{s*}(\bar{r})> \quad (1)$$

We multiply (31), Section 3.3, by its complex conjugate, take the ensemble average, and then subtract from it the coherent intensity by using (27), Section 7.2. Next, we apply the distorted Born approximation to the result equation by letting

$$E_{jl}(w_m^{(j)} w_n^{(l)}) \simeq E_j(w_m^{(j)}) E_l(w_n^{(l)}) \quad (2)$$

The incoherent intensity is then given by the following expression

$$<\overline{\mathcal{E}}^s(\bar{r}) \cdot \overline{\mathcal{E}}^{s*}(\bar{r})> = n_o \int_{z_j<0} d\bar{r}_j \sum_m \sum_{m'} \Big[T_m a_m^E e^{i\overline{K}_d \cdot \bar{r}_j} \overline{\psi}_m(k\overline{r}\bar{r}_j)$$
$$\cdot T_{m'}^* a_{m'}^{E*} e^{-i\overline{K}_d^* \cdot \bar{r}_j} \overline{\psi}_{m'}^*(k\overline{r}\bar{r}_j) \Big]$$
$$+ n_o^2 \mathrm{Re} \int_{z_l, z_j < 0} d\bar{r}_j \, d\bar{r}_l [g(\bar{r}_j - \bar{r}_l) - 1]$$
$$\times \sum_m \sum_{m'} \Big[T_m a_m^E e^{i\overline{K}_d \cdot \bar{r}_j} \overline{\psi}_m(k\overline{r}\bar{r}_j)$$
$$\cdot T_{m'}^* a_{m'}^{E*} e^{-i\overline{K}_d^* \cdot \bar{r}_l} \overline{\psi}_{m'}^*(k\overline{r}\bar{r}_l) \Big] \quad (3)$$

where $\overline{\psi}_m$ stands for $\overline{M}_{\mu\nu}$ and $\overline{N}_{\mu\nu}$ vector wave functions so that $\sum_m$ represents summation of $\overline{M}_{\mu\nu}$ and $\overline{N}_{\mu\nu}$ functions over both μ and ν. Since the coherent exciting field is completely determined by the results of the previous section, the incoherent intensity can be readily calculated by evaluation of the integral in (3). To evaluate the bistatic scattering coefficients, we note that the observation point $\bar{r}$ is in the far-field in region 0. Thus, the far-field approximation can be made to facilitate the evaluation of the integrals in (3). Let $\hat{k}_s$ denote the scattered direction at angle (θ_s, ϕ_s). Then the vertically polarized scattered intensity is

$$<|\overline{\mathcal{E}}^s(\bar{r})\cdot\hat{\theta}_s|^2> = \frac{n_o}{k^2 r^2}|\overline{L}(\theta_s,\phi_s)\cdot\hat{\theta}_s|^2 W(\theta_s,\phi_s)A \qquad (4)$$

and the horizontally polarized scattered intensity is the expression in (4) with $\hat{\theta}_s$ replaced by $\hat{\phi}_s$ in the dot product with $\overline{L}(\theta_s,\phi_s)$.

In (4), A is the target area under observation, and

$$W(\theta_s,\phi_s) = \frac{A}{2\,Im(K_z)}\left\{1 + n_o\,Re\int_{-\infty}^{\infty}d\bar{r}\,[g(\bar{r})-1]\,e^{i(Re\overline{K}_d - \bar{k}_s)\cdot\bar{r}}\right\} \qquad (5)$$

Comparing (5) with (3), Section 5.3, shows that the integral in (5) is proportional to the structure factor (Waseda, 1980). As indicated in Section 5, closed-form expressions exist for the structure factor of the Percus-Yevick and Verlet-Weis pair-distribution functions.

For an incident vertically polarized wave,

$$\overline{L}(\theta_s,\phi_s) = -i\sum_{mn}(-1)^m\frac{(2n+1)}{n(n+1)}$$

$$\cdot\left\{T_n^{(M)}\overline{C}_{mn}(\theta_s,\phi_s)Y_n^{(M)}\left[\hat{\theta}_{td}\cdot\overline{C}_{-mn}(\pi-\theta_t,\phi_i)\right]\right.$$

$$\left. + T_n^{(N)}\overline{B}_{mn}(\theta_s,\phi_s)Y_n^{(N)}\left[\hat{\theta}_{td}\cdot\overline{B}_{-mn}(\pi-\theta_t,\phi_i)\right]\right\} \qquad (6)$$

For an incident horizontally polarized wave, the expression for $\overline{L}(\theta_s,\phi_s)$ is the same as (6) with $\hat{\theta}_{td}$ replaced by $\hat{\phi}_i$ in the dot products with $\overline{C}_{-mn}$ and $\overline{B}_{-mn}$. Summation over the mth index can be carried out in (6) by using the vector addition theorem in the Section 5.9, Chapter 3.

The backscattering $(\theta_s = \theta_i, \phi_s = \pi + \phi_i)$ cross sections can be calculated. Using (4) through (6) and the addition theorem of Section 5.9 (Chapter 3), the results obtained are

$$\sigma_{vv} = \frac{4\pi n_o}{k^2}W(\theta_i,\pi+\phi_i)\left|\sum_n\frac{(2n+1)}{n(n+1)}(-1)^n\right.$$

Solution in Low-Frequency Limit

$$\times \left\{ T_n^{(M)} Y_n^{(M)} \frac{P_n^1 (\cos(\theta_i - \theta_t))}{\sin(\theta_i - \theta_t)} \right.$$

$$+ T_n^{(N)} Y_n^{(N)} \left[\cot(\theta_i - \theta_t) P_n^1 (\cos(\theta_i - \theta_t)) \right.$$

$$\left. \left. + n(n+1) P_n (\cos(\theta_i - \theta_t)) \right] \right\} \Big|^2 \quad (7)$$

$$\sigma_{hh} = \frac{4\pi n_o}{k^2} W(\theta_i, \pi + \phi_i) \left| \sum_n \frac{(2n+1)}{n(n+1)} (-1)^n \right.$$

$$\times \left\{ T_n^{(N)} Y_n^{(N)} \frac{P_n^1 (\cos(\theta_i - \theta_t))}{\sin(\theta_i - \theta_t)} \right.$$

$$+ T_n^{(M)} Y_n^{(M)} \left[\cot(\theta_i - \theta_t) P_n^1 (\cos(\theta_i - \theta_t)) \right.$$

$$\left. \left. + n(n+1) P_n (\cos(\theta_i - \theta_t)) \right] \right\} \Big|^2 \quad (8)$$

and $\sigma_{vh} = \sigma_{hv} = 0$.

7.5 Solution in Low-Frequency Limit

Closed-form solutions can be obtained in the low-frequency limit. In the low-frequency limit ($ka \ll 1$ and $k_s a \ll 1$), only the electric dipole term $T_1^{(N)}$ contributes, and the effective propagation constant is as given in (5), Section 6.3. From the results of Section 7.3, we note that (5), Section 6.3, is true for both incident vertical and horizontal polarization and is also angular independent. Furthermore, the phase-matching condition of (10) through (12), Section 7.2, gives $K_z = K \cos\theta_t$.

Case A: Vertically Polarized Incidence

Solution of the generalized Ewald-Oseen extinction theorem (9), Section 7.3, gives

$$Y_1^{(N)} = -\frac{(K_z - k_{iz}) k_{iz} k}{3\pi i n_o T_1^{(N)} \cos(\theta_i - \theta_t)} \quad (1)$$

Using (10), Section 7.3, in the low-frequency limit and retaining only the electric dipole $T_1^{(N)}$ contributions, the coherent reflected field is

$$\langle \overline{E}^s(\overline{r}) \rangle = \hat{v}_i \, e^{i \overline{k}_i \cdot \overline{r}} \frac{K_z - k_{iz} \cos(\theta_i + \theta_t)}{K_z + k_{iz} \cos(\theta_i - \theta_t)}$$

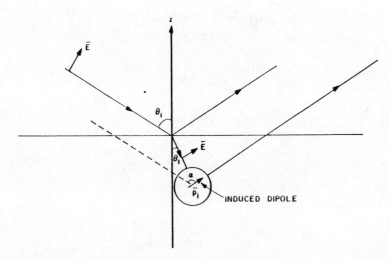

Fig. 6.24 Induced dipole radiating in the absence of other scatterers giving the Brewster angle condition of $\theta_i + \theta_t = \pi/2$.

$$= \hat{\theta}_i \, e^{i\bar{k}_i \cdot \bar{r}} \, \frac{K^2 k_{iz} - k^2 K_z}{K^2 k_{iz} + k^2 K_z} \tag{2}$$

Thus, the TM Fresnel reflection coefficient is exhibited. At $\theta_i + \theta_t = \pi/2$ the coherent reflected field is zero, giving the Brewster angle effect. The above equation for the Brewster angle is linked to the fact that after the final exciting field is determined, the induced dipole radiates in the absence of other scatterers (i.e., in the k medium) (Figure 6.24). Had the final radiation of the induced dipole been in the effective K medium, (Figure 6.25), the Brewster angle would be given by the condition of $\theta_t = \pi/4$. The backscattering cross section is, using (7), Section 7.4,

$$\sigma_{vv} = \frac{1}{2\pi n_o} |(K_z - k_{iz})k_{iz}|^2 \, \frac{1}{Im(K_z)} \times \left\{ 1 + n_o \int_{-\infty}^{\infty} d\bar{r} \, [g(\bar{r}) - 1] \right\} \tag{3}$$

Case B: Horizontally Polarized Incidence

Solution of the E-O theorem [(11), Section 7.3] gives

$$Y_1^{(N)} = -\frac{(K_z - k_{iz})k_{iz}k}{3\pi} \, \frac{1}{i n_o T_1^{(N)}} \tag{4}$$

The coherent reflected field is, on using (4) in (12), Section 7.3,

$$<\overline{E}^s(\bar{r})> = \hat{\phi}_i \, e^{i\bar{k}_i \cdot \bar{r}} \, \frac{k_{iz} - K_z}{k_{iz} + K_z} \tag{5}$$

Solution in Low-Frequency Limit

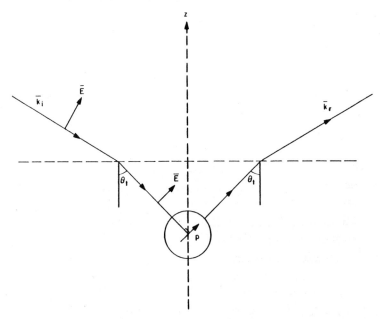

Fig. 6.25 Induced dipole radiating in the presence of other scatterers giving the Brewster angle condition of $\theta_i + \theta_t = \pi/2$.

Thus, the Fresnel TE reflection coefficient is exhibited.

From (8), Section 7.4, we have

$$\sigma_{hh} = \frac{1}{2\pi n_o}|(K_z - k_{iz})k_{iz}|^2 \frac{1}{Im(K_z)}\left\{1 + n_o\int_{-\infty}^{\infty} d\bar{r}\,[g(\bar{r}) - 1]\right\} \quad (6)$$

It is interesting to note that from (3) and (6) that $\sigma_{vv} = \sigma_{hh}$. This is also a direct consequence of the fact that the final radiation of the induced dipole is in the absence of other scatterers (in the k medium). We note from Figure 6.24 that radiation intensity in the backscattering direction for the vertically polarized case is proportional to the square of the product of the transmission coefficient of the electric field and $\sin\alpha$. Since $\alpha = \pi/2 + (\theta_i - \theta_t)$, the radiation intensity is proportional to

$$\left|\frac{k}{K}\frac{2K^2 k_{iz}}{K^2 k_{iz} + k^2 K_z}\cos(\theta_i - \theta_t)\right|^2$$

which can be shown to be equal to $|2k_{iz}/(K_z + k_{iz})|^2$. Thus the radiation intensity is identical to the backscattering radiation intensity of the horizontally polarized case. We emphasize the fact that $\sigma_{vv} = \sigma_{hh}$ is only true in the low-frequency limit and under the version of the distorted Born

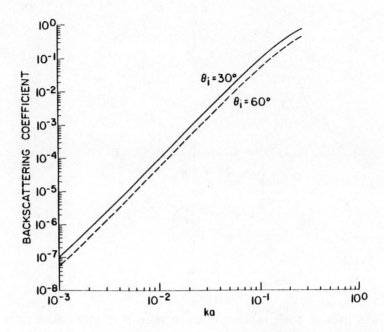

Fig. 6.26 Backscattering coefficient in the low-frequency limit as a function of ka for $\epsilon_s = 3.24(1 + i0.001)\epsilon_o$, $f = 0.2$ and $\theta_i = 30\deg$ and $60\deg$.

approximation of (2), Section 7.4. It is also to be noted that the Fresnel reflection coefficient for the coherent wave (2) and (5) is only true in the low-frequency limit and is independent of frequency if ϵ_s is nondispersive.

In Figure 6.26, we plot the backscattering coefficient in the low-frequency limit as a function of ka for $f = 0.2$. We note that the backscattering coefficient is a rapidly increasing function of frequency. The Percus-Yevick pair-distribution function is used in computing the results of Figure 6.26.

7.6 Results at Higher Frequency

For larger values of ka, the generalized Lorentz-Lorenz law and the Ewald-Oseen extinction theorem are to be solved numerically. The effective propagation constant K is calculated by solving the set of homogeneous equations (7) and (8), Section 7.3. The values of $Y_n^{(M)}$ and $Y_n^{(N)}$ are then determined by using the single inhomogeneous equation of (9), Section 7.3, for the case of vertical polarization and (11), Section 7.3, for the case of horizontal polarization. The coherent reflected waves are then calculated from (10) and (12), Section 7.3, and the backscattering coefficients from (7) and (8), Section 7.4.

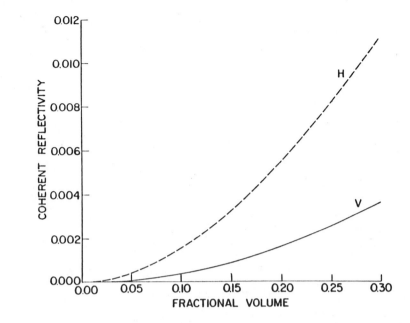

Fig. 6.27 Coherent reflectivity as a function of fractional volume of scatterers f for $\epsilon_s = 3.24(1 + i0.001)\epsilon_o$, $ka = 0.5$, $\theta_i = 30\,\text{deg}$, and for vertical and horizontal polarizations.

The pair-distribution function is calculated by a computer code that programs the analytic expressions of Wertheim (1963) and Verlet-Weis (1972), (McQuarrie, 1976). The Wigner $3j$ symbols are also calculated by a computer code. In the subsequent figures, the Percus-Yevick pair distribution function is used unless specified otherwise. In the following, we illustrate results using typical parameters of snow.

In Figure 6.27, the coherent reflectivity $= |<\overline{E}_s>|^2$ is plotted as a function of fractional volume of scatterers. The medium becomes more reflective as the number of scatterers increases. In Figures 6.28 and 6.29, we show respectively the coherent reflectivity and the backscattering coefficients as a function of incidence angle for $ka = 1$. The coherent reflectivity for the horizontal polarization increases with angle whereas that for the vertical polarization exhibits the Brewster angle effect. The backscattering coefficient decreases with incident angle and the polarization dependence of the backscattering coefficient is very small.

In Figures 6.30 and 6.31, the coherent reflectivity and backscattering coefficient are shown as a function of ka. We note that while for low frequencies, the coherent reflectivity as given in (2) and (5), Section 7.5, is non-dispersive, dispersive effects are exhibited at higher frequencies. The

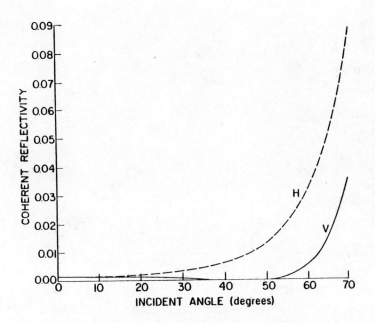

Fig. 6.28 Coherent reflectivity as a function of incident angle θ_i for $\epsilon_s = 3.24(1 + i0.001)\epsilon_o$, $ka = 1$, $f = 0.2$, and for vertical and horizontal polarizations.

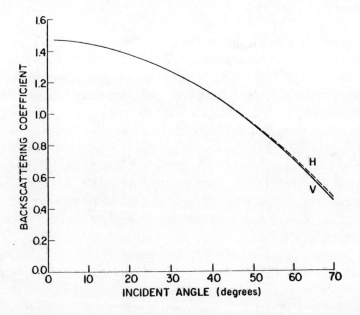

Fig. 6.29 Backscattering coefficient as a function of incident angle θ_i for $\epsilon_s = 3.24(1 + i0.001)\epsilon_o$, $ka = 1$, $f = 0.2$, and for vertical and horizontal polarizations.

Results at Higher Frequency

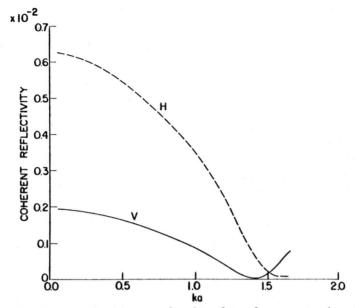

Fig. 6.30 Coherent reflectivity as a function of ka for $\epsilon_s = 3.24(1 + i0.001)\epsilon_o$, $f = 0.2$, $\theta_i = 30$ deg, and for vertical and horizontal polarizations.

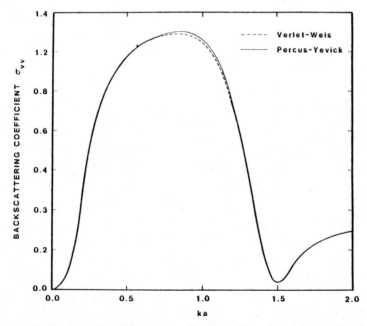

Fig. 6.31 Backscattering coefficient as a function of ka for $\epsilon_s = 3.24(1 + i0.001)\epsilon_o$, $f = 0.2$, $\theta_i = 30$ deg, and for vertical polarization. Both PY and VW pair-distribution functions are used.

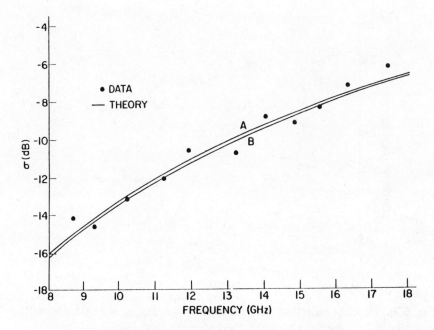

Fig. 6.32 Backscattering coefficient as a function of frequency. Comparison of experimental data (solid circles) and theory (solid lines). Two sets of parameters are used in the theoretical curves: (a) $\theta_i = 50$ deg, $\epsilon_s = 3.24(1+i0.001)\epsilon_o$, $a = 0.05$ cm, and $f = 0.25$; and (b) $\theta_i = 50$ deg, $\epsilon_s = 3.24(1+i0.03)\epsilon_o$, $a = 0.15$ cm, and $f = 0.25$. The experimental data were taken over dry snow for HH polarization. The snow depth was 48 cm, and the water equivalent was 10.5 cm (Stiles and Ulaby, 1980a).

minimum for the coherent reflectivity and for the backscattering coefficient around $ka = 1.5$ is a result of the scattering pattern of a spherical particle (Wichramasinghl, 1973). As can be seen from Figure 6.31, the difference between using the PY and VW pair-distribution functions is insignificant.

In Figure 6.32, we apply the theory to match experimental data (Stiles and Ulaby, 1980a) collected from dry snow. The data was taken over a slab of snow of finite thickness over ground. However, it seems that the data shows relatively little effect of the bottom interface for frequencies between 8 and 17 GHz. The fractional volume of ice in snow is slightly larger than the water equivalent divided by the snow depth. Thus, a value of $f = 0.25$ is chosen to match the data. We note that the theoretical results match very well with the experimental data in spite of the half-space model that was used. The two sets of parameters chosen in the theoretical curves represent typical values for dry snow.

8 NONSPHERICAL PARTICLES

8.1 Introduction

In this section, scattering of a plane electromagnetic wave obliquely incident on a half-space of non-spherical particles with prescribed orientation distribution will be studied. Coherent wave transmission and reflection are investigated with the quasicrystalline approximation. Generally, there can be two characteristic waves that are excited in the scattering medium. The propagation constants and the polarization properties of the two characteristic waves are calculated from the generalized Lorentz-Lorenz law. The amplitudes of the characteristic waves are next obtained by solving the equations of the generalized Ewald-Oseen extinction theorem. The coherent reflectivities are then calculated. In the limit of low concentration of particles, the results are identical to those of Foldy's approximation. The incoherent scattered field is calculated with the distorted Born approximation and generally gives a nonzero depolarization return in the backscattering direction. Very little study has been made on the pair-distribution functions of nonspherical particles. For the sake of illustration, we will assume that the orientation of two particles are independent of their position and are independent of each other.

8.2 Generalized Ewald-Oseen Extinction Theorem and Lorentz-Lorenz Law

Consider a plane electromagnetic wave obliquely incident on a half-space of N dielectric scatterers with permittivity ϵ_s and centered at $\bar{r}_1, \bar{r}_2, \ldots \bar{r}_N$ (Figure 6.33). The background medium has permittivity ϵ and the same background medium occupies the region $z \geq 0$. The particles are identical in shape and size. They can be aligned or follow a prescribed orientation distribution. We use $\overline{\overline{T}}^{(j)}$ to denote the T-matrix of the j th particle. Let $\overline{\overline{\hat{T}}}$ represent the T-matrix of the particles in their natural frame. Suppose the j th particle is oriented with Eulerian angles α_j, β_j, and γ_j with respect to the principal frame. Then from (25), Section 5.7, Chapter 3,

$$\overline{\overline{T}}^{(j)} = \overline{\overline{D}}(\alpha_j \beta_j \gamma_j) \overline{\overline{\hat{T}}} \, \overline{\overline{D}}^{-1}(\alpha_j \beta_j \gamma_j) \tag{1}$$

where $\overline{\overline{D}}$ is the rotation matrix. The multiple scattering equations assume

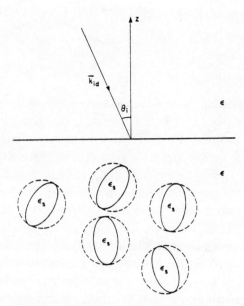

Fig. 6.33 Plane electromagnetic wave obliquely incident on a half-space of non-spherical dielectric particles.

the following form by using (40), Section 3.3,

$$\overline{w}^{(l)} = \sum_{\substack{j=1 \\ j \neq l}}^{N} \overline{\overline{\sigma}}(k\overline{r_l r_j}) \overline{\overline{T}}^{(j)} \overline{w}^{(j)} + \exp(i\overline{k}_{id} \cdot \overline{r}_l) \overline{a}^{inc} \qquad (2)$$

where $\overline{a}^{inc}$ and $\overline{w}^{(j)}$ are column matrices representing the coefficients a_j^{inc} and $w_l^{(j)}$ in (1) and (2), respectively, and $\overline{\overline{\sigma}}(k\overline{r_l r_j})$ is the transformation matrix that transforms spherical waves centered at $\overline{r}_j$ to spherical waves centered at $\overline{r}_l$ (Section 3.3). The subscript d in (2) indicates that the incident wave vector is downward going as in (1), Section 7.2. Equation (2) is true under the conditions that the particles can be enclosed by circumscribing spheres that do not overlap each other (Figure 6.33 and Section 3.3). After the exciting field in (2) is solved, the scattered field $\overline{E}^s$ can be calculated from the sum of the scattered fields from all particles as in (31), Section 3.3.

To solve for the coherent field, ensemble average can be taken over (2). To truncate the hierarchy of multiple scattering equations, the quasi-crystalline approximation is made.

$$\overline{E}_l(\overline{w}^{(l)}) = \overline{w}(\overline{r}_l) \qquad (3)$$

We assume that the orientation of two particles are independent of their positions and independent of each other. The conditional average of the exciting field coefficients then satisfies the following equation

$$\overline{w}(\overline{r}_1) = n_o \int_{z_2<0} d\overline{r}_2 g(\overline{r}_2, \overline{r}_1) \overline{\overline{\sigma}}(k\overline{r}_1\overline{r}_2) <\overline{\overline{T}}>_{st} \overline{w}(\overline{r}_2)$$

$$+ \exp(i\overline{k}_{id} \cdot \overline{r}_1)\overline{a}^{inc} \qquad (4)$$

In (4), $<\overline{\overline{T}}>_{st}$ represents the T-matrix averaged over the orientation distribution.

$$<\overline{\overline{T}}>_{st} = \int_0^{2\pi} d\alpha \int_0^{\pi} d\beta \int_0^{2\pi} d\gamma\, p(\alpha, \beta, \gamma) \overline{\overline{D}}(\alpha\beta\gamma) \overline{\overline{T}}\,\overline{\overline{D}}^{-1}(\alpha\beta\gamma) \qquad (5)$$

where $p(\alpha, \beta, \gamma)$ is the probability density function for the Eulerian angles of rotation α, β and γ. Since the exclusion volumes are spherical, we shall use spherical pair-distribution functions (Percus and Yevick, 1958; Verlet and Weis, 1972) which are in terms of the exclusion sphere radius $b/2$ and the fractional exclusion volume $f_e = n_o \pi b^3/6$ of the particles. The general problem of calculating the pair-distribution function for non-spherical particles with prescribed orientation distribution has not been done. For aligned particles, the cases of elliptical and ellipsoidal correlations have been considered (Twersky, 1978).

To solve (4), let

$$\overline{w}(\overline{r}_1) = \sum_\tau c_\tau \overline{a}_\tau^E \exp(i\overline{K}_{\tau d} \cdot \overline{r}_1) \qquad (6)$$

where the index τ indicates the τ characteristic wave, c_τ is the amplitude and $\overline{K}_{\tau d}$ is the downward-going effective wavevector for the τ characteristic wave. We have

$$\overline{K}_{\tau d} = K_{\tau x}\hat{x} + K_{\tau y}\hat{y} - K_{\tau z}\hat{z} \qquad (7)$$

and the effective wavenumber K_τ for the τ characteristic wave is

$$K_\tau = (K_{\tau x}^2 + K_{\tau y}^2 + K_{\tau z}^2)^{1/2} \qquad (8)$$

Substituting (6) into (4), the integrals can be evaluated as in Section 7.2. Balancing the wave dependences of the terms gives rise to the generalized Ewald-Oseen extinction theorem, the generalized Lorentz-Lorenz law, the phase-matching conditions of

$$K_{\tau x} = k_{ix} \qquad (9)$$

$$K_{ry} = k_{iy} \tag{10}$$

and Snell's law

$$K_r \sin\theta_{rt} = k \sin\theta_i \tag{11}$$

The effective propagation constant K_r is generally angular dependent. The transmitted angle θ_{rt} is slightly complex since K_r is complex due to a combination of absorption and scattering.

The generalized Eward-Oseen extinction theorem is as follows

$$n_o \sum_r c_r \overline{\overline{U}}^{(k)}(k, K_r) <\overline{\overline{T}}>_{st} \bar{a}_r^E = -\bar{a}^{inc} \tag{12}$$

where

$$\overline{\overline{U}}^{(k)}(k, K_r) = \begin{bmatrix} \overline{\overline{X}}^{(k)}(k, K_r) & \overline{\overline{Z}}^{(k)}(k, K_r) \\ \overline{\overline{Z}}^{(k)}(k, K_r) & \overline{\overline{X}}^{(k)}(k, K_r) \end{bmatrix} \tag{13}$$

with $\overline{\overline{X}}^{(k)}(k, K_r)$ and $\overline{\overline{Z}}^{(k)}(k, K_r)$ being $L_{max} \times L_{max}$ matrices with $L_{max} = N_{max}(N_{max} + 2)$ and N_{max} being the highest multipole retained in the computations. The quantities $X^{(k)}_{mn\mu\nu}(k, K_r)$ and $Z^{(k)}_{mn\mu\nu}(k, K_r)$ are given, respectively, in (25) and (26), Section 7.2, with K replaced by K_r and K_z replaced by K_{rz}. The generalized Ewald-Oseen extinction theorem is a system of inhomogeneous equations with dimensions $2N_{max}(N_{max} + 2)$ for the unknowns c_r. The system of equations in (13) can be reduced to two independent equations.

Consider the incident plane wave to be a linear combination of vertically and horizontally polarized waves

$$\overline{E}^{inc} = (E_v^{inc} \hat{\theta}_{id} + E_h^{inc} \hat{\phi}_{id}) e^{i\bar{k}_{id} \cdot \bar{r}} \tag{14}$$

where $\hat{\theta}_{id}$ and $\hat{\phi}_{id}$ are the vertically and horizontally polarized vectors associated with the downward-going incident wave (Section 7.2). Then

$$a_{mn}^{inc\,(M)} = (-1)^m \frac{1}{\gamma_{mn}} \frac{(2n+1)}{n(n+1)} i^n (E_v^{inc} \hat{\theta}_{id} + E_h^{inc} \hat{\phi}_{id}) \cdot \overline{C}_{-mn}(\pi - \theta_i, \phi_i) \tag{15}$$

$$a_{mn}^{inc\,(N)} = (-1)^m \frac{1}{\gamma_{mn}} \frac{(2n+1)}{n(n+1)} i^n (E_v^{inc} \hat{\theta}_{id} + E_h^{inc} \hat{\phi}_{id}) \cdot \frac{\overline{B}_{-mn}(\pi - \theta_i, \phi_i)}{i} \tag{16}$$

Substituting (15) and (16) into (12) and simplifying, the system reduces to two independent scalar equations by canceling $\overline{C}_{-mn}(\pi - \theta_i, \phi_i)$ and

$\overline{B}_{-mn}(\pi - \theta_i, \phi_i)$ on both sides of the system of equations (Tsang, 1984b; Problem 30)

$$-E_v^{inc} = \frac{2\pi n_o}{k_{iz}k} \sum_{\tau,\mu,\nu} c_\tau \gamma_{\mu\nu} \frac{(-1)^\nu i^{\nu+1}}{K_{\tau z} - k_{iz}} \left[<h_{\tau\mu\nu}^{(M)}> (\hat{\theta}_{id} \cdot \overline{C}_{\mu\nu}(\pi - \theta_i, \phi_i)) \right.$$

$$\left. + <h_{\tau\mu\nu}^{(N)}> (\hat{\theta}_{id} \cdot i\overline{B}_{\mu\nu}(\pi - \theta_i, \phi_i)) \right] \quad (17)$$

$$-E_h^{inc} = \frac{2\pi n_o}{k_{iz}k} \sum_{\tau,\mu,\nu} c_\tau \gamma_{\mu\nu} \frac{(-1)^\nu i^{\nu+1}}{K_{\tau z} - k_{iz}} \left[<h_{\tau\mu\nu}^{(M)}> (\hat{\phi}_{id} \cdot \overline{C}_{\mu\nu}(\pi - \theta_i, \phi_i)) \right.$$

$$\left. + <h_{\tau\mu\nu}^{(N)}> (\hat{\phi}_{id} \cdot i\overline{B}_{\mu\nu}(\pi - \theta_i, \phi_i)) \right] \quad (18)$$

where

$$<\overline{h}_\tau> = \begin{bmatrix} <\overline{h}_\tau^{(M)}> \\ <\overline{h}_\tau^{(N)}> \end{bmatrix} = <\overline{\overline{T}}>_{st} \overline{a}_\tau^E \quad (19)$$

The $\overline{a}_\tau^E$ and K_τ are, respectively, the eigenvectors and effective propagation constants calculated from the Lorentz-Lorenz law

$$\overline{a}_\tau^E = n_o \overline{\overline{U}}^{(K)}(k, K_\tau) <\overline{\overline{T}}>_{st} \overline{a}_\tau^E \quad (20)$$

where $\overline{\overline{U}}^{(K)}(k, K_\tau)$ is a matrix of dimensions $2N_{max}(N_{max}+2) \times 2N_{max}(N_{max}+2)$ given in (15) through (19), Section 7.2, with the replacement of K by K_τ and θ_t by $\theta_{\tau t}$. The generalized Lorentz-Lorenz law is a homogeneous system of equations for the coefficients $\overline{a}_\tau^E$. Setting the determinant of (20) to zero gives the dispersion relation for the effective propagation constant K_τ and $\overline{a}_\tau^E$ is the corresponding eigenvector for the τ characteristic wave. Then $\overline{h}_\tau$ is calculated according to (19). The generalized Ewald-Oseen extinction theorem of (17) and (18) furnishes two independent equations for the unknown amplitudes, c_τ, of the characteristic waves. Hence, in general there can only be two characteristic waves, $\tau = 1, 2$, associated with the generalized Lorentz-Lorenz law and Ewald-Oseen extinction theorem.

8.3 Coherent Reflected Wave and Incoherent Scattered Field

The coherent reflected wave can be calculated from taking the ensemble average of (31), Section 3.3, and using (6), Section 8.2.

$$<\overline{E}^s(\overline{r})> = n_o \sum_{mn\tau} c_\tau \int_{z_j<0} d\overline{r}_j \exp(i\overline{K}_{\tau d} \cdot \overline{r}_j)$$

$$\cdot \left\{ \overline{M}_{mn}(k\overline{\overline{r}}\overline{r}_j) <h_{\tau mn}^{(M)}> + \overline{N}_{mn}(k\overline{\overline{r}}\overline{r}_j) <h_{\tau mn}^{(N)}> \right\} \quad (1)$$

where $<h^{(M)}_{\tau mn}>$ and $<h^{(N)}_{\tau mn}>$ are defined in (19), Section 8.2. The $d\bar{r}_j$ integration can be carried out readily, and (1) gives

$$<\overline{E}^s(\bar{r})> = \frac{2\pi n_o}{kk_{iz}} e^{i\bar{k}_i \cdot \bar{r}} \sum_{mn\tau} \frac{c_\tau}{K_{\tau z} + k_{iz}} \gamma_{mn}(-i)^{n-1}$$
$$\cdot \left\{ <h^{(M)}_{\tau mn}> \overline{C}_{mn}(\theta_i, \phi_i) + <h^{(N)}_{\tau mn}> i\overline{B}_{mn}(\theta_i, \phi_i) \right\} \quad (2)$$

where $\bar{k}_i$ denotes the upward-propagating wavevector that is specularly related to the downward incident wavevector $\bar{k}_{id}$.

To calculate the incoherent scattered intensity, the distorted Born approximation that corresponds to single scattering of the coherent wave is applied. Since the coherent exciting field is completely determined by the results of Section 8.2, the incoherent intensity can be calculated and is simplified by making far-field approximations as in Section 7.4

$$<\overline{\mathcal{E}}^s(\bar{r})\overline{\mathcal{E}}^{s^*}(\bar{r})> = \frac{n_o A}{k^2 r^2} \sum_{\tau,\tau_1} \frac{1}{i(K^*_{\tau_1 z} - K_{\tau z})}$$

$$\cdot \left\{ \int_0^{2\pi} d\alpha \int_0^\pi d\beta \int_0^{2\pi} d\gamma\, p(\alpha\beta\gamma)\, \bar{c}^{(s)}_\tau(\alpha\beta\gamma)\, \bar{c}^{(s)^*}_{\tau_1}(\alpha\beta\gamma) \right.$$

$$+ Re\left[<\bar{c}^{(s)}_\tau> <\bar{c}^{(s)^*}_{\tau_1}> n_o \int_{-\infty}^\infty d\bar{r}'\, [g(\bar{r}') - 1] \right.$$

$$\left. \left. \times \exp\left(i(k_{ix} - k_{sx})x' + i(k_{iy} - k_{sy})y' \right.\right.\right.$$

$$\left.\left.\left. - \frac{i(K_{\tau z} + K^*_{\tau_1 z} + 2k_{sz})z'}{2} \right) \right] \right\} \quad (3)$$

where $\bar{k}_s$ is the wavevector in the scattered direction

$$\bar{c}^{(s)}_\tau(\alpha\beta\gamma) = c_\tau \sum_{mn} \gamma_{mn} i^{-n-1} \left[h^{(M)}_{\tau mn}(\alpha\beta\gamma)\, \overline{C}_{mn}(\theta_s, \phi_s) \right.$$
$$\left. + h^{(N)}_{\tau mn}(\alpha\beta\gamma)\, i\overline{B}_{mn}(\theta_s, \phi_s) \right] \quad (4)$$

and

$$\begin{bmatrix} \bar{h}^{(M)}_\tau \\ \bar{h}^{(N)}_\tau \end{bmatrix} = \bar{h}_\tau(\alpha\beta\gamma) = \overline{\overline{T}}(\alpha\beta\gamma)\, \bar{a}^E_\tau \quad (5)$$

In (3) through (5), $p(\alpha\beta\gamma)$ is the probability density function of particle orientation and $<\bar{c}_r^{(s)}>$ is the expression of (4) with $h_{rmn}^{(M)}(\alpha\beta\gamma)$ and $h_{rmn}^{(N)}(\alpha\beta\gamma)$ replaced respectively by $<h_{rmn}^{(M)}>$ and $<h_{rmn}^{(N)}>$ as given by (19), Section 8.2. The integral in the second term in (3) is proportional to the structure factor (Waseda, 1980) which can be evaluated exactly for both the Percus-Yevick and Verlet-Weis pair-distribution functions. Using (3), the vertically polarized scattered intensity $\hat{\theta}_s \cdot <\overline{\mathcal{E}}^s(\bar{r})\overline{\mathcal{E}}^{s*}(\bar{r})> \cdot \hat{\theta}_s$ and the horizontally polarized scattered intensity $\hat{\phi}_s \cdot <\overline{\mathcal{E}}^s(\bar{r})\overline{\mathcal{E}}^{s*}(\bar{r})> \cdot \hat{\phi}_s$ can be readily calculated. The bistatic and backscattering coefficients can also be calculated.

8.4 Sparse Concentration of Particles

It will be shown that in the limit of low concentration of particles, the results of the Section 8.2 reduce to those of Foldy's approximation. The results for the effective propagation constants in the low concentration limit also serve as good initial guesses for numerically solving the homogeneous system of equations of the Lorentz-Lorentz law.

In the limit of low concentration of particles, $K \simeq k$ and $g(r) = 1$ for $r > b$. Hence, making the substitution in (19), Section 7.2,

$$S_p(k, K) = \frac{-i}{2k_{iz}k(K_z - k_{iz})} \tag{1}$$

Substituting (1) into the Lorentz-Lorenz law, and making use of (20) and (21), Section 7.2, it follows that

$$X_{mn\mu\nu}^{(K)}(k, K) = -X_{mn\mu\nu}^{(k)}(k, K) \tag{2}$$

$$Z_{mn\mu\nu}^{(K)}(k, K) = -Z_{mn\mu\nu}^{(k)}(k, K) \tag{3}$$

To solve the Lorentz-Lorenz law, assume the following trial solution with unknown amplitudes E_v and E_h.

$$a_{mn}^{E(M)} = (-1)^m \frac{1}{\gamma_{mn}} \frac{(2n+1)}{n(n+1)} i^n (E_v \hat{\theta}_{id} + E_h \hat{\phi}_{id}) \cdot \overline{C}_{-mn}(\pi - \theta_i, \phi_i) \tag{4}$$

$$a_{mn}^{E(N)} = (-1)^m \frac{1}{\gamma_{mn}} \frac{(2n+1)}{n(n+1)} i^n (E_v \hat{\theta}_{id} + E_h \hat{\phi}_{id}) \cdot \frac{\overline{B}_{-mn}(\pi - \theta_i, \phi_i)}{i} \tag{5}$$

Substituting (2) through (5) into the Lorentz-Lorenz law of (20), Section 8.2, and simplifying, the system of equations reduces to two linear independent equations governing the unknowns E_v and E_h

$$(E_v\hat{\theta}_{id}+E_h\hat{\phi}_{id}) = \frac{2\pi n_o}{k_{iz}(K_z-k_{iz})}\overline{\overline{F}}(\pi-\theta_i,\phi_i;\pi-\theta_i,\phi_i)\cdot(E_v\hat{\theta}_{id}+E_h\hat{\phi}_{id}) \tag{6}$$

where $\overline{\overline{F}}$ is the average far-field scattering amplitude dyad and is related to the T-matrix elements by

$$\overline{\overline{F}}(\theta,\phi;\theta',\phi') = \frac{4\pi}{k}\sum_{n,m,n',m'}(-1)^{m'}i^{n'-n-1}$$

$$\cdot\left\{\left[<T^{(11)}_{mnm'n'}>_{st}\gamma_{mn}\overline{C}_{mn}(\theta,\phi)\right.\right.$$

$$\left.+<T^{(21)}_{mnm'n'}>_{st}i\gamma_{mn}\overline{B}_{mn}(\theta,\phi)\right]\gamma_{-m'n'}\overline{C}_{-m'n'}(\theta',\phi')$$

$$+\left[<T^{(12)}_{mnm'n'}>_{st}\gamma_{mn}\overline{C}_{mn}(\theta,\phi)\right.$$

$$\left.\left.+<T^{(22)}_{mnm'n'}>_{st}i\gamma_{mn}\overline{B}_{mn}(\theta,\phi)\right]\gamma_{-m'n'}\frac{\overline{B}_{-m'n'}(\theta',\phi')}{i}\right\} \tag{7}$$

In deriving (6), we have also made use of the approximation $K^2-k^2\simeq 2k_{iz}(K_z-k_{iz})$. Let

$$\overline{\overline{M}}(\theta,\phi) = \frac{i2\pi n_o}{k}\overline{\overline{F}}(\theta,\phi;\theta,\phi) \tag{8}$$

and

$$\overline{E}(\theta,\phi) = (E_v\hat{\theta}+E_h\hat{\phi}) \tag{9}$$

Then the Lorentz-Lorenz law reduces to the following 2×2 eigenvalue problem.

$$i(K-k)\overline{E}(\pi-\theta_i,\phi_i) = \overline{\overline{M}}(\pi-\theta_i,\phi_i)\cdot\overline{E}(\pi-\theta_i,\phi_i) \tag{10}$$

Results of (10) are identical to those derived under Foldy's approximation (16), Section 4.2, and extensively used in wave propagation in rain and other hydrometeors (Oguchi, 1973, 1983; Ishimaru, 1983; Ishimaru and Yeh, 1984). There are two propagation constants, K_τ, $\tau=1,2$ and two characteristic wave polarizations associated with the solution of (10).

Let $[E_v^\tau, E_h^\tau]$ be the 2 x 1 eigenvector associated with the τ characteristic wave in (10). The Ewald-Oseen extinction theorem of (17) and (18), Section 8.2, then gives

$$\begin{bmatrix}E_v^{inc}\\E_h^{inc}\end{bmatrix} = \sum_{\tau=1}^{2}c_\tau\begin{bmatrix}E_v^\tau\\E_h^\tau\end{bmatrix} \tag{11}$$

which provides two simultaneous equations for the unknown amplitude coefficients c_τ, $\tau = 1, 2$, of the characteristic waves.

8.5 Low-Frequency Limit

The solution of the Ewald-Oseen theorem and the Lorentz-Lorenz law in the low-frequency limit for spheroids uniformly distributed in γ will be examined. For small spheroids, the $\hat{\bar{\bar{T}}}$ matrix in the natural frame of the particle is obtained by choosing $N_{max} = 1$, and retaining only the $\hat{\bar{\bar{T}}}^{(22)}$ matrix. The $\hat{\bar{\bar{T}}}^{(22)}$ matrix is

$$\hat{\bar{\bar{T}}}^{(22)} = \begin{bmatrix} T_1 & 0 & 0 \\ 0 & T_o & 0 \\ 0 & 0 & T_1 \end{bmatrix} \tag{1}$$

where T_o and T_1 are given in (11) and (12), Section 5.6, Chapter 3. If the spheroid is oriented with angle $\alpha\beta\gamma$ with respect to the principal frame, then the relation (1), Section 8.2, can be used to calculate the T-matrix in the principal frame. Since $N_{max} = 1$, therefore (Problem 31)

$$\bar{\bar{D}} = \begin{bmatrix} e^{-i(\alpha+\gamma)}\cos^2\frac{\beta}{2} & -\frac{1}{\sqrt{2}}e^{-i\gamma}\sin\beta & e^{-i\gamma}\sin^2\frac{\beta}{2}e^{i\alpha} \\ \frac{1}{\sqrt{2}}e^{-i\alpha}\sin\beta & \cos\beta & -\frac{1}{\sqrt{2}}e^{i\alpha}\sin\beta \\ e^{i\gamma}\sin^2\frac{\beta}{2}e^{-i\alpha} & \frac{1}{\sqrt{2}}e^{i\gamma}\sin\beta & e^{i(\alpha+\gamma)}\cos^2\frac{\beta}{2} \end{bmatrix} \tag{2}$$

Because the spheroid is axisymmetric, the T-matrix $\bar{\bar{T}}^{(22)}$ is independent of the Eulerian angle α.

To simplify calculations, we allow the orientation probability density function to depend on β only. Let $p(\beta)$ be the orientation distribution in β normalized by the condition $\int_0^\pi d\beta \sin\beta p(\beta)/2 = 1$. Then

$$<\bar{\bar{T}}^{(22)}>_{st} = \begin{bmatrix} T_{a1} & 0 & 0 \\ 0 & T_{ao} & 0 \\ 0 & 0 & T_{a1} \end{bmatrix} \tag{3}$$

where

$$T_{a1} = \int_0^\pi d\beta \frac{\sin\beta}{2} p(\beta) \left[T_1 + (T_o - T_1) \frac{\sin^2\beta}{2} \right] \tag{4}$$

$$T_{ao} = \int_0^\pi d\beta \frac{\sin\beta}{2} p(\beta) \left[(T_o - T_1)\cos^2\beta + T_1 \right] \tag{5}$$

Substituting (3) into (20), Section 8.2, and retaining only the electric dipole interaction terms give

$$\overline{a}_E^{(N)} = n_o \overline{\overline{X}}^{(K)} <\overline{\overline{T}}^{(22)}>_{st} \overline{a}_E^{(N)} \tag{6}$$

where $\overline{a}_E^{(N)}$ is a 3×1 column matrix. Making the low-frequency approximations as done in Sections 6.3 and 7.5, we find two dispersion relations in (6) with two associated eigenvectors.

Eigen-Solution (A)

The dispersion relation corresponds to that of vertical polarization and is $K = K_v$ with

$$\frac{K_v^2}{k^2} = \frac{1}{\dfrac{\epsilon}{\epsilon_{effz}} \sin^2 \theta_t + \dfrac{\epsilon}{\epsilon_{efft}} \cos^2 \theta_t} \tag{7}$$

with associated eigenvector

$$\overline{a}_E^{(N)} = c_v \begin{bmatrix} -r_B\, e^{i\phi_i} \\ 1 \\ r_B\, e^{-i\phi_i} \end{bmatrix} \tag{8}$$

where c_v is an arbitrary constant to be determined by the E-O theorem. In (7) and (8)

$$r_B = \frac{i3}{2\sqrt{2}} \frac{\cos\theta_t}{\sin\theta_t} \frac{4\pi n_o T_{ao}}{k^3} \frac{\epsilon_{effz}}{\left[1 - \dfrac{4\pi n_o}{k^3} T_{a1}(i + W_o)\right](\epsilon_{effz} - \epsilon)} \tag{9}$$

$$\frac{\epsilon_{efft}}{\epsilon} = \frac{1 - (4\pi n_o/k^3)T_{a1}(i + W_o)}{1 + (4\pi n_o/k^3)T_{a1}(i/2 - W_o)} \tag{10}$$

$$\frac{\epsilon_{effz}}{\epsilon} = \frac{1 - (4\pi n_o/k^3)T_{ao}(i + W_o)}{1 + (4\pi n_o/k^3)T_{ao}(i/2 - W_o)} \tag{11}$$

where

$$W_o = k^3 \int_0^\infty dr\, r^2\, [g(r) - 1] \tag{12}$$

In view of the phase-matching conditions (9) through (11), Section 8.2, let

$$K_z^v = \sqrt{K_v^2 - k^2 \sin^2 \theta_i} \tag{13}$$

Low-Frequency Limit

which, in view of (7), satisfies the equation

$$\frac{1}{k^2} = \frac{1}{\frac{\epsilon}{\epsilon_{\mathit{efft}}} k^2 \sin^2 \theta_i + \frac{\epsilon}{\epsilon_{\mathit{efft}}}(K_z^v)^2} \tag{14}$$

Hence, K_z^v is first calculated via (14), then K_v solved through (13), and then θ_t through (11), Section 8.2.

Eigen-Solution (B)

The dispersion relation corresponds to that of horizontal polarization and is $K = K_h$ with

$$\frac{K_h^2}{k^2} = \frac{\epsilon_{\mathit{efft}}}{\epsilon} \tag{15}$$

with associated eigenvector

$$\bar{a}_E^{(N)} = c_h \begin{bmatrix} e^{i\phi_i} \\ 0 \\ e^{-i\phi_i} \end{bmatrix} \tag{16}$$

where c_h is an arbitrary constant to be determined by the E O theorem.

Thus the dispersion relations correspond to an effective uniaxial medium with the effective permittivity dyad $\bar{\bar{\epsilon}}_{\mathit{eff}}$ given by

$$\bar{\bar{\epsilon}}_{\mathit{eff}} = \begin{bmatrix} \epsilon_{\mathit{efft}} & 0 & 0 \\ 0 & \epsilon_{\mathit{efft}} & 0 \\ 0 & 0 & \epsilon_{\mathit{effz}} \end{bmatrix} \tag{17}$$

In using the pair-distribution function $g(r)$, we shall let b be the diameter of the smallest circumscribing sphere enclosing the spheroid. To solve the constants c_v and c_h, we need to solve the Ewald-Oseen extinction theorem [(17) and (18), Section 8.2]. Consider separately the cases of vertically and horizontally polarized incidence.

A. Vertically Polarized Incidence

Using the eigenvectors (8) and (16) in (17) and (18), Section 8.2, gives $c_h = 0$ and

$$c_v = \sqrt{\frac{2}{3\pi}} \frac{(K_z^v - k_{iz})k_{iz}k}{in_o \left[-T_{a1}r_B\sqrt{2}\cos\theta_i + T_{ao}\sin\theta_i\right]} \tag{18}$$

Next we calculate the coherent reflected field by using (2), Section 8.3, and find that

$$<\overline{E}_s(\overline{r})> = \hat{\theta}_i\, e^{i\overline{k}_i\cdot\overline{r}} \frac{(K_z^v - k_{iz})(T_{a1}r_B\cos\theta_i + \sin\theta_i T_{ao}/\sqrt{2})}{(K_z^v + k_{iz})(T_{a1}r_B\cos\theta_i - \sin\theta_i T_{ao}/\sqrt{2})} \quad (19)$$

Using (9) in (19), it follows, after some algebra, that

$$<\overline{E}_s(\overline{r})> = \hat{\theta}_i\, e^{i\overline{k}_i\cdot\overline{r}} \frac{\epsilon_{efft} k_{iz} - \epsilon K_z^v}{\epsilon_{efft} k_{iz} + \epsilon K_z^v} \quad (20)$$

which is the expected result for the vertically polarized reflected wave for a uniaxial medium with vertical optic axis (Kong, 1975).

B. Horizontally Polarized Incidence

In this case $c_v = 0$, and

$$c_h = -\frac{(K_z^h - k_{iz})k_{iz}k}{\sqrt{3}\pi n_o T_{a1}} \quad (21)$$

where

$$K_z^h = \sqrt{K_h^2 - k^2\sin^2\theta_i} \quad (22)$$

$$<\overline{E}_s(\overline{r})> = \hat{\phi}_i\, e^{i\overline{k}_i\cdot\overline{r}} \frac{k_{iz} - K_z^h}{k_{iz} + K_z^h} \quad (23)$$

which is also the expected result for the horizontally polarized reflected wave. We can calculate the backscattering coefficients by applying the distorted Born approximation in a manner similar to that performed in Section 7 (Problem 32). In the backscattering direction $\hat{k}_s = -\hat{k}_{id}$, depolarization is exhibited as a first-order effect.

In Figure 6.34, we plot the effective permittivities ϵ_{efft} and ϵ_{effz} for oblate spheroids as a function of fractional volume $f = n_o 4\pi a^2 c/3$ occupied by the particles. The orientation probability $p(\beta)$ is equal to $2/[1 - \cos(\Delta\beta)]$ for $0 \le \beta \le \Delta\beta$ and equal to zero otherwise. The quantity $\Delta\beta$ is chosen to be 30 deg so that the axes of symmetry of the oblate spheroids are more inclined toward the vertically direction. We note that horizontally polarized waves suffer more scattering than vertically polarized waves because of the ability of the tangential electric field to penetrate into the dielectric particles. Thus, we see from the figure that both the real and the imaginary parts of ϵ_{efft} are, respectively, larger than the real and imaginary parts of ϵ_{effz}.

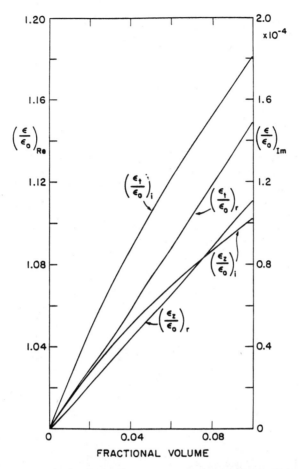

Fig. 6.34 Effective permittivities $\epsilon_{eff t}$ and $\epsilon_{eff z}$ in the low-frequency limit as a function of fractional volume occupied by dielectric oblate spheroids with $\epsilon_s = 3.2(1+i0.001)\epsilon_0$, $ka = 0.1847$, $a/c = 2$, and $\Delta\beta = 30\,\text{deg}$. The fractional volume is $n_0 4\pi a^2 c/3$.

8.6 Numerical Illustrations

For higher frequencies, the effective propagation constants are calculated by solving numerically the Lorentz-Lorenz law of (20), Section 8.2, and the amplitudes of the two characteristic waves are then calculated from the two equations (17) and (18), Section 8.2, of the Ewald-Oseen extinction theorem. The coherent reflected wave is then solved by using (2), Section 8.3. In solving the roots of the Lorentz-Lorenz law of (20), Section 8.2, the solution for low concentrations in Section 8.4 and the solutions for low frequencies in Section 8.5 provide good initial guesses. The Percus-Yevick pair function

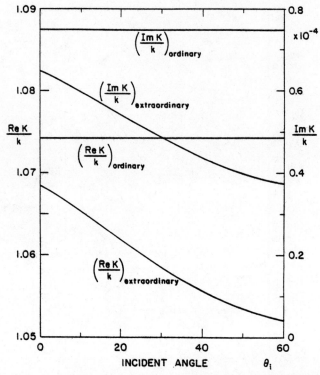

Fig. 6.35 Effective propagation constants for ordinary and extraordinary waves as a function of incidence angle θ_i for $ka = 0.1, \epsilon_s = 3.24(1 + i0.001)\epsilon_o$, $a/c = 2$, $\phi_i = 0$, $f = 0.1$, and Eulerian angles $\beta = \pi/6$ and $\gamma = 5\pi/6$.

is used in computing the results.

We shall consider numerical results for aligned oblate spheroids using parameters of ice particles in snow. The orientations of the oblate spheroids are described by Eulerian angles β and γ with respect to the principal frame. The axis of symmetry $\hat{z}_b$ of the oblate spheroids is in the direction $(-\cos\gamma \sin\beta \hat{x} + \sin\gamma \sin\beta \hat{y} + \cos\beta \hat{z})$. In Figures 6.35, 6.36, and 6.37, we plot, respectively, the effective propagation constants, coherent reflectivity, and backscattering coefficients for $ka = 0.1$, $\beta = \pi/6$, and $\gamma = 5\pi/6$ as a function of incidence angle.

We note from Figure 6.35 that the effective propagation constant of the ordinary wave does not vary with angle because the dispersion relation of the ordinary wave at low frequency is a sphere. Ordinary waves suffer more scattering than extraordinary waves so that the effective wavenumbers are larger than that of the extraordinary wave. From Figure 6.36, we note that the coherent reflectivities of incident horizontally and vertically polarized waves exhibit results resembling that of Fresnel reflection coefficients. Because

Numerical Illustrations

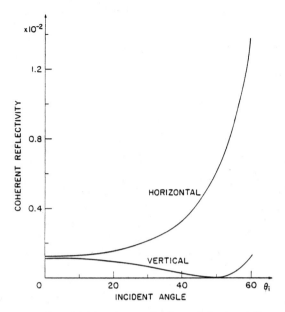

Fig. 6.36 Coherent reflectivities for vertically and horizontally polarized incident waves as a function of incidence angle θ_i. Other parameters are as in Fig. 6.35.

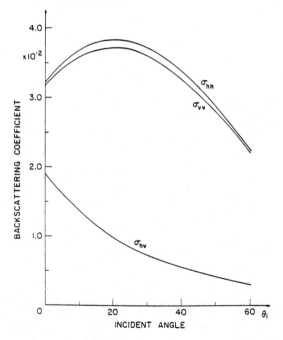

Fig. 6.37 Backscattering coefficients as a function of incident angle θ_i. Other parameters are as in Figure 6.35.

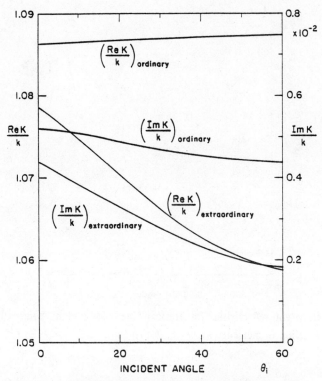

Fig. 6.38 Effective propagation constants for ordinary and extraordinary waves as a function of incidence angle θ_i for $ka = 1.0$, $\epsilon_s = 3.24(1 + i0.001)\epsilon_o$, $a/c = 2$, $\phi_i = 0$, $f = 0.1$, $\beta = \pi/6$, and $\gamma = 5\pi/6$.

of the absence of azimuthal symmetry, the reflectivities of vertically and horizontally polarized waves are not equal at $\theta_i = 0$. From Figure 6.37, we note that backscattering coefficients contain significant depolarized return which is quite common in the microwave remote sensing of geophysical media.

In Figures 6.38, 6.39, and 6.40 we plot, respectively, the effective propagation constants, the coherent reflectivity, and the backscattering coefficients for the case $ka = 1$. We note that the effective wave numbers of the ordinary wave also have angular variations. When particle size is comparable to wavelength, the dispersion relation for the ordinary wave is no longer a sphere. Due to the orientation of the aligned oblate spheroids, the geometric cross section intercepting the incident wave decreases with the incident angle. Hence, $Im(K/k)$ decreases with the incidence angle. We note from Figure 6.40 that the backscattering coefficients are much larger than those of Figure 6.37 because of the increase of the size parameter ka.

The frequency dependence of coherent reflectivity is shown in Figure

Numerical Illustrations

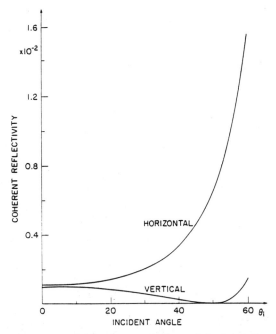

Fig. 6.39 Coherent reflectivities for vertically and horizontally polarized incident waves as a function of incidence angle θ_i. Other parameters are as Figure 6.38.

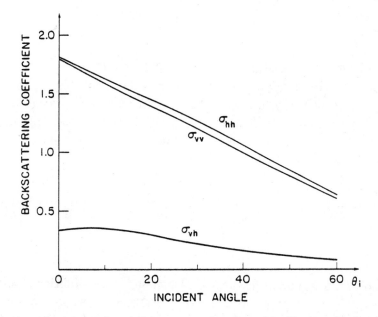

Fig. 6.40 Backscattering coefficients as a function of incident angle θ_i. Other parameters are as in Figure 6.38.

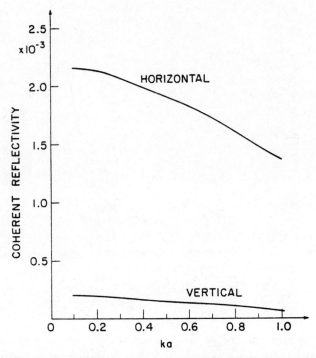

Fig. 6.41 Coherent reflectivities for vertically and horizontally polarized incident waves as a function of ka for $\epsilon_s = 3.24(1 + i0.001)\epsilon_o$, $a/c = 2$, $\theta_i = 30\,\text{deg}$, $\phi_i = 0\,\text{deg}$, $f = 0.1$, $\beta = 90\,\text{deg}$, and $\gamma = 0$.

6.41. The oblate spheroids have their axes of symmetry pointing horizontally in the $\hat{x}$ direction. We note that if the particle permittivity ϵ_s does not change with frequency, then the Fresnel reflectivity is also invariant with frequency. However, the coherent reflectivity does vary with frequency in Figure 6.41. This implies that the Fresnel reflectivity formulas are only valid in the low-frequency limit.

9 DISPERSION RELATIONS BASED ON COHERENT POTENTIAL

9.1 Introduction

Multiple scattering problems are also of interest because of their application to the understanding of alloys, liquid metals, amorphous semiconductors, etc. In the latter application, the introduction of the coherent potential gives better results. In this section, the method of coherent potential (Section 4.4) is applied to treat electromagnetic scattering by discrete scatterers. In the

T-matrix formalism, the propagator of the background medium is used in formulating the multiple scattering equations. However, the coherent potential approach leads to the use of the propagator of the effective medium. The coherent potential dispersion relations are cast in momentum representation through the solutions of integral equations in Section 4.4. The low-frequency solutions of the propagation constants for EFA-CP and QCA-CP were given in Section 4.5.

In general, a straightforward iteration of the integral equations, corresponding to the Born series, converges only for weak scatterers; this is not the case for geological materials. For cases of strong dielectric contrast, other methods of solving the integral equations are required. Functional properties of the unknown variables are used to convert the integral equations to algebraic equations which are valid for arbitrary frequencies (Tsang and Kong, 1980a). Expressions for the effective propagation constants K for spherical scatterers under both the EFA-CP and QCA-CP in the low-frequency limit were obtained in Section 4.5 and will be illustrated in Section 9.3 by using the Percus-Yevick pair-distribution function.

9.2 Dispersion Relation for Dielectric Spheres

The dispersion equation as derived in Section 4.4 for coherent potential is

$$\det\left[\overline{\overline{G}}_o^{-1}(\bar{p}) - n_o \overline{\overline{\hat{C}}}_p(\bar{p},\bar{p})\right] = 0 \tag{1}$$

where

$$\overline{\overline{\hat{C}}}_p(\bar{p}_1,\bar{p}_2) = \overline{\overline{U}}_p(\bar{p}_1,\bar{p}_2)$$

$$+ \int \frac{d\bar{p}_3}{(2\pi)^3} \overline{\overline{U}}_p(\bar{p}_1,\bar{p}_3)\overline{\overline{G}}_c(\bar{p}_3) q_p(\bar{p}_3 - \bar{p}_2)\overline{\overline{\hat{C}}}_p(\bar{p}_3,\bar{p}_2) \tag{2}$$

and the coherent potential propagator $\overline{\overline{G}}_c(\bar{p})$ is

$$\overline{\overline{G}}_c(\bar{p}) = \left[\overline{\overline{G}}_o^{-1}(\bar{p}) - n_o \overline{\overline{\hat{C}}}_p(\bar{p},\bar{p})\right]^{-1} \tag{3}$$

and $q_p(\bar{p})$ depends on the pair-correlation function of particle position $g(\bar{r})$. From (10), Section 4.3,

$$q_p(\bar{p}) = 1 + n_o \int d\bar{r}\, e^{-i\bar{p}\cdot\bar{r}}[g(\bar{r}) - 1] \tag{4}$$

For effective field approximation–coherent potential (EFA-CP), $g(\bar{r}) = 1$. For quasicrystalline approximation–coherent potential (QCA-CP), a more accurate pair-distribution, e.g., the Percus-Yevick result, will be used.

We consider the problem of scattering by dielectric spheres of radius a. To solve (2), we first expand the scattering potential $\overline{\overline{U}}_p(\bar{p}_1, \bar{p}_2)$ in terms of spherical harmonics vectors

$$\overline{\overline{U}}_p(\bar{p}_1, \bar{p}_2) = \sum_{n,m,\alpha,\beta} U_{mn}^{\alpha\beta} \overline{V}_{mn}^{\alpha}(\Omega_{p_1}) \overline{V}_{-mn}^{\beta}(\Omega_{p_2}) \tag{5}$$

where $\overline{V}_{mn}^{\alpha}(\Omega_p)$ with $\alpha = 1, 2$, and 3 are the vector spherical harmonics [(9) through (11), Section 5.1, Chapter 3] with angular variables θ_p and ϕ_p.

$$U_{mn}^{11}(p_1, p_2) = \frac{u_{mn} a^2}{p_1^2 - p_2^2} \left[p_1 j_n(p_1 a) j_n'(p_2 a) - p_2 j_n(p_2 a) j_n'(p_1 a) \right] \tag{6}$$

$$U_{mn}^{12}(p_1, p_2) = u_{mn} \frac{a j_n(p_1 a) j_n(p_2 a)}{p_1 p_2} \tag{7}$$

$$U_{mn}^{21}(p_1, p_2) = U_{mn}^{12}(p_1, p_2) \tag{8}$$

$$U_{mn}^{22}(p_1, p_2) = \frac{u_{mn}}{n(n+1)} \frac{1}{p_1 p_2} \left\{ a j_n(p_1 a) j_n(p_2 a) \right.$$

$$\left. + \frac{p_1 p_2 a^2}{p_1^2 - p_2^2} \left[p_1 j_n(p_1 a) j_n'(p_2 a) - p_2 j_n(p_2 a) j_n'(p_1 a) \right] \right\} \tag{9}$$

$$U_{mn}^{13}(p_1, p_2) = U_{mn}^{23}(p_1, p_2) = U_{mn}^{31}(p_1, p_2) = U_{mn}^{32}(p_1, p_2) = 0 \tag{10}$$

$$U_{mn}^{33}(p_1, p_2) = \frac{u_{mn}}{n(n+1)} \frac{a^2}{p_1^2 - p_2^2}$$

$$\times \left[p_2 j_n(p_1 a) j_n'(p_2 a) - p_1 j_n(p_2 a) j_n'(p_1 a) \right] \tag{11}$$

with

$$u_{mn} = 4\pi(-1)^m z(2n+1) \tag{12}$$

$$z = k_s^2 - k^2 \tag{13}$$

To solve (2), let the unknown $\overline{\overline{C}}_p$ assume the following form

$$\overline{\overline{C}}_p(\bar{p}_1, \bar{p}_2) = \sum_{n,n',m,\alpha,\beta} W_{mnn'}^{\alpha\beta}(p_1, p_2) \overline{V}_{mn}^{\alpha}(\Omega_{p_1}) \overline{V}_{-mn'}^{\beta}(\Omega_{p_2}) \tag{14}$$

where $|m| < \min(n, n')$.

Also expand $q_p(\bar{p}_3 - \bar{p}_2)$ in (4) in terms of spherical harmonics

$$q_p(\bar{p}_3 - \bar{p}_2) = \sum_{n,m} (-1)^m q_n(p_3, p_2) Y_n^m(\Omega_{p_3}) Y_n^{-m}(\Omega_{p_2}) \quad (15)$$

with

$$q_n(p_1, p_2) = \delta_{no} + g_n^{(1)}(p_1, p_2) + g_n^{(2)}(p_1, p_2) \quad (16)$$

where δ_{no} is the Kronecker delta, and

$$g_n^{(1)}(p_1, p_2) = -\frac{4\pi n_o (2n+1) b^2}{p_1^2 - p_2^2}$$

$$\times \left[p_2 j_n(p_1 b) j_n'(p_2 b) - p_1 j_n(p_2 b) j_n'(p_1 b) \right] \quad (17)$$

$$g_n^{(2)}(p_1, p_2) = n_o \int_b^\infty dr \, r^2 j_n(p_1 r) j_n(p_2 r) S_n(r) \quad (18)$$

$$S_n(r) = \sigma_1 4\pi (2n+1)[g(\bar{r}) - 1] \quad (19)$$

In (19), when $\sigma_1 = 0$, the results correspond to EFA-CP. For $\sigma_1 = 1$, we have the results corresponding to QCA-CP.

A set of equations for the unknowns $W_{mnn'}^{\alpha\beta}(p_1, p_2)$ is obtained by substituting (5) through (16) into (2), making use of the orthogonal relations of vector spherical harmonics, and using the addition theorems expressing the product of spherical harmonics and vector spherical harmonics in terms of vector spherical harmonics [(22), Section 7.2; Tsang and Kong, 1980a]. Solutions of these equations for arbitrary frequencies have not been obtained. The low-frequency results for EFA-CP and QCA-CP are given respectively in (23) and (24) and (27) and (28), Section 4.5. By using the results for the Percus-Yevick pair function of (7), Section 5.3, the result of QCA-CP becomes

$$K^2 = k^2 + \frac{f(k_s^2 - k^2)}{1 + \frac{k_s^2 - k^2}{3K^2}(1-f)}$$

$$\times \left\{ 1 + i \frac{2(k_s^2 - k^2) K a^3 (1-f)^4}{9\left[1 + \frac{k_s^2 - k^2}{3K^2}(1-f)\right](1+2f)^2} \right\} \quad (20)$$

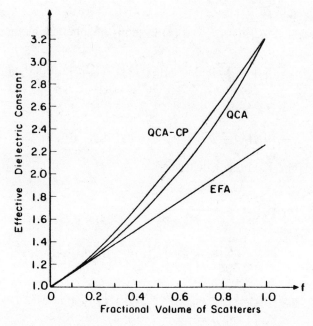

Fig. 6.42 Effective dielectric constant $(K^2/k^2)_r$ as a function of fractional volume of scatterers for $\epsilon = \epsilon_o$, $\epsilon_s = 3.2\epsilon_o$, $a = 0.1$ cm, and frequency = 5 GHz. The results of EFA, QCA, and QCA-CP are compared.

9.3 Numerical Illustrations

In Figure 6.42, we plot the effective dielectric constant $\epsilon_{eff} = (K^2/k^2)_r$, with r standing for real part, as a function of fractional volume of scatterers for fixed scatterer size. The size of the scatterer is small compared with wavelength. The frequency of excitation and the medium parameters chosen in Figures 6.42 through 6.44 represents typical values encountered in microwave remote sensing of snow. In Figure 6.42, the results are shown for f from 0 to 1. The physical limit of $f = 1$ is actually not attainable for identical discrete spherical scatterers of the same sizes. Nevertheless, it is interesting to note that both QCA and QCA-CP give $\epsilon_{eff} = \epsilon_s/\epsilon_o = 3.2$ in the limit $f = 1$. Because K_r differs appreciably from k, the numerical results of QCA-CP are different from that of QCA.

In Figure 6.43, we plot the attenuation rate $2K_i$ as a function of fractional volume of scatterers f. We note that the result for EFA increases rapidly with f. The results of QCA and QCA-CP predict a saturation trend as f increases. In Figure 6.44, we plot $(2K_i)_{ind}/2K_i$, the ratios of attenuation rate due to independent scattering to those due to EFA, QCA, and QCA-CP. With f in the range of 0.1 to 0.4, the ratio of QCA-CP is

Numerical Illustrations

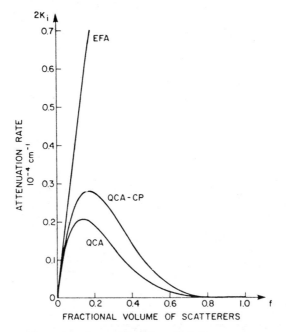

Fig. 6.43 Attenuation rate $2K_i$ as a function of f for $\epsilon = \epsilon_o$, $\epsilon_s = 3.2\epsilon_o$, $a = 0.1$ cm, and frequency = 5 GHz. The results of EFA, QCA, and QCA-CP are compared.

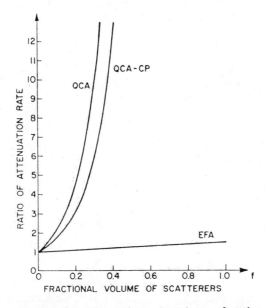

Fig. 6.44 Ratios of attenuation rate $(2K_i)_{ind}/(2K_i)$ as a function of f for $\epsilon = \epsilon_o$, $\epsilon_s = 3.2\epsilon_o$, $a = 0.1$ cm, and frequency = 5 GHz. The results of EFA, QCA, and QCA-CP are compared.

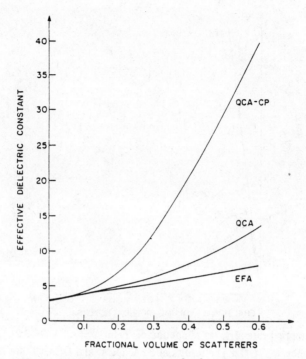

Fig. 6.45 Effective dielectric constant $(K^2/k^2)_r$ as a function of f for $\epsilon = (3 + i0.01)\epsilon_o$, $\epsilon_s = (79 + i6)\epsilon_o$, and frequency = 1.5 GHz. The results of EFA, QCA, and QCA-CP are compared.

between 2 to 12. In Figure 6.45, the effective dielectric constant is plotted as a function of the fractional volume of scatterers using typical values in soil moisture studies.

10 MULTIPLE SCATTERING OF SECOND MOMENT

10.1 Introduction

In Sections 3.4 and 3.5, Chapter 5, multiple scattering equations for the second moment of the field are derived by including all the ladder terms and the cyclical terms of Bethe-Salpeter equation. These are examined by using the model of uncorrelated scatterers which is equivalent to setting the pair distribution function $g(\bar{r}) = 1$ everywhere. Summation of the ladder terms leads to the Schwarzschild-Milne integral equation that is equivalent to the radiative transfer equations. Summation of the cyclical terms leads to a cyclical transfer equation. Both radiative transfer and cyclical transfer equations include multiple scattering effects of the second moment.

Introduction

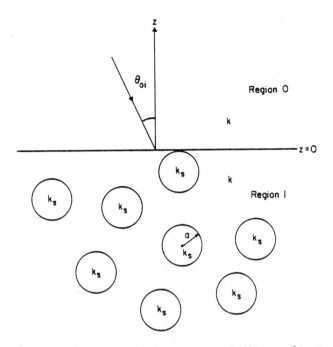

Fig. 6.46 Plane scalar wave obliquely incident on a half-space of scatterers.

For the case of correlated scatterer position, the Percus-Yevick and Verlet-Weis pair functions were used to represent the pair function. In Section 7.4 and 8.3, the second moment of the field is calculated by using the distorted Born approximation which corresponds to single scattering of the coherent field. Thus it is desirable to have multiple scattering equations for the second moment of the field that take into account correlations of particle positions. In Section 4.6, it was shown that a modified ladder approximation for the second moment can be used with the QCA-CP approximation for the first moment that will obey energy conservation. These two approximations will be used in Section 10.3 to derive the radiative transfer equations for dense non-tenuous particles. Because of the possible significant difference between the effective propagation constant calculated by QCA and QCA-CP, it will be shown that the use of QCA will lead to radiative transfer equations that violate energy conservation appreciably for dense non-tenuous particles. By extending the diagrammatic technique, the cyclical equation is also derived for dense medium.

The scalar wave equation will be used in this section for the sake of simplicity. Consider a plane wave incident onto a half-space of densely distributed particles with wavenumber k_s (Figure 6.46). The background medium in region 1 and the medium in region 0 have the same wavenumber

k. The difference between k_s and k can be significant. The development of the multiple scattering equations in this section will be parallel to that of Sections 3.4 through 3.6, Chapter 5, with the following major differences in approach and results: (i) The QCA-CP and modified ladder approximations will be used to include the effect of pair correlation functions of particle positions. (ii) The QCA-CP approximation for the first moment produces an effective propagation constant K with a real part that can be significantly different from k. (iii) The results of the first and second moment indicate that the propagation directions in region 0 and 1 are related by Snell's law. (iv) The contrast between K and k produces a reflecting interface at $z = 0$ and the result of the mean Green's function contains a reflection term. (v) Nonzero correlation exists between the upward and downward going waves that are specularly related. Thus, the field covariance is no longer the Fourier transform of the specific intensity. It is modified in a manner similar to that of Section 5, Chapter 5. (vi) The dense media radiative transfer equations derived from the QCA-CP and the modified ladder approximation assume a form that is identical to the classical transport equations. However, the extinction coefficient, scattering coefficient, albedo, and phase functions have been modified to include the effects of dense media.

10.2. First Moment for Half-Space Medium

Consider a plane wave incident in the direction θ_{oi} on a half-space of densely distributed discrete scatterers of radius a (Figure 6.46). The scalar wave equation will be used. The background medium in region 1 and the medium in region 0 has the same wavenumber k. The wavenumber for the scatterers is k_s. Low frequency assumption will be used so that $ka \ll 1$ and $k_s a \ll 1$. The wavenumber k_s can be significantly different from k as in the case of geophysical media. The fractional volume $f = 4\pi n_o a^3/3$ can be appreciable. The QCA-CP will be used in this section to solve for the mean Green's function and the mean field. In the next section, the first moment solution will be used with the modified ladder approximation for the second moment to derive the radiative transfer equations for the half-space medium.

The scalar wave equation is

$$-(\nabla^2 + k^2)\psi(\bar{r}) = \sum_{j=1}^{N} U_j(\bar{r})\psi(\bar{r}) \qquad (1)$$

where $U_j(\bar{r}) = U(\bar{r} - \bar{r}_j)$ and

$$U(\bar{r}) = \begin{cases} k_s^2 - k^2 & r \leq a \\ 0 & \text{otherwise} \end{cases} \quad (2)$$

The Green's function obeys the equation

$$-(\nabla^2 + k^2)G(\bar{r},\bar{r}') = \delta(\bar{r} - \bar{r}') + \sum_{j=1}^{N} U_j(\bar{r}) G(\bar{r},\bar{r}') \quad (3)$$

The operator form of the QCA-CP equations are, from (30) through (33), Section 4.4,

$$\hat{t} = U + UG_c \hat{t} \quad (4)$$

$$\hat{C}_j = \hat{t}_j + \hat{t}_j n_o \int_{z_l < 0} d\bar{r}_l \, h(\bar{r}_l - \bar{r}_j) G_c \hat{C}_l \quad (5)$$

$$E(G) \equiv G_c = G_o + n_o G_o \int_{z_j < 0} d\bar{r}_j \, \hat{C}_j G_c \quad (6)$$

The integration domain of (5) and (6) are confined to the lower half-space. To solve (4) through (6), we assume a trial solution of the form

$$G_c(\bar{r},\bar{r}') = \frac{e^{iK|\bar{r}-\bar{r}'|}}{4\pi|\bar{r}-\bar{r}'|}$$

$$+ \frac{i}{8\pi^2} \int_{-\infty}^{\infty} d\bar{k}_\perp \frac{e^{i\bar{k}_\perp \cdot (\bar{r}_\perp - \bar{r}'_\perp)}}{K_z} R(\bar{k}_\perp) \, e^{-iK_z(z+z')} \quad (7)$$

for $\bar{r}$ and $\bar{r}'$ in region 1. In (7), K is the effective propagation constant, $K_z = (K^2 - k_\perp^2)^{1/2}$, and $R(\bar{k}_\perp)$ is the reflection coefficient for the transverse direction $\bar{k}_\perp$. Both K and $R(\bar{k}_\perp)$ are to be determined by substituting (7) into (4) through (6). For dense non-tenuous discrete scatterers, K can be substantially different from k. The contrast between K and k creates a reflecting boundary at the interface $z = 0$. The reflection has been taken into account by including the second term in (7).

The quantity $\hat{t}$ is short ranged since it is only nonzero in the vicinity of the particle. If $\bar{r}$ is at least several diameters deep inside the half-space medium, thus neglecting boundary layer effect as in Section 6, the reflection term of the Green's function G_c plays a negligible role. Thus the solution of (4) proceeds in the same manner as Section 4.5. Note that G_c is a scalar

Green's function and does not have the Dirac delta function singularity as the dyadic Green's function $\overline{\overline{G}}_c$ in Section 4.5. The solution of (4) is

$$\hat{t}(\overline{r},\overline{r}') = \hat{t}_m \, v_o \, \delta(\overline{r}) \, \delta(\overline{r}') \tag{8}$$

where

$$\hat{t}(\overline{r},\overline{r}') \equiv <\overline{r}|\hat{t}|\overline{r}'> \tag{9}$$

$$\hat{t}_m = (k_s^2 - k^2)\left[1 + i\frac{(k_s^2 - k^2)}{3}Ka^3\right] \tag{10}$$

Since h in equation (5) is also short-ranged, the reflection term of G_c can again be neglected in (5) and the solution can proceed in a manner similar to that of Section 4.5. Hence,

$$\hat{C}_j(\overline{r},\overline{r}') = \hat{c}\,\delta(\overline{r} - \overline{r}_j)\,\delta(\overline{r} - \overline{r}_j) \tag{11}$$

where

$$\hat{c} = v_o \hat{t}_m \left[1 + if\,\hat{t}_m K \int_0^\infty dr\, r^2\, h(r)\right] \tag{12}$$

Combining (9) and (12), it follows that

$$\hat{c} = v_o(k_s^2 - k^2)\left\{1 + i\frac{(k_s^2 - k^2)Ka^3}{3}\left[1 + 4\pi n_o \int_0^\infty dr\, r^2 h(r)\right]\right\} \tag{13}$$

The mass operator $n_o \int d\overline{r}_j\, \hat{C}_j$ is not short ranged. Hence the reflection term in G_c plays a role in (6). Substitute (7) and (11) in (6) and make use of the plane wave representation of $G_o(\overline{r},\overline{r}') = \exp(ik|\overline{r} - \overline{r}'|)/4\pi|\overline{r} - \overline{r}'|$. Without loss of generality, let $z > z'$ in (6) when taking the space representation. After some algebra, we get

$$\frac{1}{K_z}\left\{e^{iK_z(z-z')} + R(\overline{k}_\perp)\,e^{-iK_z(z+z')}\right\}$$

$$= \frac{1}{k_z}e^{ik_z(z-z')} - \frac{n_o\hat{c}}{2k_zK_z}\left[\frac{2}{K_z^2 - k_z^2}\left(K_z e^{ik_z(z-z')} - k_z e^{iK_z(z-z')}\right)\right.$$

$$\left. - \frac{1}{K_z + k_z}e^{-ik_z z - iK_z z'}\right]$$

$$- \frac{n_o \hat{c}\, R(\overline{k}_\perp)}{2k_z K_z}\left[\frac{e^{-ik_z z - iK_z z'}}{K_z - k_z} - \frac{2k_z e^{-iK_z(z+z')}}{K_z^2 - k_z^2}\right] \tag{14}$$

where $k_z = [k^2 - k_\perp^2]^{1/2}$. There are four kinds of terms in (14) with dependence $\exp[iK_z(z - z')]$, $\exp[ik_z(z - z')]$, $\exp[-iK_z(z - z')]$ and

$\exp(-ik_z z - iK_z z')$. They should be balanced as in the Ewald-Oseen extinction theorem and the Lorentz-Lorenz law. Balancing the first three dependences result in the same equation

$$1 = \frac{n_o \hat{c}}{K_z^2 - k_z^2} \tag{15}$$

Balancing the dependence $\exp(-ik_z z - iK_z z')$ gives

$$R(\bar{k}_\perp) = \frac{K_z - k_z}{K_z + k_z} \tag{16}$$

which is recognized to be the Fresnel reflection coefficient. Since $K_z^2 - k_z^2 = K^2 - k^2$, using (13) in (15) gives

$$K^2 = k^2 + f(k_s^2 - k^2)\left\{1 + i(k_s^2 - k^2)\frac{Ka^3}{3}\left[1 + 4\pi n_o \int_0^\infty dr\, r^2 h(r)\right]\right\} \tag{17}$$

Next consider a plane wave incident onto the half-space medium (Figure 6.46). The incident wave is

$$\psi_{inc}(\bar{r}) = e^{ik_{ix}x + ik_{iy}y - ik_{iz}z} \tag{18}$$

with $k_{ix} = k\sin\theta_{oi}\cos\phi_i$, $k_{iy} = k\sin\theta_{oi}\sin\phi_i$ and $k_{iz} = k\cos\theta_{oi}$.

By using (6) and (11), the mean field in region 1 obeys the equation

$$E[\psi(\bar{r})] = \psi_{inc}(\bar{r}) + n_o \hat{c} \int_{z_j < 0} d\bar{r}_j\, G_o(\bar{r}, \bar{r}_j) E[\psi(\bar{r}_j)] \tag{19}$$

Manipulations can be performed as in the calculation of G_c giving the Ewald-Oseen theorem and the Lorentz-Lorenz law as well as the phase matching condition as in Section 7 and Problem 6.34. The final result is

$$E[\psi(\bar{r})] = T_i\, e^{ik_{ix}x + ik_{iy}y - iK_{iz}z} \tag{20}$$

where

$$K_{iz} = (K^2 - k_{ix}^2 - k_{iy}^2)^{1/2} \tag{21}$$

in accordance with phase matching and

$$T_i = \frac{2k_{iz}}{k_{iz} + K_{iz}} \tag{22}$$

is the transmission coefficient for the incident wave. Since $K'_{iz} \gg K''_{iz}$, the coherent transmitted wave is in the θ_i direction with θ_i related to θ_{oi}

by Snell's law $\theta_i = \sin^{-1}(k \sin \theta_{oi}/K')$, and the ϕ angle is the same for regions 0 and 1.

If QCA is used instead of QCA-CP, then the G_c in (4) and (5) are to be replaced by G_o. The effective propagation constant, mean Green's function, and mean field can be calculated by the same procedure. The result of mean Green's function and mean field assume the same form as in (7), (16), and (20) through (22) with the following different expression for K^2. Under QCA

$$K^2 = k^2 + f(k_s^2 - k^2)\left\{1 + i(k_s^2 - k^2)\frac{ka^3}{3}\left[1 + 4\pi n_o \int_0^\infty dr\, r^2 h(r)\right]\right\} \quad (23)$$

The second term inside the curly bracket contains the factor k instead of K in (17). The result is significantly different from that of QCA-CP when K' deviates appreciably from k which can be the case in dense non-tenuous particles.

The mean Green's function $G_c(\bar{r}, \bar{r}')$ with $\bar{r}$ in region 0 and $\bar{r}'$ in region 1 can also be calculated. Taking the coordinate representation of (6) and using (11) gives

$$G_c(\bar{r}_1, \bar{r}_2) = G_o(\bar{r}_1, \bar{r}_2) + n_o \hat{c} \int_{z_j < 0} d\bar{r}_j\, G_o(\bar{r}_1, \bar{r}_j)\, G_c(\bar{r}_j, \bar{r}_2) \quad (24)$$

Let $\bar{r}_1$ be in region 0 and $\bar{r}_2$ in region 1. Substituting (7) in the integrand of (24), performing the integration, and using the results of (15) and (16) (Problem 6.35) gives

$$G_c(\bar{r}_1, \bar{r}_2) = \frac{i}{8\pi^2}\int_{-\infty}^{\infty} d\bar{k}_\perp \frac{e^{i\bar{k}_\perp \cdot (\bar{r}_{1\perp} - \bar{r}_{2\perp})}}{K_z}\left[1 + R(\bar{k}_\perp)\right] e^{ik_z z_1 - iK_z z_2} \quad (25)$$

for $z_1 \geq 0$ and $z_2 \leq 0$.

The mean field in region 0 can also be calculated by setting $\bar{r}$ in region 0 in equation (19). The result is (Problem 6.35)

$$E[\psi(\bar{r})] = \psi_{inc}(\bar{r}) + R_{01i}\, e^{ik_{ix} x + ik_{iy} y + ik_{iz} z} \quad (26)$$

where

$$R_{01i} = \frac{k_{iz} - K_{iz}}{k_{iz} + K_{iz}} \quad (27)$$

10.3 Second Moment Radiative Transfer Equation and Cyclical Transfer Equation

The second moment obeys the equation

$$E[\psi(\bar{r})\psi^*(\bar{r}')] = E[\psi(\bar{r})]E[\psi^*(\bar{r}')]$$
$$+ \int d\bar{r}_1 \int d\bar{r}_2 \int d\bar{r}'_1 \int d\bar{r}'_2\, E[G(\bar{r},\bar{r}_1)]E[G^*(\bar{r}',\bar{r}'_1)]$$
$$\cdot I(\bar{r}_1,\bar{r}_2;\bar{r}'_1,\bar{r}'_2)\, E[\psi(\bar{r}_2)\psi^*(\bar{r}'_2)] \tag{1}$$

where I is the intensity operator. Using the modified ladder approximation of (16), Section 4.6, and the results of $\hat{C}_j$ in Section 10.2 gives the following result for the intensity operator for small scatterers.

$$I(\bar{r}_1,\bar{r}_2;\bar{r}'_1,\bar{r}'_2)$$
$$= n_o \int_{z_j<0} d\bar{r}_j\, |\hat{c}|^2\, \delta(\bar{r}'_1-\bar{r}_j)\,\delta(\bar{r}'_2-\bar{r}_j)\,\delta(\bar{r}_1-\bar{r}_j)\,\delta(\bar{r}_2-\bar{r}_j)$$
$$+ n_o^2 \int_{z_j<0} d\bar{r}_j \int_{z_l<0} d\bar{r}_l\, h(\bar{r}_l,\bar{r}_j)\,|\hat{c}|^2\,\delta(\bar{r}'_1-\bar{r}_j)\,\delta(\bar{r}'_2-\bar{r}_j)$$
$$\cdot \delta(\bar{r}_1-\bar{r}_l)\,\delta(\bar{r}_2-\bar{r}_l) \tag{2}$$

Next define the incoherent field as

$$\tilde{\psi}(\bar{r}) = \psi(\bar{r}) - <\psi(\bar{r})> \tag{3}$$

Using (2) and (3) in (1) gives the second moment equation as

$$<\tilde{\psi}(\bar{r}_1)\tilde{\psi}^*(\bar{r}_2)> = n_o|\hat{c}|^2 \int_{z_j<0} d\bar{r}_j\, E[G(\bar{r}_1,\bar{r}_j)]E[G^*(\bar{r}_2,\bar{r}_j)]E[|\psi(\bar{r}_j)|^2]$$
$$+ n_o^2|\hat{c}|^2 \int_{z_j<0} d\bar{r}_j \int_{z_l<0} d\bar{r}_l\, h(\bar{r}_l,\bar{r}_j)E[G(\bar{r}_1,\bar{r}_l)]$$
$$\cdot E[G^*(\bar{r}_2,\bar{r}_j)]E[\psi(\bar{r}_l)\psi^*(\bar{r}_j)] \tag{4}$$

Comparing with the case of sparse concentration of scatterers as given by (5), Section 3.6, Chapter 5, it can be seen that (4) contains an additional term that includes the correlation of particle positions. Since $h(\bar{r}_l,\bar{r}_j)$ is only nonzero for $|\bar{r}_l - \bar{r}_j| = O(b)$ where b is the diameter of the particle and because of the low frequency approximation of $ka \ll 1$ and $k_s a \ll 1$,

$\bar{r}_l$ can be set equal to $\bar{r}_j$ for the rest of the integrand in (4). Hence equation (4) simplifies to

$$<\tilde{\psi}(\bar{r}_1)\tilde{\psi}^*(\bar{r}_2)> = n_o|\hat{c}|^2\omega \int_{z_j<0} d\bar{r}_j\, E[G(\bar{r}_1,\bar{r}_j)]E[G^*(\bar{r}_2,\bar{r}_j)]E[|\psi(\bar{r}_j)|^2] \quad (5)$$

where

$$\omega = 1 + 4\pi n_o \int_0^\infty dr\, r^2\, h(r) \quad (6)$$

Equation (5) is similar in form to that of (5), Section 3.6, Chapter 5. However, equation (5) for dense non-tenuous medium contains a boundary reflection term that is absent in the case of sparse concentration of particles. The development of radiative transfer equation for half-space random medium including the effect of interface was done in Tsang and Kong (1979). A similar procedure will be followed here.

We next make a spectral decomposition of the incoherent field. Let $d\bar{k}_\perp = dk_x dk_y$ and

$$\tilde{\psi}(\bar{r}) = \int d\bar{k}_\perp\, e^{i\bar{k}_\perp \cdot \bar{r}_\perp} \left\{ \tilde{\psi}_u(z,\bar{k}_\perp) e^{iK'_z z} + \tilde{\psi}_d(z,\bar{k}_\perp) e^{-iK'_z z} \right\} \quad (7)$$

where $\bar{k}_\perp = k_x \hat{x} + k_y \hat{y}$, $K_z = (K^2 - k_x^2 - k_y^2)^{1/2}$ and $K'_z = Re\, K_z$. In (7) $\tilde{\psi}_u$ and $\tilde{\psi}_d$ are, respectively, upward-going and downward-going fields in the direction $\bar{k}_\perp$. Because of the reflection term in $<G>$, we seek solutions of the form,

$$<\tilde{\psi}_u(z_1,\bar{k}_{\perp 1})\tilde{\psi}_u^*(z_2,\bar{k}_{\perp 2})> = \delta(\bar{k}_{\perp 1} - \bar{k}_{\perp 2})J_u(z_1,z_2,\bar{k}_{\perp 1}) \quad (8)$$

$$<\tilde{\psi}_d(z_1,\bar{k}_{\perp 1})\tilde{\psi}_d^*(z_2,\bar{k}_{\perp 2})> = \delta(\bar{k}_{\perp 1} - \bar{k}_{\perp 2})J_d(z_1,z_2,\bar{k}_{\perp 1}) \quad (9)$$

$$<\tilde{\psi}_u(z_1,\bar{k}_{\perp 1})\tilde{\psi}_d^*(z_2,\bar{k}_{\perp 2})> = \delta(\bar{k}_{\perp 1} - \bar{k}_{\perp 2})J_c(z_1,z_2,\bar{k}_{\perp 1}) \quad (10)$$

The assumption in (8) through (10) is that incoherent fields with different transverse directions are uncorrelated. However, for the same transverse direction, there exists a correlation between the upward and the downward propagating waves because of reflection at the boundary $z = 0$. The correlation is represented by J_c in (10). Further let

$$I_\alpha(z,\bar{k}_\perp) = J_\alpha(z,z,\bar{k}_\perp) \qquad (\alpha = u,d,c) \quad (11)$$

represent intensities.

Second Moment Transfer Equations

In multiple scattering problems, generally $K' \gg K''$. Thus there are two distance scales: the extinction distance $l_e = 1/2K''$ and the wavelength $= 2\pi/K'$. The extinction distance is orders of magnitude larger than the wavelength. The quantities $\tilde{\psi}_u$ and $\tilde{\psi}_d$ in (7) represent envelop that vary on the extinction scale. Hence the quantities J_u, J_d, J_c, I_u, I_d and I_c are all slowly varying on the wavelength scale. For $|\bar{r}_1 - \bar{r}_2| \ll l_e$, the covariance of the incoherent field can be simplified to

$$<\tilde{\psi}(\bar{r}_1)\tilde{\psi}^*(\bar{r}_2)> = \int_{-\infty}^{\infty} d\bar{k}_\perp \, e^{i\bar{k}_\perp \cdot (\bar{r}_{1\perp} - \bar{r}_{2\perp})} \left\{ I_u\left(\frac{z_1 + z_2}{2}, \bar{k}_\perp\right) e^{iK'_z(z_1 - z_2)} \right.$$

$$+ I_d\left(\frac{z_1 + z_2}{2}, \bar{k}_\perp\right) e^{-iK'_z(z_1 - z_2)}$$

$$\left. + 2\,Re\left[I_c\left(\frac{z_1 + z_2}{2}, \bar{k}_\perp\right) e^{iK'_z(z_1 + z_2)}\right] \right\} \quad (12)$$

Substitute (12) into (5), use the mean Green's function and mean field given respectively by (7) and (20) of Section 10.2, and perform the integration over $d\bar{r}_{j\perp}$. The result is

$$I_u\left(\frac{z_1 + z_2}{2}, \bar{k}_\perp\right) e^{iK'_z(z_1 - z_2)} + I_d\left(\frac{z_1 + z_2}{2}, \bar{k}_\perp\right) e^{-iK'_z(z_1 - z_2)}$$

$$+ 2\,Re\left[I_c\left(\frac{z_1 + z_2}{2}, \bar{k}_\perp\right) e^{iK'_z(z_1 + z_2)}\right]$$

$$= \frac{n_0 \omega |\hat{c}|^2}{16\pi^2 |K_z|^2} \int_{-\infty}^{0} dz_j \left\{ e^{iK_z|z_1 - z_j|} + R(\bar{k}_\perp) e^{-iK_z(z_1 + z_j)} \right\}$$

$$\cdot \left\{ e^{-iK^*_z|z_2 - z_j|} + R^*(\bar{k}_\perp) e^{iK^*_z(z_2 + z_j)} \right\}$$

$$\cdot \left\{ |T_i|^2 e^{2K''_{iz} z_j} + \int_{-\infty}^{\infty} d\bar{\gamma}_\perp \left[I_u(z_j, \bar{\gamma}_\perp) + I_d(z_j, \bar{\gamma}_\perp) \right. \right.$$

$$\left. \left. + 2\,Re\left(I_c(z_j, \bar{\gamma}_\perp) e^{2iK'_{z\gamma} z_j}\right) \right] \right\} \quad (13)$$

where $K_{z\gamma} = (K^2 - \gamma_\perp^2)^{1/2}$.

Without loss of generality, let $z_1 > z_2$ in (13). In performing integration over dz_j on the right hand side of (13), consider a typical integral of the form

$$\int_{z_1}^{0} dz_j \, e^{iK_z(z_j - z_1)} e^{-iK^*_z(z_j - z_2)} = \frac{e^{-iK_z z_1 + iK^*_z z_2}}{2K''_z} \left[e^{-2K''_z z_1} - 1 \right] \quad (14)$$

The integrand in (13) is a product of the field and the complex conjugate of the field. When the phases of the two factors balance each other, then the integrand has little phase variation over the range of integration. Thus the result of the typical integral in (14) is of order $O(1/K'')$. Next consider the following typical integral

$$\int_{z_1}^{0} dz_j\, e^{iK_z(z_j - z_1)}\, e^{iK_z^*(z_2 + z_j)} = \frac{e^{-iK_z z_1 + iK_z^* z_2}}{2iK_z'}\left[1 - e^{2iK_z' z_1}\right] \quad (15)$$

The phases of the two factors in the integrand of (15) do not balance each other. The integrand has phase variation over the range of integration. The integral is of order $O(1/K')$ and is much smaller than that of (14). Physically this corresponds to the fact that scattering due to products of fields traveling with different phases will average out to a small number after several wavelengths. Thus terms of the type in (14) are called constructive interference and are much larger than that of destructive interference terms of the type in (15). Destructive interference terms on the right-hand side of (13) will be discarded. Thus there are four types of phase dependences in equation (13): $\exp[iK_z'(z_1 - z_2)]$, $\exp[-iK_z'(z_1 - z_2)]$, $\exp[iK_z'(z_1 + z_2)]$ and $\exp[-iK_z'(z_1 + z_2)]$. The terms of the same phase dependence are equated giving four equations. We then set $z_1 = z_2$ in the four equations. Further, let $\tau = 2K''z$ to denote the scale of optical thickness. These result in the following equations. The result of equating terms with dependence $\exp[iK_z'(z_1 - z_2)]$ and then setting $z_1 = z_2$ gives

$$I_u(\tau, \overline{k}_\perp) = \frac{n_o \omega |\hat{c}|^2}{16\pi^2 |K_z|^2} \left\{ \frac{|T_i|^2}{2K_z'' + 2K_{iz}''} e^{K_{iz}'' \tau / K''} \right.$$
$$\left. + \frac{1}{2K''}\int_{-\infty}^{\tau} d\tau_j\, e^{K_z''(\tau_j - \tau)/K''} J(\tau_j) \right\} \quad (16)$$

where

$$J(\tau_j) = \int_{-\infty}^{\infty} d\overline{\gamma}_\perp \left[I_u(\tau_j, \overline{\gamma}_\perp) + I_d(\tau_j, \overline{\gamma}_\perp)\right] \quad (17)$$

The result of equating terms with dependence $\exp[-iK_z'(z_1 - z_2)]$ gives

$$I_d(\tau, \overline{k}_\perp) = \frac{n_o \omega |\hat{c}|^2}{16\pi^2 |K_z|^2} \left\{ \frac{|T_i|^2}{2K_{iz}'' - 2K_z''}\left(e^{K_z'' \tau / K''} - e^{K_{iz}'' \tau / K''}\right) \right.$$
$$\left. + \frac{|R(\overline{k}_\perp)|^2 |T_i|^2}{2K_{iz}'' + 2K_z''} e^{K_z'' \tau / K''}\right.$$

Second Moment Transfer Equations

$$+ \frac{1}{2K''} \int_{-\infty}^{\tau} d\tau_j \, |R(\bar{k}_\perp)|^2 \, e^{K_z''(\tau+\tau_j)/K''} J(\tau_j)$$

$$+ \frac{1}{2K''} \int_{\tau}^{0} d\tau_j \left(e^{K_z''(\tau-\tau_j)/K''} \right.$$

$$\left. + |R(\bar{k}_\perp)|^2 \, e^{K_z''(\tau+\tau_j)/K''} \right) J(\tau_j) \right\} \quad (18)$$

The results of equating terms with dependence $\exp[iK'_z(z_1 + z_2)]$ and of equating terms with dependence $\exp[-iK'_z(z_1 + z_2)]$ and then setting $z_1 = z_2$ gives the same equation as follows

$$I_c(\tau, \bar{k}_\perp) = \frac{n_o \omega |\hat{c}|^2}{16\pi^2 |K_z|^2} \left\{ |T_i|^2 \, R^*(\bar{k}_\perp) \frac{e^{(K_z'' + K_{iz}'')\tau/K''}}{2K_z'' + 2K_{iz}''} \right.$$

$$\left. + \frac{R^*(\bar{k}_\perp)}{2K''} \int_{-\infty}^{\tau} d\tau_j \, e^{K_z'' \tau_j / K''} J(\tau_j) \right\} \quad (19)$$

Consider only radiating waves in the spectral representation, by transforming $\bar{k}_\perp$ to (θ, ϕ) direction. Let $k_x = K' \sin\theta \cos\phi$ and $k_y = K' \sin\theta \sin\phi$ so that

$$\int dk_\perp = \int_0^{\pi/2} d\theta \sin\theta \cos\theta \int_0^{2\pi} d\phi \, K'^2$$

The attenuation rates in the z direction can be expressed as follows

$$K_z'' = \frac{K''}{\mu} = \frac{\kappa_e}{2\mu} \quad (20)$$

$$K_{iz}'' = \frac{K''}{\mu_i} = \frac{\kappa_e}{2\mu_i} \quad (21)$$

where $\mu = \cos\theta$ and $\mu_i = \cos\theta_i$. The factor $|K_z|^2$ in the denominator of (16), (18), and (19) can be approximated by $K'^2 \mu^2$. Following Tsang and Kong (1979) and (19), Section 3.6, Chapter 5, define specific intensity I for incoherent wave in region 1 as

$$I_\alpha(\tau, \theta, \phi) = \frac{K'^3 \cos\theta}{k} I_\alpha(\tau, \bar{k}_\perp) \quad (\alpha = u, d, c) \quad (22)$$

and the albedo $\tilde{\omega}$ as

$$\tilde{\omega} = \frac{n_o \omega |\hat{c}|^2}{8\pi K''} \quad (23)$$

It then follows from (16) through (19) that

$$I_u(\tau,\theta,\phi) = \frac{\tilde{\omega}}{4\pi\mu}\left\{\frac{K'}{k}|T_i|^2\frac{e^{\tau/\mu_i}}{1/\mu+1/\mu_i} + \int_{-\infty}^{\tau}d\tau_j\,e^{(\tau_j-\tau)/\mu}J(\tau_j)\right\} \quad (24)$$

$$I_d(\tau,\theta,\phi) = \frac{\tilde{\omega}}{4\pi\mu}\left\{\frac{K'}{k}|T_i|^2\left[\frac{(e^{\tau/\mu}-e^{\tau/\mu_i})}{1/\mu_i-1/\mu} + |R(\theta)|^2\frac{e^{\tau/\mu}}{1/\mu+1/\mu_i}\right]\right.$$

$$+ \int_{-\infty}^{\tau}d\tau_j\,|R(\theta)|^2\,e^{(\tau+\tau_j)/\mu}J(\tau_j)$$

$$\left. + \int_{\tau}^{0}d\tau_j\left[e^{(\tau-\tau_j)/\mu} + |R(\theta)|^2\,e^{(\tau+\tau_j)/\mu}\right]J(\tau_j)\right\} \quad (25)$$

$$I_c(\tau,\theta,\phi) = e^{\tau/\mu}R^*(\theta)\,I_u(\tau,\theta,\phi) \quad (26)$$

where

$$J(\tau) = \int_0^{\pi/2}d\theta\sin\theta\int_0^{2\pi}d\phi\,[I_u(\tau,\theta,\phi)+I_d(\tau,\theta,\phi)] \quad (27)$$

The integral equation in (24) through (27) can be readily converted into differential equation (Problem 6.36)

$$\frac{dI_u(\tau,\theta,\phi)}{d\tau} = -\frac{I_u(\tau,\theta,\phi)}{\mu} + \frac{\tilde{\omega}}{4\pi\mu}\left\{|T_i|^2\,e^{\tau/\mu_i}\frac{K'}{k}+J(\tau)\right\} \quad (28a)$$

$$-\frac{dI_d(\tau,\theta,\phi)}{d\tau} = -\frac{I_d(\tau,\theta,\phi)}{\mu} + \frac{\tilde{\omega}}{4\pi\mu}\left\{|T_i|^2\,e^{\tau/\mu_i}\frac{K'}{k}+J(\tau)\right\} \quad (28b)$$

The boundary condition for the differential equation in (28) can be deduced by setting $\tau=0$ in the integral equations in (24) through (27). It is

$$I_d(\tau=0,\theta,\phi) = |R(\theta)|^2\,I_u(\tau=0,\theta,\phi) \quad (29)$$

By using (13), Section 10.2, in (23), it follows that the albedo of the scattering medium is

$$\tilde{\omega} = \frac{fa^3}{3}|k_s^2-k^2|^2\left\{1+4\pi n_o\int_0^{\infty}dr\,r^2\,h(r)\right\}\Big/2K'' \quad (30)$$

Second Moment Transfer Equations

The imaginary part of the effective propagation constant K'', under QCA-CP, is calculated by using (17), Section 10.2. It is easily verified from (30) and (17), Section 10.2, that

$$\tilde{\omega} = 1 \quad \text{(real } k_s \text{ and QCA-CP)} \tag{31}$$

From Section 10.2, the specific intensity for the coherent wave is in the downward direction and can be calculated. We shall denote it by $I_d^{(c)}(\tau, \theta, \phi)$. It is

$$I_d^{(c)}(\tau, \theta, \phi) = \frac{K'}{k}|T_i|^2 e^{\tau/\mu_i} \delta(\mu - \mu_i) \delta(\phi - \phi_i) \tag{32}$$

Energy conservation can be tested by calculating the divergence of the flux (Ishimaru 1978) and using (28) and (32). It is

$$\int_0^{\pi/2} d\theta \sin\theta \int_0^{2\pi} d\phi \left\{ \mu \frac{dI_u(\tau, \theta, \phi)}{d\tau} - \mu \frac{dI_d(\tau, \theta, \phi)}{d\tau} - \mu \frac{dI_d^{(c)}(\tau, \theta, \phi)}{d\tau} \right\}$$

$$= -(1 - \tilde{\omega})\left[J(\tau) + |T_i|^2 \frac{K'}{k} e^{\tau/\mu_i} \right] \tag{33}$$

The square bracket on the right hand side of (33) is equal to the integration of total specific intensity over solid angles. In the case when the particles are non-absorptive, k_s is real and, from (31), $\tilde{\omega} = 1$ based on QCA-CP. Hence, the right hand side of (33) vanishes and energy conservation is satisfied. It is interesting to note that if QCA is used instead of QCA-CP with K^2 as given by (23), Section 10.2, then

$$\tilde{\omega} = \frac{K'}{k} \quad \text{(real } k_s \text{ and QCA)} \tag{34}$$

Hence the albedo is not equal to 1 and is approximately equal to 1 for medium with tenuous particles or medium with sparse concentration of particles. The radiative transfer equations satisfy energy conservation approximately for these two restrictive cases and can violate energy conservation substantially for dense non-tenuous particles.

The incoherent field in region 0 can be calculated by setting $\bar{r}_1$ and $\bar{r}_2$ in region 0 in equation (5) (Problem 6.37). Evaluate the integral by using the half-space Green's function $E(G)$ as given by (25), Section 10.2, and the intensity in region 1 is given by (20), Section 10.2, and (12). Retain only constructive interference term and compare with the result of $I_u(0, \bar{k}_\perp)$ as given by (16). The result is, for $\bar{r}_1$ and $\bar{r}_2$ in region 0

$$\langle \tilde{\psi}(\bar{r}_1) \tilde{\psi}^*(\bar{r}_2) \rangle = \int_{-\infty}^{\infty} d\bar{k}_\perp \, e^{i\bar{k}_\perp \cdot (\bar{r}_{1\perp} - \bar{r}_{2\perp})} e^{ik_z z_1 - ik_z^* z_2} I_u^{(0)}(\bar{k}_\perp) \tag{35}$$

where

$$I_u^{(0)}(\overline{k}_\perp) = |1 + R(\overline{k}_\perp)|^2 \, I_u(\tau = 0, \overline{k}_\perp) \tag{36}$$

Following (22), specific intensity in region 0 is defined by

$$\mathcal{I}_u^{(0)}(\theta_o, \phi_o) = k^2 \cos\theta_o I_u^{(0)}(\overline{k}_\perp) \tag{37}$$

where (θ_o, ϕ_o) are the angular variables in region 0 corresponding to $\overline{k}_\perp$. Since $K_z' \gg K_z''$, we have, from (16), Section 10.2,

$$R(\overline{k}_\perp) = \frac{K' \cos\theta - k \cos\theta_o}{K' \cos\theta + k \cos\theta_o} \tag{38}$$

where (θ_o, ϕ_o) and (θ, ϕ) are related by Snell's law. Hence using (22), (37), and (38), we have

$$\mathcal{I}_u^{(0)}(\theta_o, \phi_o) = \frac{k^2}{K'^2} \left[1 - |R(\overline{k}_\perp)|^2\right] \mathcal{I}_u(\tau = 0, \theta, \phi) \tag{39}$$

It is interesting to note that (39) contains the divergence of beam factor as required for specific intensity when passing from one (effective) dielectric medium to another (Ishimaru 1978).

We note from (15), Section 4.6, that the modified ladder approximation is a generalization of the ladder approximation that includes scattering from particles that are correlated in positions. Thus, by extending the diagrammatic techniques of Section 3.4 and 3.5, Chapter 5, the cyclical equation can also be derived for dense media. Let

$$\otimes^j \;=\; \hat{C}_j \tag{40}$$

$$\begin{aligned}
\overset{\otimes^j}{\underset{\otimes_l}{\mid}} &= n_o \int d\overline{r}_j \int d\overline{r}_l \, q(\overline{r}_l, \overline{r}_j) \, \hat{C}_j \, \hat{C}_l^* \\
&= n_o \int d\overline{r}_j \, \hat{C}_j \, \hat{C}_j^* + n_o^2 \int d\overline{r}_j \int d\overline{r}_l \, h(\overline{r}_l, \overline{r}_j) \, \hat{C}_j \, \hat{C}_l^*
\end{aligned} \tag{41}$$

Thus, $\hat{C}_j$ operator is used instead of the transition operator. The notation in (41) is that when two $\otimes$'s are joined by a solid line, then it includes integration over the same particle as well as other particles weighted by $h(\overline{r}_l, \overline{r}_j)$. Following Sections 3.4 and 3.5 of Chapter 5, let

$=\!=\!=\!=$ = mean Green's function $E[G(\overline{r}, \overline{r}')]$ with both observation and source points $\bar{r}$ and $\bar{r}'$ in region 1 as given by equation (7), Section 10.2

⁓ = mean field $E[\psi(\bar{r})]$ in region 1

≈ = mean Green's function $E[G(\bar{r},\bar{r}')]$ with observation point $\bar{r}$ in region 0 and source $\bar{r}'$ in region 1 as given by (25), Section 10.2

The ladder intensity in region 0, $I_L(\bar{r})$ is

$$I_L(\bar{r}) = \boxed{L} \qquad (42)$$

with the operator equation

$$\boxed{L} = | + \boxed{\quad L} \qquad (43)$$

The total intensity is

$$I(\bar{r}) = I_L(\bar{r}) + I_C(\bar{r}) \qquad (44)$$

where $I_C(\bar{r})$ is the cyclical intensity given by

$$I_C(\bar{r}) = \boxed{C} \qquad (45)$$

and the operator obeys the diagrammatic equation

$$\boxed{C} = \times + \times\!\!\times + \times\!\!C\!\!\times \qquad (46)$$

as in Section 3.5, Chapter 5.

PROBLEMS

6.1 Show that the polarizability α of a dielectric sphere with permittivity ϵ_s and radius a in a background medium of ϵ is given by $\alpha =$

$v_o 3(\epsilon_s - \epsilon)\epsilon/(\epsilon_s + 2\epsilon)$ where $v_o = 4\pi a^3/3$ is the volume of the sphere.

6.2 Show that (3), Section 2.1, is the same as (5), Section 2.1.

6.3 In Section 2, we have used the relation $\overline{E}^e = \overline{E} + \overline{P}/3\epsilon$. Prove this relation. What will be the change in the result if an ellipsoid instead of a sphere of smooth polarization is subtracted?

6.4 Derive (11), Section 2.3, from (9), Section 2.3 by making use of the relations given below (9), Section 2.3.

6.5 The simple model in Section 2 is done for spherical particles. Extend the model to treat ellipsoidal scatterers with axes a, b and c. Calculate (1) effective permittivity (2) effective attenuation rate, and (3) backscattering coefficients for a half-space medium.

6.6 In treating the simple model in Section 2, backscattering coefficients are calculated for a half-space medium. Extend the model to calculate backscattering due to a slab of small spherical scatterers over a homogeneous medium of permittivity ϵ_2.

6.7 Derive (12), Section 3.2, for case B with source point $\bar{r}'$ in region V_j. Your analysis should be similar to that of case A.

6.8 The free-space Green's dyadic function is given in (2), Section 3.1, Chapter 2. Show that the free-space dyadic Green's operator is given by (17), Section 3.2.

6.9 Show that (2), Section 3.2, is the coordinate representation of the operator equation (19), Section 3.2.

6.10 Make translation addition theorem substitutions in (4), Section 3.3, to have a common center at $\bar{r}_l$ for the spherical wave functions that has an observation point at $\bar{r}'$. Next, by balancing coefficients of $Rg\overline{M}_{mn}(k\overline{r'r_l})$ and $Rg\overline{N}_{mn}(k\overline{r'r_l})$, show that the following two sets of equations are obtained

$$0 = ik \int_{S_l} dS (-1)^m \overline{M}_{-mn}(k\overline{rr_l}) \cdot i\omega\mu\, \hat{n}(\bar{r}) \times \overline{H}_l(\bar{r})$$

Problems

$$+ ik^2 \int_{S_l} dS (-1)^m \overline{N}_{-mn}(k\overline{r}\overline{r}_l) \cdot \hat{n}(\overline{r}) \times \overline{E}_l(\overline{r})$$

$$+ ik \sum_{\substack{j \neq l \\ j=1}}^{N} \sum_{\nu\mu} \int_{S_j} dS (-1)^\mu A_{mn\mu\nu}(k\overline{r}_l\overline{r}_j)$$

$$\cdot Rg\overline{M}_{-\mu\nu}(k\overline{r}\overline{r}_j) \cdot i\omega\mu\,\hat{n}(\overline{r}) \times \overline{H}_j(\overline{r})$$

$$+ ik \sum_{\substack{j \neq l \\ j=1}}^{N} \sum_{\nu\mu} \int_{S_j} dS (-1)^\mu B_{mn\mu\nu}(k\overline{r}_l\overline{r}_j)$$

$$\cdot Rg\overline{N}_{-\mu\nu}(k\overline{r}\overline{r}_j) \cdot i\omega\mu\,\hat{n}(\overline{r}) \times \overline{H}_j(\overline{r})$$

$$+ ik^2 \sum_{\substack{j \neq l \\ j=1}}^{N} \sum_{\nu\mu} \int_{S_j} dS (-1)^\mu B_{mn\mu\nu}(k\overline{r}_l\overline{r}_j)$$

$$\cdot Rg\overline{M}_{-\mu\nu}(k\overline{r}\overline{r}_j) \cdot \hat{n}(\overline{r}) \times \overline{E}_j(\overline{r})$$

$$+ ik^2 \sum_{\substack{j \neq l \\ j=1}}^{N} \sum_{\nu\mu} \int_{S_j} dS (-1)^\mu A_{mn\mu\nu}(k\overline{r}_l\overline{r}_j)$$

$$\cdot Rg\overline{N}_{-\mu\nu}(k\overline{r}\overline{r}_j) \cdot \hat{n}(\overline{r}) \times \overline{E}_j(\overline{r})$$

$$+ \sum_{\mu\nu} \left[a_{\mu\nu}^{(M)} Rg A_{mn\mu\nu}(k\overline{r}_l) + a_{\mu\nu}^{(N)} Rg B_{mn\mu\nu}(k\overline{r}_l) \right] \quad (1)$$

$$0 = ik \int_{S_l} dS (-1)^m \overline{N}_{-mn}(k\overline{r}\overline{r}_l) \cdot i\omega\mu\hat{n}(\overline{r}) \times \overline{H}_l(\overline{r})$$

$$+ ik^2 \int_{S_l} dS (-1)^m \overline{M}_{-mn}(k\overline{r}\overline{r}_l) \cdot \hat{n}(\overline{r}) \times \overline{E}_l(\overline{r})$$

$$+ ik \sum_{\substack{j \neq l \\ j=1}}^{N} \sum_{\nu\mu} \int_{S_j} dS (-1)^\mu B_{mn\mu\nu}(k\overline{r}_l\overline{r}_j)$$

$$\cdot Rg\overline{M}_{-\mu\nu}(k\overline{r}\overline{r}_j) \cdot i\omega\mu\,\hat{n}(\overline{r}) \times \overline{H}_j(\overline{r})$$

$$+ ik \sum_{\substack{j \neq l \\ j=1}}^{N} \sum_{\nu\mu} \int_{S_j} dS (-1)^\mu A_{mn\mu\nu}(k\overline{r}_l\overline{r}_j)$$

$$\cdot Rg\overline{N}_{-\mu\nu}(k\overline{r}\overline{r}_j) \cdot i\omega\mu\,\hat{n}(r) \times \overline{H}_j(\overline{r})$$

$$+ ik^2 \sum_{\substack{j \neq l \\ j=1}}^{N} \sum_{\nu\mu} \int_{S_j} dS (-1)^\mu A_{mn\mu\nu}(k\overline{r}_l\overline{r}_j)$$

$$\cdot Rg\overline{\overline{M}}_{-\mu\nu}(k\overline{rr_j}) \cdot \hat{n}(\bar{r}) \times \overline{E}_j(\bar{r})$$

$$+ ik^2 \sum_{\substack{j\neq l \\ j=1}}^{N} \sum_{\nu\mu} \int_{S_j} dS(-1)^\mu B_{mn\mu\nu}(k\overline{r_l r_j})$$

$$\cdot Rg\overline{\overline{N}}_{-\mu\nu}(k\overline{rr_j}) \cdot \hat{n}(\bar{r}) \times \overline{E}_j(\bar{r})$$

$$+ \sum_{\mu\nu} \left[a^{(M)}_{\mu\nu} Rg B_{mn\mu\nu}(k\bar{r}_l) + a^{(N)}_{\mu\nu} Rg A_{mn\mu\nu}(k\bar{r}_l) \right] \quad (2)$$

for $l = 1, 2, \ldots, N$. Next make the tangential surface field expansions as indicated in (17) and (18), Section 3.3, and derive (19) and (20), Section 3.3.

6.11 Using the appropriate expansion of the dyadic Green's function and expansion of the surface fields in (30), Section 3.3, and noting that $|\overline{r'r_j}| > |\overline{rr_j}|$ for $\bar{r}$ lying on S_j, derive (31) and (33) from (30), Section 3.3.

6.12 Prove that (39), Section 3.3, is true.

6.13 Show that for a particle centered at $\bar{r}_j$, $\overline{\overline{T}}_j = e^{-i\bar{p}\cdot\bar{r}_j} \overline{\overline{T}} e^{i\bar{p}\cdot\bar{r}_j}$ where $\bar{p}$ is momentum operator, and $\overline{\overline{T}}_j$ and $\overline{\overline{T}}$ are transition operators centered at $\bar{r}_j$ and 0 respectively.

6.14 1. Calculate the dyadic Green's function $\overline{\overline{G}}_s(\bar{r}_1, \bar{r}_2)$ for a single spherical dielectric particle (Tai, 1971; Tsang and Kong, 1980a). Label the regions outside and inside the particle respectively as 1 and 2. Use the following definitions

$$\overline{\overline{G}}_s = \begin{cases} \overline{\overline{G}}_{11}(\bar{r}_1, \bar{r}_2) & \text{if } \bar{r}_1 \text{ and } \bar{r}_2 \text{ are outside the particle} \\ \overline{\overline{G}}_{12}(\bar{r}_1, \bar{r}_2) & \text{if } \bar{r}_1 \text{ is outside the particle} \\ & \text{and } \bar{r}_2 \text{ is inside the particle} \\ \overline{\overline{G}}_{21}(\bar{r}_1, \bar{r}_2) & \text{if } \bar{r}_1 \text{ is inside the particle} \\ & \text{and } \bar{r}_2 \text{ is outside the particle} \\ \overline{\overline{G}}_{22}(\bar{r}_1, \bar{r}_2) & \text{if } \bar{r}_1 \text{ and } \bar{r}_2 \text{ are inside the particle} \end{cases} \quad (3)$$

Give expressions for $\overline{\overline{G}}_{11}, \overline{\overline{G}}_{12}, \overline{\overline{G}}_{21}$ and $\overline{\overline{G}}_{22}$.

2. Calculate the transition operator in momentum representation for a spherical particle by making use of the relation of (13), Section 4.1

(Tsang and Kong, 1980a).

3. Verify the relation (17), Section 4.1, between scattering dyad $\overline{\overline{F}}$ and $\overline{\overline{T}}_p(\bar{p}_1, \bar{p}_2)$ for spherical particles.

6.15 In calculating the coherent reflected field in (8), Section 2.3, the far-field approximation is made for $\overline{W}$ in obtaining (7), Section 2.3. Repeat the coherent reflected field calculation by using the exact field integral representation of a dipole.

6.16 By taking the statistical average of (3), Section 4.3, and using (11) and (12), Section 4.3, show that the dispersion relation under the quasicrystalline approximation is given by (14) and (15), Section 4.3. By using the Lippmann-Schwinger equation of (27), Section 3.2, derive (16), Section 4.3. In a similar manner, show that (26) and (27), Section 4.4, is equivalent to (24), Section 4.4.

6.17 The pair-distribution function $g(x)$ for Percus-Yevick approximation is given by (20) through (23), Section 5.2. Derive explicit expressions for $g_1(x)$ and $g_2(x)$ by calculating the residues at t_o, t_1, and t_2 which are the three roots of the polynomial $S(t)$ of (23), Section 5.2. Express your answers in terms of t_o, t_1, and t_2.

6.18 Use the results of $c(x)$ in (15), Section 5.2, to derive (6), Section 5.3.

6.19 Derive (7), Section 5.3, using (4) through (6), Section 5.3, and (15) through (19), Section 5.2.

6.20 Calculate the structure factor for the Verlet-Weis pair-distribution function.

6.21 Derive (2), Section 6.2, by taking conditional average of equation (40), Section 3.3, with particle $\bar{r}_l$ fixed.

6.22 Derive Kasterin's representation of $h_p(kr)P_p(\cos\theta)$ as given by (12), Section 6.2.

6.23 Use the definitions in (12) and (13), Section 3.3, to show that (17) and (18), Section 6.2, are true.

6.24 1. Prove the two summation identities in (21), Section 6.2.
2. Show that by using (21) and (22), Section 6.2, the generalized

E-O theorem of (19) and (20), Section 6.2, reduces to a single scalar equation of (23), Section 6.2.

6.25 In this problem, we shall derive the generalized Lorentz-Lorenz law and the generalized Ewald-Oseen extinction theorem for oblique incidence. On substituting (6) into (5), Section 7.2, and using $\overline{\overline{\sigma}}(k\overline{r_1 r_2})$ as given in (10) and (11) and (24), Section 3.3, the key integral to be evaluated is of the following form

$$I_{mn\mu\nu p} = \int_{\substack{z_2 < 0 \\ |\overline{r}_2 - \overline{r}_1| > b}} d\overline{r}_2\, g(\overline{r}_2 - \overline{r}_1)\, e^{i\overline{K}_d \cdot \overline{r}_2} \tag{4}$$

$$\times h_p(k|\overline{r_1 r_2}|) P_p^{\mu-m}(\cos\theta_{r_1 r_2})\, e^{i(\mu-m)\phi_{r_1 r_2}}$$

Write
$$g(\overline{r}) = 1 + (g(\overline{r}) - 1) \tag{5}$$

as in the case of normal incidence, so that

$$I_{mn\mu\nu p} = I^{(1)}_{mn\mu\nu p} + I^{(2)}_{mn\mu\nu p} \tag{6}$$

where $I^{(1)}_{mn\mu\nu p}$ corresponds to 1 in (5) and $I^{(2)}_{mn\mu\nu p}$ corresponds to $g(\overline{r}) - 1$ in (5).

1. Show that

$$I^{(1)}_{mn\mu\nu p} = \frac{(-1)^p e^{i\overline{K}_d \cdot \overline{r}_1}}{K^2 - k^2} \left\{ e^{-ik_{iz}z_1 + iK_z z_1} L^{(d)}_{p(\mu-m)} - b^2 L^{(e)}_{p(\mu-m)} \right\} \tag{7}$$

where

$$L^{(d)}_{nm} = \frac{2\pi i^{n+1}}{k}(-1)^{n+m} P_n^m\left(\frac{k_{iz}}{k}\right) e^{im\phi_i} \frac{k_{iz} + K_z}{k_{iz}} \tag{8}$$

$$L^{(e)}_{nm} = 4\pi [kj_n(Kb)h'_n(kb) - Kj'_n(Kb)h_n(kb)]$$
$$\times i^n P_n^m(-\cos\theta_i) e^{im\phi_i} \tag{9}$$

The solutions in (7) through (9) have made use of the phase matching conditions of (10) through (13), Section 7.2.

2. Show that

$$I^{(2)}_{mn\mu\nu p} = e^{i\overline{K}_d \cdot \overline{r}_1} (-1)^p L^{(g)}_{p(\mu-m)} \tag{10}$$

where

$$L_{nm}^{(g)} = 4\pi i^n P_n^m(-\cos\theta_t) e^{im\phi_i} \int_b^\infty dr\, r^2 [g(r) - 1] j_n(Kr) h_n(kr) \tag{11}$$

3. Show that using the above results in (5), Section 7.2, one obtains the generalized Lorentz-Lorenz law of (14), Section 7.2, and the generalized Ewald-Oseen extinction theorem of (23), Section 7.2.

6.26 Using (22), Section 7.2, and the completeness relation of spherical harmonics

$$\sum_{nm}(-1)^m \frac{2n+1}{4\pi} Y_n^{-m}(\theta', \phi') Y_n^m(\theta, \phi) \tag{12}$$

$$= \delta(\cos\theta' - \cos\theta)\delta(\phi' - \phi)$$

and the orthogonality property of vector spherical harmonics in (11), Section 5.1 (Chapter 3), derive (20) and (21), Section 7.2.

6.27 Beginning with the relation (Friedman and Russek, 1954),

$$P_\nu^\mu(x) P_n^{-m}(x) = \sum_p a(\mu, \nu | -m, n | p) P_p^{\mu-m}(x) \tag{13}$$

derive the addition theorem for vector spherical harmonics in (22), Section 7.2.

6.28 The orthogonality relation of Wigner $3j$ symbols is

$$\sum_{m_2} \begin{pmatrix} j_1 & j_2 & j_3 \\ -m+m_2 & -m_2 & m \end{pmatrix} \begin{pmatrix} j_1 & j_2 & j_3' \\ -m+m_2 & -m_2 & m \end{pmatrix}$$

$$= \frac{1}{2j_3+1} \delta_{j_3 j_3'} \tag{14}$$

Use it to derive the orthogonality relations in (4) through (6), Section 7.3.

6.29 Substitute (1) through (3), Section 7.3 into the Ewald-Oseen Theorem of (23), Section 7.2, and make use of the addition theorem in Section 5.9, Chapter 3. Show that (23), Section 7.2 can be reduced to a single scalar equation as given by (9), Section 7.3.

6.30 Substitute (15) and (16) into (12), Section 8.2, and use the expressions for $\overline{\overline{U}}^{(k)}(k, K_r)$ in (25) and (26), Section 7.2, to derive the two independent scalar equations in (17) and (18), Section 8.2, for the generalized Ewald-Oseen extinction theorem.

6.31 By using the results of Section 5.7, Chapter 3, show that $\overline{\overline{D}}$ for $N_{max} = 1$ is given by (2), Section 8.5.

6.32 Show that in the low-frequency limit, the backscattering cross sections under the distorted Born approximation for oblate spheroids uniformly distributed in γ are given by the following

A. Vertically polarized incidence:

$$\sigma_{vv} = \frac{3n_o}{16k^2 Im(K_z^v)} \int_0^\pi d\beta \frac{\sin \beta}{2} p(\beta) |c_v|^2$$

$$\times \left\{ 2 \Big| - 2T_1 r_B \cos \theta_i + \cos \theta_i \sin^2 \beta \, r_B (T_1 - T_0) \right.$$

$$\left. - \sqrt{2}(T_1 - T_0) \sin \theta_i \cos^2 \beta + \sqrt{2} T_1 \sin \theta_i \Big|^2 \right.$$

$$\left. + \sin^2 \beta \cos^2 \beta \, |T_1 - T_0|^2 \Big| \sqrt{2} \cos \theta_i - 2 \sin \theta_i r_B \Big|^2 \right.$$

$$\left. + \cos^2 \theta_i \, |r_B|^2 \sin^4 \beta \, |T_1 - T_0|^2 \right\}$$

$$+ \frac{3n_o^2}{4k^2 Im(K_z^v)} Re \int d\bar{r} [g(\bar{r}) - 1] e^{i(Re\overline{K}_d - \bar{k}_s) \cdot \bar{r}}$$

$$\times |c_v|^2 \Big| \sqrt{2} T_{a1} r_B \cos \theta_i - T_{ao} \sin \theta_i \Big|^2 \quad (15)$$

$$\sigma_{hv} = \frac{3n_o}{16k^2 Im(K_z^v)} \int_0^\pi d\beta \frac{\sin^3 \beta}{2} p(\beta) |c_v|^2 |T_1 - T_0|^2$$

$$\times \left\{ 2 \cos^2 \beta + \sin^2 \beta \, |r_B|^2 \right\} \quad (16)$$

B. Horizontally polarized incidence:

$$\sigma_{hh} = \frac{3n_o}{16k^2 Im(K_z^h)} \int_0^\pi d\beta \frac{\sin \beta}{2} p(\beta) |c_h|^2$$

$$\times \left\{ 2\left| 2T_1 - (T_1 - T_o)\sin^2\beta \right|^2 + \sin^4\beta |T_1 - T_o|^2 \right\}$$

$$+ \frac{3n_o}{2k^2 Im(K_z^h)} Re \int d\bar{r}[g(\bar{r}) - 1]$$

$$\times e^{i(Re\overline{K}_d - \overline{k}_s)\cdot\bar{r}} |c_h|^2 |T_{a1}|^2 \qquad (17)$$

Thus, depolarization is exhibited as a first-order effect.

6.33 Show that (24), Section 9.2, is the same as (23), Section 9.2.

6.34 The concept of Ewald-Oseen extinction theorem and Lorentz-Lorenz law is also applicable to propagation in random media. Consider bilocal approximation to Dyson's equation as given by (5) through (8), Section 3.2, Chapter 5. For the case of short correlation lengths, show that the mean field $<\psi(\bar{r})>$ obeys the equation

$$<\psi(\bar{r})> = \psi_{inc}(\bar{r}) + W \int_{z'<0} d\bar{r}'\, G^{(0)}(\bar{r}, \bar{r}') <\psi(\bar{r}')> \qquad (18)$$

where

$$W = \int d\bar{r}\, G^{(0)}(\bar{r}) C(\bar{r}) \qquad (19)$$

and $C(\bar{r})$ is the correlation function. Equation (18) is the result of a plane wave incident on a half-space random medium that occupies the region $z < 0$. The incident wave is thus

$$\psi_{inc}(\bar{r}) = \exp(ik_{ix}x + ik_{iy}y - ik_{iz}z) \qquad (20)$$

To solve (18) and (20), assume a trial solution in a manner analogous to Section 7,

$$<\psi(\bar{r})> = T\exp(iK_{ix}x + iK_{iy}y - iK_{iz}z) \qquad (21)$$

where $K = (K_{ix}^2 + K_{iy}^2 + K_{iz}^2)^{1/2}$ is the unknown effective wavenumber. By using (21) and the plane wave representation of $G^{(0)}(\bar{r}, \bar{r}')$, the integral in (18) can be carried out. Show that (18) then becomes

$$T e^{iK_{ix}x + iK_{iy}y - iK_{iz}z}$$

$$= e^{ik_{ix}x + ik_{iy}y - ik_{iz}z}$$

$$- WT e^{iK_{ix}x + iK_{iy}y} \left\{ -\frac{e^{-ik_z z}}{2(k_z - K_{iz})} + \frac{e^{-iK_{iz}z}}{k_z^2 - K_{iz}^2} \right\} \qquad (22)$$

where
$$k_z = (k^2 - K_{ix}^2 - K_{iy}^2)^{1/2} \tag{23}$$

In order to balance the phases of the terms in (22), we must have

$$K_{ix} = k_{ix} \tag{24}$$

$$K_{iy} = k_{iy} \tag{25}$$

which are the phase matching conditions. Show from (23) through (25) that

$$k_z = k_{iz} \tag{26}$$

Thus there are two types of waves in (22). First type is of dependence

$$\exp(ik_{ix}x + ik_{iy}y - ik_{iz}z) \tag{27}$$

and the second type is

$$\exp(ik_{ix}x + ik_{iy}y - iK_{iz}z) \tag{28}$$

Show that balancing terms of the type in (28) gives the Lorentz-Lorenz law

$$k_{iz}^2 - K_{iz}^2 = -W \tag{29}$$

Since $k_{iz}^2 - K_{iz}^2 = k^2 - K^2$, show that (29) is equivalent to

$$K^2 = k^2 + W \tag{30}$$

which is the same as (8), Section 3.2, Chapter 5. Show that balancing terms of the type in (27) gives the Ewald-Oseen extinction theorem

$$1 + \frac{WT}{2k_{iz}(k_{iz} - K_{iz})} = 0 \tag{31}$$

Show that solving (29) and (31) gives

$$T = \frac{2k_{iz}}{k_{iz} + K_{iz}} \tag{32}$$

which is the Fresnel transmission coefficient.

6.35 Perform the integration on the right hand side of (24), Section 10.2, for $\bar{r}_1$ in region 0 and $\bar{r}_2$ in region 1 to derive the result in (25),

Problems

Section 10.2. In a similar manner, derive the result of the mean field in region 0 as given by (26), Section 10.2.

6.36 By differentiating the integral equations of (24) and (25), Section 10.3, with respect to τ, show that the integral equation can be converted to differential equations in (28), Section 10.3. Next set $\tau = 0$ in (24) and (25), Section 10.3, and derive the boundary condition as given by (29), Section 10.3.

6.37 Derive the results in (35) and (39), Section 10.3, by using equation (5), Section 10.3, and following the steps as indicated in the text.

BIBLIOGRAPHY

Abramowitz, M. and J. A. Stegun (1965), *Handbook of Mathematical Functions*, Dover Publications, New York.

Adler, B. J. and C. E. Hecht (1969), Studies in molecular dynamics. VII Hard sphere distribution functions and an augmented van der Waals Theory, *J. Chem. Phys.*, 50, 2032-2037.

Agarwal, G. S. (1977), Interaction of electromagnetic waves at rough dielectric surfaces, *Phys. Rev. B*, 15, 2371-2383.

Ambach, W. and A. Denoth (1980), The dielectric behavior of snow: a study versus liquid water content, *NASA Workshop on the Microwave Remote Sensing of Snowpack Properties*, Ft. Collins, Colorado, NASA CP-2153.

Attema, E. P. W. and F. T. Ulaby (1978), Vegetation modeled as a water cloud, *Radio Science*, 13, 357-364.

Axline, R. M. and A. K. Fung (1978), Numerical computation of scattering from a perfectly conducting random surface, *IEEE Trans. Ant. Prop.*, AP-26, 482-488.

Bahar, E. (1978), Full wave and physical optics solutions for scattered radiation fields by rough surfaces – energy and reciprocity relationships, *IEEE Trans. Ant. Prop.*, AP-26, 603-614.

Bahar, E. and D. E. Barrick (1983), Scattering cross sections for composite surfaces that cannot be treated as perturbed-physical optics problems, *Radio Science*, 18, 129-137.

Barabanenkov, Y. N. (1973), Wave corrections to the transfer equation for backscattering, *Izv. Vyssh. Uchebn. Zaved. Radiofiz.*, 16, 88-94.

Barabanenkov, Y. N. and V. M. Finkelberg (1968), Radiative transport equation for correlated scatterers, *Sov. Phys. JETP.*, 26, 587-591.

Barabanenkov, Y. N., Y. A. Kravtsov, S. M. Rytov, and V. I. Tatarski (1971), Status of the theory of propagation of wave in a randomly inhomogeneous medium, *Sov. Phys. Usp.*, 13, 551-580.

Barber, P. and C. Yeh (1975), Scattering of electromagnetic waves by arbitrary-shaped dielectric bodies, *Appl. Optics*, **14**, 2864–2872.

Barker, J. A. and D. Henderson (1971), Molecular Monte Carlo values for the radial distribution function of a system of fluid hard spheres, *Molecular Physics*, **21**, 187–191.

Barrett, E. C. and D. W. Martin (1981), *The Use of Satellite Data in Rainfall Monitoring*, Academic Press, New York.

Barrick, D. E. (1968), Relationship between slope probability density function and the physical optics integral in rough surface scattering, *Proc. IEEE*, **56**, 1728–1729.

Barrick, D. E. (1970), Unacceptable height correlation coefficients and the quasispecular component in rough surface scattering, *Radio Science*, **5**, 647–654.

Bass, F. G. and I. M. Fuks (1979), *Wave Scattering from Statistically Rough Surface*, translated by C. B. Vesecky and J. F. Vesecky, Pergamon Press, Oxford.

Beckmann, P. and A. Spizzichino (1963), *The Scattering of Electromagnetic Waves from Rough Surfaces*, MacMillan, New York.

Beckmann, P. (1965), Shadowing of random rough surfaces, *IEEE Trans. Ant. Prop.*, **AP-13**, 384–388.

Beckmann, P. (1968), *The Depolarization of Electromagnetic Waves*, Golem Press, Boulder, Colorado.

Beckmann, P. (1975), Scattering by non-Gaussian surfaces, *IEEE Trans. Ant. Prop.*, **AP-23**, 169–175.

Bellman, R. and G. M. Wing (1975), *An Introduction to Invariant Imbedding*, John Wiley & Sons, New York.

Beran, M. J. (1968), *Statistical Continuum Theories*, Wiley-Interscience, New York.

Besieris, I. M. and W. E. Kohler (1981), Two-frequency radiative transfer equation for statistically inhomogeneous and anisotropic absorptive medium, in *Multiple Scattering and Waves in Random Media*, edited by P. L. Chow, W. E. Kohler, and G. C. Papanicolaou, North-Holland Publishing Company, New York.

Blanchard, D. C. (1972), Bentley and Lenard: Pioneers in cloud physics, *Am. Sci.*, **60**, 746.

Blanchard, A. J. and J. W. Rouse, Jr. (1980), Depolarization of electromagnetic waves scattered from an inhomogeneous half space bounded by a rough surface, *Radio Science*, **15**, 773–780.

Blanchard, A. J. and B. R. Jean (1983), Antenna effects in depolarization measurements, *IEEE Trans. Geosci. Rem. Sens.*, **GE-21**, 113–117.

Blinn, III, J. C., J. E. Conel, and J. G. Quade (1972), Microwave emission from geological materials: Observations of interference effects, *J. Geophys. Res.*, **77**, 4366–4378.

Boerner, W. M., M. B. El-Arini, C. Y. Chan, and P. M. Mastoris (1981), Polarization dependence in electromagnetic inverse problems, *IEEE Trans. Ant. Prop.*, **AP-29**, 262–271.

Bolomey, J. C. and J. A. Wirgin (1974), Numerical comparison of the Green's function and the Waterman and Rayleigh theories of scattering from a cylinder with arbitrary cross section, *Proc. IEEE*, **121**, 794–804.

Bolotovskii, B. M. and A. N. Lebedev (1968), On threshold phenomena in classical electrodynamics, *Sov. Phys. JETP*, **26**, 784.

Booker, H. G. and W. E. Gordon (1950), Outline of a theory of radio scatterings in the troposphere, *J. Geophys. Res.*, **55**, 241–246.

Born, M. and E. Wolf (1975), *Principles of Optics*, Fifth Edition, Pergamon Press, New York.

Bottcher, C. J. F. (1952), *Theory of Electric Polarization*, Elsevier, Amsterdam.

Bowman, J. J., T. B. A. Senior, and P. L. E. Uslenghi (1969), *Electromagnetic and Acoustic Scattering by Simple Shapes*, North Holland, Amsterdam.

Bremmer, H. (1964), Random volume scattering, *J. Res. Nat. Bur. Std.*, **68D**, 967–981.

Bringi, V. N., V. V. Varadan, and V. K. Varadan (1982), The effects on pair correlation function of coherent wave attenuation in discrete random media, *IEEE Trans. Ant. Prop.*, **AP-30**, 805–808.

Brown, G. S. (1978), Backscattering from a Gaussian-distributed perfectly conducting rough surface, *IEEE Trans. Ant. Prop.*, **AP-26**, 472–482.

Brown, G. S. (1980), Coherent wave propagation through a sparse concentration of particles, *Radio Science*, **15**, 705–710.

Brown, G. S. (1982), A stochastic Fourier transform approach to scattering from perfectly conducting randomly rough surfaces, *IEEE Trans. Ant.*

Prop., AP-30, 1137-1144.

Brown, G. S. (1984), The validity of shadowing correction in rough surface scattering, *Radio Science*, 19, 1461-1468.

Brown, W. P. (1971), Second moment of a wave propagating in a random medium, *J. Opt. Soc. Am.*, 61, 1051-1059.

Bryan, M. L. and J. R. W. Larson (1975), The study of fresh-water lake ice using multiplexed imaging radar, *J. Glaciology*, 14.

Bugnolo, D. (1960), Transport equation for the spectral density of a multiple scattered electromagnetic field, *J. Appl. Phys.*, 31, 1176-1182.

Burke, H.-H. K. and T. J. Schmugge (1982), Effects of varying soil moisture contents and vegetation canopies on microwave emissions, *IEEE Trans. Geosci. Rem. Sens.*, GE-20, 268-274.

Burke, W. J., T. Schmugge, and J. F. Paris (1979), Comparison of 2.8 and 1 cm microwave radiometer observations over soils with emission model calculations, *J. Geophys. Res.*, 84, 287-294.

Burrows, M. L. (1973), On the composite model for rough surface scattering, *IEEE Trans. Ant. Prop.*, AP-21, 241-243.

Bush, T. F. and F. T. Ulaby (1976), Radar return from a continuous vegetation canopy, *IEEE Trans. Ant. Prop.*, AP-24, 269-276.

Bush, T. F. and F. T. Ulaby (1978), An evaluation of radar as a crop classifier, *Remote Sensing of Environment*, 7, 15-36.

Callen, H. B. and T. A. Welton (1951), Irreversibility and generalized noise, *Phys. Rev.*, 83, 34.

Campbell, W. J., P. Gloersen, H. J. Zwally, R. O. Ramseier, and C. Elachi (1980), Simultaneous passive and active microwave observations of near-shore Beaufort Sea Ice, *J. Petrol. Technol.*, 21, 1105-1112.

Carver, K. R. (1977), Radiometric recognition of coherence, *Radio Science*, 12, 371-379.

Case, K. M. and P. F. Zweifel (1967), *Linear Transport Theory*, Addison-Wesley, Reading, Mass.

Chandrasekhar, S. (1960), *Radiative Transfer*, Dover, New York.

Chang, S. K. and K. K. Mei (1976), Application of the unimoment method to electromagnetic scattering of dielectric cylinders, *IEEE Trans. Ant. Prop.*, AP-24, 35-42.

Chang, T. C., P. Gloersen, T. Schmugge, T. T. Wilheit, and H. J. Zwally (1976), Microwave emission from snow and glacier ice, *J. Glaciology*, **16**, 23–29.

Chan, H. L. and A. K. Fung (1978), A numerical study of the Kirchhoff approximation in horizontally polarized backscattering from a random surface, *Radio Science*, **13**, 811–818.

Cheung, R. L-T. and A. Ishimaru (1982), Transmission backscattering, and depolarization of waves in randomly distributed spherical particles, *Appl. Optics*, **21**, 3792–3798.

Chew, W. C. and P. N. Sen (1982), Dielectric enhancement due to electrochemical double layer: thin double layer approximation, *J. Chem. Phys.*, **77**, 4683–4693.

Choudhury, J. B., T. J. Schmugge, A. Chang, and R. W. Newton (1979), Effect of surface roughness on the microwave emission from soils, *J. Geophys. Res.*, **84**.

Chow, P. L., W. E. Kohler, and G. C. Papanicolaou, eds. (1981), *Multiple Scattering and Waves in Random Media*, North-Holland Publishing Company, Amsterdam.

Chu, T. S. (1974), Rain-induced cross-polarization at centimeter and millimeter wavelengths, *Bell System Tech. J.*, **53**, 1557–1579.

Chu, T. S. and D. C. Hogg (1968), Effects of precipitation on propagation at 0.63, 3.5 and 10.6 microns, *Bell System Tech. J.*, **47**, 723–759.

Chuang, S. L. and J. A. Kong (1981), Scattering of waves from periodic rough surface, *Proc. IEEE*, **69**, 1132–1144.

Chuang, S. L., J. A. Kong, and L. Tsang (1980), Radiative transfer theory for passive microwave remote sensing of two-layer random medium with cylindrical structures, *J. Appl. Phys.*, **51**, 5588–5593.

Colbeck, S. (1972), A theory of water percolation in snow, *J. Glaciology*, **11**, 369–385.

Colbeck, S. C. (1979), Grain clusters in wet snow, *J. Colloid Interface Sci.*, **72**, 371–384.

Colbeck, S. (1982), The geometry and permittivity of snow at high frequencies, *J. Appl. Phys.*, **53**, 4495–4500.

Cole, J. D. (1968), *Perturbation Methods in Applied Math*, Blaisdell, Waltham, Mass.

Cox, D. C. (1981), Depolarization of radio waves by atmospheric hydrometeors in earth-space paths: A review, *Radio Science*, 16, 781–812.

Crane, R. K. (1974), Bistatic scatter from rain, *IEEE Trans. Ant. Prop.*, AP-22, 312–320.

Cruzan, O. R. (1961), Translational addition theorems for spherical vector wave functions, TR-906, Diamond Ordinance Fuse Laboratories, Department of the Army, Washington DC.

Cruzan, O. R. (1962), Translational addition theorems for spherical vector wave functions, *Quart. J. Appl. Math.*, 20, 33–40.

Cumming, W. A. (1952), The dielectric properties of ice and snow at 3.2 cm, *J. Appl. Phys.*, 23, 768–773.

Dashen, R. (1979), Path integrals for waves in random media, *J. Math. Phys.*, 20, 894–920.

Davenport, W. B., Jr. and W. L. Root (1958), *An Introduction to the Theory of Random Signals and Noise*, McGraw-Hill, New York.

Davison, B. (1958), *Neutron Transport Theory*, Oxford University Press, London.

Deirmendjian, D. (1969), *Electromagnetic Scattering on Spherical Polydispersions*, Elsevier, New York.

Dence, D. and J. E. Spence (1973), Wave propagation in random anisotropic media, in *Probabilistic Methods in Applied Mathematics*, 3, edited by A. T. Bharucha-Reid, Academic Press, NY.

Devaney, A. J. (1980), Multiple scattering theory for discrete elastic random media, *J. Math. Phys.*, 21, 2603–2611.

De Loor, G. P. (1968), Dielectric properties of heterogeneous mixtures containing water, *J. Microwave Power*, 3, 67–73.

De Loor, G. P., A. A. Jurrins, and H. Gravesteijn (1974), The radar backscatter from selected agricultural crops, *IEEE Trans. Geosci. Electron.*, GE-12, 70–77.

DeSanto, J. A. (1974), Green's function for electromagnetic scattering from a random rough surface, *J. Math. Phys.*, 15, 283–288.

DeSanto, J. A. (1975), Scattering from a sinusoid: Derivation of linear equations for the field amplitudes, *J. Acoust. Soc. Am.*, 57, 1195–1197.

DeSanto, J. A. (1983), Scattering of scalar waves from a rough interface using a single integral equation, *Wave Motion*, **73**, 125–135.

DeSanto, J. A., A. W. Saenz, and W. W. Zachary, ed., (1980), *Mathematical Methods and Applications of Scattering Theory*, Springer-Verlag, New York.

deWolf, D. A. (1971), Electromagnetic reflection from an extended turbulent medium: cumulative forward-scatter single backscatter approximation, *IEEE Trans. Ant. Prop.*, **AP-19**, 254–262.

deWolf, D. A. (1972), Discussion of radiative transfer methods applied to electromagnetic reflection from turbulent plasma, *IEEE Trans. Ant. Prop.*, **AP-20**, 805–807.

Djermakoye, B. and J. A. Kong (1979), Radiative transfer theory for the remote sensing of layered random media, *J. Appl. Phys.*, **50**, 6600–6604.

Dobson, C., H. Stiles, D. Brunfeldt, T. Metzler, and S. McMeeking (1977), Data documentation: 1975 MAS 1-8 and MAS 8-18 vegetation experiments, *Rep. RSL TR 264-15*, Remote Sensing Lab., Univ. of Kansas Center for Res., Inc., Lawrence, Kansas.

Dunbar, M. (1975), Interpretation of solar imagery of sea ice in Nares Strait and the Arctic Ocean, *J. Glaciology*, **15**.

Edgerton, A. T., A. Stogryn, and G. Poe (1971), Microwave radiometric investigations of snowpacks, Final Report 1285R-4, Aerojet-General Corp., El Monte, CA.

Edmonds, A. R. (1957), *Angular Momentum in Quantum Mechanics*, Princeton University, Princeton, NJ.

Elachi, C., M. L. Bryan, and W. F. Weeks (1976), Imaging radar observations of frozen Arctic lakes, *Rem. Sens. Environ.*, **5**, 169–175.

England, A. W. (1974), Thermal microwave emission from a half-space containing scatterers, *Radio Science*, **9**, 447–454.

England, A. W. (1975), Thermal microwave emission from a scattering layer, *J. Geophys. Res.*, **80**, 4484–4496.

Evans, S. (1965), Dielectric properties of ice and snow – a review, *J. Glaciology*, **5**, 773–792.

Fante, R. L. (1973), Propagation of electromagnetic waves through turbulent plasma using transport theory, *IEEE Trans. Ant. Prop.*, **AP-21**, 750–755.

Fante, R. L. (1981), Relation between radiative-transport theory and Maxwell's equations in dielectric media, *J. Opt. Soc. Am.*, **71**, 460–468.

Faulkner, J. S. (1970), Electronic states of a liquid metal from the coherent-potential approximation, *Phys. Rev. B*, **1**, 934–936.

Faulkner, J. S. (1979), Multiple-scattering approach to band theory, *Phys. Rev. B*, **19**, 6186–6206.

Felsen, L. B. and N. Marcuvitz (1973), *Radiation and Scattering of Waves*, Prentice-Hall, Englewood Cliffs, New Jersey.

Feynman, R. P., R. B. Leighton, and M. Sand (1964), *The Feynman Lectures on Physics*, Addison-Wesley, Reading, MA.

Fikioris, J. G. and P. C. Waterman (1964), Multiple scattering of waves, II. Hole corrections in the scalar case, *J. Math. Phys.*, **5**, 1413–1420.

Finney, J. L. (1970), Random packing and the structure of simple liquids, II. The molecular geometry of simple liquids, *Proc. Royal Soc. Lond.*, **A.319**, 495–507.

Flock, W. L. (1979), *Electromagnetics and the Environment: Remote Sensing and Telecommunications*, Prentice-Hall, Englewood Cliffs, New Jersey.

Foldy, L. L. (1945), The multiple scattering of waves, *Phys. Rev.*, **67**, 107–119.

Fraser, K. S., N. E. Gaut, E. C. Geifenstein, II, and H. Sievering (1975), Interaction mechanisms – within the atmosphere, in *Manual of Remote Sensing*, I, R. G. Reeves, ed., American Society of Photogrammetry, Falls Church, Virginia, Chapter 5, 207–210.

Friedman, B. and J. Russek (1954), Addition theorems for spherical waves, *Quart. Appl. Math.*, **12**, 13–23.

Frisch, V. (1968), Wave propagation in random medium, *Probabilistic Methods in Applied Mathematics*, **1**, edited by Bharuch-Reid, Academic Press.

Fung, A. K. (1979), Scattering from a vegetation layer, *IEEE Trans. Geosci. Electro.*, **GE-17**, 1–6.

Fung, A. K. and H. L. Chan, (1969), Backscattering of waves by composite rough surfaces, *IEEE Trans. Ant. Prop.*, **AP-17**, 590–597.

Fung, A. K. and M. F. Chen (1981a), Emission from an inhomogeneous layer with irregular interfaces, *Radio Science*, **16**, 289–298.

Fung, A. K. and M. F. Chen (1981b), Scattering from a Rayleigh layer with an irregular interface, *Radio Science*, **16**, 1337–1347.

Fung, A. K. and H. J. Eom (1981), A theory of wave scattering from an inhomogeneous layer with an irregular interface, *IEEE Trans. Ant. Prop.*, **AP-29**, 899–910.

Fung, A. K. and H. S. Fung (1977), Application of first order renormalization method to scattering from a vegetation-like half space, *IEEE Trans. Geosci. Electron.*, **GE-15**, 189–195.

Fung, A. K. and F. T. Ulaby (1978), A scatter model for leafy vegetation, *IEEE Trans. Geosci. Electron.*, **GE-16**, 281–285.

Furutsu, K. (1975), Multiple scattering of waves in a medium of randomly distributed particles and derivation of the transport equation, *Radio Science*, **10**, 29–44.

Furutsu, K. (1983), Statistical theory of scattering and propagation over a random surface, *IEE Proc.*, **130**, 601–622.

Garcia, N. and N. Cabrera (1978), New method for solving the scattering of waves from a periodic hard surface: Solutions and numerical comparisons with the various formalisms, *Phys. Rev. B*, **18**, 576–589.

Garcia, N., V. Celli, N. R. Hill, and N. Cabrera (1978), Ill-conditioned matrices in the scattering of waves from hard corrugated surfaces, *Phys. Rev. B*, **18**, 5184–5189.

Glisson, A. W. and D. R. Wilton (1980), Simple and efficient numerical methods for problems of electromagnetic radiation and scattering from surfaces, *IEEE Trans. Ant. Prop.*, **AP-28**, 593–603.

Gloersen, P., W. Nordberg, T. J. Schmugge, T. T. Wilheit, and W. J. Campbell (1973), Microwave signatures of first-year and multiyear sea ice, *J. Geophys. Res.*, **78**, 3564–3572.

Goodman, F. O. (1977), Scattering of atoms by a scattering sinusoidal hard wall: Rigorous treatment in $(n+1)$ dimensions and comparison with the Rayleigh method, *J. Chem. Phys.*, **66**, 976–982.

Goodman, J. W. (1968), *Introduction to Fourier Optics*, McGraw-Hill, New York.

Goodman, J. W. (1976), Some fundamental properties of speckle, *J. Opt. Soc. Am.*, **66**, 1145–1149.

Gradshteyn, I. S., and I. M. Ryzhik (1965), *Table of Integrals, Series and Products*, Academic Press, New York.

Gray, E. P., R. W. Hart, and R. A. Farrell (1978), An application of a variational principle for scattering by random rough surfaces, *Radio*

Science, **13**, 333–343.

Green, H. S. (1969), *The Molecular Theory of Fluids*, Dover, New York.

Grody, N. C. (1976), Remote sensing of atmospheric water content from satellites using microwave radiometry, *IEEE Trans. Ant. Prop.*, **AP-24**, 155–162.

Gurvich, A. S. and V. I. Tatarskii (1975), Coherent and intensity fluctuations of light in the turbulent atmosphere, *Radio Science*, **10**, 3–14.

Gurvich, A. S., V. L. Kalinin, and D. T. Matveyer (1973), Influence of the internal structure of glaciers on their thermal radio emission, *Atm. Oceanic Phys. USSR*, **9**, 713–717.

Gyorffy, B. L. (1970), Electronic states in liquid metals: A generalization of the coherent-potential approximation for a system with short-range order, *Phys. Rev. B*, **1**, 3290–3299.

Hagfors, T. (1964), Backscattering from an undulating surface with applications to radar return from the moon, *J. Geophys. Res.*, **69**, 3779–3784.

Harrington, R. F. (1968), *Field Computation by Moment Methods*, MacMillan, New York.

Havelka, U. D. (1971), *The effect of leaf type, plant density, and row spacing on canopy architecture and plant morphology in grain sorghum*, Doctoral Dissertation, Texas A & M University, College Station, Texas.

Hawley, S. W., T. H. Kays, and V. Twersky (1967), Comparison of distribution function from scattering data on different sets of spheres, *IEEE Trans. Ant. Prop.*, **AP-15**, 118–135.

Hildebrand, F. B. (1956), *Introduction to Numerical Analysis*, McGraw-Hill, New York.

Hipp, J. E. (1974), Soil electromagnetic parameters as a function of frequency, soil density and soil moisture, *Proc. IEEE*, **62**, 98–103.

Hoekstra, P. and A. Delaney (1974), Dielectric properties of soils at UHF and microwave frequencies, *J. Geophys. Res.*, **79**, 1699–1708.

Hofer, R. and E. Shanda (1978), Signatures of snow in the 5 to 94 GHz range, *Radio Science*, **13**, 365–369.

Hofer, R. and W. Good (1979), Snow parameter determination by multichannel microwave radiometry, *Rem. Sens. Environ.*, **8**, 211–224.

Holt, A. R. (1982), Electromagnetic wave scattering by spheroids: A comparison of experimental and theoretical results, *IEEE Trans. Ant. Prop.*, AP-30, 758–760.

Holzer, J. A. and C. C. Sung (1978), Scattering of electromagnetic waves from a rough surface. II, *J. Appl. Phys.*, **49**, 1002–1011.

Hunka, J. F. and K. K. Mei (1981), Electromagnetic scattering by two bodies of revolution, *Electromagnetics*, **1**, 329–347.

Hutley, M. C. and V. M. Bird (1973), A detailed experimental study of the anomalies of a sinusoidal diffraction grating, *Opt. Acta.*, **20**, 771–782.

Ishimaru, A. (1975), Correlation function of a wave in a random distribution of stationary and moving scatterers, *Radio Science*, **10**, 45–52.

Ishimaru, A. (1978), *Wave Propagation and Scattering in Random Media*, **1** and **2**, Academic Press, New York.

Ishimaru, A. and R., L-T. Cheung (1980), Multiple scattering effects on wave propagation due to rain, *Ann. Telecommunication*, **35**, 373–379.

Ishimaru, A. and Y. Kuga (1982), Attenuation constant of a coherent field in a dense distribution of particles, *J. Opt. Soc. Am.*, **72**, 1317–1320.

Ishimaru, A. and K. J. Pinter (1980), Backscattered pulse shape due to small-angle multiple scattering in random media, *Radio Science*, **15**, 87–93.

Ishimaru, A., D. Lesselier, and C. Yeh (1984), Multiple scattering calculations for nonspherical particles based on the vector radiative transfer theory, *Radio Science*, **19**, 1356–1366.

Ishimaru, A., R. Woo, J. W. Armstrong, and D. C. Backman (1982), Multiple scattering calculation of rain effects, *Radio Science*, **17**, 1425–1433.

Ishimaru, A. and C. W. Yeh (1984), Matrix representations of the vector radiative transfer equations for randomly distributed nonspherical particles, *J. Opt. Soc. Am. A*, **1**, 359–364.

Iskander, M. F., A. Lakhtakia, and C. H. Durney (1983), A new procedure for improving the solution stability and extending the frequency range of the EBCM, *IEEE Trans. Ant. Prop.*, AP-31, 317–324.

Jackson, J. D. (1975), *Classical Electrodynamics*, John Wiley & Sons, New York.

Jackson, T. J. and T. J. Schmugge (1981), Aircraft active microwave measurements for estimating soil moisture, *Photogrammetr. Eng. Rem. Sens.*,

47, 801–805.

Johnson, J. D. and L. D. Farmer (1971), Use of side-looking airborne radar for sea ice identification, *J. Geophys. Res.*, 76, 2138–2155.

Karal, F. C. and J. B. Keller (1964), Elastic electromagnetic and other waves in random medium, *J. Math. Phys.*, 5, 537–547.

Karam, M. A. and A. K. Fung (1982), Vector forward scattering theorem, *Radio Science*, 17, 752–756.

Karam, M. A. and A. K. Fung (1983), Scattering from randomly oriented circular discs with applications to vegetation, *Radio Science*, 18, 557–565.

Kattawar, G. W., G. N. Press, and F. E. Catchings (1973), Matrix operator theory of radiative transfer, *Appl. Opt.*, 12, 1071–1084.

Keller, J. B. (1964), Stochastic equations and wave propagation in random media, *Proc. Symp. Appl. Math.*, 16, 145–170 (Am Math. Soc., Providence, RI.).

Keller, J. B. and F. C. Karal (1966), Effective dielectric constant, permeability and conductivity of a random medium and the velocity and attenuation coefficient of coherent waves, *J. Math. Phys.*, 7, 661–670.

Kerker, M. (1969), *The Scattering of Light and Other Electromagnetic Radiation*, Academic Press, NY.

Ketchum, R. D., Jr., and S. G. Tooma, Jr. (1973), Analysis and interpretation of air-borne multifrequency side-looking radar sea ice imagery, *J. Geophys. Res.*, 3, 520–538.

Kodis, R. D. (1966), A note on the theory of scattering from an irregular surface, *IEEE Trans. Ant. Prop.*, AP–14, 77–82.

Kohler, W. E. and G. C. Papanicolaou (1973), Power statistics for wave propagation in one dimension and comparison with radiative transport theory, *J. Math. Phys.*, 14, 1733–1745.

Kohler, W. E. and G. C. Papanicolaou (1981), Some application of the coherent potential approximation, in *Multiple Scattering and Waves in Random Media*, edited by P. L. Chow, W. E. Kohler, and G. C. Papanicolaou, pp. 199–223, North-Holland Publishing Company, New York.

Kong, J. A. (1975), *Theory of Electromagnetic Waves*, Wiley-Interscience, New York.

Kong, J. A., ed. (1981), *Research Topics in Electromagnetic Wave Theory*, Wiley-Interscience, New York.

Kong, J. A., S. L. Lin, and S. L. Chuang (1984), Microwave thermal emission from periodic surfaces, *IEEE Trans. Geosci. Rem. Sens.*, **GE-22**, 377–382.

Kong, J. A., R. Shin, J. C. Shiue, and L. Tsang (1979), Theory and experiment for passive remote sensing of snowpacks, *J. Geophys. Res.*, **84**, 5669–5673.

Korringa, J. and R. L. Mills (1972), Coherent-potential approximation for random systems with short-range correlations, *Phys. Rev. B*, **5**, 1654–1655.

Kraus, J. D. (1966), *Radio Astronomy*, McGraw-Hill, New York.

Kravtsov, Y. A. and A. I. Saichev (1982), Effects of double passage of waves in randomly inhomogeneous media, *Sov. Phys. Usp.*, **25**, 494–508.

Kritikos, H. N. and J. Shiue (1979), Microwave remote sensing from orbit, *IEEE Spectrum*, 34–41.

Kuga, Y. and A. Ishimaru (1984), Retroreflectance from a dense distribution of spherical particles, *J. Opt. Soc. Am.*, A, **1**, 831–835

Kunzi, K. F., A. D. Fisher, D. H. Staelin, and J. W. Waters (1976), Snow and ice surfaces measured by the Numbus-5 microwave spectrometer, *J. Geophys. Res.*, **81**, 4965–4980.

Landau, L. and E. Lifshitz (1960), *Electrodynamics of Continuous Media*, Pergamon Press, Oxford.

Lane, J. and J. Saxton (1952), Dielectric dispersion in pure liquids at very high radio frequencies, *Proc. Roy. Soc.*, **A213**, 400–408.

Lang, R. H. and J. S. Sidhu (1983), Electromagnetic backscattering from a layer of vegetation: A discrete approach, *IEEE Trans. Geosci. Rem. Sens.*, **GE-21**, 62–71.

Lang, R. H. (1981), Electromagnetic backscattering from a sparse distribution of lossy dielectric scatterers, *Radio Science*, **16**, 15–33.

Laws, J. O. and D. A. Parsons, (1943), The relationship of raindrop size to intensity, *Trans. Am. Geophys. Union*, **24**, 452–460.

Lax, M. (1951), Multiple scattering of waves, *Rev. Modern Phys.*, **23**, 287–310.

Lax, M. (1952), Multiple scattering of waves II. The effective field in dense systems, *Phys. Rev.*, **85**, 261–269.

Lax, M. (1973), Wave propagation and conductivity in random media, *SIAM-AMS Proc.*, **VI**, 35–95.

Leader, J. C. (1971), Bidirectional scattering of electromagnetic waves from rough surfaces, *J. Appl. Phys.*, **42**, 4808–4816.

Lettau, H. (1971), Antarctic atmosphere as a test tube for meteorological theories, in *Research in the Antarctic*, pp. 443–375, Am. Assn. for the Adv. of Sci., Washington, DC.

Linlor, W. (1980), Permittivity and attenuation of wet snow between 4 and 12 GHz, *J. Appl. Phys.*, **23**, 2811–2816.

Liszka, E. G. and J. J. McCoy (1982), Scattering at a rough boundary – extensions of the Kirchhoff approximation, *J. Acoust. Soc. Am.*, **71**, 1093–1100.

Long, M. W. (1975), *Radar Reflectivity of Land and Sea*, Lexington Books, Lexington, MA.

Lundien, J. R. (1971), Terrain analysis by electromagnetic means, *Technical Report 3-727*, U.S. Army Engineer Waterways Experiment Station, Vicksbury, Miss.

Lynch, P. J. and R. J. Wagner (1968), Energy conservation for rough surface scattering, *J. Acoust. Soc. Am.*, **47**, 819.

Lynch, P. J. and R. J. Wagner (1970), Rough-surface scattering: Shadowing, multiple scatterer and energy conservation, *J. Math. Phys.*, **11**, 3032–3042.

MacDonald, H. C. and W. P. Waite (1971), Soil moisture detection with imaging radars, *Water Resources Res.*, **7**, 100–110.

Maradudin, A. A. (1983), Iterative solutions for electromagnetic scattering by gratings, *J. Opt. Soc. Am.*, **73**, 759–764.

Marshall, J. S. and J. W. M. Palmer (1948), The distribution of rain drops with size, *J. Meteorol.*, **5**, 165–166.

Marvin, A. M. (1980), Kirchhoff approximation and closed-form expressions for atom-surface scattering, *Phys. Rev. B*, **22**, 5759–5767.

Masel, R. I., R. P. Merrill and W. H. Miller (1975), Quantum scattering from a sinusoidal hard wall: Atomic diffraction from solid surface, *Phys. Rev. B*, **12**, 5545–5551.

Mathur, N. C. and K. C. Yeh (1964), Multiple scattering of electromagnetic waves by random discrete scatterers of finite size, *J. Math. Phys.*, **5**, 1619–1628.

Matzler, C., E. Schanda, and W. Good (1982), Towards the definition of optimum sensor specifications for microwave remote sensing of snow, *IEEE Trans. Geosci. Rem. Sens.*, **GE-20**, 57–66.

Maxwell-Garnett, J. C. (1904), *Trans. Roy. Soc.*, London, **203**, 385.

McQuarrie, D. A. (1976), *Statistical Mechanics*, Harper and Row, New York.

Medhurst, R. G. (1965), Rainfall attenuation of centimeter waves: comparison of theory and measurement, *IEEE Trans. Ant. Prop.*, **AP-13**, 550–564.

Mei, K. K. (1974), Unimoment method of solving antenna and scattering problems, *IEEE Trans. Ant. Prop.*, **AP-22**, 760–766.

Meier, M. F. and A. T. Edgerton, (1971), Microwave emission from snow – A progress report, *Proc. of the 7th International Symp. Rem. Sens. Environ.*, **2**, University of Michigan, Ann Arbor, MI.

Meier, M. F. (1975), Application of remote sensing techniques to the study of seasonal snow cover, *J. Glaciol.*, **15**, 251–265.

Merzbacher, E. (1970), *Quantum Mechanics*, John Wiley & Sons, Second Edition, New York.

Millar, R. F. (1973), The Rayleigh hypothesis and a related least squares solution to scattering problems for periodic surfaces and other scatterers, *Radio Science*, **8**, 785–796.

Mo, T., B. J. Choudhury, T. J. Schmugge, J. R. Wang, and T. J. Jackson (1982), A model for microwave emission from vegetation-covered fields, *J. Geophys. Res.*, **87**, 11229–11237.

Mo, T., T. J. Schmugge, and T. J. Jackson (1984), Calculation of radar backscattering coefficient of vegetation-covered soils, *Rem. Sens. Environ.*, **15**, 119–133.

Moore, R. K. (1966), Radar scatterometry – an active remote sensing tool, *Proc. 4th Symp. Rem. Sens. Environ.*, pp. 339–375, University of Michigan, Ann Arbor, MI.

Morgan, M. A. (1980), Finite element computation of microwave scattering by raindrops, *Radio Science*, **15**, 1109–1119.

Morgan, M. A. and K. K. Mei (1979), Finite element computation of scattering by inhomogeneous penetrable bodies of revolution, *IEEE*

Trans. Ant. Prop., AP-27, 202-214.

Morrison, J., G. C. Papanicolaou, and J. B. Keller (1971), Mean power transmission through a slab of random medium, *Commun. Pure Appl. Math.*, 24, 473-489.

Morse, P. M. and H. Feshbach (1953), *Methods of Theoretical Physics*, Part II, McGraw-Hill, New York.

Nakayama, J., H. Ogura, and M. Sakata (1981), A probabilistic theory of electromagnetic wave scattering from a slightly random rough surface, 1. Horizontal polarization, *Radio Science*, 16, 831-845.

Nakayama, J., M. Sakata, and H. Ogura (1981), A probabilistic theory of electromagnetic wave scattering from a slightly random surface, 2. Vertical polarization, *Radio Science*, 16, 847-853.

Neviere, M., D. Maystone, and W. R. Hunter (1978), On the use of classical and conical diffraction mountings for XUV gratings, *J. Opt. Soc. Am.*, 68, 1106-1113.

Newton, R. G. (1966), *Scattering Theory of Waves and Particles*, McGraw-Hill, New York.

Newton, R. W., Q. R. Black, S. Makanvaud, A. J. Blanchard, and B. R. Jean (1982), Soil moisture information and microwave emission, *IEEE Trans. Geosci. Rem. Sens.*, GE-20, 275-281.

Newton, R. W. and J. W. Rouse, Jr. (1980), Microwave radiometer measurements of soil moisture constant, *IEEE Trans. Ant. Prop.*, AP-28, 680-686.

Newton, R. W. (1976), Microwave remote sensing and its application to soil moisture detection, Technical Report RSC-81, Remote Sensing Center, Texas A & M University, College Station, Texas.

Nieto-Vesperinas, M. (1982), Depolarization of electromagnetic waves scattered from slightly rough random surfaces: A study by means of the extinction theorem, *J. Opt. Soc. Am.*, 72, 539-547.

Njoku, E. G. and J. A. Kong (1977), Theory of passive sensing of near-surface soil moisture, *J. Geophys. Res.*, 82, 3108-3118.

Njoku, E. G. and P. E. O'Neill (1981), Multifrequency microwave radiometer measurements of soil moisture, *IEEE Trans. Geosci. Rem. Sens.*, GE-20, 468-475.

Oguchi, T. (1973), Attenuation and phase rotation of radio waves due to rain: Calculation at 19.3 and 34.8 GHz, *Radio Science*, **8**, 32–38.

Oguchi, T. (1983), Electromagnetic wave propagation and scattering in rain and other hydrometeors, *Proc. IEEE*, **71**, 1029–1078.

Ogura, H. (1975), Theory of waves in a homogeneous random medium, *Phys. Rev. A*, **11**, 942–956.

Onstott, R. G., R. K. Moore, and W. F. Weeks (1979), Surface-based scatterometer results of Arctic sea ice, *IEEE Trans. Geosci. Electron.*, **GE-17**, 78–85.

Ozisik, M. N. (1973), *Radiative Transfer*, Wiley-Interscience, New York.

Papanicolaou, G. C., and J. B. Keller (1971), Stochastic differential equations with applications to random harmonic oscillators and wave propagation in random media, *SIAM J. Appl. Math.*, **21**, 287–305.

Parashar, S. K., R. M. Haralick, R. K. Moore, and A. W. Biggs (1977), Radar scatterometer discrimination of sea-ice types, *IEEE Trans. Geosci. Electron.*, **GE-15**, 83–87.

Peake, W. H. (1959), Interaction of electromagnetic waves with some natural surfaces, *IEEE Trans. Ant. Prop.*, AP-7, Special Supplement, S324–S329.

Peake, W. H., D. E. Barrick, A. K. Fung, and H. L. Chan (1970), Comments on 'backscattering of waves by composite rough surfaces,' *IEEE Trans. Ant. Prop.*, **AP-18**, 716–726.

Percus, J. K. and G. J. Yevick (1958), Analysis of classical statistical mechanics by means of collective coordinates, *Phys. Rev.*, **110**, 1–13.

Peterson, B. and S. Strom (1973), T matrix for electromagnetic scattering from an arbitrary number of scatterers and representation of E(3), *Phy. Rev. D*, **8**, 3661–3678.

Polder, D. and J. H. van Santern (1946), The effective permeability of mixture of solids, *Physica*, **12**, 257–271.

Pruppacher, H. R. and R. L. Pitter (1971), A semi-empirical determination of cloud and rain drops, *J. Atmos. Sci.*, **28**, 86–94.

Ralston, A. (1965), *A First Course in Numerical Analysis*, McGraw-Hill, New York.

Rango, A., A. T. C. Chang, and J. L. Foster (1979), The utilization of spaceborne microwave radiometers for monitoring snowpack properties, *Nordic Hydrol.*, **10**, 25–40.

Ray, P. S. (1972), Broadband complex refractive indices of ice and water, *Applied Optics*, **11**, 1836–1844.

Rayleigh, L. (1892), On the influence of obstacles arranged in rectangular order on the properties of a medium, *Phil. Mag.*, **34**, 481–502.

Rice, S. O. (1963), Reflection of EM waves by slightly rough surfaces, in *The Theory of Electromagnetic Waves*, edited by M. Kline, Interscience, New York.

Rosenbaum, J. S. (1969), On the coherent motion in bounded, randomly fluctuating regions, *Radio Science*, **4**, 709–719.

Rosenbaum, S. (1971), The mean Green's function: A nonlinear approximation, *Radio Science*, **6**, 379–386.

Rotenberg, M., R. Bivins, N. Metroplis, and J. K. Wooten, Jr. (1959), *The 3-j and 6-j Symbols*, Technology Press, Cambridge, MA.

Roth, L. (1974), Effective-medium approximation for liquid metals, *Phys. Rev. B*, **9**, 2476–2484.

Rouse, J. W., Jr. (1969), Arctic ice type identification by radar, *Proc. IEEE*, **57**, 605–611.

Rowlinson, J. S. (1965), Self-consistent approximations for molecular distribution functions, *Molecular Physics*, **9**, 217–227.

Ruck, G. T., D. E. Barrick, W. D. Stuart, and C. K. Krichbaum (1970), *Radar Cross-Section Handbook*, **1,2**, McGraw-Hill, New York.

Ryzhov, Y. A., V. V. Tamoikin and V. I. Tatarskii (1965), Spatial dispersion of inhomogeneous media, *Sov. Phys. JETP*, **21**, 433–438.

Ryzhov, Y. A. and V. V. Tamoikin (1970), Radiation and propagation of electromagnetic waves in randomly inhomogeneous media, *Radiophysics Quantum Electron.*, **13**, 273–300.

Sancer, M. I. (1969), Shadow-corrected electromagnetic scattering from a randomly rough surface, *IEEE Trans. Ant. Prop.*, **17**, 577–585.

Saxton, J. A. and J. A. Lane (1952), Electrical properties of sea water, *Wireless Engineer*, **29**, 269–275.

Schmugge, T. J. (1980), Effect of soil texture on the microwave emission from soils, *IEEE Trans. Geosci. Rem. Sens.*, **GE-18**, 353–361.

Schmugge, T. J. (1983), Remote sensing of soil moisture: Recent advances, *IEEE Trans. Geosci. Rem. Sens.*, **GE-21**, 336–344.

Schmugge, T. J. and B. J. Choudhury, (1981), A comparison of radiative transfer models for predicting the microwave emission from soils, *Radio Science*, **16**, 927–938.

Schmugge, T. J., P. Gloerson, T. Wilheit, and F. Geiger (1974), Remote sensing of soil moisture with microwave radiometers, *J. Geophys. Res.*, **79**, 317–323.

Schuster, A. (1905), Radiation through a foggy atmosphere, *Astrophys. J.*, **21**, 1–22.

Schwartz, L. and A. Bansil (1974), Total and component density of states in substitutional alloys, *Phys. Rev. B*, **10**, 3261–3272.

Schwartz, L. and D. L. Johnson (1984), Long-wavelength acoustic propagation in ordered and disordered suspensions, *Phys. Rev. B*, **30**, 4302–4313.

Schwartz, L. and T. J. Plona (1984), Ultrasonic propagation in close-packed disordered suspensions, *J. Appl. Phys.*, **55**, 3971–3977.

Sekera, Z. (1966), Scattering matrices and reciprocity relationships for various representation of the state of polarization, *J. Opt. Soc. Am.*, **56**, 1732–1740.

Semenov, B. I. (1965), Scattering of electromagnetic waves by bounded regions of rough surfaces having finite conductivity, *Radio Eng. Electron. Phys.*, **10**, 1666–1673.

Semenov, B. I. (1966), Approximate computation of scattering of electromagnetic waves by rough surface contours, *Radio Eng. Electron. Phys.*, **11**, 1179–1187.

Sen, P. N., C. Scala, and M. H. Cohen (1981), A self-similar model for sedimentary rocks with application to the dielectric constant of fused glass beads, *Geophysics*, **46**, pp. 781–795.

Shen, J. and A. A. Maradudin (1980), Multiple scattering of waves from random rough surfaces, *Phys. Rev. B*, **22**, 4234–4240.

Shin, R. T. and J. A. Kong (1981), Radiative transfer theory for active remote sensing of homogeneous layer containing spherical scatterers, *J. Appl. Phys.*, **52**, 4221–4230.

Shin, R. T. and J. A. Kong, Theory for thermal microwave emission from a homogeneous layer with rough surfaces containing spherical scatterers, *J. Geophys. Res.*, **87**, 5566–5576.

Shiue, J. C., A. T. C. Chang, H. Boyne, and D. Ellerbruch (1978), Remote sensing of snowpack with microwave radiometers for hydrologic applications, *Proc. 12th International Symp. Rem. Sens. Environ.*, **2**, 877–886, Univ. of Michigan, Ann Arbor, MI.

Singh, V. A. (1981), Analytic (unitary-preserving) approximation for the electronic structure of amorphous systems, *Phys. Rev. B*, **24**, 4852–4854.

Skolnik, M. I. (1962), *Radar Systems*, McGraw-Hill, New York.

Skolnik, M. I. (1970), *Radar Handbook*, McGraw-Hill, New York.

Smith, B. G. (1967), Geometrical shadowing of a random rough surface, *IEEE Trans. Ant. Prop.*, **AP-15**, 668–671.

Sobolev, V. V. (1963), *A Treatise on Radiative Transfer*, Van Nostrand, Princeton, NJ.

Soven, P. (1966), Approximate calculations of electronic structure of disordered alloys – Application to alpha brass, *Phys. Rev.*, **151**, 539–550.

Soven, P. (1967), Coherent-potential model of substitutional disordered alloys, *Phys. Rev.*, **156**, 809–813.

Soven, P. (1969), Contributions to the theory of disordered alloys, *Phys. Rev.*, **178**, 1136–1144.

Staelin, D. H. (1969), Passive remote sensing at microwave wavelengths, *Proc. IEEE*, **57**, 427–459.

Stiles, W. H. and F. T. Ulaby (1980a), Radar observation of snowpacks, in NASA Conference Publication 2153 *Microwave Remote Sensing of Snowpack properties*, Workshop at Fort Collins, Colorado, May 20–22, 1980.

Stiles, W. H. and F. T. Ulaby (1980b), The active and passive microwave response to snow parameters: 1. Wetness, *J. Geophys. Res.*, **85**, 1037–1044.

Stogryn, A. (1967), Electromagnetic scattering from rough finitely conducting surfaces, *Radio Science*, **2**, 415–428.

Stogryn, A. (1970), The brightness temperature of a vertically structured medium, *Radio Science*, **5**, 1397–1406.

Stogryn, A. (1974), Electromagnetic scattering by random dielectric constant fluctuations in a bounded medium, *Radio Science*, **9**, 509–518.

Stogryn, A. (1984), Correlation functions for random granular media in strong fluctuation theory, *IEEE Trans. Geosci. Rem. Sens.*, **GE-22**, 150–154.

Stott, P. (1968), Transport theory for multiple scattering of electromagnetic waves by turbulent plasma, *J. Phys. A*, **1**, 675–689.

Stratton, J. A. (1941), *Electromagnetic Theory*, McGraw-Hill, New York.

Sung, C. C. and J. A. Holzer (1976), Scattering of electromagnetic waves from a rough surface, *Appl. Phys. Letters*, **28**, 429–431.

Sung, C. C. and W. D. Ekerhardt (1978), Scattering of an electromagnetic wave from a very rough semi-infinite dielectric plane (exact treatment of the boundary conditions), *J. Appl. Phys.*, **49**, 994–1001.

Swift, C. T. (1974), Microwave radiometer measurements of the Cape Cod Canal, *Radio Science*, 641–655.

Tai, C. T. (1971), *Dyadic Green's Function in Electromagnetic Theory*, International Textbook, Scranton, PA.

Tai, C. T. (1973), On the eigenfunction expansion of dyadic Green's functions, *Proc. IEEE*, **61**, 480–481.

Tamoikin, V. V. (1971), The average field in a medium having strong anisotropic inhomogeneities, *Radiophysics Quantum Electron.*, **14**, 228–233.

Tan, H. S., A. K. Fung, and H. Eom (1980), A second order renormalization theory for cross-polarized backscatter from a half space random medium, *Radio Science*, **15**, 1059–1065.

Tatarskii, V. I. (1961), *Wave Propagation in a Turbulent Medium*, translated from Russian by R. Silverman, McGraw-Hill, New York.

Tatarskii, V. I. (1964), Propagation of electromagnetic waves in a medium with strong dielectric constant fluctuations, *Sov. Phys. JETP*, **19**, 946–953.

Tatarskii, V. I. (1971), The effects of turbulent atmosphere on wave propagation, National Tech. Information Service, 472, Springfield, VA.

Tatarskii, V. I. and M. E. Gertsenshtein (1963), Propagation of waves in a medium with strong fluctuations of the refractive index, *Sov. Phys. JETP*, **17**, 458–463.

Thiele, E. (1963), Equation of state for hard spheres, *J. Chem. Phys.*, **39**, 474–479.

Throop, G. J. and R. J. Bearman (1965), Numerical solutions of the Percus-Yevick equation for the hard-sphere potential, *J. Chem. Phys.*, **42**,

2408-2411.

Tiuri, M. E. (1982), Theoretical and experimental studies of microwave emission signatures of snow, *IEEE Trans. on Geoscience and Remote Sensing*, **GE-20**, 51-57.

Toigo, F., A. Marvin, and N. R. Hill (1977), Optical properties of rough surfaces: General theory and the small roughness limit, *Phys. Rev. B*, **15**, 5618-5626.

Tomiyasu, K. (1974), Remote sensing of the earth by microwaves, *Proc. IEEE*, **62**, 86-92.

Tomiyasu, K. (1983), Computer simulation of speckle in a synthetic aperture radar image pixel, *IEEE Trans. Geosci. Rem. Sens.*, **GE-21**, 357-363.

Tooma, S. G., R. A. Mennella, J. P. Hollinger, and R. D. Ketchum, Jr. (1975), Comparison of sea-ice type identification between airborne dual-frequency passive microwave radiometry and standard laser infrared techniques, *J. Glaciology*, **15**, 225-239.

Tsang, L. (1984a), Thermal emission of nonspherical particles, *Radio Science*, **19**, 966-974.

Tsang, L. (1984b), Scattering of electromagnetic waves from a half space of nonspherical particles, *Radio Science*, **19**, 1450-1460.

Tsang, L. and A. Ishimaru (1984), Backscattering enhancement of random discrete scatterers, *J. Opt. Soc. Am. A*, **1**, 836-839.

Tsang, L. and J. A. Kong (1976a), Thermal microwave emission from a half-space random media, *Radio Science*, **11**, 599-610.

Tsang, L. and J. A. Kong (1976b), Microwave remote sensing of a two-layer random medium, *IEEE Trans. Ant. Prop.*, **AP-24**, 283-287.

Tsang, L. and J. A. Kong (1977), Thermal microwave emission from a random homogeneous layer over a homogeneous medium using the method of invariant imbedding, *Radio Science*, **12**, 185-194.

Tsang, L. and J. A. Kong (1978), Radiative transfer theory for active remote sensing of half-space random media, *Radio Science*, **13**, 763-773.

Tsang, L. and J. A. Kong (1979), Wave theory for microwave remote sensing of a half-space random medium with three-dimensional variations, *Radio Science*, **14**, 359-369.

Tsang, L. and J. A. Kong (1980a), Multiple scattering of electromagnetic waves by random distribution of discrete scatterers with coherent potential and quantum mechanical formulism, *J. Appl. Phys.*, **15**, 3465–3485.

Tsang, L. and J. A. Kong (1980b), Energy conservation for reflectivity and transmissivity at a very rough surface, *J. Appl. Phys.*, **51**, 673–680.

Tsang, L. and J. A. Kong (1981), Scattering of electromagnetic waves from random media with strong permittivity fluctuations, *Radio Science*, **16**, 303–320.

Tsang, L. and J. A. Kong (1982), Effective propagation constant for coherent electromagnetic waves in media embedded with dielectric scatterers, *J. Appl. Phys.*, **53**, 7162–7173.

Tsang, L. and J. A. Kong (1983), Scattering of electromagnetic waves from a half space of densely distributed dielectric scatterers, *Radio Science*, **18**, 1260–1272.

Tsang, L., E. Njoku, and J. A. Kong (1975), Microwave thermal emission from a stratified medium with nonuniform temperature distribution, *J. Appl. Phys.*, **46**, 5127–5133.

Tsang, L., J. A. Kong, and R. W. Newton (1982a), Application of strong fluctuation random medium theory to scattering of electromagnetic waves from a half space of dielectric mixture, *IEEE Trans. Ant. Prop.*, **AP-30**, 292–302.

Tsang, L., J. A. Kong, and T. Habashy (1982b), Multiple scattering of acoustic waves by random distribution of discrete spherical scatterers with the quasicrystalline and Percus-Yevick approximation, *J. Acoust. Soc. Am.*, **71**, 552–558.

Tsang, L., J. A. Kong, E. Njoku, D. H. Staelin, and J. W. Waters (1977), Theory for microwave thermal emission from a layer of cloud or rain, *IEEE Trans. Ant. Prop.*, **AP-25**, 650–657.

Tsang, L., J. A. Kong, and R. T. Shin (1984), Radiative transfer theory for active remote sensing of a layer of nonspherical particles, *Radio Science*, **19**, 629–642.

Tsang, L., M. C. Kubacsi, and J. A. Kong (1981), Radiative transfer theory for active remote sensing of a layer of small ellipsoidal scatterers, *Radio Science*, **16**, 321–329.

Tsang, L. and R. W. Newton (1982), Microwave emissions from soils with rough surfaces, *J. Geophys. Res.*, **87**, 9017–9024.

Twersky, V. (1962), On scattering of waves by random distributions I. Free space scatterer formalism, *J. Math. Phys.*, **3**, 700–715.

Twersky, V. (1964), On propagation in random media of scatterers, *Proc. Symp. Appl. Math.*, **16**, 84–116 (Am. Math. Soc., Providence, RI.).

Twersky, V. (1967), Multiple scattering of electromagnetic waves by arbitrary configurations, *J. Math. Phys.*, **8**, 589–610.

Twersky, V. (1977), Coherent scalar field in pair-correlated random distributions of aligned scatterers, *J. Math. Phys.*, **18**, 2468–2486.

Twersky, V. (1978), Coherent electromagnetic waves in pair-correlated random distributions of aligned scatterers, *J. Math. Phys.*, **19**, 215–230.

Twersky, V. (1979), Propagation in pair-correlated distributions of small-spaced lossy scatterers, *J. Opt. Soc. Am.*, **69**, 1567–1572.

Twersky, V. (1983), Reflection and scattering of sound by correlated rough surfaces, *J. Acoust. Soc. Am.*, **73**, 85–94.

Twersky, V. (1983), Propagation in correlated distributions of large-spaced scatterers, *J. Opt. Soc. Am.*, **73**, 313–320.

Twomey, S., H. Jacobowitz, and H. B. Howell (1966), Matrix methods for multiple-scattering problems, *J. Atmos. Sci.*, **23**, 289–296.

Ulaby, F. T. (1975), Radar response to vegetation, *IEEE Trans. Ant. Prop.*, **AP-23**, 36–45.

Ulaby, F. T. and P. P. Batlivala (1976), Diurnal variations of radar backscatter from a vegetation canopy, *IEEE Trans. Ant. Prop.*, **AP-24**, 11–17.

Ulaby, F. T. and W. H. Stiles (1980), The active and passive microwave response to snow parameters: 2. Water equivalent of dry snow, *J. Geophys. Res.*, **85**, 1045–1049.

Ulaby, F. T., R. K. Moore, and A. K. Fung (1981), *Microwave Remote Sensing: Active and Passive*, **1** and **2**, Addison-Wesley, Reading, MA.

Valenzuela, G. R. (1967), Depolarization of EM waves by slightly rough surfaces, *IEEE Trans. Ant. Prop.*, **AP-15**, 552–557.

Valenzuela, G. R. (1968), Scattering of EM waves from a tilted slightly rough surface, *Radio Science*, **3**, 1057–1064.

Vallese, F. and J. A. Kong (1981), Correlation function studies for snow and ice, *J. Appl. Phys.*, **52**, 4921–4925.

Bibliography

Van Bladel, J. (1961), Some remarks on Green's dyadic for infinite space, *IRE Trans. Ant. and Prop.*, **AP-9**, 563–566.

Van den Berg, P. M. (1971), Diffraction theory of a reflection grating, *Appl. Sci. Res.*, **24**, 261–293.

Van de Hulst, H. C. (1957), *Light Scattering by Small Particles*, John Wiley & Sons, New York.

Varadan, V. K. (1980), Multiple scattering of acoustic, electromagnetic and elastic waves, in *Acoustic Electromagnetic and Elastic Wave Scattering – Focus on the T-Matrix Approach*, edited by V. K. Varadan and V. V. Varadan, Pergamon Press, New York.

Varadan, V. K., V. N. Bringi, V. V. Varadan, and A. Ishimaru (1983), Multiple scattering theory for waves in discrete random media and comparison with experiments, *Radio Science*, **18**, 321–327.

Varadan, V. K., V. V. Varradan, and Y. H. Pao (1978), Multiple scattering of elastic waves by cylinders of arbitrary cross section, I. SH waves, *J. Acoust. Soc. Am.*, **63**, 1310–1319.

Velicky, B., S. Kirkpatrick, and H. Ehrenreich (1968), Single-site approximations in the electronic theory of simple binary alloys, *Phys. Rev.*, **175**, 747–766.

Verlet, L. and J. J. Weis (1972), Equilibrium theory of simple liquids, *Phys. Rev. A.* **5**, 939–952.

Vezzetti, D. J. and J. B. Keller (1967), Refractive index, attenuation, dielectric constant and permeability of waves in a polarizable medium, *J. Math. Phys.*, **8**, 1861–1870.

Von Hippel, A. R. (1954), *Dielectric Materials and Applications*, M.I.T. Press, Cambridge, MA.

Wagner, R. J. (1967), Shadowing of randomly rough surfaces, *J. Acoust. Soc. Am.*, **41**, 138–147.

Waite, W. P. and H. C. McDonald (1969–70), Snow-field mapping with K-band radar, *Rem. Sens. Environ.*, **1**, 143–150.

Wait, J. R. (1970), *Electromagnetic Waves in Stratified Media*, 2nd ed., Pergamon Press, New York.

Wang, J. R. (1980), The dielectric properties of soil-water mixtures at microwave frequencies, *Radio Science*, **15**, 977–985.

Wang, J. R. and T. J. Schmugge (1980), An empirical model for the complex permittivity of soils as a function of water content, *IEEE Trans. Geosci. Rem. Sens.*, **GE-18**, 288–295.

Wang, J. R., P. E. O'Neill, T. J. Jackson, and E. T. Engman (1983), Multifrequency measurements of the effects of soil moisture, soil texture and surface roughness, *IEEE Trans. Geosci. Rem. Sens.*, **GE-21**, 44–51.

Wang, J. R., R. W. Newton, and J. W. Rouse (1980), Passive microwave remote sensing of soil moisture: The effect of tilled row structure, *IEEE Trans. Geosci. Rem. Sens.*, **GE-18**, 296–302.

Wang, J. R., J. C. Shiue, S. L. Chuang, R. T. Shin and M. Dombrowski (1984), Thermal microwave emission from vegetated fields; A comparison between theory and experiment, *IEEE Trans. on Geosci. Rem. Sens.*, **GE-22**, 143–150.

Waseda, Y. (1980), *The Structure of Non-Crystalline Materials, Liquids and Amorphous Solids*, McGraw-Hill, New York.

Waterman, P. C. and R. Truell (1961), Multiple scattering of waves, *J. Math. Phys.*, **2**, 512–537.

Waterman, P. C. (1965), Matrix formulation of electromagnetic scattering, *Proc. IEEE*, **53**, 805–811.

Waterman, P. C. (1968), New formulation of acoustic scattering, *J. Acoust. Soc. Am.*, **45**, 1417–1429.

Waterman, P. C. (1971), Symmetry, unitarity and geometry in electromagnetic scattering, *Phys. Rev. D*, **3**, 825–839.

Waterman, P. C. (1975), Scattering by periodic surfaces, *J. Acoust. Soc. Am.*, **57**, 791–802.

Waters, J. W. (1976), Absorption and emission of microwave radiation by atmospheric gases, in *Methods of Experimental Physics*, M. L. Meeks, ed., 12, Part B, Radio Astronomy, Academic Press.

Watson, K. M. (1969), Multiple scattering of electromagnetic waves in an underdense plasma, *J. Math. Phys.*, **10**, 688–702.

Watson, K. M. (1970), Electromagnetic wave scattering within a plasma in the transport approximation, *Phys. Fluids*, **13**, 2514–2523.

Watson, J. G. and J. B. Keller (1983), Reflection, scattering, and absorption of acoustic waves by rough surfaces, *J. Acoust. Soc. Am.*, **74**, 1887–1894.

Weil, H. and C. M. Chu (1976), Scattering and absorption of electromagnetic radiation by thin dielectric discs, *Appl. Optics*, **15**, 1832–1836.

Wertheim, M. S. (1963), Exact solution of the Percus-Yevick integral equation for hard spheres, *Phys. Rev. Lett.*, **20**, 321–323.

Wertheim, M. S. (1964), Analytical solution of the Percus-Yevick equation, in *Equilibrium Theory of Classical Fluids*, edited by H. L. Frisch and J. L. Lebowitz, W. A. Benjamin, Inc., New York.

Whitman, G. M. and F. Schwering (1977), Scattering by periodic metal surfaces with sinusoidal height profiles – a theoretical approach, *IEEE Trans. Ant. Prop.*, **AP-25**, 869–876.

Wichramasinghl, M. C. (1973), *Light Scattering Functions for Small Particles*, John Wiley & Sons, New York.

Wiener, N. (1958), *Nonlinear Problems in Random Theory*, MIT Press, Cambridge, Mass.

Wilheit, T. T. (1978), Radiative transfer in a plane stratified dielectric, *IEEE Trans. Geosci. Electron.*, **GE-16**, 138–143.

Wilheit, T. T., A. T. C. Chang, M. S. V. Rao, E. B. Rodgers, and J. S. Theon (1977), A satellite technique for quantitatively mapping rainfall rate over the oceans, *J. Appl. Meteorology*, **16**, 551–560.

Wu, T. K. and L. L. Tsai (1977), Scattering from arbitrary-shaped lossy dielectric bodies of revolution, *Radio Science*, **12**, 709–718.

Yang, C. C. and K. C. Yeh (1984), Effects of multiple scattering and Fresnel diffraction on random volume scattering, *IEEE Trans. Ant. Prop.*, **AP-32**, 347–355.

Yeh, K. C. (1983), Mutual coherence functions and intensities of backscattered signals in a turbulent medium, *Radio Science*, **18**, 159–165.

Zavody, A. M. (1974), Effect of scattering by rain on radiometer measurements at millimeter wavelengths, *Proc. IEE*, **121**, 257–263.

Ziman, J. M. (1979), *Models of Disorder*, Cambridge University Press, New York.

Zipfel, G. C. Jr., and J. A. DeSanto (1972), Scattering of a scalar wave from a random rough surface: A diagrammatic approach, *J. Math. Phys.*, **13**, 1903–1911.

Zuniga, M. and J. A. Kong (1980), Active remote sensing of random media,

J. Appl. Phys., **51**, 74–79.

Zuniga, M. and J. A. Kong (1981), Mean dyadic Green's function for a two-layer random medium, *Radio Science*, **16**, 1255–1270.

Zuniga, M. A. and J. A. Kong (1980), Modified radiative transfer for a two-layer random medium, *J. Appl. Phys.*, **51**, 5228–5244.

Zuniga, M., J. A. Kong, and L. Tsang (1980), Depolarization effects in the active remote sensing of random media, *J. Appl. Phys.*, **51**, 2315–2325.

Zuniga, M., T. M. Habashy, and J. A. Kong (1979), Active remote sensing of layered random media, *IEEE Trans. Geosci. Electron.*, **GE-17**, 296–302.

Zwally, H. J. (1977), Microwave emissivity and accumulation rate of polar firn, *J. Glaciology*, **18**, 195–215.

INDEX

Absorption, 149, 220, 234
  of atmospheric gases, 306
  attenuation, 381
  cross section, 138, 146, 162
  modified, 148, 150
  profile, 221, 223, 297
  scale, 394
  by water droplets, 222, 306, 308
Absorption coefficient, 131, 221, 223, 234, 254
  small ellipsoids, 162
  nonspherical particles, 147
  spheres, 230
Absorptive cross section, 155, 158, 162
Absorptivity, 10, 15
Active remote sensing, 1
Addition of intensities, 330, 336
Addition theorem, 189, 199, 213
  spherical harmonics, 199
  vector spherical harmonics, 189, 212, 513, 569
Aerosols, 20
Albedo, 246, 251, 276, 357, 373, 559, 561
  iterative parameter, 254
  large, 311, 367
  small, 220, 257, 258, 384
Aligned scatterers, 245, 527
Amorphous semiconductors, 542
Amorphous solid, 480, 481
Anisotropic medium, 10, 22
Antarctic, 318
Antarctica, 260, 261
Associated Legendre polynomial, 169, 172, 192, 211, 492
Atmosphere, 20-21
Atmospheric gaseous absorption, 280
  coefficients, 284
  profile, 306
Atmospheric gaseous emission, 285
Atmospheric gases, 222
Attenuation rate, 139, 354, 373, 546, 559
Axisymmetric objects, 197, 211

Azimuthal indices, 197
Azimuthal symmetry, 228, 246, 249, 258, 286

Background absorption coefficient, 327
Background absorption profile, 308
Background Green's operator, 465
Background medium absorption, 273
Backscattering coefficient, 381, 383, 438, 515, 538
  for three-layer random medium, 410
Backscattering cross sections, 88, 108, 111, 254, 403
Backscattering enhancement, 228, 235, 328-330, 334-336, 364, 392
  for random discrete scatterers, 358-370
  for reflective boundary, 336
Backscattering matrices, 302
Backward scattering coefficient, 292
Backward scattering phase matrices, 266
Basaltic rock chips, 273
Bayes' rule, 433, 455, 456
Bessel functions, 64, 81, 115, 375
Bethe-Salpeter equation, 112, 318, 337, 347-351, 352, 424, 548
  for covariance, 414
  diagrammatic representation, 349
  for dyadic field covariance, 399
Bilocal approximation, 342-347, 378, 393, 422, 476
  to Dyson's equation, 352, 376, 571
  to scalar wave equation, 343-344, 411
  validity of, 380
  to vector wave equation, 344-345
Bilocal diagrams, 343
Bistatic scattering coefficients, 152, 227, 361, 438, 506, 516
  for cyclical intensity, 360
  for ladder intensity, 356, 358
  definition, 3, 82, 254, 356
  of half-space spherical particles, 237-240

of isotropic point scatterers, 357
of rough surface, 88, 89, 94, 95, 108
Bistatic scattering intensity, 481
Black body, 10
Boltzmann probability distribution, 6
Boltzmann's constant, 6, 46, 130, 143, 482, 486
Born approximation, 155, 163, 318, 319, 321, 392, 410
  Nth order, 338
  second order, 359
Born series, 318, 319, 543
Bose-Einstein statistics, 4
Boundary conditions, 60-61, 150, 152, 200-207, 267, 395, 401, 442
Boundary layer effect, 382, 492, 493, 494, 508, 551
Boundary reflection, 556
Boundary value problem, 220, 291
Brewster angle, 29, 50
  effect, 506, 518, 521
Brightening effects, 281
  scattering induced, 234, 284, 285
Brightness temperature, 10, 41, 116, 121, 153, 222, 231, 296, 303
  ice over water, 305
  of stratified medium, 45-53
Brine pockets, 18

Canonical ensemble, 486
Cauchy's theorem, 484
Characteristic correlation function, 330
Characteristic function, 79
Characteristic impedance, 143
Characteristic wave, 139, 141, 525, 527, 537
  polarization, 460, 532
Christoffel weighting functions, 264, 301
Classical mixture formulas, 429
Classification of random discrete scatterers, 429
Clausius-Mossoti formula, 431, 506
Cloud, 20, 23, 223, 232-235, 280, 291, 306
  droplets, 21
  over ocean, 283
Cluster expansion, 325, 339
Coherent potential, 464-469, 475, 542, 543
  dispersion relations, 543
  operator, 465
  propagator, 543
Coherent propagation, 243
Coherent reflection, 435, 506, 529-531, 537, 538
Coherent Stokes vector, 407

Coherent wave decay distance, 400
Coherent wave-like effects, 404
Complementary error function, 95
Complex conjugate vector space, 348
Complex transmitted angle, 528
Composite rough surfaces, 112
Conditional average, 456, 491, 567
  of exciting field, 507, 511, 513, 526
Conditional probability, 433, 455, 462, 464
Configurational average, 433, 436, 462, 465
  of multiple scattering equations, 455-458
Conical diffraction, 57, 166
Conjugate field, 327
Continuous random medium, 476
  with small variance of permittivity, 346, 350
Coordinate representation, 443, 444, 478, 554, 564
Correlation coefficient, 79
Correlation function, 107, 165, 166, 321, 384, 421
  anisotropic, 376
  determination of, 420
  direct computation of, 166
  exponential, 323, 344, 345
  Gaussian, 87
  nonspherical, 377
  spherically symmetric, 318, 344, 345, 376-378, 381
  time and space, 42
  two-point, 321, 325, 339, 397
Correlation length, 81, 87, 107, 321, 330, 343, 392, 400, 571
Correlation of particle positions, 337, 430, 432, 549, 555
Corrugation depth, 65
Covariance function, 16
Covariance matrix, 117
Covariance of field, 337, 370, 390
Critical angle, 203
Cyclical:
  bistatic coefficient, 362, 363
  diagrams, 319, 337, 359
  equation, 319, 361, 363, 548, 555-563
  integral equation, 362
  intensity, 563
  operator, 360
  terms, 359, 360, 365, 367
Cylinders, 19
Cylindrical structure, 166-168, 314

Darkening effect, 281, 285
  caused by ice, 311

Index

scattering induced, 234, 263, 270, 273
Debye equation, 17
Dense medium, 18, 279, 429, 550, 556
Dense non-tenuous particles, 476, 506, 549, 551, 561
Depolarization, 19, 88, 108, 111, 237, 252, 324, 514, 525, 536, 571
  effects, 325-327
  return, 384, 540
  backscattering cross sections, 256
  strong, 241
Deterministic symmetry, 258
Diagrammatic technique, 112, 337, 347, 378, 478, 549, 562
  equation, 360, 563
  notation, 352, 360, 444
  strongly connected, 340, 341, 348, 349, 413, 422
  weakly connected, 340, 341, 343, 348, 349, 413
Dielectric constant, 17
  ice, 427
  rocks, 376
  snow, 376
Dielectric enhancement effect, 420
Dielectric oblate spheroids, 537
Dielectric response of rocks, 418
Diffraction integral, 72
Diffuse scattering, 270
Dirac delta function, 33, 56, 80, 103, 151, 165, 253, 344, 471
Dirac's notation, 443
Discrete ordinate-eigenanalysis, 220, 234, 257, 258-291
Discrete spherical particles, 259
Discrete spherical scatterer model, 275
Disks, 19, 249
Dispersion relation, 459, 463, 468, 497, 543
  for coherent effective propagation constant, 490
  of coherent wave, 457
  for composite medium, 497
  for dielectric spheres, 543-546
  for effective propagation, 509, 529
  for horizontal polarization, 535
  for ordinary wave, 538, 540
  for vertical polarization, 534
  under EFA, 472
  under EFA-CP, 467, 474
  under QCA, 464, 473, 479, 567
  under QCA-CP, 474
Dispersive effects, 521
Displacement relation, 431, 432
Distance scales, 391, 400, 557

Distorted Born approximation, 382, 515, 530, 536, 570
Divergence of beam factor, 202, 562
Divergence theorem, 145, 434
Double resonant scattering, 332
Double scattering, 227, 252, 327, 328
Doubling method, 112
Drop-size distribution, 222, 232, 233, 280, 307, 310
Duality, 27, 28
Dyad transition operator, 444
Dyadic Green's function, 32-41, 71, 183, 321, 397
  singularity of, 319, 345, 376, 377, 412, 470
  symmetry relation, 143
Dyadic Green's operator, 443
Dyadic notation, 476
Dyson's equation, 112, 337-342, 352, 393, 397
  diagrammatic representation, 342
  for mean field, 347, 414, 476
  for mean scalar Green's function, 421

Earth terrain, 16-21
Effective conductivity, 420
Effective field approximation (EFA), 458-461, 465, 470-472, 475, 479, 546
Effective field approximation-coherent potential (EFA-CP), 466-467, 473-474, 544
Effective loss tangent, 344, 498, 500
Effective medium theory, 376, 379
Effective optical thickness, 281
Effective permittivity, 19, 319, 377, 428, 430-431, 535
  bilocal approximation, 377, 379-381
  mixture of coated spheres, 419
  for oblate spheroids, 536
  very low frequency limit, 377
Effective propagation constants, 140, 343, 353, 396, 422, 460, 463, 498, 538
  for mean Green's function, 343
  for TE and TM polarizations, 398
  for two-layer random medium, 406
Effective propagation vector, 459
Effective uniaxial medium, 535
Effective wavenumber, 508
  for characteristic wave, 527
Eigenanalysis, 220
Eigenmatrix, 141, 224
Eigensolutions, 140
Eigenvalues, 140, 243, 286

Eigenvectors, 141, 286, 460, 529
Electric dipole, 190, 497, 517
　interaction terms, 534
Electric dipole moment, 44
Electron diffraction studies, 481
Electrostatic field, 411
Ellipsoidal correlation, 527
Ellipsoidal scatterers, 160, 564
Ellipsoids, 133, 139, 208, 412, 564
Elliptical correlation, 527
Elliptical polarization, 121-125, 208
Emission:
　from bare agricultural field, 67
　from ice, 299
　from ocean surface, 285
　from water, 299, 305
Emission vector, 121, 129, 131, 142- 148, 149, 158, 303
Emissivity, 10, 16, 22, 29, 41, 129, 149, 204
　lower limit, 204
　of rough surface, 96
　of periodic surfaces, 65-70
　upper limit, 98, 204
Energetically self-consistent formulation, 414
Energy conservation, 65, 90-93, 95, 178, 350, 414-417, 475-479, 549, 561
Energy density, 46
　spectrum, 4
Energy flux vector, 414, 476
Ensemble average, 46, 80, 87, 320, 347, 352, 455, 515, 526
　of fields, 511
　over particle positions, 329
Equation of state, 486, 487
Euler's constant, 364
Eulerian angles of rotation, 133, 158-160, 208, 245, 525, 533, 538
Evanescent mode, 67
Ewald-Oseen extinction theorem, 494, 511, 514, 535
　generalized, 490-497
　for random medium, 506, 571
Exciting field, 431, 439, 446, 452, 492, 500
Exclusion volume, 412, 434, 484, 527
　nonspherical, 471
　shape of, 344, 377, 411
　spherical, 345, 377, 413, 420, 471, 493
Expectation value, 46, 456
Exponential integral, 357, 359, 375, 424
Extended boundary condition (EBC), 53, 65, 99, 100, 181-188, 446, 447, 455
Extinction coefficient, 158, 197, 230, 234, 270, 550

Extinction cross section, 137, 138, 155, 177, 191
　sphere, 189-190
Extinction distance, 371, 392, 557
Extinction matrix, 121, 129, 131, 148, 158, 177, 224, 242, 253, 302, 401
　modified, 149
　for nonspherical particles, 138-142
　for radiative transfer equations, 461
Extinction rate, 351, 354, 427
Extinction theorem, 55, 99, 113, 138, 434
Extraordinary wave, 538
Extreme order, 481

Far-field approximation, 37, 72, 132, 163, 177, 321, 355, 458, 516, 530
　dyadic Green's function, 321, 331, 410
Far-field scattering dyad, 458, 532
Feynman diagrams, 112, 339, 347, 355
Field correlation, 350
Field covariance, 350, 550
First-order smoothing approximation, 342
Floquet modes, 57-60, 66, 69
Fluctuation-dissipation theory, 42-53, 129, 142
Fog, 20, 241
Foldy's approximation, 139, 149, 458-461, 500, 531-532
Forest terrain, 330
Forward propagation matrix, 40
Forward scattering coefficient, 292
Forward scattering phase matrices, 266
Forward scattering theorem, 138
Fourier series expansion, 53, 56, 62, 286
Fourier transform, 32, 42, 104, 107, 165, 214, 343, 351, 462, 481
　of correlation functions, 326
　of direct correlation function, 485
　of pair-distribution function, 481, 485
　of specific intensity, 323, 370, 390, 550
　of total correlation function, 485
　of two-point correlation function, 407
　three-dimensional, 380
　two-dimensional, 356
Fourth-order moment of the field, 351
Fractional exclusion volume, 485, 527
Fractional volume, 17, 158, 209, 223, 378, 536
　effective, 279, 280
　of ice in dry snow, 427
　of scatterers, 472, 498, 521, 546
Fresnel emissivity, 299

Index

Fresnel reflection coefficient, 16, 73, 104, 242, 437, 518, 553
Fresnel reflectivity, 262, 267, 542
Fresnel transmission coefficient, 332, 382, 436, 572
Fresnel transmissivity, 268, 327

Gas, 480, 481
Gaseous absorption, 308
Gaussian random process, 117, 338
Gaussian random variable, 71
Gaussian statistics, 325
Gaussian-Legendre quadrature, 188, 263, 270, 282, 286, 287, 300
Generalized Ewald-Oseen extinction theorem, 490-497, 508-513, 525-529, 568, 570
Generalized Lorentz-Lorenz law, 490-497, 508-513, 525-529, 568
Generalized susceptibility, 42
Geometrical optics approximation, 77, 85, 203, 217
Glory effect, 235, 239, 240, 328
Grand canonical ensemble, 486
Graphical representations, 340
Green's function:
 infinite space, 477
 periodic, 57
 two-dimensional, 55
 unperturbed, 351
Green's operator, 444, 463, 465
Green's theorem, 65, 493
 scalar, 55
 vector, 115, 182, 184, 211, 441
Ground, 256, 257
Ground truth measurement, 279

Half-space:
 laminar structure, 318
 identical spherical scatterers, 490
 spherical scatterers, 273
 random medium, 319
Hamiltonian, 42
Hankel functions, 169, 178
Hard sphere potential, 483, 486, 488, 496
 radially symmetric, 489
Haze, 20, 291
Hermitian adjoint of operator, 477
Hermitian conjugate, 180
High absorption, 24

Hole-correction (HC) approximation, 479, 481
Hole-correction (HC) pair function, 505
Homogeneous profiles, 220, 258
Huygen's principle, 55, 71, 99, 136
Hydrometeors, 20, 461, 532

Ice, 18, 241, 297, 375
 measurements, 290
 layer in atmosphere, 305
Ice particles, 279
 in snow, 538
 in upper atmosphere, 306
Ill-conditioned matrix, 65
Illumination function, 93
Incident field state, 446
Incident flux vector, 301
Incoherent intensities, 78
Incoherent scattered field, 529-531
Incoherent scattered intensity, 515-517
Incoherent scattering, 435
Independent particle position, 352, 464, 479
Independent position approximation, 479, 481
Independent scattering, 280, 427, 546
Induced dipole, 156, 430-432, 519
Inhomogeneous slab, 300-303
Inhomogeneous profiles, 28, 220, 291
 of absorption and temperature, 285, 305-311
Initial value problem, 220, 291, 302
Integral equation, 193, 224, 226, 231, 319, 338, 359
Integral representation, 32-34, 211
 of dyadic Green's function, 410
 of vector spherical harmonics, 172
 of vector spherical waves, 211
Integral variable transformation, 375
Integrated optical relation, 416, 476
Integrated water content, 309
Intensity operator, 348-350, 352, 414, 477, 555
 exact, 478
 for small scatterers, 555
Interference effects, 52, 319, 390
 constructive, 31, 228, 400, 401, 404, 558, 561
 destructive, 31, 558
Interpenetration of particles, 439, 483
Invariant imbedding, 220, 285, 291-311
Inverse of rotation group, 195

Isotropic point scatterers, 224-228, 240, 250-252, 351, 359
Isotropic radiated field, 8
Iterative method, 220, 411

Jacobi polynomial, 195, 212
Joint probability density, 79
  for particle positions, 352

Kasterin's representation, 493, 567
Kirchhoff approximation, 70-98, 203, 205, 217
Kirchhoff's law, 10, 14-16
Kronecker delta, 83, 545

Ladder approximation, 319, 337, 350, 352, 355, 359, 374, 399, 476, 478, 562
  of Bethe-Salpeter equation, 318, 371, 390, 401
  of Dyson's equation, 352
  of intensity operator, 416
  for isotropic point scatterer, 351-358
  limitation of, 392
  modified, 476, 478, 479, 549, 550, 555, 562
  second moment of field, 475
Ladder:
  bistatic scattering coefficient, 358
  equation, 363
  intensity, 356, 360, 563
  series, 356
Ladder terms, 355, 356, 548
  Nth order, 365
  second-order, 359, 365
  summation, 359
Laminar structure, 165-166, 259-263, 297, 314
Laplace equation, 138, 155, 156, 192, 208
Layered random medium, 319-336
Layered structure, 318
Legendre polynomial, 200, 213, 237, 264, 286, 287, 492, 493
Lippmann-Schwinger equation, 445, 463, 466, 469, 470, 567
Liquid, 480, 481
  metals, 542
  molecules, 489
  structures, 485
Liquid water content, 21
Local thermodynamic equilibrium, 43
Lorentz-Lorenz formula, 431

Lorentz-Lorenz law, 511, 514, 529, 531, 532, 533, 537, 553, 571, 572
  generalized, 490-497
  for random medium, 506
Loss tangent, 239, 297, 390
Lossless scatterer, 180 .
Lossy scatterers, 190, 239
Low concentration of particles, 354, 525
Low-absorption areas, 323
Low-frequency approximation, 223, 246, 376, 555
  brightness temperatures, 262
  effective propagation constant, 376
Low-frequency limit, 211, 432, 470, 472, 474, 497, 517-520, 533-537
  for spheroids, 533
Low-frequency range, 234
Lunar regolith, 273

Macroscopic field, 431, 432, 436
Magnetic dipole moment, 44
Marshall and Palmer distribution, 21
Mass density, 223
Mass operator, 341, 352, 413, 422, 477
  coordinate representation, 477
  dyadic, 342
  for discrete scatterers, 464
  under QCA-CP, 469
Mass profile, 223
Maxwell-Garnett mixing formula, 420, 431
Mean dyadic Green's function, 341, 342
Mean field, 78, 337
Mean Green's function, 355, 393, 476, 551
Mean square surface slope, 87, 95, 206
Mean wavelength, 392
Mean wavenumber, 338
Meteorology, 223
Method of asymptotics, 86
Method of moments, 364
Mie extinction coefficient, 234
Mie scattering, 138, 212, 234, 275, 307
  pattern, 235, 239, 328
  phase functions, 273, 274
  tabulation of, 239
Mixed representation, 470, 472
Mixture formulas, 19, 431, 475
Mixture, 16, 375, 378
  of three constituents, 385, 388
  of two constituents, 384, 420
Modified Bessel functions, 97, 110
Modified physical optics expansion, 62
Modified radiative transfer (MRT) equations, 319, 370, 390-409, 422-423

Index

Moisture content, 18
Molecular polarizability of particle, 431
Momentum eigenstate, 443, 458
Momentum operator, 443, 465
Momentum representation, 443, 457, 459, 462, 466, 472, 475, 543
Monostatic coefficient, 3
Monte Carlo simulation, 488
Mueller's method, 500
Multilayer model, 314
Multiple resonances, 273, 332
Multiple scattering, 76, 207, 243, 328, 359
  equations, 445-446
  by large spheres, 290
Multiply-scattered amplitude formalism, 505
Mutual coherence function, 121

Nadir, 51
Natural axes, 158, 193
Natural frame, 193, 245, 525, 533
Natural light, 125-127
Near-surface region, 24
Neumann matrices, 62
Neumann series, 338, 341, 350, 413
  averaged, 347
  for covariance, 347
  for random dyadic Green's function, 347
Neutron diffraction studies, 481
Neutron diffusion, 120
Non-absorptive, 561
Non-emissive slab, 301
Non-Gaussian statistics, 339
Non-spherical particles, 19, 525
Non-tenuous medium, 429
Nonabsorptive scatterers, 273
Nondispersive, 520
Nonlinear approximation, 337, 342, 346-347, 351, 392, 397, 470, 475
  in random medium theory, 464
  of mean field, 350, 475
  to Dyson's equation, 346, 393, 414, 422
  to mass operator, 416
Nonlinear equation, 470, 474
Nonlinear integral equation, 469
Nonspherical particles, 223, 241-250
Nontenuous dense media, 351
Nonuniform, 24
  scattering profiles, 297
  temperature profiles, 260
Normalized covariance function, 164, 214
Normalized phase velocity, 498, 500
Numerical simulations, 112

Numerical solution, 234, 256, 257
Numerical techniques, 193

Oblate spheroids, 161, 211, 244-246, 538
  aligned, 538, 540
  axes of symmetry, 536
  small, 312, 316
Operator equation, 444, 457, 464, 563, 564
  for mean Green's operator, 469
Operator form, 454, 464
  of QCA-CP, 551
Operator formalism, 475
Opposition effect, 235
Optic axes, 28
Optical distance, 358
Optical gratings, 53
Optical theorem, 135-138, 155, 177, 190, 191, 212
Optical thickness, 225, 234, 281, 357, 359, 367, 558
Ordinary wave, 538
Orientation distribution, 245, 525, 533
Orientation of particle, 455, 526
Orientation probability density function, 198, 533
Ornstein-Zernike equation, 482, 485
Orthogonality relation, 170, 196
  for momentum eigenstates, 443
  of rotation group representation, 196
  of vector spherical harmonics, 510, 545, 569
  of Wigner 3j symbols, 512, 569
Orthonormal system 72, 102
Over-estimation of scattering effect, 280

Pair-correlation function, 464, 543, 550
Pair-distribution function, 433, 462, 479-489, 521, 567
  for nonspherical particles, 525, 527
  of particle positions, 435, 479
  Percus-Yevick, 489, 498, 500, 505, 520, 521, 537, 543
  short range effect, 479, 552
  spherical, 527
Partial order, 481
Partial polarization, 121, 125-127
Particle size distributions, 275
Passive remote sensing, 4
Percus-Yevick, 489, 505, 531, 549
Percus-Yevick approximation, 567
  defects of, 486
  improvement of, 486-489

Percus-Yevick equation, 482-485, 487
Percus-Yevick hard-sphere pair function, 485-486
Periodic surface, 18, 53
  coupled matrix equations, 60-65
  Dirichlet problem, 62
  hybrid matrix, 63
Permeability tensors, 28
Permittivity, 28
  of fresh water, 31
  of ice, 29
  of ocean, 283
  of vegetation, 19
  of water droplets, 280
  tensors, 14
Permittivity fluctuation, 165
  large, 319, 345
  small, 155, 319, 476
  variance of, 164, 319, 323
Perturbation expansion, 444
Perturbation orders, 220
Perturbation series, 226, 393
Phase function, 225, 550
Phase matching, 35, 508, 517, 527, 534, 553, 568, 572
Phase matrix, 131-135, 197
  modified, 149
  for random media, 162-168, 269
  Rayleigh, 155-158
  for small ellipsoids, 160-162
Phase velocity, 505
Planck's constant, 4, 143
Planck's radiation law, 4-8
Plane wave expansion, 212
Plane wave representation, 552, 571
  of Green's function, 357, 361, 374
Planetary soil surfaces, 112
Poincare sphere, 124, 125, 127, 208
Point matching, 364
Point scatterers, 225, 272
  analytic wave theory, 351
Polarizability of sphere, 431, 563
Polarizable medium, 411
Polarization charge, 431
Polarization current, 413
Polarization dependence, 285
Polarization ellipse, 123
Polder and van Santern mixing formula, 319, 376, 379
Position eigenstate, 443
Position operator, 443
Potential energy between two particles, 482
Potential function, 443
Potential operator, 443, 465

Power flux density, 130
Power series, 487
Power series expansion, 81
Poynting power density, 82, 83
Poynting vector, 130, 135, 179, 202
Poynting's theorem, 44
Precipitation rate, 20
Pressure equation of canonical ensemble, 486
Principal frame, 133, 193, 525, 533, 538
Principal value, 344, 377, 471
Principal value integral, 412, 413
Principle of superposition, 294
Probability density function, 86, 241, 455
  for Eulerian angle, 527
  of particle orientation, 531
  for N-particles, 433
Probability distribution, 244
Probability occurence of slopes, 85
Propagating mode, 67
Propagation matrix, 24, 40, 46, 48, 331
Propagator, 378, 469
  of background medium, 543
  of effective medium, 543

Quadrature, 220, 263
Quadrature formula, 264
Quantum mechanical potential scattering problems, 321
Quasi-monochromatic, 125
Quasicrystalline approximation (QCA), 461-464, 472- 473, 505, 526, 546, 549, 554, 561
Quasicrystalline approximation-coherent potential (QCA-CP), 467-469, 474-475, 477, 542-548, 554, 561
Quasistatic limit, 420

Radar cross section, 2
Radar equation, 1
Radially symmetric problems, 433
Radiation condition, 12
Radiative transfer theory:
  limitations of, 318
  relation to ladder approximation, 351
  vector, 249
Radiative transfer equations:
  alternative derivation, 370
  for dense media, 550
  for isotropic point scatterers, 424
  for isotropic scatterers, 358
  second moment, 555-563

Index

vector, 128-155, 129-132
Radiometer, 10, 45, 53, 67, 69, 275
Rain, 241, 280, 461, 532
   layer over ocean, 283
Rain droplets, 20
Rain layer, 234, 235
Rainfall, 223, 306
   droplets, 232
   layers, 291
Random bosses, 112
Random height distribution, 71, 204
Random medium, 259
   one dimensional, 337
   three dimensional, 259, 269-272, 393
   cylindrical sturcture, 314
Random rough surfaces, 18, 70-113
Rayleigh absorption, 223
Rayleigh approximations, 272
Rayleigh hypothesis, 99
Rayleigh limit, 498
Rayleigh mixing formula, 431
Rayleigh scattering, 138, 155, 265, 429
   phase function, 155-158, 209, 273, 274, 288
Rayleigh-Jean's approximation, 6
Rayleigh-Jean's law, 7
Reciprocity, 10-14, 178, 303, 329, 336
   rough surface, 90-93, 207
   scattering amplitude, 149
   for scattering function matrix, 133-134
Recurrence relation, 28, 48, 173, 211
Reflected temperature, 296, 303
Reflection coefficient, 28, 48, 88, 89, 201, 396, 398, 551
Reflective surface, 334
Reflectivity, 29, 41, 202
   function, 293, 294, 301
   matrix, 242
   of coherent wave, 429
   of moon, 235
Regular wave functions, 169, 171, 196
Renormalization methods, 403, 406, 424
Renormalized mass operator, 346
Representation of rotation group, 195
Resonance, 234, 273, 275
Resonant scattering, 270, 500
Retro-reflection, 235
Rock, 420, 427
   porosity, 419, 427
Root mean square (rms) slope, 77
Rotation matrix, 525
Rotation of T-matrix, 193-197
Rotational symmetry, 198, 399
Rough dielectric interface, 10, 200, 203-207

Rough surface effects, 249
Row structure, 18, 53, 67
Row-structured plowed field, 65

Saddle point method, 321, 410
Saturation, 234
Scattered fields, uncorrelated, 330
Scattering:
   amplitude, 130, 137, 149, 351
   dyad, 134, 177, 458, 567
   attenuation, 381, 430
   coefficient, 90, 157, 164, 209, 230, 269, 326, 550
   from particles correlated in position, 562
   function matrix, 131-135, 156, 157, 178
   potential, 544
   processes, 227
   profile, 297, 298
   rate, 373, 398
   scale, 394
Scattering cross section, 177, 190, 191
   lower bound, 252
   for sphere, 189-190
Schwarzschild-Milne integral equation, 318, 359, 370, 424, 548
Schwarzschild-Milne problem, 358
Sea ice, 18
Secular terms, 394
Self-patch contribution, 364
Self-similar model, 418, 420
   for sedimentary rocks, 379
Shadowing effect, 76, 93-96, 203, 207
Shadowing function, 95
Single scattering, 227, 256
   of coherent wave, 530
Sinusoidal profiles, 53
Sinusoidal surface, 64-65, 116
Size distribution, 21
Slab of point scatterers, 354, 365
Slightly rough surface, 99
Small perturbation method (SPM), 98-113
Small-scale fluctuations, 343
Small-space lossy-scatterers, 505
Smog, 20
Smoke, 20
Snell's law, 84, 89, 202, 327, 436, 508, 528, 554
Snow, 18, 256, 275-280, 470, 500, 521, 546
   afternoon data, 334
   density, 18
   depth, 389, 524
   diurnal change, 332, 334
   dry, 375, 384, 427, 506, 524

grain sample, 166, 167
grain size, 389
morning data, 334
measurements, 275-280, 290
wet, 375, 384, 389, 427
wetness, 18, 389
Soil, 17-18, 20, 290
  clay, 17
  moisture, 17, 67, 69, 548
  sand, 17
  silt, 17
  texture, 17, 53
  Yuma sand, 17
Sommerfeld integral identity, 375
Source point, 319
Source state, 446
Sources of radiation, 376
Space harmonics, 59
Space representation, 470
Sparse concentration of particles, 458, 464, 481, 500, 531-533
  isotropic, 476
  nonspherical, 241
Sparse medium, 429
Spatial dispersion effects, 380
Specific intensity, 6, 7, 8-10, 46, 221, 225
  diffuse, 358, 424
  invariance, 9
  outgoing, 251
Specific water content, 232, 307
Spectral decomposition, 371, 556
Spectral density, 107, 214, 324
Spectral functions, 371
Spectral representation, 559
  of mean Green's function, 372, 417
  radiating waves, 373
Specular, 255, 326
  direction, 514
  reflection, 77, 90
  surface, 90
  transmission, 77, 89, 90
Spheres, 138, 188-191
  coated, 419
  single sized, 276
Spherical Bessel function, 169, 215, 493
Spherical coordinate system, 457
Spherical Hankel functions, 169, 177, 178
Spherical harmonics, 170, 492, 510, 545
  completeness relation, 510, 569
Spherical harmonics vectors, 544
Spherical pair-distribution functions, 527
Spherical particles, 138, 228-235, 272-285, 470, 479, 564, 566
  scattering pattern, 524

small, 190-191, 252-258, 286-291, 564
Spherical Rayleigh scatterers, 288, 314
Spherical wave functions, 183, 447
  combined index, 175
Spherical wave transformations, 446
Spheroids, 161, 191-193
  aligned, 223, 249
  small, 192-193, 209, 215, 224, 316, 533
States of particle, 455
Stationary Gaussian process, 79
Stationary point, 37
Stationary random processes, 42
Stationary-phase method, 37, 85-90, 146, 215
Stationary-phase point, 85, 88, 146, 205
Statistical mechanics of liquids, 489
Statistical symmetry, 258
Statistically homogeneous, 79, 321, 462
Statistically isotropic, 79
Stepping forward procedure, 291, 202, 308
Stochastic Fourier Stieltjes integral, 105
Stochastic Fourier transform approach, 112
Stokes matrix, 131-135, 157, 198, 208
  averaged, 427
  modified, 149, 152
Stokes parameters, 121-128, 149, 221, 223
  incoherent addition, 127-128
  modified, 122
  transformation of, 128
Stratified medium, 26, 38-41
Stratus clouds, 233, 280
Strong dielectric contrast, 543
Strong permittivity fluctuation theory, 319, 375-389
Strong permittivity mixing formula, 419
Structure factor, 481, 485-486, 488, 506, 516, 531, 567
  of pair-distribution function, 516
Subsurface:
  boundary, 272, 319
  emission, 262
  temperature distributions, 261
  temperature profile, 260
Surface field expansion, 62-64, 186, 453
Surface harmonics, 213
Surface integrals, 493
Surface slopes, 76
Symmetric matrices, 266
Symmetry of particle positions, 492
Symmetry, 150, 192, 195, 266, 364, 478
  of dyadic Green's function, 36, 40, 116, 143, 394, 398
System of extreme disorder, 481
System transfer operator approach, 168

Index

T-matrix, 452, 506, 525, 532
   approach, 113, 168
   averaged over orientation distribution, 527
   for axisymmetric objects, 186-188
   coefficients, 176
   formalism, 446, 455, 475, 490, 543
   multiple scattering equation, 507
   in principal frame, 533
   vector spherical wave expansion, 440
Tangent plane approximation, 70, 72-77
Tangential surface field expansions, 184, 566
Taylor expansion method, 422
Temperature equilibrium, 15
Temperature profiles, 221, 297, 306
Tensor product, 347
   of Neumann series, 348
Tenuous medium, 346, 350, 429, 561
Thermal emission, 4, 142, 153, 221-224
   of half-space laminar medium, 260
Thermal radiation, 4
Thermodynamic equilibrium, 16, 43
Threshold angles, 69
Total field state, 446
Transformation matrix, 210, 452
Transition moisture, 18
Transition operator, 360, 439, 444-445, 463, 470, 473, 479, 562, 566
   modified, 466, 473
   in momentum representation, 458, 566
   for point scatterers, 352, 356
   single particle, 439
Translation addition theorem, 449, 450, 454, 564
   for vector spherical, 448-454
Translational invariance, 356, 361, 422, 447, 493
Translational property, 462
Translational theorem of operators, 457
Transmission, 25-32
Transmission coefficients, 201, 398
Transmissivity, 202, 324
Transmissivity function, 293, 301
Transverse electric (TE), 25, 55
Transverse gradient operator, 54
Transverse magnetic (TM), 26, 55
Turbulent medium, 337, 351, 359, 476
Two particle probability density function, 433
Two-distance scales, 371
Two-point distribution function, 433

Two-variable expansion method, 392-394, 397, 422
Two-variable integral equation, 363
   cyclical-transfer, 359

Uncorrelated scatterers, 337, 548
Uniaxial medium, 28, 536
Uniform particle distributions, 433
   outside the hole, 479
Uniform profile, 298
Uniform-sized particles, 275
Unimoment method, 193
Unit dyad operator, 443
Unitarity, 178-180
Unitary operator, 195
Upper hemisphere, 14, 91

Variation of parameters, 243
Vector spherical harmonics, 170, 510, 544
Vector spherical wave functions, 196, 446
Vegetation, 19-20, 166, 290, 330, 334, 461
   alfalfa, 19
   corn, 19
   leaves, 19, 241
   sorghum, 19
   soybeans, 19, 249
   stalks, 19
   trees, 330
   wheat, 19
Verlet-Weis pair-distribution function, 488, 505, 531, 549, 567
Very low frequency limit, 475
Virial equation, 486

Water, 17, 233, 297, 375
Water droplets, 222, 306
Water equivalent, 19, 389, 524
Wave equation, 155, 375, 476, 550
Wavelength, 371, 400, 557
Wavenumber, 351, 354
Weak permittivity fluctuation, 345, 375
Weak scatterers, 543
Wiener-Hermite nonlinear functional representation, 112
Wigner 3j symbols, 450, 500, 521

X-ray, 481
   diffraction study, 485